ANNUAL REVIEW OF
NEUROSCIENCE

ANNUAL REVIEW OF NEUROSCIENCE

VOLUME 16, 1993

W. MAXWELL COWAN, *Editor*
Howard Hughes Medical Institute

ERIC M. SHOOTER, *Associate Editor*
Stanford University School of Medicine

CHARLES F. STEVENS, *Associate Editor*
Salk Institute for Biological Studies

RICHARD F. THOMPSON, *Associate Editor*
University of Southern California

ANNUAL REVIEWS INC   4139 EL CAMINO WAY   P.O. BOX 10139   PALO ALTO, CALIFORNIA 94303-0897

ANNUAL REVIEWS INC.
Palo Alto, California, USA

International Standard Serial Number: 0147-006X
International Standard Book Number: 0-8243-2416-1

TYPESET BY BPCC-AUP GLASGOW LTD., SCOTLAND
PRINTED AND BOUND IN THE UNITED STATES OF AMERICA

Annual Review of Neuroscience
Volume 16, 1993

CONTENTS

vi   CONTENTS *(continued)*

SOME RELATED ARTICLES IN OTHER *ANNUAL REVIEWS*

Theodore H. Bullock

Annu. Rev. Neurosci. 1993. 16:1–15

INTEGRATIVE SYSTEMS RESEARCH ON THE BRAIN: Resurgence and New Opportunities

Theodore H. Bullock

Neurobiology Unit, Scripps Institution of Oceanography and Department of Neurosciences, University of California, San Diego, La Jolla, California 92093-0201

KEY WORDS: neuronal integration, coding circuitry, EEG, evoked potentials, neuroethology, evolution

Introduction

The explosive growth of new information in neuroscience is an exciting development in an inherently exciting field of inquiry. We should remind ourselves, however, that the vast increase in knowledge still leaves us fundamentally ignorant of how brain cells work together—to achieve recognition, evaluation of sensory input, selection, and coordination of response. We are rich in knowledge of mechanisms at the level of components: molecular, intracellular, even some circuits. We are still poor in understanding higher levels, for example, cerebellum, striatum and cerebral cortex or feature extraction, repertoire selection, and motor control, and still poorer in respect to volition, cognition, and learning, in spite of spectacular advances in each of these areas.

A major part of the task of neuroscience will only be approached as we devote new attention to the difficult questions of organization at the higher levels—from the intact limbic system to the forms of signal processing that result in expectation, recognition, motivation, and choice. With those examples, I intend to highlight the span of intermediate integrative levels between the single cell and the cognitive achievement, and to imply the several levels defined by structure and function, as well as those defined by behavior. A great deal, of course, has already been discovered, and a formidable body of knowledge accumulated about these levels, including

1

0147–006X/93/0301–0001$02.00

important, though incomplete, information on anatomical localization, relevant cellular components, and parameters of modulation. These are the familiar aspects of neural systems research and will, no doubt, continue to be a prime arena of advance while genuine understanding of how neural assemblies operate slowly unfolds. The complex levels of the brain, in its manifold integrative tiers, will doubtless be elucidated less rapidly than the cellular componentry and will emerge in a disorderly, piecemeal fashion.

This is the scope of the remarks that follow. I believe the best use I can make of this space and opportunity is to illustrate the magnificent possibilities now available for insightful investigation of integrative brain physiology. While molecular and cellular componentry forges ahead, I hear from all sides a resurgence of interest in integrative systems level research. The upswing is already under way on many fronts. I am well aware that this broad-based interest is not adequately represented by the remarks that follow; they are samples selected to show a range of issues where major advance is available in the near future.

I formulate some major issues as questions to emphasize that an important part of the challenge, as in the old game of "20 Questions," is to formulate carefully what needs to be answered to narrow the uncertainty — and to do this over some range of puzzles beyond one's immediate research goals. The selection of issues and question is personal and appears to undervalue by neglect major parts of the spectrum of approaches. To convert my limited list of examples into an agenda, they must be complemented by others to represent more widely the whole armamentarium of strategic and tactical means for attaining new perspectives on operations in the brain.

New Parameters of Neural Integration Are Likely to be Discovered at Several Levels

The integrative function of synapses is universally appreciated, but I find it generally overlooked that this function incorporates many variables, at least partly independent, even if we consider only the elementary neuronal level. It is important not to oversimplify models that aim at understanding how the brain works. The combinations of these quasi-independent variables mean that there are not a few dozen or hundreds of kinds of neurons, but many orders of magnitude more, at least in higher animals.

At the neuronal level, do we know all the variables that contribute to integration?
The list is already long, but we have no assurance that it is complete. Well known are a number of dynamic, as well as static, variables that determine output of the neuron as a function of input. Familiar examples are membrane potential; space constants; threshold; accommodation; adaptation;

facilitation and its opposite; regular, irregular, or bursty firing; and a wide range of slopes of the curve relating depolarization to firing frequency. In 1984, I enumerated more than 40 variables, and new discoveries have added to the list. Bennett et al (1985) have elaborated on the earlier finding of an influence of chemicals upon electrical transmission in gap junctions and of electrical events upon chemically transmitting junctions (Bennett 1974). More recently, the action of specific transmitters upon particular ion channels and the actions of modulators upon transmitter efficacy have been documented (Barrio et al 1991; Christenson & Grillner 1991; Gardner 1991), as well as actions of chemicals at a distance (Vizi & Lábos 1991). My bet is that important parameters remain to be discovered, in addition to those already known, but not generally incorporated into models presumed to be fairly realistic.

One reason for calling attention to the formidable battery of variables is that permutations of properties or proclivities along the numerous spectra of variables tend to be characteristic of a neuron or class of neurons. One class, with many subclasses, may be brisk, whereas another is sluggish; one may tend to fire with quite irregular intervals, others more and more regularly (over several orders of magnitude of coefficient of variation); one with a slowly decaying after-effect, others with rebound— and so on for dozens of variables. Although each neuron tends toward a certain "personality" profile, this can vary from time to time, with ontogenetic stage, brain state, and evolutionary level. The personality profile of dynamic functional characteristics is as important in defining cell types and roles as shape, transmitters, and modulators.

Most of the known variables have been discovered serendipitously, a few per year for decades. No end is in sight. Only occasionally are they found by deliberate search or logical process. The recognition of additional neuronal integrative variables is as heuristic and insightful as the recognition of new channels and messengers—the material substrata for the dynamic traits. The need to uncover any remaining variables is an intellectual and technical challenge worthy of our best effort.

At higher levels, what additional integrative parameters operate?
In my opinion, the operations of organized assemblies and circuits of neurons (overlapping but not equivalent terms!) are not adequately described merely by extrapolating the numbers and concatenations of the neuronal variables. Emergent parameters become important because of cooperativity among neurons, at least in more advanced brains, levels, stages, and states. Abeles (1991) has shown that when a cell receives convergent input from several sources, their combined effect is greater than the sum of their isolated effects. He further claims that diverging and converging chains of neurons "are bound to transmit activity synchron-

ously." I am not convinced that it is necessarily true that diverging and converging chains of neurons are bound to act synchronously, but I would not be surprised to learn that a marked evolution has occurred between the major animal groups in the degree of development of synchronous activity. Synchrony, in fact, must be one of the prime emergent variables, as between states of the brain, stages in ontogeny, and different parts of the brain, as well as grades in evolution (Bullock 1992). Yet, synchrony is rarely evaluated quantitatively. I believe that it is not a unitary variable with a single spectrum from low to high, but a heterogeneous class; distinct mechanisms of several kinds operate for different frequencies in different parts of the brain.

Toward the eventual recognition of a list of integrative variables at the multiunit level, I would like to mention a few other candidates to highlight the opportunities for new research and the need for conceptual synthesis. Probably related to the synchrony variables are (i) the degree of rhythmicity or peakiness of the power spectrum versus a broad spectrum with inconstant bumps, and (ii) a tendency in some places and conditions for adequate stimuli to cause a phase shift or "reordering" of certain frequencies (Başar 1992). Another dimension of system options is (iii) the degree to which subsystems are plastic or "hard-wired," as in canary song versus the innately emitted and recognized calls of chickens. (iv) The degree to which some rhythms are timed by circuit time constants versus intracellular oscillations of pacemaker cells would seem to be an important distinction, with the possibility of intermediate, modulating influences. In practice, they are difficult to distinguish. (v) Alternative forms of time series, with different implications as to causal relations among underlying functions, are limit cycles versus chaotic versus thoroughly stochastic activity: Each of these appears to be realized in particular situations in the brain. One of the major challenges in the burgeoning field of neural computation is to provide means to make this distinction meaningfully, quantitatively, and efficiently on data samples of a few seconds or fractions of a second.

Many other integrative variables and system options will be identified in both spike trains and compound field potential time series, particularly with multiple electrode recording that reveals spatiotemporal activity patterns. Each such identification will represent a breakthrough in discovering new relationships and clues to derivative ones.

What correlations, rules, or trends hold between these variables and the kind of behavior, the part of the brain, the state or stage of the nervous system, or the animal group?
This nearly virgin form of question becomes accessible and interesting

with such higher level integrative variables, whereas one does not expect many general correlations with the elementary neuronal variables. To me, an important implication of this perspective is that it does not ask which one of the alternatives actually exists in the brain, as though they were mutually exclusive. It assumes that this organ is eclectic and employs different alternatives in its various subsystems and states and in different taxa. I find it heuristic that this perspective calls for searching among species and situations, rather than inferring a universal answer to how the brain works.

A Wide Variety of Candidate Spike and Nonspike Codes Appears to Operate, in Series and Parallel

I hope it is obvious where I am going. The aim is to imply, as well as to make explicit, some areas of research that appear particularly pregnant to me. Each bases great promise on new techniques, attitudes, and effort, and the likelihood of exciting insights that depart from prevailing paradigms of brain operation.

Which set of the many candidate spike codes and nonspike codes are used in given places and times?

In a canvass of available codes, Perkel & Bullock (1968) systematized many of the numerous plausible candidate codes. We may still emphasize the candidate status, pending demonstration that both the sending and the receiving aspects are verified in the same system—the encoding and the decoding by the candidate parameter—a requirement that is still almost as seldom met as it was then.

The means by which information is represented should be a prime item on the research agenda for a communication system. It is an attractive target for neural research, because we now have good reason to conclude there is not just one code; instead, several means of representation clearly operate, although this is not likely to be a high number. The classical spike frequency code is actually a class with distinct members, including spikes per 200 ms, spikes per 2 s, weighted integral over a forgetting time constant, and instantaneous interval. Distribution of interval fluctuation is another class, including simple variance and shorter or longer patterned sequences. Coding without change of intervals can utilize phase relative to other channels and is well established in sensory systems that follow rhythmic stimuli, even into hundreds of cycles per second, thus manifesting sensitivity to fractions of a microsecond. Nonspike codes are a diverse class of transmitters released, subthreshold membrane voltage, and impedance changes.

Another reason coding is an attractive target for study is the compound cryptographic challenge: Clearly, there are multiple, simultaneous codes among the many parallel channels. A given neuron, or even a given axon, probably conveys information at the same time in two or more distinct codes for different recipient cells or arrays. One postsynaptic cell, for example, may read the number of spikes in the last 0.5 s, with a certain weighting for the most recent intervals, while another postsynaptic cell integrates over several seconds, and a third is sensitive to so-called trophic factors or to irregularity of intervals or to collaterals where impulses do not reach the terminals and only decremented, graded, slow potentials carry information from this input channel. The wide range of species, preparations, subsystems, and approaches available for study should stimulate imaginative investigators.

Here, I want to call special attention to the need and opportunity for work on multichannel codes, those that depend on some kind of spatiotemporal pattern or cooperative activity among sets of cells. These include the assessment of correlation of spikes, bursts, and slow waves among populations of units as information-bearing parameters. Perhaps more than most other codes thus far suggested, spatiotemporal correlation codes may differ significantly from region to region and among animal groups, as a consequence of evolution. The evidence for this inference is the fall-off of coherence with distance between recording electrodes. Coherence is approximately a cross-correlation for each frequency. We have estimated its decline with distance for each of five or six frequency bands between 1 and 50 Hz (Bullock & McClune 1989).

The results of preliminary comparisons of ongoing, unstimulated activity of the highest centers in the gastropod, *Aplysia*, the cephalopod, *Octopus*, elasmobranchs, teleost fish, reptiles, and mammals suggest that, in spite of wide variance, the tendency for a spatially defined population of cells to be coherent increases in the order of the taxa named, that is, coherence falls more slowly with distance. Brain size, at least within mammals, appears to have little effect on this measure. Curiously, such an evolutionary spread in respect to cross-correlation does not seem to apply for evoked responses time-locked to stimuli; they differ more widely, depending on such factors as region of the brain, modality, and temporal pattern of stimulation, but not so much on animal group.

Because the measure of rate of decline of coherence with distance is an estimate of synchrony in the population, and perhaps the best one we have so far, it is now possible, given enough sampling, to test such propositions. Synchrony, at least in certain frequency bands, e.g. the gamma band (ca. 30–50 Hz in mammals), may increase with advancement of the brain among classes of vertebrates, with higher cognitive functions, with higher levels of the brain, or in some way in ontogeny. My bet is that some evolutionary trend will likely occur in a broad or rough way, with notable exceptions, but against a close correlation with cognition. This does not mean that I undervalue cross-correlation as a mediator of integrative achievements. I see it employed here and there, at lower as well as higher levels, and not during many higher cognitive processes. It is poorly correlated with higher levels of cognition, if we consider a wide sample of appropriate situations.

What rules or correlations could give meaning to the wide range of apparent reliability of spike codes?
Reproducibility of spiking to the same conditions varies among neurons from unequivocally high to apparently very low. A common mistake is to speak of neurons as inherently noisy in the sense of sloppy or unreliable. In the best studied cases, where precisely the same stimulus elicits varying responses, the unjustified assumptions are two: (i) that our measure of response is relevant to the normal code, e.g. number or distribution of spikes in an arbitrary segment of time, and (ii) that the neuron under study and the system of which it is a part are in a steady, unchanging state. Unexplained variation in response should be called just that, not noise or slop.

We know that some neurons are extremely reliable, and that some examples of apparent noise in spike intervals are actually useful jitter (Bullock 1970). Arriving volleys of impulses from diverse sources naturally create a stochastic time series of synaptic potentials, and these may interact with the threshold and recovery processes in the responding neuron to generate more or less stochastic output series, without any necessary consequent noise in the proper sense of unwanted or antisignal activity. The serious question remains: What rules or correlations may explain the wide range of *apparent* reliability of single neuron response? No satisfactory answers are available, as far as I know, if we consider a fair sample of cases, not restricted to a select few. This is another frontier on which I expect interesting advance by a combination of experiments with living systems and conceptually imaginative queries.

Working Out the Circuit Is Not Enough

The great change dates from the 1960s, when the conviction gradually grew that, at least in favorable invertebrates, whole circuits could be unraveled, cell by cell. This triggered a vast effort, which has realized this goal to a significant extent in a good many species. The hope has been that this would amount to explaining or understanding the system. The prevailing paradigm of neural operation is circuits and spikes. The sometime analogy to digital computers is well discredited, there being no evidence of the time-slicing required for a digital device. Even today, however, a view of circuitry seems to prevail that only occasionally recognizes "local circuits," with dendro-dendritic and nonspike signaling.

What must be noted is that widespread features of neural operation and organization are not implicit in the concept of circuitry, even that of local circuits, as generally understood (Rakic 1975). I mention only three: 1. Field effects, both chemical and electrical, clearly operate over many micrometers, taking us far beyond classical synaptic transmission. Under chemical field effects, I include the nonsynaptic interactions at a distance, sometimes called volume transmission (Vizi & Lábos 1991). It remains to evaluate how important these avenues of communication may be in each part of the brain, species, stage, and state. 2. The characteristic three-dimensional geometry of terminal arbors and dendritic ramifications must play a role in integrating activity across large arrays of endings and synapses over many tens of micrometers, taking us beyond the textbook synapse defined by electron microscopy of specialized appositions. 3. There is an enormous diversity of cell types, which go beyond shapes, transmitters, and modulators, to include the above-mentioned permutations of scores of integrative parameters (I call them "personality traits" to underscore at once that they tend to characterize a neuron and, yet, are labile).

The diversity of classes of cells is clearer when we remember the non-equivalence of cells with different receptive and projective fields. I have elsewhere estimated that there may be tens of millions of *kinds* of neurons in the human brain, on all criteria of nonequivalence, and far fewer in goldfish and lobster (Bullock 1980). Circuit diagrams can show connectivity in simple systems, but if they try to include the personality traits, three-dimensional geometry, and field effects, they quickly become more of a fat book of specifications. The questions I find hopeful and exciting for future discovery can be subsumed under the following.

How can we formulate hypotheses for the roles of field effects, geometry of arbors, and the diversity of specified cell types in our thinking about circuits? One can claim, with some reason, that we now need another model, paradigm, or frame of reference that preserves the extensive knowledge of circuitry and the established capacity of spike trains to convey information, but takes account of additional essentials. These include the diversity of personalities, the three-dimensional spatial distribution of dendritic and axonal arbors, and the possibility that nonclassical forms of signaling between cells are employed. It is all the more important to recognize the diversity of units and parameters of communication today, because of the heavy influence of "neural network" research, which largely ignores these realities. The next question is similar, but focuses on specific regions or centers.

How can we think of more adequate formulations of the functions or operations performed by parts of the brain?
We are no longer content with the vague wordings we were taught: The cerebellum "coordinates" motor patterns. . . . Striate cortex is a "primary sensory area" for vision. The intellectual challenge stretches the imagination how best to investigate and then name or summarize adequately the functional role of a region or system (e.g. motor, limbic, extra-pyramidal), or the operations it performs upon its input, or even without current input. The evidence is quite uneven and points to a nonequivalence, in a logical sense, of functions of different parts of the brain. It also warns us not to assume engineering or common sense organization. One reason this kind of challenge is stimulating is that answers cannot be expected solely from deductive analysis, but call for creative leaps followed by vigorous efforts to disprove each hypothesis.

Available Windows into the System in Operation Are Each too Narrow; Spike and Slow Wave Windows Are Nonredundant

We do not soon expect to understand how the brain works at its higher levels of operation, in a comprehensive sense, and we must not be discouraged by finding ever-increasing degrees of complexity. Without denigrating the popular search for basic cellular components and fundamental common denominators, a true understanding of brain operation in higher animals has to include the emergent levels of system complexity. For this, our technical armamentarium, wide as it is diverse, and exciting in its new developments, still severely limits our windows into the operating system. We need all the windows we have—and more.

I must comment on the unfortunate dichotomy among brain researchers who work on the systems aspect of brian physiology, between those who concentrate on nerve impulse firing in cellular units and those who concentrate on compound field potentials, mostly nonspike, slow waves. Each group tends to rationalize its preferred approach and derogate the other. One can understand the attractions of the two, quite asymmetrical positions, and still regret the lack of a more common synergetic relationship.

I will not recapitulate the case for either view, but only note that the unit approach, exemplified by Abeles (1991), has led to elegant quantitative analyses based on known components (which are not necessarily representative). It interfaces with similarly based models that also oversimplify by overlooking, for example, the summation of slow fields, the three-dimensional structure of neuropile, and the possibility of subthreshold codes other than prespike synaptic potentials. As exemplified by Başar (1992), the compound field approach, in its turn, although based on less understood components, including the three just mentioned, has led to robust correlations with subtle stimuli, states, and cognitive events and to sophisticated descriptors of traits, dynamics, and cooperativity. It has also led to maps, which are rapidly becoming more elaborate and include some deep recording sites. The unit spike approach is chiefly based on nonhuman animal subjects, the slow potentials chiefly on human recordings. One of the promising opportunities is to work on bridge building between these two approaches. The severe limitations of each approach are only partly complemented by the other. Besides the mutual disparagement, a vast loss of valuable data is regrettable when spike studies high-pass the recorded potentials before storing the activity picked up by microelectrodes.

Slow, compound-field potentials may, indeed, be "soft," uninterpretable in cellular terms, and "merely" by-products of operation. But, we cannot be sure; even if so, they are phenomena deserving study in their own right and likely to yield at least clues to new properties and levels of organization. They are, generally speaking, not predictable from our knowledge of unit spike activity.

Even apart from the dichotomy just discussed, important differences in viewpoint among workers in the compound, slow potential area make communication and progress more difficult than in the more cellular levels of investigation. I illustrate with a few questions upon which agreement has not been reached and which I predict will lead to new tools and effort, thus significantly altering our view of the microstructure of brain activity.

Is the EEG a mixture of a few rhythms or is it closer to a stochastic or chaotic series?
Some workers are impressed by or interested in the one or two, rarely three, peaks of the power spectrum visible in some electroencephalogram (EEG) samples. For the sake of counterbalance, I underline the lack of consistent peaks in most species, apart from a gentle maximum. Only in particular states, such as walking in the rat or eyes closed, awake, and not concentrating in the human, are peaks—usually only one—consistently present. Even during these states, much of the power is broadband. The EEG is a broadband time series in most species most of the time. Without additional information, we cannot choose between the possibilities that most EEG activity is a mixture of several sinusoidal oscillators or of quasistochastic, broadband events, e.g. sawtooth postsynaptic potentials (Lopes da Silva 1992). Phase spectra and coherence spectra, especially when computed from intracranial, direct cortical recordings, are typically only gently sloping or nearly flat over several octaves, incompatible with a model of discrete oscillators. I have elsewhere commented on the pre-liminary reports of a chaotic, as distinct from a limit cycle or a stochastic,

nature of the EEG (Bullock 1990); it is too early to wax enthusiastic, but it may well turn out that some measure or form of display based on the nonlinear dynamics will reveal patterns in time and space and discriminate among regions, states, and species. Here, again, one can hope for elucidation that might markedly alter our prevailing views.

Is synchrony a steadily maintained state for particular frequencies (multicellular oscillator model), is it episodic and rapidly fluctuating (stochastic model), or is it nearly uniform across a wide band of frequencies (event-driven model)?
Synchrony among a population of neurons is commonly identified by visual inspection of the EEG. ["This record shows a synchronized (or a desynchronized) state."] Such a statement generally means simply that the record is characterized by high-voltage, low-frequency activity (or low-voltage, fast activity), without any actual measurement of synchronization. Synchrony is considered to be the principal basis of high-voltage episodes and of spectral peaks, and each frequency is believed to characterize certain brain states. So little quantitative information is at hand about population synchrony, other than the simple amplitude itself, that we cannot really decide whether synchrony is episodic, rapidly fluctuating, or sustained for certain frequencies! The best method for evaluating this property of neural assemblies is probably the rate of decline of coherence as a function of distance, measured for each frequency band, as mentioned above (Bullock & McClune 1989). Our data for subdural, surface recordings in humans, rabbits, rats, turtles, and rays agree most with the third alternative in the above question, but intracortical microelectrodes give a more complex picture. I expect that clarification with more work will substantially enlighten the picture and, possibly, turn our thinking to new directions.

Are evoked potentials basically phase shifted, "reordered" background oscillations, or are they largely transient synchronized activity added to the ongoing background?
Included in the latter alternative would be phase shifts between subsets of the cell population. A negative answer to the latter alternative is the proposal that evoked potentials are reset cycles of the EEG (Başar 1980, 1992). This proposal can apply only to those records in which the evoked potential does not rise appreciably above the EEG. It may, therefore, be an appropriate analysis of some situations, but not of the general case, because many evoked potentials, recorded in the best locus, are large in amplitude and show a characteristic shape, including early, fast waves and, later, fast and slow waves. One reason to expect important elucidation, comparing time-locked stimulus or event related responses in various species and levels of the brain, is that quite heterogeneous cellular and

assembly operations underlie these responses in different situations, even during the time course of a response. Some apparently oscillatory event-related potentials turn out to be caused by successive events in spatially separate populations (Mangun 1992).

These few examples suffice to indicate that basic biology remains to be done and that insight will depend on comparison of numerous samples representing different species, as well as states, stages, and levels of the brain.

Neuroethological and Macroevolutionary Approaches Are Distinct and Asymmetrical

These two approaches are not widely familiar although each has a small coterie of enthusiasts. My concern here is the following:

Which aims of these two approaches have general neurobiological interest? Neuroethology has led to advances in general neurobiology not likely to have occurred without this slant. In the crustacean optic lobe (Waterman & Wiersma 1963; Wiersma 1967, 1974) and in retinal ganglion cells of the frog (Lettvin et al 1959; Maturana et al 1960) classes of cells that are especially responsive to complex, natural stimuli opened a new field of research into cells that recognize combinations of features. Suga's (1988a–c, 1989) stepwise unraveling of the higher levels of auditory processing of stimulated bat cries and echos led to a new appreciation of parallel processing and topographic representation of derived functions. Species-specific song-selective and face-selective neurons are the most advanced examples of complex recognition cells so far (Margoliash 1986; Konishi 1989; Perrett et al 1989; Perrett & Rolls 1983; Rolls et al 1982). Heiligenberg (1988, 1991) and colleagues, working on electric fish social responses, have given us the most complete analysis to date of a piece of normal behavior followed through more than 14 orders of neurons from receptors to effectors, characterizing each one anatomically and physiologically. Our understanding of the neural organization and control of behavior, including walking, feeding, swimming, vocalization, and startle responses in leeches, insects, crustaceans, snails, lampreys, bony fish, anurans, and a few birds and mammals, has chiefly depended on studies of nonstandard laboratory animals, based on their species-characteristic activities. I have detailed the relevance of studies on fish and electroreception to various aspects of general neurobiology elsewhere (Bullock 1983, 1986). Findings of broad interest concern, for example, ultrastructural changes in ribbon synapses with physiological state, resonance of receptors at their best excitatory frequency, and plastic alteration of such tuning, diversity of spike codes, and lifelong addition of sense cells to a fixed number of afferent axons (Bullock & Heiligenberg 1986).

The evolutionary approach is commonly understood to overlap virtually completely with the neuroethological, because most comparative studies deal with adaptive specialization among species and taxa at about the same grade of complexity. Macroevolution of the brain has, however, led to great differences in grade of complexity (cf. flatworms, earthworms, insects, cephalopods, fish, and mammals)—a span of levels of complexity surpassing any other. Complexity in this context can be measured by the number of kinds of component anatomical structures, physiological interrelationships, and behavioral choices, including sensory discriminations and response repertoire. This great span of complexity, a salient feature of biological evolution, has been relatively neglected as a topic of research, but raises a new class of questions.

What changes in structure, physiology, or chemistry characterize more advanced levels of complexity?
A deep asymmetry with the foregoing approach lies in the issue of levels. How much can be attributed to simple increase in number of cells and connections, and how much depends on novelties, such as new cell types, spines, recurrent collaterals, number of intrinsic cells and cell types, reciprocal connections, new physiological processes, chemical messengers, and emergent properties? These have been little studied across major taxa of clearly distinct levels of complexity, because most effort is to find basic, common mechanisms. Perhaps a factor has been reluctance to compare species in different classes or phyla, or even to admit some groups are more complex than others. The concept of "higher" animals has mistakenly been thought to require an anthropocentric, linear assumption of evolution. Neurally more advanced higher categories (phyla, classes, some orders), however, clearly evolved—sporadically—long before primates, mammals, or even vertebrates appeared. I, for one, cannot assume that this fact is trivial by being an obvious consequence of adaptive pressures, because survival value, adaptiveness, and reproductive potential do not necessarily increase with neural complexity.

These approaches are complementary and not equivalent. They differ mainly in the level of analysis and in the comparison of closely related versus well separated taxa of distinct grades of complexity. Neuroethology, in the explanation of special adaptations, is more likely to turn up basic, common mechanisms. Macroevolutionary comparative anatomy, physiology, and chemistry are more likely to turn up the novelties that permitted more advanced brains (Bullock 1992).

Coda

The formulation of pregnant approaches I have chosen is not based entirely on logic or current interests, but is partly autobiographical. I have enjoyed struggling with each of these major issues as the themes underlying most of my published work. This formulation manifests my belief that by hitching one's vehicle to a high enough objective in the galaxy of questions about nature, one can at least glean extra inner rewards, in spite of the transience of explanations and hypotheses. To me, the measure of value of a hypothesis, as of a model, is not its plausibility or compatibility with a subset of facts, or its presumed validity, but is heuristic potential—how much it suggests for the next stage of investigation. These five areas, among others, have particularly high heuristic potential toward the goal of disclosing, bit by bit, how the brain works in its wondrous array of internal levels and species differences.

A major component of this potential that deserves its own emphasis is the power of an unorthodox approach—descriptive natural history. It has been inherent in many of the questions and examples selected that the likelihood of major findings by this approach is still strong today. Unexpected relationships and dynamic properties are noticed by the trained eye of a naturalist and become the grist for the reductionist mill in the future. I, for one, believe that the brain still conceals a host of unrecognized qualitative and quantitative traits, including basic principles, still awaiting description and, only then, analysis.

Literature Cited

Abeles, M. 1991. *Corticonics*. Cambridge, Engl: Cambridge Univ. Press
Barrio, L. C., Suchyna, T., Bargiello, T., Xu, L. X., Roginski, R. S., et al. 1991. Gap junctions formed by connexins 26 and 32 alone and in combination are differently affected by applied voltage. *Proc. Natl. Acad. Sci. USA* 88: 8410–14
Başar, E. 1992. Brain natural frequencies are causal factors for resonances and induced rhythms. In *Induced Rhythms in the Brain*, ed. E. Başar, T. Bullock. Boston, Mass: Birkhäuser
Başar, E. 1980. *EEG-Brain Dynamics*. Amsterdam: Elsevier
Bennett, M. V. L. 1974. *Synaptic Transmission and Neuronal Interaction*. New York: Raven
Bennett, M. V. L., Zimering, M. B., Spira, M. E., Spray, D. C. 1985. Interaction of electrical and chemical synapses. In *Gap Junctions*, ed. M. Bennett, D. Spray, pp.

355–66. Cold Spring Harbor, NY: Cold Spring Harbor Lab.
Bullock, T. H. 1992. How are more complex brains different? One view and an agenda for comparative neurobiology. *Brain Behav. Evol.* In press
Bullock, T. H. 1990. An agenda for research on chaotic dynamics. In *Chaos in Brain Function*, ed. E. Başar, pp. 31–41. Berlin: Springer
Bullock, T. H. 1986. Significance of findings on electroreception for general neurobiology. In *Electroreception*, ed. T. Bullock, W. Heiligenberg, pp. 651–74. New York: Wiley
Bullock, T. H. 1984. A framework for considering basic levels of neural integration. In *Cortical Integration: Basic, Archicortical and Association Levels of Integration*, ed. F. Reinoso-Suarez, C. Ajmone-Marsan, pp. 27–36. New York: Raven

Bullock, T. H. 1983. Why study fish brains? Some aims of comparative neurology today. In *Fish Neurobiology*, ed. R. Davis, R. Northcutt, 2: 361–68. Ann Arbor: Univ. Mich. Press

Bullock, T. H. 1980. Reassessment of neural connectivity and its specification. In: *Information Processing in the Nervous System*, ed. H. Pinsker, W. Willis, pp. 199–220. New York: Raven

Bullock, T. H. 1970. The reliability of neurons. *J. Gen. Physiol.* 55: 565–84

Bullock, T. H., Heiligenberg, W., ed. 1986. *Electroreception*. New York: Wiley

Bullock, T. H., McClune, M. C. 1989. Lateral coherence of the electrocorticogram: a new measure of brain synchrony. *Electroencephalogr. Clin. Neurophysiol.* 73: 479–98

Christenson, J., Grillner, S. 1991. Primary afferents evoke excitatory amino acid receptor-mediated EPSPs that are modulated by presynaptic GABA$_B$ receptors in lamprey. *J. Neurophysiol.* 66: 2141–49

Gardner, D. 1991. Presynaptic transmitter release is specified by postsynaptic neurons of *Aplysia* buccal ganglia. *J. Neurophysiol.* 66: 2150–54

Heiligenberg, W. 1991. The neural basis of behavior: a neuroethological view. *Annu. Rev. Neurosci.* 14: 247–67

Heiligenberg, W. 1988. The neuronal basis of electrosensory perception and its control of a behavioral response in a weakly electric fish. In *Sensory Biology of Aquatic Animals*, ed. J. Atema, R. Fay, A. Popper, W. Tavolga, pp. 851–68. New York: Springer-Verlag

Konishi, M. 1989. Birdsong for neurobiologists. *Neuron* 3: 541–49

Lettvin, J. Y., Maturana, H. R., McCulloch, W. S., Pitts, W. H. 1959. What the frog's eye tells the frog's brain. *Proc. Inst. Radio Eng.* 47: 1940–51

Lopes da Silva, F. 1991. Neural mechanisms underlying brain waves: from neural membranes to networks. *Electroencephalogr. Clin. Neurophysiol.* 79: 81–93

Mangun, G. R. 1992. Human visual evoked potentials: induced rhythms or separable components? In *Induced Rhythms in the Brain*, ed. E. Başar, T. H. Bullock, pp. 217–31. Boston: Birkhäuser

Margoliash, D. 1986. Preference for autogenous song by auditory neurons in a song system nucleus of the white-crowned sparrow. *J. Neurosci.* 6: 1643–61

Maturana, H. R., Lettvin, J. Y., McCulloch, W. S., Pitts, W. H. 1960. Anatomy and physiology of vision in the frog (*Rana pipiens*). *J. Gen. Physiol.* 43: 129–75

Perkel, D. H., Bullock, T. H. 1968. Neural coding. *Neurosci. Res. Program Bull.* 6: 221–348

Perrett, D. I., Harries, M. H., Bevan, R., Thomas, S., Benson, P. J., et al. 1989. Frameworks of analysis for the neural representation of animate objects and actions. *J. Exp. Biol.* 146: 87–113

Perrett, D. I., Rolls, E. T. 1983. Neural mechanisms underlying the visual analysis of faces. In *Advances in Vertebrate Neuroethology*, NATO ASI Ser. A: Life Sci., ed. J. Ewert, R. Capranica, D. Ingle, 56: 543–68. New York: Plenum

Rakic, P. 1975. Local circuit neurons. *Neurosci. Res. Program. Bull.* 13: 289–446

Rolls, E. T., Perrett, D. I., Caan, A. W., Wilson, F. A. W. 1982. Neuronal responses related to visual recognition. *Brain* 105: 611–46

Suga, N. 1989. Principles of auditory information-processing derived from neuroethology. *J. Exp. Biol.* 146: 277–86

Suga, N. 1988a. Auditory neuroethology and speech processing: complex-sound processing by combination-sensitive neurons. In *Auditory Function*, ed. W. Edelman, W. Gall, W. Cowan, pp. 679–720. New York: Wiley

Suga, N. 1988b. Parallel-hierarchical processing of biosonar information in the mustached bat. In *Animal Sonar*, ed. P. Nachtigall, P. Moore, pp. 149–59. New York: Plenum

Suga, N. 1988c. What does single-unit analysis in the auditory cortex tell us about information processing in the auditory system? In *Neurobiology and Neocortex*, ed. P. Rakic, W. Singer, pp. 331–49. New York: Wiley

Vizi, E. S., Lábos, E. 1991. Non-synaptic interactions at presynaptic level. *Prog. Neurobiol.* 37: 145–63

Waterman, T., Wiersma, C. A. G. 1963. Electrical responses in decapod crustacean visual systems. *J. Cell. Comp. Physiol.* 61: 1–16

Wiersma, C. A. G. 1974. Behavior of neurons. In *The Neurosciences: Third Study Program*, ed. F. Schmitt, F. Worden, pp. 419–31. Cambridge, Mass: MIT Press

Wiersma, C. A. G. 1967. Visual central processing. In *Invertebrate Nervous Systems, Their Significance for Mammalian Neurophysiology*, ed. C. Wiersma, pp. 269–84. Chicago: Univ. Chicago Press

Annu. Rev. Neurosci. 1993. 16:17–29

TRANSSYNAPTIC CONTROL OF GENE EXPRESSION

Robert C. Armstrong and Marc R. Montminy

The Clayton Foundation Laboratories for Peptide Biology,
The Salk Institute, La Jolla, California 92037

KEY WORDS: immediate early genes, neuronal gene expression, phosphory-
              lation, CREB, cAMP

INTRODUCTION

Poised for relay or storage of synaptic information, the postsynaptic neuron must assimilate a variety of incoming synaptic signals and then integrate these into short- and long-term responses. Neuronal signals not only regulate the quantal release of transmitters and peptides acutely, but also dictate the form and content of genetic program that the postsynaptic neuron will express. Indeed, the activity-dependent release of neuro-transmitters begins a cascade of events, which culminates in postsynaptic changes in gene expression. Upon binding to postsynaptic receptors, a neurotransmitter evokes specific changes in postsynaptic second messenger pathways that ultimately trigger the rapid induction of postsynaptic genes. Many of these rapidly induced genes then feed-forward to activate genetic programs, which are important for neuronal cell function. Such activity-dependent changes may underlie complex phenomena, like long-term memory and synaptic plasticity.

This review traces our current understanding of activity-regulated gene expression, from model systems in which neuronal activity alters post-synaptic gene expression, to receptor systems and second messengers that participate in the induction, and finally to the molecular mechanisms that may mediate these changes at the transcriptional level.

NEURONAL ACTIVITY AND GENE EXPRESSION

Synaptic activity alters postsynaptic gene expression in two phases. The first consists of immediate early genes (IEGs), which are rapidly induced

17

0147–006X/93/0301–0017$02.00

in response to neuronal stimulation. The second contains late-onset genes, which encode differentiated neuronal products, like neuropeptides and neurotransmitter biosynthetic enzymes.

Regulation of Immediate Early Genes

First detected in growth factor-stimulated fibroblasts and, subsequently, in neuronal cell lines (Curran & Morgan 1987; Lau & Nathans 1987), the IEGs share several distinguishing features: They are generally expressed at very low levels in nonstimulated cells. They are rapidly induced by cellular stimuli. Their transcriptional induction is very transient (30–60 minutes) and independent of new protein synthesis. Subsequent transcriptional termination requires new protein synthesis. And, their corresponding mRNAs are rapidly degraded. As most IEGs encode predominantly nuclear proteins, several groups have speculated that such proteins might serve as transcription factors (Morgan & Curran 1989), which could in turn stimulate late-onset gene expression in response to neuronal signals (Christy et al 1988; Curran & Franza 1988; Milbrandt 1987).

Perhaps the best characterized IEG to date is the c-*fos* gene (Curran 1988). First characterized as a viral oncogene in feline osteosarcoma cells, the cellular proto-oncogene counterpart, c-*fos*, appears to function largely as a transcription factor. In response to neuronal stimulation, c-*fos* expression is rapidly activated and then turned off within one hour. C-*fos* activates transcription of target genes by heterodimerizing with yet another IEG, c-*jun*, to form a transcription factor known as AP-1.

The induction of IEGs in response to synaptic activity was first demonstrated with seizure-inducing agents. Within one hour of metrazole administration, for example, c-*fos* expression was dramatically induced in the brain (Morgan et al 1987; Saffen et al 1988; Sonnenberg et al 1989a,b). This induction was inhibited by anticonvulsant drugs, which suggests that neuronal activity was required to stimulate c-*fos* expression. Other IEGs such as c-*jun*, *Jun*B (Sonnenberg et al 1989a,b), NGFI-A (Saffen et al 1988; Sonnenberg et al 1989c), and NGFI-B (Watson & Milbrandt 1989) were coincidentally activated by such seizure-evoked activity. The IEG response appeared to be independent of the seizure-inducing method employed, as such treatments as kainic acid (Sonnenberg et al 1989b), electrical lesion of the dentate gyrus hilus (White & Gall 1987), and electroconvulsive ear shock (Daval et al 1989) were equipotent.

These initial studies provided strong evidence to support the hypothesis that synaptic activity modulates postsynaptic gene expression, with the caveat that such stimuli were somewhat extreme. More recent studies have demonstrated that IEGs can also be induced by physiologically relevant neuronal stimuli. Hunt et al (1987), for example, observed the induction of c-*fos* immunoreactivity in spinal cord dorsal horn neurons following stimulation of peripheral sensory neurons. This finding was confirmed and extended by Bullitt (1989), who showed that noxious stimuli would not only induce c-*fos* in spinal cord neurons, but also thalamic nocioceptive neurons in the afferent pathway. Indeed, under similar conditions, other IEGs, including NGFI-A, NGFI-B, and c-*jun*, were also stimulated in the spinal cord (Wisden et al 1991).

Recent studies with IEG stimulation in cortical structures have been even more provocative. Electrical stimulation of rodent hindlimb neocortex, for example, elicits metabolic changes in postsynaptic neurons within the cerebellum, as measured by 2-deoxyglucose uptake. Remarkably, c-*fos* expression is also induced in the same postsynaptic regions that had been metabolically mapped (Sharp et al 1989). In hippocampus, moreover, stimulation of the perforant path-granule cell synapse [which can induce long-term potentiation (LTP)] markedly increases the levels of NGFI-A, c-*fos*, c-*jun*, and *jun*B mRNAs (Cole et al 1989).

Taken together, these studies suggest that physiologically relevant synaptic activity in the central nervous system leads to the induction of cellular IEGs. These IEGs have been proposed to mediate transcriptional activation of late-onset genes in response to neuronal activity, although a direct link has yet to be established (Sonnenberg et al 1989c).

Regulation of Late Onset Genes

Neuropeptide genes are among the best studied neuronal late-onset genes. Released synaptically in response to neuronal activity, these peptides appear to function as neuromodulators. And, current evidence suggests that transcriptional control is indeed important in maintaining intracellular peptide levels (Black et al 1987). Dopaminergic inputs into the striatum, for instance, are critical for expression of proenkephalin and other neuropeptide genes. Lesioning the major nigrostriatal dopaminergic pathway with 6-hydroxydopamine (6-OHDA) substantially reduces substance P mRNA and peptide levels in the striatum. In contrast, proenkephalin mRNA and peptide levels are actually increased by this treatment (Gerfen et al 1990; Sivam et al 1987; Voorn et al 1987; Young et al 1986), thus demonstrating the divergent effects of synaptic activity.

In the peripheral nervous system, pharmacologic agents have been employed to examine the importance of neuronal activity on neuropeptide expression. Reserpine, a catecholamine reuptake blocking agent, has been used extensively to examine the effect of sympathetic inputs on several neuropeptides. Indeed, chronic reserpine treatment appears to increase splanchnic nerve activity, which in turn stimulates postsynpatic expression of neuropeptide Y (Higuchi et al 1990) and tyrosine hydroxylase (Tank et al 1985) mRNAs in the adrenal gland. As transection of the splanchnic nerve inhibits the reserpine-mediated induction of these mRNAs, these results demonstrate that such regulation probably occurs through a trans-synaptic mechanism. Other regions of the autonomic nervous system demonstrate similar transsynaptic regulation. Reserpine treatment stimulates neuropeptide Y mRNA accumulation in the rat superior cervical ganglion (SCG) (Hanze et al 1991), whereas denervation of the SCG decreases the expression of several SCG synaptic vesicle antigens (Greif & Trenchard 1988).

NEUROTRANSMITTER RECEPTOR SYSTEMS WHOSE ACTIVATION REGULATES GENE EXPRESSION

Transsynaptic activation requires, as its initial step, the binding of a neurotransmitter to its receptor. Several receptor systems mediate postsynaptic changes in gene transcription through second messenger systems, which are activated upon ligand binding. These second messengers link the neurotransmitter receptors to the molecular regulators of gene transcription.

The dopamine receptors constitute one of the most thoroughly studied receptor systems involved in the regulation of postsynaptic gene expression. Two dopamine receptor subtypes (D_1 and D_2) that exhibit distinct pharmacological properties have been identified (Creese & Fraser 1987). Best characterized in the striatum, they differentially activate or inhibit adenylate cyclase through interactions with specific GTP-binding proteins (G-proteins) (Albert et al 1990). In the striatum, the two major efferent pathways express dopamine receptors, with D_1 receptors primarily on striatonigral and D_2 receptors on striatopallidal neurons (Le Moine et al 1990). Dopamine released from afferent nigrostriatal neurons can differentially regulate efferent striatonigral and striatopalladal neurons through activation of D_1 and D_2 receptors, respectively. Indeed, many of the changes in neuropeptide expression noted following chemical ablation of nigrostriatal dopamine neurons appear to be mediated by these two receptor subtypes (Gerfen et al 1990). Thus, the 6-OHDA-induced elevation of proenkephalin mRNA in striatopalladal neurons was reversed by continuous administration of the D_2 agonist quinpirole; reductions in substance P mRNA were reversed by administering the D_1 agonist SKF-38393. These results demonstrate that neuronal signals can elicit very different cellular responses, depending on receptor subtypes that are expressed in target cells.

The expression of multiple receptor subtypes that evoke different cellular responses to a single neurotransmitter may be particularly important in the hippocampus. Indeed, glutamate receptor expression in these cells appears to be important for LTP formation (Mody & Heinemann 1987), a process that may involve changes in gene expression (Madison et al 1991). To date, three glutamate receptor subtypes have been characterized, based on specific agonist binding properties (Cotman et al 1987): N-methyl-d-aspartate (NMDA), quisqualate (Q), and kianate (K).

Glutamate receptor activation, either by glutamate or NMDA, induces c-*fos* expression in primary cultures of rat cerebellar neurons (Szekely et al 1987). This induction is blocked by NMDA receptor-specific antagonists. Surprisingly, the induction of c-*fos* and other IEGs by metrazole-induced seizures in whole animals is also effectively blocked by NMDA receptor antagonists (Nakabeppu et al 1988; Sonnenberg et al 1989b), which suggests that the NMDA receptor pathway may be preferentially activated by these agents.

At least two types of glutamate receptors appear to induce IEG expression. Sonnenberg et al (1989b) observed that both kianate and NMDA administration stimulated c-*fos* expression in hippocampal cells. The time course of kainate-induced expression, however, was significantly prolonged compared with NMDA, which suggests that the two receptor subtypes may employ different second-messenger pathways to transmit their signals.

In addition to IEGs, the glutamate receptors appear to regulate late-onset genes. Glutamate treatment of primary cerebellar granule cells in culture, for example, stimulates GABA receptor subunit mRNA expression (Memo et al 1991). NMDA receptor antagonists reduce GABA receptor mRNA levels. Thus, although the mechanism of NMDA-receptor action has not been elucidated, these observations provide a very exciting prospect for future studies on LTP and transcriptional control.

NEUROTRANSMITTER RECEPTOR-GENERATED SECOND MESSENGER SYSTEMS

Neurotransmitter receptors can be divided into two general classes: those linked to GTP binding proteins, and those that form ligand gated ion channels. Although each receptor type can stimulate gene expression in response to ligand binding, this transcriptional effect is ultimately dependent upon changes in second messenger production.

G-Protein Interactions

Many G-proteins have been characterized, and each possesses the ability to effect distinct cellular second messengers (Birnbaumer 1990). For example, the D_1 and D_2 dopamine receptor subtypes interact with G_s and G_i, respectively. These interactions regulate adenylate cyclase activity, either positively (G_s) or negatively (G_i) (Creese & Fraser 1987). Thus, D_1 and D_2 receptors may correspondingly regulate neuronal gene expression through cAMP-dependent mechanisms.

The cAMP pathway is not the sole, or even primary, second messenger system in neuronal cells. For example, many receptors are coupled to G_0, which regulates the activity of phospholipase C. These include the muscarinic acetylcholine (type M1 and M2), adrenergic (α1), and serotonin (5HT2 and 5HT1c) receptors (Boyer et al 1989; Fisher & Agranoff 1987). Phospholipase C acts on membrane phospholipids to generate two second messengers, diacylglycerol (DAG) and inositol-triphosphate (IP_3) (Martin 1991). Diacylglycerol can directly activate protein kinase C (PKC), which mediates the induction of several IEGs, including c-*fos* (Gilman 1988; Greenberg et al 1986). Generation of IP_3 leads to the liberation of intracellular stores of Ca^{2+}. Calcium may then bind to calmodulin and activate the calmodulin-dependent kinase (CAM-kinase). Indeed, CAM-kinase activity is necessary for the induction of c-*fos* in response to depolarization of PC12 cells (Morgan & Curran 1986).

Gated Ion Channels

The nicotinic acetylcholine and the NMDA/glutamate receptors are the only ion channel receptors currently implicated in the regulation of gene expression. Both of these channels regulate the flow of cations across the membrane, with the nicotinic acetylcholine receptor allowing passage of Na^+, and the NMDA-receptor channel fluxing Ca^{2+}. Indeed, both of these receptors may ultimately stimulate gene expression through Ca^{2+}-dependent mechanisms (Morgan & Curran 1986). Although the NMDA receptor can directly increase intracellular Ca^{2+} levels, nicotinic acetylcholine receptors appear to affect intracellular Ca^{2+} levels more indirectly. The flow of Na^+ through these channels depolarizes the cell, thereby activating voltage-gated Ca^{2+} channels and allowing increased flow of Ca^{2+}. The rise in intracellular Ca^{2+} produced by each of these channels acts as a second messenger in the same manner as Ca^{2+} released from intracellular stores.

MOLECULAR BIOLOGY OF ACTIVITY-DEPENDENT GENE REGULATION

As with other cellular inducers, synaptic signals appear to stimulate specific target genes through *cis*-acting regulatory sequences. The molecular mechanisms by which neuronal signals regulate these genes are probably not unique, but appear to be shared with other stimuli that can employ similar second messenger pathways. Indeed, work on these pathways in non-neuronal cells has been critical in understanding the mechanisms by which neuronal activity regulates gene expression.

Cyclic-AMP and Gene Regulation

Extracellular signals that activate second messenger pathways cause rapid cellular responses and consistently regulate gene expression through pathways that do not require new protein synthesis. As a result, such changes in gene activity require intrinsic changes in pre-existing regulatory proteins and in the ways they interact with their cognate promoter sequences. Covalent modifications, like phosphorylation, are especially appealing candidates, because such changes are both rapid and easily reversible.

Cyclic-AMP has long been implicated in the regulation of genes in the nervous system (Eiden et al 1984; Murdoch et al 1982). All of the known cellular effects of cAMP occur via the catalytic subunit of the cAMP-dependent protein kinase (PKA), which suggests that this second messenger might correspondingly regulate transcription through the reversible phosphorylation of specific transcription factors.

The rat somatostatin gene has been widely used as a model system to dissect this pathway. First described as a 14-amino acid peptide in hypothalamic cells, somatostatin was characterized by its ability to inhibit release of growth hormone from somatotrophs (Brazeau et al 1973). Early experiments showing that forskolin (an activator of adenyl cyclase) could stimulate somatostatin mRNA accumulation in fetal rat hypothalamic cells prompted us to characterize this response in the neuroendocrine cell line PC12 (Montminy et al 1986a).

In PC12 cells, cAMP had profound effects on somatostatin expression, inducing the activity of a transfected somatostatin promoter-chloramphenicol acetyltransferase (CAT) fusion gene 15–20-fold. Promoter deletion studies revealed a 30-base cAMP response element (CRE), which was both necessary and sufficient for cAMP inducibility. Remarkably, comparison of this CRE with other cAMP responsive genes revealed a short-core palindrome, 5′-TGACGTCA-3′, which was conserved among several neuronally expressed genes, including VIP (Tsukada et al 1987) and tyrosine hydroxylase (Lewis et al 1987). Furthermore, cAMP-dependent protein kinase activity appeared to be critical for CRE-mediated transcriptions, as a PK-A-deficient mutant PC12 cell line A126-1B2 was transcriptionally unresponsive to cAMP (Montminy et al 1986b).

The conserved nature of the CRE suggested that cAMP would regulate gene expression through a common DNA-binding protein, whose activity would be modulated by cAMP-dependent protein kinase. This protein was first detected in nuclear extracts of PC12 cells by DNAse I footprinting assay (Montminy & Bilezikjian 1987) and later purified as a single CRE-binding (CREB) protein of 43 kD from both brain and PC12 extracts by sequence-specific DNA affinity chromatography. By using primary sequence information from the purified protein, we obtained a CREB cDNA that encoded a 341-amino acid protein. Hoefler et al (1988) also identified a CREB cDNA by screening a bacterial cDNA expression library with a CRE oligonucleotide.

The most striking feature of the CREB sequence is a single consensus PK-A phosphorylation site (Arg-Arg-x-Ser). This site is efficiently phosphorylated in vitro by PKA, and phosphorylation of the PK-A site in situ is stimulated 20–30-fold within 15 minutes of treatment with forskolin. As forskolin causes no detectable changes in CREB protein levels, these results suggest that the increase in phosphorylation is, indeed, independent of new protein synthesis.

Mutagenesis studies with the cloned CREB cDNA have been useful in demonstrating the regulatory importance of the PK-A phosphorylation site. When cotransfected with a somatostatin-CAT reporter plasmid, an RSV-CREB expression plasmid can direct high level expression of this reporter gene in a cAMP-dependent manner. Mutagenesis of the PK-A phosphoacceptor site (Ser133) completely abrogates CREB activity, thus demonstrating that Ser133 serves as a central control point for cAMP action. The ability of PK-A to regulate CREB activity without apparently affecting its DNA binding properties suggests that phosphorylation may trigger a limited conformational change in the molecule that promotes new interactions between CREB and other proteins in the RNA polymerase II complex.

CRE-binding does not appear to be the only factor capable of mediating transcriptional responses to cAMP. Such proteins as ATF-1 (Rehfuss et al 1991) and *Jun*D (Kobierski et al 1991) can also stimulate transcription in a cAMP dependent manner. Promoter sequences in the proenkephalin gene may mediate the cAMP response by binding *Jun*D (Kobierski et al 1991).

Transcriptional Activation by Ca^{2+}

Many neuronal stimuli activate transcription through Ca^{2+}-dependent mechanisms. PC12 cells have also been used extensively to dissect the components of this pathway. Both K^+-depolarization and nicotinic acetylcholine receptor activation of PC12 cells stimulate c-*fos* expression in a Ca^{2+}-dependent manner (Greenberg et al 1986; Morgan & Curran 1986). And, CAM-kinase inhibitors block c-*fos* expression in response to these stimuli. As promoter deletion analysis reveals that Ca^{2+} may partly stimulate c-*fos* expression through the CRE element (Sheng et al 1990), it appears that both Ca^{2+} and cAMP responses may converge on a single calcium/cyclic AMP responsive element (CaRE/CRE). Two-dimensional tryptic phosphopeptide mapping experiments suggest that depolarization and Ca^{2+} may stimulate the phosphorylation of CREB at the Ser133 phosphoacceptor site (Sheng et al 1991). Indeed, CREB can be directly phosphorylated by the Ca^{2+}-dependent CAM-kinase in vitro, which suggests that the Ca^{2+} and cAMP pathways may converge on the same regulatory protein through distinct kinases. As elevated Ca^{2+} did not appear to stimulate cAMP accumulation, Sheng et al proposed that Ca^{2+} would regulate CREB activity in a PK-A independent manner.

Other laboratories have disputed this finding, however, by showing that functional PKA activity is indeed required for Ca^{2+} to induce c-*fos* expression (Ginty et al 1991). By using the PKA deficient line A-126, the authors observed that the transcriptional response to Ca^{2+} in these cells was severely attenuated in face of normal CAM-kinase activity. These conflicting reports may be reconciled by two possible factors: PC12 cells may express a Ca^{2+}-sensitive adenyl cyclase that activates CREB through a cAMP-dependent route. Or, the PKA deficient A-126 cell line may have additional, as yet uncharacterized, deficiencies that prevent a normal Ca^{2+} response. This work is also complicated by the presence of additional Ca^{2+}-responsive elements on the c-*fos* promoter that are distinct from the CaRE/CRE site (Sheng et al 1990). Future work with different Ca^{2+}-regulated genes in better defined cell systems will be critical to understand this response. Indeed, potassium depolarization of PC12 cells induces other IEGs (Bartel et al 1989), and several of these genes contain CaRE/CRE-like sites in their promoters (Changelian et al 1989; Christy et al 1988; Watson & Milbrandt 1989). However, the molecular basis for their induction in response to cAMP or Ca^{2+} remains to be elucidated.

Transcriptional Activation by PKC

Identification of genes that are regulated by PKC has been facilitated by the use of phorbol esters, a group of membrane-permeant compounds that can substitute for DAG in activating PKC. As with the study of transcription mediated by other pathways, the study of phorbol ester stimulated transcription has focused on the c-*fos* gene. Promoter mapping studies have revealed that TPA stimulates c-*fos* transcription through a dyad symmetrical sequence previously characterized as the serum response element (SRE) (Gilman 1988; Treisman 1986). The SRE appears to mediate several responses through a cellular protein termed the serum response factor (SRF) (Greenberg et al 1987). Although the precise role of SRF in the PKC-mediated transcriptional activation of c-*fos* is not understood, current evidence showing that SRF is a phosphoprotein suggests that this factor may be regulated in a manner analogous to CREB (Norman et al 1988).

Two additional SRE-associated factors have recently been identified. One protein, p62^{DFB}, binds directly to the sequence just 5′ of the c-*fos* SRE (Ryan et al 1989). The other, p62^{TCF}, associates with the SRE/SRF complex in a DNA-independent manner (Shaw et al 1989). Although the precise role that these accessory factors perform has not been characterized, recent evidence suggests that they are important for both phorbol ester (Graham & Gillman 1991) and serum responsive (Shaw et al 1989) transcription. Indeed, several other IEGs contain SRE-like sequences, including β-actin (Ortia et al 1989), NGFI-A (Christy & Nathans 1989), and krox-20 (Chavrier et al 1989), which suggests that the SRF and associated factors may be critical for neurotransmitter-activated transcription.

Regulation of Genes by AP-1: A Concerted Cellular Response

Many of the IEGs encode transcription factors and might act as nuclear "third" messengers in a cellular response to external stimuli (Morgan & Curran 1989). The best studied IEGs are those whose products contribute to AP-1 binding activity: c-*fos*, the *Fra*s, and members of the *Jun* family (Morgan & Curran 1989). Indeed, metrazole-induced seizures appear to induce AP-1 binding activity, as determined by gel retardation assay. The protein composition of this AP-1 binding activity also appears to be modulated in response to metrazole. Thus, Fos and several Fras sequentially appear and disappear over time (Sonnenberg et al 1989a). The increased AP-1 binding activity in response to metrazole suggests that new target genes may be activated as a consequence of its changing molecular composition.

CONCLUSIONS

Although neuronal activity can clearly regulate gene expression, it remains to be determined whether and in which manner such regulation is important for neuronal function. For example, are dopamine-induced changes in striatal proenkephalin expression critical for striatal function? And, are NMDA-induced changes in gene expression important to LTP formation and short-term memory? The availability of cDNA clones encoding receptors, kinases, and transcription factors may at last provide us with the means to interfere with specific signaling pathways in the brain. And, transgenic technology may prove invaluable in assessing the consequences of such manipulations on behavioral processes as well.

ACKNOWLEDGMENTS

The authors gratefully acknowledge helpful discussions with other members of Dr. Montminy's laboratory and the secretarial assistance of Bethany Coyne. This work was supported by National Institutes of Health grants GM13786 and GM37828, National Cancer Institute grant CA-14195, the McKnight Foundation, and the Foundation for Medical Research, Inc. (FMR). Marc Montminy is an FMR investigator.

Literature Cited

Albert, P. R., Neve, K. A., Bunzow, J. R., Civelli, O. 1990. Coupling of a cloned rat dopamine-D_2 receptor to inhibition of adenylyl cyclase and prolactin secretion. *Biol. Chem.* 265: 2098–2104

Bartel, D. P., Sheng, M., Lau, L. F., Greenberg, M. E. 1989. Growth factors and membrane depolarization activate distinct programs of early response gene expression : dissociation of *fos* and *jun* induction. *Genes Dev.* 3: 304–13

Birnbaumer, L. 1990. G proteins in signal transduction. *Annu. Rev. Pharmacol. Toxicol.* 30: 675–705

Black, I. B., Adler, J. E., Dreyfus, C. F., Friedman, W. F., Lagamma, E. F., et al. 1987. Biochemistry of information storage in the nervous system. *Science* 236: 1263–68

Boyer, J. L., Hepler, J. R., Harden, T. K. 1989. Hormone and growth factor receptor-mediated regulation of phospholipase C activity. *Trends Pharmacol. Sci.* 10: 360–64

Brazeau, P., Vale, W., Burgus, R., Ling, N., Butcher, M., et al. 1973. Hypothalamic polypeptide that inhibits the secretion of immunoreactive pituitary growth hormone. *Science* 179: 77–79

Bullitt, E. 1989. Induction of c-*fos*-like protein within lumbar spinal cord and thalamus of the rat following peripheral stimulation. *Brain Res.* 493: 391–97

Changelian, P. S., Feng, P., King, T. C., Milbrandt, J. 1989. Structure of the NGFI-A gene and detection of upstream sequences responsible for its transcriptional induction by nerve growth factor. *Proc. Natl. Acad. Sci. USA* 86: 377–81

Chavrier, P., Janssen-Timmen, U., Mattei, M. G., Zerial, M., Bravo, B., et al. 1989. Structure, chromosome location, and expression of the mouse zinc finger gene Krox-20: multiple gene product and coregulation with the proto-oncogene c-*fos. Mol. Cell. Biol.* 9: 787–97

Christy, B. A., Lau, L. F., Nathans, D. 1988. A gene activate in mouse 3T3 cells by serum growth factors encodes a protein with "zinc finger" sequences. *Proc. Natl. Acad. Sci. USA* 85: 7857–61

Christy, B. A., Nathans, D. 1989. Functional serum response elements upstream of the growth factor-inducible gene *zif*268. *Mol. Cell. Biol.* 9: 4889–95

Cole, A. J., Saffen, D. W., Baraban, J. M., Worley, P. F. 1989. Rapid increase of an immediate early gene messenger RMA in hippocampal neurons by synaptic NMDA receptor activation. *Nature* 340: 474–76

Cotman, C. W., Monaghan, D. T., Ottersen, O. P., Storm-Mathisen, J. 1987. Anatomical organization of excitatory amino acid receptors and their pathways. *Trends Neurosci.* 7: 273–79

Creese, I., Fraser, C. M. 1987. *Receptor Biochemistry and Methodology: Dopamine Receptors*, Vol. 8. New York: Liss

Curran, T. 1988. The *fos* oncogene. In *The Oncogene Handbook*, ed. E. P. Reddy, A. M. Skalka, pp. 307–25. Amsterdam: Elsevier

Curran, T., Franza, B. R. 1988. *Fos* and *Jun*: the AP-1 connection. *Cell* 55: 395–97

Curran, T., Morgan, J. I. 1987. Memories of *fos. Bioessays* 7: 255–58

Daval, J. L., Nakajima, T., Gleiter, C. H., Post, R. M., Marangos, P. J. 1989. Mouse brain c-*fos* mRNA distribution following a single electroconvulsive shock. *J. Neurochem.* 52: 1954–57

Eiden, L. E., Girand, P., Affolter, H. U., Herbert, E., Hotchkiss, A. J. 1984. Alternate modes of enkephalin biosynthesis regulation by reserpine and cyclic AMP in cultured chromaffin cells. *Proc. Natl. Acad. Sci. USA* 81: 3949–53

Fisher, S. K., Agranoff, B. W. 1987. Receptor activation and inositol lipid hydrolysis in neural tissues. *J. Neurochem.* 48: 999–1017

Gerfen, C. R., Engber, T. M., Mahan, L. C., Susel, Z., Chase, T. N., et al. 1990. D1 and

D2 dopamine receptor-regulated gene expression of striatonigral and striatopallidal neurons. *Science* 250: 1429–31

Gilman, M. J. 1988. The c-*fos* serum response element responds to protein kinase C-dependent and -independent signals. *Genes Dev.* 2: 394–402

Ginty, D. D., Glowacka, D., Bader, D. S., Hidaka, H., Wagner, J. A. 1991. Induction of immediate early genes by Ca^{2+} influx requires cAMP-dependent protein kinase in PC12 cells. *J. Biol. Chem.* 266: 17454–58

Graham, R., Gillman, M. 1991. Distinct protein targets for signals acting at the c-*fos* serum response element. *Science* 251: 189–92

Greenberg, M. E., Siegfried, Z., Ziff, E. B. 1987. Mutation of the c-*fos* gene dyad symmetry element inhibits serum inducibility of transcription in vivo and the nuclear regulatory factor binding in vitro. *Mol. Cell. Biol.* 7: 1217–25

Greenberg, M. E., Ziff, E. B., Greene, L. A. 1986. Stimulation of neuronal acetylcholine receptors induces rapid gene transcription. *Science* 234: 80–83

Greif, K. F., Trenchard, H. 1988. Neonatal deafferentation prevents normal expression of synaptic vesicle antigens in the developing rat superior cervical ganglion. *Synapse* 2: 1–6

Hanze, J., Kummer, W., Haas, M., Lang, R. E. 1991. Neuropeptide Y mRNA regulation in rat sympathetic ganglia: effect of reserpine. *Neurosci. Lett.* 124: 119–21

Higuchi, H., Iwasa, A., Yoshida, H., Miki, N. 1990. Long lasting increase in neuropeptide Y gene expression in rat addrenal gland with reserpine treatment: positive regulation of transsynaptic activation and membrane depolarization. *Mol. Pharmacol.* 38: 614–23

Hoefler, J. P., Meyer, T. E., Yun, Y., Jameson, J. L., Habener. J. F. 1988. Cyclic-AMP-responsive DNA-binding protein: structure based on a cloned placental cDNA. *Science* 242: 1430–32

Hunt, S. P., Pini, A., Evan, G. 1987. Induction of c-*fos*-like protein in spinal cord neurons following sensory stimulation. *Nature* 328: 632–34

Kobierski, L. A., Chu, H. M., Tan, Y., Comb, M. J. 1991. cAMP-dependent regulation of proenkephalin by *JunD* and *JunB*: positive and negative effects of AP-1 proteins. *Proc. Natl. Acad. Sci. USA* 88: 10222–26

Lau, L. F., Nathans, D. 1987. Expression of a set of growth related immediate early genes in BALB/c 3T3 cells: coordinate regulation with c-*fos* and c-*myc*. *Proc. Natl. Acad. Sci. USA* 84: 1182–86

Le Moine, C., Normand, E., Guitteny, A. F., Fouque, B., Teoule, R., et al. 1990. Dopamine receptor gene expression by enkephalin neurons in rat forebrain. *Proc. Natl. Acad. Sci. USA* 87: 230–34

Lewis, E. J., Harrington, C. A., Chikaraishi, D. M. 1987. Transcriptional regulation of the tyrosine hydroxylase gene by glucocorticoids and cyclic AMP. *Proc. Natl. Acad. Sci. USA* 84: 3550–54

Madison, D. V., Malenka, R. C., Nicoll, R. A. 1991. Mechanism underlying long term potentiation of synaptic transmission. *Annu. Rev. Neurosci.* 14: 379–97

Martin, T. F. 1991. Receptor regulation of phosphoinisitidase C. *Pharmacol. Ther.* 49: 329–45

Memo, M., Bovolin, P., Costa, E., Grayson, D. R. 1991. Regulation of GABA receptor subunit expression by activation of N-Methyl-d-aspartate-selective glutamate receptors. *Mol. Pharmacol.* 39: 599–603

Milbrandt, J. 1987. A Nerve Growth Factor-induced gene encodes a possible transcriptional regulatory factor. *Science* 238: 797–99

Mody, I., Heinemann, U. 1987. NMDA receptors of dentate gyrus granule cells participate in synaptic transmission following kindling. *Nature* 326: 701–4

Montminy, M. R., Bilezikjian, L. M. 1987. Binding of a nuclear protein to the cyclic-AMP response element of the somatostatin gene. *Nature* 328: 175–78

Montminy, M. R., Low, M. J., Tapia-Arancibia, L., Reichlin, S., Mandel, G., et al. 1986a. Cyclic AMP regulates somatostatin mRNA accumulation in primary diencephalic cultures and in transfected fibroblast cells. *J. Neurosci.* 6: 1171–76

Montminy, M. R., Sevarino, K. A., Wagner, J. A., Mandel, G., Goodman, R. H. 1986b. Identification of cyclic-AMP response element within the rat somatostatin gene. *Proc. Natl. Acad. Sci. USA* 83: 6682–86

Morgan, J. I., Cohen, D. R., Hempstead, J. L., Curran, T. 1987. Mapping patterns of c-*fos* expression in the central nervous system after seizure. *Science* 237: 192–96

Morgan, J. I., Curran, T. 1989. Stimulus-transcription coupling in neurons: role of cellular immediate-early genes. *Trends Neurosci.* 12: 459–62

Morgan, J. I., Curran, T. 1986. The role of ion flux in the control of c-*fos* expression. *Nature* 322: 552–55

Murdoch, G. H., Rosenfeld, M. G., Evans, R. M. 1982. Eukaryotic transcriptional regulation and chromatin associated protein phosphorylation by cyclic AMP. *Science* 218: 1315–17

Nakabeppu, Y., Ryder, K., Nathans, D. 1988. DNA binding activities of three murine *Jun* proteins: Stimulation by *fos.* *Cell* 55: 907–15

Norman, C., Runswick, M., Pollack, R. M., Treisman, R. 1988. Isolation and characterization of cDNA clones encoding SRF, a transcription factor that binds the c-*fos* serum response element. *Cell* 55: 989–1003

Ortia, S., Makino, K., Kawamoto, T., Niwa, H., Sugiyama, H., et al. 1989. Identification of a site that mediates transcriptional response of the human beta-actin gene to serum factors. *Gene* 75: 13–19

Rehfuss, R. P., Walton, K. M., Loriaux, M. M., Goodman, R. H. 1991. The cAMP-regulated enhancer-binding protein ATF-1 activates transcription in response to cAMP-dependent protein kinase A. *J. Biol. Chem.* 266: 18431–34

Ryan, W. A., Franza, B. R., Gilman, M. Z. 1989. Two distinct cellular phosphoproteins bind to the c-*fos* serum response element. *EMBO J.* 8: 1785–92

Saffen, D. W., Cole, A. J., Worley, P. F., Christy, B. A., Ryder, K., et al. 1988. Convulsant-induced increase in transcription factor messenger RNA's in rat brain. *Proc. Natl. Acad. Sci. USA* 85: 7795–99

Sharp, F. R., Gonzalez, M. P., Sharp, J. W., Sagar, S. M. 1989. c-*fos* expression and (^{14}C) 2-deoxyglucose uptake in the caudal cerebellum of the rat during motor/sensory cortex stimulation. *J. Comp. Neurol.* 284: 621–36

Shaw, P. E., Schroter, H., Nordheim, A. 1989. The ability of a ternary complex to form over the serum response element correlates with serum inducibility of the human c-*fos* promoter. *Cell* 56: 563–72

Sheng, M., McFadden, G., Greenberg, M. E. 1990. Membrane depolarization and calcium induce c-*fos* transcription via phosphorylation of transcription factor CREB. *Neuron* 4: 571–82

Sheng, M., Thompson, M. A., Greenberg, M. E. 1991. CREB: A Ca+2-regulated transcription factor phosphorylated by calmodulin-dependent kinases. *Science* 252: 1427–30

Sivam, S. P., Breese, G. R., Krause, J. K., Napier, T. C., Mueller, R. A., et al. 1987. Neonatal and adult 6-hydroxydopamine-induced lesions differentially alter tachykinin and enkephalin gene expression. *J. Neurochem.* 49: 1623–33

Sonnenberg, J. L., Macgregor-Leon, P. F., Curran, T., Morgan, J. I. 1989a. Dynamic alterations occur in the levels and composition of transcription factor AP-1 complexes after seizure. *Neuron* 3: 359–65

Sonnenberg, J. L., Mitchelmore, C., Macgregor-Leon, P. F., Hempstead, J., Morgan, J. I., et al. 1989b. Glutamate receptor agonists increase the expression of *Fos*, *Fra*, and AP-1 DNA binding activity in the mammalian brain. *J. Neurosci. Res.* 24: 72–80

Sonnenberg, J. L., Rauscher, F. J., Morgan, J. I., Curran, T. 1989c. Regulation of proenkephalin by *Fos* and *Jun*. *Science* 246: 1622–25

Szekely, A. M., Baraccia, M. L., Costa, E. 1987. Activation of specific glutamate receptor subtypes increases c-*fos* proto-oncogene expression in primary cultures of neonatal rat cerebellar granule cells. *Neuropharmacology* 26: 1779–82

Tank, A. W., Lewis, E. J., Chikaraishi, D. M., Weiner, N. 1985. Elevation of RNA coding for tyrosine hydroxylase in rat adrenal gland by reserpine treatment and exposure to cold. *J. Neurochem.* 45: 1030–33

Treisman, R. 1986. Identification of a protein-binding site that mediates transcriptional response of the c-*fos* gene to serum factors. *Cell* 46: 567–74

Tsukada, T., Fink, J. S., Mandel, G., Goodman, R. H. 1987. Identification of a region in the human vasoactive intestinal polypeptide gene responsible for regulation by cyclic AMP. *J. Biol. Chem.* 262: 8743–47

Voorn, P., Roest, G., Groeneweggen, H. J. 1987. Increase of enkephalin and decrease of substance P immunoreactivity in the dorsal and ventral striatum of the rat after midbrain 6-hydroxydopamine lesions. *Brain Res.* 412: 391–96

Watson, M. A., Milbrandt, J. 1989. The NGFI-B gene, a transcriptionally inducible member of the steroid receptor gene superfamily: genomic structure and expression in rat brain after seizure induction. *Mol. Cell. Biol.* 9: 4213–19

White, J. D., Gall, C. M. 1987. Differential regulation of neuropeptide and proto-oncogene mRNA content in the hippocampus following recurrent seizures. *Mol. Brain Res.* 3: 21–29

Wisden, W., Errington, M. L., Williams, S., Dunnett, S. B., Waters, C., et al. 1991. Differential expression of immediate early genes in the hippocampus and spinal cord. *Neuron* 4: 603–14

Young, W. S., Bonner, T. I., Brann, M. R. 1986. Mesencephalic dopamine neurons regulate the expression of neuropeptide mRNAs in the rat forebrain. *Proc. Natl. Acad. Sci. USA* 83: 9827–31

Annu. Rev. Neurosci. 1993. 16:31–46

MOLECULAR MECHANISMS OF DEVELOPMENTAL NEURONAL DEATH

E. M. Johnson, Jr. and T. L. Deckwerth

Department of Molecular Biology and Pharmacology,
Washington University School of Medicine, St. Louis, Missouri 63110

KEY WORDS: cell death, trophic factor, macromolecular synthesis, apoptosis, cell calcium

INTRODUCTION

Neuronal cell death is an important phenomenon in the series of steps involved in the development of the mature vertebrate nervous system. Neurogenesis produces about twice as many neurons in a given structure as survive in the adult organism. This initial excess of neurons is pruned within a narrow time span that differs among structures in both the peripheral and central nervous systems. This seemingly wasteful process is conceived to be a mechanism whereby the size of a neuronal pool is matched to the amount of target tissue to be innervated. This general interpretation is supported by decades of work in which naturally occurring cell death can, in most situations, be reduced by increasing the mass of the target tissue or exacerbated by removing target tissue. Our objective is not to review the literature on the general phenomenon of cell death during the development of the nervous system; this has already been well described in detail (Oppenheim 1991). Rather, our charge is to focus more narrowly on recent studies of possible molecular mechanisms involved in developmental cell death and its probable pathological analogues (axotomy, target removal) and the mechanistic insights that might be obtained from studies of other physiologically appropriate cell deaths seen in other tissues and invertebrates.

31

0147–006X/93/0301–0031$02.00

A TWO-STEP VIEW OF NEURONAL DEATH

Neuronal death can be viewed as a two-step process: First, neuronal cell-surface receptors signal the presence of ligands for these receptors in the extracellular environment to the neuron. Based upon this information and other factors (e.g. afferent electrical input), the neuron "decides" whether to live or die. Then, the decision is executed by maintaining cellular processes that ensure viability or trigger intracellular molecular events that lead to death.

Step One: The Decision to Live or Die

The molecules critical to the first step are target derived or locally acting neurotrophic factors. Nerve growth factor (NGF), the prototype of such factors, has been studied for decades (Levi-Montalcini & Angeletti 1966), and considerable data are available regarding its chemistry, biosynthesis, and regulation. Furthermore, its physiological targets have been characterized extensively in respect to the timing of acquisition and loss of the dependence upon NGF for survival (for recent reviews, see Barde 1989; Snider & Johnson 1989). Nerve growth factor is but one member of a family of very similar factors that act via a family of tyrosine protein kinase receptors to exert the trophic effects on various neuronal populations (concisely reviewed in Bothwell 1991). In addition to the NGF or "neurotrophin" family of factors that has been defined in the context of the biology of neuronal function and survival, many other factors (e.g. fibroblast growth factor) have also been shown to exert survival-promoting or neurotrophic effects on neuronal populations. However, despite the myriad effects exerted by the factors in in vitro and in vivo paradigms, NGF is currently the only factor whose physiological function is reasonably well defined. Nerve growth factor mediates the critical role of the target in determining the survival of NGF-dependent cell types that include sympathetic neurons and some neural crest-derived sensory neurons. Nerve growth factor is a polypeptide constitutively synthesized and released in minute quantities by targets of these neurons. It then binds to specific receptors on neuronal processes and is subsequently retrogradely transported to the cell body. Although the structural nature of the critical transported message is uncertain, that message induces a complex set of responses in the cell that includes morphological, biochemical, and functional alterations. In the developing neuron and, less acutely, in the mature neuron, a continual access to neurotrophic factor is required for survival of at least some neurons (Gorin & Johnson 1979, 1980). The role of NGF on dependent cells has been demonstrated by experiments in which NGF is either supplemented (Hamburger et al 1981; Hendry & Campbell

1976) or removed in vivo (Gorin & Johnson 1979; Levi-Montalcini & Booker 1960). Exogenously administered NGF prevents naturally occurring neuronal death during development or following axotomy. Conversely, greater numbers of sympathetic and sensory neurons die if deprived of endogenous NGF by neutralizing antibodies. Such results are consistent with the concept that NGF availability is a critical determinant of neuronal survival. Thus, what is known from three decades of work with NGF defines the general paradigm through which other neurotrophic factors, whether members of the NGF family or chemically distinct, are generally thought to act. The search for neurotrophic factors and corresponding receptors and for the definition of cell types responsive to an individual factor represents a rapidly expanding area of research. Interest in identifying these factors is motivated by both their basic biological importance and their potential pharmacological utility in treating neurodegenerative conditions.

Step Two: Executing the Decision to Die

The considerations above describe the process by which extracellular neurotrophic factors interact with specific cell-surface receptors to produce neurotrophic effects and secure the survival of the neuron that, at least during development, is acutely dependent on the factor. However, few previous studies of neurotrophic factors and their receptors have addressed the question of the mechanism by which the developing neuron dies when deprived of trophic factor and which changes occur upon maturation that make neurons less acutely dependent upon trophic factor. We focus on these questions in the remainder of this review.

TWO GENERAL SCHEMES Two general schemes may be envisioned to account for the death of neurons when trophic support becomes insufficient. One possibility is that neurons require trophic factor for sustained general metabolic activity and that absence of these factors leads to a loss of these activities and, ultimately, to cellular degeneration and death. Cell death by this scheme is a passive process. Alternatively, death caused by trophic factor deprivation may be a metabolically active process, and the role of neurotrophic factors may be to repress a "suicide" response. That is, NGF and other trophic factors promote survival not by stimulating life, but by suppressing death.

The former scheme is inferred by the term "trophic factor" (meaning to nourish or sustain) and generally seems to be the way the survival-promoting actions of neurotrophic factors were previously viewed. Two general kinds of observations, one genetic and one cell biological, have recently focused on the latter view, which suggests that specific cellular mechanisms exist, by which the cell commits suicide. Stated another way, neuronal death during development is physiologically appropriate and represents an alternate, albeit terminal, pathway of differentiation. As with other differentiation pathways, death presumably results from specific intra-cellular events that probably involve the regulation of the expression of specific gene products.

THE EVIDENCE   The genetic experiments implicating specific gene products involved in cell death have been performed in the nematode *Caenorhabditis elegans*, in which the divisions and death of individual cells can be observed visually. Studies over the last 15 years have described the life history of all cells in the organism. During the development of the adult hermaphroditic organism, 1090 somatic cells are formed and 131 undergo a precisely timed and cell-autonomous programed cell death. In contrast to neuronal death in developing vertebrates, cell death in *C. elegans* appears to be "hard-wired" into the pattern of development and not to be governed by cell-cell interactions (Yuan & Horvitz 1990) or competitive phenomena (reviewed in Ellis et al 1991). However, the morphological appearance of *C. elegans* neurons undergoing programed cell death resembles that observed in ver-tebrate neurons dying physiologically appropriate death, and similarities might exist between the suicide mechanisms in both cases. Horvitz et al have identified several genes in *C. elegans* that alter various aspects of some or all of the cells that normally die. Particularly relevant to the current discussion are the two genes *ced-3* and *ced-4*; mutations in either of these genes cause survival of almost all cells that would normally die (Ellis & Horvitz 1986). In genetically mosaic animals made of cells expressing either functional or nonfunctional *ced-3*, only those cells expressing the functional *ced-3* gene die. These results suggest that *ced-3* and *ced-4* genes encode proteins that act within the dying cell or a close ancestor and can either kill the cell themselves or interact with other intracellular molecules to produce cell death. A third gene, *ced-9*, appears to repress the expression of *ced-3* and *ced-4*. Mutations that inactivate *ced-9* are lethal, but only if *ced-3* and *ced-4* are active. The death of the affected cells is speculated to result from inappropriate cell death caused by the unregulated expression of *ced-3* and *ced-4*. Taken together, these data suggest that a highly regulated genetic program is responsible for physio-logically appropriate programed cell death in *C. elegans* (Ellis et al 1991).

Importantly, in *ced-3* and *ced-4* mutants, the cells that would normally die survive and differentiate into recognizable phenotypes. This infers that expression of *ced-3* and *ced-4* is not critical to normal functioning of cells and that the primary or only function of these genes is in the expression of one differentiated phenotype, which is degeneration leading to death.

What is the evidence that a similar genetic program is responsible for neuronal death in vertebrate neurons during development or other situations of trophic factor deprivation, as after axotomy? The DNA sequences of *ced-3* and *ced-4* have not been reported, and whether there are clear homologues in vertebrate, or even other invertebrate, species is not known. In dealing with vertebrate neurons, the elegant genetic approaches available in *C. elegans* are virtually impossible. However, cell biological methods not readily available in the nematode can be used in vertebrates. The most provocative result consistent with, but not proving, that trophic factor deprivation-induced neuronal death is caused by an active process involving specific gene products is the demonstration that such neuronal death requires both RNA and protein synthesis. This was first demonstrated in established cultures of dissociated sympathetic neurons derived from embryonic rat. During an initial culture period of one week, the neurons were maintained in the presence of NGF, which is required for their survival. Nerve growth factor was then removed, a manipulation designed to mimic axotomy or target removal in producing acute trophic factor deprivation. Under these conditions, the vast majority of the neurons died within 48 hours. Neuronal death was prevented by the addition of inhibitors of RNA or protein synthesis at the time of, or for many hours subsequent to, NGF removal from the cells (Martin et al 1988). Thus, when unable to synthesize new RNA or protein, these neurons do not undergo trophic factor deprivation-induced death. This initial demonstration was extended to other neuronal types (Scott & Davies 1990) sustained by other trophic factors in vitro: Chicken trigeminal mesencephalic neurons maintained by brain-derived neurotrophic factor, dorsomedial trigeminal ganglion cells maintained by NGF, and ciliary neurons maintained by ciliary neurotrophic factor (CNTF). In an in vivo paradigm in chick embryos, Oppenheim and colleagues (1990) demonstrated that naturally occurring neuronal death in both sensory and motor neurons was prevented by inhibitors of macromolecular synthesis. In addition, the augmented neuronal death caused by limb removal or axotomy was reduced significantly. Taken together, the results obtained in both in vitro and in vivo experiments indicate that active gene expression is required for trophic factor deprivation-induced neuronal death and suggest that neurotrophic factors suppress the expression and/or function of specific cellular activities used for the production of a specific differentiation pathway, i.e. cell death.

Utilization of inhibitors of macromolecular synthesis to mimic the survival-promoting action of neurotrophic factors is a crude experiment. The prevention of death by inhibitors of macromolecular synthesis is consistent with several possible molecular mechanisms discussed below. The demon-

stration of the requirement for macromolecular synthesis for this physiologically appropriate mode of cell death relates neuronal cell death more
closely with death of various other cell types in response to both positive
and negative signals in developing and adult animals. In many cases, the
death of other cell types offers logistical advantages for study and has been
the subject of much investigation into molecular mechanism. Thus, we
now briefly review what is known about cell death in other systems and
our current understanding, or lack thereof, of the possible similarities and
differences among the mechanisms of cell death.

COMMON CHARACTERISTICS OF PHYSIOLOGICALLY APPROPRIATE CELL DEATH OF OTHER CELL TYPES

Cells of different tissue origin display a wide variety of morphologies when
dying during development or following a toxic insult. Attempts have been
made to classify different forms of death by using morphological criteria.
Such a classification appears useful when it is assumed that cells demonstrating a common morphology of death die by similar underlying molecular mechanisms. The subcellular localization of such mechanisms is suggested by pathologically altered organelles for which certain aspects of
their normal functions are known. Generally, an alteration of this function
is then believed to be involved in or responsible for the demise of the cell.

For nonneuronal cells, a binary classification scheme that has received
much attention attempts to distinguish between a mechanism for physiologically appropriate death termed "apoptosis" and a pathologic mechanism termed "necrosis," which occur upon nonphysiologic intoxication
(Arends & Wyllie 1991; Duvall & Wyllie 1986). Morphologically, apoptosis
is characterized by condensation of the chromatin, segmentation of the
nucleus, and the convolution of the plasma membrane into bulbous
appendices that constrict at their base and detach as plasma membrane-
enclosed fragments of the cytoplasm and nucleus. Such apoptotic bodies,
as well as the cytoplasm of the remainder of the cell, contain largely intact
organelles and, in intact tissue, are phagocytosed by macrophages or other
cells. Necrosis, on the other hand, may be caused by compromise of cellular
ionic homeostasis, which leads to dilation of the endoplasmic reticulum,
alteration of the mitochondria, swelling of the cell, and rupture of the
plasma membrane.

Biochemically, the hallmark of apoptosis is the degradation of the
nuclear genomic DNA into oligonucleosomal fragments (multiples of
approximately 180 basepairs) caused by a Ca^{++}/Mg^{++}-dependent endonuclease. DNA fragmentation precedes lysis and may be the critical event

leading to death. In support of this notion, aurintricarboxylic acid, a mixture of monomers and polymers of acidic hydroxytriphenylmethane dyes displaying a wide spectrum of inhibitory and toxic effects (Batistatou & Greene 1991; Bina-Stein & Tritton 1976; Gonzáles et al 1979; McConkey et al 1989), and Zn^{++} (Cohen & Duke 1984) prevent both DNA fragmentation and apoptotic death. DNA fragmentation occurs in cells displaying apoptotic nuclei (Appleby & Modak 1977; Umansky et al 1981). Morphology interpreted as similar to apoptotic nuclei has been produced by digestion of nuclei with micrococcal nuclease (Arends et al 1990). Thus, evidence exists for the correlation of DNA fragmentation, nuclear morphology, and death. However, the mechanism by which DNA fragmentation might lead to death is not understood. An element of this uncertainty relates to the short time interval between observed DNA fragmentation and the death of the cell. Clearly, DNA fragmentation cannot kill the cell merely by disabling RNA and protein synthesis (Clarke 1990). In cell types showing DNA fragmentation, other parameters have been often, but not uniformly, observed and are, therefore, less diagnostic for apoptotic death. These parameters include a moderate increase of $[Ca^{++}]_i$; a decrease of protein and RNA synthesis; exposure of hidden glycan groups, which allows recognition of the apoptotic cell by phagocytic cells; and the induction of transglutaminase (Arends & Wyllie 1991). In necrotic death, a depletion of ATP, which enables massive calcium influx, activates Ca^{++}-dependent phospholipases and leads to irreversible membrane disruption and lysis of the cell. Lysis causes the degradation of genomic DNA into random-size fragments. Extensive lists of cells dying of apoptosis can be found in recent reviews (Arends & Wyllie 1991; Clarke 1990).

Despite the existence of many cell types displaying apoptotic characteristics, apoptosis alone is insufficient to describe the wide varieties of physiologically appropriate cell death. Alternate schemes of classification have been proposed (Beaulaton & Lockshin 1982; Clarke 1990; Schwartz et al 1991; Schweichel & Merker 1973). DNA fragmentation is not observed in the death of intersegmental muscles of the moth *Manduca sexta* upon deprivation of 20-hydroxyecdysone (Schwartz et al 1991). Specifically regarding physiologically appropriate neuronal death, a greater diversity of morphologies has been described than is compatible with the current definition of apoptosis (Clarke 1990). Environmental conditions, as well as the mode of triggering death, appear to influence the morphology of dying neurons. For example, chick motoneurons and ciliary neurons dying during development and upon target removal show two morphological modes of death (Cho-Wang & Oppenheim 1978; Pilar & Landmesser 1976). The morphology of dying neurons of the avian isthmooptic nucleus changes upon the mode of deprivation of the presumed neurotrophic factor (Clarke 1982; Hornung et al 1989). The morphological characteristics of death of sympathetic neurons dying of NGF-deprivation differ between in vivo and in vitro situations (Levi-Montalcini et al 1969; Martin et al 1988; Wright et al 1983). However, in all these studies at least some of the neurons show a nuclear morphology reminiscent of apoptosis. This suggests a possible role for DNA fragmentation in neuronal death. Extensive damage to the nuclear DNA is observed upon endocytic death of neurons of the chick isthmooptic nucleus, where a large fraction of the nuclear DNA is found in autophagic vacuoles (Clarke & Hornung 1989). DNA fragmentation has been seen in some models of excitotoxin- and neurotoxin-induced death (Dipasquale et al 1991; Kure et al 1991), whereas other such studies failed to detect DNA fragmentation (Masters et al 1989). DNA fragmentation has also been suggested to occur upon NGF-deprivation of sympathetic neurons, based upon the observation that aurintricarboxylic acid maintains somal integrity (Batistatou & Greene 1991). Recent studies in sympathetic neurons have demonstrated that DNA fragmentation does occur after NGF deprivation (Edwards et al 1991; Deckwerth and Johnson, unpublished). The role of DNA fragmentation, if any, in the process of killing the neuron is unclear.

THE ROLE OF MACROMOLECULAR SYNTHESIS IN CELL DEATH

Continued RNA and protein synthesis is required in multiple forms of physiologically appropriate death. For example, death of palatal epithelial cells (Pratt & Greene 1976), the resorption of tadpole tails (Tata 1966), and the death of the intersegmental muscles in *Manduca sexta* upon decline of 20-hydroxyecdysone levels (Schwartz et al 1990a) are prevented by inhibition of macromolecular synthesis. A similar dependence is seen in the death of glucocorticoid-exposed thymocytes (Cohen & Duke 1984), a model for the physiologically appropriate death of immature thymocytes during negative selection. Such data are based upon the ability of general inhibitors of protein and RNA synthesis to prevent physiologically appropriate death before the inhibitors' action kills the cell. This phenomenon appears to be more easily demonstrable in neurons than in nonneuronal cells, because neurons tolerate longer periods of inhibition of macromolecular synthesis without irreversible damage (Martin et al 1988).

The requirement for macromolecular synthesis is subject to several possible explanations. Such mechanisms may require the involvement of a specific gene product conferring death to the cell. We have referred to such a putative product of a killer gene as "thanatin" (Johnson et al 1989). Without trying to be comprehensive, we envision five general models that might explain the ability of inhibitors to macromolecular synthesis to prevent neuronal death. The first three invoke the existence of one or perhaps several thanatins; the last two do not.

1. Thanatin is normally not expressed or is expressed at only a very low level in the cells. Upon the decision to die, thanatin is synthesized de novo, a process requiring macromolecular synthesis.

2. Thanatin is synthesized constitutively and is turned over rapidly, encoded by a labile mRNA species. The degradation of protein and mRNA is maintained by the trophic factor. Trophic factor deprivation leads to inhibition of degradation and to accumulation of thanatin. Blocking macromolecular synthesis also prevents accumulation of thanatin.

3. An inactive form of the thanatin is present constitutively and turned over very slowly; thus, it is not affected by inhibition of macromolecular synthesis. Trophic factor deprivation results in the activation of thanatin by a posttranslational mechanism via one or several gene products, which themselves are either newly synthesized (as in model 1) or stabilized (as in model 2), thus making the process of activation dependent upon macromolecular synthesis.

Remark: These three possibilities are not mutually exclusive. One might envision, particularly in the case of a postmitotic, irreplaceable neuron, that both macromolecular synthesis-dependent and posttranslational mechanisms exist to produce an active form of thanatin. The presence of a posttranslational switch would also allow for a rapid inactivation of an activated thanatin, thus enabling a last-minute abortion of the death program.

4. Upon trophic factor deprivation, a process involved in macromolecular synthesis is altered, which causes the generation of a potentially lethal by-product (not a specific gene product). Blocking macromolecular synthesis prevents the synthesis of this material, thereby preventing death.

5. Trophic factor deprivation leads to a general breakdown of cellular regulatory mechanisms. Inappropriate levels of many RNA and protein species are produced, which causes a general disorganization of the cell followed by loss of integrated function and death. Inhibiting macromolecular function prevents this disorganization by creating a state of "suspended animation," in which the levels of cellular macromolecular are locked in a state compatible with life for an extended period of time.

Several observations in the literature lead us to favor one of the first three possibilities that invoke a thanatin as a specific killer of the cell. Such a model is strongly suggested by the genetic evidence derived from the study of programed cell death in *C. elegans* (Ellis et al 1991). In addition, several specific mRNA species have been isolated from cells subjected to a manipulation leading to cell death. Selected examples include TRPM-2, originally isolated from prostate epithelium dying of androgen deprivation (Léger et al 1987) but later found to be expressed under a variety of degenerative conditions, and polyubiquitin expression in dying intersegmental muscles of *Manduca sexta* after eclosion (Schwartz et al 1990b). Other messages induced in dying prostate epithelium include *c-fos*, *c-myc*, and *hsp-70* (Buttyan et al 1988). However, no direct evidence exists that such genes are required for vertebrate cell death. In addition, we have found that these mRNA species (except for *hsp-70*, which was not investigated) are not increased in sympathetic neurons dying of NGF-deprivation (A. Ito and K. Horigome, unpublished). In apoptotic death, the Ca^{++}/Mg^{++}-dependent endonuclease may be a candidate for a thanatin. In thymocytes, this endonuclease turns over rapidly, consistent with model 2 (McConkey et al 1990). A combination of models 1 or 2 with model 3 would be consistent with the finding that there appears to be an interval during which sympathetic neurons deprived of NGF cannot be rescued from death within inhibitors of macromolecular synthesis, but NGF is still active as a saving agent (Edwards et al 1991).

PHARMACOLOGY OF TROPHIC FACTOR DEPRIVATION-INDUCED DEATH

In an effort to provide insight into molecular mechanisms of neuronal death and to develop means of manipulating the phenomenon in vitro and in vivo, many drugs have been examined for their ability to inhibit neuronal death induced by trophic factor deprivation. In addition to the inhibitors of macromolecular synthesis described above, only a few agents have been found that prevent death caused by trophic factor removal (reviewed in Martin & Johnson 1991), and, in all cases, the mechanistic implications of the pharmacological data are uncertain. The first agent shown to prevent neuronal death in a well-defined trophic factor-deprivation paradigm was potassium ion (K^+), acting by inducing depolarization (Wakade et al 1983). Indeed, it has been known for more than two decades (Scott & Fisher 1970) that elevated K^+ enhances the survival of a variety of peripheral and central nervous system neurons in vitro. This fact is often used empirically to maintain neurons in culture. At least in sympathetic neurons, K^+ is the most effective "saver" in a variety of agents that act by elevating intracellular Ca^{++} level (see below). Although phorbol esters, presumably acting by activation of protein kinase C, maintain chick embryo sympathetic neurons in vitro (Wakade et al 1988), similar experiments conducted with rat sympathetic neurons fail to demonstrate survival (Martin & Johnson 1991). A variety of agents that mimic or elevate cAMP in sympathetic neurons prevent neuronal death after NGF removal (Edwards et al 1991; Rydel & Greene 1988). As mentioned above, aurintricarboxylic acid also prevents death of sympathetic neurons in vitro (Batistatou & Greene 1991). The precise mechanism by which these agents exert this pharmacological effect is not clear, nor is there any direct evidence that activation of protein kinase C or protein kinase A is immediately involved in the mechanism by which trophic factors suppress death.

Other than inhibitors of macromolecular synthesis, the only agent that prevents trophic factor deprivation-induced neuronal death in vitro that has been extrapolated to the in vivo situation is flunarizine (Rich & Hollowell 1990). This agent, generally classified as a diphenylalkylamine Ca^{++} channel antagonist, enhances survival of dissociated dorsal root ganglion (DRG) neurons deprived of NGF in vitro. In an analogous in vivo paradigm, administration of flunarizine to postnatal rats significantly reduced neuronal death of DRG sensory neurons after sciatic nerve lesion. The mechanism by which flunarizine acts to maintain neuronal survival is not known. This agent exerts many pharmacological actions in addition to Ca^{++} channel blockade. Ca^{++} channel blockade appears not to be the operative mechanism, because the concentration of flunarizine required to exert survival-promoting effects in vitro is well above those required for Ca^{++} channel blockade. And, Ca^{++} channel blockers belonging to other chemical classes exert no survival-promoting activity (Rich & Hollowell 1990). Irrespective of the mechanism of action of flunarizine, it is encouraging that a survival-promoting agent detected in an in vitro paradigm acts similarly in neuronal injury models in vivo, where the physiological trophic factor (i.e. NGF) also exerts a saving effect (Yip et al 1984). Such demonstrations provide further impetus to discover drugs that mimic the survival-promoting effect of trophic factor.

ENHANCEMENT OF NEURONAL SURVIVAL BY ELEVATING INTRACELLULAR CALCIUM

Depolarization induced by elevation of extracellular K^+ promotes neuronal survival (Scott & Fisher 1970). This effect is mediated by an influx of extracellular Ca^{++} ion via dihydropyridine-sensitive Ca^{++} channels in several cell types. This conclusion is based upon the ability of K^+-induced cell rescue to be prevented by dihydropyridine Ca^{++} channel blockers and intracellular Ca^{++} chelators. Additionally, Ca^{++} channel agonists augment K^+ rescue (Collins & Lile 1989; Gallo et al 1987; Koike et al 1989). A "Ca^{++} set-point hypothesis" of neuronal trophic factor dependence (Koike et al 1989) posits that the concentration of free cytosolic Ca^{++} ($[Ca^{++}]_i$) or some specific kinetic Ca^{++} pool, which generally correlates well with intracellular free Ca^{++}, determines the degree to which neurons require trophic factor to suppress the mechanism responsible for neuronal death. This hypothesis makes certain predictions that may explain several important phenomena associated with neuronal death during development (Johnson et al 1992). For example, it predicts that maintaining neurons in the presence of elevated K^+, which saves neurons independent of neurotrophic factor, will result in a sustained elevation of $[Ca^{++}]_i$. A sustained two- to threefold elevation of $[Ca^{++}]_i$ is found for days in sympathetic (Koike & Tanaka 1991) and ciliary (Collins et al 1991) neurons. Neuronal survival generally correlates with $[Ca^{++}]_i$. Similarly, other drugs that cause sustained elevations of $[Ca^{++}]_i$, such as thapsigargin (Thastrup et al 1990), also promote survival (P. A. Lampe, unpublished). Thus, the data available to date are consistent with the idea that modest elevations of $[Ca^{++}]_i$ enhance neuronal survival and decrease trophic factor dependence. That such elevations in $[Ca^{++}]_i$ enhance neuronal survival may be related to important developmental observations. Blocking afferent electrical input by lesioning or pharmacological interruption increases naturally occurring neuronal death in several situations (reviewed in Oppenheim 1991). Afferent electrical input leads to episodic depolarization of the neurons and causes at least transient elevations in $[Ca^{++}]_i$ that are predicted to reduce the dependence upon target-derived trophic factor; conversely, decreasing afferent activity results in reduced $[Ca^{++}]_i$, increased trophic factor dependence, and, ultimately, a greater likelihood that a neuron will fail to gain access to sufficient trophic factor, thus endangering its survival. Such a model is consistent with the observation that episodic K^+ depolarization of as little as ten minutes per day is sufficient to enhance rat myenteric neuron survival in vitro (Thigpen et al 1989).

The apparent inverse relationship between $[Ca^{++}]_i$ and trophic factor dependence may explain the decrease in trophic factor dependence observed both in vitro and in vivo as neurons mature. During development, as targets become innervated by a surplus of neurons, these neurons must be acutely dependent for survival upon target-derived trophic factor to allow the innervating neuronal population to match the target in size. However, once cell death has determined the final number of postmitotic, irreplaceable neurons, the neurons must be brought back from this brink of acute trophic factor dependence; otherwise, the mature nervous system would be a fragile apparatus. Koike et al (1989), and Johnson et al (1992) have suggested that a developmentally associated increase in $[Ca^{++}]_i$ may underlie the decrease in trophic factor dependence associated with maturation. Such a prediction is not readily testable in vivo. However, a similar decrease in NGF dependence is seen in sympathetic and sensory neurons in vitro. In both types of neurons, $[Ca^{++}]_i$ rise two- to threefold from acutely dependent younger cells to mature trophic factor-independent cells; the time course of $[Ca^{++}]_i$ correlates inversely with the time course of trophic factor dependence, as predicted by the Ca^{++}-set point hypothesis (Eichler et al 1992; Koike & Tanaka 1991). Future work will be required to determine the generality of these observations and whether the critical parameter in these phenomena is simply cytosolic free Ca^{++} or some other related parameter that correlates well with $[Ca^{++}]_i$.

SUMMARY AND CONCLUSION

Data derived from several experimental approaches demonstrate that naturally occurring neuronal death during development has many parallels with the physiologically appropriate death seen in nonneuronal cells. Physiologically appropriate death in different cell types may share some common mechanisms. These general notions must remain vague and tentative, because details of the mechanisms by which cells die in response to physiological positive or negative signals are poorly understood in any cell type. Current thinking focuses on the idea that cells possess a mechanism, which involves specific gene products, that are designed to kill the cell in response to appropriate physiological signals.

Genetic studies of cell death in *C. elegans* and the demonstrations of increased expression of specific genes temporally associated with death in nonneuronal cells are consistent with this view. However, in the latter studies, there is no direct evidence that such temporally related genes are critical to the process of cell death or whether such gene expression may be related to some other aspect of the response to the hormonal manipulations that produce the death of the cell under study. Therefore, the mechanism of death of any cell type is not understood, and whether neuronal death during development or after experimental manipulation results from the same mechanism is unknown.

Several approaches are currently being pursued in a number of laboratories to address this general problem. These include pharmacological studies, such as described above, and studies aimed at analyzing biochemical and morphological changes associated with death. Attempts to find mRNAs or proteins whose increased expression is associated with neuronal death can be addressed by subtractive and differential hybridization strategies, by two-dimensional protein gel electrophoresis, and by examining genes whose increased expression is temporally correlated with cell death. Success in these various strategies will provide an understanding of neuronal death and relate to it cell death in other cell types.

If future work provides direct evidence for a genetic program acting physiologically to produce death in the developing nervous system, an obvious question becomes the possible role that loss of transcriptional control of such a program plays in the adult in responses to mechanical or chemical trauma, neurodegenerative disease, or neuronal attrition associated with aging. Studies addressing the basic developmental process of trophic factor deprivation-induced death should provide molecular markers of and pharmacological approaches to these pathological processes in the adult.

ACKNOWLEDGMENTS

The authors thank Drs. E. B. Cornbrooks, S. Estus, J. L. Franklin, and R. S. Freeman for careful reading of and valuable comments on the manuscript, as well as Ms. P. A. Osborne and Ms. C. Pupillo for excellent editorial assistance in preparing this review. Work cited from our own laboratory was supported by National Institutes of Health grant NS-24679, the Washington University Alzheimer Disease Research Center AG05861, and the American Paralysis Association.

Literature Cited

Appleby, D. W., Modak, S. P. 1977. DNA degradation in terminally differentiating lens fiber cells from chick embryos. *Proc. Natl. Acad. Sci. USA* 74: 5579–83

Arends, M. J., Morris, R. G., Wyllie, A. H. 1990. Apoptosis. The role of the endonuclease. *Am. J. Pathol.* 136: 593–608

Arends, M. J., Wyllie, A. H. 1991. Apoptosis: Mechanisms and roles in pathology. *Int. Rev. Exp. Pathol.* 32: 223–54

Barde, Y.-A. 1989. Trophic factors and neuronal survival. *Neuron* 2: 1525–34

Batistatou, A., Greene, L. A. 1991. Aurintricarboxylic acid rescues PC12 cells and sympathetic neurons from cell death caused by nerve growth factor deprivation: Correlation with suppression of endonuclease activity. *J. Cell Biol.* 115: 461–71

Beaulaton, J., Lockshin, R. A. 1982. The relation of programmed cell death to development and reproduction: Comparative studies and an attempt at classification. *Int. Rev. Cytol.* 79: 215–35

Bina-Stein, M., Tritton, T. R. 1976. Aurintricarboxylic acid is a nonspecific enzyme inhibitor. *Mol. Pharmacol.* 12: 191–93

Bothwell, M. 1991. Keeping track of neurotrophin receptors. *Cell* 65: 915–18

Buttyan, R., Zakeri, Z., Lockshin, R., Wolgemuth, D. 1988. Cascade induction of c-*fos*, c-*myc*, and heat shock 70 K transcripts during regression of the rat ventral prostate gland. *Mol. Endocrinol.* 2: 650–57

Chu-Wang, I.-W., Oppenheim, R. W. 1978. Cell death of motoneurons in the chick embryo spinal cord. I. A light and electron microscopic study ·of naturally occurring and induced cell loss during development. *J. Comp. Neurol.* 177: 33–58

Clarke, P. G. H. 1990. Developmental cell death: Morphological diversity and multiple mechanisms. *Anat. Embryol.* 181: 195–213

Clarke, P. G. H. 1982. Labelling of dying neurones by peroxidase injected intravascularly in chick embryos. *Neurosci. Lett.* 30: 223–28

Clarke, P. G. H., Hornung, J. P. 1989. Changes in the nuclei of dying neurons as studied with thymidine autoradiography. *J. Comp. Neurol.* 283: 438–49

Cohen, J. J., Duke, R. C. 1984. Glucocorticoid activation of a calcium-dependent endonuclease in thymocyte nuclei leads to cell death. *J. Immunol.* 132: 38–42

Collins, F., Lile, J. D. 1989. The role of dihydropyridine-sensitive voltage-gated calcium channels in potassium-mediated neuronal survival. *Brain Res.* 502: 99–108

Collins, F., Schmidt, M. F., Guthrie, P. B., Kater, S. B. 1991. Sustained increase in intracellular calcium promotes neuronal survival. *J. Neurosci.* 11: 2582–87

Dispasquale, B., Marini, A. M., Youle, R. J. 1991. Apoptosis and DNA fragmentation induced by 1-methyl-4-phenylpyridinium in neurons. *Biochem. Biophys. Res. Commun.* 181: 1442–48

Duvall, E., Wyllie, A. H. 1986. Death and the cell. *Immunol. Today* 7: 115–19

Edwards, S. N., Buckmaster, A. E., Tolkovsky, A. M. 1991. The death programme in cultured sympathetic neurones can be suppressed at the posttranslational level by nerve growth factor, cyclic AMP, and depolarization. *J. Neurochem.* 57: 2140–43

Eichler, M. E., Dubinsky, J. M., Rich, K. M. 1992. Relationship of intracellular calcium to dependence on nerve growth factor in dorsal root ganglion neurons in cell culture. *J. Neurochem.* 58: 263–69

Ellis, H. M., Horvitz, H. R. 1986. Genetic control of programmed cell death in the nematode *C. elegans. Cell* 44: 817–29

Ellis, R. E., Yuan, J., Horvitz, H. R. 1991. Mechanisms and functions of cell death. *Annu. Rev. Cell Biol.* 7: 663–98

Gallo, V., Kinsbury, A., Balázs, R., Jørgensen, R. O. 1987. The role of depolarization in the survival and differentiation of cerebellar granule cells in culture. *J. Neurosci.* 7: 2203–13

Gonzáles, R. G., Blackburn, B. J., Schleich, T. 1979. Fractionation and structural elucidation of the active components of aurintricarboxylic acid, a potent inhibitor of protein nucleic acid interactions. *Biochim. Biophys. Acta* 562: 534–45

Gorin, P. D., Johnson, E. M. 1980. Effects of long-term nerve growth factor deprivation on the nervous system of the adult rat: an experimental approach. *Brain Res.* 198: 27–42

Gorin, P. D., Johnson, E. M. 1979. Experimental autoimmune model of nerve growth factor deprivation: Effect on developing peripheral sympathetic and sensory neurons. *Proc. Natl. Acad. Sci. USA* 76: 5382–86

Hamburger, V., Brunso-Bechthold, J. K., Yip, J. 1981. Neuronal death in the spinal ganglia of the chick embryo and its reduction by nerve growth factor. *J. Neurosci.* 1: 60–71

Hendry, I. A., Campbell, J. 1976. Morphometric analysis of rat superior cervical ganglion after axotomy and nerve growth factor treatment. *J. Neurocytol.* 5: 351–60

Hornung, J. P., Koppel, H., Clarke, P. G. H. 1989. Endocytosis and autophagy in dying neurons: An ultrastructural study in chick embryos. *J. Comp. Neurol.* 283: 425–37

Johnson, E. M. Jr., Chang, J. Y., Koike, T., Martin, D. P. 1989. Why do neurons die when deprived of trophic factor? *Neurobiol. Aging* 10: 549–52

Johnson, E. M. Jr., Koike, T., Franklin, J. 1992. A "calcium set-point hypothesis" of neuronal dependence on neurotrophic factor. *Exp. Neurol.* 115: 163–66

Koike, T., Martin, D. P., Johnson, E. M. Jr. 1989. Role of Ca^{2+} channels in the ability of membrane depolarization to prevent neuronal death induced by trophic-factor deprivation: Evidence that levels of internal Ca^{2+} determine nerve growth factor dependence of sympathetic ganglion cells. *Proc. Natl. Acad. Sci. USA* 86: 6421–25

Koike, T., Tanaka, S. 1991. Evidence that nerve growth factor dependence of sympathetic neurons for survival in vitro may be determined by levels of cytoplasmic free Ca^{2+}. *Proc. Natl. Acad. Sci. USA* 88: 3892–96

Kure, S., Tominaga, T., Yoshimoto, T., Tada, K., Narisawa, K. 1991. Glutamate triggers internucleosomal DNA cleavage in neuronal cells. *Biochem. Biophys. Res. Commun.* 179: 39–45

Léger, J. G., Montpetit, M. L., Tenniswood, M. P. 1987. Characterization and cloning of androgen-repressed mRNAs from rat ventral prostate. *Biochem. Biophys. Res. Commun.* 147: 196–203

Levi-Montalcini, R., Angeletti, P. U. 1966. Immunosympathectomy. *Pharmacol. Rev.* 18: 619–28

Levi-Montalcini, R., Booker, B. 1960. Destruction of the sympathetic ganglia in mammals by an antiserum to the nerve-growth promoting factor. *Proc. Natl. Acad. Sci. USA* 46: 384–91

Levi-Montalcini, R., Caramia, F., Angeletti, P. U. 1969. Alterations in the fine structure of nucleoli in sympathetic neurons following NGF-antiserum treatment. *Brain Res.* 12: 54–73

Martin, D. P., Johnson, E. M. Jr. 1991. Programmed cell death in the peripheral nervous system. In *Apoptosis: The Molecular Basis of Cell Death*, ed. L. D. Tomei, F. O. Cope, pp. 247–61. Cold Spring Harbor, NY: Cold Spring Harbor Lab. Press

Martin, D. P., Schmidt, R. E., DiStefano, P. S., Lowry, O. H., Carter, J. G., Johnson, E. M. Jr. 1988. Inhibitors of protein synthesis and RNA synthesis prevent neuronal death caused by nerve growth factor deprivation. *J. Cell Biol.* 106: 829–44

Masters, J. N., Finch, C. E., Sapolsky, R. M. 1989. Glucocorticoid endangerment of hippocampal neurons does not involve deoxyribonucleic acid cleavage. *Endocrinology* 124: 3083–88

McConkey, D. J., Hartzell, P., Nicotera, P., Orrenius, S. 1989. Calcium-activated DNA fragmentation kills immature thymocytes. *FASEB J.* 3: 1843–49

McConkey, D. J., Hartzell, P., Orrenius, S. 1990. Rapid turnover of endogenous endonuclease activity in thymocytes: Effects of inhibitors of macromolecular synthesis. *Arch. Biochem. Biophys.* 278: 284–87

Mesner, P. W., Winters, T. R., Green, S. H. 1991. PC12 and sympathetic neuronal apoptosis do not involve DNA fragmentation. *Soc. Neurosci. Abstr.* 21: 1124

Oppenheim, R. W. 1991. Cell death during development of the nervous system. *Annu. Rev. Neurosci.* 14: 453–501

Oppenheim, R. W., Prevette, D., Tytell, M., Homma, S. 1990. Naturally occurring and induced cell death in the chick embryo in vivo requires protein and RNA synthesis: Evidence for the role of cell death genes. *Dev. Biol.* 138: 104–13

Pilar, G., Landmesser, L. 1976. Ultrastructural differences during embryonic cell death in normal and peripherally deprived ciliary ganglia. *J. Cell Biol.* 68: 339–56

Pratt, R. M., Greene, R. M. 1976. Inhibition of palatal epithelial cell death by altered protein synthesis. *Dev. Biol.* 54: 135–45

Rich, K. M., Hollowell, J. P. 1990. Flunarizine protects neurons from death after axotomy or NGF deprivation. *Science* 248: 1419–21

Rydel, R. E., Greene, L. A. 1988. cAMP analogs promote survival and neurite outgrowth in cultures of rat sympathetic and sensory neurons independently of nerve growth factor. *Proc. Natl. Acad. Sci. USA* 85: 1257–61

Schwartz, L. M., Kosz, L., Kay, B. K. 1990a. Gene activation is required for developmentally programmed cell death. *Proc. Natl. Acad. Sci. USA* 87: 6594–98

Schwartz, L. M., Myer, A., Kosz, L., Engelstein, M., Maier, C. 1990b. Activation of polyubiquitin gene expression during developmentally programmed cell death. *Neuron* 5: 411–19

Schwartz, L. M., Smith, S., Jones, M. E., Osborne, B. A. 1991. Two distinct molecular mechanisms mediate programmed cell death. *Soc. Neurosci. Abstr.* 21: 228

Schweichel, J. U., Merker, H. J. 1973. The morphology of various types of cell death in prenatal tissues. *Teratology* 7: 253–66

Scott, B. S., Fisher, K. C. 1970. Potassium concentration and number of neurons in cultures of dissociated ganglia. *Exp. Neurol.* 27: 16–22

Scott, S. A., Davies, A. M. 1990. Inhibition of protein synthesis prevents cell death in sensory and parasympathetic neurons deprived of neurotrophic factor in vitro. *J. Neurobiol.* 21: 630–38

Snider, W. D., Johnson, E. M. Jr. 1989. Neurotrophic molecules. *Ann. Neurol.* 26: 489–506

Tata, J. R. 1966. Requirement for RNA and protein synthesis for induced regression of the tadpole tail in organ culture. *Dev. Biol.* 13: 77–94

Thastrup, O., Cullen, P. J., Drøbak, B. K., Hanley, M. R., Dawson, A. P. 1990. Thapsigargin, a tumor promotor, discharges intracellular Ca^{2+} stores by specific inhibition of the endoplasmic reticulum Ca^{2+}-ATPase. *Proc. Natl. Acad. Sci. USA* 87: 2466–70

Thigpen, J. C., Franklin, J. L., Willard, A. L. 1989. Calcium-dependent effects of acute potassium depolarization on survival of rat myenteric neurons in culture. *Soc. Neurosci. Abstr.* 15: 438

Umansky, S. R., Korol, B. A., Nelipovich, P. A. 1981. In vivo DNA degradation in thymocytes of γ-irradiated or hydrocortisone-treated rats. *Biochim. Biophys. Acta* 655: 9–17

Wakade, A. R., Edgar, D. A., Thoenen, H. 1983. Both nerve growth factor and high K^+ concentrations support the survival of chick embryo sympathetic neurons. *Exp. Cell Res.* 144: 377–84

Wakade, A. R., Wakade, T. D., Malhorta, R. K., Bhave, S. V. 1988. Excess K^+ and phorbol ester activate protein kinase C and support the survival of chick sympathetic neurons in culture. *J. Neurochem.* 451: 975–83

Wright, L. L., Cunningham, T. J., Smolen, A. J. 1983. Developmental neuronal death in the rat superior cervical sympathetic ganglion: Cell counts and ultrastructure. *J. Neurocytol.* 12: 727–38

Yip, H. K., Rich, K. M., Lampe, P. A., Johnson, E. M. Jr. 1984. The effects of nerve growth factor and its antiserum on the postnatal development and survival after injury of sensory neurons in rat dorsal root ganglia. *J. Neurosci.* 4: 2986–92

Yuan, J., Horvitz, H. R. 1990. The *Caenorhabditis elegans* genes ced-3 and ced-4 act cell autonomously to cause programmed cell death. *Dev. Biol.* 138: 33–41

Annu. Rev. Neurosci. 1993. 16:47–71

GENETIC AND CELLULAR ANALYSIS OF BEHAVIOR IN *C. ELEGANS*

Cornelia I. Bargmann

Department of Anatomy, Programs in Developmental Biology, Genetics, and Neuroscience, University of California, San Francisco, California 94143-0452

KEY WORDS: neuronal development, axon guidance, mechanosensation, neurogenetics, olfaction

INTRODUCTION

Behavior arises through the interplay of innate properties of the nervous system, environmental stimuli, and experience. An opportunity to integrate neuronal and genetic approaches to study behavior is provided by the soil nematode *Caenorhabditis elegans*. *C. elegans* is attractive for study because of the simplicity and accessibility of its nervous system. The adult hermaphrodite is 1 mm long, and its nervous system is composed of only 302 neurons (White et al 1986). The nucleus of each neuron can be identified in live animals by differential interference microscopy (Brenner 1974), and the cell lineage that gives rise to each of these neurons has been described in its entirety (Sulston & Horvitz 1977; Sulston et al 1983). *C. elegans* develops to adulthood in about three days at 25°C, which facilitates observation of its development and genetic analysis (Brenner 1974).

Despite its small size, the *C. elegans* nervous system generates and regulates many behaviors. Mechanosensory neurons mediate an escape response to light touch (Chalfie & Sulston 1981). *C. elegans* chemotaxes to attractive chemicals, avoids repellent chemicals, or thermotaxes to the temperature at which it was raised (Dusenbery 1974; Hedgecock & Russell 1975; Ward 1973). Male worms locate and mate with hermaphrodites (Hodgkin 1983; Sulston & Horvitz 1977). Food availability regulates feed-

47

0147–006X/93/0301–0047$02.00

ing, defecation, and egg-laying (Avery & Horvitz 1990; Horvitz et al 1982; Thomas 1990).

Neurons that generate many of these behaviors have been identified and characterized. The defined neuronal circuits in *C. elegans* are compact: Egg-laying requires only one pair of motor neurons; five neurons sense light touch; and a single motor neuron is required for feeding. This information can be used to characterize genes that affect those neurons. The nature of these genes has shed light on the genetic and molecular mechanisms that operate in nervous system development and function. Chalfie & White (1988) have provided a detailed review of the *C. elegans* nervous system, and Hedgecock et al (1987) have reviewed its neural development.

THE STRUCTURE OF THE *C. ELEGANS* NERVOUS SYSTEM

The anatomy of the *C. elegans* nervous system has been described in great detail through the work of White and colleagues, who examined the entire *C. elegans* nervous system in serial section electron micrographs of individual worms (Albertson & Thomson 1976; White et al 1976, 1986). From these micrographs, the position and morphology of each neuron have been deduced. In addition, putative synapses were defined, based on ultrastructural specializations of neurons that are characteristic of synapses in other organisms. The total wiring diagram was deduced to contain about 5000 chemical synapses, 600 gap junctions, and 2000 neuromuscular junctions.

By comparing morphologies, dendritic specializations, and connectivities of neurons, the 302 *C. elegans* neurons could be subdivided into 118 classes of neurons, with 1 to 13 neurons in each class. Based on their structures, about half of the classes of neurons are interneurons, about one third are sensory neurons, and about one quarter are motor neurons (a few neurons are classified as mixed, such as sensory/motor neurons or interneurons/motor neurons). Most sensory neurons and interneurons belong to bilaterally symmetric pairs, with one member of each pair on the left and right sides of the animal. Motor neurons in the body are arranged in repeating groups along the anterior to posterior axis, whereas motor neurons in the head have a four- or sixfold radial symmetry around the nose. About three quarters of the classes of neurons send processes into the nerve ring, a large neuropil that circles the pharynx in the head region.

In general, *C. elegans* neurons are simple in structure. Most neurons have one or two unbranched processes (although some have more complex process morphologies). Most sensory neurons have cilia or other sensory structures on specialized dendrites and synapses on separate axonal processes. Nonneuronal support cells are associated with many sensory structures. Most motor neurons have defined presynaptic and postsynaptic regions. In an unusual arrangement, the muscles send arms to meet the motor neurons within the nerve cords. Interneurons generally make and receive synapses en passant within the nerve bundles, without clear axonic or dendritic specializations.

Inspection of the *C. elegans* neuroanatomy suggests that different neurons have intrinsic properties that mediate adhesion or association among neurons. Within the nerve ring and in other process bundles, the position of a particular process is reproducible: A given process adheres to the same partners for much of its length. Cell adhesion molecules expressed on subsets of axons might mediate the bundling (fasciculation) of processes, as occurs in insects and vertebrates (Grenningloh et al 1990; Jessell 1988). A given neuron has synaptic connections with only about 15% of the neurons it contacts, so some information in addition to process adjacency directs synapse formation between two cells (Durbin 1987; White et al 1986).

Several classes of *C. elegans* neurons position the axons of other neurons. These include the AVG neurons, which pioneer the ventral nerve cord that runs the length of the worm (Durbin 1987); the PVPR neuron, which pioneers the left subcord of the ventral nerve cord (Durbin 1987); the PVQ neurons, which pioneer the lumbar ganglion in the tail (Durbin 1987); and the BDU neurons, which guide the AVM sensory neuron into the nerve ring (Walthall & Chalfie 1988). In each case, killing the pioneer or guide neuron causes disorganization or failure of the following processes. These pioneer neurons persist throughout life, but as yet they are not known to have any function in the adult. Perhaps, they are similar to the neurons in vertebrates and insects that function transiently to guide development of subsequent neurons (Kutsch & Bentley 1987; McConnell et al 1989).

Electron micrographs of at least two sectioned animals were examined for most of the nervous system. The synapses made by homologous neurons in the two two animals were highly reproducible, but not identical (Durbin 1987; White et al 1986). For any synapse between two neurons in one animal, there was a 75% chance that a similar synapse would be found in the second animal (Durbin 1987). If two neurons were connected by more than two synapses, the chances that they would be interconnected in the other animal increased greatly (92% identity). Most synapses fall into this highly reproducible set.

The wiring diagram constructed through these heroic efforts has inspired and aided in the interpretation of almost all *C. elegans* neurobiological experiments. However, neuronal functions cannot yet be predicted purely from the neuroanatomy. The electron micrographs do not indicate whether a synapse is excitatory, inhibitory, or modulatory. Nor do the morphologically defined synapses necessarily represent the complete set of physiologically relevant neuronal connections in this highly compact nervous system. Traditionally, electrophysiological recording from neurons would be used to determine their connectivity. Because of the small size and close packing of *C. elegans* cells, the electrophysiological analysis of neuronal properties is just starting.

Thus, the neuroanatomy suggests roles of neurons in circuits and behaviors, but is a structure in search of a function, rather than an end to *C. elegans* neurobiology. The challenge represented by the wiring diagram is to integrate it with other information to understand how the nervous system develops and how neurons act together to generate coherent behaviors. These problems have been attacked by using laser ablations, genetic analysis, pharmacology, and behavioral analysis.

THE NEURONAL BASIS OF BEHAVIORS IN *C. ELEGANS*

Movement

C. elegans moves forward or backward across solid substrates (in the laboratory, on an agar surface) by making sinusoidal body waves. The animal lies on its side, so that dorsal-ventral head bends initiate forward movement. Surface tension from a water film over the animal presses it into a groove in the agar. Muscle contraction in an undulatory pattern maintains most of the worm's body within the groove and propels it longitudinally either forward or backward (Niebur & Erdos 1991). In addition, turning or reversal movements can be made from the head or anterior body region. The body moves only within the plane of the agar, whereas the head can generate turns and move up, down, or from side to side in more complex movements (Croll 1975a).

Most of the *C. elegans* nervous system is present at hatching, but the motor system in the body changes substantially during postembryonic development. Four of the seven classes of motor neurons in the ventral nerve cord are born after hatching (Sulston & Horvitz 1977), and one of the embryonic classes of motor neurons alters its synaptic connectivity during the first postembryonic larval stage (White et al 1978). Nonetheless, the overall sinusoidal pattern of worm movement remains similar through development.

ELECTROPHYSIOLOGY OF MOVEMENT IN *ASCARIS*   The number and structure of neurons in the parasitic nematode *Ascaris lumbricoides* are remarkably similar to those in *C. elegans* (Angstadt et al 1989; Stretton et al 1985). *Ascaris* grows to 15–20 cm long as an adult, and its large neurons make it accessible to electrophysiological and biochemical analysis of single neurons. The *Ascaris* motor neurons that have been examined lack classical all-or-none action potentials, so electrical activity is transmitted through the neurons passively (Davis & Stretton 1989a,b).

The ventral nerve cord of *Ascaris* contains the motor neurons that drive movement and the processes of the interneurons that regulate the motor neurons. There are seven classes of *Ascaris* motor neurons, each of which corresponds in structure to a class of *C. elegans* motor neuron (Stretton et al 1978). Each class of neuron innervates either dorsal or ventral muscles. Five classes of motor neurons excite the body wall muscles and probably contain acetylcholine (del Castillo et al 1963; Johnson & Stretton 1985; Walrond et al 1985), and two classes of motor neurons are inhibitory and contain the neurotransmitter GABA (del Castillo et al 1964; Johnson & Stretton 1987). The inhibitory neurons that innervate the ventral side receive innervation from the dorsal excitatory motor neurons, and dorsal inhibitory neurons receive innervation from the ventral excitatory motor neurons (Walrond & Stretton 1985). These GABAergic neurons probably act as cross-inhibitors that ensure that the dorsal side of the animal is relaxed when the ventral side is contracted and vice versa, which is important for sinusoidal movement. Connections between other classes of *Ascaris* motor neurons are also described (Stretton et al 1985). These connections are probably part of a system by which sinusoidal movement is propagated through interactions between stimulated motor neurons and muscles.

LASER KILLING OF MOTOR NEURONS AND INTERNEURONS   One way to infer the function of a neuron is to remove it from a circuit and examine the behavior of the resulting animal. This sort of experiment is feasible in *C. elegans*, because of the transparency of the nematode and the reproducibility of cell lineage and cell position in different animals. A laser microbeam is focused through the objective of a microscope onto the nucleus of a single cell, which is then killed by photoablation (Avery & Horvitz 1987; Chalfie & Sulston 1981; Sulston & White 1980). This technique has been used extensively to probe interactions between cells in *C. elegans* development (e.g. Kimble 1981; Sternberg & Horvitz 1986; Sulston & White 1980). When a sufficiently accurate laser is used, this technique can also be used to kill single neurons. In several cases, both functional and structural assays have confirmed loss of function for neurons killed in young animals (Avery & Horvitz 1989; Bargmann & Horvitz 1991a; Chalfie & Sulston 1981). However, if nuclei are killed in later larval stages, the functions of some cells can apparently persist (Avery & Horvitz 1987; Chalfie & Sulston 1981).

Some limitations apply to this experimental approach. Neurons with subtle effects on a behavior may be overlooked. Further, under some circumstances in development, killing a cell that normally takes some function allows a second cell to take its place. This sort of regulation is probably rare among mature neurons, but could be more common early in development. In addition, the requirement for neurons in some behavior need not be direct. For example, some neurons could produce factors necessary for the survival of other neurons, or instruct other neurons' development.

Ablation experiments were used to define the roles of neurons in movement. Surprisingly, forward and backward movement are largely independently controlled at the neuronal level. Throughout the body, one system of neurons controls forward movement in parallel with a similar system that controls backward movement. Four interneurons appear to coordinate movement by activating sets of motor neurons throughout the ventral nerve cord (Chalfie et al 1985). The interneurons AVB and PVC drive forward movement: Killing AVB and PVC results in animals that fail to generate forward waves of movement. The interneurons AVA and AVD drive backward movement. Killing AVA and AVD results in animals that cannot back. AVA, AVB, PVC, and AVD have the largest axon diameters in the ventral nerve cord and are reminiscent of the giant interneurons that drive movement in other invertebrates (Daley & Camhi 1988).

The motor neurons required for forward or backward movement are also distinct. Only three classes of motor neurons are present in the body of young animals. One class of neurons (DB-type) is required for coordinated forward movement, and one (DA-type) for backward movement (Chalfie et al 1985). These neurons correspond to excitatory motor neurons in *Ascaris* (Johnson & Stretton 1985; Walrond et al 1985). The third class of neurons (DD-type) is essential for normal forward and backward movement (Chalfie et al 1985). The DD-like neurons in *Ascaris* are inhibitory and may coordinate sinusoidal movement (Johnson & Stretton 1987; Walrond et al 1985).

These and other experiments suggest that two motor neuron classes (DB and VB) direct forward movement, two or three classes (DA, VA, and perhaps AS) direct backward movement, and two classes (DD and VD) coordinate sinusoidal movement on the dorsal and ventral sides of the animal for both forward and backward movement.

GENES REQUIRED FOR COORDINATED MOVEMENT   *C. elegans* requires little in the way of neuronal function for viability in the laboratory, so mutants with widespread defects in the nervous system can be propagated as fertile strains (Avery & Horvitz 1989). Many such mutants fell among the first *C. elegans* mutants isolated, including the uncoordinated, or *unc* mutants (Brenner 1974). 120 *unc* genes have been defined, and mutations in many other genes also cause defective movement (Edgley & Riddle 1990). Movement requires the function of muscles, as well as multiple neuronal types, so the cause and site of action of mutant defects are not necessarily obvious. However, various techniques permit *unc* genes to be assigned to particular functions.

Muscle uncs   The muscle structure of *C. elegans* can be visualized in live animals by polarized light or in electron micrographs of fixed animals. More than 25 *unc* genes encode structural components of the muscles, including myosin, actin, and numerous accessory proteins [reviewed in Waterston (1988) and Epstein (1990)]. The functions of additional *unc* genes have been mapped to the muscles by using genetic mosaic analysis, in which the behavior of animals composed of known mixtures of wild-type and mutant cells is examined (Herman 1984, 1989).

Neurotransmitter and neurotransmitter receptor genes   Like vertebrate and *Ascaris* neuromuscular junctions, neuromuscular junctions in *C. elegans* appear to utilize acetylcholine as an excitatory neurotransmitter and GABA as an inhibitory transmitter. Enzymatic assays have demonstrated the presence of choline acetyltransferase (ChAT), which synthesizes acetylcholine (Rand & Russell 1985), and acetylcholinesterase, which degrades it (Johnson & Russell 1983) in *C. elegans*. The analysis of mutants with decreased levels of these enzymes indicates that normal acetylcholine levels are important for normal movement. Thus, mutations (in the *cha-1* gene) that reduce levels of ChAT result in uncoordinated movement (Rand & Russell 1984); mutations that eliminate ChAT entirely are lethal (Rand 1989). Mutations that decrease acetylcholinesterase also result in uncoordinated movement; elimination of acetylcholinesterase is lethal (Johnson et al 1981, 1988).

GABA-like immunoreactivity has been described in *C. elegans* motor neurons (McIntire et al 1992), and the GABA agonist muscimol causes a flaccid paralysis (Chalfie & White 1988). These results suggest that GABA acts as an inhibitory neurotransmitter at *C. elegans* neuromuscular junctions and may regulate dorsal/ventral cross-inhibition of muscles, as it does in *Ascaris*.

The functions of other neurotransmitters in *C. elegans* have been less clear. Dopamine and serotonin-like material has been observed in *C. elegans* neurons, but mutants with greatly decreased levels of these neurotransmitters are apparently normal in viability, movement, and other behaviors (Avery & Horvitz 1990; Desai et al 1988; Sulston et al 1975). FMRFamide-like peptides are also present in some *C. elegans* neurons (Li & Chalfie 1991).

Pharmacological approaches have identified genes that encode an acetyl-choline receptor in *C. elegans*. Exogenously applied cholinergic agonists, including the antihelmintic levamisole, cause hypercontraction of body muscles (Brenner 1974). Levamisole is thought to act as an agonist of nicotinic acetylcholine receptors in the body muscles (Brenner 1974; Lewis et al 1980b). Mutations in the genes *unc-29*, *unc-38*, *lev-1*, *unc-63*, *unc-50*, and *unc-74* all result in levamisole resistance, accompanied by a slow-moving uncoordinated phenotype (Brenner 1974; Lewis et al 1980a). Mutations in *unc-29*, *unc-50*, and *unc-74* eliminate binding to nicotinic agonists (Lewis et al 1980b). The congruence of resistance to the agonists, loss of agonist binding, and a characteristic uncoordinated phenotype support the notion that these genes are either structural subunits of an acetylcholine receptor or are required for its assembly.

Some *unc* genes probably have roles in the process of synaptic transmission. Animals mutant for the *unc-104* gene are severely uncoordinated or dead, with a greatly reduced number of synaptic vesicles in their neuronal processes (Hall & Hedgecock 1991). *unc-104* encodes a member of the kinesin family of microtubule motors and could be a motor protein involved in the transport of synaptic vesicles to presynaptic regions (Otsuka et al 1991). Mutations in several genes lead to aberrant acetyl-choline accumulation, perhaps by preventing presynaptic release of acetyl-choline (Hosono & Kamiya 1991; Hosono et al 1987). One of these genes, *unc-13*, encodes a gene product with homology to the calcium- and phos-pholipid-binding domain (but not the kinase domain) of protein kinase C (Maruyama & Brenner 1991).

Genes required for normal axon morphology   A few *unc* mutants have been examined for defects in nervous system structure by electron microscopy (Hedgecock et al 1990; McIntire et al 1992). More restricted screens for neuronal defects have been done by staining *unc* mutants with antisera that recognize subsets of neurons in *C. elegans*. The antisera used for these experiments have used antiserotonin antisera, which recognize about 10 of the 302 neurons in a hermaphrodite (Desai et al 1988), anti-GABA antisera (26 neurons) (McIntire et al 1992), antihorseradish peroxidase antisera (27 neurons) (Siddiqui & Culotti 1991), and antitubulin antisera (5 neurons) (Siddiqui 1990; Siddiqui et al 1989). Sixteen chemosensory neurons can also be visualized in live animals soaked in the fluorescent dye FITC (Hedgecock et al 1985). In all cases, the neuronal cell bodies and their processes are stained by the antisera or fluorescent dye.

Mutations in more than 30 *unc* or *mig* genes cause morphological abnormalities in the structures of neurons visualized by these reagents. The defects lead to process abnormalities that may arise from failures of cell migration, axon outgrowth, axon guidance, or axon fasciculation (bundling).

Strikingly, most of the genes affect in multiple types of neurons, even including cells outside the nervous system. The function of some genes can be inferred by comparing their effects on different cells (Hedgecock et al 1987). For example, the *unc-5* gene is required for dorsal axon growth and cell migration of numerous cell types; the *unc-40* gene, for ventral axon growth and cell migration; and the *unc-6* gene, for both dorsal and ventral axon growth (Hedgecock et al 1990). All three of these genes only affect axons or cells that migrate on the epidermis, not those that elongate along pre-existing processes, and may encode components that participate in cell-substrate interactions (Desai et al 1988; Hedgecock et al 1990; McIntire et al 1992). Genetic mosaic analysis indicates that *unc-6* must be expressed by the epidermal substrate for normal axon growth (Hedgecock et al 1987, 1990).

Other genes, including *unc-34*, *unc-71*, and *unc-76*, cause failures of axon elongation and fasciculation within the larger neuronal bundles of the worm, especially the ventral nerve cord that runs from the head to the tail (Desai et al 1988; Hedgecock et al 1985; McIntire et al 1992). Such genes could be primarily important in axon-axon interactions. Additional genes, including *unc-33*, *unc-44*, and *unc-51*, are required for normal axon elongation in any context and may be involved in the mechanics of axon growth (Desai et al 1988; Hedgecock et al 1985; McIntire et al 1992).

A few *unc* mutants have defects restricted to a particular cell type. Mutations in the *unc-4* gene, which encodes a homeobox-containing protein, cause the VA motor neurons, which drive backward movement, to have synapses appropriate to the VB neurons, which drive forward movement (Miller et al 1992; White et al 1992). *unc-4* may regulate the transcription of genes that determine VA synaptic specificity.

Mechanosensation

A touch avoidance response can be elicited either by stroking *C. elegans* with a hair or by prodding it with a platinum wire (Chalfie & Sulston 1981). Touch to the anterior body causes the worm to move backward; touch to the tail induces or accelerates forward movement (Chalfie & Sulston 1981). Light touch also inhibits pharyngeal pumping (feeding) (Chalfie et al 1985) and alters the timing of the defecation cycle (Thomas 1990).

Vibration of the substrate, caused by tapping the plate that contains the worms, also elicits either reversed movement or accelerated forward movement (Chalfie & Sulston 1981; Rankin et al 1990). Because the same mutations eliminate both the touch response and the response to vibration (see below), vibration is thought to be sensed as a delocalized form of touch (Rankin 1991). Interestingly, the response to vibration varies depending on developmental stage: A tap to the plate causes mostly reversals in adult animals and mostly accelerations in second larval stage animals, with mixed responses in the other larval stages (Chiba & Rankin 1990).

The touch avoidance response fails after repeated touching of the animal (Chalfie et al 1985; Rankin et al 1990). Repeated tapping with a mechanical device can attenuate the response to less than 5% of the initial response (Rankin et al 1990). The animals are not merely exhausted, because they immediately recover about 50% of the original response following electric shock. This plasticity fits the classical criteria for habituation, a form of nonassociative learning. In addition, the touch response can be sensitized by taps shortly before the test stimulus, so that about a twofold increase in the response to touch is observed (Rankin et al 1990).

THE TOUCH AVOIDANCE CIRCUIT   A set of neurons called microtubule cells is required for touch responses in the body (Chalfie & Sulston 1981). The microtubule cells are attached to the epidermis immediately below the outer cuticle of the worm and contain large-diameter microtubules composed of 15 protofilaments, unlike the 11-protofilament microtubules present in most nematode cells (Chalfie & Thomson 1979).

When all of the microtubule cells are killed with a laser microbeam, the resulting animals fail to respond to stroking with a hair (Chalfie & Sulston 1981). They still respond to prodding with a wire, however, which indicates that the microtubule cells are required for the response to light, but not heavy touch. Two microtubule cells in the tail (PLMR and PLML) mediate the response to tail touch, whereas three anterior microtubule cells (ALML, ALMR, and AVM) are required for the response to touch in the anterior part of the body. The AVM neurons are born postembryonically; either these neurons or the postembryonically derived motor neurons may account for the changes in the frequency of accelerations or reversals to tap seen at different developmental stages (Chiba & Rankin 1990; Sulston & Horvitz 1977).

An alternate method for identifying candidate neurons responsible for a behavior is to isolate mutants with behavioral abnormalities and identify cells affected by the mutations. Mutations in *mec-3*, a gene that determines touch cell identity (see below), lead to abnormalities in the response to both light and heavy touch (Way & Chalfie 1989). In transgenic animals containing fusions of the *mec-3* gene to the bacterial *lacZ* gene, β-galactosidase is expressed in the microtubule cells and in two other putative sensory neurons, PVD and FLP. This result suggested, and subsequent laser experiments confirmed, that the PVD sensory neurons mediate the residual response to heavy tail touch after the microtubule cells are killed (Way & Chalfie 1989).

A sensory neuron-interneuron-motor neuron reflex circuit generates each touch response (Chalfie et al 1985). The anterior touch cells synapse onto interneurons that initiate a wave of backward movement. The posterior touch cells synapse onto interneurons that accelerate forward movement. Each of these interneurons synapses onto an appropriate set of motor neurons. Laser killing of sensory neurons, interneurons, or motor neurons in the circuit disrupts the touch response.

Light touch can be sensed in the head after all of the microtubule cells are killed, so additional touch-sensitive neurons in this region remain to be identified (Chalfie & Sulston 1981).

TOUCH-INSENSITIVE MUTANTS   Mutants with defects in mechanosensation were identified as animals that did not respond to stroking with a hair, but did respond to prodding with a wire. Four hundred seventeen independent mutations that confer these properties have been isolated; these mutations fall in 18 genes (Chalfie & Au 1989; Chalfie & Sulston 1981). Most of these genes are called *mec* genes, for *mec*hanosensory-defective. Because multiple mutations have been identified in all but one of the *mec* genes, these mutations probably define all or nearly all of the genes that can mutate to give this specific phenotype (i.e. the screen has been saturated).

In some of the touch-insensitive mutants, no recognizable microtubule cells are present. In animals mutant for either *unc-86* or *lin-32*, the microtubule cells are never generated, because of defective cell lineages during development (Chalfie & Au 1989; Chalfie et al 1981; C. Kenyon and E. Hedgecock, personal communication). The *mec-3* gene is also required for normal touch-cell development, but seems to act in the specification of the microtubule cell identity after the lineages are complete (Way & Chalfie 1988). Both *unc-86* and *mec-3* encode putative DNA-binding transcription factors of the homeobox class (Finney et al 1988; Way & Chalfie 1988). In these neuronal cell lineages, as in the development of other cell types, transcription factors appear to specify the differences between cell types (Kessel & Gruss 1990; Levine & Hoey 1988).

In the other touch-insensitive mutants, clearly recognizable microtubule cells are present. Four genes, *mec-7, 12, 17*, and *mec(u455)*, are necessary for characteristic large-diameter microtubules that give the cells their names (Chalfie & Au 1989). *mec-7* has been cloned and shown to encode a β-tubulin isoform (Savage et al 1989). Two genes, *mec-1* and *mec-5*, are required for the formation of the mantle, an extracellular matrix that attaches the microtubule cells to the epidermis under the worm's cuticle (Chalfie & Au 1989). Mutations in the ten remaining genes, *mec-2, 4, 6, 8, 9, 10, 14, 15, 18,* and *egl-5*, do not affect the microtubule cell structure, as visualized in electron micrographs.

The touch-insensitive mutants have been screened for other behavioral or structural phenotypes, and eight genes were found to have effects in other cell types. *unc-86, lin-32,* and *egl-5* are pleiotropic genes with effects on numerous cell lineages (Chalfie & Au 1989; Chalfie et al 1981; Chisholm 1991; C. Kenyon and E. Hedgecock, personal communication). *mec-3* affects the touch cells that sense heavy touch, as well as the microtubule cells (Way & Chalfie 1989). *mec-1, mec-2,* and *mec-8* are required for normal morphology or function of chemosensory neurons (Lewis & Hodgkin 1977; Perkins et al 1986), and *mec-6* can have effects in the PVC interneurons (Driscoll & Chalfie 1991). Animal mutants for the remaining ten *mec* genes are not known to have any defects other than touch-insensitivity; therefore, these genes may be specifically required in only the microtubule cells. Of these, the *mec-4* gene has been cloned; it encodes a novel transmembrane protein that has been proposed to be involved in signal transduction in the microtubule cells (Driscoll & Chalfie 1991).

Mutants with behavioral abnormalities may be defective either in the development of neurons required for the behavior or for the behavior itself. Four of the five microtubule neurons are born embryonically, but function throughout life, so the development and function of the neurons are separated in time. For some *mec* genes, temperature-sensitive mutations can be used to ask when a gene product is required. *mec-2* mutations are reversibly touch-insensitive and can shift between sensitivity and insensitivity at any time during larval growth (Chalfie & Au 1989). Thus, *mec-2* is probably directly required for touch sensation or signal propagation by the microtubule cells. *mec-7* and *mec-12*, which affect the characteristic 15-protofilament microtubules, are required during the middle larval stages while the microtubule cells are growing (Chalfie & Au 1989). *mec-4, 6, 8,* and *15* also function through much of larval life.

Egg-Laying

The hermaphrodite *C. elegans* lays fertilized eggs through the vulva, by using a specific set of vulval muscles. Defects in either the vulva, the vulval muscles, or the HSN motor neurons, which innervate the vulval muscles, can lead to the retention of eggs, as can killing these cell types with the laser (Trent et al 1983). Food deprivation also causes retention (Horvitz et al 1982).

GENES REQUIRED FOR NORMAL EGG-LAYING   Egg-laying defective mutants can be easily identified based on the retention of late-stage eggs, which may even hatch inside their mother and eat her. More than 50 genes necessary for normal egg-laying (*egl* genes) have been identified by screening for animals containing late-stage eggs or larvae (Desai & Horvitz 1989; Trent et al 1983).

Antisera against the neurotransmitter serotonin recognize the HSN motor neurons, and exogenous serotonin causes vulval muscle contraction and egg-laying (Croll 1975b; Desai et al 1988; Horvitz et al 1982). These observations suggest that serotonin may be an endogenous HSN transmitter that drives egg-laying, and allowed the development of a pharmacological screen for mutants that affect HSN function. Egg-laying defective mutants were soaked in serotonin; mutants with abnormalities in the vulva or vulval muscles could not lay eggs in response to exogenous serotonin, but those with HSN defects could (Trent et al 1983). The mutants were also treated with imipramine, which potentiates endogenous serotonin; HSN-defective mutants should not have a source of endogenous serotonin and, therefore, should not respond to imipramine.

Sixteen *egl* genes with HSN defects were identified by using the pharmacological criteria (Desai & Horvitz 1989; Trent et al 1983). Further analysis of these mutations was facilitated by the ability to visualize the HSN neurons and processes with antiserotonin antisera. The antisera were also used to screen other mutants (e.g. *unc* mutants) for HSN defects. Thirty-eight genes with abnormal HSNs were identified by using these combined approaches, and other unidentified genes probably also affect these neurons (Desai et al 1988).

The HSNs were obviously abnormal, as visualized with antiserotonin antibodies, in animals mutant for 32 of the 38 genes (Desai et al 1988). Nine genes affected axonal outgrowth (*mig-2*, *unc-6*, *33*, *34*, *40*, *51*, *71*, *73*, *76*; see above), two were necessary for serotonin expression (*cat-1*, *4*), and nine were required for a long-range migration of the HSN neurons to their final position during embryonic development (*egl-18*, *20*, *27*, *mig-1*, *2*, *10*, *12*, *unc-71*, *73*).

Twelve additional genes affected multiple properties of the HSNs (for example, both axon outgrowth and serotonin expression) [*egl-1*, *5*, *41*, *44*, *45*, *46*, *egl(n1653)*, *ham-1*, *sem-4*, *unc-86*, *her-1*, *tra-2*]. Although in all cases the cells had some properties of HSNs, these genes appear to be required for the HSNs to fulfill their developmental potential (Desai et al 1988). *unc-86* encodes a homeobox transcription factor that also influences microtubule cell lineages and other neuronal lineages (Finney et al 1988).

Several other *egl* genes that do not fit the pharmacological criteria for affecting the HSN neurons might also act in the nervous system. Egg-laying is regulated by food, and disruption of this regulation could lead to an egg-laying defective (or egg-laying constitutive) phenotype (Desai et al 1988; Trent et al 1983).

Chemosensation and Thermosensation

C. elegans chemotaxes to several salts, cAMP, biotin, and some amino acids in the micromolar to millimolar concentration range (Bargmann & Horvitz 1991a; Dusenbery 1974; Ward 1973). It is possible to saturate the response to an attractant by assaying chemotaxis in the presence of a high uniform level of the attractant (Ward 1973). This property has been used to divide these water-soluble attractants into at least four classes; the attractants within one class cross-saturate each others' responses, but not responses to other classes (Ward 1973). Many volatile organic molecules are also attractive to C. elegans (Bargmann et al 1990).

High osmotic strength repels C. elegans, as do acid, D-tryptophan, garlic, copper, and a substance released by dead worms (Bargmann et al 1990; Culotti & Russell 1978; Dusenbery 1975; Ward 1973).

The chemosensory nervous system also influences the overall development of the animal. C. elegans regulates its development based on environmental stimuli, chiefly a pheromone that reflects nematode density [reviewed in Riddle (1988)]. In the presence of high pheromone levels, animals arrest in a specialized larval stage called a dauer larva (Golden & Riddle 1982). Food and temperature affect this decision, and food availability also permits dauer larvae to recover and continue growth (Golden & Riddle 1984).

Although chemotaxis has not been studied in as much detail as the mechanosensory response, it also seems to fall off after continuous stimulation in a form of adaptation or habituation. Thus, worms that chemotax to an attractant leave after about ten minutes and return again later (Ward 1973). In addition, experiments with tethered worms have shown that switching a worm into a less attractive chemical environment leads to a round of movement reversals that ends within about one minute (Dusenbery 1980b).

In a thermal gradient, worms migrate across the plate to the temperature at which they were raised (Hedgecock & Russell 1975). Animals placed in a radial thermal gradient make isothermal circular tracks at their preferred temperature. Thermotaxis is suppressed or inhibited by starvation; within six hours of food deprivation, animals disperse, rather than accumulate to a particular temperature (Hedgecock & Russell 1975). C. elegans has a galvanotaxis response that causes it to orient itself in electric fields (Sukul & Croll 1978), a reversal response to light (Burr 1985) and a preference for higher oxygen concentrations (Dusenbery 1980a). The nature of these responses is not well understood.

CHEMOSENSORY NEURONS Eleven classes of putative chemosensory neurons in *C. elegans* were identified based on electron micrographs, because they had endings that were exposed to the environment through specialized sensory structures in the cuticle (Ward et al 1975; Ware et al 1975; White et al 1986). The neurons required for chemotaxis, chemical avoidance, and normal development have been identified by laser killing. All of these neurons have ciliated sensory endings in the amphid sensory organ.

The ASH neurons are important in avoidance of high osmotic strength (Bargmann et al 1990; J. Thomas, personal communication). Some other neurons mediate a weak residual avoidance of high osmotic strength after the ASH neurons are killed.

A single neuron pair, the ASE neurons, are required for normal chemotaxis to cAMP, biotin, Na+, or Cl− ions (Bargmann & Horvitz 1991a). Interestingly, these attractants are discriminated by the animal in saturation assays (see above), yet all are recognized by a single sensory neuron class. Additional chemosensory neurons direct an inefficient chemotaxis response to the same chemicals. Different amphid chemosensory neurons sense volatile odorants (Bargmann et al 1990).

Some of the minor neurons in the chemotaxis response, the ADF, ASG, and ASI neurons, are also required to regulate the developmental decision between dauer and nondauer development (Bargmann & Horvitz 1991b). If they are killed, dauer development always occurs. The ASJ neurons are most important in recovery from the dauer stage (Bargmann & Horvitz 1991b).

The unique properties of each type of sensory neuron are likely to arise as a consequence of both its chemosensory specificity and its connections to different interneurons. Different amphid neurons have distinct synaptic partners, but the functions of the interneurons are not yet described.

In both the chemosensory and mechanosensory nervous systems, there are indications that multiple sensory neuron types act partly in parallel. Although ASH and ASE are the primary sensory neurons in avoidance and chemotaxis, respectively, each is assisted by other chemosensory neurons with partly overlapping functions (Bargmann & Horvitz 1991a; Bargmann et al 1990). The ADF, ASI, and ASG neurons are also at least partly redundant in their ability to promote normal development (Bargmann & Horvitz 1991b). Similarly, in the mechanosensory system, several neuronal types act to sense touch with different thresholds. The ALM neurons sense light touch in the anterior body; the AVM neuron senses light touch but is either less touch sensitive or habituates more quickly than the ALMs; and the PVD neurons sense heavy touch to the same part of the body (Chalfie et al 1985; Way & Chalfie 1989).

CHEMOTAXIS AND THERMOTAXIS-DEFECTIVE MUTANTS   Chemosensory mutants have been generated by using behavioral screens for chemotaxis- or avoidance-defective mutants (Culotti & Russell 1978; Dusenbery et al 1975; Lewis & Hodgkin 1977). Other mutants that were first identified by their effects on dauer formation also have chemosensory defects (Albert et al 1981). The processes and cell bodies of some exposed chemosensory neurons take up the fluorescent dye FITC, which allows their direct visualization in live animals; this assay has also been used to isolate mutants in the chemosensory system (Perkins et al 1986).

The most common class of chemotaxis-defective mutants are pleiotropically defective in chemosensation. At least 14 genes can be mutated to produce animals that are chemotaxis-defective, defective in avoiding chemical repellents, and dauer formation-defective (Albert et al 1981; Culotti & Russell 1978; Lewis & Hodgkin 1977; Perkins et al 1986). Animals mutant in these genes have structural abnormalities in the chemosensory neurons that can be seen in electron micrographs. Eight of these genes (che-2, 3, 13, daf-10, 19, osm-1, 5, 6) are required for normal morphology of all sensory cilia in the animal (Albert et al 1981; Lewis & Hodgkin 1977; Perkins et al 1986). Six additional genes (che-10, 11, 12, 14, daf-6, osm-3) are required for normal amphid chemosensory anatomy, but not for all ciliated sensory neurons (Albert et al 1981; Perkins et al 1986). Some of these genes appear to affect the neurons themselves, whereas others affect the nonneuronal support cells associated with sensory structures (Herman 1984; Perkins et al 1986). The structural and behavioral defects in these mutants are consistent with the laser ablation experiments, which indicates that ciliated amphid neurons are important in these chemosensory responses. Thermotaxis is normal in these mutants, so the thermosensory neurons do not require this sort of ciliated sensory ending for function (Perkins et al 1986).

A second class of chemotaxis-defective mutant has normal osmotic avoidance and dauer formation, so only subsets of chemosensory functions are deficient (che-1, 5, 6, 7, tax-2, 3, 4, 6) (Culotti & Russell 1978; Dusenbery et al 1975; Lewis & Hodgkin 1977). In all of these mutants, the sensory cilia of the neurons are exposed normally through the amphid opening (Lewis & Hodgkin 1977; Perkins et al 1986). tax-2, 3, 4, and tax-6 mutants are also thermotaxis defective; this genetic overlap between chemotaxis and thermotaxis could be caused by common gene products in different sensory neurons, or by a common motor pathway in thermotaxis and chemotaxis (Hedgecock & Russell 1975).

Thermotaxis-defective mutations can result in either a thermophilic, cryophilic, or athermotactic phenotype (Hedgecock & Russell 1975). At least one thermotaxis gene, *ttx-1*, is not essential for normal chemotaxis.

In addition to the genes that affect both chemotaxis and dauer formation, a genetic pathway for dauer formation includes many *daf* genes that do not lead to defects in other sensory processes (Riddle et al 1981; Vowels & Thomas 1992). These mutants may have abnormalities in sensory neurons that affect only dauer development, or they might act in other cells downstream of the sensory neurons.

Feeding and Defecation

C. elegans ingests bacteria, its food, through a muscular pharynx that concentrates and grinds up bacteria (Croll 1978). Pharyngeal motil function consists of two components: pumping, a nearly synchronous contraction of the pharyngeal muscles, and peristalsis, or locomotion of food toward the gut.

Defecation occurs in three steps: a posterior body contraction and relaxation, followed by anterior body contraction and relaxation, and finally opening of the anus and defecation (Croll 1978). These three steps occur in this fixed order about once every minute in feeding worms. This defecation cycle is responsive to external stimuli; it slows if food is limited, and the cycle "clock" is reset by light touch to the animal (Thomas 1990).

The pharynx, which ingests and grinds up food, contains 20 neurons that are connected to the remainder of the *C. elegans* nervous system by only a single pair of neurons (Albertson & Thomson 1976). Surprisingly, after all 20 neurons of the pharynx are killed with the laser, pharyngeal pumping continues, although peristalsis stops and the worm starves (Avery & Horvitz 1989). Based on this result, pumping is proposed to be generated by a myogenic rhythm analogous to the beating of a vertebrate heart. A single neuron, called M4, is necessary and sufficient for peristalsis (Avery & Horvitz 1987). The other 19 pharyngeal neurons are not required for viability of *C. elegans*. However, the two MC pharyngeal neurons are necessary for normal regulation of pumping by food, and some of the other neurons increase the efficiency and coordination of feeding (Avery & Horvitz 1989).

At least in the pharynx, most neurons function in coordinating and regulating a behavior, rather than generating it. In fact, very few *C. elegans* neurons are essential for viability. Two factors contribute to their dispensability: First, the ability of behaviorally abnormal, even paralyzed, animals to survive under laboratory conditions; second, the redundancy of neuronal functions (Avery & Horvitz 1989; Bargmann & Horvitz 1991a,b;

Chalfie et al 1981). Only M4, the neuron that regulates pharyngeal peristalsis, and the CAN neurons are known to be required for viability in *C. elegans*. Animals lacking CAN arrest development during larval growth and die, perhaps because of defects in osmoregulation or excretion (J. Sulston, personal communication).

RESPONSES TO STARVATION   Food availability regulates feeding, movement, defecation, and egg-laying directed by the nervous system (Avery & Horvitz 1990; Horvitz et al 1982; Thomas 1990). Animals in the presence of bacteria feed regularly, lay eggs, and are relatively inactive. If food is removed, they become more active, retain eggs, and stop pharyngeal pumping. Their defecation cycle also slows down in the absence of food, to the extent that the animals become constipated (Thomas 1990). After long-term food deprivation, other behavioral changes occur. Animals starved for several hours cease to thermotax to the temperature at which they were raised (Hedgecock & Russell 1975), and their pharyngeal pumping becomes hyperresponsive to bacteria (Avery & Horvitz 1990). It is not known whether these alterations are coordinately regulated by a single starvation signal, or whether each is induced independently. Dauer larva development (see previous section) is also partly regulated by starvation. Interestingly, most dauer-constitutive mutants are egg-laying defective if they grow to adulthood, which suggests a common genetic step in two starvation responses (Riddle et al 1981).

GENES THAT AFFECT THE DEFECATION CYCLE   Defecation-defective mutants in 18 genes have been identified by direct inspection for constipated mutants, and by screening the *unc* mutants for phenotypes (Thomas 1990). Single mutations can eliminate any one of the three motor components of defecation: anterior body contraction (*unc-16, 33, 44, 101*), posterior body contraction (*pbo-1, egl-8*), and expulsion (*exp-1, 2, unc-25, 47*). Mutations in the six *aex* genes affect both anterior body contraction and expulsion. Because any unaffected motor response continues to cycle regularly in each of these mutants, no step of the defecation cycle depends entirely on the completion of another step.

The genes that affect motor components of defecation do not necessarily affect the timing of the defecation cycle (Thomas 1990). In fact, the cycle time is not set at any one step, because a regular cycle can occur when any step is absent. The generator of the cycle must be responsive to feeding and touch, but its nature is unknown. Mutation of the gene encoding choline acetyltransferase (*cha-1*) lengthens the cycle, which suggests that acetylcholine may influence its periodicity. In addition, one mutant with a specific defect in cycle timing has been isolated (*dec-1*) (Thomas 1990).

Male Mating

The most complex behavior known in *C. elegans* is mating of the male with the hermaphrodite (Hodgkin 1983). During male mating, a series of events with both chemosensory and mechanosensory feedback must occur in a stereotypic order. The male chemotaxes to the hermaphrodite, circles her, locates the vulva with his tail, opens the vulva, and deposits sperm there. The hermaphrodite does not display obvious behavioral changes during this process.

The male contains about 80 additional neurons that presumably help it to locate hermaphrodites and mate with them (Hodgkin 1983; Sulston & Horvitz 1977). The connectivity of these male-specific neurons is not fully known. Among the many male mating-defective mutants, some have defective development of male-specific neurons, whereas others lack obvious structural defects (Hodgkin 1983).

CONCLUSIONS: GENETIC ANALYSIS OF IDENTIFIED NEURONS

Genetic analysis can be a powerful tool for identifying novel or unexpected molecules that participate in a process of interest. In an organism as genetically tractable as *C. elegans*, generating mutants is simple; the challenge is discovering how each gene product affects behavior, which arises from groups of cells acting together.

Understanding the roles of individual neurons in behavioral responses provides one crucial link between the gene and the behavior. Where neuronal functions are defined by laser ablation, mutant phenotypes can be correlated with the properties of a single cell type. Chemosensation and mechanosensation have been particularly amenable to such analysis. The effects of pharmacological agents on wild-type and mutant animals allowed hypotheses to be formulated about genes that affect egg-laying or movement. Structural analysis of mutants by using antisera or electron microscopy identified numerous genes that may direct neuronal development.

In each case, models for gene action were suggested when behavioral defects could be interpreted at the level of the cell. The validity of these models is being tested by molecular analysis of the relevant genes. The sequence and expression pattern of several genes have confirmed (and extended) expectations derived from their genetic and phenotypic characterization.

To what extent are the same genes used by different cell types, with the particular properties of a given cell determined by the combination of genes it expresses, and to what extent do cell-type specific genes exist? The best-characterized neurons are the HSN neurons (egg-laying) and the microtubule neurons (touch). Of the 38 genes that affect the HSNs, 34 also have effects in other cell types (Desai et al 1988). By contrast, 10 of 18 *mec* genes are only known to affect the microtubule cells, which suggests that these neurons are more genetically specialized than the HSNs (Chalfie & Au 1989). The HSNs are motor neurons that innervate a single muscle type and may be genetically similar to other motor neurons in the body. The touch neurons, with their specialized microtubules, may require a number of dedicated genes for mechanosensation and signal transduction.

A distinction can be made between genes that act primarily during neuronal development and those that disrupt specific neuronal functions. Gene products used in neuronal development are often shared by multiple neuronal cell types. All the genes required for HSN identity or microtubule cell identity are utilized in other cell types, as well (Chalfie & Au 1989; Chisholm 1991; Desai et al 1988). Most notable, the *unc-86* gene is required for normal development of the HSNs and the microtubule cells and is also expressed in several additional neuronal cell types (Finney & Ruvkun 1990).

Almost all of the genes that affect axon guidance and cell migration also affect numerous cell types. Similarly, many genes required for neuronal function-neurotransmitters, neurotransmitter receptors, and the like act in multiple cells. Following this line of logic, there could be neuronal cell types whose unique properties are entirely encoded by combinations of shared gene products.

The pleiotropy of genes that affect neuronal development points out one limitation of a genetic approach. A screen for behavioral mutants tends to overlook broadly acting genes essential for viability or movement. Indeed, many mutants with abnormal axon guidance or cell migration of the microtubule cells were not recognized as touch insensitive, either because they were paralyzed or because misplaced microtubule cell processes can still mediate a touch response (Siddiqui 1990).

On the other hand, pleiotropic mutations can provide valuable information about cellular or molecular components that act in several processes. The large class of chemotaxis-defective, dauer-defective, avoidance-defective mutants reflects the requirement for ciliated sensory neurons in all three of these responses. Similarly, the congruence of dauer-constitutive and egg-laying defective phenotypes, or anterior-body contraction/expulsion defects in defecation-defective mutants, suggests that a common neuron, neurotransmitter, or process may operate in both responses.

Insights from the behavior, neuroanatomy, cell ablation studies, and molecular genetics of *C. elegans* illuminate its nervous system development and function. Prospects for the future include characterizing the molecules encoded by behavioral genes, applying circuit analysis and genetic techniques to more complex behaviors, studying the functions of groups of interacting neurons, and elucidating plasticity of behavioral responses.

ACKNOWLEDGMENTS

Many thanks to Leon Avery, Mike Finney, Bob Horvitz, Carl Johnson, Cynthia Kenyon, Bruce Kimmel, Michel Labouesse, Chris Li, Kent Nybakken, Steve Salser, Piali Sengupta, and Jim Thomas for discussions and comments on this manuscript. C. I. Bargmann is a Lucille P. Markey Scholar.

Literature Cited

Albert, P. S., Brown, S. J., Riddle, D. L. 1981. Sensory control of dauer larva formation in *Caenorhabditis elegans*. *J. Comp. Neurol.* 198: 435–51

Albertson, D. G., Thomson, J. N. 1976. The pharynx of *Caenorhabditis elegans*. *Philos. Trans. R. Soc. London Ser. B* 275: 299–325

Angstadt, J. D., Donmoyer, J. E., Stretton, A. O. 1989. Retrovesicular ganglion of the nematode *Ascaris. J. Comp. Neurol.* 284: 374–88

Avery, L., Horvitz, H. R. 1990. Effects of starvation and neuroactive drugs on feeding in *Caenorhabditis elegans. J. Exp. Zool.* 253: 263–70

Avery, L., Horvitz, H. R. 1989. Effects of killing identified pharyngeal neurons on feeding behavior in *Caenorhabditis elegans. Neuron* 3: 473–85

Avery, L., Horvitz, H. R. 1987. A cell that dies during wild-type *C. elegans* development can function as a neuron in a *ced-3* mutant. *Cell* 51: 1071–78

Bargmann, C. I., Horvitz, H. R. 1991a. Chemosensory neurons with overlapping functions direct chemotaxis to multiple chemicals in *C. elegans. Neuron* 7: 729–42

Bargmann, C. I., Horvitz, H. R. 1991b. Control of larval development by chemosensory neurons in *Caenorhabditis elegans. Science* 251: 1243–46

Bargmann, C. I., Thomas, J. H., Horvitz, H. R. 1990. Chemosensory cell function in the behavior and development of *Caenorhabditis elegans. Cold Spring Harbor Symp. Quant. Biol.* 55: 529–38

Brenner, S. 1974. The genetics of *Caenorhabditis elegans. Genetics* 77: 71–94

Burr, A. H. 1985. The photomovement of *Caenorhabditis elegans*, a nematode which

lacks ocelli. Proof that the response is to light not radiant heating. *Photochem. Photobiol.* 41: 577–82

Chalfie, M., Au, M. 1989. Genetic control of differentiation of the *Caenorhabditis elegans* touch receptor neurons. *Science* 243: 1027–33

Chalfie, M., Horvitz, H. R., Sulston, J. E. 1981. Mutations that lead to reiterations in the cell lineages of *C. elegans. Cell* 24: 59–69

Chalfie, M., Sulston, J. 1981. Developmental genetics of the mechanosensory neurons of *Caenorhabditis elegans. Dev. Biol.* 82: 358–70

Chalfie, M., Sulston, J. E., White, J. G., Southgate, E., Thomson, J. N., Brenner, S. 1985. The neural circuit for touch sensitivity in *Caenorhabditis elegans. J. Neurosci.* 5: 956–64

Chalfie, M., Thomson, J. N. 1979. Organization of neuronal microtubules in the nematode *Caenorhabditis elegans. J. Cell Biol.* 82: 278–89

Chalfie, M., White, J. 1988. The nervous system. In *The Nematode* Caenorhabditis elegans, ed. W. B. Wood, pp. 337–91. Cold Spring Harbor, NY: Cold Spring Harbor Labs. 667 pp.

Chiba, C. M., Rankin, C. H. 1990. A developmental analysis of spontaneous and reflexive reversals in the nematode *Caenorhabditis elegans. J. Neurobiol.* 21: 543–54

Chisholm, A. 1991. Control of cell fate in the tail region of *C. elegans* by the gene *egl-5. Development* 111: 921–32

Croll, N. A. 1978. Integrated behavior in the feeding phase of *Caenorhabditis elegans. J. Zool.* 184: 507–17

Croll, N. A. 1975a. Components and patterns in the behavior of the nematode

Caenorhabditis elegans. J. Zool. 176: 159–76

Croll, N. A. 1975b. Indolealkylamines in the coordination of nematode behavioral activities. *Can. J. Zool.* 53: 894–903

Culotti, J. G., Russell, R. L. 1978. Osmotic avoidance defective mutants of the nematode *Caenorhabditis elegans. Genetics* 90: 243–56

Daley, D. L., Camhi, J. M. 1988. Connectivity pattern of the cercal to giant interneuron system of the American cockroach. *J. Neurophys.* 60: 1350–68

Davis, R. E., Stretton, A. O. W. 1989a. Passive membrane properties of motorneurons and their role in long distance signaling in the nematode *Ascaris. J. Neurosci.* 9: 403–14

Davis, R. E., Stretton, A. O. W. 1989b. Signalling properties of *Ascaris* motor neurons: graded synaptic transmission and tonic transmitter release. *J. Neurosci.* 9: 415–25

del Castillo, J., deMello, W. C., Morales, T. 1964. Inhibitory action of GABA on *Ascaris* muscle. *Experientia* 20: 141

del Castillo, J., deMello, W. C., Morales, T. 1963. The physiological role of acetylcholine in the neuromuscular system of *Ascaris lumbricoides. Arch. Int. Physiol. Biochim.* 71: 741–57

Desai, C., Garriga, G., McIntire, S. L., Horvitz, H. R. 1988. A genetic pathway for the development of the *Caenorhabditis elegans* HSN motor neurons. *Nature* 336: 638–46

Desai, C., Horvitz, H. R. 1989. *Caenorhabditis elegans* mutants defective in the functioning of the motor neurons responsible for egg laying. *Genetics* 121: 703–21

Driscoll, M., Chalfie, M. 1991. The *mec-4* gene is a member of a family of *Caenorhabditis elegans* genes that can mutate to induce neuronal degeneration. *Nature* 349: 588–93

Durbin, R. M. 1987. Studies on the development and organisation of the nervous system of *Caenorhabditis elegans.* PhD thesis, Univ. Cambridge, England

Dusenbery, D. B. 1980a. Appetitive response of the nematode *Caenorhabditis elegans* to oxygen. *J. Comp. Physiol.* 136: 333–36

Dusenbery, D. B. 1980b. Responses of the nematode *Caenorhabditis elegans* to controlled chemical stimulation. *J. Comp. Physiol.* 136: 327–31

Dusenbery, D. B. 1975. The avoidance of D-tryptophan by the nematode *Caenorhabditis elegans. J. Exp. Zool.* 193: 413–18

Dusenbery, D. B. 1974. Analysis of chemotaxis in the nematode *Caenorhabditis elegans* by countercurrent separation. *J. Exp. Zool.* 188: 41–47

Dusenbery, D. B., Sheridan, R. E., Russell, R. L. 1975. Chemotaxis-defective mutants of the nematode *Caenorhabditis elegans. Genetics* 80: 297–309

Edgley, M. L., Riddle, D. L. 1990. The nematode *Caenorhabditis elegans. Genet. Maps* 5: 3.111–3.133

Epstein, H. F. 1990. Genetic analysis of myosin assembly in *Caenorhabditis elegans. Mol. Neurobiol.* 4: 1–25

Finney, M., Ruvkun, G. 1990. The *unc-86* gene couples cell lineage and cell identity in *C. elegans. Cell* 63: 895–905

Finney, M., Ruvkun, G., Horvitz, H. R. 1988. The *C. elegans* cell lineage and differentiation gene *unc-86* encodes a protein with a homeodomain and extended similarity to transcription factors. *Cell* 55: 757–69

Golden, J. W., Riddle, D. L. 1984. The *Caenorhabditis elegans* dauer larva: developmental effects of pheromone, food, and temperature. *Dev. Biol.* 102: 368–78

Golden, J. W., Riddle, D. L. 1982. A pheromone influences larval development in the nematode *Caenorhabditis elegans. Science* 218: 578–80

Grenningloh, G., Bieber, A., Rehm, J., Snow, P. M., Tarquina, Z., et al. 1990. Molecular genetics of neuronal recognition in *Drosophila*: evolution and function of immunoglobulin superfamily cell adhesion molecules. *Cold Spring Harbor Symp. Quant. Biol.* 55: 327–40

Hall, D. H., Hedgecock, E. M. 1991. Kinesin-related gene *unc-104* is required for axonal transport of synaptic vesicles in *C. elegans. Cell* 65: 837–47

Hedgecock, E. M., Culotti, J. G., Hall, D. H. 1990. The unc-5, unc-6, and unc-40 genes guide circumferential migrations of pioneer axons and mesodermal cells on the epidermis in *C. elegans. Neuron* 4: 61–85

Hedgecock, E. M., Culotti, J. G., Hall, D. H., Stern, B. D. 1987. Genetics of cell and axon migrations in *Caenorhabditis elegans. Development* 100: 365–82

Hedgecock, E. M., Culotti, J. G., Thomson, J. N., Perkins, L. A. 1985. Axonal guidance mutants of *Caenorhabditis elegans* identified by filling sensory neurons with fluorescein dyes. *Dev. Biol.* 111: 158–70

Hedgecock, E. M., Russell, R. L. 1975. Normal and mutant thermotaxis in the nematode *Caenorhabditis elegans. Proc. Natl. Acad. Sci. USA* 72: 4061–65

Herman, R. K. 1989. Mosaic analysis in the nematode *Caenorhabditis elegans* [erratum, in *J. Neurogenet.* 1989. 5(2): 159]. *J. Neurogenet.* 5: 1–24

Herman, R. K. 1984. Analysis of genetic mosaics of the nematode *Caenorhabditis elegans. Genetics* 108: 165–80

Hodgkin, J. 1983. Male phenotypes and mating efficiency in *Caenorhabditis elegans. Genetics* 103: 43–64

Horvitz, H. R., Chalfie, M., Trent, C., Sulston, J. E., Evans, P. D. 1982. Serotonin and octopamine in the nematode *Caenorhabditis elegans. Science* 216: 1012–14

Hosono, R., Kamiya, Y. 1991. Additional genes which result in an elevation of acetylcholine levels by mutations in *Caenorhabditis elegans. Neurosci. Lett.* 128: 243–44

Hosono, R., Sassa, T., Kuno, S. 1987. Mutations affecting acetylcholine levels in the nematode *Caenorhabditis elegans. J. Neurochem.* 49: 1820–23

Jessell, T. M. 1988. Adhesion molecules and the hierarchy of neuronal development. *Neuron* 1: 3–13

Johnson, C. D., Duckett, J. G., Culotti, J. G., Herman, R. K., Meneely, P. M., Russell, R. L. 1981. An acetylcholinesterase-deficient mutant of the nematode *Caenorhabditis elegans. Genetics* 97: 261–79

Johnson, C. D., Rand, J. B., Herman, R. K., Stern, B. D., Russell, R. L. 1988. The acetylcholinesterase genes of *C. elegans*: identification of a third gene (*ace-3*) and mosaic mapping of a synthetic lethal phenotype. *Neuron* 1: 165–73

Johnson, C. D., Russell, R. L. 1983. Multiple molecular forms of acetylcholinesterase in the nematode *Caenorhabditis elegans. J. Neurochem.* 41: 30–46

Johnson, C. D., Stretton, A. O. W. 1985. Localization of choline acetyltransferase within identified motoneurons of the nematode *Ascaris. J. Neurosci.* 5: 1984–92

Johnson, C. D., Stretton, A. O. W. 1987. GABA-immunoreactivity in inhibitory motor neurons of the nematode *Ascaris. J. Neurosci.* 7: 223–35

Kessel, M., Gruss, P. 1990. Murine developmental control genes. *Science* 249: 374–79

Kimble, J. 1981. Alterations in cell lineage following laser ablation of cells in the somatic gonad of *Caenorhabditis elegans. Dev. Biol.* 87: 286–300

Kutsch, W., Bentley, D. 1987. Programmed cell death of peripheral pioneer neurons in the grasshopper embryo. *Dev. Biol.* 123: 517–25

Levine, M., Hoey, T. 1988. Homeobox proteins as sequence-specific transcription factors. *Cell* 55: 537–40

Lewis, J. A., Hodgkin, J. A. 1977. Specific neuroanatomical changes in chemo-sensory mutants of the nematode *Caenorhabditis elegans. J. Comp. Neurol.* 172: 489–510

Lewis, J. A., Wu, C. H., Berg, H., Levine, J. H. 1980a. The genetics of levamisole resistance in the nematode *Caenorhabditis elegans. Genetics* 95: 905–28

Lewis, J. A., Wu, C. H., Levine, J. H., Berg, H. 1980b. Levamisole-resistant mutants of the nematode *Caenorhabditis elegans* appear to lack pharmacological acetylcholine receptors. *Neuroscience* 5: 967–89

Li, C., Chalfie, M. 1991. Organogenesis in *C. elegans*: positioning of neurons and muscles in the egg-laying system. *Neuron* 4: 681–95

Maruyama, I. N., Brenner, S. 1991. A phorbol ester/diacylglycerol-binding protein encoded by the *unc-13* gene of *Caenorhabditis elegans. Proc. Natl. Acad. Sci. USA* 88: 5729–33

McConnell, S. K., Ghosh, A., Shatz, C. J. 1989. Subplate neurons pioneer the first axon pathways from the cerebral cortex. *Science* 245: 978–82

McIntire, S. L., Garriga, G., White, J. G., Jacobson, D., Horvitz, H. R. 1992. Genes necessary for directed axonal elongation or fasciculation in *Caenorhabditis elegans. Neuron* 8: 307–22

Miller, D. M., Shen, M. M., Shamu, C. E., Burglin, T. R., Ruvkun, D., et al. 1992. *C. elegans unc-4* gene encodes a homeodomain that determines the pattern of synaptic input to specific motor neurons. *Nature* 355: 841–45

Niebur, E., Erdos, P. 1991. Theory of the locomotion of nematodes. *Biophys. J.* 60: 1132–46

Otsuka, A. J., Jeyaprakash, A., Garcia-Anoveros, J., Tang, L. Z., Fisk, G., et al. 1991. The *C. elegans unc-104* gene encodes a putative kinesin heavy chain-like protein. *Neuron* 6: 113–22

Perkins, L. A., Hedgecock, E. M., Thomson, J. N., Culotti, J. G. 1986. Mutant sensory cilia in the nematode *Caenorhabditis elegans. Dev. Biol.* 117: 456–87

Rand, J. B. 1989. Genetic analysis of the cha-1-unc-17 gene complex in *Caenorhabditis. Genetics* 122: 73–80

Rand, J. B., Russell, R. L. 1985. Properties and partial purification of choline acetyltransferase from the nematode *Caenorhabditis elegans. J. Neurochem.* 44: 189–200

Rand, J. B., Russell, R. L. 1984. Choline acetyltransferase-deficient mutants of the nematode *Caenorhabditis elegans. Genetics* 106: 227–48

Rankin, C. H. 1991. Interactions between two antagonistic reflexes in the nematode *Caenorhabditis elegans. J. Comp. Physiol.* 169: 59–67

Rankin, C. H., Beck, C. D., Chiba, C. M. 1990. *Caenorhabditis elegans*: a new model system for the study of learning and memory. *Behav. Brain Res.* 37: 89–92

Riddle, D. L. 1988. The dauer larva. In *The Nematode* Caenorhabditis elegans, ed. W. B. Wood, pp. 393–412. Cold Spring Harbor, NY: Cold Spring Harbor Lab. 667 pp.

Riddle, D. L., Swanson, M. M., Albert, P. S. 1981. Interacting genes in nematode dauer larva formation. *Nature* 290: 668–71

Savage, C., Hamelin, M., Culotti, J. G., Coulson, A., Albertson, D. G., Chalfie, M. 1989. mec-7 is a beta-tubulin gene required for the production of 15-protofilament microtubules in *Caenorhabditis elegans*. *Genes Dev.* 3: 870–81

Siddiqui, S. S. 1990. Mutations affecting axonal outgrowth and guidance of motor neurons and mechanosensory neurons in the nematode *Caenorhabditis elegans*. *Neurosci. Res. (Suppl.)* 13: 171–90

Siddiqui, S. S., Aamodt, E., Rastinejad, F., Culotti, J. 1989. Anti-tubulin monoclonal antibodies that bind to specific neurons in *Caenorhabditis elegans*. *J. Neurosci.* 9: 2963–72

Siddiqui, S. S., Culotti, J. G. 1991. Examination of neurons in wild type and mutants of *Caenorhabditis elegans* using antibodies to horseradish peroxidase. *J. Neurogenet.* 7: 193–211

Sternberg, P. W., Horvitz, H. R. 1986. Pattern formation during vulval development in *C. elegans*. *Cell* 44: 761–72

Stretton, A. O. W., Davis, R. E., Angstadt, J. D., Donmoyer, J. E., Johnson, C. D. 1985. Neural control of behavior in *Ascaris*. *Trends Neurosci.* 8: 294–300

Stretton, A. O. W., Fishpool, R. M., Southgate, E., Donmoyer, J. E., Walrond, J. P., et al. 1978. Structure and physiological activity of the motoneurons of the nematode *Ascaris*. *Proc. Natl. Acad. Sci. USA* 75: 3493–97

Sukul, N. C., Croll, N. A. 1978. Influence of potential difference and current on the electrotaxis of *Caenorhabditis elegans*. *J. Nematol.* 10: 314–17

Sulston, J., Dew, M., Brenner, S. 1975. Dopaminergic neurons in the nematode *Caenorhabditis elegans*. *J. Comp. Neurol.* 163: 215–26

Sulston, J. E., Horvitz, H. R. 1977. Postembryonic cell lineages of the nematode, *Caenorhabditis elegans*. *Dev. Biol.* 56: 110–56

Sulston, J. E., Schierenberg, E., White, J. G., Thomson, J. N. 1983. The embryonic cell lineage of the nematode *Caenorhabditis elegans*. *Dev. Biol.* 100: 64–119

Sulston, J. E., White, J. G. 1980. Regulation and cell autonomy during postembryonic development of *Caenorhabditis elegans*. *Dev. Biol.* 78: 577–97

Thomas, J. H. 1990. Genetic analysis of defecation in *Caenorhabditis elegans*. *Genetics* 124: 855–72

Trent, C., Tsung, N., Horvitz, H. R. 1983. Egg-laying defective mutants of the nematode *Caenorhabditis elegans*. *Genetics* 104: 619–47

Vowels, J. J., Thomas, J. H. 1992. Genetic analysis of chemosensory control of dauer formation in *Caenorhabditis elegans*. *Genetics* 130: 105–23

Walrond, J. P., Kass, I. S., Stretton, A. O. W., Donmoyer, J. E. 1985. Identification of excitatory and inhibitory motoneurons in the nematode *Ascaris* by electrophysiological techniques. *J. Neurosci.* 5: 1–8

Walrond, J. P., Stretton, A. O. W. 1985. Reciprocal inhibition in the motor nervous system of the nematode *Ascaris*: direct control of ventral inhibitory motoneurons by dorsal excitatory motoneurons. *J. Neurosci.* 5: 9–15

Walthall, W. W., Chalfie, M. 1988. Cell-cell interactions in the guidance of late-developing neurons in *Caenorhabditis elegans*. *Science* 239: 643–45

Ward, S. 1973. Chemotaxis by the nematode *Caenorhabditis elegans*: identification of attractants and analysis of the response by use of mutants. *Proc. Natl. Acad. Sci. USA* 70: 817–21

Ward, S., Thomson, N., White, J. G., Brenner, S. 1975. Electron microscopical reconstruction of the anterior sensory anatomy of the nematode *Caenorhabditis elegans*. *J. Comp. Neurol.* 160: 313–37

Ware, R. W., Clark, D., Crossland, K., Russell, R. L. 1975. The nerve ring of the nematode *Caenorhabditis elegans*: sensory input and motor output. *J. Comp. Neurol.* 162: 71–110

Waterston, R. H. 1988. Muscle. In *The nematode* Caenorhabditis elegans, ed. W. B. Wood, pp. 281–335. Cold Spring Harbor, NY: Cold Spring Harbor Lab. 667 pp.

Way, J. C., Chalfie, M. 1989. The mec-3 gene of *Caenorhabditis elegans* requires its own product for maintained expression and is expressed in three neuronal cell types. *Genes Dev.* 3: 1823–33

Way, J. C., Chalfie, M. 1988. mec-3, a homeobox-containing gene that specifies differentiation of the touch receptor neurons in *C. elegans*. *Cell* 54: 5–16

White, J. G., Albertson, D. G., Arness, M. A. R. 1978. Connectivity changes in a class of motoneurone during the development of a nematode. *Nature* 271: 764–66

White, J. G., Southgate, E., Thomson, J. N. 1992. Mutations in the *Caenorhabditis elegans unc-4* gene alter the synaptic input to ventral cord motor neurons. *Nature* 355: 838–41

White, J. G., Southgate, E., Thomson, J. N., Brenner, S. 1986. The structure of the nervous system of the nematode *Caenorhabdidit elegans*. *Philos. Trans. R. Soc. London Ser. B* 314: 1–340

White, J. G., Southgate, E., Thomson, J. N., Brenner, S. 1976. The structure of the ventral nerve cord of *Caenorhabditis elegans*. *Philos. Trans. R. Soc. London Ser. B* 275: 327–48

Annu. Rev. Neurosci. 1993. 16:73–93

NEUROTRANSMITTER TRANSPORTERS: Recent Progress[1]

Susan G. Amara

Vollum Institute for Advanced Biomedical Research, Oregon Health
Sciences University, Portland, Oregon 97201

Michael J. Kuhar

Neurosciences Branch, Addiction Research Center, National Institute on
Drug Abuse, Baltimore, Maryland

KEY WORDS: neurotransmitter reuptake, synaptic signal termination, uptake

INTRODUCTION

It has been known for many years that neurons and glia can accumulate
neurotransmitters by a sodium dependent cotransport process that is, in
many respects, similar to systems present in most cells for concentrating
metabolites. By cotransporting a solute with sodium, the energy stored in
transmembrane electrochemical gradients can be used to drive the solute
into the cell (reviewed in Kanner & Schuldiner 1987; Trendelenburg 1991).
A special appreciation of the neurotransmitter cotransport systems has
developed from studies, which indicate the existence of multiple uptake
systems, each relatively selective for a specific neurotransmitter. Table 1
presents a list of brain uptake systems that have been identified. These
transport activities are localized within the synaptic membranes of neurons
that use the same transmitter and are probably the most important mechanism for terminating synaptic transmission. Many of these transporters

---

[1] The US government has the right to retain a nonexclusive, royalty-free license in and to
any copyright covering this paper.

73

Table 1   Neurotransmitter (candidates)
have high affinity transport systems

Dopamine	GABA
Norepinephrine	Glycine
Serotonin	Taurine
Glutamate	Proline
Aspartate	Adenosine

or reuptake systems have been implicated as important sites for drug action. The augmentation of synaptic activity by selective inhibitors of specific sodium-dependent monoamine transport forms the basis for the mechanism of action of clinically important antidepressant drugs. Recently, interest in these uptake systems has peaked, because these same transporters are the primary site of action of cocaine and other stimulant drugs of abuse. Many of these topics have been reviewed (Amara & Pacholczyk 1991; Horn 1990; Iversen 1975; Kanner & Schuldiner 1987; Snyder 1970; Trendelenburg 1991).

Most recently, several of these plasma membrane transporters have been cloned with the resulting promise of rapid new advances in understanding the mechanisms of neurotransmitter reuptake. In this article, we review this and other topics. We focus on more recent developments on neurotransmitter transporters and provide some historical data. This review does not deal with uptake of neurotransmitters into intracellular vesicles, which has been reviewed elsewhere (Kanner & Schuldiner 1987; Rudnick 1986; Winkler et al 1986).

Neurotransmitter Uptake Is a Mechanism to Inactivate Released Transmitter

The large body of experimental data that has led to the conclusion that transporters are a major mechanism for terminating synaptic transmission (see Iversen 1975) is beyond the scope of this review. As an overview, we present some of the classic data of norepinephrine.

Burn (1932) was one of the first to suggest that norepinephrine might be taken up by tissues. Many subsequent reports supported this notion. Axelrod and coworkers (Axelrod et al 1959; Whitby et al 1961) showed that tritiated norepinephrine and tritiated epinephrine were removed from circulation in animals after injections of small doses. These radiolabeled catecholamines accumulated in various peripheral tissues. The fact that the tritiated norepinephrine was found in tissues with the highest degree of sympathetic innervation suggested that the catecholamine was accumulating in sympathetic neurons. In support of this latter notion, a decrease in norepinephrine uptake was found after destruction of sympathetic neurons in various peripheral organs. For example, in rodents that had been given nerve growth factor antiserum to hinder the development of the sympathetic nervous system, the accumulation of tritiated norepinephrine was reduced in peripheral organs (Iversen et al 1966; Sjöqvist et al 1967; Zaimis et al 1965). Also, lesions of ascending noradrenergic pathways in brain resulted in a reduction of tritiated norepinephrine uptake by synaptosomal factions prepared from areas receiving these pathways (Kuhar 1973). Taken together, these and many other findings suggest the existence of an uptake mechanism in sympathetic neurons for accumulating catecholamines. This evidence was strengthened by additional autoradiographic and histochemical studies.

Several sources provide evidence that the reuptake process terminates synaptically released norepinephrine. Especially important is the observation that inhibitors of norepinephrine metabolism do not potentiate noradrenergic neurotransmission, whereas drugs that inhibit the noradrenergic uptake system can have substantial enhancing effects on neurotransmission (Iversen 1967, 1975). Also, little or no norepinephrine overflows into perfusing fluids from various organs during sympathetic nerve stimulation; however, if inhibitors of norepinephrine uptake are present, a large increase in norepinephrine overflow occurs after sympathetic nerve stimulation (Brown & Gillespie 1957). In tissues with dense sympathetic innervation, as much as 70–80% of released neurotransmitter may be recaptured under ordinary conditions (Iversen 1975).

The above evidence strongly supports the uptake hypothesis; over the years, similar experiments have been carried out with other neurotransmitter systems. In general, the results support this hypothesis for many neurotransmitters that have high affinity uptake systems in nerve terminals. Table 1 lists the various neurotransmitter candidates that have an associated high affinity transport system. Interestingly, acetylcholine is not present on the list; it is inactivated by enzymatic breakdown. There is, instead, a high affinity transport system for the precursor, choline (Kuhar & Murrin 1978; Yamamura & Snyder 1973).

Transporters as Drug Receptors

Powerful physiologic effects result from the administration of drugs that block neurotransmitter reuptake. This observation not only supports the functional importance of transporters as described above, but has led to the exploration of transporter mechanisms and pharmacology by the drug industry to develop potentially useful therapeutic drugs. For example, a strong current interest in the serotonin transporter stems from the potent blocking actions that many antidepressant drugs, such as fluoxetine (Fuller

& Wong 1990) and imipramine (Langer & Briley 1981), have at the serotonin transporter in brain. Imipramine binding sites are also interesting because their number is reduced in the platelets of depressed patients relative to controls (Langer & Raisman 1983; Paul et al 1981). Other antidepressant drugs, such as the tricyclic antidepressant desipramine, are also inhibitors of norepinephrine uptake (Rehavi et al 1982). Although the precise role of transporters in the antidepressant action of these drugs is not yet known, the fact that so many drugs are active at these sites has resulted in a focus on transporters as a potential mechanism of action of these drugs.

Interest in the dopamine transporter also has recently increased, because of evidence that it may be the primary site of action or the "receptor" for psychostimulant drugs, such as cocaine (Bergman et al 1989; Ritz et al 1987; Spealman et al 1989). The data in support of this concept have been reviewed (Kuhar et al 1991).

The major classes of brain reuptake systems (see Figure 1) are defined by substrate and antagonist specificities, kinetic properties, and ionic requirements. Transporter antagonists have also provided evidence for

Figure 1 Model of the transmembrane topology of the dopamine transporter. Solid dots denote amino acid residues conserved between rat dopamine (DAT) (Kilty et al 1991; Shimada et al 1991a), human norepinephrine (NET) (Pacholczyk et al 1991), rat GABA (GAT) (Guastella et al 1990), rat serotonin (5-HTT) (Blakely et al 1991a), and canine betaine (BGT-1) (Yamauchi et al 1992) transporters.

functionally distinct transporter subtypes within a class, particularly among the amino acid neurotransmitter transporters. At least two γ-aminobutyric acid (GABA) transport systems have been described, based upon differential sensitivity to cis-1,3-aminocyclohexane carboxylic acid (ACHC) (Bowery et al 1976) and β-alanine (Iversen & Kelly 1975). It has been proposed that these subtypes represent distinct neuronal and glial transporters, but some studies contradict this hypothesis (Cummins et al 1982; Larsson et al 1986). Glial norepinephrine, serotonin, and glutamate uptake activities have also been described (Kimelberg & Katz 1985, 1986; Kimelberg & Pelton 1983; Roberts & Watkins 1975), thus supporting the notion that glia may be important regulators of extracellular neurotransmitter concentrations.

Solubilization, Purification, and Characterization of Transporters

A number of transporters have been solubilized, (partially) purified, and reconstituted in different ways. Several transporter molecules with available binding ligands can be purified by using a relatively simple ligand-binding technique. For example, the choline transporter has been partially purified from rat brain by using tritiated hemicholinium-3 as a binding ligand (Yamada et al 1988). Similarly, the dopamine and neuronal serotonin transporters have been solubilized and partially purified by using ligand binding as an assay (Berger et al 1990; Biessen et al 1990; Graham et al 1991; Grigoriadis et al 1989; Lew et al 1991a; Sallee et al 1989). Transporters that do not have high-affinity binding ligands require more difficult reconstitution assays for purification. The glutamate transporter (Gordon & Kanner 1988) and the sodium and chloride coupled glycine transporter (Lopez-Corcuera et al 1989) have been solubilized and partially purified in this way. In addition, the GABA transporter has been solubilized and reconstituted in proteoliposomes (Kanner et al 1989; Radian & Kanner 1986). In general, studies have shown that neurotransmitter transporters are glycoproteins with apparent molecular weights ranging from 60,000 to 85,000 Da.

Although the availability of high-affinity ligands has greatly facilitated the study of the monoamine and choline carriers, none have been purified to homogeneity. This may partly be caused by their low abundance relative to the transporters for GABA and glutamate, the apparent major inhibitory and excitatory neurotransmitters in the central nervous system.

The ACHC-sensitive GABA transporter, which to date has been the most thoroughly studied transport activity, has generated evidence for a variety of biochemical, structural, and functional characteristics. GABA transporter has been immunocytochemically localized to both neuronal and glial processes in rat brain by using polyclonal antibodies generated against the purified carrier (Radian et al 1990). In addition, limited papain digestion of the purified 80 kD transporter can generate one or more oligosaccharide-containing bands with apparent molecular weights of approximately 60 kD (Kanner et al 1989). When repurified by lectin chromatography, these proteolytic fragments can reconstitute GABA transport in liposomes, which suggests that the deleted domains are not critical to the fundamental transport activity of the protein. Reconstitution experiments that use synthetic liposomes have also explored the role of brain lipids in determining the basal transport activity; the results suggest that cholesterol must associate with the transporter in the membrane for optimal transport function (Shouffani & Kanner 1990).

Ionic Dependence and Electrogenic Properties of Transport

Studies of transport mechanisms have demonstrated that the plasma membrane transporters are cotransporters (or symporters), which transport sodium and frequently chloride ions (Kuhar & Zarbin 1978) along with the neurotransmitter (reviewed in Johnstone 1990). The accumulation of neurotransmitter is coupled to the cotransport of sodium ions down a concentration gradient, a process analogous to the active transport of many other substrates. For the ACHC-sensitive GABA carrier, kinetic experiments suggest an apparent stoichiometry in which the zwitterionic form of GABA is cotransported with 2.5 sodium ions and one chloride ion (Keynan & Kanner 1988). Although the L-glutamate transporter is also sodium-coupled, chloride is not required. In addition, glutamate transport has an absolute dependence on internal potassium; accumulated evidence suggests that potassium ions are counter-transported as part of a mechanism for recycling carriers to the extracellular side of the membrane. This overall topic has been reviewed in detail (Kanner & Schuldiner 1987) and is not reviewed further here.

An important aspect of the neurotransmitter transport process is that it is potentially electrogenic. For the sodium-dependent GABA and glutamate transporters, a net translocation of charges across the membrane and the process generates a current. These inward currents have been directly measured by using whole cell patch clamp analysis of glutamate uptake, in salamander retinal glial cells (Schwartz & Tachibana 1990; Szatkowski et al 1990) and in skate horizontal cells (Malchow & Ripps 1990). The technique has also enabled studies of internal and external ionic dependence, stoichiometry of transport, voltage dependence of transport, and the conditions that lead to reversal of the uptake (reviewed in Nicholls & Attwell 1990). Interestingly, norepinephrine (Trendelenburg 1991), GABA (Schwartz 1982, 1987), and glutamate (Szatkowski et al 1990) carriers can function in reverse, thus transporting substrates out of the cell when external potassium concentration rises and the cell is depolarized. This reversal of uptake may explain observed instances of calcium-independent, nonvesicular glutamate and GABA release (for example, see Pin & Bockaert 1989).

MOLECULAR BIOLOGY OF NEUROTRANSMITTER TRANSPORTERS

Recent molecular biologic studies of neurotransmitter transporters have greatly increased our understanding of the structure of these important synaptic proteins and provided insights into the mechanisms of their function.

Expression in Xenopus Oocytes

Xenopus oocytes have proven to be a useful heterologous expression system for characterizing neurotransmitter transport activities. They have been used to study endogenous brain mRNAs and have provided a basis for expression cloning strategies. The oocyte expression system remains an important analytical tool for characterizing the activities encoded by cloned cDNAs.

Sodium-dependent transport activity for GABA, glutamate, glycine, choline, serotonin, and dopamine has been expressed in *Xenopus* oocytes injected with mRNA isolated from discrete rat brain regions (Blakely et al 1988; Sarthy 1986). The expressed transport activities exhibit the same characteristics and pharmacology as the high-affinity, sodium-dependent transporters observed in brain slices and synaptosomal preparations. More recently, mRNA from PC12 cells, as well as transcripts from a rat midbrain cDNA library, have been used to direct the expression of cocaine-sensitive dopamine uptake in oocytes (Uhl et al 1991). Oocytes injected with human substantia nigra mRNA also demonstrate similar sodium-dependent dopamine uptake activity (Bannon et al 1990).

Expression cloning strategies that use oocytes offer a powerful approach to the isolation of genes encoding low abundance membrane proteins. This approach is most useful when the gene product of interest is encoded by a single RNA species. Several sodium cotransporters, including the bacterial Na^+/proline (Nakao et al 1987) and the mammalian Na^+/glucose (Hediger et al 1987, 1989) had previously been shown to contain single subunits. Experimental support for the idea that neurotransmitter transporters are encoded by single RNAs was obtained from oocyte expression studies that demonstrated that discrete size classes of mRNA direct the synthesis of different amino acid neurotransmitter transporters. Sucrose gradient fractionation experiments using oocytes also addressed the structural diversity of mRNAs that encode the amino acid neurotransmitter transporters (Blakely et al 1991a). The transporters for glutamate, GABA, and glycine are encoded by different size classes of RNA in different brain regions and, together with pharmacologic data, suggested a greater complexity of carrier subtypes than had been anticipated.

Purification and Cloning of a Brain GABA Transporter

A rat brain GABA transporter purified to apparent homogeneity (Radian et al 1986; Radian & Kanner 1986) provided an important reagent for the molecular cloning of a GABA transporter cDNA. Partial protein sequence information from cyanogen bromide fragments generated peptide sequences for oligonucleotide probes that were used to isolate a GABA transporter cDNA by standard plaque hybridization techniques (Guastella et al 1990). When expressed in *Xenopus* oocytes, this cDNA clone encoded a GABA transporter that was sodium- and chloride-dependent and had the pharmacologic sensitivity of the purified GABA transporter. Hydropathy analysis contributes clues to the structure of the transporter and suggests that the predicted 67 kDa protein contains 11–13 potential membrane-spanning domains. The amino-terminus lacks an identifiable signal sequence (von Heijne 1983) and is thus likely to be retained in the cytoplasm. Although several variations can be proposed, the GABA transporter structure is most readily described by a model with 12 transmembrane domains, intracellular amino- and carboxyl-termini, and a large extracellular loop with three glycosylation sites between transmembrane domains 3 and 4. (See Figure 1.)

Expression Cloning of the Norepinephrine Transporter

A COS cell expression system served as the basis for isolating a clone encoding a human norepinephrine transporter (Pacholczyk et al 1991). The SK-N-SH human neuroblastoma cell line expresses a homogeneous population of sodium-dependent catecholamine transporters with the pharmacologic characteristics of a norepinephrine transporter (Richards & Sadee 1986). In addition to catecholamines, this carrier can also transport meta-iodobenzylguanidine (mIBG), a compound structurally similar to the neuronal blocking agents guanethidine and bretylium. This analogue can be labeled with [^{125}I] and remains stable when accumulated intracellularly. A cDNA library constructed by using RNA from SK-N-SH cells was transfected into COS-1 cells; those cells that expressed specific uptake activity were identified by autoradiography by using [^{125}I]mIBG. Episomal DNA from areas of cells corresponding to autoradiographic spots was rescued by Hirt lysis and transformed into bacteria. This process led to a single, purified clone encoding a novel carrier with the substrate-specificity and pharmacologic properties, including antidepressant- and cocaine-sensitivity, expected for a norepinephrine transporter.

RNA hybridization studies with the clone confirm the presence of transporter mRNA in tissues and cell lines that synthesize the norepinephrine transporter. Interestingly, both human SK-N-SH and rat PC-12 cells contain two RNAs, 3.6 kb and 5.8 kb in length, which hybridize at high stringency to probes specific for the norepinephrine transporter cDNA. The 5.8 kb species has the expected pattern of distribution of the norepinephrine transporter, with highest expression in the adrenal gland and brainstem regions containing the locus coeruleus. The 3.6 kb RNA, which appears to arise from the same gene, is more diffusely expressed in neural tissues and could, perhaps, encode a norepinephrine carrier in glial cells.

The sequence of the norepinephrine transporter cDNA predicts a protein of 617 amino acids and a molecular weight of approximately 69 kD. The hydrophobicity plot of the norepinephrine sequence is virtually superimposable upon that of the GABA, which suggests similar membrane topographies for the two. The GABA and norepinephrine transporters also show significant primary amino acid sequence identity (46%), particularly within and adjacent to the putative transmembrane domains (see Figure 2). In fact, in one stretch of conserved residues overlapping transmembrane domains 1 and 2, 17 of 18 residues are identical in the two transporters. Perhaps surprisingly, the two proteins show little sequence similarity with other known ion-coupled cotransporters or neurotransmitter receptors; thus, they define a novel gene family of neurotransmitter cotransport proteins.

Cloning of Dopamine and Serotonin Transporters by Sequence Similarity

The significant homology between the norepinephrine and GABA transporters became the basis for the isolation of cDNA clones encoding rat (Kilty et al 1991; Shimada et al 1991a) and bovine (Usdin et al 1991) dopamine transporters, as well as a serotonin transporter (Blakely et al 1991a; Hoffman et al 1991). Degenerate oligonucleotides corresponding to regions of high sequence identity between the norepinephrine and GABA transporters served as probes or were the basis of polymerase chain reactions (PCR) to generate probes for screening cDNA libraries. Figure 2 compares the amino acid sequences predicted for the members of this gene family. The high degree of amino acid identity among the norepinephrine, dopamine, and serotonin transporters is not surprising in light of their

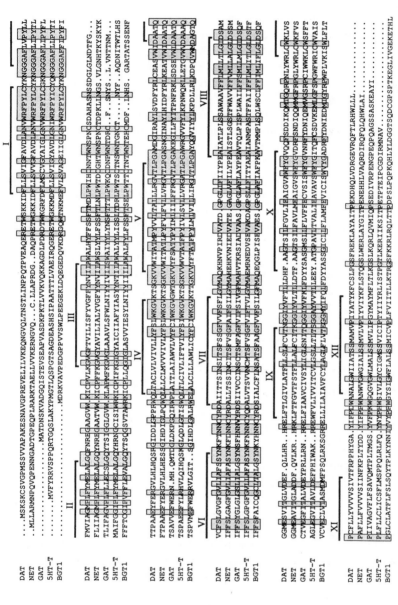

Figure 2   Alignment of the predicted amino acid sequences of the rat dopamine (DAT) (Kilty et al 1991; Shimada et al 1991a), human norepinephrine (NET) (Pacholczyk et al 1991), rat GABA (GAT) (Guastella et al 1990), rat serotonin 5-HTT (Blakely et al 1991a), and canine betaine (BGT-1) (Yamauchi et al 1992) transporters. Shaded areas represent regions of amino acid identity among all five transporters. The location of 12 potential transmembrane helices are noted.

overlapping pharmacologic sensitivities. However, several regions appear to be subject to different selective pressures. Among all members of the family, the amino termini, large extracellular loops, and carboxyl-termini show much less sequence similarity than the putative membrane-spanning domains and their immediate boundaries.

In all brain regions and cell lines studied thus far, the dopamine and serotonin transporters appear to be encoded by single 3.6 kb and 3.7 kb mRNAs, respectively. However, the existence of minor splicing variants or very closely related genes that alter the sequence of the transporter, but not the overall size or hybridization stringency of the mRNA, cannot be categorically ruled out.

Both transporters displayed an appropriate pharmacology and have been localized to brain pathways specific to each neurotransmitter. In situ hybridization studies that use a cloned dopamine transporter cDNA show intense hybridization in the substantia nigra, with more caudal sections containing an additional region of hybridization corresponding to dopaminergic cell bodies present in the ventral tegmental area (Kilty et al 1991; Shimada et al 1991a). There was also hybridization in the periphery of the olfactory bulb and in discrete regions of the hypothalamus, both sites of dopaminergic cell bodies. Within the rat brain, serotonin transporter RNA was observed within dorsal and medial subdivisions of the midbrain Raphe complex (Blakely et al 1991a).

A Sodium- and Chloride-Dependent Transporter Gene Family

Additional members of the neurotransmitter transporter gene family have been identified by using PCR primers based on regions of significant amino acid sequence similarity between the norepinephrine and GABA transporters (Clark et al 1991; Peek et al 1991; Shimada et al 1991b). These experiments have generated cDNA probes encoding many novel products with clear structural similarities to norepinephrine and GABA carriers. The regional distribution of RNAs encoding these putative transporters (Clark et al 1991) suggests that the expression of members of the family is not restricted to the nervous system. Compelling evidence for this idea is provided by the expression cloning in *Xenopus* oocytes of the transporter for the osmolyte glycine betaine from MDCK cells (Yamauchi et al 1992). This transporter, which is expressed predominantly in the renal medulla, has a predicted amino acid sequence that is remarkably similar to members of the neurotransmitter transporter gene family. This carrier catalyzes the sodium- and chloride-dependent accumulation of betaine and GABA. Table 2 summarizes some of the basic structural and pharmacologic properties of currently characterized members of this gene family.

Table 2 Properties of cloned transporters

Transporter	rat GABA[a]	hum GABA[b]	hum NE[c]	rat DA[d]	bov DA[e]	rat 5-HT[f]	dog betaine[g]
Amino acids	599 aa	599 aa	617 aa	619 aa	693 aa	607, 653 aa	614 aa
Relative molecular mass	67 kD	67 kD	69 kD	69 kD	77 kD	69, 73 kD	69 kD
Glycosylation sites[h]	3	3	3	4	3	2	2
Phosphorylation sites[h]	yes	yes	yes	yes	yes	yes	yes
Substrates	GABA	GABA	NE, DA	DA	DA	5-HT	betaine, GABA
Na$^+$ and Cl$^-$ dependent	yes	n.d.	yes	yes	yes	yes	yes
Selective inhibitors	ACHC, 2,4-DABA	n.d.	desipramine	GBR12909	GBR12909	paroxetine	yes
Other inhibitors	nipecotic acid	n.d.	cocaine, amphetamines, tricyclic antidepressants	cocaine, amphetamines	cocaine, amphetamines	cocaine, amphetamines, tricyclic antidepressants	quinidine, phloretin, beta-alanine
RNA size	4.2 kb	4.4 kb	3.6, 5.8 kb	3.6 kb	3.0 kb	3.7 kb	2.4, 3.0 kb
Tissue distribution	brain	brain	locus coeruleus, adrenal, PC12 cells, SKNSH cells	SN, VTA, hypothalamus, olfactory bulb	SN, VTA, hypothalamus, olfactory bulb	Raphe nuclei, lung, adrenal, RBL cells	kidney, MDCK cells

[a] Guastella et al 1990.
[b] Nelson et al 1990.
[c] Pacholczyk et al 1991.
[d] Kilty et al 1991; Shimada et al 1991.
[e] Usdin et al 1991. The carboxyl terminus of the bovine dopamine transporter diverges and is much longer than the three reported rat sequences for catecholamine transporters. This may be the result of a base change or sequencing error that shifts the reading frame.
[f] Blakely et al 1991; Hoffman et al 1991. The two reported rat serotonin transporter cDNAs have isolated base differences that cause the protein sequences to diverge at their carboxyl termini. Their amino-termini diverge completely. Whether these reflect actual differences or cloning artifacts remains to be determined.
[g] Yamauchi et al 1992.
[h] Predicted from consensus sequences apparent within the primary amino acid sequence.

Common Structural Motifs

The striking similarities in the sequences of the cloned neurotransmitter carriers begin to define the structural features common to this transporter gene family (Figure 2). Conserved features may point to the structural elements important in generic functions, such as sodium and chloride cotransport. Regions of divergence may suggest potential determinants of specificity for each transporter, including substrate and inhibitor binding sites. Mutagenesis studies should prove useful in defining these regions. A few aspects of the predicted structures are worth discussing.

Although the neurotransmitter transporters have no significant sequence homology with other families of carriers, the presence of 11–13 transmembrane domains appears to be a recurrent theme among both bacterial and mammalian symporters (Hediger et al 1987; Kaback 1988; Nakao et al 1987). In the neurotransmitter carrier family, transmembrane domains 1, 2 and 4–8 show the highest degree of sequence identity. The conservation of sequence in these membrane-spanning regions argues for their functional importance in transport activity, but details of their function is not yet known. Sequence identity is most pronounced at points where the polypeptide would be expected to enter or exit the membrane. This degree of conservation may reflect the importance of these residues in determining the folding pattern and orientation of helices within the membrane.

Four leucine residues arranged in a pattern resembling a leucine zipper are present in transmembrane domain 2 in norepinephrine and GABA (Pacholczyk et al 1991), but are less well conserved in the other cloned transporters. Such motifs could be the basis for dimerization of subunits or interactions with other membrane-spanning domains. Although a single gene product suffices to reconstitute transport activity in mammalian expression systems, data do not distinguish whether any of these transporters are functionally active as monomers [as suggested for the lac permease protein (Costello et al 1987)], as homomultimeric complexes [as suggested for the Na^+/glucose transporter protein (Stevens et al 1990)], or as heteromultimeric complexes. Interestingly, radiation inactivation studies on the dopamine transporter predict a size for the carrier complex that is significantly larger than the cloned transporter protein and may suggest the presence of additional modulatory subunits (Berger et al 1991).

Several discrete charged and polar residues within otherwise hydrophobic transmembrane domains are conserved between the transporters (e.g. glutamate residues in transmembrane domains 2 and 10), and at least one charged residue is unique to the monoamine carriers. For the G-protein coupled receptors, ligand-binding domains appear to reside in multiple regions largely within the hydrophobic core of the proteins. Residues implicated as part of ligand-binding domains in adrenergic receptors have been examined in site-directed mutagenesis studies, which suggest that a conserved aspartate residue in transmembrane domain 3 is critical for the binding of both agonists and antagonists with protonated amine groups (Horstman et al 1990; Strader et al 1987). By analogy, specific residues common to the catecholamine transporters, but absent in the GABA transporter, could combine to form a three-dimensional catecholamine binding site. The aspartate residue in transmembrane domain 1 (Asp 79 in the dopamine transporter) is conserved in dopamine, norepinephrine, and serotonin transporters, but is substituted by a glycine in the GABA and betaine transporters, which suggests that this residue may be selectively important for monoamine transporter function. In support of this idea, studies on the structure activity relationships of tricyclic antidepressants have led to the proposal that these agents, like norepinephrine, form bonds from a terminal ammonium group to a negatively charged residue in the norepinephrine transporter (DePaulis et al 1978; Maxwell et al 1969).

Serine residues have been proposed in the adrenergic receptors to act as a hydrogen bond donor for interactions with the catechol groups of catecholamines. Putative transmembrane domain 7, although hydrophobic in character, has a very different character with a high percentage of serine, threonine, and cysteine residues throughout the segment. In the dopamine transporter transmembrane, domains 6 and 8 also have several serine and threonine residues. Thus, this portion of the transporter could also be functionally significant in the transport of the catechol moiety of the catecholamines.

Charged residues are frequently positioned at the boundaries between hydrophilic loops and putative membrane-embedded domains. An analysis by von Heijne (1986) revealed that integral membrane proteins frequently have charged residues bordering on hydrophobic regions and that these residues may play a role in determining the correct positioning of the protein within the membrane. Focal clusters of 4–5 charged residues in the cytoplasmic loop immediately preceding transmembrane helix 9 are found in all cloned neurotransmitter transporters. Similar clusters of charged residues are evident in analogous locations in the bacterial Na^+/proline and mammalian Na^+/glucose transport proteins (Hediger et al 1989).

Membrane-buried proline residues serve an important role in transport proteins. A survey of integral membrane protein sequences (Brandl & Deber 1986) provides evidence that, although proline residues are largely excluded from membrane-spanning segments in nontransport proteins, they are evenly distributed between aqueous and membrane domains in transport proteins. Prolines have been suggested to have three potential roles within membrane-spanning helices. Local structural changes arising from regulated *cis-trans* isomerization of membrane-buried X-proline peptide bonds (X = unspecified amino acid) of transport proteins could result in the realignment of amphipathic helices to provide the conformational changes required for substrate translocation, such that a hydrophilic pore becomes accessible to a substrate ready for transport across the membrane (Brandl & Deber 1986). Prolines also introduce a bend in transmembrane helices, possibly allowing the formation of a pocket that could accommodate substrates or ligands. Finally, the more negative character of carbonyl groups of X-proline peptide bonds make them more likely to participate in hydrogen bonding or interact with cations within the membrane. The neurotransmitter transporter family has conserved proline residues in 5 of 12 (1, 2, 5, 11, 12) transmembrane domains.

The large putative extracellular loop between transmembrane domains 3 and 4 in both transporters has three predicted N-glycosylation sites, but otherwise has little sequence homology. The significance of the oligosaccharides to transport, ligand-binding, or protein targeting has not yet been explored. However, a study in which neuraminidase treatment of synaptosomes resulted in a 40% decrease in Vmax of dopamine transport (Zalecka & Erecinska 1987) suggests that carbohydrate moieties may serve an important role in transporter function. The presence of multiple (2–4) N-linked sites in the different transporters also raises the possibility that variations in glycosylation patterns could contribute to transporter heterogeneity. Biochemical studies support this idea. The apparent molecular weight of the dopamine transporter from rat nucleus accumbens is higher than that in the striatum (Lew et al 1991b), but after deglycosylation of the transporters from the two regions, the difference in apparent molecular weight is negligible (Lew et al 1992).

Although the large extracellular loop is less well conserved in the neurotransmitter transporters, two cysteines, 9 residues apart, are common to all members of the family. Dixon et al (1987) and Fraser (1989) have addressed the role of conserved cysteine residues in adrenergic receptor function. Site-directed mutagenesis of two cysteine residues within different extracellular loops of the β-2 adrenergic receptor (residues 106 and 184) characteristically alters the ligand-binding properties (Dixon et al 1987) and has led to the proposal that these conserved cysteine residues form a disulfide-linked pair. Similarly, the cysteine residues that are highly conserved within neurotransmitter transporters could form disulfide bonds and impart tertiary structure to the extracellular loop.

It is tempting to assign the regions of sequence homology common to all the transporters to functions involved in the transport of sodium. A putative "sodium-binding" domain has been proposed for a bacterial Na^+/glutamate transporter, based on a short, loosely defined region of amino acid sequence similarity present in several sodium symporters (Deguchi et al 1990). This sequence motif is not present in the neurotransmitter transporter family. However, such common motifs might not be expected, because although all the neurotransmitter transporters co-transport sodium, the stoichiometry of sodium to transmitter, the codependence on extracellular chloride, and the role of intracellular K differ between the various members of the family, and from other symporters, as well.

Regulation of Neurotransmitter Transport

The presence of serine and threonine residues that could be substrates for phosphorylation in the cytoplasmic amino- and/or carboxyl-termini of the neurotransmitter transporters cloned to date (Blakely et al 1991a; Guastella et al 1990; Hoffman et al 1991; Kilty et al 1991; Nelson et al 1990; Pacholczyk et al 1991; Shimada et al 1991a; Usdin et al 1991; Yamauchi et al 1992) supports the hypothesis that second messengers may dynamically regulate the function of these transporters. In the nervous system, membrane depolarization and second messengers, such as calcium and arachidonic acid, modulate activity or expression of choline (Boksa et al 1988; Murrin & Kuhar 1976; Saltarelli et al 1988, 1990; Simon et al 1976) and glutamate (Barbour et al 1989; Murrin et al 1978; Yu et al 1986) transport, which suggests that synaptic activity itself may influence neurotransmitter reuptake. Studies of primary astrocytes have indicated that activation of β-adrenergic receptors (Hansson & Ronnback 1989) or protein kinase C (Gomeza et al 1991) may regulate glial GABA transport. Additional studies support a role for second messenger pathways in modulating monoamine uptake activities. Both serotonin transport in a choriocarcinoma cell line (JAR cells) (Cool et al 1991) and dopamine transport in primary hypothalamic cells (Kadowaki et al 1990) are stimulated by cAMP. Inhibitors of arachidonic acid metabolism are relatively potent dopamine uptake blockers in rat striatal slices (Cass et al 1991), which suggests that arachidonic acid metabolites may be important for the regulation of dopamine transport activity.

There are likely to be other mechanisms that regulate the activities of neurotransmitter transporters. These include the direct influence of transmembrane electrochemical gradients that determine the driving force for transport, and in some instances, when altered, can lead to reversed transport. In addition, some reports have suggested that an important mechanism for regulating transport activity of facilitated glucose transporters is the subcellular redistribution of transporters from cryptic intracellular pools to the plasma membrane (reviewed in Simpson & Cushman 1986). For neurotransmitter transporters, such a mechanism could contribute to observed instances of regulated changes in the Vmax of transport. Developmental and trans-synaptic events are also likely to control the expression of brain transporter genes and lead to changes in transporter mRNA abundance (for example, Blakely et al 1991b). The availability of cloned probes should now make gene expression studies more feasible.

CONCLUSIONS

Although much emphasis has been placed on the molecular biology of proteins, such as ion channels and receptors that mediate synaptic signaling, less attention has been directed at the gene products important for the process of signal termination. Recent progress in molecular cloning and the elucidation of carrier structure establishes a new gene family encoding Na^+-dependent neurotransmitter transporters. The availability of new tools for studying reuptake processes at the molecular level should enable rapid progress in the field.

Literature Cited

Amara, S. G., Pacholczyk, T. 1991. Sodium-dependent neurotransmitter reuptake systems. *Curr. Opin. Neurobiol.* 1: 84–90

Axelrod, J., Weil-Malherbe, H., Tomchick, R. 1959. The physiological disposition of H^3-epinephrine and its metabolite metanephrine. *J. Pharmacol. Exp. Ther.* 127: 251–56

Bannon, M. J., Zue, C. H., Shibata, K., Dragovic, L. J., Kapatos, G. 1990. Expression of a human cocaine-sensitive dopamine transporter in *Xenopus laevis* oocytes. *J. Neurochem.* 54: 706–8

Barbour, B., Szatkowski, M., Ingledew, N., Attwell, D. 1989. Arachidonic acid induces a prolonged inhibition of glutamate uptake into glial cells. *Nature* 342: 918–20

Berger, P., Kempner, E. S., Paul, S. M. 1991. Radiation inactivation studies of the dopamine transporter protein. *Soc. Neurosci. Abstr.* 17: 190

Berger, P., Martenson, R., Laing, P., Thurcauf, A., DeCosta, B., et al. 1990. Photoaffinity labeling of the dopamine reuptake carrier protein using a novel high affinity ayidoderivative of GBR-12935. *Soc. Neurosci. Abstr.* 16: 13

Bergman, J., Madras, B. K., Johnson, S. E., Spealman, R. D. 1989. Effects of cocaine and related drugs in nonhuman primates III. Self-administration by squirrel monkeys. *J. Pharmacol. Exp. Ther.* 251: 150–55

Biessen, E. A., Horn, A. S., Robillard, G. T. 1990. Partial purification of the 5-hydroxytryptophan reuptake system from human blood platelets using a citalopram-derived affinity resin. *Biochemistry* 29: 3349–54

Blakely, R. D., Berson, H. E., Fremeau, R. T., Caron, M. G., Peek, M. M., et al. 1991a. Cloning and expression of a functional serotonin transporter from rat brain. *Nature* 354: 66–70

Blakely, R. D., Clark, J. A., Pacholczyk, T., Amara, S. G. 1991b. Distinct, developmentally regulated brain and mRNAs direct the synthesis of neurotransmitter transporters. *J. Neurochem.* 56: 860–71

Blakely, R. D., Robinson, M. B., Amara, S. G. 1988. Expression of neurotransmitter transpot from rat brain mRNA in *Xenopus laevis* oocytes. *Proc. Natl. Acad. Sci. USA* 85: 9846–50

Boksa, P., Mykitta, S., Collier, B. 1988. Arachidonic acid inhibits choline uptake and depletes acetylcholine content in rat cerebral cortical synaptosomes. *J. Neurochem.* 50: 1309–18

Bowery, N. G., Jones, G. P., Neal, M. J. 1976. Selective inhibition of neuronal

GABA uptake by cis-1,3-aminocyclohexane carboxylic acid. *Nature* 264: 281–84

Brandl, C. J., Deber, C. M. 1986. Hypothesis about the function of membrane-buried proline residues in transport proteins. *Proc. Natl. Acad. Sci. USA* 83: 917–21

Brown, G. L., Gillespie, J. S. 1957. The output of sympathetic transmitter from the spleen of the cat. *J. Physiol.* 138: 81–102

Burn, J. H. 1932. The action of tyramine and ephedrine. *J. Pharmacol. Exp. Ther.* 46: 75–95

Cass, W. A., Larson, G., Fitzpatrick, F. A., Zahniser, N. R. 1991. Inhibitors of arachidonic acid metabolism: effects on rat striatal dopamine release and uptake. *J. Pharmacol. Exp. Ther.* 257: 990–96

Clark, J. A., Fluet, A. A., Amara, S. G. 1991. Isolation of new members of a transporter gene family by sequence homology-based PCR. *Soc. Neurosci. Abstr.* 17: 1183

Cool, D. R., Leibach, F. H., Bhalla, V. K., Mahesh, V. B., Ganapathy, V. 1991. Expression and cAMP regulation of a high affinity serotonin transporter in human placental choriocarcinoma cell line (JAR). *J. Biol. Chem.* 266: 15750–57

Costello, M. J., Escaig, J., Matsushita, K., Viitanen, P. V., Menick, D. R., Kaback, H. R. 1987. Purified *lac* permease and cytochrome o oxidase are functional as monomers. *J. Biol. Chem.* 262: 17072–82

Cummins, C. J., Glover, R. A., Sellinger, O. Z. 1982. Beta-alanine is not a marker for brain astroglia in culture. *Brain Res.* 239: 299–302

Deguchi, Y., Yamato, I., Anraku, Y. 1990. Nucleotide sequence of *glt*S, the Na^+/glutamate symport carrier gene of *Escherichia coli* B*. *J. Biol. Chem.* 265: 21704–8

DePaulis, T., Kelder, D., Ross, S. B. 1978. On the topology of the norepinephrine transport carrier in rat hypothalamus: the site of action of tricyclic uptake inhibitors. *Mol. Pharmacol.* 14: 596–606

Dixon, R. A. F., Sigal, I. S., Candelore, M. R., Register, R. B., Scattergood, W., et al. 1987. Structural features required for ligand binding to the β-adrenergic receptor. *EMBO J.* 6: 3269–75

Fraser, C. M. 1989. Site-directed mutagenesis of β-adrenergic receptors. *J. Biol. Chem.* 264: 9266–70

Fuller, R. W., Wong, D. T. 1990. *Ann. NY Acad. Sci.* 600: 68–80

Gomeza, J., Casado, M., Gimenez, C., Aragon, C. 1991. Inhibition of high-affinity γ-aminobutyric acid uptake in primary astrocyte cultures by phorbol esters and phospholipase C. *Biochem. J.* 275: 435–39

Gordon, A. M., Kanner, B. I. 1988. Partial purification of the sodium- and potassium-coupled L-glutamate transport glycoprotein from rat brain. *Biochim. Biophys. Acta* 944: 90–96

Graham, D., Esnaud, H., Langer, S. Z. 1991. Characterization and purification of the neuronal sodium-ion-coupled 5-hydroxytryptamine transporter. *Biochem. Soc. Trans.* 19: 99–102

Grigoriadis, D. E., Wilson, A. A., Lew, R., Sharkey, J. S., Kuhar, M. J. 1989. Dopamine transport sites selectively labeled by a novel photoaffinity probe: ^{125}I-DEEP. *J. Neurosci.* 9: 2664–70

Guastella, J., Nelson, N., Nelson, H., Czyzk, L., Keynan, S., et al. 1990. Cloning and expression of a rat brain GABA transporter. *Science* 249: 1303–6

Hansson, E., Ronnback, L. 1989. Regulation of glutamate and GABA transport by adrenoceptors in primary astroglial cell cultures. *Life Sci.* 44: 27–34

Hediger, M. A., Coady, M. J., Ikeda, T. S., Wright, E. M. 1987. Expression cloning and cDNA sequencing of the Na^+/glucose cotransporter. *Nature* 330: 379–81

Hediger, M. A., Turk, E., Wright, E. M. 1989. Homology of the human intestinal Na^+/glucose and *Escherichia coli* Na^+/proline cotransporters. *Proc. Natl. Acad. Sci. USA* 86: 5748–52

Hoffman, B. J., Mezby, E., Brownstein, M. J. 1991. Cloning of a serotonin transporter affected by antidepressants. *Science* 254: 79–80

Horn, A. S. 1990. Dopamine uptake: A review of progress in the last decade. *Prog. Neurobiol.* 34: 387–400

Horstman, D. A., Brandon, S., Wilson, A. L., Guyer, C. A., Cragoe, E. J., Limbird, L. E. 1990. An aspartate conserved among G-protein receptors confers allosteric regulation of a2-adrenergic receptors by sodium. *J. Biol. Chem.* 265: 21590–95

Iversen, L. L. 1975. Uptake process of biogenic amines. In *Handbook of Psychopharmacology*, ed. L. L. Iversen, S. D. Iversen, S. H. Snyder, pp. 381–442. New York: Plenum

Iversen, L. L. 1967. *The Uptake and Storage of Noradrenaline in Sympathetic Nerves.* London: Cambridge Univ. Press

Iversen, L. L., Glowinski, J., Axelrod, J. 1966. The physiological disposition and metabolism of norepinephrine in immunosympathectomized animals. *J. Pharmacol. Exp. Ther.* 151: 273–84

Iversen, L. L., Kelly, J. A. 1975. Uptake and metabolism of γ-aminobutyric acid by neurones and glial cells. *Biochem. Pharmacol.* 24: 933–38

Johnstone, R. 1990. Ion-coupled cotransport. *Curr. Opin. Cell. Biol.* 2: 735–41

Kaback, H. R. 1988. Site-directed muta-genesis and ion-gradient driven active transport: on the path of the proton. *Annu. Rev. Physiol.* 50: 243–56

Kadowaki, K., Hirota, K., Koike, K., Ohmichi, M., Kiyama, H., et al. 1990. Adenosine 3′,5′-cyclic monophosphate enhances dopamine accumulation in rat hypothalamic cell culture containing dopaminergic neurons. *Neuroendocrinology* 52: 256–61

Kanner, B. I., Keynan, S., Radian, R. 1989. Structural and functional studies on the sodium- and chloride-coupled γ-amino-butyric acid transporter: deglycosylation and limited proteolysis. *Biochemistry* 28: 3722–28

Kanner, B. I., Schuldiner, S. 1987. Mech-anism of transport and storage of neuro-transmitters. *CRC Crit. Rev. Biochem.* 22: 1–38

Keynan, S., Kanner, B. I. 1988. γ-amino-butyric acid transport in reconstituted preparations from rat brain: coupled sodium and chloride fluxes. *Biochemistry* 27: 12–17

Kilty, J. E., Lorang, D., Amara, S. G. 1991. Cloning and expression of a cocaine-sen-sitive rat dopamine transporter. *Science* 254: 578–79

Kimelberg, H. K., Katz, D. M. 1986. Regional differences in 5-hydroxytryp-tamine and catecholamine uptake in pri-mary astrocytic cultures. *J. Neurochem.* 47: 1647–52

Kimelberg, H. K., Katz, D. M. 1985. High affinity uptake of serotonin into immuno-cytochemically identified astrocytes. *Science* 228: 889–95

Kimelberg, H. K., Pelton, E. W. 1983. High-affinity uptake of [3H]norepinephrine by primary astrocyte cultures and its inhi-bition by tricyclic antidepressants. *J. Neurochem.* 40: 1265–70

Kuhar, M. J. 1973. Neurotransmitter up-take: a tool for identifying neurotrans-mitter-specific pathways. *Life Sci.* 33: 1623–34

Kuhar, M. J., Murrin, L. C. 1978. Sodium-dependent, high affinity choline uptake. *J. Neurochem.* 30: 15–21

Kuhar, M. J., Ritz, M. C., Boja, J. W. 1991. The dopamine hypothesis of the rein-forcing properties of cocaine. *Trends Neurosci.* 14: 299–302

Kuhar, M. J., Zarbin, M. A. 1978. Synap-tosomal transport: a chloride dependence for choline, GABA, glycine, and several other compounds. *J. Neurochem.* 31: 251–56

Langer, S. Z., Briley, M. 1981. High affinity ³H-imipramine binding: a new biological tool for studies in depression. *Trends Neurosci.* 4: 28–31

Langer, S. Z., Raisman, R. 1983. Binding of [³H]imipramine and [³H]desipramine as biochemical tools for studies in de-pression. *Neuropharmacology* 22: 407–13

Larsson, O. M., Griffiths, R., Allen, I. C., Schousboe, A. 1986. Mutual inhibition kinetic analysis of γ-aminobutyric acid, taurine, and β-alanine high-affinity trans-port into neurons and astrocytes: evidence for similarity between taurine and β-ala-nine carriers in both cell types. *J. Neuro-chem.* 47: 426–32

Lew, R., Grigoriadis, D. E., Wilson, A., Boja, J. W., Simantov, R., Kuhar, M. J. 1991a. Dopamine transporter: deglyco-sylation with exo- and endoglycosidases. *Brain Res.* 539: 239–46

Lew, R., Patel, A., Vaughn, R. A., Wilson, A., Kuhar, M. J. 1992. Microhetero-geneity of dopamine transporters in rat striatum and nucleus accumbens. *Brain Res.* 584: 266–71

Lew, R., Vaughn, R., Simantov, R., Wilson, A., Kuhar, M. J. 1991b. Dopamine trans-porters in the nucleus accumbens and the striatum have different apparent molec-ular weights. *Synapse* 8: 152–53

Lopez-Corcuera, B., Kanner, B. I., Aragon, C. 1989. Reconstitution and partial puri-fication of the sodium and chloride-coupled glycine transporter from rat spinal cord. *Biochim. Biophys. Acta* 983: 247–52

Malchow, R. P., Ripps, H. 1990. Effects of gamma-aminobutyric acid on skate retinal horizontal cells: evidence for an elec-trogenic uptake mechanism. *Proc. Natl. Acad. Sci. USA* 87: 8945–49

Maxwell, R. A., Keenan, P. D., Chaplin, E., Roth, B., Eckhardt, S. B. 1969. Molecular features affecting the potency of tricyclic antidepressants and structurally-related compounds as inhibitors of the uptake of tritiated norepinephrine by rabbit aortic strips. *J. Pharmacol. Exp. Ther.* 166: 320–29

Murrin, L. C., Kuhar, M. J. 1976. Activation of high affinity choline uptake in vitro by depolarizing agents. *Mol. Pharmacol.* 12: 1082–90

Murrin, L. C., Lewis, M. S., Kuhar, M. J. 1978. Amino acid transport: alterations due to synaptosomal depolarization. *Life Sci.* 22: 2009–16

Nakao, T., Yamato, I., Anraku, Y. 1987. Nucleotide sequence of *put P*, the proline carrier gene of *Escherichia coli* K12. *MGG* 208: 70–75

Nelson, H., Mandiyan, S., Nelson, N. 1990. Cloning of the human brain GABA trans-porter. *FEBS Lett.* 269: 181–84

Nicholls, D., Attwell, D. 1990. The release

and uptake of excitatory amino acids. *Trends Pharmacol. Sci.* 11: 462–68

Pacholczyk, T., Blakely, R. D., Amara, S. G. 1991. Expression cloning of a cocaine- and antidepressant-sensitive human noradrenaline transporter. *Nature* 350: 350–53

Paul, S. M., Rehavi, M., Skolnick, P., Ballenger, J. C., Goodwin, F. K. 1981. Depressed patients have decreased binding of ^{3}H-imipramine to the platelets serotonin "transporter." *Arch. Gen. Psychiatry* 38: 1315–17

Peek, M. M., Fremeau, R. T., Caron, M. G., Blakely, R. D. 1991. Identification of multiple members of the neurotransmitter transporter gene family. *Soc. Neurosci. Abstr.* 17: 904

Pin, J.-P., Bockaert, J. 1989. Two distinct mechanisms, differentially affected by excitatory amino acids, trigger GABA release from fetal mouse striatal neurons in primary culture. *J. Neurosci.* 9: 648–56

Radian, R., Bendahan, A., Kanner, B. I. 1986. Purification and identification of the functional sodium- and chloride-coupled γ-aminobutyric acid transport glycoprotein from rat brain. *J. Biol. Chem.* 261: 15437–41

Radian, R., Kanner, B. I. 1986. Reconstitution and purification of the sodium- and chloride-coupled γ-aminobutyric acid transport glycoprotein from rat brain. *J. Biol. Chem.* 260: 11859–65

Radian, R., Ottersen, O. P., Storm-Mathisen, J., Castel, M., Kanner, B. I. 1990. Immunocytochemical localization of the GABA transporter in rat brain. *J. Neurosci.* 10: 1319–30

Rehavi, M., Skolnick, P., Brownstein, M. J., Paul, S. M. 1982. High affinity binding of [^{3}H] desipramine to rat brain: A presynaptic marker for noradrenergic uptake sites. *J. Neurochem.* 38: 889–95

Richards, M. L., Sadee, W. 1986. Human neuroblastoma cell lines as models of catechol uptake. *Brain Res.* 384: 132–37

Ritz, M. C., Lamb, R. J., Goldberg, S. R., Kuhar, M. J. 1987. Cocaine receptors on dopamine transporters are related to self-administration of cocaine. *Science* 237: 1219–23

Roberts, P. J., Watkins, J. C. 1975. Structural requirements for the inhibition of L-glutamate uptake by glia and nerve endings. *Brain Res.* 85: 120–25

Rudnick, G. 1986. ATP-driven H$^+$ pumping into intracellular organelles. *Annu. Rev. Physiol.* 48: 403

Sallee, F. R., Fogel, E. L., Schwartz, E., Choi, S. M., Curran, D. P., Niznik, H. B. 1989. Photoaffinity labeling of the mammalian dopamine transporter. *FEBS Lett.* 256: 219–24

Saltarelli, M. D., Lopez, J., Lowenstein, P. R., Coyle, J. T. 1988. The role of calcium in the regulation of [^{3}H]Hemicholinium-3 binding sites in rat brain. *Neuropharmacology* 27: 1301–8

Saltarelli, M. D., Yamada, K., Coyle, J. T. 1990. Phospholipase-A$_2$ and ^{3}H-Hemicholinium-3 binding site in rat brain: a potential second messenger role for fatty acids in the regulation of high-affinity choline uptake. *J. Neurosci.* 10: 62–72

Sarthy, V. 1986. γ-aminobutyric acid (GABA) uptake by *Xenopus* oocytes injected with rat brain mRNA. *Mol. Brain Res.* 1: 97–100

Schwartz, E. 1987. Depolarization without calcium can release GABA from a retinal neuron. *Science* 238: 350–55

Schwartz, E. 1982. Calcium-independent release of GABA from isolated horizontal cells of the toad retina. *J. Physiol.* 323: 211–27

Schwartz, E., Tachibana, M. 1990. Electrophysiology of glutamate and sodium cotransport in a glial cell of the salamander retina. *J. Physiol.* 426: 43–80

Sjöqvist, F., Taylor, P. W. Jr., Titus, E. 1967. The effects of immunosympathectomy on the retention and metabolism of noradrenaline. *Acta Physiol. Scand.* 69: 13–22

Shimada, S., Kitayama, S., Lin, C.-L., Patel, A., Nanthakumar, E., et al. 1991a. Cloning and expression of a cocaine-sensitive dopamine transporter complementary DNA. *Science* 254: 576–78

Shimada, S., Kitayama, S., Lin, C.-L., Patel, A., Nanthakumar, E., et al. 1991b. Diversity of neurotransmitter transporter cDNAs in brain libraries. *Soc. Neurosci. Abstr.* 17: 1182

Shouffani, A., Kanner, B. I. 1990. Cholesterol is required for the reconstitution of the sodium-coupled and chloride-coupled gamma-aminobutyric acid transporter from rat brain. *J. Biol. Chem.* 265: 6002–8

Simon, J. R., Atweh, S., Kuhar, M. J. 1976. Sodium-dependent high affinity choline uptake: a regulatory step in the synthesis of acetylcholine. *J. Neurochem.* 26: 909–22

Simpson, I. A., Cushman, S. W. 1986. Hormonal regulation of mammalian glucose transport. *Annu. Rev. Biochem.* 55: 1059–89

Snyder, S. H. 1970. Putative neurotransmitter in the brain: selective neuronal uptake, subcellular localization and interactions with centrally acting drugs. *Biol. Psychiatry* 2: 367–89

Spealman, R. D., Madras, B. K., Bergman,

J. 1989. Effects of cocaine and related drugs in nonhuman primates. II. Stimulant effects on schedule-controlled behavior. *J. Pharmacol. Exp. Ther.* 261: 142–49

Stevens, B. R., Fernandez, A., Hirayama, B., Wright, E. M., Kempner, E. S. 1990. Intestinal brush border membrane Na+/glucose cotransporter functions in situ as a homotetramer. *Proc. Natl. Acad. Sci. USA* 87: 1456–60

Strader, C. D., Sigal, I. S., Register, R. B., Candelore, M. R., Rands, E., Dixon, R. A. F. 1987. Identification of residues required for ligand binding to the β-adrenergic receptor. *Proc. Natl. Acad. Sci. USA* 84: 4384–88

Szatkowski, M., Barbour, B., Attwell, D. 1990. Non-vesicular release of glutamate from glial cells by reversed electrogenic glutamate uptake. *Nature* 348: 443–45

Trendelenburg, U. 1991. The TIPS lecture: Functional aspects of the neuronal uptake of noradrenaline. *Trends Pharmacol. Sci.* 12: 334–37

Uhl, G. R., O'Hara, B., Shimada, S., Zaczek, R., DiGiorgianni, J., Nishimori, T. 1991. Dopamine transporter: expression in *Xenopus* oocytes. *Mol. Brain Res.* 9: 23–29

Usdin, T. B., Mezey, E., Chen, C., Brownstein, M. J., Hoffman, B. J. 1991. Cloning of the cocaine-sensitive bovine dopamine transporter. *Proc. Natl. Acad. Sci. USA* 88: 11168–71

von Heijne, G. 1983. Patterns of amino acids near signal-sequence cleavage sites. *Eur. J. Biochem.* 133: 17–21

von Heijne, G. 1986. The distribution of positively charged residues in bacterial inner membrane proteins correlates with the trans-membrane topology. *EMBO J.* 5: 3021–27

Whitby, L. G., Axelrod, J., Weil-Malherbe, H. 1961. The fate of H^3-norepinephrine in animals. *J. Pharmacol. Exp. Ther.* 132: 193–201

Winkler, H., Apps, D. K., Fisher-Collrie, R. 1986. The molecular functions of adrenal chromaffin granules: established facts and unresolved topics. *Neuroscience* 18: 261–90

Yamada, K., Saltarelli, M. D., Coyle, J. T. 1988. Solubilization and characterization of a [^{3}H]Hemicholinium-3 binding site in rat brain. *J. Neurochem.* 50: 1759–64

Yamamura, H. I., Snyder, S. H. 1973. High affinity transport of choline into synaptosomes of rat brain. *J. Neurochem.* 21: 1355–74

Yamauchi, A., Uchida, S., Kwon, H. M., Preston, A. S., Robey, R. B., et al. 1992. Cloning of a Na^+- and Cl^--dependent betaine transporter that is regulated by hypertonicity. *J. Biol. Chem.* 267: 649–52

Yu, A. C. H., Chan, P. H., Fishman, R. A. 1986. Effects of arachidonic acid on glutamate and γ-aminobutyric acid uptake in primary cultures of rat cerebral cortical astrocytes and neurons. *J. Neurochem.* 47: 1181–89

Zaimis, E., Berk, L., Callingham, B. 1965. Morphological, biochemical and functional changes in the sympathetic nervous system of rats treated with NGF-antiserum. *Nature* 206: 1221–22

Zaleska, M. M., Erecinska, M. 1987. Involvement of sialic acid in high-affinity uptake of dopamine by synaptosomes from rat brain. *Neurosci. Lett.* 82: 107–12

Annu. Rev. Neurosci. 1993. 16:95–127

PROTEIN TARGETING IN THE NEURON

Regis B. Kelly and Eric Grote

Department of Biochemistry and Biophysics and Hormone Research Institute, University of California, San Francisco, California 94143-0534

KEY WORDS: polarity, sorting, synaptic vesicle biogenesis, axonal transport, membrane cytoskeleton interactions

INTRODUCTION

The morphology of the neuron is essential to its function. Each morphological specialization of the neuron, the dendrite, the dendritic spines, the axon, the axon hillock, the nodes of Ranvier, the active zone, and the synaptic vesicle appears to have a unique protein composition. To explain the architecture of the neuron in molecular terms, therefore, we need to know how unique proteins achieve their distributions within the cell.

Because neurons have such morphological complexity, we may wonder whether they have evolved unique mechanisms of protein sorting, or use the same mechanisms found in nonneuronal cells. Although scanty, the existing data on sorting of proteins in neuronal cells can still be adequately explained by mechanisms that exist in other cell types. What may be surprising is the diversity of sorting mechanisms that are used. In this review, we summarize evidence for selective targeting, selective retention, selective stabilization, selective exclusion, and RNA targeting.

Protein sorting is conveniently divided into two issues: how proteins are targeted to organelles and how organelles reach their cellular destination. A second major subdivision is between sorting mechanisms that are inherent to cells and occur in dissociated cells and those that are induced by a cell's interaction with other cells or with extracellular matrix. In this review, we attempt to organize the available data into these conventional categories. It is also conventional to review membrane protein targeting or cytoskeletal assembly as two separate topics. Because the morphology

95

of neurons depends so heavily on the interactions between membrane and cytoskeleton, we attempt to integrate the fields. We hope that the insights achieved from a unified picture compensate for the somewhat superficial treatment that is sometimes necessary.

PROTEIN TARGETING IN NONNEURONAL CELLS

Before examining sorting in neurons, it is useful to review some of the concepts that have evolved to explain protein sorting in cells that are simple to study.

Membrane vesicles bud from one organelle within a cell and fuse with another (for review, see Rothman & Orci 1992). Despite the constant membrane traffic between cellular compartments, organelles maintain their unique membrane protein composition. Two mechanisms by which this is achieved are selective budding and selective retention (Figure 1). Selective budding accounts, for example, for the selective removal of the mannose 6-phosphate receptor from the trans Golgi network (TGN) or for the endocytosis of the transferrin receptor from the cell surface. Selective

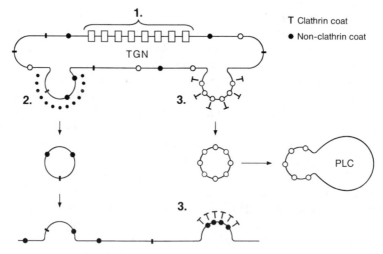

Figure 1 Membrane sorting by selective retention and selective budding in traffic from TGN to cell surface. Selective retention in the TGN (1) occurs because proteins have signals that prevent them from moving through the default pathway (2). Selective budding occurs (3) in at least two places: exit from the TGN of proteins destined for the prelysosomal compartment (PLC) and endocytosis from the cell surface. Coating mechanisms are required for budding events. Nonselective budding (2) involves coatamers and selective budding requires clathrin-coated structures. The two examples of clathrin-mediated budding (3) use different adaptins, proteins that link the clathrin coat to the membrane protein cargo.

retention keeps protein with the appropriate carboxyterminus in the rough endoplasmic reticulum, or a protein with the correct transmembrane domain in the Golgi complex. To be selectively budded, a membrane protein must have a sorting domain in its cytoplasmic sequences. The sorting domain is recognized by cytoplasmic machinery, which induces the budding of a vesicle with the correct cargo. The only cytoplasmic machinery currently known to account for selective budding is the clathrin-coated vesicle. There are two major sites of selective budding in the cell: the TGN and the cell surface (Figure 1). Each site is associated with its own type of coated vesicle. The TGN has vesicles that use the HA1 (or AP1) adaptin, a protein that is thought to link the sorting domains on the cargo proteins to the clathrin coat. Plasma membrane-coated vesicles use a similar, but different, adaptin (HA2 or AP2).

If a membrane protein is neither retained nor selectively removed from a donor organelle, it is carried by bulk flow to acceptor organelles. Intracellular membrane transport that goes by bulk flow and requires no sorting information is called a default pathway (Figure 1). Nonspecific budding from a donor compartment has been associated with a different class of coating machinery, one that involves a coat made of "coatamers" (for review, see Rothman & Orci 1992).

After a transport vesicle buds from a donor compartment, it must fuse with the correct acceptor. A vesicle leaving the TGN, for example, fuses with the cell surface. Although some form of selective docking must occur (Figure 2), little is known about the recognition events that are involved. In yeast mutants defective in small *ras*-like GTP binding proteins, the transport vesicles cannot fuse with the acceptor compartment; therefore, they accumulate in the cytoplasm. Different cellular organelles have different types of small GTP-binding proteins, now called *rab* proteins. At present, rab proteins offer perhaps the best experimental access to the selective organelle docking that must precede fusion (for review, see Pfeffer 1992).

After leaving the TGN, a post-TGN transport vesicle must recognize the appropriate docking proteins on the acceptor membrane (Figure 2). The selective docking of post-TGN vesicles with the cell surface can be facilitated by moving them along the microtubule network that radiates, in most cells, from a microtubule-organizing center near the TGN (for review, see Kelly 1990). When the microtubule network is polarized with an abundance of microtubules pointing at one region of the cell surface, preferential docking and fusion of post-TGN vesicles occur at that cell surface region (Figure 2*B*). Organelle accumulation caused by microtubule transport occurs in a wide variety of cells (for review, see Singer & Kupfer 1986). We use the term location-specific targeting to describe the type

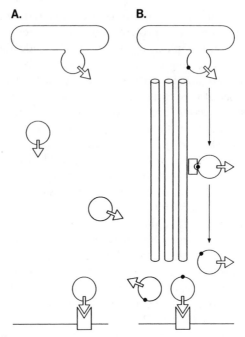

Figure 2 Organelle targeting mechanisms. In (*A*), a vesicle diffuses from the TGN until it encounters a docking site on the plasma membrane. If the docking sites are concentrated at one point, there will be a high concentration of membrane insertion at that point. Alternatively in (*B*), the plus ends of microtubules that transport the vesicles are concentrated at one region of the plasma membrane. Such location-specific targeting results in selective insertion at the tips of microtubules. If only some organelles are transported along the microtubules, then organelle-specific, location-specific targeting (not shown) can occur, in principle.

of vesicle accumulation that requires no unique property of membrane proteins, but only an asymmetry of microtubule distribution. Selectivity can be achieved if only some intracellular organelles do not move on microtubules or move in only one direction. Such selectivity is believed to explain why the Golgi complex accumulates at minus ends of microtubules near the organizing center, whereas the endoplasmic reticulum is spread throughout the cytoplasm (for review, see Kelly 1990).

Location-specific targeting requires that asymmetric microtubule distributions be generated. Microtubules grow constantly from the organizing center with their minus ends at the center and plus or growing ends distal. They are unstable and can suddenly switch from a growing state to a state of very rapid depolymerization from the plus end. If, however, the plus end makes contact with a stabilizing influence, such as a chromosomal kinetochore, it loses the capacity for catastrophic collapse and becomes stable (Kirschner & Mitchison 1986). Presumably, microtubules that are oriented toward one region of the cell surface (Figure 2) can be selectively stabilized and, thus, generate more efficient membrane transport to that region.

The plasma membrane at the surface of a cell can be subdivided into domains of different protein composition. In general, such specializations arise when cells contact either the substratum or other cells. For example, when epithelial cells are grown in culture without contact, little evidence of polarity exists. When cell-to-cell contact is made via the adhesion molecule E-cadherin, epithelium formation begins. The face of the epithelium that contacts the extracellular matrix becomes the basal side of the cell, and the other face the apical. An actin-fodrin cytoskeleton develops beneath the basolateral plasma membranes. The membrane composition of the two epithelial membranes becomes polarized. The basolateral membrane is enriched in housekeeping proteins, such as the sodium pump, and is experimentally identified as the site of insertion of the vesicular stomatitis virus glycoprotein. The apical membrane has proteins that are required for apical function, such as isosucrase maltase, and proteins with a glycolipid anchor, such as Thy-1, and is experimentally identified as the site of insertion of the viral hemagglutinin proteins. The asymmetric protein distribution can be generated by targeting newly synthesized proteins from the TGN to the correct surface, either directly or by transcytosis. After targeting, the membrane proteins are inhibited from diffusion from the correct membrane by a ring of tight junctions that separate basolateral from apical membranes. Some basolateral proteins are not selectively targeted, but instead are selectively stabilized by interaction with the cytoskeleton that lies under the basolateral surface. Selective targeting and selective stabilization both appear to require cell-to-cell contact (for review, see Nelson et al 1990).

The adhesion molecules of the cadherin and integrin family do not cause cell sticking merely by their avidity for each other. For true cell-cell attachment to occur, the cytoskeleton underlying the point of contact must be rearranged. If the cytoplasmic regions of these adhesion molecules are removed, they do not lose their extracellular binding capacity, but can no longer cause cell adhesion. The cytoplasmic tails bind to the actin-based cytoskeleton.

The interaction of a T lymphocyte with an antigen-bearing target cell illustrates the importance of cytoskeleton rearrangements in cell-cell contact and the regulation of targeting by phosphorylation. When a T-cell receptor recognizes a target cell bearing the correct antigen protein, a cascade of signaling events, including phosphorylations, is initiated. This leads to the formation of an actin-based cytoskeletal array under the T-cell plasma membrane at the site of contact and the selective delivery of the contents of secretory vesicles to the contact site. Protein kinase C-catalyzed phosphorylation leads to the strengthening of the adhesivity of integrin molecules already on the surface of the T-cell, thus cementing the interaction between T cell and target (Dustin & Springer 1991). The T cell illustrates that the two steps in cell-cell contact, the specific recognition of target, and adhesion to the target may be handled by two different types of protein, in this case the T-cell receptor and an integrin. It also illustrates how localized phosphorylation events induced by contact can cause adhesion without increasing the number of adhesion molecules at a site. Phosphorylation can give localization of a function to a site without requiring localization of the protein.

In the following sections, we discuss how these principles of sorting are pertinent to neuronal polarization, process extension, synapse formation, and synaptic vesicle biogenesis.

TARGETING OF PROTEINS INTO CELLULAR PROCESSES

Although dendrites and axons of adult neurons are readily distinguishable, neuronal cells and cell lines in culture often extend processes that are clearly neither. These processes are generally referred to as neurites. Because sorting of proteins into the three types of process—neurites, axons, and dendrites—can use different mechanisms, each process is considered separately.

Sorting into Neurites

When dissociated neurons are placed in culture, they extend short processes of undefined morphology. In early stages of hippocampal cell culture (Stage II), the microtubules in the processes have their plus ends near the periphery, which is an axonal characteristic (Baas et al 1989). However, the dendritic marker, MAP2, is present in all processes (Dotti et al 1987). Neurites, in fact, may share properties with the processes of nonneuronal cells, such as follicular dendritic cells and oligodendrocytes. The pituitary cell line, AtT-20, which has been extensively studied, can extend long processes in appropriate culture conditions. Secretory granules accumulate at the process tips by microtubule-based transport, but there appears to be little selectivity in what is transported. Constitutive secretory vesicles, endosomes, and ER markers, such as immunoglobulin heavy chain binding protein (BiP), are also found in the growing tips (Matsuuchi et al 1988;

Tooze et al 1989). As has been noted by Letourneau & Shattuck (1989), such process extension by addition of membrane at the tips is highly reminiscent of membrane addition at the growing edge of fibroblasts. The latter is an example of location-specific targeting (Figure 2). There is, as yet, no evidence for preferential insertion of some categories of membrane proteins at neurite tips. Accumulation at the tips may therefore result from location-specific targeting by microtubules and involve no selective sorting of membranes.

Two major proteins that are identified in growth cones are GAP43 (van Lookeren Campagne et al 1989) and pp60^{c-src} (Maness et al 1988). Both proteins have the properties of membrane-bound regulatory proteins. GAP-43 is the target of protein kinase C; binds calmodulin when dephosphorylated; regulates PIP kinase; interacts with G$_o$, thus stimulating GTP exchange; and associates with the cytoskeleton (Meiri & Gordon-Weeks 1990). It certainly has the hallmarks of a protein that regulates the actin-based cytoskeleton of the growth cone lamellipodium (Forscher 1989). pp60^{c-src} is the target of many kinases, including protein kinase C (Bjelfmen et al 1990); executes tyrosine phosphorylation; and had a well-documented interaction with the actin-based cortical cytoskeleton (Sobue 1990; Tsukita et al 1991). Both proteins have fatty acids on their amino terminus and require their amino terminus for targeting to the membrane (Resh 1989; Zuber et al 1989b).

There is good evidence that pp60^{c-src} and GAP43 can influence neurite extension. Expression of GAP43 by transfection in fibroblasts induces process extention (Zuber et al 1989a), and overexpression in PC12 cells promotes process extension (Yankner et al 1990). Expression of pp60^{v-src} in PC12 cells also promotes neurite extension (Alema et al 1985). On the other hand, neither GAP43 nor pp60^{c-src} is essential. Lamellipodia in nonneuronal cells are functional, despite the absence of GAP43. Furthermore, process extension is normal in PC12 cells that lack GAP43 (Baetge & Hammang 1991). A transgenic mouse lacking pp60^{src} has no dramatic abnormalities in its nervous system (Soriano et al 1991).

The target protein to which GAP43 binds is unknown, but is likely to be nonneural, as GAP43 accumulates in lamellipodial regions in transfected fibroblasts (Zuber et al 1989a). GAP43 has the properties of a protein that binds to lamellipodial structures and regulates them without being essential.

How are we to explain the concentration of microtubules in processes? To ensure that sufficient microtubules are directed from the Golgi complex to the growth cone, microtubules with plus ends near the growth cone must be stabilized, according to the hypothesis of dynamic instability (Kirschner & Mitchison 1986). There is now some evidence in favor of selective stabilization of microtubules in nerve cell processes. By studying fluorescent tubulin in living frog neurons, it was possible to observe microtubules extending into and retracting from the growth cone in real time. The observed kinetics were consistent with those predicted from other measurements of dynamic instability in frog (Tanaka & Kirschner 1991). A key observation is that microtubules do not just follow the lamellipodia, but guide it. The direction a process will take can be predicted from the turning of its microtubules, both in cultured cells (Tanaka & Kirschner 1991) and in pioneer cells of the grasshopper in situ (Sabry et al 1991). Microtubule turning appears to involve selective stabilization, at least in part, but the molecular basis for stabilization is unknown.

A plausible scenario to account for process extension is that a local increase occurs in the concentration of calcium or some other signaling factor, thus initiating neurite extension by triggering assembly of actin into a lamellipodial structure. Microtubules that grow toward a lamellipodium are stabilized. Membranes containing newly synthesized proteins are transported to the plus end of the stabilized microtubules, where they accumulate between the ends of the microtubules and the actin filaments of the lamellipodia. Such accumulation could reflect inefficient retrograde transport to the cell body, rather than any specific targeting event. Process extension requires fusion of the membranous structures with the growth cone of plasma membrane. Peripheral and integral membrane proteins that interact with the lamellopodial elements tend to be selectively retained at the growth cone. It is, therefore, not surprising that proteins that seem to show the most marked localization at the growth cones (Letourneau & Shattuck 1989) are those known to interact with the actin cytoskeleton, such as N-cadherin, GAP43, or pp60^{c-src}. In conclusion, the key targeting mechanisms associated with neurite extension appear to be location-specific, microtubule-based targeting, but may include selective retention, via interaction with the actin-based cytoskeleton of the growth cone.

Extension of Axonal Processes

In culture, the axon and the neurite differ most dramatically in their relative rate of growth. At Stage III of culture in vitro, a single neurite of a hippocampal cell changes to become an axonal process. Apparently, the key factor that determines which neurite becomes an axon is length. When the length of one neurite exceeds that of others by more than 10–15 μm, it acquires the rapid extension rate characteristic of an axon, while suppressing the capacity to other neurites to become axons (Goslin & Banker 1989). Associated with an enhanced rate of extension of the axonal process in Stage III of development is the accumulation of GAP43 at the axonal tip (Goslin et al 1990), and its disappearance from the tips of other processes. The synaptic vesicle proteins synapsin and synaptophysin also move out of the cell body to accumulate in the growth cones and distal axons (Fletcher et al 1991).

To explain how one neurite is selected to be the axon, Goslin & Banker (1989) suggest that a regulatory protein, such as GAP43, is present in limiting amounts that depend on axonal length. Because GAP43 can enhance lamellipodial extension and accumulate selectively at axonal tips, it is suspected of playing a major role in the neurite to axon conversion. Unfortunately, it is not yet possible to know whether the change in GAP43 targeting causes the conversion, or is caused by it. Because other proteins, such as synapsin and synaptophysin, also accumulate at axonal growth cones, it may be more likely that the conversion is caused by a sudden increase in the efficiency of transport down one neurite, or selective retention at one growing tip. How such changes might occur, however, is completely mysterious.

The microtubule-associated proteins, tau and MAP1, are restricted to axons. They are induced before process extension in PC12 cells (Drubin et al 1985, 1988). Data from transfected cells implicate tau directly in axon extension. Introduction of tau by transfection causes dispersed microtubules in fibroblast cell lines to bundle together, separated by an approximately 20 nm spacing (Kanai et al 1989; Lewis et al 1989). When tau is expressed to high levels in insect cells by using a baculovirus expression system, long processes, thin and cylindrical, are extended (Knops et al 1991). Studies in which tau expression is suppressed by using an antisense RNA oligonucleotide have confirmed the role of tau in axonal extension. Caceres & Kosik (1990) studied rat cerebellar neurons in culture, which also extend neurites, only one of which develops into a neuron. When antisense oligonucleotides are added to the culture medium, tau synthesis is inhibited and extension of the stable axon is suppressed, whereas neurite extension is unaffected (Caceres & Kosik 1990). These results are consistent with a role of tau in either initiating conversion of a neurite to an axon, or stabilizing a newly converted process. Arguing, however, against conversion is the observation that tau expression appears to come after the conversion in dissociated rate cerebral cells (Kosik & Finch 1987). The apparent conflict between systems may arise because dissociated cells are regenerating processes lost in isolation, not generating them de novo.

Expression or activation of a microtubule bundling factor, such as tau, could account for both the conversion of the neurite to axonal morphology and the acquisition of a higher rate of growth. Axons are cylindrical and lack ribosomes. Depolymerization of microtubules allows ribosomes to enter the axon (Baas et al 1987). Bundling of microtubules could, therefore, cause ribosome exclusion from the axon. The presence of bundled microtubules in growth cones precedes the constriction of the axonal membrane; thus, bundling may be involved in converting an amorphously shaped growth cone into a cylindrical axon (Sabry et al 1991; Tanaka & Kirschner 1991). Bundling may also stimulate microtubule extension by enhancing the net microtubule growth rate (Tanaka & Kirschner 1991) and by facilitating microtubule translocation (Reinsch et al 1991). It would be surprising, nonetheless, if tau were the only factor required for neurite to axon conversion.

Although much is known about neurite to axon conversion in dissociated hippocampal cells, the information may be more pertinent to regenerating processes than axon extension in vitro. In some cells in culture (Bruckenstein & Higgins 1988a,b) and in pioneer neurons in vivo (Lefcort & Bentley 1989), a single axonal process is extended. In the simplest model of axon extension, the expression of such proteins as tau could cross-link the microtubules, which could in turn convert the first neurite extended into an axon of uniform diameter and fast growth rate. If cross-linked microtubules lead to more efficient transport, then axonal accumulation of such proteins as GAP43, synaptophysin, and synapsin could also be explained. Any process that is subsequently extended would become a dendrite.

Targeting of Proteins to Dendrites

Because dendritic morphology is highly variable and subject to such an array of influences, it is particularly important for neurobiologists to understand their biogenesis.

Dendrites differ from axons in both their formation and their composition. Axons form in the absence of protein synthesis and in serum-free medium, whereas dendritic development requires protein synthesis and the presence of growth factors or supporting cells (Bruckenstein & Higgins 1988a,b; Lein & Higgins 1991; Rousselet et al 1990). During development, the axons form first, followed considerably later by dendritic proliferation (Bruckenstein et al 1989; Ramoa et al 1988; Voyvodic 1987). Dendrites have the machinery for protein synthesis and are enriched in specific cytoskeletal proteins, such as MAP2 (for review, see Matus 1990), dystrophin (Lidov et al 1990), ankyrin (Kordeli & Bennett 1991), and spectrin isoforms (Lazarides et al 1984), which are absent from axons. They also have the very unusual property of antiparallel microtubules (Baas et al 1988). Dendrites and axons should, therefore, be considered as quite different morphological specializations with different targeting mechanisms.

The two isoforms of MAP2, MAP2A, and MAP2B are closely associated with dendritic arborization. Like the tau protein, MAP2 is a microtubule cross-linking protein, especially when it is correctly phosphorylated (Brugg & Matus 1991). Inhibition of its expression by antisense oligonucleotides can prevent retinoic acid-induced neurite extension in embryonal carcinoma cells in culture (Dinsmore & Solomon 1991). Inhibition may, however, be indirect and result from effects on cell division. The appearance of MAP2 does not correlate with dendrite extension. MAP2 is detectable in neuritic and axonal extensions in immature neurons and PC12 cells (Dotti et al 1987; Fischer et al 1991; Rousselet et al 1990). It may therefore be involved in neurite extension, but not in the conversion of neurites to dendrites.

Which factors might cause the restriction of MAP2 to the dendrite? The first possibility is that MAP2 is targeted to the dendrites and excluded from the axon. The high packing density of axonal cytoplasm could, for example, prevent the access of selected proteins (Rousselet et al 1990). Arguing against an exclusion mechanism is the observation that, although ribosomes are excluded from PC12 axonal-like processes, MAP2 is not (Fischer et al 1991). In a direct demonstration that MAP2 was not excluded from axons, Okabe & Hirokawa (1989) injected labeled MAP2 into dissociated neurons. The MAP2 was initially recovered in both dendritic and axonal processes, but disappeared from axons with time in culture. The possibility that MAP2 might bind to some dendrite-specific location is supported by evidence that MAP2 was readily extracted from the axon, but not from the dendrite. Selective stabilization of proteins that assemble into the cytoskeleton is a well-established phenomenon (for review, see Kelly 1991a). If MAP2 binds to the dendritic cytoskeleton and acquires a slower turnover time, its relative concentration in the dendrite would increase.

The asymmetric distribution of MAP2 within the cell may result from selective binding to dendritic elements. As yet unknown is what those dendritic elements might be. Alternatively, because MAP2 needs to be phosphorylated correctly to bind (Brugg & Matus 1991), dendritic binding might be caused by dendritic kinases, rather than by specific binding proteins. MAP2 is phosphorylated when the NMDA receptor is activated (Halpain & Greengard 1990). Activation of dendritic receptors by synaptic input cannot, however, account for the development of dendrites, because this is an inherent property of neurons, albeit sensitive to external regulation. Dissociated hippocampal and cerebral cells express MAP2-rich processes (Dotti et al 1987; Kosik & Finch 1987; Rousselet et al 1990) that have antiparallel microtubules (Baas et al 1987) without cell contact.

Another possible mechanism for the accumulation of MAP2 in dendrites is very unusual, namely mRNA targeting to dendrites. There is good evidence that mRNAs, especially those encoding cytoskeletal proteins, are not uniformly distributed throughout the cell (for review, see Lawrence & Singer 1991). Such asymmetries in mRNA distribution have been seen in muscle and epithelial cells, fibroblasts, and frog and Drosophila oocytes. The targeting of mRNA seems to depend on an intact cytoskeleton (Peter et al 1991; Sundell & Singer 1991; Yisraeli et al 1990). The levels of MAP2 mRNA are enriched in dendrites compared with tubulin and neuro-filament-68 mRNAs (Garner et al 1988; Kleiman et al 1990). Dendritic targeting of mRNA gives the dendrite the capacity for local protein synthesis, which is probably advantageous in allowing local dendritic remodeling, at sites far removed from the cell body. Still left unresolved is how RNA targeting occurs. One possibility is movement toward the minus ends of microtubules.

The other striking characteristic of the dendrite is the polarity of its microtubules. Whereas all axonal microtubules have plus or growing ends distal to the cell body, dendrites have microtubules of both polarities (Baas et al 1987). This unconventional arrangement of microtubules is also found in mitral neurons (Burton 1988). In the conventional arrangement, the microtubules splay out from a microtubule organizing center, which is located near the Golgi complex (for review, see Kelly 1990). The conventional arrangement is lost in at least one major cell type, the epithelial cell. As epithelial cells make contact, the centrioles migrate to the apical surface, and apical sites now nucleate the growth of microtubules that run perpendicular to the plane of the epithelium, with minus ends apical (Bacallao et al 1989; Mogensen et al 1989). The presence of antiparallel microtubules in hippocampal dendrites tells us that the dendrites of inter-neurons must also contain microtubule organizing elements. It is perhaps noteworthy that epithelial cells accumulate actin mRNA near the minus ends of their microtubules (Cheng & Bjerknes 1989) as do neurons, consistent with mRNA transport to the minus ends.

Because of the currently popular association between basolateral membranes of epithelial and dendritic membranes (Dotti et al 1991; Dotti & Simons 1990), it is worth noting that in two properties, the selective accumulation of mRNA and the ability to nucleate microtubule assembly, dendrites resemble apical membranes of epithelial cells, not basolateral ones. Further, when neurons develop in vivo, they express in their dendrites Thy-1, a protein often associated with apical epithelial membranes (Xue et al 1991). This contrasts with what was found in dissociated hippocampal neurons (Dotti et al 1991). To what extent polarity in epithelial cells resembles polarity in neurons is, therefore, not clear. The key issue is whether the preferential association of viral hemagglutinin proteins with axonal membranes and vesicular stomatitis virus protein with dendritic ones (Dotti & Simons 1990) is caused by the operation of the same sorting mechanisms that gives protein targeting in epithelial cells or to different mechanisms. Viral protein sorting in epithelial cells requires cell-cell contact and a barrier to membrane diffusion. Because neither is required for the type of polarized distribution seen in dissociated hippocampal neurons, the sorting mechanism involved may differ from that involved in generating the viral protein asymmetries. The viral proteins may interact, for example, with cytoskeletal components that are asymmetrically distributed in the dissociated neurons.

PROTEIN REDISTRIBUTION DURING SYNAPTOGENESIS

Synapses form when a growth cone touches its target. Because the conversion of growth cone to synapse can be very fast (Buchanan et al 1989; Haydon et al 1990), growth cones are expected to contain many of the elements needed to form a synapse. The growth cone proteins GAP43 and pp60$^{c\text{-src}}$ are also present in the synapse and may function during transmitter release (Barnekow et al 1990; Dekker et al 1990). Synapsin and synaptic vesicles are also present in the immature axon (Fletcher et al 1991) and accumulate at foci only when synaptic contact is made. All the correct components must be present to allow synapse formation. NG-108 cells can form synapses in culture only after the levels of synapsin IIb are elevated by transfection (Han et al 1991). Because synaptic vesicle components are already present in the growth cone, synapse formation involves a retargeting of already existing components.

It is possible to mimic the transition from a growth cone to a synapse at least partially in vitro. When cAMP is elevated in an Aplysia growth cone (Forscher et al 1987), it undergoes a dramatic, reversible rearrangement, reminiscent of the morphology changes during synaptogenesis. The membranes normally in the center of the growth cone migrate to the periphery, thus replacing the actin sheets of the lamellipodium. Such endings resemble "transitional elements," quiescent growth cones seen in neuronal cell cultures in which the GAP43-labeled subcortical cytoskeleton disintegrates and synaptic vesicles accumulate (Burry 1991).

There are many potential signaling mechanisms that might trigger growth cone metamorphosis (for reviews, see Heidemann & Buxbaum 1991; Strittmatter & Fishman 1991). One major protein in the growth cone that could well regulate this transition is a member of the trimeric G protein family, G_o. Initially, trimeric G proteins were thought to respond only to external signals, but they are not known to play an intracellular role in regulating membrane traffic (Donaldson et al 1991; Stow et al 1991). It is not known what regulates G_o in vivo, but in vitro the rate of binding of GTPγs to G$_o$ is enhanced by GAP43 (Strittmatter et al 1990). GAP43 could be a regulator of G$_o$, or its substrate. We also know that GAP43, a major substrate of protein kinase C, can be phosphorylated in growth cones. Phosphorylation is a late event that occurs approximately at the time that axons reach their targets and only near the growth cone, not in the proximal axons or cell body (Meiri et al 1991). An appealing possibility is that protein kinase C phosphorylation of GAP43 is associated with the adhesion of a growth cone to its postsynaptic target. A lack of GAP43 has been associated with reduced adhesivity of PC12 cells (Baetge & Hammang 1991). Two of the factors that influence growth cones, carbachol and nerve growth factor, can stimulate GAP43 phosphorylation (Meiri & Burdick 1991; van Hooff et al 1989). This may, therefore, be another case in which local function is achieved by local phosphorylation.

Cell-cell contact must also induce postsynaptic specializations. Although dendritic growth is an inherent property of neurons, stabilizing the dendritic morphology clearly requires cell-cell contact (Ramoa et al 1988). We know most about the formation of postsynaptic specializations at the neuromuscular junction. The protein agrin is transported to nerve terminals and secreted into the contact zone between nerve and muscle, thus causing the aggregation of acetylcholine receptors (Campanelli et al 1991). Aggregation of receptors also involves a rearrangement of the subplasmalemmal cytoskeleton. A 43 kd protein and a dystrophin-like protein cocluster with the receptor (Froehner 1991). Cell surface zones enriched in membrane proteins almost always correlate with specializations of a subplasmalemma cytoskeleton, even in nonsynaptic regions. The accumulation of sodium channels and pumps at nodes of Ranvier and the axon hillock seems to be defined by both axoglial junctions and interaction with portions of the cytoskeleton, particularly an isoform of ankyrin R (Kordeli et al 1990). A second isoform of ankyrin Ro, initially identified in erythrocytes, is also present in the dendrites and cell bodies of some neurons where its patchy distribution suggests association with postsynaptic specializations (Kordeli & Bennett 1991). Another cytoskeletal protein, dystrophin, also has a patchy, somatodendritic pattern in some central nervous system neurons (Lidov et al 1990). Ankyrin binds spectrin to membranes, and dystrophin is in the same protein family as spectrin (Dubreuil 1991). These findings suggest that postsynaptic specializations require the targeting of both receptors and cytoskeletal elements to the point of cell-cell contact. The cytoskeletal elements appear to play a crucial role in neural function, as mice deficient in either ankyrin Ro or dystrophin have strong neurological aberrations (Kordeli & Bennett 1991; Lidov et al 1990).

Targeting of membrane proteins in the neuron, therefore, has the same two requirements as other cases of cell surface specialization. The region of cell surface must contact another cell or matrix component, and the subcortical cytoskeleton must be reorganized. The universal, but somewhat unexpected, requirement for cytoskeletal arrangement is not yet explained satisfactorily. The cytoskeleton might aggregate the adhesion molecules, prevent their degradation, change their adhesivity, or allow the assembly of a diverse population of membrane proteins at one site.

In conclusion, synapse formation involves assembly from locally available materials that is triggered by cell-cell contact and assisted by cytoskeletal rearrangements. Local control of synapse formation makes sense, given the distances of some dendrites and axons from the cell body. Local control can include localized translation of dendritic mRNA and localized activation of protein kinases.

GENERATION OF SYNAPTIC VESICLES

A defining characteristic of neurons with chemical synapses is the presence of secretory vesicles, which release their contents in response to calcium influx into the presynaptic nerve terminal. The protein composition of the synaptic vesicle is now very well defined (Sudhof & Jahn 1991). Some proteins, such as the proton pump, are shared with other organelles; some, such as the transmitter transporters, are present only in subclasses of vesicles; and others, such as synaptotagmin and cytochrome b561, are also present in dense core synaptic vesicles. A final class of proteins, almost unique to synaptic vesicles, is typified by synaptophysin and synaptobrevin. From the enrichment of such proteins as synaptophysin and synaptic vesicles, we can infer that there are synaptic vesicle-specific sorting domains and the cellular machinery to segregate them from other cellular proteins. In this section, we discuss the possible sites at which synaptic vesicle protein sorting may occur. Subsequently, we consider potential mechanisms of sorting.

Sorting of Synaptic Vesicle Proteins at the Nerve Terminal

To release neurotransmitter by exocytosis, the membrane of the synaptic vesicle must fuse with the plasma membrane of the nerve terminal. Considerable experimental evidence has shown that after exocytosis, the membrane of the synaptic vesicle is recovered and refilled with neurotransmitter from cytoplasmic pools (reviewed in Heuser 1989). Recycling of synaptic vesicle proteins requires that they be sorted away from the plasma membrane proteins. There are three basic mechanisms by which the segregation may occur.

The first mechanism (Figure 3*A*) proposes that a pore forms between vesicle and plasma membranes, but the vesicle does not flatten out into the membrane. Although there is evidence for pore formation between protein-containing secretory vesicles and the cell surface (Chow et al 1992; Spruce et al 1990), there is no direct evidence for pore formation with synaptic vesicles. The predicted structure of the major synaptic vesicle protein synaptophysin does, however, bear some resemblance to that of channel formers. As Linstedt & Kelly (1991a) have argued, it is difficult to reconcile the pore-forming model with the extensive data that show that exocytosis depletes the nerve terminal of synaptic vesicles, especially when endocytosis is inhibited. If pore formation occurs, it cannot be the exclusive mechanism of neurotransmitter release.

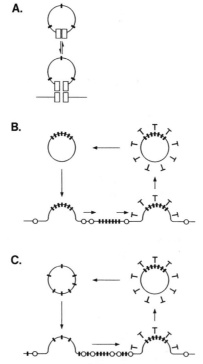

Figure 3   Three models of exocytosis. (*A*) The vesicles do not flatten out and so no coat is necessary. (*B*) Vesicles flatten out; thus, a coat is necessary to recognize synaptic vesicle proteins and internalize them. (*C*) The proteins of the synaptic vesicle mix with the plasma membrane. Recycling requires three steps: segregation of the vesicle and plasma membrane proteins, recognition by a coat, and internalization.

In the second model (Figure 3*B*), the synaptic vesicle flattens into the plane of the plasma membrane, but the vesicle proteins do not mix with the other membrane proteins. Because it requires energy to curve a membrane, it is generally believed that budding from a donor organelle requires cytoplasmic machinery of the clathrin or coatamer type (see Introduction). To recycle synaptic vesicle membrane in this scheme, one predicts that cytoplasmic domains of the aggregated synaptic vesicle proteins are recognized, thus triggering coat-mediated endocytosis. In this model, therefore, two steps are needed: recognition of the aggregate of synaptic vesicle proteins and internalization. For secretory vesicle exocytosis in endocrine cells, the secretory granule membrane proteins apparently remain patchy on the cell surface and do not mix freely (De Camilli et al 1976; Patzak & Winkler 1986). No such evidence is currently available for synaptic vesicle exocytosis.

In the third model (Figure 3*C*), synaptic vesicle proteins mix with the plasma membrane after exocytosis. To recover the synaptic vesicle membranes, some mechanism must exist to reassemble them into a cluster before internalization. In this model, each time a synaptic vesicle recycles, a step is required in addition to recognition and internalization. Dispersed synaptic vesicle proteins must be resorted from plasma membrane proteins and clustered into an aggregate. Although the extra step appears wasteful, it is not without precedent. Recycling LDL receptors are thought to mix freely with the other membrane proteins before clustering into a coated pit. The only direct evidence for mixing comes from the classic experiments of Heuser & Reese (1973), who noted in freeze-fracture experiments that large intramembranous particles, abundant in synaptic vesicles, dispersed from the active zone after exocytosis and then reassembled at the site of endocytosis.

A major difficulty of the third model is to explain how synaptic vesicle proteins are segregated from other membrane proteins, such as the LDL receptor, that are also undergoing clathrin-mediated endocytosis. The strongest evidence that such segregation occurs was generated by studying neuroendocrine cells. Cells such as the pheochromocytoma cell line PC12 have small, clear vesicles with a size and density similar to rat brain synaptic vesicles (Wiedenmann et al 1988). These vesicles, which we call endocrine synaptic vesicles, contain at least five synaptic vesicle membrane proteins. They arise by endocytosis, but exclude the receptors for transferrin and LDL (Cameron et al 1991; Clift-O'Grady et al 1990; Linstedt & Kelly 1991b). Although the function of the endocrine synaptic vesicles needs further examination, current evidence suggests a role in paracrine communication between endocrine cells (Reetz et al 1991) that is mechanistically similar to synaptic transmission. Despite the uncertainty regarding function, the presence in PC12 cells of synaptic vesicles that exclude endocytotic markers allows us to study in detail sorting events during synaptic vesicle biogenesis.

If mixing of membrane proteins occurs after exocytosis, sorting of synaptic vesicle membrane proteins might not occur at the cell surface, but in nerve terminal endosomes. In epithelial cells, both the transferrin receptor and the polyIg receptor are taken up by endocytosis from the basolateral surface and delivered to the endosome. In the endosome, the polyIg receptor is segregated away from the transferrin receptor into transcytotic vesicles (for review, see Apodaca et al 1991). There is circumstantial evidence that endosomes are involved in synaptic vesicle biogenesis. In PC12 cells, at least 80% of the endosomes labeled by the uptake of labeled transferrin contain synaptophysin (E. Grote and R. B. Kelly, unpublished). Immunofluorescence microscopy has also shown considerable overlap between the straining patterns for synaptophysin and internalized transferrin (Johnston 1989). Synaptophysin is also found in endosomes when transfected into nonneuronal cells (Cameron et al 1991; Johnston et al 1989; Linstedt et al 1991b). The presence of synaptic vesicle proteins in endosomes does not prove a role in biogenesis, however. Targeting to endosomes may be a default pathway taken by synaptic vesicle proteins when the capacity to make synaptic vesicles is inadequate or absent.

Direct evidence for a major role of endosomes in synaptic vesicle biogenesis remains elusive. Endosomes can be observed in some presynaptic nerve terminals, especially following a variety of treatments, including intense transmitter release and low temperature (Schaeffer & Raviola 1978). Coated vesicles can be seen budding from intraterminal membranous compartments in *Drosophila shibire* mutants maintained at nonpermissive temperatures (Koenig & Ikeda 1989), but the vacuoles were not identified. Studies with such tracers as horseradish peroxidase have suggested a role of endosomes in vesicle recycling (Miller & Heuser 1984), as has treatment of frog retinal photoreceptors with the weak base ammonium chloride (Sulzer & Holtzman 1989).

In the model presented in Figure 3C, all the machinery for collecting synaptic vesicle proteins and budding off a synaptic vesicle is present in the nerve terminal and is used during each recycling event. In the other two models, the segregation of synaptic vesicle proteins from other membrane proteins has already occurred at another site. Given the role of TGN in sorting, the most obvious prediction of the first two models is that the sorting of vesicle proteins occurs there first, a prediction examined in the next section.

Sorting of Synaptic Vesicle Proteins in the TGN

Newly synthesized synaptic vesicle proteins are likely to follow the conventional secretory pathway through the endoplasmic reticulum and Golgi apparatus to the TGN, where sorting at membrane proteins usually occurs. In the TGN, synaptic vesicle proteins may be sorted directly into newly formed synaptic vesicles or transported to the plasma membrane in vesicles that carry other nerve terminal membrane proteins. Biogenesis of the vesicle then occurs for the first time at the nerve terminal (Figure 3C). If synaptic vesicles are generated from the TGN, newly synthesized synaptic vesicle proteins should be detected in synaptic vesicles undergoing axonal transport to presynaptic nerve terminals. Synaptic vesicles were not identified in early electron micrographs of vesicles that accumulate proximal to a blockade of axonal transport (Tsukita & Ishikawa 1980).

A class of organelles, termed VP_0, isolated from the electric organ of Torpedo, has been proposed to be newly synthesized synaptic vesicles (Stadler & Kiene 1987). These VP_0 vesicles are similar in size to the recycled VP_1 and VP_2 synaptic vesicles, but are less dense, and have been isolated both from the nerve terminal and from axons. In support of the proposition that they are de novo synaptic vesicles, they contain synaptic vesicle antigens and exclude neurosecretory granule contents. Although they do not contain neurotransmitters, they are capable of acetylcholine uptake (Agoston et al 1989; Stadler & Kiene 1987). The possibility that VP_0 vesicles are constitutive secretory vesicles has not yet been rigorously excluded. To demonstrate that these vesicles are authentic de novo synaptic vesicles, it is necessary to demonstrate that they exclude such markers as acetylcholine esterase and agrin. It is also important to demonstrate that the fusion of these putative synaptic vesicles is regulated.

Two other recent reports are consistent with axonal transport of synaptic vesicles. Newly synthesized proteins undergoing fast axonal transport have been isolated from retinal optic nerve axons in a light density component that is composed primarily of 80 nm vesicles, slightly larger than the expected size of synaptic vesicles (Morin et al 1991). Synaptophysin, kinesin, and the glucose transporter, a plasma membrane marker, are all found in vesicles of this size. It was claimed, without presenting data, that antibodies against the cytoplasmic domain of synaptophysin did not bind to vesicles containing the glucose transporter. If documented, this result would imply sorting of synaptic vesicle and plasma membrane proteins in the cell body. Another important discovery was made by using the *unc-104* mutant of *C. elegans* (Hall & Hedgecock 1991; Otsuka et al 1991). Although axons form in this mutant, they are devoid of synaptic vesicles. Structures with the dimensions of synaptic vesicles, although with denser cores than usual, accumulate in the cell body. What is needed to prove that synaptic vesicles can form in the neuronal somata of *C. elegans* is evidence that the putative synaptic vesicles lack other nerve terminal membrane markers.

An alternative model of synaptic vesicle biogenesis proposed that synaptic vesicle membrane proteins would be delivered to the presynaptic plasma membrane as components of dense core granule membranes. Although synaptophysin is a component, albeit minor, of dense core granules (Lowe et al 1988; Obendorf et al 1988), recent experiments have conclusively demonstrated that synaptic vesicles are not generated from components recycled from dense core granule membranes (Cutler & Cramer 1990). Pulse-chase analysis revealed that synaptophysin can be detected in the synaptic vesicles of PC12 cells before secretion from dense core granules. Furthermore, a variant of the AtT20 cell line, which is defective in dense core granule biogenesis, still contains endocrine synaptic vesicles (Matsuuchi & Kelly 1991).

A third possibility is that synaptic vesicle membrane proteins are delivered to the presynaptic plasma membrane as a component of constitutive secretory vesicles. Evidence favoring this model was obtained by following the fate of synaptophysin labeled in the TGN of PC12 cells with $[^{35}S]SO_4$ (Regnier-Vigouroux et al 1991). Because protein components of the regulated secretory pathway are readily labeled with $[^{35}S]SO_4$, it was assumed that any synaptophysin taking the regulated route would be labeled. $[^{35}S]SO_4$ labeled synaptophysin exited the TGN in vesicles of the size and density of constitutive secretory vesicles, but distinct from dense core granules. Synaptophysin appeared at the cell surface after a 10-minute chase, but could not be detected in synaptic vesicles until after a 60-minute chase. These data show that in PC12 cells, the majority of the newly synthesized membrane proteins enters synaptic vesicles via the recycling pathway.

Mechanisms for Targeting Proteins to Synaptic Vesicles

To create a synaptic vesicle, their membrane proteins must be sorted away from other membrane proteins and concentrated into a small region of membrane, before budding. Whatever the intracellular site of sorting, TGN or nerve terminal, the mechanism probably involves both the inherent tendency of vesicle proteins to self-associate and cellular machinery, such as coated pits, which can interact with cytoplasmic domain of vesicle proteins. To what extent coat proteins might drive the self-association is not currently known. In this section, we summarize the evidence for protein aggregation, coated vesicle formation, and targeting domains.

AGGREGATION   In detergent extracts from synaptosomes, synaptic vesicle membrane proteins remain associated with others in large aggregates. Complexes containing a mixture of synaptic vesicle proteins, but excluding a plasma membrane marker, can be identified by immunoprecipitation from detergent solubilized preparations of rat brain synaptosomes (Bennett et al 1992). Aggregates contained different combinations of vesicle proteins, depending on the detergent used. Dialysis from octylglucoside to CHAPs allowed the association of a larger complex from smaller ones, thus demonstrating that these complexes are not native to synaptic vesicles, but rather reflect the inherent tendency of synaptic vesicle proteins to self-associate. Self-association could prevent mixing of synaptic vesicle proteins after fusion (Figure 3B) or promote vesicle formation either at the TGN, at the plasma membrane, or in endosomes. These data must be interpreted cautiously, however, until we have evidence that the protein associations detected in detergent are pertinent in vivo.

COATED VESICLES   Clathrin coated vesicles are the only cytoplasmic structures with a well-characterized role in the sorting of membrane proteins (Figure 1). The discovery that approximately 30% of brain-coated vesicles contain synaptic vesicle proteins (Pfeffer & Kelly 1985) supports the morphological evidence that suggests a role of clathrin-coated vesicles in synaptic vesicle recycling. Clathrin-mediated synaptic vesicle recycling is strongly supported by the morphology of *Drosophila shibire* nerve terminals, in which endocytosis is blocked by maintaining the flies at nonpermissive temperatures (Koenig & Ikeda 1989). Knowing that one class of adaptins (HA1) is used in lysosome biogenesis in the TGN, and another class (HA2) in receptor-mediated endocytosis, it makes sense to look for a third class, one used in neurons for the segregation of synaptic vesicle proteins into clathrin-coated structures away from lysosomal proteins or endocytotic receptors. Possible candidates are the neural-specific assembly protein (AP3) (Murphy et al 1991) and auxilin (Ahle & Ungewickell 1990). Although these proteins are not themselves adaptin components, they may interact with the HA1 or HA2 adaptins to modify their specificities.

If endosomes are the site of biogenesis of synaptic vesicles, clathrin-coated vesicles might carry both synaptic vesicle proteins and endocytotic receptors together from the cell surface to the endosome (Figure 4). A new coating mechanism, not involving clathrin, could cause the segregation of synaptic vesicle proteins from the others in the endosome. How recycling vesicles or transcytotic vesicles bud selectively from the endosome is not yet understood. Clearly, they become concentrated in the tubules that extend from endosomes, but it is not known how.

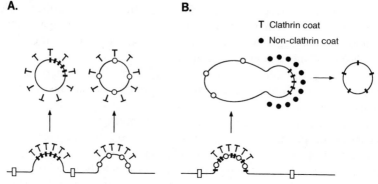

Figure 4   Endocytosis of synaptic vesicle proteins from the nerve terminal plasma membrane. (*A*) Synaptic vesicle proteins and other proteins undergoing endocytosis, such as the transferrin receptor, are segregated into two different types of coated pit. (*B*) They are internalized together by a clathrin-coated vesicle and segregated in the endosome. The budding mechanism associated with the endosomes is not known.

TARGETING DOMAINS OF VESICLE PROTEINS   Recovery of synaptic vesicle membrane after exocytosis requires internalization sequences that are recognized by the budding mechanism. If the vesicle proteins form one or more aggregates (Bennett et al 1992), then only one protein in the aggregate need have internalization domains. A protein in the aggregate that lacks internalization domains must have information that specifies selective association. All proteins in the synaptic vesicle, therefore, must have information for either selective budding or selective association with a protein that does.

Although we expect that recombinant DNA technology will reveal the location and nature of the budding and association domains, all that is currently available is information on endocytotic targeting of synaptophysin. If synaptophysin missing its carboxy-terminal cytoplasmic domain is expressed in fibroblasts, it is no longer targeted to endosomes (Linstedt & Kelly 1991c). Conversely, if this cytoplasmic region of synaptophysin replaces the cytoplasmic tail of the LDL receptor, the chimera recycles through early endosomes with great efficiency (Y. Kaneda and R. B. Kelly, unpublished).

In the amino acid sequence of synaptophysin, there must be information targeting it to synaptic vesicles, in addition to its endosomal targeting signal. Multiple targeting signals have been identified on the cytoplasmic tails of the mannose 6-phosphate receptor and the polymeric immunoglobulin receptor (Breitfeld et al 1990; Lobel et al 1989). We must await the in vitro mutagenesis studies that will identify the sorting domains of synaptophysin, synaptobrevin, and the other sequenced synaptic vesicle proteins.

TARGETING OF SYNAPTIC VESICLES

In addition to understanding how proteins are targeted to synaptic vesicles, it is important to know what causes synaptic vesicle targeting within the neuron. Synaptic vesicle proteins leaving the TGN first show sorting into axons at Stage III of hippocampal development before synapse development. After synapse development, they become localized more tightly to the points of synaptic contact (Fletcher et al 1991). In the mature synapse, there are two intracellular destinations for synaptic vesicles. A small fraction of the vesicles must be docked at relax sites on the plasma membrane, to allow rapid response to stimulation. The majority often form a reserve pool in the vicinity of the active zone. Before undergoing exocytosis, the reserve pool vesicles must dock at the active zone. The three issues in synaptic vesicle targeting are, therefore, targeting to axons, docking sites, and the reserve pool.

Axonal Transport of Synaptic Vesicle Components

We think conventionally of synaptic vesicle components transported to the nerve terminal by fast axonal transport. Might synaptic vesicles also be transported into dendrites? Because dendrodendritic synapses are frequently observed, dendritic transport of synaptic vesicles must be possible, at least sometimes. Accumulation at the synaptic terminals might, therefore, reflect selective retention, rather than selective targeting.

To reach the nerve terminal, synaptic vesicle proteins are transported through the axon (reviewed in Sheetz et al 1989). Although the vesicles in which transport occurs have not been identified, synaptic vesicle membrane proteins undergo fast axonal transport in both the anterograde and retrograde directions (Booj et al 1989). Calcitonin gene-related protein (CGRP), which is secreted from neurosecretory granules, is transported only in the anterograde direction (Booj et al 1989).

Anterograde axonal transport is directed toward the distal or plus ends of microtubules. Microtubules in dendrites, by contrast, are oriented in both directions (Baas et al 1989). If a vesicle carrying newly synthesized synaptic vesicle proteins had a plus end-directed motor, it would only be transported toward the terminal of an axon, whereas it could travel in both the anterograde and retrograde directions in dendrites. Such a vesicle would preferentially accumulate in axons. In a developing neuron, however, other mechanisms must exist for targeting, because antiparallel microtubules are not observed until after synaptic vesicles are selectively localized in axons (Fletcher et al 1991).

Two classes of motor proteins, which are proposed to move organelles along microtubules, have been purified: kinesin and cytoplasmic dynein (reviewed in Brady 1991). In vitro, kinesin moves organelles toward the plus end of microtubules, whereas dynein transports organelles to minus ends. Thus, kinesin was expected to be the motor for anterograde transport to nerve terminals. Antibodies to kinesin block anterograde and retrograde transport, but inhibitor studies have failed to identify kinesin definitively as the motor for anterograde transport. No synaptic vesicle binding site has yet been identified for a microtubule motor.

Kinesin was the first identified member of a family of motor proteins. Although mutants in the *C. elegans* kinesin heavy-chain homologue *unc-116* had effects on microtubule dependent structures in cell bodies, synaptic vesicle targeting was normal (Hall et al 1991). As mentioned earlier, a mutant in the kinesin-like gene *unc-104* develops axons, but accumulates synaptic vesicle-like structures in the cell body (Hall & Hedgecock 1991; Otsuka et al 1991).

Another way to target vesicles to axons selectively is by exclusion from dendrites. Microtubule bundling proteins may regulate the accessibility of microtubules as substrates for organelle transport. By using in vitro kinesin motility assays, MAP2, but not tau, was found to inhibit motility (Heins et al 1991). Because MAP2 is found in dendrites and tau in axons of mature neurons, axonal microtubules may carry synaptic vesicles to their plus ends more efficiently than dendritic microtubules.

Targeting of Synaptic Vesicles Within the Nerve Terminal

There are two pools of synaptic vesicles near each synapse: an active pool, which is docked to the presynaptic membrane, and a reserve pool held in place by interactions with the actin-based cytoskeleton.

Two synaptic vesicle membrane proteins, synaptophysin and synapto-tagmin, interact with components of the presynaptic plasma membrane, thus making them candidates for involvement in vesicle docking. Physo-phillin was identified by its ability to bind to immobilized synaptophysin and then localized to the presynaptic plasma membrane (Thomas & Betz 1990). Interestingly, mild protease treatment has shown that physophillin binds to a region of synaptophysin other than the carboxy-terminal cyto-plasmic tail, which contains the signal for endocytosis.

Synaptotagmin binds specifically to the α-latrotoxin receptor immo-bilized on α-latrotoxin affinity columns, thus implicating the latrotoxin receptor as a synaptotagmin binding protein (Petrenko et al 1991). The massive stimulation of synaptic vesicle exocytosis by α-latrotoxin suggests a role for the α-latrotoxin receptor/synaptotagmin complex in fusion, perhaps by docking synaptic vesicles at the active zone. However, the ω-conotoxin receptor, which is a calcium channel, also binds synaptotagmin (Leveque et al 1992) and is an alternative candidate for a docking receptor.

PC12 variants lacking synaptotagmin are normal in their exocytosis mechanisms (Shoji-Kasai et al 1992), however, which suggests that synaptotagmin docking may be essential.

The use of nonhydrolysable GTP analogues has led to the realization that GTP binding proteins are important regulators of many aspects of membrane traffic (for review, see Pfeffer 1992). As mentioned in the Introduction, the rab family of GTPase has been associated with regulating vesicle transport. rab3a is one of several GTP binding proteins that bind to synaptic vesicles. In cultured cells, rab3a is associated with only those synaptophysin positive structures found in the tips of processes (Matteoli et al 1991). Consistent with a role in recycling and docking, rab3a remains with the plasma membrane (Matteoli et al 1991) and does not dissociate into the cytoplasm.

Synaptic vesicles that are not docked at the active zone, but are in the reserve pool, are associated with a dense actin/fodrin cytoskeletal meshwork (Hirokawa et al 1989). Synapsins form the link between synaptic vesicles and the cytoskeleton. The interaction of synapsins with synaptic vesicles and cytoskeletal elements is negatively regulated by phosphorylation at identified sites mediated by Ca/calmodulin-dependent protein kinase II and cAMP-dependent protein kinase (De Camilli et al 1990). Within the axon, cytoskeletal proteins, including synapsins in their phosphorylated form, are transported via the slow component of axonal transport and do not appear to interact with synaptic vesicle membrane proteins, which are transported more rapidly (Petrucci et al 1991). The importance of synapsin for the localization of synaptic vesicles was demonstrated by the induction of varicosities-containing synaptic vesicle upon transfection of synapsin IIb into cultured neurons (Han et al 1991).

TARGETING OF PROTEINS TO DENSE CORE SECRETORY VESICLES

In addition to small, clear synaptic vesicles, neurons have neuropeptide-containing dense core secretory vesicles or granules. The dense core secretory granules contain proteins, which are released exocytotically. Neuropeptide-secreting cells respond relatively sluggishly to cell stimulation (Chow et al 1992). After release, neuropeptides affect cells at some distance away from themselves, over a time-scale of seconds to minutes. In contrast, synaptic vesicles contain no protein, release their contents in milliseconds, and regulate in a paracrine fashion only cells with which they are in contact. Some membrane proteins, such as synaptotagmin and cytochrome b561 and components of the proton pump, appear to be targeted to both types of vesicle. To our knowledge, no membrane protein has yet been found that is unique to all dense core secretory granules that is absent in synaptic vesicles. The peripheral membrane proteins pp60$^{c\text{-}src}$ and rab3A appear to be targeted to both classes of vesicle (Darchen et al 1990; Fischer von Mollard et al 1990; Grandori & Hanafusa 1988; Linstedt et al 1992; Parsons & Creutz 1986). Neuropeptides are, of course, unique to secretory granules.

Most of the newly synthesized peptide hormones arriving in the TGN are segregated from constitutively secreted proteins into immature secretory granules. Because the area has already been extensively reviewed (Kelly 1991b), we summarize only the major points and focus on new developments.

Although Chung et al (1989) report that sorting involves a hormone-specific carrier protein, the more favored model of sorting at this moment is that the soluble proteins destined to be secretory granule content selectively associated with each other, and then the aggregate binds selectively to secretory granule membrane proteins (Chanat & Huttner 1991). Transfection experiments between cell types have revealed that the sorting mechanism is remarkably conserved across tissue types. It is reasonable to assume, therefore, that if a neuropeptide is expressed in a neuron, it will be targeted to a dense core secretory granule. In Aplysia, two different neuropeptide fragments from the same precursor protein are targeted to different secretory granules (Fisher et al 1988). Expression of the same precursor protein in mouse pituitary AtT-20 gives only one polypeptide in secretory granules. The other exited constitutively (Jung & Scheller 1991).

The ability to observe sorting into secretory granules in vitro has revealed some details of the sorting process. The immature secretory granule is light in density. Maturation of the secretory granule is either by fusion of small granules (Tooze et al 1991), by the removal of excess material (Grimes & Kelly 1992), or by both. If the excess material that is removed includes constitutively secreted proteins, then sorting is not an instantaneous event, but continues during maturation. This idea has already been invoked to explain sorting in exocrine cells (van Zastrow & Castle 1987).

Although the sorting domains that target secretory granule content proteins have not been defined, there is recent progress in the targeting of secretory granule membrane proteins. A membrane protein of platelet α granules and endothelial Weibel-Palade bodies, P-selectin, is targeted to endocrine secretory granule membranes (Disdier et al 1992; Koedam et al 1992). By making chimeric proteins, the cytoplasmic tail can be shown to contain the required sorting domains (Disdier et al 1992). It is not known how the secretory granule membrane proteins are segregated away from other TGN membrane proteins. Immature secretory granules are, however, associated with clathrin-coated structures (Orci et al 1987; Tooze & Tooze 1986).

After exocytosis in neuroendocrine cells, the secretory granule proteins are recycled and used again to make a new generation of secretory granules. One possibility is that secretory granule proteins have sorting information that first targets to the secretory granule and then targets their selective retrieval from the cell surface and return to the Golgi complex. An alternative possibility is that the secretory granule membrane proteins recycle by a generic recycling path for all membrane proteins. Recent measurements of membrane protein recycling rates in ricin-resistant PC12 cells have revealed the existence of a major generic recycling path (Green & Kelly 1992), and so favor the latter model.

CONCLUSION

A neuron acquires its distinctive morphology by targeting different proteins to different locations within the cell. Although the field of neuronal protein targeting is in its infancy, the basic outlines can be discerned. The neuron uses many of the sorting mechanisms familiar in other cell types: microtubule stabilization, microtubule-dependent organelle transport, membrane protein self-association, enhanced resistance to turnover, clathrin-coated vesicles, mRNA targeting, local phosphorylation, and cytoskeletal rearrangements at points of cell-cell contact. The impressive morphology of the neuron seems to result from utilizing all of these sorting mechanisms, with small modifications. These modifications include neural-specific elements of the clathrin-coated vesicle, neural-specific microtubule-associated proteins, and sophisticated regulation of growth cone movement.

ACKNOWLEDGMENTS

We owe a great deal of thanks to members of our lab, especially Mani Ramaswami, Sam Green, and Frank Bonzelius. We also thank Leslie Spector for preparing and editing the manuscript. The authors acknowledge the financial support from the National Institutes of Health (grants NS15927 and NS09878).

Literature Cited

Agoston, D., Dowe, G., Whittaker, V. 1989. Isolation and characterization of secretory granules storing a vasoactive intestinal polypeptide-like peptide in Torpedo cholinergic electromotor neurones. *J. Neurochem.* 52: 1729–40

Ahle, S., Ungewickell, E. 1990. Auxilin, a newly identified clathrin-associated protein in coated vesicles from bovine brain. *J. Cell Biol.* 111: 19–29

Alema, S., Casalbore, P., Agostini, E., Tato, F. 1985. Differentiation of PC12 pheochromocytoma cells induced by v-src oncogene. *Nature* 316: 557–59

Apodaca, G., Bomsel, M., Arden, J., Breit-feld, P., Tang, K., Mostov, K. 1991. The polymeric immunoglobulin receptor. A model protein to study transcytosis. *J. Clin. Invest.* 87: 1877–82

Baas, P., Black, M., Banker, G. 1989. Changes in microtubule polarity orientation during the development of hippocampal neurons in culture. *J. Cell Biol.* 109: 3085–94

Baas, P., Deitch, J., Black, M., Banker, G. 1988. Polarity orientation of microtubules in hippocampal neurons: uniformity in the axon and nonuniformity in the dendrite. *Proc. Natl. Acad. Sci. USA* 85: 8335–39

Baas, P., Sinclair, G., Heidemann, S. 1987. Role of microtubules in the cytoplasmic compartmentation of neurons. *Brain Res.* 420: 73–81

Bacallao, R., Antony, C., Dotti, C., Karsenti, E., Stelzer, E., Simons, K. 1989. The subcellular organization of Madin-Darby canine kidney cells during the formation of a polarized epithelium. *J. Cell Biol.* 109: 2817–32

Baetge, E., Hammang, J. 1991. Neurite outgrowth in PC12 cells deficient in GAP-43. *Neuron* 6: 21–30

Barnekow, A., Jahn, R., Schartl, M. 1990. Synaptophysin: a substrate for the protein tyrosine kinase pp60c-src in intact synaptic vesicles. *Oncogene* 5: 1019–24

Bennett, M., Calakos, N., Kreiner, T., Scheller, R. 1992. Synaptic vesicle membrane proteins interact to form a multimeric complex. *J. Cell Biol.* 116: 761–74

Bjelfman, C., Meyerson, G., Cartwright, C., Mellstrom, K., Hammerling, U., Pahlman, S. 1990. Early activation of endogenous pp60src kinase activity during neuronal differentiation of cultured human neuroblastoma cells. *Mol. Cell. Biol.* 10: 361–70

Booj, S., Goldstein, M., Fischer-Colbrie, R., Dahlstrom, A. 1989. Calcitonin gene-related peptide and chromogranin a: presence and intra-axonal transport in lumbar motor neurons in the rat, a comparison with synaptic vesicle antigens in immunohistochemical studies. *Neuroscience* 30: 479–501

Brady, S. T. 1991. Molecular motors in the nervous system. *Neuron* 7: 521–33

Breitfeld, P., Casanova, J., McKinnon, W., Mostov, K. 1990. Deletions in the cytoplasmic domain of the polymeric immunoglobulin receptor differentially affect endocytotic rate and postendocytotic traffic. *J. Biol. Chem.* 265: 13750–57

Bruckenstein, D., Higgins, D. 1988a. Morphological differentiation of embryonic rat sympathetic neurons in tissue culture. I. Conditions under which neurons form axons but not dendrites. *Dev. Biol.* 128: 324–36

Bruckenstein, D., Higgins, D. 1988b. Morphological differentiation of embryonic rat sympathetic neurons in tissue culture. II. Serum promotes dendritic growth. *Dev. Biol.* 128: 337–48

Bruckenstein, D., Johnson, M., Higgins, D. 1989. Age-dependent changes in the capacity of rat sympathetic neurons to form dendrites in tissue culture. *Brain Res.* 46: 21–32

Brugg, B., Matus, A. 1991. Phosphorylation determines the binding of microtubule-associated protein 2 (MAP2) to microtubules in living cells. *J. Cell Biol.* 114: 735–43

Buchanan, J., Sun, Y.-A., Poo, M.-M. 1989. Studies of nerve-muscle interactions in Xenopus cell culture: fine structure of early functional contacts. *J. Neurosci.* 9: 1540–54

Burry, R. 1991. Transitional elements with characteristics of both growth cones and presynaptic terminals observed in cell cultures of cerebellar neurons. *J. Neurocytol.* 20: 124–32

Burton, P. 1988. Dendrites of mitral cell neurons contain microtubules of opposite polarity. *Brain Res.* 473: 107–15

Caceres, A., Kosik, K. 1990. Inhibition of neurite polarity by tau antisense oligonucleotides in primary cerebellar neurons. *Nature* 343: 461–63

Cameron, P., Sudhof, T., Jahn, R., De Camilli, P. 1991. Colocalization of synaptophysin with transferrin receptors: implications for synaptic vesicle biogenesis. *J. Cell Biol.* 115: 151–64

Campanelli, J., Hoch, W., Rupp, F., Kreiner, T., Scheller, R. 1991. Agrin mediates cell contact-induced acetylcholine receptor clustering. *Cell* 67: 909–16

Chanat, E., Huttner, W. 1991. Milieu-induced, selective aggregation of regulated secretory proteins in the trans-Golgi network. *J. Cell Biol.* 115: 1505–20

Cheng, H., Bjerknes, M. 1989. Asymmetric distribution of actin mRNA and cytoskeletal pattern generation in polarized epithelial cells. *J. Mol. Biol.* 210: 541–49

Chow, R. H., von Rudin, L. 1992. Delay in vesicle fusion revealed by electrochemical monitoring of single secretory events in adrenal chromaffin cells. *Nature* 356: 60–63

Chung, K., Walter, P., Aponte, G., Moore, H. 1989. Molecular sorting in the secretory pathway. *Science* 243: 192–97

Clift-O'Grady, L., Linstedt, A., Lowe, A., Grote, E., Kelly, R. 1990. Biogenesis of synaptic vesicle-like structures in a pheochromocytoma cell line PC-12. *J. Cell Biol.* 110: 1693–1703

Cutler, D., Cramer, L. 1990. Sorting during transport to the surface of PC12 cells: divergence of synaptic vesicle and secretory granule proteins. *J. Cell Biol.* 110: 721–30

Darchen, F., Zahraoui, A., Hammel, F., Monteils, M.-P., Tavitian, A., Scherman, D. 1990. Association of the GTP-binding protein Rab3A with bovine adrenal chromaffin granules. *Proc. Natl. Acad. Sci. USA* 87: 5692–96

De Camilli, P., Benfenati, F., Valtorta, F., Greengard, P. 1990. The synapsins. *Annu. Rev. Cell Biol.* 6: 433–60

De Camilli, P., Peluchetti, D., Meldolesi, J. 1976. Dynamic changes of the luminal plasmalemma in stimulated parotid acinar cells. *J. Cell Biol.* 70: 59–74

Dekker, L., De Graan, P., De Wit, M., Hens, J., Gispen, W. 1990. Depolarization-induced phosphorylation of the protein kinase C substrate B-50 (GAP-43) in rat cortical synaptosomes. *J. Neurochem.* 54: 1645–52

Dinsmore, J., Solomon, F. 1991. Inhibition of MAP2 expression affects both morphological and cell division phenotypes of neuronal differentiation. *Cell* 64: 817–26

Disdier, M., Morrissey, J. H., Fugate, R. D., Baintain, D. F., McEver, R. P. 1992. Cytoplasmic domain of P-selectin (CD62) contains the signal for sorting into the regulated secretory pathway. *Mol. Cell. Biol.* 3: 309–21

Donaldson, J. G., Kahn, R. A., Lippincott-Schwartz, J., Klausner, R. D. 1991. Binding of ARF and β-COP to Golgi membranes: possible regulation by a trimeric G protein. *Science* 254: 1197–99

Dotti, C., Banker, G., Binder, L. 1987. The expression and distribution of the microtubule-associated proteins Tau and microtubule-associated protein 2 in hippocampal neurons in the rat in situ and in cell culture. *Neuroscience* 23: 121–30

Dotti, C., Parton, R., Simons, K. 1991. Polarized sorting of glypiated proteins in hippocampal neurons. *Nature* 349: 158

Dotti, C., Simons, K. 1990. Polarized sorting of viral glycoproteins to the axon and dendrites of hippocampal neurons in culture. *Cell* 62: 63–72

Drubin, D., Feinstein, S., Shooter, E., Kirschner, M. 1985. Nerve growth factor-induced neurite outgrowth in PC12 cells involves the coordinate induction of microtubule assembly and assembly promoting factors. *J. Cell Biol.* 101: 1788–1807

Drubin, D., Kobayashi, S., Kellogg, D., Kirschner, M. 1988. Regulation of micro-

tubule protein levels during cellular morphogenesis in nerve growth factor-treated PC12 cells. *J. Cell Biol.* 106: 1583–91

Dubreuil, R. 1991. Structure and evolution of the actin crosslinking proteins. *Bio-Essays* 13: 219

Dustin, M., Springer, T. 1991. Role of lymphocyte adhesion receptors in transient interactions and cell locomotion. *Annu. Rev. Immunol.* 9: 27–66

Fischer, I., Richter-Landsberg, C., Safaei, R. 1991. Regulation of microtubule associated protein 1 (MAP2) expression by nerve growth factor in PC12 cells. *Exp. Cell Res.* 194: 195–201

Fischer von Mollard, G., Sudhof, T., Jahn, R. 1991. A small GTP-binding protein dissociates from synaptic vesicles during exocytosis. *Nature* 349: 79–81

Fisher, J., Sossin, W., Newcomb, R., Scheller, R. 1988. Multiple neuropeptides derived from a common precursor are differentially packaged and transported. *Cell* 54: 813–22

Fletcher, T., Cameron, P., De Camilli, P., Banker, G. 1991. The distribution of synapsin I and synaptophysin in hippocampal neurons developing in culture. *J. Neurosci.* 11: 1617–26

Forscher, P. 1989. Calcium and polyphosphoinositide control of cytoskeletal dynamics. *Trends Neurosci.* 12: 468–74

Forscher, P., Kaczamarek, L., Buchanan, J., Smith, S. 1987. Cyclic AMP induces changes in distribution and transport of organelles within growth cones of Aplysia bag cell neurons. *J. Neurosci.* 7: 3600–11

Froehner, S. 1991. The submembrane machinery for nicotinic acetylcholine receptor clustering. *J. Cell Biol.* 114: 1–7

Garner, C., Tucker, R., Matus, A. 1988. Selective localization of mRNA for cytoskeletal protein MAP2 in dendrites. *Nature* 336: 674–77

Goslin, K., Banker, G. 1989. Experimental observations on the development of polarity by hippocampal neurons in culture. *J. Cell Biol.* 108: 1507–16

Goslin, K., Schreyer, D., Skene, J., Banker, G. 1990. Changes in the distribution of GAP-43 during the development of neuronal polarity. *J. Neurosci.* 10: 588–602

Grandori, C., Hanafusa, H. 1988. pp60c-src is complexed with a cellular protein in subcellular compartment involved in exocytosis. *J. Cell Biol.* 107: 2125–35

Green, S. A., Kelly, R. B. 1992. Transport of rapidly internalized membrane proteins from the cell surface to the Golgi apparatus is efficient and non-selective in PC12 cells. *J. Cell Biol.* 117: 47–55

Grimes, M., Kelly, R. B. 1992. Intermediates in the constitutive and regulated secretory pathways formed in vitro from semi-intact cells. *J. Cell Biol.* 117: 531–38

Hall, D. H., Hedgecock, E. M. 1991. Kinesin-related gene *unc*-104 is required for axonal transport of synaptic vesicles in *C. elegans. Cell* 65: 837–47

Hall, D. H., Plenefisch, J., Hedgecock, E. M. 1991. Ultrastructural abnormalities of kinesin mutant *unc*-116. *J. Cell Biol.* 115: 389a

Halpain, S., Greengard, P. 1990. Activation of NMDA receptors induces rapid dephosphorylation of the cytoskeletal protein MAP2. *Neuron* 5: 237–46

Han, H.-Q., Nichols, R. A., Rubin, M. R., Bahler, M., Greengard, P. 1991. Induction of formation of presynaptic terminals in neuroblastoma cells by synapsin IIb. *Nature* 349: 697–99

Haydon, P., Zoran, M., Man-son-Hing, H., Sievers, E., Doyle, R. 1990. A relation between synaptic specificity and the acquisition of presynaptic properties. *J. Physiol. (Paris)* 84: 111–20

Heidemann, S., Buxbaum, R. 1991. Growth cone motility. *Curr. Opin. Neurobiol.* 1: 339–45

Heins, S., Song, Y. H., Wille, H., Mandelkow, E., Mandelkow, E. M. 1991. Effect of MAP2, MAP2c and tau on kinesin-dependent microtubule motility. *J. Cell. Sci.* 14: 121–24

Heuser, J. 1989. Review of electron microscopic evidence favouring vesicle exocytosis as the structural basis for quantal release during synaptic transmission. *Q. J. Exp. Physiol.* 74: 1051–69

Heuser, J., Reese, T. 1973. Evidence for recycling of synaptic vesicle membrane during transmitter release at the frog neuromuscular junction. *J. Cell Biol.* 57: 315–44

Hirokawa, N., Sobue, K., Kanda, K., Harada, A., Yorifuji, H. 1989. The cytoskeletal architecture of the presynaptic terminal and molecular structure of synapsin 1. *J. Cell Biol.* 108: 111–26

Johnston, P., Cameron, P., Stukenbrok, H., Jahn, R., De Camilli, P., Sudhof, T. 1989. Synaptophysin is targeted to similar microvesicles in CHO and PC12 cells. *EMBO J.* 8: 2863–72

Jung, L., Scheller, R. 1991. Peptide processing and targeting in the neuronal secretory pathway. *Science* 251: 1330–35

Kanai, Y., Takemura, R., Oshima, T., Mori, H., Ihara, Y., et al. 1989. Expression of multiple tau isoforms and microtubule bundle formation in fibroblasts transfected with a single tau cDNA. *J. Cell Biol.* 109: 1173–84

Kelly, R. B. 1991a. Neurobiology: A system for synapse control. *Nature* 349: 650–51

Kelly, R. B. 1991b. Secretory granule and synaptic vesicle formation. *Curr. Opin. Cell Biol.* 3: 654–60

Kelly, R. B. 1990. Microtubules, membrane traffic and cell organization. *Cell* 61: 5–7

Kirschner, M., Mitchison, T. 1986. Beyond self-assembly: from microtubules to morphogenesis. *Cell* 45: 329–42

Kleiman, R., Banker, G., Stewart, O. 1990. Differential subcellular localization of particular mRNAs in hippocampal neurons in culture. *Neuron* 5: 821–30

Knops, J., Kosik, K., Lee, G., Pardee, J., Cohen-Gould, L., McConlogue, L. 1991. Overexpression of tau in a nonneuronal cell induces long cellular processes. *J. Cell Biol.* 114: 725–33

Koedam, J., Cramer, E., Briend, E., Furie, B., Wagner, D. 1992. P-selectin, a granule membrane protein of platelets and endothelial cells, follows the regulated secretory pathway in AtT-20 cells. *J. Cell Biol.* 116: 617–25

Koenig, J., Ikeda, K. 1989. Disappearance and reformation of synaptic vesicle membrane upon transmitter release observed under reversible blockage of membrane retrieval. *J. Neurosci.* 9: 3844–60

Kordeli, E., Bennett, V. 1991. Distinct ankyrin isoforms at neuron cell bodies and nodes of Ranvier resolved using erythrocyte ankyrin-deficient mice. *J. Cell Biol.* 114: 1243–59

Kordeli, E., Davis, J., Trapp, B., Bennett, V. 1990. An isoform of ankyrin is localized at nodes of Ranvier in myelinated axons of central and peripheral nerves. *J. Cell Biol.* 110: 1341–52

Kosik, K., Finch, E. 1987. MAP2 and tau segregate into dendritic and axonal domains after the elaboration of morphologically distinct neurites: an immunocytochemical study of cultured rat cerebrum. *J. Neurosci.* 7: 3142–53

Lawrence, J., Singer, R. 1991. Spatial organization of nucleic acid sequences within cells. *Semin. Cell Biol.* 2: 83–101

Lazarides, E., Nelson, W., Kasamatsu, T. 1984. Segregation of two spectrin forms in the chicken optic system: a mechanism for establishing restricted membrane-cytoskeletal domains in neurons. *Cell* 36: 269–78

Lefcort, F., Bentley, D. 1989. Organization of cytoskeletal elements and organelles preceding growth cone emergence from an identified neuron in situ. *J. Cell Biol.* 108: 1737–49

Lein, P., Higgins, D. 1991. Protein synthesis is required for the initiation of dendritic growth in embryonic rat sympathetic neurons in vitro. *Dev. Brain Res.* 60: 187–96

Letourneau, P., Shattuck, T. 1989. Distribution and possible interactions of actin-associated proteins and cell adhesion molecules of nerve growth cones. *Development* 105: 505–19

Leveque, C., Hoshino, T., David, P., Shoji-Kasai, Y., Leys, K., et al. 1992. The synaptic vesicle protein synaptotagmin associates with calcium channels and is a putative Lambert-Eaton myasthenic syndrome antigen. *Proc. Natl. Acad. Sci. USA* 89: 3625–29

Lewis, S., Ivanov, I., Gwo-Hwa, L., Cowan, N. 1989. Organization of microtubules in dendrites and axons is determined by a short hydrophobic zipper in microtubule-associated proteins MAP2 and tau. *Nature* 342: 498–505

Lidov, H., Byers, T., Watkins, S., Kunkel, L. 1990. Localization of dystrophin to postsynaptic regions of central nervous system cortical neurons. *Nature* 348: 725–28

Linstedt, A. D., Kelly, R. B. 1991a. Molecular architecture of the nerve terminal. *Curr. Opin. Neurobiol.* 1: 382–92

Linstedt, A. D., Kelly, R. 1991b. Synaptophysin is sorted from endocytotic markers in neuroendocrine PC12 cells but not transfected fibroblasts. *Neuron* 7: 309–17

Linstedt, A., Kelly, R. 1991c. Endocytosis of the synaptic vesicle protein, synaptophysin, requires the COOH-terminal tail. *J. Physiol. (Paris)* 85: 90–96

Linstedt, A. D., Vetter, M., Bishop, J. M., Kelly, R. B. 1992. Specific association of the proto-oncogene product pp60c-src with an intracellular organelle, the PC12 synaptic vesicle. *J. Cell Biol.* In press

Lobel, P., Fujimoto, K., Ye, R., Griffiths, G., Kornfeld, S. 1989. Mutations in the cytoplasmic domain of the 275 dd mannose 6-phosphate receptor differentially alter lysosomal enzyme sorting and endocytosis. *Cell* 57: 787–96

Lowe, A. W., Madeddu, L., Kelly, R. B. 1988. Endocrine secretory granules and neuronal synaptic vesicles have three integral membrane proteins in common. *J. Cell Biol.* 106: 51–59

Maness, P., Aubry, M., Shores, C., Frame, L., Pfenninger, K. 1988. c-src gene product in developing rat brain is enriched in nerve growth cone membranes. *Proc. Natl. Acad. Sci. USA* 85: 5001–5

Matsuuchi, L., Buckley, K., Lowe, A., Kelly, R. 1988. Targeting of secretory vesicles to cytoplasmic domains in AtT-20 and PC-12 cells. *J. Cell Biol.* 106: 239–51

Matsuuchi, L., Kelly, R. 1991. Constitutive and basal secretion from endocrine cell line, AtT-20. *J. Cell Biol.* 112: 843–52

Matteoli, M., Takei, K., Cameron, R., Hurlbut, P., Johnston, P., et al. 1991. Association of rab3A with synaptic vesicles at late stages of the secretory pathway. *J. Cell Biol.* 115: 625–33

Matus, A. 1990. Microtubule-associated proteins and the determination of neuronal form. *J. Physiol.* 84: 134–37

Meiri, K., Gordon-Weeks, P. 1990. GAP-43 in growth cones is associated with areas of membrane that are tightly bound to substrate and is a component of a membrane skeleton subcellular fraction. *J. Neurosci.* 10: 256–66

Meiri, K., Bickerstaff, L., Schwob, J. 1991. Monoclonal antibodies show that kinase C phosphorylation of GAP-43 during axonogenesis is both spatially and temporally restricted in vivo. *J. Cell Biol.* 112: 991–1005

Meiri, K., Burdick, D. 1991. Nerve growth factor stimulation of GAP-43 phosphorylation in intact isolated growth cones. *J. Neurosci.* 11: 3155–64

Miller, T., Heuser, J. 1984. Endocytosis of synaptic vesicle membrane at the frog neuromuscular junction. *J. Cell Biol.* 98: 685–98

Mogensen, M., Tucker, J., Stebbings, H. 1989. Microtubule polarities indicate that nucleation and capture of microtubules occurs at cell surfaces in Drosophila. *J. Cell Biol.* 108: 1445–52

Morin, P., Liu, N., Johnson, R., Leeman, S., Fine, R. 1991. Isolation and characterization of rapid transport vesicle subtypes from rabbit optic nerve *J. Neurochem.* 56: 415–27

Murphy, J.-E., Pleasure, I., Puszkin, S., Prasad, K., Keen, J. 1991. Clathrin assembly protein AP-3. *J. Biol. Chem.* 266: 4401–8

Nelson, W., Hammerton, R., Wang, A., Shore, E. 1990. Involvement of the membrane-cytoskeleton in development of epithelial cell polarity. *Semin. Cell Biol.* 1: 359–71

Obendorf, D., Schwarzenbrunner, U., Fischer-Colbrie, R., Laslop, A., Winkler, H. 1988. In adrenal medulla synaptophysin (protein 38) is present in chromaffin granules and in a special vesicle population. *J. Neurochem.* 51: 1573–80

Okabe, S., Hirokawa, N. 1989. Rapid turnover of microtubule-associated protein MAP2 in the axon revealed by microinjection of biotinylated MAP2 into cultured neurons. *Proc. Natl. Acad. Sci. USA* 86: 4127–31

Orci, L., Ravazzola, M., Storch, M., Anderson, R., Vassalli, J., Perrelet, A. 1987. Proteolytic maturation of insulin is a post-Golgi event which occurs in acidifying clathrin-coated secretory vesicles. *Cell* 87: 865–68

Otsuka, A. J., Jeyaprakash, A., Garcia-anoveros, J., Lan, Z. T., Fisk, G., et al. 1991. The *C. elegans unc*-104 gene encodes a putative kinesin heavy chain-like protein. *Neuron* 6: 113–22

Parsons, S., Creutz, C. 1986. pp60c-src activity detected in the chromaffin granule membrane. *Biochem. Biophys. Res. Commun.* 134: 736–42

Patzak, A., Winkler, H. 1986. Excytotic exposure and recycling of membrane antigens of chromaffin granules: ultrastructural evaluation after immuno-labeling. *J. Cell Biol.* 102: 510–15

Peter, A., Schittny, J., Niggli, V., Reuter, H., Sigel, E. 1991. The polarized distribution of poly(A+)-mRNA-induced functional ion channels in the Xenopus oocyte plasma membrane is prevented by anti-cytoskeletal drugs. *J. Cell Biol.* 114: 455–64

Petrenko, A., Perin, M., Davletov, B., Ushkaryov, Y., Geppert, M., Sudhof, T. 1991. Binding of synaptotagmin to the a-latrotoxin receptor implicates both in synaptic vesicle exocytosis. *Nature* 353: 65–68

Petrucci, T., Macioce, P., Paggi, P. 1991. Axonal transport kinetics and post-translational modification of synapsin I in mouse retinal ganglion cells. *J. Neurosci.* 11: 2938–46

Pfeffer, S. 1992. GTP-binding proteins in intracellular transport. *Trends Cell Biol.* 2: 41–46

Pfeffer, S., Kelly, R. 1985. The subpopulation of brain coated vesicles that carries synaptic vesicle proteins containing two unique polypeptides. *Cell* 40: 949–57

Ramoa, A., Campbell, G., Shatz, C. 1988. Dendritic growth and remodeling of cat retinal ganglion cells during fetal and postnatal development. *J. Neurosci.* 8: 4239–61

Reetz, A., Solimena, M., Matteoli, M., Folli, F., Takei, K., DeCamilli, P. 1991. GABA and pancreatic B-cells: colocalization of glutamic acid decarboxylase (GAD) and GABA with synaptic-like microvesicles suggests their role in GABA storage and secretion. *EMBO J.* 10: 1275–84

Regnier-Vigouroux, A., Tooze, S., Huttner, W. 1991. Newly synthesized synaptophysin is transported to synaptic-like microvesicles via constitutive secretory vesicles and the plasma membrane. *EMBO J.* 10: 3589–3601

Reinsch, S., Mitchison, T., Kirschner, M. 1991. Microtubule polymer assembly and transport during axonal elongation. *J. Cell Biol.* 115: 365–79

Resh, M. 1989. Specific and saturable binding of pp60v-*src* to plasma membranes: evidence for a myristyl-*src* receptor. *Cell* 58: 281–86

Rothman, J., Orci, L. 1992. Molecular dissection of the secretory pathway. *Nature* 355: 409–15

Rousselet, A., Autillo-Touati, A., Araud, D., Prochiantz, A. 1990. In vitro regulation of neuronal morphogenesis and polarity by astrocyte-derived factors. *Dev. Biol.* 137: 33–45

Sabry, J., O'Connor, T., Evans, L., Toroian-Raymond, A., Kirschner, M., Bentley, D. 1991. Microtubule behavior during guidance of pioneer neuron growth cones in situ. *J. Cell Biol.* 115: 381–95

Schaeffer, S., Raviola, E. 1978. Membrane recycling in the cone cell endings of the turtle retina. *J. Cell Biol.* 79: 802–25

Sheetz, M., Steuer, E., Schroer, T. 1989. The mechanism and regulation of fast axonal transport. *Trends Neurosci.* 12: 474–78

Shoji-Kasai, Y., Yoshida, A., Sato, K., Hoshino, T., Ogura, A., et al. 1992. Neurotransmitter release from synaptotagmin (p65)-deficient clonal variants of PC12 cells. *Science.* In press

Singer, S., Kupfer, A. 1986. The directed migration of eukaryotic cells. *Annu. Rev. Cell Biol.* 2: 337–65

Sobue, K. 1990. Involvement of the membrane cytoskeletal proteins and the *src* gene product in growth cone adhesion and movement. *Neurosci. Res.* 13: S80–S91

Soriano, P., Montgomery, C., Geske, R., Bradley, A. 1991. Targeted disruption of the c-*src* proto-oncogene leads to osteopetrosis in mice. *Cell* 64: 693–702

Spruce, A., Breckenridge, L., Lee, A., Almers, W. 1990. Properties of the fusion pore that forms during exocytosis of a mast cell secretory vesicle. *Neuron* 4: 643–54

Stadler, H., Kiene, M. 1987. Synaptic vesicles in electromotoneurones. II. Heterogeneity of populations is expressed in uptake properties; exocytosis and insertion of a core proteoglycan into the extracellular matrix. *EMBO J.* 6: 2217–21

Stow, J. L., de Almeida, J. B., Narula, N., Holtzman, E. J., Ercolani, L., Ausiello, D. A. A. 1991. A heterotrimeric G protein, G α i-3, on Golgi membranes regulates the secretion of a heparan sulfate proteoglycan in LLC-PK1 epithelial cells. *J. Cell Biol.* 1143: 1113–24

Strittmatter, S., Fishman, M. 1991. The neuronal growth cone as a specialized transduction system. *BioEssays* 13: 127–34

Strittmatter, S., Valenzuela, D., Kennedy, T., Neer, E., Fishman, M. 1990. Go is a major growth cone protein subject to regulation by GAP-43. *Nature* 344: 836–41

Sudhof, T., Jahn, R. 1991. Proteins of synaptic vesicles involved in exocytosis and membrane recycling. *Neuron* 6: 665–77

Sulzer, D., Holtzman, E. 1989. Acidification and endosome-like compartments in the presynaptic terminals of frog retinal photoreceptors. *J. Neurocytol.* 18: 529–40

Sundell, C., Singer, R. 1991. Requirement of microfilaments in sorting of actin messenger RNA. *Science* 253: 1275–77

Tanaka, E., Kirschner, M. 1991. Microtubule behavior in the growth cones of living neurons during axon elongation. *J. Cell Biol.* 115: 345–63

Thomas, L., Betz, H. 1990. Synaptophysin binds to physophilin, a putative synaptic plasma membrane protein. *J. Cell Biol.* 111: 2041–52

Tooze, J., Hollinshead, M., Fuller, S., Tooze, S., Huttner, W. 1989. Morphological and biochemical evidence showing neuronal properties in AtT-20 cells and their growth cones. *Eur. J. Cell Biol.* 49: 259–73

Tooze, J., Tooze, S. 1986. Clathrin-coated vesicular transport of secretory proteins during the formation of AC-FH-containing secretory granules in AtT-20 cells. *J. Cell Biol.* 103: 839–50

Tooze, S., Flatmark, T., Tooze, J., Huttner, W. 1991. Characterization of the immature secretory granule, and intermediate in granule biogenesis. *J. Cell Biol.* 115: 1491–4

Tsukita, S., Ishikawa, H. 1980. The movement of membranous organelles in axons: Electron microscopic identification of anterogradely and retrogradely transported organelles. *J. Cell Biol.* 84: 513–30

Tsukita, S., Oishi, K., Akiyama, T., Yamanashi, Y., Yamamoto, T., Tsukita, S. 1991. Specific proto-oncogenic tyrosine kinases of *src* family are enriched in cell-to-cell adherens junctions where the level of tyrosine phosphorylation is elevated. *J. Cell Biol.* 113: 867–79

van Hooff, C., De Graan, P., Oestreicher, A., Gispen, W. 1989. Muscarinic receptor activation stimulates B-50/GAP43 phosphorylation in isolated nerve growth cones. *J. Neurosci.* 9: 3753–59

van Lookeren Campagne, M., Oestreicher, A., Van Bergen en Henegouwen, P., Gispen, W. 1989. Ultrastructural immunocytochemical localization of B-50/GAP43, a protein kinase C substrate, in isolated presynaptic nerve terminals and neuronal growth cones. *J. Neurocytol.* 18: 479–89

von Zastrow, M., Castle, J. 1987. Protein sorting among two distinct export pathways occurs from the content of maturing exocrine storage granules. *J. Cell Biol.* 105: 2675

Voyvodic, J. 1987. Development and regulation of dendrites in the rat superior cervical ganglion. *J. Neurosci.* 7: 904–12

Wiedenmann, B., Rehm, H., Knierim, M., Becker, C. 1988. Fractionation of synaptophysin-containing vesicles from rat brain and cultured PC12 pheochromocytoma cells. *FEBS Lett.* 240: 71–77

Xue, G., Rivero, B., Morris, R. 1991. The surface glycoprotein Thy-1 is excluded from growing axons during development: a study of the expression of Thy-1 during axogenesis in hippocampus and hindbrain. *Development* 112: 161–76

Yanker, B., Benowitz, L., Villa-Komaroff, L., Neve, R. 1990. Transfection of PC12 cells with the human GAP-43 gene: effects on neurite outgrowth and regeneration. *Mol. Brain Res.* 7: 39–44

Yisraeli, J., Sokol, S., Melton, D. 1990. A two-step model for the localization of maternal mRNA in Xenopus oocytes: involvement of microtubules and microfilaments in the translocation and anchoring of Vg1 mRNA. *Development* 108: 289–98

Zuber, M., Goodman, D., Karns, L., Fishman, M. 1989a. The neuronal growth-associated protein GAP-43 induces filopodia in nonneuronal cells. *Science* 244: 1193–95

Zuber, M., Strittmatter, S., Fishman, M. 1989b. A membrane-targeting signal in the amino terminus of the neuronal protein GAP-43. *Nature* 341: 345–48

Annu. Rev. Neurosci. 1993. 16:129–58

MOLECULAR CONTROL OF CELL FATE IN THE NEURAL CREST: The Sympathoadrenal Lineage

David J. Anderson

Howard Hughes Medical Institute, Division of Biology, California Institute of Technology, Pasadena, California 91125

KEY WORDS: nerve growth factor, glucocorticoids, cell lineage, peripheral nervous system, neurogenesis

INTRODUCTION

A central problem in developmental neurobiology is understanding the cellular and molecular mechanisms that generate the diversity of cell types found in the nervous system. Although the general problem of cell type specification can be addressed in many non-neuronal tissues, it is particularly challenging in the nervous system, because of the enormous variety of cell types that exist, and the phenotypic plasticity they display. In recent years, several technical breakthroughs have permitted an intensive analysis of cell lineage relationships in various parts of the vertebrate nervous system. These have included the development of recombinant retroviruses for genetic marking of cell fate (Sanes et al 1986; Turner & Cepko 1987) and membrane-impermanent lineage tracers for micro-injection of single progenitor cells (Holt et al 1988; Wetts & Fraser 1988). These techniques have revealed that in many (but not all) systems, individual neural precursor cells give rise to a variety of different cell types, thus demonstrating that they are multipotent. In the case of the neural crest, application of both techniques has indicated that many neural crest cells are multipotent, before (Bronner-Fraser & Fraser 1988, 1989; Frank

129

& Sanes 1991) or shortly after (Fraser & Bronner-Fraser 1991) they migrate from the neural tube.

If neural precursors are multipotent, how do they choose their fates? In principle, fates could be determined stochastically, be governed by cell-autonomous developmental programs, or be controlled by environmental signals. In the case of the neural crest, transplantation of cell populations has indicated that the fate(s) of these populations can be altered by changing their environment (for reviews, see Le Douarin 1980, 1982). However, there have been relatively few cases in which it has been possible to identify the specific signals that control neural cell fate and study their actions on identified cells. In the optic nerve, studies of a bipotential glial progenitor cell, the O2A cell, have suggested that one fate, the oligodendrocyte, is the "default" pathway for the cell, whereas the other fate, the type 2 astrocyte, is dependent upon instructive signals (which include CNTF and extracellular matrix-associated factors) (for reviews, see Lillien & Raff 1990; Raff 1989). Although this system has proven excellent for analysis at the cellular level, it has been less accessible for study at the molecular level.

One neural crest lineage that has been investigated in detail is the sympathoadrenal (SA) lineage (Landis & Patterson 1981). This lineage derives from neural crest cells that migrate ventrally from the apex of the neural tube to the dorsal aorta, where they aggregate and differentiate to form sympathetic neurons, or to the adrenal gland primordia, where they differentiate to form chromaffin cells (Figure 1). Chromaffin cells are round secretory cells that lack the axons and dendrites characteristic of sympathetic neurons; their secretory vesicles are also larger than those of neurons (Doupe et al 1985a). A third and minor SA cell type, the small intensely fluorescent (SIF) cell, has an intermediate morphology with short processes (Eranko 1975). A striking feature of these cell types is that they can be phenotypically interconverted by specific environmental signals (Patterson 1978). This plasticity has been interpreted to reflect developmental history, i.e. sympathetic neurons, adrenal chromaffin cells, and SIF cells can be interconverted because they may develop from a common embryonic progenitor cell (Anderson & Axel 1986; Doupe et al 1985b; Unsicker et al 1989). Moreover, the ability of glucocorticoids (GC) and nerve growth factor (NGF) to interconvert or maintain these phenotypes has suggested that these factors may be important environmental determinants of cell fate in vivo (Aloe & Levi-Montalcini 1979; Anderson 1988; Doupe et al 1985a,b; Seidl & Unsicker 1989a,b; Unsicker et al 1978).

Recently, it has become possible to isolate SA progenitors from rat embryos by using monoclonal antibodies, and derive immortal cell lines from these progenitors (Anderson 1988; Birren & Anderson 1990;

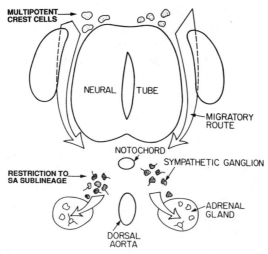

Figure 1   Schematic cross-section through a midgestational rat embryo at the caudal thoracic level, showing migratory route taken by neural crest cells that give rise to the sympatho-adrenal lineage. For clarity, other crest derivatives, such as sensory neurons, melanocytes, and glia, are omitted. Cells at several different stages of development are shown in the same figure.

Carnahan & Patterson 1991b). These advances have permitted a critical test of the "common progenitor" hypothesis and an examination of the influences of environmental signals on this embryonic cell. This review summarizes new insights into the biology of the SA lineage gained over the last several years as a consequence of these studies. A minireview on this topic has recently appeared elsewhere (Patterson 1990).

HISTORICAL OVERVIEW: PLASTICITY AND DEVELOPMENTAL HISTORY

Experimental evidence for a close developmental relationship between chromaffin cells and sympathetic neurons initially came from the observation that transplantation of postnatal adrenal medullary tissue into the anterior chamber of the eye produced outgrowth of neuritic processes (Olson 1970). Unsicker et al (1978) subsequently demonstrated that neurite outgrowth could be induced from dissociated postnatal chromaffin cells by NGF in vitro. This induction of neurite outgrowth could be blocked or delayed by GC, which suggests that the (presumably) high local concentration of steroids produced by the adrenal cortex (Roos 1967) is important for maintaining the endocrine phenotype of the medullary cells.

Nerve growth factor also promoted neurite outgrowth from a clonal cell line, PC12, derived from a rat adrenal medullary tumor (Greene & Tischler 1976). Moreover, injections of NGF into rat fetuses caused the replacement of chromaffin cells by ganglionic neurons in situ (Aloe & Levi-Montalcini 1979). Taken together, these data revealed that the chromaffin phenotype is plastic and identified NGF and GC as environmental signals of potential importance in controlling the fate of these cells in vivo.

Although these early in vitro studies documented short-term neurite outgrowth from chromaffin cells in response to NGF, they did not determine whether these endocrine cells could undergo a complete phenotypic conversion into bona fide sympathetic neurons. This was resolved by Doupe et al (1985a), who demonstrated that individual chromaffin cells could convert into cells that are morphologically and antigenically indistinguishable from true sympathetic neurons, when exposed to NGF for several weeks. Furthermore, such a transdifferentiation occurred in single cells, in the absence of DNA synthesis or cell division (Doupe et al 1985b) [however, in the absence of mitotic inhibitors, NGF exerted a mitogenic effect on chromaffin cells (Lillien & Claude 1985)]. The conversion to a neuronal phenotype also involved the de novo induction of expression of neuron-specific genes (Anderson & Axel 1985). In these studies, GC not only inhibited neuronal transdifferentiation, but also acted as a survival factor for the chromaffin cells (Doupe et al 1985a).

Based on these and the preceding observations, it was hypothesized that chromaffin cells and sympathetic neurons derive from a common embryonic progenitor (Doupe et al 1985b; Landis & Patterson 1981). A cell with the properties expected of such a progenitor was identified in cultures of neonatal sympathetic ganglia (Doupe et al 1985b) (although the lack of markers for this cell made it difficult to establish its relationship to embryonic precursor cells in vivo). This cell could differentiate into a SIF-like phenotype in moderate concentrations of GC (10^{-8} M) or to an adrenergic chromaffin cell in micromolar GC. Once generated, such SIF-like cells could be converted to sympathetic neurons by exposure to NGF (Doupe et al 1985b). This observation, coupled with the fact that the conversion of chromaffin cells to sympathetic neurons appeared to occur through a SIF-like intermediate, led to the proposal that SIF cells represent the embryonic progenitor of both adrenal medullary cells and sympathetic neurons (Doupe et al 1985b). High GC would favor the differentiation of SIF cells into chromaffin cells. Nerve growth factor would promote their differentiation into sympathetic neurons. And, a combination of NGF and moderate concentrations of GC would maintain or stabilize the SIF phenotype. This idea was appealing in that it provided a role for SIF cells (Eranko 1975), whose function had previously been controversial.

Fetal SA Progenitors are Bipotential, but Distinct from SIF Cells

A direct test of the bipotential progenitor hypothesis was made possible by the development of immunologic methods to isolate fetal SA progenitors. These cells have been isolated from E14.5 rat adrenal glands (Anderson 1988; Anderson & Axel 1986) by using monoclonal antibody HNK-1 (Abo & Balch 1981), and from sympathetic ganglia (Carnahan & Patterson 1991b) by using novel monoclonal antibodies generated by an immuno-suppression technique (Carnahan & Patterson 1991a). Progenitors from fetal adrenal glands have also been isolated by density gradient centrifugation (Seidl & Unsicker 1989a). Quantitative analysis of such purified cell populations (Anderson & Axel 1986; Carnahan & Patterson 1991b), as well as serial observations of identified cells (Anderson & Axel 1986; Michelsohn & Anderson 1992) has confirmed that many such progenitors are bipotential, able to develop into either chromaffin cells or sympathetic neurons depending upon the culture conditions. Nevertheless, the SA progenitor appears to have lost the ability to give rise to other crest derivatives, such as glia (D. L. Stemple and D. J. Anderson 1991, unpublished), and therefore represents a relatively late stage in neural crest lineage diversification. The idea that the glia and sympathoadrenal lineages diverge early in sympathetic gangliogenesis is supported by retroviral lineage-tracing experiments in fetal superior cervical ganglion (SCG) (Hall & Landis 1991a).

Fetal SA progenitors can be demonstrated to be bipotential in vitro, but do such cells exist in vivo? Individual cells coexpressing neuron-specific and chromaffin-specific antigenic markers have been observed in early (E12.5) sympathetic ganglia primordia (Anderson et al 1991; Carnahan & Patterson, 1991a). Subsequently, the chromaffin-specific markers are lost by the cells that remain in the ganglia (Anderson et al 1991; Carnahan & Patterson 1991a), whereas the neuron-specific markers are lost by progenitors that continue migrating to the adrenal gland (Anderson 1988; Anderson & Axel 1986; Vogel & Weston 1990). The observation of such transiently dual phenotype cells provides circumstantial evidence for the existence of a bipotential SA progenitor in vivo. Is this cell a SIF cell? Although the morphology and high catecholamine content of the embryonic SA progenitor is similar to that of a postnatal SIF cell, SA progenitors and SIF cells are distinct by antigenic criteria (Carnahan & Patterson 1991a), and SIF cells develop later, not earlier, than principal neurons in sympathetic ganglia (Hall & Landis 1991b). These data suggest that the embryonic SA progenitor is different from a SIF cell. However, SIF cells have a similar developmental potential as the SA progenitor (Doupe et al

1985b) and might represent a postnatal/adult form of the SA progenitor (Carnahan & Patterson 1991b), as has been described in the glial 02A lineage (Wolswijk & Noble 1989).

ROLE OF POLYPEPTIDE GROWTH FACTORS IN THE NEURONAL DIFFERENTIATION OF SA PROGENITOR CELLS

Fetal SA Progenitors are Initially Responsive to Fibroblast Growth Factor (FGF) but not to NGF

The ability of NGF to trigger neurite outgrowth from postnatal chromaffin cells, SIF cells, and PC12 cells initially suggested that NGF is an important determinant of neuronal fate in the SA lineage. Indeed, neonatal post-mitotic sympathetic neurons are absolutely dependent upon NGF for survival both in vitro (Chun & Patterson 1977; Levi-Montalcini & Angeletti 1963) and in vivo (Levi-Montalcini & Booker 1960). But, freshly isolated SA progenitors from fetal adrenal glands are initially unresponsive to NGF, by criteria of neurite outgrowth, mitotic rate, and survival (Anderson & Axel 1986; S. J. Birren and D. J. Anderson 1991, unpublished). Similar observations have been made for neuronal precursors from chick (Ernsberger et al 1989a) and mouse (Coughlin & Collins 1985) sympathetic ganglia. These data raise several new questions: Why are SA progenitors unresponsive to NGF? Which factors promote the acquisition of NGF-responsiveness and NGF-dependence? Are there other factors that substitute for NGF in the early stages of neuronal differentiation?

Answers to some of these questions have come, in part, from studies of an SA progenitor cell line, called MAH (Myc-infected, Adrenal-derived, HNK-1[+]) cells. MAH cells were produced by retroviral transduction of the avian *v-myc* oncogene into SA progenitors isolated from fetal adrenal glands by fluorescence-activated cell sorting; their morphology and antigenic phenotype are similar to those of their primary counterparts (Birren & Anderson 1990). Like primary SA progenitors, MAH cells fail to respond to NGF by criteria of neurite outgrowth, survival, or induction of neuron-specific genes (Birren & Anderson 1990). The large number of homogeneous cells provided by the MAH line has permitted a molecular analysis of NGF receptor expression in SA progenitors. Northern blot analysis has indicated that MAH cells grown in the absence of dexamethasone express neither mRNA encoding p75 (Birren & Anderson 1990), the low affinity NGF receptor (Johnson et al 1986; Radeke et al 1987), nor p140^{trk} (Birren et al 1992), recently identified as a signal-transducing component of the NGF receptor (Kaplan et al 1991a; Klein et al 1991). Thus, the lack of NGF responsiveness appears due, at least in part, to a lack of NGF receptor expression. However, it may also reflect the absence of some other component(s) of the NGF signal-transduction pathway.

If NGF does not promote the initial neuronal differentiation of SA progenitors, do other factors play this role? Experiments in PC12 cells have identified bFGF (Rydel & Greene 1987; Togari et al 1985), interleukin-6 (Satoh et al 1988), and "pleiotrophin" (Kuo et al 1990; Li et al 1990) as factors able to mimic the ability of NGF to induce neurite outgrowth and, in the case of bFGF, neuron-specific gene expression (Leonard et al 1987; Stein et al 1988). In postnatal adrenal chromaffin cells, both bFGF (Stemple et al 1988) and aFGF (Claude et al 1988) promote neuronal differentiation and increase cell proliferation. Unlike NGF, however, FGF cannot act as a long-term survival factor for sympathetic neurons (Stemple et al 1988). Similarly, bFGF can induce proliferation and neurite outgrowth from both MAH cells (Birren & Anderson 1990) and primary SA progenitor cells (S. J. Birren and D. J. Anderson 1991, unpublished), but does not support their long-term survival. Taken together, these results suggest that FGF, rather than NGF, may promote the proliferation and initial neuronal differentiation of embryonic SA progenitors, whereas NGF acts as a survival factor for mature neurons. In chromaffin cells, FGF can induce neuronal differentiation independent of its ability to induce proliferation (Stemple et al 1988); it is not yet clear whether this is also true for embryonic SA progenitors.

Other factors besides bFGF also stimulate the proliferation or survival of immature sympathetic neuroblasts, including depolarization, insulin, insulin-like growth factor I (IGF-I), and vasoactive intestinal peptide (DiCicco-Bloom & Black 1988, 1989; DiCicco-Bloom et al 1990; Pincus et al 1990; Wolinsky et al 1985); for review see Rohrer (1990). In such studies, it is often difficult to distinguish between the mitogenic effect of a growth factor and its ability to enhance simply the survival of proliferating cells without affecting their rate of cell division. Two other "growth" factors, ciliary neurono-trophic factor (CNTF) and cholinergic differentiation factor/leukemia inhibitory factor (CDF/LIF), inhibit the proliferation of sympathetic neuroblasts and MAH cells (Ernsberger et al 1989; Ip et al 1992). With the exception of NGF (Levi-Montalcini & Booker 1960; Rohrer et al 1988), the requirement for any of these growth factors in sympathetic neuronal proliferation or differentiation in vivo has not yet been assessed. However, embryonic chick sympathetic ganglia contain bFGF immunoreactivity at early stages of gangliogenesis (Kalcheim & Neufeld 1990).

Induction of NGF Receptors and NGF-Responsiveness in SA Progenitors

The survival of sympathetic and other neurons is independent of NGF before their axons reach the sources of NGF in the periphery (Davies et al 1987; Korsching & Thoenen 1988). How do differentiating sympathetic neurons acquire their responsiveness to, and ultimately their trophic dependence upon, NGF? Because SA progenitors initially do not appear to express NGF receptors, a critical early step must be the induction of such receptors. In MAH cells, treatment with both FGF and NGF (but not with either factor alone) leads to a small population of cells (0.5–10%), which differentiate to postmitotic neurons. Such cells are NGF-dependent, which implies that they express functional NGF receptors (Birren & Anderson 1990). This suggested that FGF might induce the expression of NGF receptors, and a low-level induction of p75 (LNGFR) mRNA was observed in MAH cells exposed to FGF for several days (Birren & Anderson 1990).

Experiments with cloned NGF receptor genes suggest that the expression of p75 is probably not sufficient for NGF responsiveness [and may not be necessary, either (Weskamp & Reichardt 1991)], and that p140^{trk} is required to form functional, high-affinity NGF receptors (Hempstead et al 1991). In MAH cells, p140^{trk} mRNA is not directly induced by FGF (or any other growth or neurotrophic factors tested); however, it is induced by membrane depolarization, which also induces a functional NGF-response (Birren et al 1992). Depolarization does not induce neurite outgrowth, however, which indicates that the induction of *trk* can be experimentally uncoupled from neuronal differentiation. The ability of FGF to induce NGF-responsiveness in a small proportion of MAH cells (Birren & Anderson 1990) may reflect an indirect induction of *trk* expression, perhaps via synaptic activity. In primary chick sympathetic neuroblasts, depolarization augmented cell survival in the presence of NGF (Ernsberger et al 1989a), although the mechanism of this effect was not investigated. Taken together, however, these results suggest that the electrical excitation of developing sympathetic neuroblasts may stimulate their acquisition of trophic-factor responsiveness. Such a mechanism could coordinate the formation of stable presynaptic and postsynaptic connections by a developing neuron.

Other mechanisms may also contribute to the acquisition of trophic factor responsiveness and dependence in developing neuroblasts. In chromaffin cells, FGF can induce an NGF-dependence in transdifferentiating neurons that are already NGF-responsive (Stemple et al 1988). Seemingly spontaneous acquisition of trophic factor dependence has been observed for primary sensory neurons in vitro (Vogel & Davies 1991); however, it was not determined whether these cells were already responsive to the neurotrophic factor [brain-derived neurotrophic factor (BDNF)] at the time of their initial isolation. Retionic acid induces both NGF-dependence and the expression of high-affinity NGF receptors in chick sympathetic neuroblasts that bear low affinity NGF receptors (Rodriguez-Tébar & Rohrer 1991). In the rat SA progenitor, however, retinoic acid does not appear to induce NGF responsiveness or expression of p140^{trk} mRNA (Birren et al 1992); this difference may reflect either a species difference or differences in the relative stages of development examined in the two systems. The availability of cloned probes for both p75 and p140^{trk} should help resolve these issues.

ROLE OF GC IN CHROMAFFIN CELL DIFFERENTIATION

Positive and Negative Action of GC on the SA Progenitor

The migration of SA progenitors to the adrenal gland primordium brings them to a microenvironment that contains a high concentration of GC hormones, synthesized by the adrenal cortex. Earlier studies have established two distinct influences of GC on the chromaffin phenotype, one positive and one negative. Glucocorticoids act positively to upregulate the expression of phenylethanolamine-N-methyl transferase (PNMT), the epinephrine-synthesizing enzyme (Bohn et al 1981; Jiang et al 1989; Pohorecky & Wurtman 1971; Teitelman et al 1982; Wurtman & Axelrod 1966). This gene is expressed by a majority of adrenal chromaffin cells, but not by sympathetic neurons (Bohn et al 1982). Glucocorticoids also act negatively to inhibit neuronal differentiation induced by both NGF (Unsicker et al 1978) and FGF (Stemple et al 1988).

The ability of GC to exert both positive and negative effects on the SA progenitor reflects the capacity of the GC receptor (GCR) to function as both a positive- and negative-acting transcriptional regulatory molecule (for review, see Beato 1989). In the case of the SA progenitor, some of the targets of these positive and negative actions of the GCR are known. The sequence of the PNMT gene contains several consensus glucocorticoid-response elements (GREs) (Batter et al 1988), and at least one of these is functional in cell transfection assays (Ross et al 1990). These data suggest that the PNMT gene is a direct target of positive regulation by GC. Glucocorticoids also repress the expression of several neuron-specific genes, including peripherin (Leonard et al 1987), SCG10 (Stein et al 1988), and GAP-43 (Federoff et al 1988), in PC12 cells. Negative GREs have not yet been identified for these genes in functional assays; therefore, it is not clear whether the action of GC on these genes is direct. However, a promoter fragment from the metalloprotease gene transin (stromelysin) contains elements for both positive regulation by NGF, and negative regulation by GC, in transfected PC12 cells (Machida et al 1989). Further analysis of this promoter should shed light on the inhibitory actions of GC.

Glucocorticoids exert both positive and negative influences on embryonic SA progenitors, as well as on PC12 cells, but these influences occur on different developmental schedules. Primary E14.5 SA progenitors are competent to respond to GC by inhibition of neuronal differentiation within the first 15–24 hours of culture, but do not express PNMT (Michelsohn & Anderson 1992). Although GCs are absolutely required for PNMT expression (Seidl & Unsicker 1989b), competence to express PNMT does not develop until after several days in culture (Michelsohn & Anderson 1992), which reflects the schedule of PNMT appearance in vivo (Bohn et al 1981; Ehrlich et al 1989; Teitelman et al 1982). In vivo, PNMT expression is also preceded by the inhibition of neuronal differentiation, as indicated by the extinction of neuron-specific markers (Anderson & Axel 1986; Anderson et al 1991; Anderson & Michelsohn 1989; Vogel & Weston 1990).

Both the early and late effects of GC appear mediated by the type II GCR (Anderson & Michelsohn 1989; Michelsohn & Anderson 1992). How can the same receptor control two different developmental events in the same cell at two different times? One clue comes from a pharmacological analysis of GC actions. A fivefold higher dose of GC is required for half-maximal induction of PNMT, than for half-maximal inhibition of neurite outgrowth. This suggests that the induction of PNMT transcription may require a higher concentration of ligand-bound GCR than does the repression of neuron-specific genes. The idea that different GC-responsive genes within the same cell may require different levels of ligand-bound receptor is supported by recent studies of model promoters (Diamond et al 1990; Simons et al 1989). The timing of PNMT induction might then be determined by the accumulation of GCR to a threshold level (Anderson & Michelsohn 1989). Consistent with this idea, GCR levels in fetal chromaffin cells increase in parallel with PNMT expression (Seidl & Unsicker 1989b), and PNMT$^+$ chromaffin cells appear to have higher levels of GCR immunoreactivity than PNMT$^-$ chromaffin cells (Ceccatelli et al 1989).

SEQUENTIAL STEPS IN CHROMAFFIN DIFFERENTIATION ARE DEPENDENT EVENTS
Why should chromaffin cell development involve two sequential GC-dependent steps? If SA progenitors are cultured for two days in the absence of GC, many cells extend neurites; these cells subsequently fail to express PNMT in response to GC, as if committed to neuronal differentiation

(Michelsohn & Anderson 1992). If neuronal differentiation is suppressed by progesterone during this initial period, then an increased number of precursors acquire competence to express PNMT (Michelsohn & Anderson 1992). Thus, the first step in chromaffin differentiation (the inhibition of neuronal commitment) is a prerequisite for the second step (the decision to express PNMT). Neuronal commitment correlates with the expression of an antigenic marker, B2 in vivo (Anderson & Axel 1986; Anderson et al 1991) (Figure 2). Consistent with the in vitro data, a lack of B2 expression by adrenal medullary cells precedes their expression of PNMT in vivo (Anderson et al 1991). Such a two-step mechanism may ensure that epinephrine is synthesized only by those progenitors that have first migrated

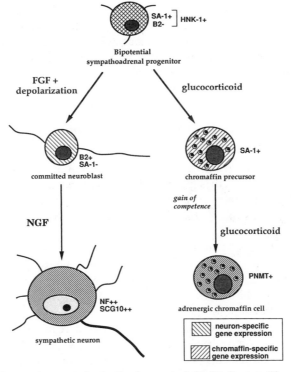

Figure 2 Progressive stages in the development of the SA lineage. Changes in marker expression that correlate with these stages are shown. Cross-hatched lines in bipotential SA progenitor indicate that both neuron-specific and chromaffin-specific genes are coexpressed. Committed neuroblasts have lost competence to respond to GC by expression of a chromaffin phenotype. Acquisition of NGF receptors appears to occur at or after the committed neuroblast stage, but also occurs in chromaffin precursors. For further details, see Figure 6 in Carnahan & Patterson (1991b).

to the adrenal gland. Not all chromaffin cells express PNMT, however (Hillarp & Hökfelt 1953). This may reflect the fact that GC do not induce competence per se, but only increase the probability that competence will be acquired (Michelsohn & Anderson 1992).

Taken together, these recent studies indicate that the development of both chromaffin cells and sympathetic neurons is progressive and can be dissected into a series of steps (Figure 2). In vitro, these steps appear to involve changes in the responsiveness of cells to environmental signals: Developing sympathetic neurons gain responsiveness to NGF and lose sensitivity to GC-inhibition, whereas developing chromaffin cells acquire competence to express PNMT. These steps are associated with changes in antigenic phenotype (Figure 2), thus allowing them to be identified in vivo (Anderson et al 1991; Carnahan & Patterson 1991a). An analogous series of stepwise changes in factor responsiveness, which are correlated with changes in cell morphology and antigenic phenotype, has been documented for glial progenitors in the 02A lineage (Gard & Pfeiffer 1990). Such progressive mechanisms seem likely to be a general feature of vertebrate neurogenesis and have been identified genetically in invertebrate nervous systems, as well (for reviews, see Ghysen & Dambly-Chaudiere 1989; Jan & Jan 1990).

Regulatory Circuits Controlling the Chromaffin-Neuron Decision

Glucocorticoids inhibit the neuronal differentiation of the SA progenitor in two ways: They directly suppress expression of the neuronal phenotype and they block the ability of FGF and NGF to promote this phenotype. In PC12 cells, GCs inhibit the ability of FGF and NGF to up-regulate neuron-specific genes (Leonard et al 1987; Stein et al 1988b). Interestingly, this antagonism is a "two-way street": FGF and NGF can inhibit the ability of GC to upregulate chromaffin-abundant genes, such as TH (Stein et al 1988). In other words, at saturating concentrations, FGF/NGF and GC reciprocally inhibit each other's actions. What is the molecular basis of such reciprocal inhibition?

Recently, novel mechanisms of transcriptional regulation by steroid hormone receptors have been uncovered (for review, see Schüle & Evans 1991). These mechanisms may have relevance for understanding the reciprocal antagonism between FGF/NGF and GC in the SA lineage. Glucocorticoid receptors interact with members of the AP-1 family of transcriptional regulatory proteins, such as c-*fos* and c-*jun*. In some cases, this interaction occurs "off" the DNA (Schüle et al 1990; Yang-Yen et al 1990); in others it occurs "on" the DNA (Diamond et al 1990). In PC12 cells, both FGF and NGF cause an induction of such AP-1 genes as c-*jun* (for review, see Sheng & Greenberg 1990). Although it is not yet certain that AP-1 molecules are required for neuronal differentiation in the SA lineage, overexpression of c-*jun* promotes autonomous neuronal differentiation of PC12 cells (I. Verma 1991, personal communication). The ability of GCR to inhibit the activity of AP-1 could, therefore, provide a molecular mechanism for the antagonism of FGF/NGF-induced neuronal differentiation by corticosteroids. Moreover, because an "off-the-DNA" interaction between AP-1 and the GCR results in a mutual inhibition of DNA-binding activity (Schüle et al 1990; Yang-Yen et al 1990), it could explain the reciprocal antagonism observed between FGF/NGF and GC at saturating ligand concentrations (Figure 3).

Such a mechanism is particularly attractive, because it would generate a "titratable" system: Progenitors containing more ligand-bound GCR than AP-1 would favor chromaffin differentiation, whereas progenitors containing more active AP-1 than GCR would favor neuronal differentiation. These two situations would obtain when the dominant environmental signals were GC or FGF/NGF, respectively. In this way, such a regulatory circuit (Figure 3) would create a cell whose choice of fate was exquisitely sensitive to the relative concentrations of competing signals in its local environment—precisely the behavior observed for SA progenitor cells. The actual role of GCR–AP-1 antagonistic interactions in the SA progenitor remains to be investigated. However, the establishment of these mechanisms in model systems provides a promising new handle on the regulatory circuits that control the chromaffin-neuron decision.

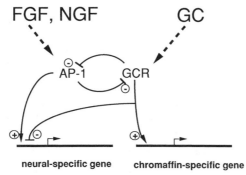

Figure 3 Hypothetical regulatory interactions between transcription factors in the AP-1 family and the GC receptor. These interactions occur in model systems, but have not yet been documented in the SA lineage. Positive and/or negative effects of AP-1 and GCR on the transcription of neuron-specific and chromaffin-specific genes are also illustrated. There is no evidence for a direct negative effect of AP-1 on chromaffin-specific gene expression, although this has not been excluded.

THE SA PROGENITOR HAS THE POTENTIAL TO GENERATE NON-CATECHOLAMINERGIC DERIVATIVES

Coexpression of Multiple Neurotransmitter Enzyme Genes in SA Progenitors

The repertoire of fates available to the SA progenitor is not restricted to catecholaminergic cell types. Sympathetic neurons and chromaffin cells can convert from a catecholaminergic to a cholinergic neurotransmitter phenotype in vitro (Doupe et al 1985b; Ogawa et al 1984; Patterson & Chun 1977). Experiments in vivo have shown that such a conversion actually occurs for the sympathetic neurons that innervate the sweat glands (Landis & Keefe 1983; Schotzinger & Landis 1988, 1990). Biochemical experiments in vitro have identified CDF/LIF (Yamamori et al 1989) and CNTF (Saadat et al 1989) as two molecules that are sufficient to induce the noradrenergic to cholinergic conversion. Whether these molecules are actually the cholinergic differentiation factors in the sweat gland is now under investigation (Rao & Landis 1990). The demonstration of cholinergic potential in sympathetic neurons and adrenal chromaffin cells raises questions regarding how early in the SA lineage this potential is established, and whether SA progenitors have other developmental potentials, as well.

Recent experiments have shed light on the answers to both of these questions. ACh synthesis has been detected in both primary SA progenitors and in MAH cells (Vandenbergh et al 1991). In addition, MAH cells transcribe low levels of ChAT mRNA. These data suggest that the cholinergic potential is established early in the SA lineage, before the choice between chromaffin and neuronal fates. In addition to ChAT mRNA, MAH cells contain low levels (approximately one copy per cell) of tryptophan hydroxylase (TpH) mRNA (Vandenbergh et al 1991). TpH transcripts were also detected in tissue from superior cervical ganglia and adrenal medulla, which indicates that the expression detected in MAH cells is not an artifact of immortalization. By contrast, no transcripts from either of the two glutamic acid decarboxylase (GAD) genes, GAD1 and GAD2, were detected in SA lineage cells or tissues (Vandenbergh et al 1991). These data suggest that SA progenitors may express a restricted repertoire of neurotransmitter biosynthetic enzyme genes.

The detection of TpH transcripts in MAH cells suggests that SA progenitors may have a serotonergic potential, as well as a cholinergic and catecholaminergic potential. Consistent with this notion, neonatal sympathetic neurons can synthesize serotonin in response to heart cell conditioned medium (Sah & Matsumoto 1987). Is this serotonergic potential actually utilized in vivo? Serotonin is one of the neurotransmitters thought to be used by enteric neurons and their endocrine counterparts, thyroid medullary "C" cells, both of which derive from the neural crest (Barasch et al 1987). This is consistent with the possibility that the SA progenitor may also give rise to enteric neurons (Figure 4), as discussed below.

SA Progenitors and Enteric Progenitors Are Similar

Circumstantial evidence supports the hypothesis that progenitors of enteric neurons may be similar, if not identical, to SA progenitors. A population of cells in the embryonic foregut transiently expresses SA lineage markers, including TH (Cochard et al 1978; Teitelman et al 1978), DBH, and high-affinity catecholamine uptake (Baetge & Gershon 1989; Jonakait et al 1985). Expression of these markers in the gut appears at E11.5-E12.5, the same time as in sympathetic ganglia. The gut cells also express SA-1 (Carnahan et al 1991) and, subsequently, B2, like neuroblasts in the sym-

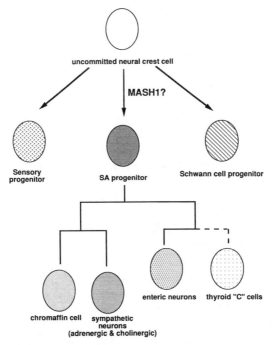

Figure 4 The SA lineage in the context of other neural crest lineages. A progenitor-progeny relationship between SA precursors and enteric neurons has not yet been formally demonstrated. Dotted line indicates that the lineage relationship between SA progenitors and thyroid C cells is speculative. MASH1 is hypothesized to play a role in the restriction of multipotent neural crest cells to the SA lineage, but the nature of this role is not established.

pathetic ganglia (Anderson et al 1991; Carnahan et al 1991). Not only are SA lineage markers expressed in gut cells, but conversely serotonin (Soinila et al 1989), a marker of enteric neurons, is transiently expressed by SA progenitors. Although shared expression of antigenic markers cannot rigorously prove a lineage relationship, the fact that so many independent gene products are expressed by both cells in the sympathetic ganglia and cells in the foregut suggests that these two populations may be closely related (Carnahan & Patterson 1991a).

The SA progenitor-like cells in the embryonic foregut are likely to be precursors of enteric neurons. This conclusion is based on the fact that neuronal "linking markers," such as neurofilament and DBH, are initially coexpressed with SA lineage markers in the foregut population and persist in enteric neurons after the SA markers have disappeared (Baetge & Gersohn 1989; Baetge et al 1990; Carnahan et al 1991). The persistent expression of such linking markers suggests that the transient SA progenitor cell-like population in the gut does not die, but rather extinguishes expression of such genes as TH and SA-1 and differentiates into enteric neurons. This implies that the environment of the gut may influence a decision between a sympathoadrenal and an enteric phenotype. Consistent with this idea, injections of NGF in vivo prolong TH expression in gut neurons (Kessler et al 1979).

The idea that the environment of the gut suppresses the expression of SA properties is also supported by in vitro experiments. Precursors of enteric neurons from the vagal neural crest express TH if cultured in NGF (Mackey et al 1988) and extinguish TH when cocultured with gut tissue (Coulter et al 1988). It is not yet known whether coculture with gut tissue can suppress TH expression in sympathetic neuroblasts. Direct tests of the postulated lineage relationships between SA progenitors and enteric neurons are clearly required before further progress on this problem can be made. But the available data, while indirect and circumstantial, suggest that the SA progenitor may have a repertoire of several developmental fates (Figure 4) (Carnahan et al 1991). This repertoire appears to be reflected in the transcriptional activation of a battery of genes that subserve several possible classical neurotransmitter phenotypes, including catecholaminergic, cholinergic, and serotonergic (Vandenbergh et al 1991), as well as multiple neuropeptide phenotypes (Nawa & Patterson 1990). The expression of a particular phenotype from this repertoire is then selected according to the various inductive and repressive signals encountered by the cell as it migrates to different embryonic environments (Anderson & Axel 1986). Following their differentiation into postmitotic cells, sympathetic neurons can undergo further phenotypic diversification with respect to their neurotransmitter and neuropeptide content, in response to target-derived instructive signals (for review, see Patterson 1990).

COMMITMENT OF EARLY NEURAL CREST CELLS TO THE SA LINEAGE

If SA progenitors have a restricted repertoire of developmental fates, how does this repertoire become established in neural crest ontogeny? Single cell lineage tracing experiments in vivo (Bronner-Fraser & Fraser 1989; Frank & Sanes 1991; Fraser & Bonner-Fraser 1991), as well as clonal analyses in vitro (Baroffio et al 1988; Dupin et al 1990; Sieber-Blum & Cohen 1980) have established that many early neural crest cells are multipotent. An individual cell may give rise to a clone containing not only catecholaminergic (SA) derivatives, but also glia, melanocytes, and sensory neurons. Because the SA progenitor seems to have lost the capacity to generate these latter three cell types, it may have become developmentally restricted. How might such a restriction event occur?

Experiments in vitro using avian neural crest cells have suggested that environmental influences, such as soluble factors (Howard & Bronner-Fraser 1985), extracellular matrix (Maxwell & Forbes 1990a,b), and the timing of cell dispersal (Vogel & Weston 1988) may influence the expression of the catecholaminergic phenotype. Ablation and rotation experiments in chick embryos have indicated that the notochord/floorplate and the dorsal aorta are required for the expression of catecholamines by neural crest cells in vivo (Stern et al 1991). The observation that SA progenitors can develop in different types of clones, which contain various subsets of other crest derivatives, has suggested that commitment may occur by a stepwise process involving "oligopotential" (i.e. partially restricted) intermediates (Anderson 1989; LeDouarin et al 1991; Sieber-Blum 1990). However, there is currently little evidence to distinguish such a progressive model from a stochastic one, a problem that has dogged the dissection of lineage commitment in the immune system for many years (Suda et al 1984). The study of lineage restriction in the neural crest has been limited by the difficulty of obtaining early markers whose expression precedes that of such differentiation products as TH (Barald 1988). Such early markers can be useful in identifying influences that might predispose cells to a particular sublineage.

A Vertebrate Homologue of Drosophila achaete-scute Is an Early Marker of SA Lineage Commitment

One recent approach that has proven productive in identifying such an early marker is to isolate vertebrate homologues of *Drosophila* genes that control early stages in neurogenic determination. The *achaete-scute* complex of *Drosophila* contains four genes whose function is required for the development of subsets of neuroblasts in appropriate positions in the fly embryo (Ghysen & Dambly-Chaudiere 1988). Molecular analysis of this complex has revealed that these genes encode nuclear regulatory proteins of the basic helix-loop-helix (bHLH) class (Alonso & Cabrera 1988; Gonzalez et al 1989; Romani et al 1989; Villares & Cabrera 1987). Included in this general class are MyoD and related proteins, encoded by mammalian genes that play a central role in myogenic determination (for review, see Weintraub et al 1991). The isolation of a *Drosophila* MyoD homologue, *nautilus*, which is specifically expressed in fly myogenic precursors (Michelson et al 1990), suggested that there has been a remarkable parallel conservation of amino acid sequence and tissue-specificity for at least some bHLH cell-type determination genes. Consistent with this idea, a recently isolated Mammalian *Achaete-Scute*-Homologous (MASH) gene, MASH1 (Johnson et al 1990), is specifically expressed by subsets of neural precursors in the rat embryo (Lo et al 1991).

A striking feature of MASH1 expression in the peripheral nervous system is that it is restricted to sympathetic ganglion primordia, but appears one day earlier than differentiation genes, such as TH (Lo et al 1991). MASH1 expression subsequently overlaps that of TH, but is eventually extinguished. MASH1 is also expressed by scattered cells in the foregut, which further supports the idea of a close developmental relationship between SA and enteric crest derivatives (see above). MASH1 is notably absent from dorsal root ganglia, glia, and melanocytes, as well as from migrating neural crest cells. It thus appears to mark neural crest cells immediately before their restriction to the SA lineage. In the CNS, MASH1 is expressed by positionally restricted subsets of precursor cells in the spinal cord and forebrain (Lo et al 1991). The fact that MASH1 is a transcription factor (Johnson et al 1992) and is homologous to a family of neural determination genes in *Drosophila* further suggests that it may have a causal role in SA lineage commitment (Figure 4). The isolation of vertebrate homologues of other *Drosophila* genes involved in neural determination (for review, see Campos-Ortega & Jan 1991), such as *daughterless* (Murre et al 1989a,b), *Notch* (Coffman et al 1990; Weinmaster et al 1991), and the RNA-binding protein *Elav* (Marusich & Weston 1992; Szabo et al 1991), suggests that this general approach should be successful in identifying molecules that will, at the very least, serve as useful markers and may at best provide handles on molecular mechanism.

Transcription Factors for Genes Expressed in SA Progenitors

Another approach to the molecular biology of SA lineage commitment is to isolate nuclear regulatory factors required for the expression of genes, such as SCG10, neurofilament, TH, and DBH, which are specifically transcribed in SA progenitors. A first step toward this goal is to delineate the *cis*-acting elements with which these factors interact. A 5.8 kb fragment of the human dopamine-β hydroxydase (DBH) gene has recently been shown to direct the expression of β-galactosidase to developing sympathetic ganglia and adrenal medulla in transgenic mice (Kapur et al 1991; Mercer et al 1991). Surprisingly, however, expression is also detected in cell types that normally express TH, but not DBH. An interesting interpretation of this observation is that the DBH promoter fragment contains regulatory elements common to all catecholamine biosynthetic enzyme genes, but lacks silencer elements that normally restrict DBH expression to noradrenergic cell types (Mercer et al 1991). The TH promoter has also been extensively analyzed. Transfection experiments in PC12 cells have defined an enhancer necessary for cell type-specific expression (Gandelman et al 1990; Harrington et al 1987). Although an intact human TH gene is specifically expressed in transgenic mice (Kaneda et al 1991), promoter fragments able to direct correct heterologous transgene expression have not yet been identified for this gene. The transcription factors that interact with these *cis*-acting control elements have yet to be identified.

Promoters for several genes expressed not only in the SA progenitor, but also in most or all neurons, have also been studied. These include the type II sodium channel (Maue et al 1990), SCG10 (a growth-associated protein) (Mori et al 1990; Vandenbergh et al 1989; Wuenschell et al 1990), and synapsin I (Sauerwald et al 1990) (for further discussion, see the article by Mandel & McKinnon in this volume). A surprising common feature of the regulation of these genes is that specificity is determined, at least in part, by selective derepression. The genes contain silencer elements that repress expression in non-neuronal cells and tissues; repression is specifically relieved in the nervous system. Recent data suggest that a common silencer mechanism may control the expression of both the type II sodium channel (Kraner et al 1992) and SCG10 (Mori et al 1992) genes. It remains to be determined when and how derepression occurs during neural crest development.

These data do not preclude a role for specifically expressed positive-acting factors in neurogenic determination. AP-2, a transcriptional activator protein, is specifically expressed in early migrating neural crest cells (Mitchell et al 1991). An insulin enhancer-binding homeodomain protein, Isl-1, is expressed in precursors to sensory and sympathetic neurons (as well as in motoneurons) (Thor et al 1991; T. Jessell 1991, personal communication). A role for this molecule as a positive-acting transcription factor is supported by its sequence similarity with *mec-3*, a homeodomain protein required for the development of mechanosensory neurons in *C. elegans* (Karlsson et al 1990; Way & Chalfie 1988).

THE SA LINEAGE IN MEDICINE AND DISEASE

The study of the SA lineage has provided insights not only into vertebrate neurogenesis, but also into the diagnosis and treatment of human disease. For example, neuroblastomas and pheochromocytomas are tumors of the sympathoadrenal lineage that afflict young children (Israel 1991). Analysis of the phenotypes of these tumors using antibody and cloned cDNA markers (Cooper et al 1990a; Trojanowski et al 1991) has suggested that different subtypes of neuroblastomas may correspond to discrete stages of differentiation in the human SA lineage (Cooper et al 1990b; Molenaar et al 1990), which have been developmentally arrested by the process of oncogenic transformation. In particular, tumors can be defined that have either a more chromaffin-like phenotype or a more neuronal phenotype. These phenotypes correspond to the SA-1$^+$B2$^-$ "chromaffin precursor" stage and the SA-1$^-$B2$^+$ "committed neuroblast" stage identified in the rat SA lineage (Figure 2) (Anderson et al 1991; Carnahan & Patterson 1991b). Interestingly, there is a correlation between tumor phenotype and prognosis. Patients whose tumors show a neuroblast phenotype have a much higher probability of spontaneous remission than those with tumors of a chromaffin phenotype (Israel 1991). The neuroblastic tumors may die because they become NGF-dependent and lack access to NGF. If so, a potential therapy for the more malignant chromaffin-like tumors might involve their in situ conversion to neuroblastic tumors, by using FGF and/or GCR antagonists. Other treatments might involve such factors as CDF/LIF and CNTF, which show an antiproliferative effect on SA progenitors (Ernsberger et al 1989b; Ip et al 1992).

Another area of medical research bearing on the SA lineage is the development of cell-replacement therapies for neurodegenerative diseases (Björklund 1991; Gage et al 1991). Adrenal chromaffin cell autografts have been used for the treatment of Parkinson's disease (Allen et al 1989), because they secrete dopamine, the missing neurotransmitter. However, such autografts have met with limited success. A more promising approach involves fetal donor tissue, which survives much better than adult tissue after grafting (Brundin et al 1986). The practical and ethical constraints on obtaining fetal donor tissue suggest that immortalized cell lines, from fetal SA progenitors as well as from other neuronal progenitors (Renfranz et al 1991; Snyder et al 1992), might prove useful in such cell-replacement therapies. Animal studies are now in progress to test this idea. The phenotypic plasticity and multipotentiality of SA lineage cells might make them useful in the treatment of other neurodegenerative diseases. For example, chromaffin cells and sympathetic neurons converted to a cholinergic phenotype by CDF/LIF have been tested for their ability to replace basal forebrain cholinergic neurons in an animal model of Alzheimer's disease (Mahanthappa et al 1990). In this way, the study of the SA lineage may have broader implications for both the understanding and treatment of human disease.

PERSPECTIVES

This review has focused on the biology of one developmentally restricted neural crest-derived progenitor cell. Although much less is known about the development of other neural crest derivatives, many of the general features of development in the SA lineage will probably apply to other crest lineages, as well. These features include the following: the generation of restricted progenitor cells; the determination of cell fate by environmental signals localized in different sites of migratory arrest; the ability of single environmental signals to promote one phenotype and repress another; the gain and loss of competence to respond to environmental signals during differentiation; and the ability of polypeptide growth factors to act sequentially as mitogens, differentiation factors, and survival factors. On the other hand, there are differences between the development of SA progenitors and that of other neural crest derivatives. Sympathoadrenal progenitors and sympathetic neuroblasts express differentiation genes (such as TH and neurofilament) while still proliferating; in other neuronal lineages, these genes are expressed subsequent to mitotic arrest (Anderson & Axel 1986; Rohrer & Thoenen 1987). And, it is not yet clear whether the phenotypic plasticity exhibited by SA derivatives is also characteristic of neurons in other crest lineages. These distinctions may reflect fundamental differences in developmental mechanisms, or they may be superficial. Further work on other neural crest sublineages will be necessary to resolve these issues.

The SA lineage has provided a well-characterized system for molecular studies of neural crest development; however, there have been recent spectacular advances in other aspects of neural crest molecular biology. Although beyond the scope of the present article, these deserve mention. The identification of the proto-oncogene p140^{trk} as a tyrosine kinase-containing NGF receptor has stimulated studies of signal transduction in PC12 cells (Kaplan et al 1991b; Kremer et al 1991; Sheng & Greenberg 1990; Vetter et al 1991) and has revealed mechanistic analogies with invertebrate systems (for reviews, see Rubin 1991; Sternberg & Horvitz 1991). The elucidation of the molecular bases of mouse mutations affecting melanocyte development has focused new attention on the role of poly-peptide growth factors and their receptors in neural crest development (for reviews, see Bowen-Pope et al 1991; Marusich & Weston 1991). And, a functional role for homeobox genes in the patterning of the cranial neural crest has been established by targeted mutagenesis of some Hox genes in mice (for review, see Hunt & Krumlauf 1991). Over the next decade, these diverse approaches should converge with cellular studies to advance our understanding of many aspects of neural crest development at the molec-ular level.

SUMMARY

Over the past five years, new insights have been gained into the biology of the SA lineage. These advances have been powered by the development of immunologic methods to isolate embryonic SA progenitors from fetal adrenal glands and sympathetic ganglia. Analysis of these embryonic pro-genitors has confirmed many of the ideas derived from earlier studies of postnatal cells, but has necessitated several revisions in our thinking, as well. First, embryonic SA progenitors appear to be distinct from mature SIF cells, a cell type initially postulated to be the central intermediate in the SA lineage. Second, FGF, not NGF, appears to be an important early influence on neuronal fate; NGF responsiveness appears relatively late in differentiation. Third, the development of both sympathetic neurons and adrenal chromaffin cells is not a one-step process, but rather involves a series of events, in which the cells change their responsiveness to growth factors and glucocorticoids. Fourth, emerging circumstantial evidence sug-gests that SA progenitors may have additional developmental potentials. Finally, new insights have been gained into the molecular mechanisms that underlie both the differentiation of SA progenitors and their determination from earlier multipotent neural crest cells. These advances have made the SA progenitor a well-defined system for studying the molecular control of cell fate in a vertebrate neurogenic precursor cell.

The analysis of the SA lineage at the cell biological level has raised several interesting molecular questions for future investigation. In the neuronal branch of the SA lineage, how is the acquisition of NGF-respon-siveness and NGF-dependence controlled, and what is the relationship of these events to the expression of $p140^{trk}$ and $p75$? In the chromaffin branch of the pathway, which molecules control the timing of PNMT expression? In the uncommitted SA progenitor, what is the molecular basis of the antagonism between the competing neuronal and chromaffin pathways of differentiation, and how does commitment to neuronal differentiation occur? Can SA progenitors differentiate to enteric neurons in vitro, and which differentiation and survival factors control this phenotype, as well as the other classical neurotransmitter and neuropeptide phenotypes expressed by SA derivatives? What are the roles of MASH1 and other regulatory genes in controlling early stages in neural crest cell determination, and how is the expression of these molecules in turn controlled? How much of the genetic regulatory network controlling neuronal differentiation in *Drosophila* has been conserved in vertebrates? These questions, and the further puzzles that their answers will inevitably create, promise fruitful territory for future investigations as the molecular biology of vertebrate neurogenesis comes of age.

ACKNOWLEDGMENTS

I thank Derek Stemple, Susan Birren, and Paul Patterson for their critical comments on this review, and Helen Walsh for her expert preparation of the manuscript. Work described in this review was supported by National Institutes of Health grant NS23476, a Pew Foundation Faculty Fellowship, and a Sloan Foundation Fellowship in Neuroscience.

Literature Cited

Abo, T., Balch, C. M. 1981. A differentiation antigen of human NK and K cells identified by a monoclonal antibody (HNK-1). *J. Immunol.* 127: 1024–29

Allen, G. S., Burns, R. S., Tulipan, N. B., Parker, R. A. 1989. Adrenal medullary transplantation to the caudate nucleus in Parkinson's Disease. *Arch. Neurol.* 46: 487–91

Aloe, L., Levi-Montalcini, R. 1979. Nerve growth factor-induced transformation of immature chromaffin cells in vivo into sympathetic neurons: Effects of antiserum to nerve growth factor. *Proc. Natl. Acad. Sci. USA* 76: 1246–50

Alonso, M. C., Cabrera, C. V. 1988. The achaete-scute complex of Drosophila melanogaster comprises four homologous genes. *EMBO J.* 7: 2585–91

Anderson, D. J. 1989. The neural crest cell lineage problem: Neuropoiesis? *Neuron* 3: 1–12

Anderson, D. J. 1988. Cell fate and gene expression in the developing neural crest. In *Neural Development and Regeneration*, ed. A. Gorio, J. R. Perez-Polo, J. D. Vellis, B. Haber. NATO ASI Ser. H, 22: 188–98. Berlin/Heidelberg: Springer-Verlag

Anderson, D. J., Axel, R. 1986. A bipotential neuroendocrine precursor whose choice of cell fate is determined by NGF and glucocorticoids. *Cell* 47: 1079–90

Anderson, D. J., Axel, R. 1985. Molecular probes for the development and plasticity of neural crest derivatives. *Cell* 42: 649–62

Anderson, D. J., Carnahan, J., Michelsohn, A., Patterson, P. H. 1991. Antibody markers identify a common progenitor to sympathetic neurons and chromaffin cells in vivo, and reveal the timing of commitment to neuronal differentiation in the sympathoadrenal lineage. *J. Neurosci.* 11: 3507–19

Anderson, D. J., Michelsohn, A. 1989. Role of glucocorticoids in the chromaffin-neuron developmental decision. *Int. J. Dev. Neurosci.* 12: 83–94

Baetge, G., Gershon, M. D. 1989. Transient catecholaminergic (TC) cells in the vagus nerves and bowel of fetal mice: relationship to the development of enteric neurons. *Dev. Biol.* 132: 189–211

Baetge, G., Pintar, J. E., Gershon, M. D. 1990. Transiently catecholaminergic (TC) cells in the bowel of the fetal rat: precursors of noncatecholaminergic enteric neurons. *Dev. Biol.* 141: 353–80

Barald, K. F. 1988. Antigen recognized by monoclonal antibodies to mesencephalic neural crest and to ciliary ganglion neurons is involved in the high affinity choline uptake mechanism in these cells. *J. Neurosci. Res.* 21: 119–34

Barasch, J. M., Mackey, H., Tamir, H., Nunez, E. A., Gershon, M. D. 1987. Induction of a neural phenotype in a serotonergic endocrine cell derived from the neural crest. *J. Neurosci.* 7: 2874–83

Baroffio, A., Dupin, E., LeDouarin, N. M. 1988. Clone-forming ability and differentiation potential of migratory neural crest cells. *Proc. Natl. Acad. Sci. USA* 85: 5325–29

Batter, D. K., D'Mello, S. R., Turzai, L. M., Hughes, H. B. III, Gioio, A. E., Kaplan, B. B. 1988. The complete nucleotide sequence and structure of the gene encoding bovine phenylethanolamine N-methyltransferase. *J. Neurosci. Res.* 19: 367–76

Beato, M. 1989. Gene regulation by steroid hormones. *Cell* 56: 335–44

Birren, S. J., Anderson, D. J. 1990. A v-myc-immortalized sympathoadrenal progenitor cell line in which neuronal differentiation is initiated by FGF but not NGF. *Neuron* 4: 189–201

Birren, S. J., Verdi, J., Anderson, D. J. 1992. Membrane depolarization induces p140^{trk} and NGF-responsiveness, but not p75^{LNGFR} in MAH cells. *Science* 257: 395–97

Björklund, A. 1991. Neural transplantation —an experimental tool with clinical possibilities. *Trends Neurosci.* 14: 319–22

Bohn, M. C., Goldstein, M., Black, I. B. 1982. Expression of phenylethanolamine N-methyltransferase (PNMT) in rat sympathetic ganglia and extra-adrenal chromaffin tissue. *Dev. Biol.* 89: 299–308

Bohn, M. C., Goldstein, M., Black, I. 1981. Role of glucocorticoids in expression of the adrenergic phenotype in rat embryonic adrenal gland. *Dev. Biol.* 82: 1–10

Bowen-Pope, D. F., van Koppen, A., Schatteman, G. 1991. Is PDGF really important? Testing the hypotheses. *Trends Genet.* 7: 413–17

Bronner-Fraser, M., Fraser, S. 1989. Developmental potential of avian trunk neural crest cells in situ. *Neuron* 3: 755–66

Bronner-Fraser, M., Fraser, S. 1988. Cell lineage analysis shows multipotentiality of some avian neural crest cells. *Nature* 335: 161–64

Brundin, P., Nilsson, O. G., Strecker, R. E., Lindvall, O., Åstedt, B., Björklund, A. 1986. Behavioural effects of human fetal dopamine neurons grafted in a rat model of Parkinson's disease. *Exp. Brain Res.* 65: 235–42

Campos-Ortega, J. A., Jan, Y. N. 1991. Genetic and molecular bases of neurogenesis in *Drosophila melanogaster*. *Annu. Rev. Neurosci.* 14: 399–420

Carnahan, J. F., Anderson, D. J., Patterson, P. H. 1991. Evidence that enteric neurons may derive from the sympathoadrenal lineage. *Dev. Biol.* 148: 552–61

Carnahan, J. F., Patterson, P. H. 1991a. Generation of monoclonal antibodies that bind preferentially to adrenal chromaffin cells and the cells of embryonic sympathetic ganglia. *J. Neurosci.* 11: 3493–3506

Carnahan, J. F., Patterson, P. H. 1991b. Isolation of the progenitor cells of the sympathoadrenal lineage from embryonic sympathetic ganglia with the SA monoclonal antibodies. *J. Neurosci.* 11: 3520–30

Ceccatelli, S., Dagerlind, A., Schalling, M., Wikstrom, A.-C., Okret, S. 1989. The glucocorticoid receptor in the adrenal gland is located in the cytoplasm of adrenaline cells. *Acta Phys. Scand.* 137: 559–60

Chun, L. L. Y., Patterson, P. H. 1977. Role of nerve growth factor in the development of rat sympathetic neurons in vitro. *J. Cell Biol.* 75: 694–704

Claude, P., Parada, I. M., Gordon, K. A., D'Amore, P. A., Wagner, J. A. 1988. Acidic fibroblast growth factor stimulates adrenal chromaffin cells to proliferate and to extend neurites, but is not a long-term survival factor. *Neuron* 1: 783–90

Cochard, P., Goldstein, M., Black, I. B. 1978. Ontogenic appearance and disappearance of tyrosine hydroxylase and catecholamines in the rat embryo. *Proc. Natl. Acad. Sci. USA* 75: 2986–90

Coffman, C., Harris, W., Kintner, C. 1990. *Xotch*, the *Xenopus* homolog of *Drosophila notch*. *Science* 249: 1438–41

Cooper, M. J., Hutchins, G. M., Cohen, P. S., Helman, L. J., Mennie, R. J., Israel, M. A. 1990a. Human neuroblastoma tumor cell lines correspond to the arrested differentiation of chromaffin adrenal medullary neuroblasts. *Cell Growth Differ.* 1: 149–59

Cooper, M. J., Hutchins, G. M., Israel, M. A. 1990b. Histogenesis of the human adrenal medulla: An evaluation of the ontogeny of chromaffin and nonchromaffin lineages. *Am. J. Pathol.* 137: 605–15

Coughlin, M. D., Collins, M. D. 1985. Nerve growth factor-independent development of embryonic mouse sympathetic neurons in dissociated cell culture. *Dev. Biol.* 110: 392–401

Coulter, D. H., Gershon, M. D., Rothman, T. P. 1988. Neural and glial phenotypic expression by neural crest cells in culture: Effects of control and presumptive aganglionic bowel from *ls/ls* mice. *J. Neurobiol.* 19: 507–31

Davies, A. M., Bandtlow, C., Heumann, R.,

Korsching, S., Rohrer, H., Thoenen, H. 1987. The site and timing of nerve growth factor (NGF) synthesis in developing skin in relation to its innervation by sensory neurons and their expression of NGF receptors. *Nature* 326: 353–58

Diamond, M. I., Miner, J. N., Yoshinaga, S. K., Yamamoto, K. R. 1990. Transcription factor interactions: selectors of positive or negative regulation from a single DNA element. *Science* 249: 1266–72

DiCicco-Bloom, E., Black, I. B. 1989. Depolarization and insulin-like growth factor-I (IGF-I) differentially regulate the mitotic cycle in cultured rat sympathetic neuroblasts. *Brain Res.* 491: 403–6

DiCicco-Bloom, E., Black, I. B. 1988. Insulin growth factors regulate the mitotic cycle in cultured rat sympathetic neuroblasts. *Proc. Natl. Acad. Sci. USA* 85: 4066–70

DiCicco-Bloom, E., Townes-Anderson, E., Black, I. B. 1990. Neuroblast mitosis in dissociated culture: regulation and relationship to differentiation. *J. Cell Biol.* 110: 2073–86

Doupe, A. J., Landis, S. C., Patterson, P. H. 1985a. Environmental influences in the development of neural crest derivatives: glucocorticoids, growth factors and chromaffin cell plasticity. *J. Neurosci.* 5: 2119–42

Doupe, A. J., Patterson, P. H., Landis, S. C. 1985b. Small intensely fluorescent (SIF) cells in culture: role of glucocorticoids and growth factors in their development and phenotypic interconversions with other neural crest derivatives. *J. Neurosci.* 5: 2143–60

Dupin, E., Baroffio, A., Dulac, C., Cameron-Curry, P., Le Douarin, N. M. 1990. Schwann-cell differentiation in clonal cultures of the neural crest, as evidenced by the anti-Schwann cell myelin protein monoclonal antibody. *Proc. Natl. Acad. Sci. USA* 87: 1119–23

Ehrlich, M. E., Evinger, M. J., Joh, T. H., Teitelman, G. 1989. Do glucocorticoids induce adrenergic differentiation in adrenal cells of neural crest origin? *Dev. Brain Res.* 50: 129–37

Eranko, O. 1975. SIF cells: structure and function of the small, intensely fluorescent sympathetic cells. *Fogarty Int. Cent. Proc. No. 30*, Dep. Health Educ. Welf. Publ. No. 76-942. Bethesda, MD: Natl. Inst. Health

Ernsberger, U., Edgar, D., Rohrer, H. 1989a. The survival of early chick sympathetic neurons in vitro is dependent on a suitable substrate but independent of NGF. *Dev. Biol.* 135: 250–62

Ernsberger, U., Sendtner, M., Rohrer, H. 1989b. Proliferation and differentiation of embryonic chick sympathetic neurons: effects of ciliary neurotrophic factor. *Neuron* 2: 1275–84

Federoff, H. J., Grabczyk, E., Fishman, M. C. 1988. Dual regulation of GAP-43 gene expression by nerve growth factor and glucocorticoids. *J. Biol. Chem.* 263: 19290–95

Frank, E., Sanes, J. R. 1991. Lineage of neurons and glia in chick dorsal root ganglia: analysis in vivo with a recombinant retrovirus. *Development* 111: 895–908

Fraser, S. E., Bronner-Fraser, M. E. 1991. Migrating neural crest cells in the trunk of the avian embryo are multipotent. *Development* 112: 913–20

Gage, F. H., Kawaja, M. D., Fisher, L. J. 1991. Genetically modified cells: applications for intracerebral grafting. *Trends Neurosci.* 14: 328–33

Gandelman, K. Y., Coker, G. T., Moffat, M., O'Malley, K. L. 1990. Species and regional differences in the expression of cell-type specific elements at the human and rat tyrosine hydroxylase gene loci. *J. Neurochem.* 55: 2149–52

Gard, A. L., Pfeiffer, S. E. 1990. Two proliferative stages of the oligodendrocyte lineage (A2B5+O4− and O4+GalC−) under different mitogenic control. *Neuron* 5: 615–25

Ghysen, A., Dambly-Chaudiere, C. 1989. Genesis of the *Drosophila* peripheral nervous system. *Trends Genet.* 5: 251–55

Ghysen, A., Dambly-Chaudiere, C. 1988. From DNA to form: the *achaete-scute* complex. *Genes Devel.* 2: 495–501

Gonzalez, F., Romani, S., Cubas, P., Modolell, J., Campuzano, S. 1989. Molecular analysis of the asense gene, a member of the *achaete-scute* complex of *Drosophila melanogaster*, and its novel role in optic lobe development. *EMBO J.* 8: 3553–62

Greene, L. A., Tischler, A. S. 1976. Establishment of a noradrenergic clonal line of rat adrenal pheochromocytoma cells which respond to nerve growth factor. *Proc. Natl. Acad. Sci. USA* 73: 2424–28

Hall, A. K., Landis, S. C. 1991a. Early commitment of precursor cells from the rat superior cervical ganglion to neuronal or nonneuronal fates. *Neuron* 6: 741–52

Hall, A. K., Landis, S. C. 1991b. Principal neurons and small intensely fluorescent (SIF) cells in the rat superior cervical ganglion have distinct developmental histories. *J. Neurosci.* 11: 472–84

Harrington, C. A., Lewis, E. J., Krzemien, D., Chikaraishi, D. M. 1987. Identification and cell-type specificity of the tyrosine hydroxylase gene promoter. *Nucl. Acids Res.* 15: 2363–84

Hempstead, B. L., Martin-Zanca, D., Kaplan, D. R., Parada, L. F., Chao, M. V.

1991. High-affinity NGF binding requires coexpression of the *trk* proto-oncogene and the low-affinity NGF receptor. *Nature* 350: 678–83

Hillarp, N. A., Hökfelt, T. 1953. Evidence of adrenaline and noradrenaline in separate adrenal medullary cells. *Acta Physiol. Scand.* 30: 55–68

Holt, C. E., Bertsch, T. W., Ellis, H. M., Harris, W. A. 1988. Cellular determination in the Xenopus retina is independent of lineage and birth date. *Neuron* 1: 15–26

Howard, M. J., Bronner-Fraser, M. 1985. The influence of neural tube-derived factors on differentiation of neural crest cells in vitro. I. Histochemical study on the appearance of adrenergic cells. *J. Neurosci.* 5: 3302–9

Hunt, P., Krumlauf, R. 1991. Deciphering the Hox code: Clues to patterning branchial regions of the head. *Cell* 66: 1075–78

Ip, N. Y., Nye, S. H., Boulton, T. G., Davis, S., Taga, T., et al. 1992. CNTP and LIF act on neuronal cells via shared signalling pathways that involve the IL-6 signal transducing receptor component gp[130]. *Cell* 69: 1121–32

Israel, M. A. 1991. Molecular origins of pediatric embryonal tumors. *Cancer Cells* 3: 193–94

Jan, Y. N., Jan, L. Y. 1990. Genes required for specifying cell fates in *Drosophila* embryonic sensory nervous system. *Trends Neurosci.* 13: 493–98

Jiang, W., Uht, R., Bohn, M. C. 1989. Regulation of phenylethanolamine N-methyltransferase (PNMT) mRNA in the rat adrenal medulla by corticosterone. *Int. J. Dev. Neurosci.* 7: 513–20

Johnson, D., Lanahan, A., Buck, C. R., Sehgal, A., Morgan, C., et al. 1986. Expression and structure of the human NGF receptor. *Cell* 47: 545–54

Johnson, J. E., Birren, S. J., Anderson, D. J. 1990. Two rat homologues of *Drosophila achaete-scute* specifically expressed in neuronal precursors. *Nature* 346: 858–61

Johnson, J. E., Birren, S. J., Saito, T., Anderson, D. J. 1992. The MASH genes encode transcriptional regulators that can activate expression of muscle creatine kinase, but do not induce myogenesis. *Proc. Natl. Acad. Sci. USA* 89: 3596–3600

Jonakait, G. M., Markey, K. A., Goldstein, M., Dreyfus, C. F., Black, I. B. 1985. Selective expression of high-affinity uptake of catecholamines by transiently catecholaminergic cells of the rat embryo: studies in vivo and in vitro. *Dev. Biol.* 108: 6–17

Kalcheim, C., Neufeld, G. 1990. Expression of basic fibroblast growth factor in the nervous system of early avian embryos. *Development* 109: 203–15

Kaneda, N., Sasaoka, T., Kobayashi, K., Kiuchi, K., Nagatsu, I., et al. 1991. Tissue-specific and high-level expression of the human tyrosine hydroxylase gene in transgenic mice. *Neuron* 6: 583–94

Kaplan, D. R., Hempstead, B. L., Martin-Zanca, D., Chao, M. V., Parada, L. F. 1991a. The *trk* proto-oncogene product: a signal transducing receptor for nerve growth factor. *Science* 252: 554–58

Kaplan, D. R., Martin-Zanca, D., Parada, L. F. 1991b. Tyrosine phosphorylation and tyrosine kinase activity of the *trk* proto-oncogene product induced by NGF. *Nature* 350: 158–60

Kapur, R. P., Hoyle, G. W., Mercer, E. H., Brinster, R. L., Palmiter, R. D. 1991. Some neuronal cell populations express human dopamine β-hydrosylase-*lacZ* transgenes transiently during embryonic development. *Neuron* 7: 717–27

Karlsson, O., Thor, S., Norberg, T., Ohlsson, H., Edlund, T. 1990. Insulin gene enhancer binding protein Isl-1 is a member of a novel class of proteins containing both a homeo- and a Cys-His domain. *Nature* 344: 879–82

Kessler, J. A., Cochard, P., Black, I. 1979. Nerve growth factor alters the fate of embryonic neuroblasts. *Nature* 280: 141–42

Klein, R., Jing, S., Nanduri, V., O'Rourke, E., Barbacid, M. 1991. The *trk* proto-oncogene encodes a receptor for nerve growth factor. *Cell* 65: 189–97

Korsching, S., Thoenen, H. 1988. Developmental changes of nerve growth factor levels in sympathetic ganglia and their target organs. *Dev. Biol.* 126: 40–46

Kraner, S. D., Shong, J. A., Tsay, H.-J., Mandel, G. 1992. Silencing the type II sodium channel gene: a model for neural-specific gene regulation. *Neuron* 9: 37–44

Kremer, N., D'Arcangelo, G., Thomas, S., Demarco, M., Brugge, J. 1991. Signal transduction by nerve growth factor and fibroblast growth factor in PC12 cells requires a sequence of *src* and *ras* actions. *J. Cell Biol.* 115: 809–19

Kuo, M. D., Oda, Y., Huang, J. S., Huang, S. S. 1990. Amino acid sequence and characterization of a heparin-binding neurite-promoting factor (P18) from bovine brain. *J. Biol. Chem.* 265: 18749–52

Landis, S. C., Keefe, D. 1983. Evidence for neurotransmitter plasticity in vivo: developmental changes in properties of cholinergic sympathetic neurons. *Dev. Biol.* 98: 349–72

Landis, S. C., Patterson, P. H. 1981. Neural

crest cell lineages. *Trends Neurosci.* 4: 172–75

LeDouarin, N. M. 1982. *The Neural Crest.* Cambridge: Cambridge Univ. Press

LeDouarin, N. M. 1980. The ontogeny of the neural crest in avian embryo chimeras. *Nature* 286: 663–69

LeDouarin, N., Dulac, C., Dupin, E., Cameron-Curry, P. 1991. Glial cell lineages in the neural crest. *Glia* 4: 175–84

Leonard, D. G. B., Ziff, E. B., Greene, L. A. 1987. Identification and characterization of mRNAs regulated by nerve growth factor in PC12 cells. *Mol. Cell. Biol.* 7: 3156–57

Levi-Montalcini, R., Angeletti, P. U. 1963. Essential role of the nerve growth factor in the survival and maintenance of dissociated sensory and sympathetic embryonic nerve cells in vitro. *Dev. Biol.* 7: 653–59

Levi-Montalcini, R., Booker, B. 1960. Destruction of the sympathetic ganglia in mammals by an antiserum to a nerve growth protein. *Proc. Natl. Acad. Sci. USA* 46: 384–91

Li, Y.-S., Milner, P. G., Chauhan, A. K., Watson, M. A., Hoffman, R. M., et al. 1990. Cloning and expression of a developmentally regulated protein that induces mitogenic and neurite outgrowth activity. *Science* 250: 1690–94

Lillien, L., Claude, P. 1985. Nerve growth factor is a mitogen for cultured chromaffin cells. *Nature* 317: 632–34

Lillien, L. E., Raff, M. C. 1990. Differentiation signals in the CNS: type-2 astrocyte development in vitro as a model system. *Neuron* 5: 111–19

Lo, L., Johnson, J. E., Wuenschell, C. W., Saito, T., Anderson, D. J. 1991. Mammalian *achaete-scute* homolog 1 is transiently expressed by spatially-restricted subsets of early neuroepithelial and neural crest cells. *Genes Dev.* 5: 1524–37

Machida, C. M., Rodland, K. D., Matrisian, L., Magun, B. E., Ciment, G. 1989. NGF induction of the gene encoding the protease transin accompanies neuronal differentiation in PC12 cells. *Neuron* 2: 1587–96

Mackey, H. M., Payette, R. F., Gershon, M. D. 1988. Tissue effect on the expression of serotonin, tyrosine hydroxylase and GABA in cultures of neurogenic cells from the neuraxis and branchial arches. *Development* 104: 205–17

Mahanthappa, N. K., Gage, F. H., Patterson, P. H. 1990. Adrenal chromaffin cells as multipotential neurons for autografts. In *Progress in Brain Research*, ed. S. B. Dunnett, S.-J. Richards, 82: 33–39

Marusich, M. F., Weston, J. A. 1992. Identification of early neurogenic cells in the neural crest lineage. *Dev. Biol.* 149: 295–306

Marusich, M. F., Weston, J. A. 1991. Development of the neural crest. *Curr. Opin. Genet. Devel.* 1: 221–29

Maue, R. A., Kraner, S. D., Goodman, R. H., Mandel, G. 1990. Neuron-specific expression of the rat brain type II sodium channel gene is directed by upstream regulatory elements. *Neuron* 4: 223–31

Maxwell, G. D., Forbes, M. E. 1990a. Exogenous basement membrane-like matrix stimulates adrenergic development in avian neural crest cultures. *Development* 101: 767–76

Maxwell, G. D., Forbes, M. E. 1990b. The phenotypic response of cultured quail trunk neural crest cells to a reconstituted basement membrane-like matrix is specific. *Dev. Biol.* 141: 233–37

Mercer, E. H., Hoyle, G. W., Kapur, R. P., Brinster, R. L., Palmiter, R. D. 1991. The dopamine β-hydroxylase gene promoter directs expression of E. coli *lacZ* to sympathetic and other neurons in transgenic mice. *Neuron* 7: 703–16

Michelsohn, A., Anderson, D. J. 1992. Changes in competence determine the timing of two sequential glucocorticoid effects on sympathoadrenal progenitors. *Neuron* 8: 589–604

Michelson, A. M., Abmayr, S. M., Bate, M., Arias, A. M., Maniatis, T. 1990. Expression of a MyoD family member prefigures muscle pattern in *Drosophila* embryos. *Genes Dev.* 4: 2086–97

Mitchell, P. J., Timmons, P. M., Hébert, J. M., Rigby, P. W. J., Tjian, R. 1991. Transcription factor AP-2 is expressed in neural crest cell lineages during mouse embryogenesis. *Genes Dev.* 5: 105–19

Molenaar, W. M., Lee, V. M.-Y., Trojanowski, J. Q. 1990. Early fetal acquisition of the chromaffin and neuronal immunophenotype by human adrenal medullary cells. An immunohistological study using monoclonal antibodies to chromogranini A, synaptophysin, tyrosine hydroxylase and neuronal cytoskeletal proteins. *Exp. Neurol.* 108: 1–9

Mori, N., Schoenherr, C., Vandenbergh, D. J., Anderson, D. J. 1992. A common silencer element in the SCG10 and type II Na⁺ channel genes binds a factor present in nonneuronal cells but not in neuronal cells. *Neuron* 9: 45–54

Mori, N., Stein, R., Sigmund, O., Anderson, D. J. 1990. A cell type-preferred silencer element that controls the neural-specific expression of the SCG10 gene. *Neuron* 4: 583–94

Murre, C., McCaw, P. S., Baltimore, D.

1989a. A new DNA binding and dimerization motif in immunoglobin enhancer binding, *daughterless, MyoD* and *myc* proteins. *Cell* 56: 777–83

Murre, C., McCaw, P. S., Vaessin, H., Caudy, M., Jan, L. Y., et al. 1989b. Interactions between heterologous helix-loop-helix proteins generate complexes that bind specifically to a common DNA sequence. *Cell* 58: 537–44

Nawa, H., Patterson, P. H. 1990. Separation and partial characterization of neuropeptide-inducing factors in heart cell conditioned medium. *Neuron* 4: 269–77

Ogawa, M., Ishikawa, I., Irimajiri, A. 1984. Adrenal chromaffin cells form functional cholinergic synapses in culture. *Nature* 307: 66–68

Olson, L. 1970. Fluorescence histochemical evidence for axonal growth and secretion from transplanted adrenal medullary tissue. *Histochemie* 22: 1–7

Patterson, P. H. 1990. Control of cell fate in a vertebrate neurogenic lineage. *Cell* 62: 1035–38

Patterson, P. H. 1978. Environmental determination of autonomic neurotransmitter functions. *Annu. Rev. Neurosci.* 1: 1–17

Patterson, P. H., Chun, L. L. Y. 1977. The induction of acetylcholine synthesis in primary cultures of dissociated rat sympathetic neurons. I. Effects of conditioned medium. *Dev. Biol.* 56: 263–80

Pincus, D. W., DiCicco-Bloom, E. M., Black, I. B. 1990. Vasoactive intestinal peptide regulates mitosis, differentiation and survival of cultured sympathetic neuroblasts. *Nature* 343: 564–67

Pohorecky, L. A., Wurtman, R. J. 1971. Adrenocortical control of epinephrine synthesis. *Pharmacol. Rev.* 23: 1–35

Radeke, M. J., Misko, T. P., Hsu, C., Herzenberg, L. A., Shooter, E. M. 1987. Gene transfer and molecular cloning of the rat nerve growth factor receptor. *Nature* 325: 593–97

Raff, M. C. 1989. Glial cell diversification in the rat optic nerve. *Science* 243: 1450–55

Rao, M. S., Landis, S. C. 1990. Characterization of a target-derived neuronal cholinergic differentiation factor. *Neuron* 5: 899–910

Renfranz, P. J., Cunningham, M. G., McKay, R. D. G. 1991. Region-specific differentiation of the hippocampal stem cell line HiB5 upon implantation into the developing mammalian brain. *Cell* 66: 713–29

Rodriguez-Tébar, A., Rohrer, H. 1991. Retinoic acid induces NGF-dependent survival response and high-affinity NGF receptors in immature chick sympathetic neurons. *Development* 112: 813–20

Rohrer, H. 1990. The role of growth factors in the control of neurogenesis. *Eur. J. Neurosci.* 2: 1005–15

Rohrer, H., Hofer, M., Hellweg, R., Korsching, S., Stehle, A. D., et al. 1988. Antibodies against mouse nerve growth factor interfere in vivo with the development of avian sensory and sympathetic neurones. *Development* 103: 545–52

Rohrer, H., Thoenen, H. 1987. Relationship between differentiation and terminal mitosis: chick sensory and ciliary neurons differentiate after terminal mitosis of precursor cells, whereas sympathetic neurons continue to divide after differentiation. *J. Neurosci.* 7: 3739–48

Romani, S., Campuzano, S., Macagno, E. R., Modolell, J. 1989. Expression of *achaete* and *scute* genes in *Drosophila* imaginal discs and their function in sensory organ development. *Genes Dev.* 3: 997–1007

Roos, T. H. 1967. Steroid synthesis in embryonic and fetal rat adrenal tissue. *Endocrinology* 81: 716–28

Ross, M. E., Evinger, M. J., Hyman, S. E., Carroll, J. M., Mucke, L., et al. 1990. Identification of a functional glucocorticoid response element in the phenylethanolamine N-methyltransferase promoter using fusion genes introduced into chromaffin cells in primary culture. *J. Neurosci.* 10: 520–30

Rubin, G. M. 1991. Signal transduction and the fate of the R7 photoreceptor in *Drosophila*. *Trends Genet.* 7: 372–77

Rydel, R. E., Greene, L. A. 1987. Acidic and basic fibroblast growth factors promote stable neurite outgrowth and neuronal differentiation in cultures of PC12 cells. *J. Neurosci.* 7: 3639–53

Saadat, S., Sendtner, M., Rohrer, H. 1989. Ciliary neurotrophic factor induces cholinergic differentiation of rat sympathetic neurons in culture. *J. Cell Biol.* 108: 1807–16

Sah, D. W. Y., Matsumoto, S. G. 1987. Evidence for serotonin synthesis, uptake and release in dissociated rat sympathetic neurons in culture. *J. Neurosci.* 7: 391–99

Sanes, J. R., Rubenstein, J. L. R., Nicolas, J. F. 1986. Use of a recombinant retrovirus to study post-implantation cell lineage in mouse embryos. *EMBO J.* 5: 3133–42

Satoh, T., Nakamura, S., Taga, T., Matsuda, T., Hirano, T., et al. 1988. Induction of neuronal differentiation in PC12 cells by B-cell stimulatory factor 2/Interleukin 6. *Mol. Cell. Biol.* 8: 3546–49

Sauerwald, A., Hoesche, C., Oschwald, R., Kilimann, M. W. 1990. The 5′-flanking region of the synapsin-I gene—a G+C-rich, TATA-less and CAAT-less, phylogenetically conserved sequence with cell type-specific promoter function. *J. Biol. Chem.* 265: 14932–37

Schotzinger, R. J., Landis, S. C. 1990. Acquisition of cholinergic and peptidergic properties by sympathetic innervation of rat sweat glands requires interaction with normal target. *Neuron* 5: 91–100

Schotzinger, R., Landis, S. C. 1988. Cholinergic phenotype developed by noradrenergic sympathetic neurons after innervation of a novel cholinergic target in vivo. *Nature* 335: 637–39

Schüle, R., Evans, R. M. 1991. Cross-coupling of signal transduction pathways: A inc finger meets leucine zipper. *Trends Genet.* 7: 377–81

Schüle, R., Rangarajan, P., Kliewer, S., Ransone, L. J., Bolado, J., et al. 1990. Functional antagonism between oncoprotein c-Jun and the glucocorticoid receptor. *Cell* 62: 1217–26

Seidl, K., Unsicker, K. 1989a. Survival and neuritic growth of sympathoadrenal (chromaffin) precursor cells in vitro. *Int. J. Dev. Neurosci.* 7: 465–73

Seidl, K., Unsicker, K. 1989b. The determination of the adrenal medullary cell fate during embryogenesis. *Dev. Biol.* 136: 481–90

Sheng, M., Greenberg, M. E. 1990. The regulation and function of c-*fos* and other immediate early genes in the nervous system. *Neuron* 4: 477–85

Sieber-Blum, M. 1990. Mechanisms of neural crest diversification. In *Comments Developmental Neurobiology*, 1: 225–49. London: Gordon & Breach

Sieber-Blum, M., Cohen, A. 1980. Clonal analysis of quail neural crest cells: they are pluripotent and differentiate in vitro in the absence of non-neural crest cells. *Dev. Biol.* 80: 96–106

Simons, S. S. J., Mercier, L., Miller, N. R., Miller, P. A., Oshima, H., et al. 1989. Differential modulation of gene induction of glucocorticoids and antiglucocorticoids in rat hepatoma tissue culture cells. *Cancer Res.* 49: 2244s–52s

Snyder, E. Y., Deitcher, D. L., Walsh, C., Arnold-Aldea, S., Hartweig, E. A., Cepko, C. L. 1992. Multipotent neural cell lines can engraft and participate in development of mouse cerebellum. *Cell* 68: 33–51

Soinila, S., Ahonen, M., Lahtinen, T., Häppölä, O. 1989. Developmental changes in 5-hydroxytryptamine immunoreactivity of sympathetic cells. *Int. J. Dev. Neurosci.* 7: 553–63

Stein, R., Orit, S., Anderson, D. J. 1988. The induction of a neural-specific gene, SCG10, by nerve growth factor in PC12

cells is transcriptional, protein synthesis dependent, and glucocorticoid inhibitable. *Dev. Biol.* 127: 316–25

Stemple, D. L., Mahanthappa, N. K., Anderson, D. J. 1988. Basic FGF induces neuronal differentiation, cell division, and NGF dependence in chromaffin cells: a sequence of events in sympathetic development. *Neuron* 1: 517–25

Stern, C. D., Artinger, K. B., Bronner-Fraser, M. 1991. Tissue interactions affecting the migration and differentiation of neural crest cells in the chick embryo. *Development* 113: 207–16

Sternberg, P. W., Horvitz, H. R. 1991. Signal transduction during *C. elegans* vulval induction. *Trends Genet.* 7: 366–71

Suda, T., Suda, J., Ogawa, M. 1984. Disparate differentiation in mouse hemopoietic colonies derived from paired progenitors. *Proc. Natl. Acad. Sci. USA* 81: 2520–24

Szabo, A., Dalmau, J., Manley, G., Rosenfeld, M., Wong, E., et al. 1991. HuD, a paraneoplastic encephalomyelitis antigen, contains RNA-binding domains and is homologous to Elav and Sex-lethal. *Cell* 67: 325–33

Teitelman, G., Joh, T. H., Park, D., Brodsky, M., New, M., Reis, D. J. 1982. Expression of the adrenergic phenotype in cultured fetal adrenal medullary cells: role of intrinsic and extrinsic factors. *Dev. Biol.* 80: 450–59

Teitelman, G., Joh, T. H., Reis, D. J. 1978. Transient expression of a noradrenergic phenotype in cells of the rat embryonic gut. *Brain Res.* 158: 229–34

Thor, S., Ericson, J., Brännström, T., Edlund, T. 1991. The homeodomain LIM protein Isl-1 is expressed in subsets of neurons and endocrine cells in the adult rat. *Neuron* 7: 1–20

Togari, A., Dickens, G., Kuzuya, H., Guroff, G. 1985. The effect of fibroblast growth factor on PC12 cells. *J. Neurosci.* 5: 307–16

Trojanowski, J. Q., Molenaar, W. M., Baker, D. L., Pleasure, D., Lee, V. M.-Y. 1991. Neural and neuroendocrine phenotype of neuroblastomas, ganglioneuroblastomas, ganglioneuromas and mature versus embryonic human adrenal medullary cells. In *Advances in Neuroblastoma Research*, 3: 335–41. New York: Wiley-Liss

Turner, D. L., Cepko, C. 1987. A common progenitor for neurons and glia persists in rat retina late in development. *Nature* 328: 131–36

Unsicker, K., Drisch, B., Otten, J., Thoenen, H. 1978. Nerve growth factor-induced fiber outgrowth from isolated rat adrenal chromaffin cells: impairment by gluco-corticoids. *Proc. Natl. Acad. Sci. USA* 75: 3498–3502

Unsicker, K., Seidl, K., Hofmann, H. D. 1989. The neuro-endocrine ambiguity of sympathoadrenal cells. *Int. J. Dev. Neurosci.* 7: 413–17

Vandenbergh, D. J., Mori, N., Anderson, D. J. 1991. Co-expression of multiple neurotransmitter enzyme genes in normal and immortalized sympathoadrenal progenitor cells. *Dev. Biol.* 148: 10–22

Vandenbergh, D. J., Wuenschell, C. W., Mori, N., Anderson, D. J. 1989. Chromatin structure as a molecular marker of cell lineage and developmental potential in neural crest-derived chromaffin cells. *Neuron* 3: 507–18

Vetter, M. L., Martin-Zanca, D., Parada, L. F., Bishop, J. M., Kaplan, D. R. 1991. Nerve growth factor rapidly stimulates tyrosine phosphorylation of phospholipase C-γ1 by a kinase activity associated with the product of the *trk* protooncogene. *Proc. Natl. Acad. Sci. USA* 88: 5650–54

Villares, R., Cabrera, C. V. 1987. The *achaete-scute* gene complex of D. melanogaster: conserved domains in a subset of genes required for neurogenesis and their homology to *myc. Cell* 50: 415–24

Vogel, K. S., Davies, A. M. 1991. The duration of neurotrophic factor independence in early sensory neurons is matched to the time course of target field innervation. *Neuron* 7: 819–30

Vogel, K. S., Weston, J. A. 1988. A subpopulation of cultured avian neural crest cells has transient neurogenic potential. *Neuron* 1: 569–77

Vogel, K. S., Weston, J. A. 1990. The sympathoadrenal lineage in avian embryos. I. Adrenal chromaffin cells lose neuronal traits during embryogenesis. *Dev. Biol.* 139: 1–12

Way, J. C., Chalfie, M. 1988. *mec-3*, a homeobox-containing gene that specifies differentiation of the touch receptor neurons in *C. elegans. Cell* 54: 5–16

Weinmaster, G., Roberts, V. J., Lemke, G. 1991. A homolog of *Drosophila Notch* expressed during mammalian development. *Development* 113: 199–205

Weintraub, H., Davis, R., Tapscott, S., Thayer, M., Krause, M., et al. 1991. The *myoD* gene family: nodal point during specification of the muscle cell lineage. *Science* 251: 761–66

Weskamp, G., Reichardt, L. F. 1991. Evidence that biological activity of NGF is mediated through a novel subclass of high affinity receptors. *Neuron* 6: 649–63

Wetts, R., Fraser, S. E. 1988. Multipotent precursors can give rise to all major types of the frog retina. *Science* 239: 1142–45

Wolinsky, E. J., Landis, S. C., Patterson, P. H. 1985. Expression of noradrenergic and cholinergic traits by sympathetic neurons cultured without serum. *J. Neurosci.* 5: 1497–1508

Wolswijk, G., Noble, M. 1989. Identification of an adult-specific glial progenitor cell. *Development* 105: 387–400

Wuenschell, C. W., Mori, N., Anderson, D. J. 1990. Analysis of SCG10 gene expression in transgenic mice reveals that neural specificity is achieved through selective derepression. *Neuron* 4: 595–602

Wurtman, R. J., Axelrod, J. 1966. Control of enzymatic synthesis of adrenaline in the adrenal medulla by adrenal cortical steroids. *J. Biol. Chem.* 241: 2301–5

Yamamori, Y., Fukada, K., Aebersold, R., Korsching, S., Fann, M.-J., Patterson, P. H. 1989. The cholinergic neuronal differentiation factor from heart cells is identical to leukemia inhibitory factor. *Science* 246: 1412–16

Yang-Yen, H. F., Chambard, J.-C., Sun, Y.-L., Smeal, T., Schmidt, T. J., et al. 1990. Transcriptional interference between *c-Jun* and the glucocorticoid receptor: mutual inhibition of DNA binding due to direct protein-protein interaction. *Cell* 62: 1205–15

Annu. Rev. Neurosci. 1993. 16:159–82

IMPLICIT MEMORY:
A Selective Review

Daniel L. Schacter, C.-Y. Peter Chiu, and Kevin N. Ochsner

Department of Psychology, Harvard University, Cambridge, Massachusetts 02138

KEY WORDS: priming, memory systems, memory disorders, skill learning

INTRODUCTION

The investigation of human memory has undergone a major transformation during the past decade, characterized by Richardson-Klavehn & Bjork (1988) as a "revolution in the way that we measure and interpret the influence of past events on current experience and behavior . . ." (pp. 476–77). It is not clear whether this transformation should be viewed as a paradigm shift, as Kuhn (1962) suggested, or, more modestly, as the emergence of a new subarea of research. What does seem clear, however, is that the "revolution" has already yielded, and will continue to yield, an outpouring of new data and ideas concerning the nature of memory.

The scientific substance of this revolution turns on a distinction between two different ways in which memory for prior experiences can be expressed, most frequently referred to as explicit and implicit memory, respectively (Graf & Schacter 1985; Schacter 1987). Explicit memory entails intentional or conscious recollection of previous experiences. In a typical explicit memory experiment, subjects are initially shown a series of words, pictures, or some other set of to-be-remembered materials. Later, they are given a recall or recognition test, in which they must think back to the study episode in order to produce or select a correct response. The vast majority of studies on human memory have, until recently, conformed to this paradigm; theoretical ideas about memory have, with few exceptions, addressed explicit remembering. Implicit memory, by contrast, entails a facilitation or change in test performance that is attributable to infor-

159

mation or skills acquired during a prior study episode, even though subjects are not required to, and may even be unable to, recollect the study episode. Implicit memory is assessed by examining the impact of study episodes on subsequent performance of tasks that do not require recollection of those episodes, such as completing a fragment of a word, choosing which of two stimuli they prefer, or reading inverted text.

Given the established tradition of research on explicit memory in cognitive psychology and neuropsychology, it took a series of striking findings to shift attention to implicit memory. By the end of the 1970s, evidence already existed that patients with organic amnesic syndromes could show substantial and, occasionally, normal levels of performance on various tests that would now be characterized as implicit memory tasks, such as motor skill learning (Milner et al 1968) and completion of fragmented words and pictures (Warrington & Weiskrantz 1974)—despite the apparent absence of any recollection for having previously performed the tasks. During the early and mid-1980s, many studies were published that confirmed, extended, and clarified these observations of spared implicit memory in amnesic patients (e.g. Cohen & Squire 1980; Graf et al 1984; Moscovitch 1982; Schacter 1985). At about the same time, several studies with normal, nonamnesic subjects revealed surprising dissociations between implicit and explicit memory: Experimental variables affected the two forms of memory in different, and even opposite, ways (Graf et al 1982; Jacoby & Dallas 1981; Tulving et al 1982), and statistical independence between implicit and explicit task performance was observed (Jacoby & Witherspoon 1982; Tulving et al 1982). These findings led to various theoretical proposals concerning the mechanisms involved in implicit memory and stimulated a virtual avalanche of research that explored numerous aspects of the phenomenon.

We should emphasize that the term "implicit memory" is a descriptive label that refers to one way in which the influence of past·experiences can be expressed in subsequent task performance—unintentionally and without conscious recollection of a learning episode. Other, less-frequently used descriptive labels with similar meanings include memory without awareness (Jacoby & Witherspoon 1982), indirect memory (Johnson & Hasher 1987), and nondeclarative memory (Squire 1987). Several different manifestations of implicit memory have been investigated in the experimental literature. Perhaps the most extensively studied implicit memory phenomenon is that of repetition, or direct priming, in which exposure to a word or object on a study list facilitates its subsequent identification when degraded perceptual cues are provided (Tulving & Schacter 1990). Various types of perceptual, motor, and cognitive skill learning can also be considered expressions of implicit memory, in that they can occur without reference to, or recollection of, prior learning episodes. At an operational level, the main differences between priming and skill learning are that the former phenomenon is typically observed after a single study trial or small number of study trials and reflects memory for specific items, whereas the latter is typically observed after numerous study trials and need not involve memory for particular items.

In addition to priming and skill learning, other manifestations of implicit memory have also been studied, including the biasing effects of previous experience on various types of perceptual, cognitive, affective, and even social judgments (e.g. Bargh 1989; Jacoby et al 1989; Kunst-Wilson & Zajonc 1980; Mandler et al 1987); classical conditioning in amnesic populations (e.g. Daum et al 1989; Weiskrantz & Warrington 1979); and learning of grammatical and other kinds of rules (Kinsbourne & Wood 1975; Knowlton et al 1992; Reber 1967). We focus here on priming, because it has been the most thoroughly investigated implicit memory phenomenon and the primary concern of most theoretical views of implicit memory. However, we consider briefly literature on skill learning, as well as the relation between priming and skill learning. The review also focuses on research from the past five years, because several reviews deal thoroughly with earlier research (cf. Richardson-Klavehn & Bjork 1988; Roediger 1990; Schacter 1987).

The principal message of the review is straightforward: Research on implicit memory has provided strong evidence for the hypothesis that memory is not a unitary entity, but is instead composed of separate, yet interacting, systems and subsystems. Some of the functional characteristics of these systems and subsystems have been delineated, and empirically based ideas about their neural basis are beginning to emerge.

PRIMING: DATA AND THEORY

We first review experimental findings concerning the nature of priming and then consider theoretical interpretation of the data. The empirical review is divided into three main sections, which correspond to three types of priming that have been most often studied: visual word priming, visual object priming, and auditory word priming. The great majority of studies have investigated visual word priming. Within each section, we first consider research on normal subjects and then turn our attention to studies of various memory-impaired populations.

Visual Word Priming

Four main experimental paradigms have been used to investigate visual word priming: stem completion, in which subjects are given the three letters with multiple possible completions and asked to provide the first word that comes to mind [e.g. for___ (forest)]; fragment completion, in which word fragments with one or two possible completions are provided with similar instructions [e.g. a–a-i-n (assassin)]; word or perceptual identification, in which target items are exposed for brief durations (e.g. 35 ms), and subjects attempt to identify them; and lexical decision, in which subjects are shown letter strings that constitute real words or nonwords (e.g. flig) and are asked to make a word/nonword decision as quickly as possible. On the first three tasks, priming is indicated when subjects complete or identify more studied than nonstudied items; on the fourth task, priming is indicated when subjects make lexical decisions about studied items more quickly than about nonstudied items.

STUDIES OF NORMAL SUBJECTS   A key finding from research in the early 1980s was that manipulating level or depth of encoding of target items during a study task had little effect on priming, despite large effects on explicit memory. That is, semantic study tasks that focus subjects' attention on meaningful properties of a word (e.g. judging the semantic category to which a word belongs) produced much higher levels of explicit memory than did nonsemantic study tasks that focus attention on a word's physical features (e.g. judging whether it has more vowels or consonants). However, the two types of study tasks yielded comparable levels of priming (Graf et al 1982; Graf & Mandler 1984; Jacoby & Dallas 1981). More recent studies have replicated and extended this phenomenon on a variety of tasks (cf. Bowers & Schacter 1990; Graf & Ryan 1990; Hashtroudi et al 1988; Roediger et al 1992; Schacter & McGlynn 1989). This finding is important because (a) it suggests that different processes are involved in priming and explicit memory and provides theoretical clues concerning the nature of those processes, and (b) it indicates that priming effects on implicit memory tasks in normal subjects should not be attributed to the "surreptitious" use of explicit memory strategies. With respect to the latter point, when college students participate in implicit memory experiments, there is always the possibility that they (in contrast to amnesic patients) may "catch on" concerning the relation between the implicit task and the prior study list, and make use of their explicit memory abilities to enhance task performance—that is, they may turn a nominally implicit task into a functionally explicit one. However, this kind of "contamination" from explicit memory is ruled out when an experimental manipulation, such as depth of study processing, has different effects on priming and explicit memory under conditions in which the same nominal cues are provided on the two tests, and only test instructions are varied: If subjects are engaging in explicit retrieval on an implicit test, then they should show higher levels of performance following semantic than nonsemantic encoding tasks (see Schacter et al 1989, for further discussion; Jacoby 1991, for another approach to the issue). Indeed, Bowers & Schacter (1990) found that subjects who were entirely unaware of the relation between the completion task and the prior study list (as assessed by a post-test questionnaire) showed robust priming effects of equivalent magnitude following semantic and nonsemantic encoding tasks. These findings indicate that it is possible to obtain dissociations between implicit and explicit memory in normal subjects that are similar to those observed in amnesic patients.

Effects of encoding processes on implicit memory have also been examined in conjunction with a phenomenon known as the generation effect—the observation that explicit memory for words that were generated at the time of study (e.g. political killer—assa–n) is typically more accurate than for words that were simply read during the study task (political killer—assassin; e.g. Slamecka & Graf 1978). A striking reversal of this generation effect on the word identification task was reported initially by Winnick & Daniel (1970) and then by Jacoby (1983): Priming effects were greater in the read than in the generate condition; in fact, significant priming was not observed in the generate condition. Although this finding was seized upon as theoretically crucial by some (Jacoby 1983; Roediger et al 1989), it is now clear that conditions exist in which priming can be higher for generate than for read items (Masson & Macleod 1992; Toth & Hunt 1990). Gardiner and colleagues have reported a series of studies in which fragment completion priming benefited from target generation during a study task (Gardiner 1988; Gardiner et al 1989), even with subjects who are "test unaware," as described earlier (Gardiner 1989). Importantly, however, such generation effects tend to be observed only when subjects are given the identical fragment at study and test (e.g. subjects generate assassin from political killer—a–a–in during the study task, and are given a–a–in at test). Generation benefits are reduced or absent when subjects are given different fragments at study and test (e.g. political killer——ss–ss—followed by a–a–in), thereby suggesting that priming can be highly specific to the perceptual form of target items.

This latter issue—the extent to which priming effects reflect the acquisition of specific perceptual information about target words—has received a great deal of experimental attention. It has been established beyond dispute that visual word priming is largely modality specific: Auditory presentation of target materials reduces and sometimes eliminates priming on stem completion (Graf et al 1985b), fragment completion (Donnelly 1988; Roediger & Blaxton 1987), word identification (Hashtroudi et al 1988; Jacoby & Dallas 1981), and lexical decision (Scarborough et al 1979) tasks. A similar pattern of reduced or absent priming is observed when subjects study pictorial equivalents of words (e.g. Weldon & Roediger 1987). And, there is little or no cross-language priming when bilingual subjects study a word in one language and are tested with same-meaning but different-form words from another language (Durgunoglu & Roediger 1987; Gerard & Scarborough 1989; Kisner et al 1984; Smith 1991, Experiment 2; but see Experiment 1).

Although the conclusion that visual word priming is largely modality specific seems inescapable, there is a good deal more uncertainty and controversy concerning the extent to which priming is specific to the precise surface form of a target word. Gardiner's (1988, 1989) work on generation effects in fragment completion shows such specificity, and Hayman & Tulving (1989) have provided related evidence for "hyperspecific" priming of fragment completion. Several studies have indicated that priming effects are reduced when the case of words is changed between study and test (e.g. uppercase to lowercase; Roediger & Blaxton 1987a; Scarborough et al 1977) or when words are studied and tested in different typographies (Jacoby & Hayman 1987). However, other experiments have failed to find evidence of such format-specific priming (e.g. Carr et al 1989; Clarke & Morton 1983; Tardif & Craik 1989; for review and discussion, see Carr & Brown 1990; Jacoby et al 1992; Kirsner et al 1989; Roediger & Blaxton 1987a; Schacter 1990; Whittlsea 1990).

Several recent studies have taken steps toward resolving the apparent inconsistencies by delineating conditions under which format-specific priming is and is not observed. In a study by Graf & Ryan (1990), for example, subjects performed different encoding tasks on target words that were presented in two Applesoft typefonts: one task focused the subjects' attention on the physical characteristics of the words (they rated the readability of the word); the other task focused on semantic attributes (they rated the pleasantness of the word). Priming was tested with a word identification task, and typefont was either held constant or changed between study and test. Graf & Ryan found less priming in the different typefont than same typefont condition following the readability task, but found no effect of changing typefont following the pleasantness task (see also Carr & Brown 1990). Marsolek et al (1992) examined the contributions of the left and right hemispheres to format-specific priming effects. Subjects studied a list of target words under full-field viewing conditions. They were then given a completion test in which half of the stems were briefly exposed in the left visual field and half were briefly exposed in the right. Case of words was either same or different at study and test. Marsolek et al found that changing case between study and test reduced priming when stems were exposed in the left visual field, but had no effect on the magnitude of priming when stems were exposed in the right, thereby suggesting a right hemisphere locus for the perceptually specific component of priming. It is thus conceivable that in the Graf & Ryan (1990) study, the readability encoding task preferentially engaged the right hemisphere and, hence, yielded format specific effects. More generally, these studies indicate that visual word priming is neither entirely specific nor entirely abstract; both kinds of priming can be observed in different conditions, and the differential contributions of the two hemispheres may play a role in determining which kind of priming is observed.

Another feature of visual word priming that appears to depend on particular experimental conditions is the time course of the phenomenon. For example, early evidence indicated that stem completion priming is a relatively transient phenomenon that disappears after a two-hour retention interval (Graf & Mandler 1984), whereas priming on the perceptual identification task persisted across a 24-hour delay (Jacoby & Dallas 1981). Priming on the fragment completion task lasted for seven days in one study (Tulving et al 1982) and over a year in another (Sloman et al 1988). However, because of initial failures to observe long-lasting fragment completion priming in amnesic patients (Squire et al 1987; see below), some investigators concluded that priming is a short-lived effect that dissipates within hours (Graf et al 1984; Squire 1987; Squire et al 1987). This conclusion appears to be incorrect, because, as discussed below, long-lasting priming has now been observed in amnesic patients. Also, recent evidence indicates that the apparent rapid decay of stem completion priming is not a fundamental feature of such priming, but rather a particular feature of the materials used by Graf & Mandler (1984) and Squire et al (1987). Roediger et al (1992) found that when word fragments and stems are constructed from the same set of target materials and tested under identical experimental conditions, long-lasting priming that persisted across one week was observed on both fragment and stem completion tasks. The longevity of visual word priming may be related to another observed feature of the phenomenon: Priming on fragment and stem completion tasks is not affected by manipulations of proactive and retroactive interference under conditions in which explicit memory is impaired by interference (Graf & Schacter 1987; Sloman et al 1988). However, conditions do exist under which some priming effects are sensitive to interference manipulations (Booker 1991; Mayes et al 1987), although the nature of the interference effects are poorly understood.

In the studies reviewed thus far, target materials have consisted of familiar words that have preexisting representations in memory. An important question concerns whether priming can also be observed for novel items that do not have preexisting memory representations. Some evidence favoring the latter possibility has been provided by studies in which subjects who were initially exposed to nonword letter strings, subsequently exhibited significant priming effects on tests of lexical decision (Bentin & Moscovitch 1988; Kersteen-Tucker 1991; Scarborough et al 1977) and perceptual identification (Feustal et al 1983; Rueckl 1990; Salaso et al 1985; Whittlsea & Cantwell 1987). However, nonword priming on the lexical decision task is exceedingly transient. For example, Bentin & Moscovitch (1988) only found nonword priming when the test trial immediately followed the initial exposure of the target; when even a single item intervened, priming effects disappeared. By contrast, longer-lasting effects have been observed with the perceptual identification test (e.g. Salasoo et al 1985), thus suggesting a possible role for task differences.

A second approach to examining priming of novel verbal information has been to test implicit memory for newly acquired associations between unrelated word pairs. For example, in a paradigm developed by Graf & Schacter (1985), subjects initially study normatively unrelated word pairs (e.g. ship-castle) and are then given a stem completion task in which the target stem is paired with its study list cue (ship-cas___; same context condition) or with some other unrelated word (officer-cas___; different context condition). Priming of new associations, as indicated by higher completion performance in the same context condition than in the different context condition, has been documented on this task. The observed priming—in contrast to explicit memory—exhibits modality specificity (Schacter & Graf 1989), insensitivity to proactive and retroactive interference (Graf & Schacter 1987), and little effect of elaborative and organizational encoding manipulations (Graf & Schacter 1989; Micco & Masson 1991; Schacter & Graf 1986a). In contrast to priming of familiar words, however, priming of new associations appears to involve some minimal degree of semantic study processing (Graf & Schacter 1985; Schacter & Graf 1986a; but see Miccio & Masson 1991) and is not observed readily in test-unaware subjects (Bowers & Schacter 1990; see also Howard et al 1991).

Little is known about the neural basis of priming in normal subjects, although the evidence suggesting a role for the right hemisphere in perceptual specificity effects provides some initial clues. Further evidence comes from a recent neuroimaging study that used positron emission tomography (PET) to investigate stem completion priming. Squire et al (1992) found that priming was associated with a reduction of blood flow to right extrastriate cortex. They also found some evidence for hippocampal activation during priming, but suggested that this effect was attributable to test awareness and subjects' use of explicit retrieval processes under the specific conditions of this experiment. Explicit memory performance (cued recall), however, was associated with marked activation in the right hippocampus.

In all the studies reviewed thus far, priming has been assessed with so-called data-driven implicit memory tasks, in which performance is primarily guided by physical properties of test cues (cf. Jacoby 1983; Roediger & Blaxton 1987b). But, it is also possible to assess priming with conceptually driven implicit tasks, in which semantic-level processing is required, such as producing a category instance in response to a category name. Indeed, priming has been observed on such tasks (Hamann 1990) and has been dissociated from perceptual priming effects on data-driven tests (e.g. Blaxton 1989; Srinivas & Roediger 1990).

STUDIES OF MEMORY IMPAIRED POPULATIONS   By the mid-1980s, it was well established that amnesic patients show intact priming of familiar words and word pairs on data-driven implicit tests, such as stem completion, perceptual identification, and lexical decision (cf. Cermak et al 1985; Graf et al 1984; Moscovitch 1982; Warrington & Weiskrantz 1974), as well as on implicit tests that involve some conceptual processing, such as category instance production and free association (Gardner et al 1973; Graf et al 1985; Schacter 1985; Shimamura & Squire 1984). Numerous authors have considered these findings as evidence that the limbic regions that are typically damaged in amnesia, including the hippocampus and related structures, are not necessary for priming of familiar verbal information (for review, see Mayes 1988; Shimamura 1986; Squire 1987, 1992). However, as alluded to earlier, questions have been raised regarding the longevity of priming in amnesia, based principally on the finding that amnesic patients did not show priming on a fragment completion test when tested at a long delay (Squire et al 1987). However, Tulving et al (1991) have recently demonstrated normal and extremely long-lasting fragment completion priming in a profoundly amnesic head-injured patient (K. C.) who exhibits no explicit memory. Indeed, K. C. showed little or no reduction in priming across a 12-month retention interval. An earlier study by MacAndrews et al (1987) revealed robust priming in K. C. and another severely amnesic patient across a one-week delay on a conceptual priming task that involved solving sentence puzzles (for other findings of very long-lasting implicit memory effects in K. C. on more complex learning tasks, see Glisky & Schacter 1988). Although the issue is not yet resolved, conditions clearly exist in which priming of words and other kinds of verbal information can persist over long delays in densely amnesic patients.

During the past several years, a great deal of attention has been paid to the question of whether amnesic patients show normal priming for novel, as well as familiar, verbal materials; results to-date are mixed (for detailed review, see Bowers & Schacter 1992). Several studies have reported impaired priming of nonwords in Korsakoff patients on tests of perceptual identification (Cermak et al 1985) and lexical decision (Smith & Oscar-Berman 1990). Diamond & Rozin (1984) had earlier found no priming of nonwords in a mixed group of memory-impaired patients. But, the Diamond-Rozin study entailed several methodological problems, and in the Cermak et al (1985) study, Korsakoff patients showed substantial nonword priming effects that were, nevertheless, smaller than those exhibited by control subjects. There are reasons to suspect, however, that control subjects' performance on the perceptual identification test was inflated by the use of explicit memory strategies (see Bowers & Schacter 1992; Haist et al 1991). Cermak et al (1988a) found near-normal priming of nonwords on a perceptual identification task in a densely amnesic encephalitic patient. Gabrieli & Keane (1988) observed what appeared to be intact nonword priming on a similar task in the well-known patient H. M. And, Haist et al (1991) found entirely normal nonword priming on a modified version of the perceptual identification task in a mixed group of amnesic patients. Similarly, Musen & Squire (1991) found that when reading and rereading lists of nonwords, amnesic patients showed normal decreases in reading time. Thus, although not all of the negative evidence can be explained, it now seems clear that amnesic patients can show intact priming of nonwords (for additional mixed outcomes, see Cermak et al 1991; Gordon 1988; Verfaillie et al 1991).

Several studies have also examined whether amnesic patients show priming for newly acquired associations, by using the Graf-Schacter (1985) cued stem completion paradigm described earlier. Graf and Schacter (1985; Schacter & Graf 1986b) observed priming of new associations in patients with relatively mild memory disorders, but not in patients with severe amnesia. Cermak and colleagues (1988b) found no evidence for associative effects in Korsakoff patients, but they (1988a) did observe priming of new associations in their encephalitic amnesic patient. Shimamura & Squire (1989) observed impaired priming of new associations in a mixed group of amnesic patients. Closer analysis of their data revealed an absence of such priming in the Korsakoff patients and a trend for associative effects in patients with amnesia attributable to anoxia and ischemia. Impaired priming of new associations in a mixed group of amnesic patients has also been reported by Mayes & Gooding (1989). Thus, although some positive trends exist, the bulk of the evidence from the Graf-Schacter paradigm suggests that newly formed associations between normatively unrelated words do not influence stem completion priming effects in densely amnesic patients (see also Moscovitch et al 1986; Tulving et al 1991).

Although research concerning priming and memory disorders has focused on amnesic patients, other populations and conditions have also been investigated. For example, several studies have reported that patients with dementia of the Alzheimer's type (AD), who typically exhibit extensive pathology in cortical association areas, show impaired word priming on a standard stem completion task (Salmon et al 1988; Shimamura et al 1987). By contrast, patients with dementia attributable to Huntington's disease (HD), which is typically associated with pathology in basal ganglia, show intact stem completion priming (Shimamura et al 1987). There have also been reports of normal word priming in AD patients when tests of lexical decision (Ober & Shenaut 1988) and word identification (Keane et al 1991) are used, and several alternative hypotheses have been offered to account for the variable results (cf. Keane et al 1991; Martin 1992). Word priming has also been investigated in normal elderly subjects, who typically exhibit explicit memory deficits. Although entirely normal priming has been observed in elderly subjects (compared with young) on stem completion, fragment completion, perceptual identification, and lexical decision tests (cf. Light & Singh 1987; Mitchell et al 1990), reports of impaired priming on some of the same tasks have also been published (e.g. Chiarello & Hoyer 1988; Davis et al 1990). However, there are reasons to suspect that the latter findings may be attributable to the use of explicit memory strategies by test-aware young subjects (see Graf 1990; Schacter et al 1992, for discussion).

Several investigators have adopted a pharmacological approach to the investigation of priming by administering drugs that typically produce explicit memory deficits to normal subjects. Although few studies have been reported, available evidence indicates that stem completion priming is spared by the benzodiazepine diazepam (Danion et al 1989), but is virtually eliminated by lorazepam, another benzodiazepine (Brown et al 1989). The anticholinergic agent scopolomine appears to reduce, but not eliminate, priming on a word fragment completion task (Nissen et al 1987).

Visual Object Priming

Although priming research has relied heavily on verbal stimuli, many investigators have developed tasks that use objects, patterns, and other nonverbal materials for studies of normal subjects and memory-impaired populations. These tasks include picture naming, in which subjects name previously presented pictures as quickly as possible (e.g. Biederman & Cooper 1991; Durso & Johnson 1979); picture fragment completion, in which subjects are given fragmented versions of pictures and asked to identify them (e.g. Snodgrass & Feenan 1990; Warrington & Weiskrantz 1968); object decision, in which subjects are shown drawings of real and nonsense objects (Kroll & Potter 1984) or structurally possible and impossible objects (Schacter et al 1990a) and are asked to make object/nonobject decisions; and dot pattern identification, in which subjects are exposed to degraded versions of dot patterns that they either copy (Musen & Treisman 1990) or complete (Gabrieli et al 1990). In all cases, priming is indicated by greater accuracy or reduced latency for studied items relative to nonstudied items (for detailed review, see Schacter et al 1990b).

STUDIES OF NORMAL SUBJECTS   As with visual word priming, a major issue in the nonverbal domain concerns the specificity of visual object priming. For example, when exposed to a picture of a table, priming effects might subsequently be observed for only that particular picture or for other pictures of tables. In an early study of picture naming, Bartram (1974) found maximal priming effects when subjects named an identical photograph of an object on two occasions, somewhat less priming when the same object was presented in two different views, and still less (but significant) priming for two different objects with the same name (e.g. two different tables). Warren & Morton (1982) and Jacoby et al (1989) subsequently reported a similar pattern of results and also found that initial study of an object's name (i.e. the word table) produced no subsequent priming of picture naming, thereby indicating that priming effects for different objects with the same name cannot be attributed to activation of the name or to activation of semantic information about the object. Specificity effects have also been observed on the fragment completion task (Snodgrass & Feenan 1990). One interpretive difficulty with these studies, however, is that they either did not include explicit memory tests or did not produce dissociations that could rule out the possible contribution of explicit memory processes (see Schacter et al 1990b). Other studies, however, have dissociated priming from explicit memory; Mitchell & Brown (1988), for example, showed that priming of picture-naming latency persisted across a six-week retention interval, despite a decline in recognition memory.

Despite the possibility that explicit memory played a role in some of the foregoing studies, they suggest rather strongly that object-specific visual information is involved in priming. Several recent experiments have provided evidence bearing on the nature of that information. For example, study-to-test changes in the size and left-right reflection (i.e. mirror image) of target objects have no effect on priming in a picture-naming paradigm (Biederman & Cooper 1992a,b) or in an object decision paradigm (Cooper et al 1992). In the latter paradigm, subjects study drawings of novel objects, half of which are structurally possible (they could exist in three dimensions) and half of which are structurally impossible (they contain various surface and edge violations that would prohibit them from existing in three dimensions). Subjects are then given brief exposures to studied and nonstudied objects and are asked to make possible/impossible decisions about them.

Importantly, performance on explicit memory tests in the Biederman & Cooper (1992a,b) and Cooper et al (1992) studies was impaired by the study/test changes in size and reflection, thereby indicating that the priming effect cannot be attributed to explicit retrieval processes. The fact that priming in these paradigms is size and reflection invariant suggests that the visual representation that supports priming does not include information about object size and reflection. By contrast, priming on the object decision task is eliminated by study-to-test changes in the picture-plane orientation of an object, thus suggesting that information about an object's orientation with respect to a principal axis may be involved in priming (Cooper et al 1991; see also Jolicouer 1985; Jolicouer & Millikan 1989).

The fact that priming is observed at all in the possible/impossible object decision paradigm highlights another important point: Priming of visual objects, like priming of verbal materials, can be observed for novel information that does not have any preexisting memory representation. Schacter et al (1990a) reported that whereas object decision priming is robust following tasks that require encoding of information about the global structure of target objects, it is not observed following tasks that require encoding of local object features or generating semantic elaborations about the objects. The latter finding is particularly important, because generating semantic elaborations about an object greatly enhances subsequent explicit memory performance (see Musen 1991, for similar findings concerning priming of novel dot patterns). Interestingly, significant priming on the object decision task is observed only for the possible objects. Impossible objects show little or no priming, perhaps because it is difficult to form an internal representation of their global three-dimensional structure (Schacter et al 1990a, 1991a; see also Kersteen-Tucker 1991, for a related finding concerning priming of novel polygons on a symmetry judgment task).

STUDIES OF MEMORY-IMPAIRED POPULATIONS   Early studies indicated that amnesic patients show some, but not normal, priming on a picture-fragment completion task (Milner et al 1968; Warrington & Weiskrantz 1968); patients' priming deficit is likely attributable to the use of explicit memory by control subjects (cf. Milner et al 1968; Schacter et al 1990b). More recently, however, evidence for normal visual object priming in amnesic patients has been provided. Cave & Squire (1992), using the picture-naming paradigm developed by Mitchell & Brown (1988), found normal priming in a mixed group of amnesic patients and further demonstrated that the priming persists over a seven-day retention interval. Intact priming of novel objects has also been demonstrated in patients with memory disorders on the possible/impossible object decision task (Schacter et al

1991b) and in patient H. M. on a dot pattern completion task (Gabrieli et al 1990). Studies of normal elderly subjects have also yielded evidence of spared priming: Mitchell et al (1990) reported long-lasting priming effects by using their picture-naming paradigm, and Schacter et al (1992) reported intact priming in the object decision paradigm. By contrast, a recent study of dementia indicates impaired priming of picture fragment completion in AD patients, together with spared priming in HD patients (Heindel et al 1990; see also Martin 1992).

Auditory Word Priming

Only a few studies have investigated priming in the auditory domain. Two main implicit tasks have been used: perceptual identification, in which spoken words are masked in white noise, and subjects attempt to identify them; and auditory stem completion, in which initial syllables are spoken, and subjects provide the first word that comes to mind (for a different task, see Jacoby et al 1988).

Several features of auditory word priming have been delineated. First, the phenomenon is largely modality specific: Cross-modal priming is weak or absent on tests of auditory identification (Gipson 1986) and stem completion (Bassili et al 1989; McClelland & Pring 1991). Second, priming on both auditory identification and completion tasks does not require semantic study processing, as indicated by the finding that priming on both tasks is either less affected or unaffected by semantic versus nonsemantic encoding manipulations that enhance explicit memory (Schacter & Church 1992). Converging evidence on this point is provided by a recent case study in which a patient who exhibited difficulties understanding spoken words as a consequence of a left-hemisphere stroke, nevertheless showed normal priming of auditory identification performance (Schacter et al 1993). Third, evidence exists that priming on the auditory stem completion task is reduced by study-to-test changes in speaker's voice (Schacter & Church 1992), whereas priming on the identification-in-noise task is unaffected by voice change (Jackson & Morton 1984; Schacter & Church 1992).

Studies of auditory priming have not yet been reported in amnesic patients, demented patients, or normal elderly. However, studies of patients undergoing surgical anesthesia have provided some evidence for priming of information presented auditorily during anesthesia, on implicit tests given postoperatively, despite a complete absence of explicit memory (e.g. Ghoneim et al 1990; Kihlstrom et al 1990). Some investigators, however, have failed to find such effects (e.g. Eich et al 1985), and the differences among studies may be related to the particular anesthetics used (e.g. Cork et al 1992; for review, see Kihlstrom & Schacter 1990). Interestingly, a recent study of implicit memory for auditory information presented during natural sleep failed to produce any evidence of priming (Wood et al 1992).

THEORETICAL ACCOUNTS OF PRIMING

Space does not permit us to provide a thorough review of the strengths and weaknesses of the various alternative theoretical accounts of priming (see Richardson-Klavehn & Bjork 1988; Schacter 1987, 1990). The main debate has involved a contrast between the multiple memory systems view, in which priming reflects the operations of a memory system (or systems) that is neurophysiologically and computationally distinct from the system that underlies explicit memory (e.g. Cohen 1984; Cohen & Eichenbaum 1992; Hayman & Tulving 1989; Keane et al 1991; Schacter 1990, 1992; Squire 1987, 1992; Tulving 1985; Tulving & Schacter 1990), and the processing view, in which priming can be understood with reference to the same principles that are used to understand explicit memory, without postulating multiple memory systems (e.g. Blaxton 1989; Jacoby 1983; Masson 1989; Roediger 1990). Various aspects of this debate have been summarized elsewhere by Roediger (1990) and Schacter (1990, 1992), who have also pointed out that the two views may serve complementary functions. We summarize one multiple memory account of priming that accommodates and integrates a good deal of the data considered here (for more detailed elaboration of this view, see Schacter 1990, 1992; Tulving & Schacter 1990; for related views, see Keane et al 1991; Squire 1992).

An adequate theory must account for at least three well-established experimental facts about perceptual priming that we have discussed in this review: 1. It can occur independently of semantic-level processing, as indicated by the weak or absent effects of semantic versus nonsemantic encoding manipulations. 2. It shows a large degree of modality specificity and, under certain circumstances, depends on highly specific perceptual information about a particular word or object. 3. It is preserved in amnesic patients. The latter finding provides particularly strong evidence that priming does not depend on the memory system that supports explicit retrieval of episodes and is tied closely to the hippocampus and other limbic structures (cf. Cohen & Eichenbaum 1992; Schacter 1987; Squire 1987, 1992). Accordingly, Schacter (1990, 1992) and Tulving & Schacter (1990) have suggested that priming effects on such data-driven implicit tests as perceptual identification, stem and fragment completion, and lexical and object decision, largely reflect experience-induced changes in a cortically based, presemantic perceptual representation system (PRS), which is in turn composed of several domain-specific subsystems. The various PRS subsystems are all dedicated to representing modality-specific information about the form and structure, but not the meaning and other associative properties, of words and objects. Independent evidence for the existence of PRS has been provided by neuropsychological studies showing that patients with impaired access to semantic knowledge of words or objects can, nevertheless, show relatively intact access to perceptual/structural knowledge of those same items (cf. Kohn & Friedman 1986; Riddoch & Humphreys 1987; Schwartz et al 1980; Warrington 1982); and PET imaging studies showing different areas of activation for perceptual and semantic processing (Peterson et al 1989).

Three PRS subsystems have been discussed with respect to priming: a visual word form system, which represents orthographic information about words and likely has a locus in extrastriate cortex (cf. Peterson et al 1989; Schacter 1990; Schacter et al 1990c); a structural description system, which computes relations among parts of objects and may be based in inferior temporal regions (cf. Plaut & Farah 1990; Schacter et al 1991b); and an auditory word form system, which handles phonological/acoustic information and is based in regions of perisylvan cortex (cf. Ellis & Young 1988; Schacter & Church 1992). The three subsystems are implicated in visual word priming, visual object priming, and auditory word priming, respectively.

This formulation does not account for all priming phenomena, and various puzzles remain. For example, it may be necessary to subdivide visual and auditory word form systems further, into abstract and form-specific subsystems that are associated with the left and right hemispheres, respectively (cf. Marsolek et al 1992; Schacter & Church 1992). Similarly, according to the PRS framework, conceptual priming occurs outside of PRS, possibly in a semantic memory system. But, the exact nature of the semantic system and its relation to episodic memory remain poorly understood. And, disagreements exist concerning the extent to which some implicit tests, such as visual stem completion, are based on perceptual versus conceptual priming (cf. Keane et al 1991; Schacter 1990). Nevertheless, the PRS view provides a reasonably coherent account of the major phenomena of priming.

SKILL LEARNING AND IMPLICIT MEMORY

As noted at the outset, although priming is the most extensively studied implicit memory effect, it is by no means the only one. Another important type of implicit memory is embodied in the phenomenon of skill learning. Since the classic studies of Milner et al (1968) on patient H. M., we have known that amnesics can acquire new perceptual/motor skills, despite absence of explicit memory for having acquired them. Subsequent work has shown that such learning can proceed normally in amnesic patients and that it extends to the acquisition of cognitive skills (e.g. Cohen & Squire 1980; Saint-Cyr et al 1988; Squire & Frambach 1990). Indeed, evidence from studies on learning of complex computer skills and knowledge indicates that severely amnesic patients can learn a great deal more than was previously thought (Glisky et al 1986; Glisky & Schacter 1988, 1989).

An important recent development concerns the double dissociation of motor skill learning and priming in patients with various forms of dementia. Specifically, investigators have shown that whereas AD patients can acquire new motor skills normally despite impaired stem completion priming, HD patients exhibit impaired motor skill learning together with intact priming (e.g. Eslinger & Damasio 1986; Heindel et al 1989; see Heindel et al 1991, for a related phenomenon). In view of evidence for abnormal corticostriatal circuits in HD patients, these findings suggest in part that motor skill learning is mediated by a corticostriatal system that is distinct from the cortical systems that subserve priming (Heindel et al 1989, 1991). There is also evidence that the cerebellum plays a significant role in learning of simple and complex sensory/motor tasks (for review, see Thach et al 1992). Additional evidence for a dissociation between priming and skill learning has been provided in studies of college students and elderly adults (Hashtroudi et al 1992; Schwartz & Hashtroudi 1991).

CONCLUDING COMMENTS

Although the systematic study of implicit memory is a recent development, it has generated an impressive amount of new information and ideas about the nature of mnemonic function. Nevertheless, much remains to be learned and many puzzles need to be solved. For example, although data on skill learning and corticostriatal function in patient populations fit well with data from animal studies (e.g. Cohen & Eichenbaum 1992; Mishkin et al 1984), animal models of priming have yet to be developed, and our knowledge of the neural basis of priming is still rather rudimentary. Given the rapid progress during the past decade, however, we suspect that these and other problems will likely be illuminated during the coming years.

ACKNOWLEDGMENTS

Supported by Air Force Office of Scientific Research grant 91-0182 and National Institute on Aging grant RO1 AG08441.

Literature Cited

Bargh, J. A. 1989 Conditional automaticity: varieties of automatic influence in social perception and cognition. In *Unintended Thoughts*, ed. J. S. Uleman, J. A. Bargh. San Francisco: Freeman

Bartram, D. J. 1974. The role of visual and semantic codes in object naming. *Cogn. Psychol.* 6: 325–56

Bassili, J. N., Smith, M. C., MacLeod, C. M. 1989. Auditory and visual word stem completion: separating data-driven and conceptually-driven processes. *Q. J. Exp. Psychol.* 41A: 439–45

Bentin, S., Moscovitch, M. 1988. The time course of repetition effects for words and unfamiliar faces. *J. Exp. Psychol. Gen.* 117: 148–60

Biederman, I., Cooper, E. E. 1992a. Evidence for complete translational and reflectional invariance in visual object priming. *Perception.* In press

Biederman, I., Cooper, E. E. 1992b. Scale invariance in visual object priming. *J. Exp. Psychol. Hum. Percept. Perform.* In press

Biederman, I., Cooper, E. E. 1991. Priming contour deleted images. Evidence for intermediate representations in visual object recognition. *Cogn. Psychol.* 23: 393–419

Blaxton, T. A. 1989. Investigating dissociations among memory measures: support for a transfer appropriate processing framework. *J. Exp. Psychol. Learn. Mem. Cogn.* 15: 657–68

Booker, J. 1991. *Interference effects in implicit memory.* PhD thesis. Univ. Arizona

Bowers, J. S., Schacter, D. L. 1992. Priming of novel information in amnesia: issues and data. In *Implicit Memory: New Directions in Cognition, Neuropsychology, and Development*, ed. P. Graf, M. E. J. Masson. New York: Academic. In press

Bowers, J. S., Schacter, D. L. 1990. Implicit memory and test awareness. *J. Exp. Psychol. Learn. Mem. Cogn.* 16: 404–16

Brown, M. W., Brown, J., Bowes, B. J. 1989. Absence of priming coupled with substantially preserved recognition in Lorazepam-induced amnesia. *Q. J. Exp. Psychol.* 41A: 599–617

Carr, T. H., Brown, J. S. 1990. Perceptual abstraction and interactivity in repeated oral reading: where do things stand? *J. Exp. Psychol. Learn. Mem. Cogn.* 16: 731–38

Carr, T. H., Brown, J. S., Charalambous, A. 1989. Repetition and reading: perceptual encoding mechanisms are very abstract but not very interactive. *J. Exp. Psychol. Learn. Mem. Cogn.* 15: 763–78

Cave, C. B., Squire, L. R. 1992. Intact and long-lasting repetition priming in amnesia. *J. Exp. Psychol. Learn. Mem. Cogn.* 18: 509–20

Cermak, L. S., Blackford, S. P., O'Connor, M., Bleich, R. P. 1988a. The implicit memory ability of a patient with amnesia due to encephalitis. *Brain Cogn.* 7: 145–56

Cermak, L. S., Bleich, R. P., Blackford, S. P. 1988b. Deficits in the implicit retention of new associations by alcoholic Korsakoff patients. *Brain Cogn.* 7: 312–23

Cermak, L. S., Chandler, K., Wolbarst, L. R. 1985. The perceptual priming phenomenon in amnesia. *Neuropsychologia* 23: 615–22

Cermak, L. S., Verfaellie, M., Milberg, W., Letourneau, L., Blackford, S. 1991. A further analysis of perceptual identification priming in alcoholic Korsakoff patients. *Neuropsychologia* 29: 725–36

Chiarello, C., Hoyer, W. J. 1988. Adult age differences in implicit and explicit memory: time course and encoding effects. *Psychol. Aging* 3: 358–66

Clarke, R., Morton, J. 1983. Cross modality facilitation in tachistoscopic word recognition. *Q. J. Exp. Psychol.* 35A: 79–96

Cohen, N. J. 1984. Preserved learning in amnesia: Evidence for multiple memory systems. In *Neuropsychology of Memory*, ed. L. R. Squire, N. Butters, pp. 83–103. New York: Guilford

Cohen, N. J., Eichenbaum, H. 1992. *Memory, Amnesia, and the Hippocampus.* Cambridge, Mass: MIT Press. In press

Cohen, N. J., Squire, L. R. 1980. Preserved learning and retention of pattern analyzing skill in amnesics: dissociation of knowing how and knowing that. *Science* 210: 207–10

Cooper, L. A., Schacter, D. L., Ballesteros, S., Moore, C. 1992. Priming and recognition of transformed three-dimensional objects: effects of size and reflection. *J. Exp. Psychol. Learn. Mem. Cogn.* 18: 43–57

Cooper, L. A., Schacter, D. L., Moore, C. 1991. *Orientation affects both structural and episodic representations of 3-D objects.* Presented at Annu. Meet. Psychon. Soc., San Francisco

Cork, R. C., Kihlstrom, J. F., Schacter, D. L. 1992. Memory with Sufentanil/nitrous oxide. *Anesthesiology.* In press

Danion, J. M., Zimmerman, M. A., Willard-Schroeder, D., Grange, D., Singer, L. 1989. Diazepam induces a dissociation between explicit and implicit memory. *Psychopharmacology* 99: 238–43

Daum, I., Channon, S., Canavar, A. 1989.

Classical conditioning in patients with severe memory problems. *J. Neurol. Neurosurg. Psychiatry* 52: 47–51

Davis, H. P., Cohen, A., Gandy, M., Colombo, P., VanDusseldorp, G., et al. 1990. Lexical priming deficits as a function of age. *Behav. Neurosci.* 104: 288–97

Diamond, R., Rozin, P. 1984. Activation of existing memories in anterograde amnesia. *J. Abnorm. Psychol.* 93: 98–105

Donnelly, R. E. 1988. *Priming across modality in implicit memory: Facilitation from auditory presentation to visual test of word fragment completion.* PhD thesis. Univ. Toronto

Durgunoglu, A. Y., Roediger, H. L. 1987. Test differences in accessing bilingual memory. *J. Mem. Lang.* 26: 377–91

Durso, F. T., Johnson, M. K. 1979. Facilitation in naming and categorizing repeated words and pictures. *J. Exp. Psychol. Hum. Learn. Mem. Cogn.* 5: 449–59

Eich, E., Reeves, J. L., Katz, R. L. 1985. Anesthesia, amnesia, and the memory/awareness distinction. *Anesth. Analg.* 64: 1143–48

Ellis, A. W., Young, A. W. 1988. *Human Cognitive Neuropsychology*, Hove, Engl: Erlbaum Assoc.

Eslinger, P. J., Damasio, A. R. 1986. Preserved motor learning in Alzheimer's disease: implications for anatomy and behavior. *J. Neurosci.* 6: 3006–9

Feustel, T. C., Shiffrin, R. M., Salasoo, A. 1983. Episodic and lexical contributions to the repetition effect in word identification. *J. Exp. Psychol. Gen.* 112: 309–46

Gabrieli, J. D. E., Keane, M. M. 1988. Priming in the patient H.M.: new findings and a theory of intact and impaired priming in patients with memory disorders. *Soc. Neurosci. Abstr.* 14: 1290

Gabrieli, J. D. E., Milberg, W., Keane, M. M., Corkin, S. 1990. Intact priming of patterns despite impaired memory. *Neuropsychologia* 28: 417–28

Gardiner, J. M. 1989. A generation effect in memory without awareness. *Br. J. Psychol.* 80: 163–68

Gardiner, J. M. 1988. Generation and priming effects in word-fragment completion. *J. Exp. Psychol. Learn. Mem. Cogn.* 14: 495–501

Gardiner, J. M., Dawson, A. J., Sutton, E. A. 1989. Specificity and generality of enhanced priming effects for self-generated study items. *Am. J. Psychol.* 102: 295–305

Gardner, H., Boller, F., Moreines, J., Butters, N. 1973. Retrieving information from Korsakoff patients: effects of cate-gorical cues and reference to the task. *Cortex* 9: 165–75

Gerard, L. D., Scarborough, D. L. 1989. Language-specific lexical access of homographs by bilinguals. *J. Exp. Psychol. Learn. Mem. Cogn.* 15: 305–15

Ghoneim, M. M., Block, R. I., Sum Ping, S. T., Ali, M. A., Hoffman, J. G. 1990. Learning without recall during general anesthesia. In *Memory and Awareness in Anesthesia*, ed. B. Bonke, W. Fitch, K. Millar, pp. 161–69. Lisse/Amsterdam: Swets & Zeitlinger

Gipson, P. 1986. The production of phonology and auditory priming. *Br. J. Psychol.* 77: 359–75

Glisky, E. L., Schacter, D. L. 1989. Extending the limits of complex learning in organic amnesia: computer training in a vocational domain. *Neuropsychologia* 27: 107–20

Glisky, E. L., Schacter, D. L. 1988. Long-term retention of computer learning by patients with memory disorders. *Neuropsychologia* 26: 173–78

Glisky, E. L., Schacter, D. L., Tulving, E. 1986. Computer learning by memory-impaired patients: acquisition and retention of complex knowledge. *Neuropsychologia* 24: 313–28

Gordon, B. 1988. Preserved learning of novel information in amnesia: evidence for multiple memory systems. *Cognition* 7: 257–82

Graf, P. 1900. Life span changes in implicit and explicit memory. *Bull. Psychon. Soc.* 28: 353–58

Graf, P., Mandler, G. 1984. Activation makes words more accessible, but not necessarily more retrievable. *J. Verbal Learn. Verbal Behav.* 23: 553–68

Graf, P., Mandler, G., Haden, P. E. 1982. Simulating amnesic symptoms in normals. *Science* 218: 1243–44

Graf, P., Ryan, L. 1990. Transfer-appropriate processing for implicit and explicit memory. *J. Exp. Psychol. Learn. Mem. Cogn.* 16: 978–92

Graf, P., Schacter, D. L. 1989. Unitization and grouping mediate dissociations in memory for new associations. *J. Exp. Psychol. Learn. Mem. Cogn.* 15: 930–40

Graf, P., Schacter, D. L. 1987. Selective effects of interference on implicit and explicit memory for new associations. *J. Exp. Psychol. Learn. Mem. Cogn.* 13: 45–53

Graf, P., Schacter, D. L. 1985. Implicit and explicit memory for new associations in normal and amnesic patients. *J. Exp. Psychol. Learn. Mem. Cogn.* 11: 501–18

Graf, P., Shimamura, A. P., Squire, L. R. 1985. Priming across modalities and priming across category levels: Extending the domain of preserved functioning in amnesia. *J. Exp. Psychol. Learn. Mem. Cogn.* 11: 385–95

Graf, P., Squire, L. R., Mandler, G. 1984. The information that amnesic patients do not forget. *J. Exp. Psychol. Learn. Mem. Cogn.* 10: 164–78

Haist, F., Musen, G., Squire, L. R. 1991. Intact priming of words and nonwords in amnesia. *Psychobiology* 19: 275–85

Hamann, S. B. 1990. Level-of-processing effects in conceptually driven implicit tasks. *J. Exp. Psychol. Learn. Mem. Cogn.* 16: 970–77

Hashtroudi, S., Chrosniak, L. D., Schwartz, B. L. 1992. A comparison of the effects of aging on priming and skill learning. *Psychol. Aging.* In press

Hashtroudi, S., Ferguson, S. A., Rappold, V. A., Chrosniak, L. D. 1988. Data-driven and conceptually-driven processes in partial-word identification and recognition. *J. Exp. Psychol. Learn. Mem. Cogn.* 14: 749–57

Hayman, C. A. G., Tulving, E. 1989. Is priming in fragment completion based on "traceless" memory system? *J. Exp. Psychol. Learn. Mem. Cogn.* 14: 941–56

Heindel, W. C., Salmon, D. P., Butters, N. 1991. The biasing of weight judgements in Alzheimer's and Huntington's disease: a priming or programming phenomenon? *J. Clin. Exp. Neuropsychol.* 13: 189–203

Heindel, W. C., Salmon, D. P., Butters, N. 1990. Pictorial priming and cued recall in Alzheimer's and Huntington's disease. *Brain Cogn.* 13: 282–95

Heindel, W. C., Salmon, D. P., Shults, C. W., Walicke, P. A., Butters, N. 1989. Neuropsychological evidence for multiple implicit memory systems: a comparison of Alzheimer's, Huntington's, and Parkinson's disease patients. *J. Neurosci.* 9: 582–87

Howard, D. V., Fry, A. F., Brune, C. M. 1991. Aging and memory for new associations: direct versus indirect measures. *J. Exp. Psychol. Learn. Mem. Cogn.* 17: 779–92

Jackson, A., Morton, J. 1984. Facilitation of auditory word recognition. *Mem. Cogn.* 12: 568–74

Jacoby, L. L. 1991. A process dissociation framework: separating automatic from intentional uses of memory. *J. Mem. Lang.* 30: 513–41

Jacoby, L. L. 1983. Remembering the data: Analyzing interactive processes in reading. *J. Verbal Learn. Verbal Behav.* 22: 485–508

Jacoby, L. L., Allan, L. G., Collins, J. C., Larwill, L. K. 1988. Memory influences

subjective experience. *J. Exp. Psychol. Learn. Mem. Cogn.* 14: 240–47

Jacoby, L. L., Baker, J. G., Brooks, L. R. 1989. Episodic effects on picture identification: implications for theories of concept learning and theories of memory. *J. Exp. Psychol. Learn. Mem. Cogn.* 15: 275–81

Jacoby, L. L., Dallas, M. 1981. On the relationship between autobiographical memory and perceptual learning. *J. Exp. Psychol. Gen.* 110: 306–40

Jacoby, L. L., Hayman, C. A. G. 1987. Specific visual transfer in word identification. *J. Exp. Psychol. Learn. Mem. Cogn.* 13: 456–63

Jacoby, L. L., Levy, B. A., Steinbach, K. 1992. Episodic transfer and automaticity: integration of data-driven and conceptually-driven processing in rereading. *J. Exp. Psychol. Learn. Mem. Cogn.* 18: 15–24

Jacoby, L. L., Witherspoon, D. 1982. Remembering without awareness. *Can. J. Psychol.* 36: 300–24

Jacoby, L. L., Woloshyn, V., Kelley, C. M. 1989. Becoming famous without being recognized: unconscious influences of memory produced by dividing attention. *J. Exp. Psychol. Gen.* 118: 115–25

Johnson, M. K., Hasher, L. 1987. Human learning and memory. *Annu. Rev. Psychol.* 38: 631–68

Jolicoeur, P. 1985. The time to name disoriented natural objects. *Brain Cogn.* 13: 289–303

Jolicoeur, P., Milliken, B. 1989. Identification of disoriented objects: effects of context and of prior presentation. *J. Exp. Psychol. Learn. Mem. Cogn.* 15: 200–10

Keane, M. M., Gabrieli, J. D. E., Fennema, A. C., Growdon, J. H., Corkin, S. 1991. Evidence for a dissociation between perceptual and conceptual priming in Alzheimer's disease. *Behav. Neurosci.* 105: 326–42

Kersteen-Tucker, Z. 1991. Long-term repetition priming with symmetrical polygons and words. *Mem. Cognit.* 19: 37–43

Kihlstrom, J. F., Schacter, D. L. 1990. Anesthesia, amnesia, and the cognitive unconscious. In *Memory and Awareness in Anesthesia*, ed. B. Bonke, W. Fitch, K. Millar, pp. 21–44. Lisse/Amsterdam: Swets & Zeitlinger

Kihlstrom, J. F., Schacter, D. L., Cork, R. C., Hurt, C. A., Behr, S. E. 1990. Implicit and explicit memory following surgical anesthesia. *Psychol. Sci.* 1: 303–6

Kinsbourne, M., Wood, F. 1975. Short-term memory processes and the amnesic syndrome. In *Short-term Memory*, ed. D.

Deutsch, J. A. Deutsch, pp. 258–91. San Diego: Academic

Kirsner, K., Dunn, J. C., Standen, P. 1989. Domain-specific resources in word recognition. In *Implicit Memory: Theoretical Issues*, ed. S. Lewandowsky, J. C. Dunn, K. Kirsner, pp. 99–122. Hillsdale, NJ: Erlbaum

Kirsner, K., Smith, M. C., Lockhart, R. S., King, M. L., Jain, M. 1984. The bilingual lexicon: Language specific units in an integrated network. *J. Verbal Learn. Verbal Behav.* 23: 519–39

Knowlton, B. J., Ramus, S. J., Squire, L. R. 1992. Intact artificial grammar learning in amnesia: dissociation of classification learning and explicit memory for specific instances. *Psychol. Sci.* 3: 172–79

Kohn, S. E., Friedman, R. B. 1986. Word-meaning deafness: a phonological-semantic dissociation. *Cogn. Neuropsychol.* 3: 291–308

Kroll, J. F., Potter, M. C. 1984. Recognizing words, pictures, and concepts: a comparison of lexical, object, and reality decisions. *J. Verbal Learn. Verbal Behav.* 23: 39–66

Kuhn, T. S. 1962. *The Structure of Scientific Revolutions*. Chicago: Univ. Chicago Press

Kunst-Wilson, W. R., Zajonc, R. B. 1980. Affective discrimination of stimuli that cannot be recognized. *Science* 207: 557–58

Light, L. L., Singh, A. 1987. Implicit and explicit memory in young and older adults. *J. Exp. Psychol. Learn. Mem. Cogn.* 13: 531–41

MacAndrews, M. P., Glisky, E. L., Schacter, D. L. 1987. When priming persists: long-lasting implicit memory for a single episode in amnesic patients. *Neuropsychologia* 25: 497–506

Mandler, G., Nakamura, Y., Van Zandt, B. J. S. 1987. Nonspecific effects of exposure on stimuli that cannot be recognized. *J. Exp. Psychol. Learn. Mem. Cogn.* 13: 646–48

Marsolek, C. J., Kosslyn, S. M., Squire, L. R. 1992. Form specific visual priming in the right cerebral hemisphere. *J. Exp. Psychol. Learn. Mem. Cogn.* 18: 492–508

Martin, A. 1992. Degraded knowledge representation in patients with Alzheimer's disease: implications for models of semantic and repetition priming. In *Neuropsychology of Memory*, ed. L. R. Squire, N. Butters. New York: Guilford. In press

Masson, M. E. J. 1989. Fluent reprocessing as an implicit expression of memory for experience. In *Implicit Memory: Theoretical Issues*, ed. S. Lewandowsky, J. C. Dunn, K. Kirsner. Hillsdale, NJ: Erlbaum

Masson, M. E. J., MacLeod, C. M. 1992. Re-enacting the route to interpretation: context dependency in encoding and retrieval. *J. Exp. Psychol. Gen.* 121: 145–76

Mayes, A. R. 1988. *Human Organic Memory Disorders*. New York: Cambridge Univ. Press

Mayes, A. R., Gooding, P. 1989. Enhancement of word completion priming in amnesics by cueing with previously novel associates. *Neuropsychologia* 27: 1057–72

Mayes, A. R., Pickering, A., Fairbairn, A. 1987. Amnesic sensitivity to proactive interference: its relationship to priming and the causes of amnesia. *Neuropsychologia* 25: 211–20

McCelland, A. G. R., Pring, L. 1991. An investigation of cross-modality effects in implicit and explicit memory. *Q. J. Exp. Psychol.* 43A: 19–33

Micco, A., Masson, M. E. J. 1991. Implicit memory for new associations: An interactive process approach. *J. Exp. Psychol. Learn. Mem. Cogn.* 17: 1105–23

Milner, B., Corkin, S., Teuber, H. L. 1968. Further analysis of the hippocampal amnesic syndrome: Fourteen year follow-up study of H.M. *Neuropsychologia* 6: 215–34

Mishkin, M., Malamut, B., Bachevalier, J. 1984. Memories and Habits: two neural systems. In *Neurobiology of Learning and Memory*, ed. G. Lynch, J. L. McGaugh, N. M. Weinberger, pp. 65–77. New York: Guilford

Mitchell, D. B., Brown, A. S. 1988. Persistent repetition priming in picture naming and its dissociation from recognition memory. *J. Exp. Psychol. Learn. Mem. Cogn.* 14: 213–22

Mitchell, D. B., Brown, A. S., Murphy, D. R. 1990. Dissociations between procedural and episodic memory: Effects of time and aging. *Psychol. Aging* 5: 264–76

Moscovitch, M. 1982. Multiple dissociations of function in amnesia. In *Human Memory and Amnesia*, ed. L. S. Cermak, pp. 337–70. Hillsdale, NJ: Erlbaum

Moscovitch, M., Winocur, G., McLachlin, D. 1986. Memory as assessed by recognition and reading time in normal and memory impaired people with Alzheimer's disease and other neurological disorders. *J. Exp. Psychol. Gen.* 115: 331–46

Musen, G. 1991. Effects of verbal labeling and exposure duration on implicity memory for visual patterns. *J. Exp. Psychol. Learn. Mem. Cogn.* 17: 954–62

Musen, G., Squire, L. R. 1991. Normal acquisition of novel verbal information in amnesia. *J. Exp. Psychol. Learn. Mem. Cogn.* 17: 1095–1104

Musen, G., Treisman, A. 1990. Implicit and explicit memory for visual patterns. *J.*

Exp. Psychol. Learn. Mem. Cogn. 16: 127–37

Nissen, M. J., Knopman, D. S., Schacter, D. L. 1987. Neurochemical dissociation of memory systems. *Neurology* 37: 789–94

Ober, B., Shenaut, G. K. 1988. Lexical decision and priming in Alzheimer's disease. *Neuropsychologia* 26: 273–86

Petersen, S. E., Fox, P. T., Posner, M. I., Mintun, M. A., Raichle, M. E. 1989. Positron Emission Tomographic studies of the processing of single words. *J. Cogn. Neurosci.* 1: 153–70

Plaut, D. C., Farah, M. J. 1990. Visual object representation: Interpreting neurophysiological data within a computational framework. *J. Cogn. Neurosci.* 2: 320–43

Reber, A. S. 1967. Implicit learning of artificial grammars. *J. Verbal Learn. Verbal Behav.* 6: 855–63

Richardson-Klavehn, A., Bjork, R. A. 1988. Measures of memory. *Annu. Rev. Psychol.* 36: 475–543

Riddoch, M. J., Humphreys, G. W. 1987. Visual object processing in optic aphasia: A case of semantic access agnosia. *Cogn. Neuropsychol.* 4: 131–86

Roediger, H. L. I. 1990. Implicit memory: Retention without remembering. *Am. Psychol.* 45: 1043–56

Roediger, H. L. I., Blaxton, T. A. 1987a. Retrieval modes produce dissociations in memory for surface information. In *The Ebbinghaus Centennial Conference*, ed. D. S. Gorfein, R. R. Hoffman, pp. 349–79. Hillsdale, NJ: Erlbaum

Roediger, H. L. I., Blaxton, T. A. 1987b. Effects of varying modality, surface features, and retention interval on priming in word fragment completion. *Mem. Cogn.* 15: 379–88

Roediger, H. L. I., Weldon, M. S., Challis,.B. H. 1989. Explaining dissociations between implicit and explicit measures of retention: A processing account. In *Varieties of Memory and Consciousness: Essays in Honor of Endel Tulving*, ed. H. L. I. Roediger, F. I. M. Craik, pp. 3–41. Hillsdale, NJ: Erlbaum

Roediger, H. L. I., Weldon, M. S., Stadler, M. L., Riegler, G. L. 1992. Direct comparison of two implicit memory tests: word fragment and word stem completion. *J. Exp. Psychol. Learn. Mem. Cogn.* In press

Rueckl, J. G. 1990. Similarity effects in word and pseudoword repetition priming. *J. Exp. Psychol. Learn. Mem. Cogn.* 16: 374–91

Saint-Cyr, J. A., Taylor, A. E., Lang, A. E. 1988. Procedural learning and neostriatal dysfunction in man. *Brain* 111: 941–59

Salasoo, A., Shiffrin, R. M., Feustal, T. C. 1985. Building permanent memory codes: Codification and repetition effects in word identification. *J. Exp. Psychol. Gen.* 114: 50–77

Salmon, D. P., Shimamura, A. P., Butters, N., Smith, S. 1988. Lexical and semantic priming deficits in patients with Alzheimer's disease. *J. Clin. Exp. Psychol.* 10: 477–94

Scarborough, D. L., Cortese, C., Scarborough, H. S. 1977. Frequency and repetition effects in lexical memory. *J. Exp. Psychol. Hum. Percept. Perform.* 3: 1–17

Scarborough, D. L., Gerard, L., Cortese, C. 1979. Accessing lexical memory: The transfer of word repetition effects across task and modality. *Mem. Cogn.* 7: 3–12

Schacter, D. L. 1992. Understanding implicit memory: a cognitive neuroscience approach. *Am. Psychol.* 47: 559–69

Schacter, D. L. 1990. Perceptual representation systems and implicit memory: Toward a resolution of the multiple memory systems debate. *Ann. NY Acad. Sci.* 608: 543–71

Schacter, D. L. 1987. Implicit memory: History and current status. *J. Exp. Psychol. Learn. Mem. Cogn.* 11: 501–18

Schacter, D. L. 1985. Priming of old and new knowledge in amnesic patients and normal subjects. *Ann. NY Acad. Sci.* 444: 44–53

Schacter, D. L., Bowers, J., Booker, J. 1989. Intention, awareness and implicit memory: The retrieval intentionality criterion. In *Implicit Memory: Theoretical Issues*, ed. S. Lewandowsky, J. C. Dunn, K. Kirsner. Hillsdale, NJ: Erlbaum

Schacter, D. L., Church, B. 1992. Auditory priming: implicit and explicit memory for words and voices. *J. Exp. Psychol. Learn. Mem. Cogn.* 18: 915–30

Schacter, D. L., Cooper, L. A., Delaney, S. M. 1990a. Implicit memory for unfamiliar objects depends on access to structural descriptions. *J. Exp. Psychol. Gen.* 119: 5–24

Schacter, D. L., Cooper, L. A., Delaney, S. M., Peterson, M. A., Tharan, M. 1991a. Implicit memory for possible and impossible objects: constraints on the construction of structural descriptions. *J. Exp. Psychol. Learn. Mem. Cogn.* 17: 3–19

Schacter, D. L., Cooper, L. A., Tharan, M., Rubens, A. B. 1991b. Preserved priming of novel objects in patients with memory disorders. *J. Cogn. Neurosci.* 3: 118–31

Schacter, D. L., Cooper, L. A., Valdiserri, M. 1992. Implicit and explicit memory for novel objects in older and younger adults. *Psychol. Aging* 7: 299–308

Schacter, D. L., Delaney, S. M., Merikle, E. P. 1990b. Priming of nonverbal information and the nature of implicit memory. In *The Psychology of Learning and Motivation*, ed. G. H. Bower, pp. 83–123. New York: Academic

Schacter, D. L., Graf, P. 1989. Modality specificity of implicit memory for new associations. *J. Exp. Psychol. Learn. Mem. Cogn.* 15: 3–12

Schacter, D. L., Graf, P. 1986a. Effects of elaborative processing on implicit and explicit memory for new associations. *J. Exp. Psychol. Learn. Mem. Cogn.* 12: 432–44

Schacter, D. L., Graf, P. 1986b. Preserved learning in amnesic patients: perspectives on research from direct priming. *J. Clin. Exp. Neuropsychol.* 8: 727–43

Schacter, D. L., Kaszniak, A. K., Kihlstrom, J. F., Valdiserri, M. 1992. On the relation between source memory and aging. *Psychol. Aging* 6: 559–68

Schacter, D. L., Kihlstrom, J. F., Kaszniak, A. K., Valdiserri, M. 1992. Preserved and impaired memory functions in elderly adults. In *Adult Information Processing: Limits on Loss*, ed. J. Cerella, W. Hoyer, J. Rybash, M. Commons. New York: Academic. In press

Schacter, D. L., McGlynn, S. 1989. Implicit memory: Effects of elaboration depend on unitization. *Am. J. Psychol.* 102: 151–81

Schacter, D. L., McGlynn, S. M., Milberg, W. M., Chuch, B. A. 1993. Spared priming despite impaired comprehension: implicit memory in a case of word meaning deafness. *Neuropsychology.* In press

Schacter, D. L., Rapcsak, S. Z., Rubens, A. B., Tharan, M., Laguna, J. M. 1990c. Priming effects in a letter-by-letter reader depend on access to the word form system. *Neuropsychologia* 28: 1079–94

Schwartz, B. L., Hashtroudi, S. 1991. Priming is independent of skill learning. *J. Exp. Psychol. Learn. Mem. Cogn.* 17: 1177–87

Schwartz, M. F., Saffran, E. M., Marin, O. S. H. 1992. Fractionating the reading process in dementia. In *Deep Dementia*, ed. M. Coltheart, K. Patterson, S. C. Marshall. London: Routledge & Kegan Paul

Shimamura, A. P. 1986. Priming effects in amnesia: Evidence for a dissociable memory function. *Q. J. Exp. Psychol.* 38A: 619–44

Shimamura, A. P., Salmon, D. P., Squire, L. R., Butters, N. 1987. Memory dysfunction and word priming in dementia and amnesia. *Behav. Neurosci.* 101: 347–51

Shimamura, A. P., Squire, L. R. 1989. Impaired priming of new associations in amnesia. *J. Exp. Psychol. Learn. Mem. Cogn.* 15: 721–28

Shimamura, A. P., Squire, L. R. 1984. Paired-associate learning and priming

effects in amnesia: a neuropsychological approach. *J. Exp. Psychol. Gen.* 113: 556–70

Slamecka, N. J., Graf, P. 1978. The generation effect: delineation of a phenomenon. *J. Exp. Psychol. Hum. Learn. Mem.* 4: 592–604

Sloman, S. A., Hayman, C. A. G., Ohta, N., Law, J. Tulving, E. 1988. Forgetting in primed fragment completion. *J. Exp. Psychol. Learn. Mem. Cogn.* 14: 223–39

Smith, M. C. 1991. On the recruitment of semantic information for word fragment completion: evidence from bilingual priming. *J. Exp. Psychol. Learn. Mem. Cogn.* 17: 234–44

Smith, M. E., Oscar-Berman, M. 1990. Repetition priming of words and pseudo-words in divided attention and amnesia. *J. Exp. Psychol. Learn. Mem. Cogn.* 16: 1033–42

Snodgrass, J. G., Feenan, K. 1990. Priming effects in picture fragment completion: support for the perceptual closure hypothesis. *J. Exp. Psychol. Gen.* 119: 276–96

Squire, L. R. 1992. Memory and the hippocampus: A synthesis from findings with rats, monkeys, and humans. *Psychol. Rev.* 99: 195–231

Squire, L. R. 1987. *Memory and Brain*. New York: Oxford Univ. Press

Squire, L. R., Frambach, M. 1990. Cognitive skill learning in amnesia. *Psychobiology* 18: 109–17

Squire, L. R., Ojemann, J. G., Miezin, F. M., Peterson, S. E., Videen, T. O., Raichle, M. E. 1992. Activation of the hippocampus in normal humans: a functional anatomical study of memory. *Proc. Natl. Acad. Sci. USA* 89: 1837–41

Squire, L. R., Shimamura, A. P., Graf, P. 1987. Strength and duration of priming effects in normal subjects and amnesic patients. *Neuropsychologia* 25: 195–210

Srinivas, K., Roediger, H. L. I. 1990. Classifying implicit memory tests: category association and anagram solution. *J. Mem. Lang.* 29: 389–412

Tardif, T., Craik, F. I. M. 1989. Reading a week later: Perceptual and conceptual factors. *J. Mem. Lang.* 28: 107–25

Thach, W. T., Goodkin, H. P., Keating, J. G. 1992. The cerebellum and the adaptive coordination of movement. *Annu. Rev. Neurosci.* 15: 403–42

Toth, J. P., Hunt, R. R. 1990. Effect of generation on a word-identification task. *J. Exp. Psychol. Learn. Mem. Cogn.* 16: 993–1003

Tulving, E. 1985. How many memory systems are there? *Am. Psychol.* 40: 385–98

Tulving, E., Hayman, C. A. G., MacDonald,

182 SCHACTER, CHIU & OCHSNER

C. 1991. Long-lasting perceptual priming and semantic learning in amnesia: A case experiment. *J. Exp. Psychol. Learn. Mem. Cogn.* 17: 595–617

Tulving, E., Schacter, D. L. 1990. Priming and human memory systems. *Science* 247: 301–6

Tulving, E., Schacter, D. L., Stark, H. 1982. Priming effects in word-fragment completion are independent of recognition memory. *J. Exp. Psychol. Learn. Mem. Cogn.* 8: 336–42

Verfaellie, M., Cermak, L. S., Letourneau, L., Zuffante, P. 1991. Repetition effects in a lexical decision task: the role of episodic memory in alcoholic Korsakoff patients. *Neuropsychologia* 29: 641–57

Warren, C., Morton, J. 1982. The effects of priming on picture recognition. *Br. J. Psychol.* 73: 117–29

Warrington, E. K. 1982. Neuropsychological studies of object recognition. *Philos. Trans. R. Soc. London Ser.* B298: 15–33

Warrington, E. K., Weiskrantz, L. 1974. The effect of prior learning on subsequent retention in amnesic patients. *Neuropsychologia* 12: 419–28

Warrington, E. K., Weiskrantz, L. 1968. New method of testing long-term retention with special reference to amnesic patients. *Nature* 217: 972–74

Weiskrantz, L., Warrington, E. K. 1979. Conditioning in amnesic patients. *Neuropsychologia* 17: 187–94

Weldon, M. S., Roediger, H. L. 1987. Altering retrieval demands reverses the picture superiority effect. *Mem. Cogn.* 15: 269–80

Whittlesea, B. W. A. 1990. Perceptual encoding mechanisms are tricky but may be very interactive: comment on Carr, Brown, and Charalambous (1989). *J. Exp. Psychol. Learn. Mem. Cogn.* 16: 727–30

Whittlesea, B. W. A., Cantwell, A. L. 1987. Enduring influence of the purpose of experiences: encoding-retrieval interactions in word and pseudoword perception. *Mem. Cogn.* 15: 465–72

Winnick, W. A., Daniel, S. A. 1970. Two kinds of response priming in tachistoscopic recognition. *J. Exp. Psychol.* 84: 74–81

Wood, J., Bootzin, R. R., Kihlstrom, J. F., Schacter, D. L. 1992. Implicit and explicit memory for verbal stimuli presented during sleep. *Psychol. Sci.* 4: 236–39

Annu. Rev. Neurosci. 1993. 16:183–205
Copyright © 1993 by Annual Reviews Inc. All rights reserved

THE NEUROFIBROMATOSIS TYPE 1 GENE

David Viskochil,[1] *Ray White,*[2] *and Richard Cawthon*[2]

[1]Department of Pediatrics and [2]Department of Human Genetics and Howard Hughes Medical Institute, University of Utah School of Medicine, Salt Lake City, Utah 84112

KEY WORDS: neurofibromin, mutations in *NF1*, alternative splice, gene mapping, *ras*-mediated differentiation

Sometimes referred to as peripheral neurofibromatosis or von Recklinghausen disease, neurofibromatosis 1 (NF1) is one of the most common medical conditions inherited in human populations. NF1, inherited as an autosomal dominant, affects approximately 1 in 3500 individuals worldwide with no apparent ethnic predilection. Its hallmark features are neurofibromas, café-au-lait spots, and Lisch nodules of the iris, although many other clinical manifestations are associated with this condition, as well.

IDENTIFICATION OF THE *NF1* GENE

Genetic mapping provided an approach to identification of the *NF1* gene. According to this paradigm, a precise chromosomal localization for a genetic trait in a family identifies candidate genes by virtue of their location in the defined region; the mutations carried by patients with the disease specify the individual gene. Investigators initially mapped the *NF1* locus to chromosome 17 by linkage with DNA markers in families (Barker et al 1987; Seizinger et al 1987). More precise mapping placed the *NF1* locus on the long arm of the chromosome, near the centromere (O'Connell et al 1989b; White et al 1987). The region so defined, however, was still too large to allow screening of individual genes for mutations; too many candidate genes were likely to be present.

Remarkably, just as the mapping results were emerging, cytogeneticists detected two independent translocation mutations in NF1 patients (Led-

183

better et al 1989; Schmidt et al 1987). The translocation chromosomes had chromosome-17 breakpoints at the exact location where the *NF1* gene had just been mapped. The simplest interpretation was that they were causing NF1 by interrupting the *NF1* gene. If the physical location of the breakpoints could be defined, the gene should be in the immediate vicinity.

The translocation breakpoints were physically localized by pulsed-field gel analysis of DNA samples from each of the two patients, probed with a high-density set of markers for the region. The probes had been developed from somatic cell hybrid lines that contained the proximal long arm of chromosome 17 as their only human complement (Leach et al 1989). These experiments identified a fragment about 600 kb long that was likely to carry the *NF1* gene (Fountain et al 1989; O'Connell et al 1989a).

Unexpectedly, however, a new opportunity arose. A mouse gene (*Evi-2*), which had been implicated in retrovirally mediated leukemogenesis, mapped to murine chromosome 11 (Buchberg et al 1988), in a region believed to be syntenic to the region of human chromosome 17 that carried both the putative *NF1* gene and a translocation breakpoint region implicated in human acute promyelocytic leukemias. Surprisingly, mapping experiments demonstrated that the human homologue of the murine gene, *EVI2*, mapped between the two *NF1*-translocation breakpoints (O'Connell et al 1990).

This oncogene mapping into the region defined by the translocation breakpoints became a strong candidate. However, the real *NF1* gene must also carry mutations in patients with the disease. *EVI2* failed this test; no mutations in this gene could be found in DNA from any NF1 patients tested (Cawthon et al 1990a).

A continued search of the region revealed that two other genes lay in the same 55-kb span of DNA between the translocation breakpoints. One of these was the gene encoding OMGP (oligodendrocyte myelin glycoprotein), a peptide previously characterized by Mikol et al (1990) as a component of the myelin sheath in the central nervous system. However, *OMGP* also failed to reveal mutations in NF1 patients (Viskochil et al 1991). The other gene was located adjacent to *EVI2*, which led to the appellations *EVI2A* and *EVI2B* (Cawthon et al 1991). Again, however, no mutations were found in patients.

Persistent probing of cDNA libraries with a DNA segment that spanned one of the translocation breakpoints and carried sequence conservation with mouse DNA identified a fourth gene. This time, mutations were found in DNA from NF1 patients, thus establishing this candidate as the gene responsible for NF1 (Cawthon et al 1990b; Viskochil et al 1990; Wallace et al 1990).

The newly identified gene, *NF1*, consists of 51 exons distributed over

350 kb of genomic DNA (Table 1). At least two of the exons are involved in alternative-splice forms of the primary transcript; with these two exons removed, the open reading frame encodes a 2818-amino-acid protein (Marchuk et al 1991; G. Xu et al, in preparation), now designated neurofibromin. Early in the DNA sequencing, a screen of peptide databases with the predicted product of partial *NF1* sequence identified an amino-acid domain related to mammalian GTPase-activating proteins (GAPs) and to *IRA1* and *IRA2* proteins, their functional homologues in yeast (Xu et al 1990b).

Table 1   Coding exons of *NF1*

Exon #	cDNA position[a]	Size (bp)	Introns (kb)	Exon #	cDNA position	Size (bp)	Introns (kb)
1	1	60	>40	26	4368	147	1.27
2	61	144	?	27	4515	258	>40
3	205	84	?	28	4773	433	1.90
4	289	366	0.22	29	5206	341	2.50
5	655	76	0.83	30	5547	203	4.10
6	731	158	0.50	31	5750	194	1.45
7	889	174	0.40	32	5944	141	0.15
8	1063	123	0.30	33	6085	280	0.40
9	1186	75	0.13	34	6365	215	0.15
10	1261	381	2.50	35	6580	62	0.06
11	1642	80	0.54	36	6642	115	0.62
12	1722	280	1.20	37	6757	102	1.80
13	2002	250	0.49	38	6859	141	6–14
14	2252	74	0.24	39	7000	127	?
15	2326	84	1.30	40	7127	132	0.96
16	2410	441	0.75	41	7259	136	2.10
17	2851	140	0.75	42	7395	158	1.70
18	2991	123	2.00	43	7553	123	0.35
19	3114	201	0.53	44	7676	131	0.20
20	3315	182	0.12	45	7807	101	1.30
21	3497	212	2.20	46	7908	143	1.70
22	3709	162	0.15	47	8051	47	0.30
23	3871	240	4.00	48	8098	217	?
(23a	4111	63	6.00)	(48a	8315	54	?)
24	4111	159	0.75	49	8315	139	
25	4270	98	1.60		3′ Noncoding		

[a] Position of the exons with respect to the coding sequence of neurofibromin. Base 1 is the A of the start codon. There are an additional 286 bp of 5′ noncoding sequence in exon 1 (G. Xu, in preparation) and approximately 2 kb of 3′ noncoding sequence for which exon boundaries have not been determined. Approximate intron sizes were determined by PCR using exon-based oligonucleotide primers and genomic DNA template. Exons 23a and 48a are insertion-exon, alternative-splice forms found in cDNAs. The GAP-related domain spans exons 20 to 27; the most highly conserved exon, FLR (Li et al 1992), is exon 24.

A single large intron encompasses all three genes previously located between the translocation breakpoints: *OMGP*, *EVI2A*, and *EVI2B*. The 5′ portion of *NF1* maps to a CpG island of hypomethylation that corresponds to the centrometric end of the 350-kb NotI fragment (Figure 1); furthermore, sequencing and primer-extension experiments at the 5′ end of the gene have defined transcription-initiation sites just proximal to the NotI site that separates the 150-kb and 350-kb fragments shown in Figure 1 (G. Xu et al 1991, personal communication). The 3′ end has not been defined, although the end of the coding sequence is contained within the 350-kb NotI fragment.

INITIAL CLUES TO THE FUNCTION OF THE *NF1* GENE

The GAP-Related Domain (NF1-GRD)

The observed similarities in amino acid sequence between a predicted peptide moiety of neurofibromin and the catalytic domains of mammalian GAP, yeast *IRA1*, and yeast *IRA2* suggested that at least one role of neurofibromin might be to stimulate ras^{GTPase} activity. This suggestion was confirmed biochemically and genetically. cDNA segments encoding the homologous catalytic domain were cloned into expression vectors for *E. coli*, baculovirus/*sf9* insect cells, and yeast (Ballester et al 1990; Martin et al 1990; Xu et al 1990a). In genetic experiments, the peptide was expressed

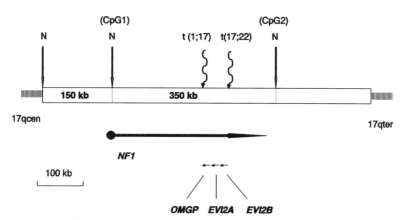

Figure 1 Pulsed-field restriction map of genomic NotI sites defining the *NF1* locus. *NF1* and three embedded genes are depicted below the map by horizontal arrows, in their orientations and relative positions (sizes of embedded genes not to scale). Translocation breakpoints in two NF1 patients are indicated. N, NotI site; CpG, island of hypomethylation.

in *ira-* yeast strains and studied by in vivo assays based on heat-shock resistance and glycogen accumulation. Transformation with the GAP-related domain of neurofibromin, NF1-GRD, reverted the *ira-* phenotype; cells survived heat shock and accumulated glycogen (Ballester et al 1990; Martin et al 1990; Xu et al 1990a). Presumably, the GRD of neurofibromin stimulates hydrolysis of GTP on yeast *RAS* proteins, thus replacing the catalytic deficiency of either *IRA1* or *IRA2* proteins in these cell lines.

In biochemical experiments, in vitro *ras*-GAP assays confirmed the hypothesis that the NF1-GRD can stimulate p21*ras* hydrolysis of GTP. Ballester et al (1990) used a cDNA construct encoding 412 amino acids that encompass the GRD to transform an *ira2-* yeast strain. Cell lysates from that transformed strain stimulated hydrolysis of *H-ras*^{GTP}, but not of the oncogenic mutant *H-ras*$_{Val12}^{GTP}$. Xu et al (1990a) purified a gluta-thione S-transferase/NF1-GRD fusion protein expressed in *E. coli* and demonstrated GTPase-stimulating activity on both yeast *RAS2*^{GTP} and *H-ras*^{GTP}. The same investigators also showed that, like the GAP and IRA2 proteins, the NF1-GRD fusion protein could not stimulate GTPase activity of the *H-ras*$_{Val12}$, *RAS2*$_{Val19}$, or *RAS2*$_{Ala42}$ mutant proteins. Martin et al (1990) purified an epitope-tagged, 474-amino-acid peptide, which encompassed the NF1-GRD from the baculovirus/*sf9* expression system. This peptide stimulated the GTPase activity of *N-ras*, but not the oncogenic mutants *N-ras*$_{Asp12}$ and *N-ras*$_{Val12}$. NG1-GRD also stimulated GTP hydrolysis of the *N-ras*$_{Ala38}$ effector mutant, but at a level less than 10% of the wild-type activity.

Recently, a full-length neurofibromin cDNA was constructed and ex-pressed in the baculovirus system (R. Clark 1991 and G. Xu 1991, personal communications). Preliminary studies show that the full-length 2818-amino acid construct has the same level of *ras*^{GTPase} stimulation as the 474-amino acid GAP-related domain peptide (G. Bollag, personal com-munication). This observation allows one to extrapolate from results of earlier experiments with the NF1-GRD peptide to interpretations of func-tion of full-length neurofibromin.

Although the initial results of in vitro assays did not biochemically distinguish NF1-GRD from p120-GAP, subsequent studies have shown some differences. At high concentrations of N-ras^{GTP}, GAP stimulates the GTPase to a much higher specific activity than does NF1-GRD; at low N-ras concentrations, the GAP and NF1-GRD activations are comparable (Martin et al 1990). This observation served as a clue that NF1-GRD might have a higher affinity for N-ras. In a competition-binding assay, NF1-GRD demonstrated a 30-fold higher affinity for N-ras^{GTP} than did GAP (Martin et al 1990); further studies demonstrated several differences between p120-GAP and NF1-GRD, both in their affinities for various *ras* proteins and in their respective activations of p21ras GTPase (Bollag & McCormick 1991).

NF1-GRD and p120-GAP also show differences with respect to their interaction with p21ras in the presence of various lipids. The activity of lipid-derived mitogenic agents depends on cellular p21 activity (Stacey et al 1988; Yu et al 1988). Arachidonic acid at 100 μg/ml inhibits stimulation of H-ras^{GTPase} by GAP catalytic fragment and NF1-GRD both, whereas high concentrations of phosphatidic acid (β-arachidonoyl-γ-stearoyl) only partially inhibited NF1-GRD and did not inhibit GAP (Golubic et al 1991). Other studies confirmed the inhibition by arachidonic acid and also demonstrated that GAP stimulation of H-ras^{GTP} was increased by prostaglandins PGF2α or PGA2 and decreased by PGI2, whereas NF1-GRD activity was not affected by the prostaglandins (Han et al 1991).

Further comparisons of p120-GAP and NF1-GAP with respect to lipid-inhibition responses on N-ras^{GTP} showed that phosphatidate, arachidonate, and phosphatidylinositol 4,5-biphosphate inhibit NF1-GAP to a much greater extent than does p120-GAP (Bollag & McCormick 1991). Furthermore, n-dodecyl β-D-Maltoside, which did not inhibit p120-GAP, was identified as a stable inhibitor of NF1-GAP activity. This selective inhibition was used to differentiate the two GAP activities in cell lysates: Inhibitable GAP activity is due to NF1-GAP, and noninhibitable GAP activity is due to p120-GAP. The potential physiological significance of the in vitro interactions of lipid moieties on GAPs raises the possibility that neurofibromin may play a key role in a signal transduction pathway within the ras-dependent mitogenic response of many cell types.

Alternative Splice in the GRD

In the course of screening for mutations in the GAP-related domain of NF1 patients, an alternate transcript previously seen in tumors (Nishi et al 1991) was independently identified in affected and unaffected individuals in RNA isolated from lymphoblastoid cell lines (R. Cawthon 1990, unpublished data). The alternate form contains an extra 63-bp exon inserted just upstream from the most conserved exon in the GAP-related domain and it encodes a basic stretch of amino acids, including six lysines out of 21 amino acids (Nishi et al 1991; Suzuki et al 1991; D. Viskochil, in preparation).

Because high lysine content could significantly alter the peptide conformation in a region presumably critical for p21ras interaction, one might expect the alternate-splice form to lack GAP function. To test this hypothesis, a baculovirus/sf9-expressed peptide (NF1-GRDII), identical to NF1-GRD (now named NF1-GRDI) except for inclusion of the 21-amino-acid insert, was purified, and its functional properties were compared with NF1-GRDI. NF1-GRDII demonstrated decreased stimulation of H-ras as compared with NF1-GRDI; however, its affinity for H-ras was about twofold greater (D. Viskochil, in preparation). Thus, the affinity of NF1-GRDII for H-ras^{GTP} is the highest yet demonstrated between wild-type p21ras^{GTP} and a GAP. This feature of the NF1-GRDII could be significant in cells where limiting concentrations of ras^{GTP} play a major role in signal transduction of physiological responses.

The biological function of the alternate-splice form of NF1-GRD has not been established, but preliminary studies have shown that a switch to higher type II/type I ratios is associated with differentiation in a neuroblastoma cell line (Nishi et al 1991). Thus, differential regulation of the neurofibromin isoforms may control the degree of influence p21ras transduction pathways have on the overall signal for cell proliferation.

GENE EXPRESSION

Tissue Distribution

Neurofibromin seems to be expressed across most human tissues; however, the level of expression in most tissues has not been quantified. Four methods have been used to ascertain expression: Northern-blot analysis, RNA-polymerase chain reaction (PCR), antibody binding, and NF1-GAP activity. The more sensitive of these assays, RNA-PCR and NF1-GAP activity, indicate that neurofibromin is present in most adult tissues.

Northern-blot analyses demonstrated an 11- to 13-kilobase transcript in murine kidney, brain, and B16 melanoma cells, but not in mouse skin, spleen, thymus, or liver (Buchberg et al 1990). A Northern signal from human tissue has proven more difficult to demonstrate, but a transcript has been seen in RNA from a choriocarcinoma cell line (Viskochil et al 1990) and from a neuroblastoma, two melanoma cell lines, brain frontal-lobe tissue, and kidney tissue (Wallace et al 1990).

Subsequent analyses of gene transcription have relied on the sensitive method of RNA-PCR: Total cellular RNA is random-primed to make tissue-specific, single-strand complementary DNA before PCR amplification. Even though such studies have not been performed quantitatively, neurofibromin transcript has been found in all tissues tested, including lymphoblastoid cell lines, skin fibroblast, spleen, muscle, brain, kidney, liver, and lung (Nishi et al 1991; Suzuki et al 1991; Wallace et al 1990).

The two alternative-splice forms in the GRD are equally represented in products of the RNA-PCR reaction from *EBV*-transformed lymphocytes, whether the cells came from NF1 patients or from unaffected individuals

(R. Cawthon, unpublished data). RNA isoforms from normal human kidney, placenta, and lung also have demonstrated equal ratios (Suzuki et al 1991). However, Nishi et al (1991) found that ratios of the two isoforms varied according to the differentiation status of a specific tissue. In addition to the observed association of the type II transcript in differentiation of HS-SYSY neuroblastoma cells, a switch in relative expression of isoforms was noted during brain development. The type I isoform was more abundant in the 20-week human fetal brain, and the ratios were reversed in the adult brain; the transition began to occur approximately 22 weeks into gestation. In contrast with these findings, a different group reported a higher type I/type II ratio in human adult brain, and, unexpectedly, this ratio appeared to be reversed in less-differentiated astrocytoma tissue from 14 samples of brain tumors that represented various stages of differentiation (Suzuki et al 1991). The respective roles the alternate-splice forms of neurofibromin play in the cellular biology of NF1 remain uncertain, but they are probably significant.

Identification of neurofibromin in cells by means of antibody has recently been reported (Daston et al 1992; DeClue et al 1991; Gutmann et al 1991). Antibodies raised against various domains of the protein have identified a 220- to 280-kilodalton protein by several techniques. By using antibodies generated against two synthetic peptides and three fusion proteins, Gutmann et al (1991) demonstrated neurofibromin as a 250-kD peptide in HeLa cells, NIH3T3 cells, and various murine tissues. These authors stated that neurofibromin was detected in all cell lines and tissues they examined. In other studies, DeClue et al (1991) used an antibody generated against an *E. coli* fusion-expression construct, which contained the GAP-related domain of neurofibromin, to immunoprecipitate a 280-kD protein from ^{35}S-labeled cell lysates from NIH3T3, HeLa, and rat Schwannoma cell lines. Daston et al (1992) used antibodies generated against two *E. coli* fusion peptides that contained approximately 310 amino acids of neurofibromin to detect a 220-kD band from total rat spinal cord on SDS/PAGE Western blots. Direct Western analysis on human spinal cord failed to detect the protein, but immunoprecipitation of human spinal cord tissue lysate with one antibody, followed by Western immunoblot analysis using the second antibody, also identified a 220-kD protein. Taken together, these studies demonstrated a 220- to 280-kD protein in cell and tissue lysates that is recognized by several different antibodies.

Immunohistocytochemical studies of neurofibromin expression in various tissues are just beginning. In rats, neurofibromin antibodies have detected the highest levels of neurofibromin in brain, spinal cord, sciatic nerve, and adrenal glands; lower levels in liver, spleen, pancreas, and cardiac tissue; and none in skeletal muscle, lung, kidney, or skin (Daston et al 1992). Furthermore, Daston et al (1992) showed that within the peripheral nervous system, the dorsal root ganglia and small axonal fibers associated with nonmyelinating Schwann cells stain intensely with anti-neurofibromin antibodies; in sections of the spinal cord, neurons and oligodendrocytes are preferentially stained.

A dichotomy is evident when one compares neurofibromin expression by the RNA and antibody methodologies. By RNA-PCR detection, neurofibromin is ubiquitously expressed; yet, Northern-blot analysis reveals restricted expression patterns. Likewise, antineurofibromin antibodies recognize a 220- to 280-kD protein from all tissue- and cell-line lysates tested; however, immunocytochemical analysis reveals a restricted pattern of expression from various rat tissues. Relative sensitivities in signal detection may explain these observations. For example, the sensitivity of RNA-PCR may allow detection of a small number of cellular *NFI* mRNAs, which represent the leaky nature of housekeeping-gene transcription, yet such a low level of expression may not provide any specific contribution to cellular phenotypes. On the other hand, like other enzymes, neurofibromin may be expressed at the lower limits of detection (e.g. antibody) in most cells, yet still contribute to the cellular phenotype. The p21ras^{GTPase}-activating function of neurofibromin provides an approach for further characterizing expression of this protein with respect to cellular physiology.

The GAP-related domain of neurofibromin is inhibited by some lipids in the rasGAP assay, whereas p120-GAP is not (Bollag & McCormick 1991). Therefore, total p21rasGAP activity from various cell lysates can be divided into p120-GAP-like activity (noninhibitable by lipids) and *NFI*-GAP-like activity (lipid inhibitable). This assumption is valid if only two rasGAP proteins are present in the cells. Bollag & McCormick (1991) found that all mammalian cells tested (NIH 3T3, SRD 3T3, PC 12, Jurkat, lymphocytes, teratocarcinoma, HL 60, stage-VI oocytes, rat brain, and human placenta) contained NF1- and p120-GAP-like activity, albeit in varying proportions. The levels of rasGAP activity cannot be compared among differing cell lines, because of the difficulty in normalizing cell numbers, DNA content, and protein content for separate preparations of cell lysates. However, one can compare the relative proportions of lipid-inhibitable and noninhibitable rasGAP activities from each cell lysate. Of note, NF1-GAP accounts for 75% of total rasGAP activity from rat pheochromocytoma cells (PC 12 cell line), whereas p120-GAP accounts for 90% of the total activity from human placental tissue. The finding of lipid-inhibitable rasGAP activity in numerous and varied cell lysates supports the conclusion from RNA-PCR experiments that neurofibromin is relatively ubiquitous. It does not, however, predict a threshold level of expression associated with a clear cellular phenotype.

Intracellular Localization

The subcellular localization of neurofibromin has not been established. One immunodetection study, which used differential centrifugation and no detergent in the cell-lysis buffer, localized neurofibromin to the non-nuclear particulate fraction of NIH 3T3 cells (DeClue et al 1991). In functional studies using 1% Nonidet P-40 in various cell lysates, Bollag & McCormick (1991) found NF1-GAP activity distributed between the post-$100,000 \times g$-spin supernatant (approximately 60% of total GAP activity) and the particulate fraction (approximately 40% of total GAP activity). It is plausible that neurofibromin is less active with the particulate fraction of the cell, and that immunologically undetectable protein in the cytosol is responsible for 60% of the detectable NF1-GAP activity. Alternatively, detergent may release loosely bound neurofibromin from the particulate fraction into the cytoplasmic fraction. DeClue et al (1991) used NP-40 detergent in the cell-lysis buffer to demonstrate that neurofibromin was immunoprecipitated from NIH 3T3, HeLa, and rat Schwannoma cell lines, both as a monomer and as a complex with an unidentified 400- to 500-kD protein. In tissue-culture cell lines, neurofibromin could be part of a complex associated with the particulate fraction of cells.

In addition to the above findings, immunostaining has demonstrated localization of neurofibromin in the cytoplasm of cells from various tissues (Daston et al 1992). In separate experiments that used different antibodies to study indirect immunofluorescence in cultured cell lines, neurofibromin sublocalized to microtubules (P. Gregory and D. Gutmann 1992, personal communication). This preliminary observation is intriguing, because the GAP-related domain of neurofibromin copurifies with tubulin from insect-cell lysates in the baculovirus expression system (R. Clark 1990, personal communication). The immunostaining and differential centrifugation studies would be consistent if the microtubule fraction of the cell lysate localizes to the particulate fraction in the absence of detergent. Until these somewhat conflicting observations are clarified, the distribution of neurofibromin activity and protein among various tissues and subcellular structures remains uncertain.

CLINICAL FEATURES OF NF1

Efforts to understand the role of neurofibromin in human physiology are intimately concerned with determining the nature and consequence of constitutional mutations in *NF1*. An appreciation of the clinical features of NF1, including complicating medical problems, is helpful both in evaluating the function of neurofibromin and in correlating abnormal expression of this protein with the underlying pathophysiology of the disease.

The clinical manifestations of NF1 are varied. Even among family members who carry the same mutation, clinical expression is as hetero-geneous as if the affected family members were not related. The severity of the disease is, therefore, unpredictable; no known relation exists between severity and age of onset, status of the affected parent, birth order, parental age, or environmental factors (Huson et al 1989).

The two most consistent clinical features of NF1 are café-au-lait spots (CLS) and neurofibromas. The CLS, which are asymptomatic, were first used as a diagnostic sign of NF1 in a population-based study of neuro-fibromatosis (Crowe et al 1956), whereby more than six on one individual served as a marker for the condition. Biopsies indicate that NF1-associated CLS contain more dopa-positive melanocytes per unit area than do com-mon "birthmarks" in relation to surrounding skin (Tong & Fitzpatrick 1990); nevertheless, the low incidence of melanomas in NF1 patients sug-gests that CLS do not undergo malignant changes (Hope & Mulvihill 1981).

Neurofibromas are the hallmark of NF1. These benign tumors arise from the peripheral nervous system and are composed of extracellular matrix (laminin, collagen, and proteoglycan), Schwann-like cells, fibro-blasts, mast cells, endothelial cells, and perineural cells. Although the cell mixture may vary from one tumor to another, Schwann-cell hyper-cellularity is a reliable finding, and proliferating Schwann cells probably lead to emergence of neurofibromas. Cutaneous neurofibromas are gener-ally small, raised nodules that are soft and often carry different pig-mentation from the surrounding skin. They commonly appear in late childhood and may increase in numbers and size during puberty and pregnancy. In contrast, plexiform neurofibromas can be quite large and usually appear either congenitally or in early childhood. They are inti-mately associated with larger nerves and extend along nerve-bundle tracts, engulfing the nerves within their sheaths.

Other clinical features are quite specific. Gliomas of the optic nerve are benign tumors seen in approximately 15% of individuals with NF1. They represent a proliferation of the glial cells and can occur at any point along the visual pathway (Hoyt & Baghdassarian 1969). Likewise, rare hamartomatous growths on the irides, called Lisch nodules (Lisch 1937) have taken on new significance as markers for NF1; more than 90% of adults with NF1 show multiple Lisch nodules bilaterally (Mulvihill et al 1990). Other NF1 clinical features are also highly specific: axillary and groin freckling (Crowe et al 1956), sphenoid wing dysplasia, and tibial pseudarthroses occur rarely, except among individuals with NF1 (Holt

1978). Taken together, these clinical signs show that the tissues most involved in the NF1 phenotype are derived from embryonic neural-crest tissue.

Neurofibromatosis 1 patients may also show any of a variety of clinical manifestations that are not specific for the condition, but occur with higher frequency among NF1-affected persons than in the general population (Riccardi & Eichner 1986). Perhaps the most significant of these associations is an eightfold increase in the prevalence of learning disabilities; as many as 50% of children with NF1 may have significant learning problems (Aron et al 1990). Frank mental retardation and epilepsy, however, are no more common than in the general population. Other manifestations may include macrocephaly, short stature, scoliosis, tibial bowing, renal artery stenosis, auditory deficits, headaches, and gastrointestinal symptoms.

Perhaps the aspect of most concern, however, is the small but significantly increased risk of malignancies associated with NF1. It is difficult to conduct unbiased epidemiological studies to determine the frequency of cancer in NF1, but some authors have reported an increased incidence of malignancy in NF1 patients followed in hospital-based clinics (Lefkowitz et al 1990). Moreover, in these studies the mean age of diagnosis of malignancy was 38 years, compared with a mean age of 65 in the general population of the United States. The latter observation may reflect the types of tumors observed in NF1 patients: optic gliomas, neurofibrosarcomas, astrocytomas, rhabdomyosarcomas, juvenile chronic myelogenous leukemia, meningiomas, and pheochromocytomas.

NF1 MUTATIONS IN CONSTITUTIONAL AND TUMOR DNA

NF1 mutations that have been detected in the constitutional DNA of NF1 patients to date include translocations, medium- to large-sized deletions, and a stop codon, all within the protein-coding regions of the gene (Cawthon et al 1990b; Estivill et al 1991; Stark 1991; Viskochil et al 1990). In one sporadic case of NF1, a de novo Alu insertion in an intron of *NF1* causes an exon to be skipped during splicing; the result is a frameshift and a premature stop (Wallace et al 1991). In another family, a point mutation leading to a Leu → Pro amino acid substitution coinherits with the NF1 disorder (Cawthon et al 1990b). Each of these mutations would be expected to inactivate neurofibromin, by truncating the protein and/or drastically changing its overall shape. Because each of these mutations lies 3′ of the NF1-GRD, stable neurofibromins with intact, functioning GRDs might still be made from the mutant *NF1* alleles.

At least two NF1 patients have been described, however, with deletions that remove the entire chromosomal region containing the NF1-GRD exons (K. Stephens 1989, personal communication; D. Viskochil 1991, unpublished data). Moreover, an NF1 patient with a "stop" mutation early in the GRD and another NF1 patient with a frameshift caused by a single-base deletion in the middle of the GRD have been found (R. Cawthon 1991, unpublished); the mutant neurofibromins predicted in these cases, even if stable, would not be expected to bind ras^{GTP} and stimulate the GTPase.

Taken together, these results provide convincing evidence that most germline *NF1* mutations lead to neurofibromatosis 1 by inactivating the gene and suggest that *NF1* is a tumor suppressor gene. If so, the remaining, normal *NF1* allele would be expected to undergo inactivation during the development of at least some of the tumors that arise in NF1 patients. Loss of an allele is one mechanism of inactivation that is relatively easy to detect.

Therefore, investigators have examined benign and malignant tumors from NF1 patients for evidence of loss of heterozygosity (LOH) on chromosome 17 in the region of the *NF1* gene. Extensive analyses have so far failed to detect LOH along chromosome 17 in neurofibromas (Skuse et al 1989, 1991). However, analyses of some malignant tumors from NF1 patients have shown LOH for DNA markers on 17p or along the entire length of chromosome 17 (Glover et al 1991; Menon et al 1990; Skuse et al 1989). These findings are somewhat ambiguous, however, because loss of a normal allele of the p53 tumor suppressor on 17p could contribute to tumor progression and drive the chromosome loss. However, other malignant tumors from NF1 patients have shown LOH specific to 17q. Skuse and colleagues (1989, 1990) observed LOH specific to 17q in one astrocytoma and several neurofibrosarcomas among patients with NF1; moreover, the alleles retained in the tumor cells were those inherited from the NF1-affected parent. Also, Ponder et al (1990) found LOH specific to 17q in some pheochromocytomas from NF1 patients. These results suggest that inactivation of both *NF1* alleles may contribute to the development of several NF1-associated malignancies. Thus, *NF1* appears to behave as a true tumor suppressor gene in these tissues. The failure to find LOH in neurofibromas suggests that events other than inactivation of the second *NF1* allele are involved in the development of these benign tumors, although inactivation of the second *NF1* allele by mechanisms more subtle than allele loss (e.g. point mutations or methylation) have not been ruled out.

Loss of marker alleles in the vicinity of *NF1* or encompassing the *NF1* locus leaves open the possibility that some other gene in the region, and not *NF1* itself, is the relevant contributor to tumor progression in NF1-

associated malignancies. The strongest evidence that directly implicates *NF1* in carcinogenesis comes from studies of neurofibrosarcomas in NF1 patients (Basu et al 1992; DeClue et al 1992). In cell lines derived from these malignant tumors, levels of neurofibromin were drastically diminished or (in some cases) barely detectable by immunoassay. Interestingly, ras^{GTP} was elevated and growth-promoting in these cells. These results suggest that a nearly complete loss of neurofibromin function contributed to development of the tumors. *NF1* apparently acts as a tumor suppressor gene in this cell type.

In the general population, sporadic cases of the tumors that have an increased incidence among NF1 patients might be expected to harbor somatic mutations in the *NF1* gene occasionally. Astrocytoma, one of the most common malignant brain tumors, appears to have an increased incidence among NF1 patients (Blatt et al 1986; Cohen & Rothner 1989; Kibirige et al 1989; Sørensen et al 1986). Initial screening of one exon in the GRD of tumor DNAs from ten non-NF1 patients with anaplastic astrocytomas (Li et al 1992) revealed a somatic point mutation that changed $codon_{1423}$ in the NF1-GRD from AAG to CAG (Lys → Gln) in one tumor. This Lys residue is normally invariant among GAP-related proteins (Wang et al 1991; Xu et al 1990b) and is, therefore, likely to be crucial for normal functioning. Indeed, the predicted mutant NF1-GRD protein from this astrocytoma binds with normal affinity to ras^{GTP}, but is severely impaired in its ability to stimulate ras GTPase (Li et al 1992). Because this somatic mutation was detected in the first *NF1* exon screened, and 50 protein-encoding *NF1* exons remain to be scanned, it is too early to assess the frequency of somatic *NF1* mutations in sporadic astrocytomas.

As neurofibromin may be an important regulator of ras^{GTP} levels in many cell types, somatic *NF1* inactivation might also result in increased concentrations of ras^{GTP} in cells. Somatic mutations that activate *ras* genes may contribute to tumorigenesis in up to 30% of human tumors (Bos 1989). Colon cancers, for example, frequently contain activated *K-ras* (Bos et al 1987), and myelodysplasias often contain activated *N-* or *K-ras* genes (Liu et al 1987; Lyons et al 1988; Padua et al 1988). Mutations of the *NF1* gene, therefore, might also promote growth in these cell types. It is interesting, therefore, that screening of the same exon found to be mutated in the astrocytoma sample mentioned above revealed a somatic mutation in 1 of 22 colonic adenocarcinomas (Li et al 1992). This mutation also affected $codon_{1423}$ by changing the normal AAG codon to a GAG (Lys → Glu). Furthermore, this same mutation was found in 1 of 28 myelodysplasias. The mutant Lys → Glu, like the mutant Lys → Gln, binds ras^{GTP} with normal affinity, but shows greatly reduced activity toward the ras GTPase.

Because colon cancers and myelodysplasias do not seem to have an increased incidence among NF1 patients, the mutant *NF1* allele encoding the neurofibromin$_{Glu1423}$ may be restricted to sporadic tumors. However, as part of a mutation screen of 80 unrelated NF1 patients, this same Lys → Glu mutation was also found in one of these individuals. The mutation coinherited with NF1 disease in the patient's family. No history of myelodysplasia, leukemia, or colon cancer was present in any of the five NF1-affected members of this family for whom clinical data were available. There is a precedent for this apparent paradox: Somatic mutations in the retinoblastoma gene are frequently found in sporadic small-cell lung cancers, yet the incidence of this tumor type is not increased among individuals with hereditary retinoblastoma (Harbour et al 1988; Yokota et al 1988). Thus, somatic mutations in the *RB* gene in lung cells may contribute to—and may even be necessary for—development of the tumors, but the mutations are not rate-limiting for tumorigenesis in this cell type.

DIVERSE ROLES OF NEUROFIBROMIN: CLUES AND SPECULATIONS

Clues

The role of neurofibromin in stimulating an increased rate of hydrolysis of ras^{GTP} to ras^{GDP} has been confirmed and detailed by in vitro biochemistry. However, the broader questions regarding the potential roles of neurofibromin as a mediator of cell growth and development through signal transduction remain unanswered. The analogies with GAP and *IRA* may extend to additional activities, e.g. effector functions, beyond the regulation of intracellular ras^{GTP} levels.

Although no direct evidence has established neurofibromin as a downstream effector of ras^{GTP}, by analogy with p120GAP it is likely to have such a role. Evidence is accumulating that supports a dual role for p120GAP interaction with ras^{GTP}; it has the capacity to function as either a negative regulator or a positive effector (Bourne et al 1990). Perhaps the most compelling circumstantial evidence that supports neurofibromin as an effector of ras^{GTP} is the clinical expression of NF1. Assuming that oncogenic ras mutants and elevated intracellular levels of wild type ras^{GTP} stimulate cellular transformation and growth via similar mechanisms, it follows that if neurofibromin acted simply as a regulator of *ras*, then inactivating mutations in patients would give rise to tumors shown to carry oncogenic *ras* mutations. In fact, cell types affected in the NF1 phenotype do not carry oncogenic *ras* mutations. Furthermore, the cell types that do carry oncogenic *ras* mutations are not represented in the clinical expression of abnormal growth in NF1 patients.

As negative regulators, both neurofibromin and p120GAP "deactivate" p21ras^{GTP}. However, as "effectors" of downstream signal propagation, these two proteins are likely to be quite different. The altered sensitivity to lipid inhibition (Bollag & McCormick 1991) is perhaps just one of many distinctions between p120GAP, neurofibromin, and other potential GAP-associated signal pathways. Further clues toward understanding which role neurofibromin may play in the transduction of ras^{GTP}-mediated signals might be found by examining cell phenotypes regulated in part by "activated" ras.

Cellular responses to increased intracellular levels of ras^{GTP} vary with respect to cell type. In many cells, e.g. mouse 3T3 or any of the many tumor-cell types in which oncogenic *ras* mutations are found, an increased level of ras^{GTP} results in a phenotype of increased growth. However, ras^{GTP} appears to be growth-inhibiting in other cell types, e.g. rat Schwann cells derived from sciatic nerve (Ridley et al 1988), pheochromocytomas (Bar-Sagi & Feramisco 1985; Noda et al 1985), human medullary thyroid carcinoma cells (Nakagawa et al 1987), 3T3-L1 cells (Benito et al 1991), F9 embryonal carcinoma cells (Yamaguchi-Iwai et al 1990), and chick embryo dorsal root ganglionic neurons (Borasio et al 1989).

Furthermore, ras may play a role in embryonic development. Ras has recently been shown to be part of a signal transduction pathway that controls development of a differentiated photoreceptor cell in *Drosophila* (Fortini et al 1992; Simon et al 1991). Likewise, determination of vulva development in *C. elegans* is regulated by the ras protein encoded by *let-60* (Beitel et al 1990; Han & Sternberg 1990). Because the meaning of an enhanced ras^{GTP} signal, whether to grow or to differentiate, differs among cell types, the interpretive mechanisms of the transduction system must lie downstream of the ras^{GTP} component.

Speculations

Does the transduction of a signal through the ras^{GTP}-neurofibromin complex carry with it the "sign" of the signal, i.e. whether to stimulate cell division or cell differentiation? We might imagine that a ras^{GTP}-neuro-fibromin-mediated signal would be essentially neutral, leaving its meaning to elements even further downstream. On the other hand, we could also imagine that the intrinsic message in a ras^{GTP}-neurofibromin complex might specify differentiation, turning off cell-cycle stimulating systems and turning on differentiation systems.

If a neurofibromin-mediated signal were differentiative in nature, the cell's interpretation of an elevated level of ras^{GTP} might be determined by the relative levels of neurofibromin and other, presumably cell growth-specific, transduction elements, such as one of the GAP proteins. This intriguing hypothesis suggests that a cell might be able to specify the meaning of an elevated ras^{GTP} level by controlling the relative activities of neurofibromin and GAPs (Figure 2). The supportive evidence for such speculation is circumstantial; nevertheless, it is worth examining such clues in greater detail.

The tissue types in which activated ras leads to differentiation are in many ways similar to the tissues involved in the clinical expression of NF1. Schwann cells, pheochromocytoma cell lines, and neurons from dorsal root ganglia have all been implicated in NF1. Thus, a reduced activity of neurofibromin as a differentiation effector may allow the abnormal growth observed as part of the clinical spectrum of this condition.

If neurofibromin sends a distinctive ras^{GTP}-mediated signal, one can speculate on the mechanisms that confer specificity in sending such a signal. Protein domains outside of the GAP-related domain of neurofibromin probably play a role in downstream signaling. Again, by analogy to p120GAP, one can envision that the interaction between the GAP-related

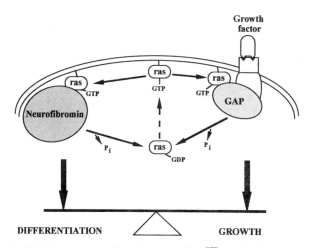

Figure 2 Model depicting neurofibromin as a p21ras^{GTP}-mediated signal transducer for differentiation and as a negative regulator of ras-mediated growth. p21ras resides at the cell membrane where it binds with GAP or neurofibromin (*solid arrows*). Ras^{GTP}/GAP complexes send growth signals, whereas ras^{GTP}/neurofibromin complexes send differentiation signals, both of which are terminated by hydrolysis to GDP. Ras^{GDP} is recycled by putative exchange proteins (*dashed arrow*). Some cells may contain other proteins that interact with activated ras to send additional signals. The lever symbolizes the concept that small changes in ras-mediated signal input may alter cell phenotype by tipping the balance in one direction or the other.

domain of neurofibromin with ras^{GTP} might "open" the protein, thereby presenting buried domains of neurofibromin to other components of the signal pathway. For example, p120GAP is an essential component in the down-regulation of potassium-channel activity in cardiac atrial cells; a model for its mechanism of action involves exposing the first SH2 or SH3 domains of GAP to the atrial membrane after the catalytic domain binds to ras^{GTP} (Martin et al 1992). Even though neurofibromin domains outside of the GRD have not been characterized, sequence homology with *IRA1* and *IRA2* may provide clues, as well as insight, in experimental approaches to further our understanding of the function of extra-GRD domains in neurofibromin. The role that neurofibromin-related domains of *IRA* play in the determination of yeast phenotypes is not known; however, recent experiments have suggested that the *IRA* gene products may directly interact with the yeast adenylate cyclase (Mitts et al 1991). If this interaction involves domains homologous to neurofibromin, experiments to evaluate potential associations between adenylate cyclase and neurofibromin may be worthwhile.

Some biochemical evidence supports a role of the NF1-GRD alternate-splice form II in transducing a ras^{GTP}-mediated differentiation signal: Activation of ras GTPase is decreased, and affinity of the NF1-GRDII for H-ras^{GTP} is increased, in comparison with NF1-GRDI (D. Viskochil, in preparation). If GTP hydrolysis is necessary for signal termination (McCormick 1989), then the combination of decreased activity with increased binding to ras^{GTP} suggests that neurofibromin isoform II might form a more stable complex with activated ras to increase the gain toward differentiation. One would expect cells with higher type II/type I ratios to be associated with a more differentiated state, as has been observed in a neuroblastoma cell line treated with retinoic acid (Nishi et al 1991). Thus, as presented in Figure 2, increasing the level of type II neurofibromin would intensify the gain of the differentiation signal and tip the balance away from cell growth.

The mutations observed in NF1 patients support two contentions regarding normal neurofibromin: It acts as an effector of ras^{GTP} and it sends a differentiation signal. Most of the mutations found so far in NF1 patients, deletions and premature stops (Cawthon et al 1990b; Estivill et al 1991; Stark et al 1991; Upadhyaya 1990; Viskochil et al 1990; Wallace et al 1991), would be expected to result in an inactive gene product, i.e. null mutations. If neurofibromin acted solely to regulate the level of ras^{GTP}, that level would be expected to rise with the reduced or eliminated neurofibromin activity resulting from the mutations. Given the differentiation-promoting activity of ras^{GTP} in neural crest-derived cells, inactivating mutations in *NF1* would not be expected to cause the abnormal proliferation of Schwann-like cells in neurofibromas. Thus, it seems unlikely that neurofibromin acts solely to regulate the level of ras^{GTP}. On the other hand, if neurofibromin has a ras-mediated effector function, reduced levels of neurofibromin in cells carrying *NF1* null mutations would decrease signal transduction, even in the presence of increased levels of ras^{GTP}. It is reasonable to suggest that, in tissue typically affected in the NF1 phenotype, activated ras enhances cell differentiation through one of its effectors, neurofibromin. According to the model depicted in Figure 2, in cells carrying *NF1* null mutations, activated ras transduces an attenuated differentiation signal, and the balance between cell proliferation and cell differentiation would be tipped toward growth.

Recent observations of the levels of ras^{GTP} in neurofibrosarcoma cell lines provide support for this idea (Basu et al 1992; DeClue et al 1992). These cells are deficient in neurofibromin and show elevated ras^{GTP}. Furthermore, reduction of ras^{GTP} in the cell line by addition of GAP catalytic domain peptide causes cell growth to cease; the elevated ras^{GTP} in these cells must have contributed to growth. Therefore, in the absence of neurofibromin, an elevated level of ras^{GTP} has the unexpected effect of contributing to the growth of these neural crest-derived cells.

It is interesting to compare these findings with the observation of ras^{GTP}-induced differentiation in the PC12 cell line derived from a rat pheochromocytoma. In these cells, increasing the level of ras^{GTP} causes differentiation (Bar-Sagi & Feramisco 1985; Noda et al 1985). There is evidence for neurofibromin GAP activity in PC12 cells (Bollag & McCormick 1991), so it remains possible that the ras^{GTP} signal is, in fact, transduced by neurofibromin in these cells. Indeed, LOH at the *NF1* locus in pheochromocytomas from NF1 patients, combined with failure to demonstrate LOH in sporadic pheochromocytomas (Ponder et al 1990), suggests that spontaneous pheochromocytomas may differ in molecular etiology from the pheochromocytomas of NF1 patients. From these observations, one could predict that enhancing activated ras in non-NF1 pheochromocytomas would lead to a more differentiated state; in NF1-associated pheochromocytomas that do not express neurofibromin, enhancement of activated ras would have no effect on differentiation and might, in fact, lead to increased proliferation.

Information gathered from mutational screens of NF1 patients and tumors will be instrumental in deciphering the various mechanisms that may lead to aberrant cell growth associated with the neurofibromin-associated signal pathway. For example, the codon$_{1423}$ mutations in NF1 and non-NF1 tumors support the model that neurofibromin can act as either a negative ras-regulator (Basu et al 1992; DeClue et al 1992) or a positive ras-effector. As this is a functional null mutation, it is difficult to reconcile the presence of the neurofibromin$_{Glu1423}$ allele in both an NF1 family and sporadic tumors not typically found in NF1. This mutation should attenuate the input to the differentiation signal and, by not catalyzing ras^{GTP} hydrolysis, increase available ras^{GTP} for the proposed growth-promoting signal pathway. This would suggest that neurofibromin plays a major role in cell determination in NF1-affected tissue through its positive-effector function and a less crucial, although still important, role for cell growth in tissue not typically affected in NF1 through its ras-regulator function. Thus, the model in Figure 2 still applies in that the balance would be tipped toward cell growth. The pathways that contribute most significantly to such growth likely vary from one tissue to another. For example, negative ras regulation by neurofibromin may be only one of several means by which non-NF1 cells control ras^{GTP} levels. In NF1-affected tissue, however, its positive-effector function might play a major, perhaps singular, role in cell determination by sending a differentiation signal.

These complexities of interpretation may stem from the intrinsic complexity of multiple, interacting systems in signal transduction. Effects and primary roles for system components may differ from one cell type to another, depending on context. Importantly, however, the several models suggested by the analysis of *NF1* mutations offer opportunities for a number of specific experiments that should shed new light on the fundamental mechanisms of genetic regulation of cell growth and differentiation.

Literature Cited

Aron, A. M., Rubenstein, A. E., Wallace, S. A., Halperin, J. C. 1990. Learning disabilities in neurofibromatosis. See Rubenstein & Korf 1990, pp. 55–58

Ballester, R., Marchuk, D., Boguski, M., Saulino, A., Letcher, R., et al. 1990. The *NF1* locus encodes a protein functionally related to mammalian GAP and yeast *IRA* proteins. *Cell* 63: 851–59

Barker, D., Wright, E., Nguyen, K., Cannon, I., Fain, P., et al. 1987. Gene for von Recklinghausen neurofibromatosis is in the pericentric region of chromosome 17. *Science* 236: 1100–2

Bar-Sagi, D., Feramisco, J. R. 1985. Microinjection of the *ras* oncogene protein into PC12 cells induces morphological differentiation. *Cell* 42: 841–48

Basu, T. N., Gutmann, D. H., Fletcher, J. A., Glover, T. W., Collins, F. S., Downward, J. 1992. Aberrant regulation of *ras* proteins in malignant tumor cells from type 1 neurofibromatosis patients. *Nature* 356: 713–15

Beitel, G. J., Clark, S. G., Horvitz, H. R. 1990. *Caenorhabditis elegans ras* gene *let-60* act as a switch in the pathway of vulval function. *Nature* 348: 503–9

Benito, M., Porras, A., Nebreda, A. R., Santos, E. 1991. Differentiation of 3T3-L1 fibroblasts to adipocytes induced by transfection of *ras* oncogenes. *Science* 253: 565–68

Blatt, J., Jaffer, R., Deutsch, M., Adkins, J. C. 1986. Neurofibromatosis and childhood tumors. *Cancer* 57: 1225–29

Bollag, G., McCormick, F. 1991. Differential regulation of rasGAP and neurofibromatosis gene product activities. *Nature* 351: 576–79

Borasio, G. D., John, J., Wittinghofer, A., Barde, Y.-A., Sendtner, M., Heumann, R. 1989. *ras* p21 protein promotes survival and fiber outgrowth of cultured embryonic neurons. *Neuron* 2: 1087–96

Bos, J. L. 1989. *Ras* Oncogenes in human cancer: a review. *Cancer Res.* 49: 4682–89

Bos, J. L., Fearon, E. R., Hamilton, S. R.,

Verlaan-de Vries, M., van Bloom, J. H., et al. 1987. Prevalence of *ras* gene mutations in human colorectal cancers. *Nature* 327: 293–97

Bourne, H. R., Sanders, D. A., McCormick, F. 1990. The GTPase superfamily: a conserved switch for diverse cell functions. *Nature* 348: 125–32

Buchberg, A., Bedigan, G., Taylor, B., Brownwell, E., Ihle, J., et al. 1988. Localization of *Evi-2* to chromosome 11: linkage to other proto-oncogene and growth factor loci using interspecific backcross mice. *Oncogene Res.* 2: 149–65

Buchberg, A. M., Cleveland, L. S., Jenkins, N. A., Copeland, N. G. 1990. Sequence homology shared by neurofibromatosis type-1 gene and *IRA-1* and *IRA-2* negative regulators of the *RAS* cyclic AMP pathway. *Nature* 347: 291–94

Cawthon, R. M., Andersen, L. B., Buchberg, A. M., Xu, G., O'Connell, P. 1991. cDNA sequence and genomic structure of *EV12B*, a gene lying within an intron of the neurofibromatosis type 1 gene. *Genomics* 9: 446–60

Cawthon, R., O'Connell, P., Buchberg, A., Viskochil, D., Weiss, R., et al. 1990a. Identification and characterization of transcripts from the neurofibromatosis 1 region: the sequence and genomic structure of *EV12* and mapping of other transcripts. *Genomics* 7: 555–65

Cawthon, R. M., Weiss, R., Xu, G., Viskochil, D., Culver, M., et al. 1990b. A major segment of the neurofibromatosis type 1 gene: cDNA sequence, genomic structure, and point mutations. *Cell* 62: 193–201

Cohen, B. H., Rothner, A. D. 1989. Incidence, types, and management of cancer in patients with neurofibromatosis. *Oncology* 3: 23–30

Crowe, F. W., Schull, W. J., Neel, J. V., eds. 1956. *A Clinical Pathological, and Genetic Study of Multiple Neurofibromatosis.* Springfield, Ill: Thomas

Daston, M. M., Scrable, H., Nordlund, M., Sturbaum, A. K., Nissen, L. M., Ratner, N. 1992. The protein product of the neurofibromatosis type 1 gene is expressed at highest abundance in neurons, Schwann cells, and oligodendrocytes. *Neuron* 8: 1–14

DeClue, J. E., Cohen, B., Lowy, D. 1991. Identification and characterization of the neurofibromatosis type 1 protein product. *Proc. Natl. Acad. Sci. USA* 88: 9914–18

DeClue, J. E., Papageorge, A. G., Fletcher, J. A., Diehl, S. R., Ratner, N., et al. 1992. Abnormal regulation of mammalian p21^{ras} contributes to malignant tumor growth in von Recklinghausen (type 1) neurofibromatosis. *Cell* 69: 265–73

Estivill, X., Lazaro, C., Casals, T., Ravella, A. 1991. Recurrence of a nonsense mutation in the *NF1* gene causing classical neurofibromatosis type 1. *Hum. Genet.* 88: 185–88

Fortini, M. E., Simon, M. A., Rubin, G. M. 1992. Signalling by the *sevenless* protein tyrosine kinase is mimicked by *Ras1* activation. *Nature* 355: 559–61

Fountain, J., Wallace, M., Bruce, M., Seizinger, B., Menon, A., et al. 1989. Physical mapping of a translocation breakpoint in neurofibromatosis. *Science* 244: 1085–87

Glover, T. W., Stein, C. K., Legius, E., Andersen, L. B., Brereton, A., et al. 1991. Molecular and cytogenetic analysis of tumors in von Recklinghausen neurofibromatosis. *Genes Chromosomes Cancer* 3: 620–70

Golubic, M., Tanaka, K., Dobrowolski, S., Wood, D., Tsai, M., et al. 1991. The GTPase stimulatory activities of the neurofibromatosis type 1 and the yeast *IRA2* proteins are inhibited by arachidonic acid. *EMBO J.* 10: 2897–2903

Gutmann, D., Wood, D., Collins, F. 1991. Identification of the neurofibromatosis type 1 gene product. *Proc. Natl. Acad. Sci. USA* 88: 9658–62

Han, J.-W., McCormick, F., Macara, I. 1991. Regulation of *Ras*-GAP and the neurofibromatosis-1 gene product by eicosanoids. *Science* 252: 576–79

Han, M., Sternberg, P. W. 1990. *let-60*, a gene that specifies cell fates during C. Elegans vulval induction, encodes a *ras* protein. *Cell* 63: 921–31

Harbour, J. W., Lai, S. L., Whang-Peng, J., Gazdar, A. F., Minna, J. D., Kaye, F. J. 1988. Abnormalities in structure and expression of the human retinoblastoma gene in SCLC. *Science* 241: 353–57

Holt, J. F. 1978. Neurofibromatosis in children. *Am. J. Radiol.* 130: 615–39

Hope, D. G., Mulvihill, J. J. 1981. Malignancy in neurofibromatosis. In *Advances in Neurology*, ed. V. Riccardi, J. Mulvihill, 29: 33–56. New York: Raven. 282 pp.

Hoyt, W. F., Baghdassarian, S. A. 1969. Optic glioma of childhood. *Br. J. Ophthalmol.* 53: 793–98

Huson, S. M., Compston, D. A. S., Clark, P., Harper, P. S. 1989. A genetic study of von Recklinghausen neurofibromatosis in south east Wales. I. Prevalence, fitness, mutation, rate, and effect of parental transmission on severity. *J. Med. Genet.* 26: 704–11

Kibirige, M. S., Birch, J. M., Campbell, R. H., Gattamaneni, H. R., Blair, V. 1989. A review of astrocytoma in childhood. *Pediatr. Hematol. Oncol.* 6: 319–29

Leach, R., Thayer, M., Schafer, A., Fournier, R. E. K. 1989. Physical mapping of human chromosome 17 using fragment-containing microcell hybrids. *Genomics* 5: 167–76

Ledbetter, D. H., Rich, D. C., O'Connell, P., Leppert, M., Carey, J. C., et al. 1989. Precise localization of *NF1* to 17q11.2 by balanced translocation. *Am. J. Hum. Genet.* 44: 20–25

Lefkowitz, I., Obringer, A., Meadows, A. 1990. Neurofibromatosis and cancer: Incidence and management. See Rubinstein & Korf 1990, pp. 99–110

Li, Y., Bollag, G., Clark, R., Conroy, L., Friedman, E., et al. 1992. Somatic mutations in the neurofibromatosis 1 gene in human tumors. *Cell* 69: 275–81

Lisch, K. 1937. Über Beteiligung der Augen, insbensondere das Vorkommen von Irisknotchen bei der Neurofibromatose (Recklinghausen). *Z. Augenheil.* 93: 137–43

Liu, E., Hjelle, B., Morgan, R., Hecht, F., Bishop, J. M. 1987. Mutation of the Kirsten *ras* proto-oncogene in human preleukemia. *Nature* 330: 186–88

Lyons, J., Janssen, J. W. G., Bartram, C., Layton, M., Mufti, G. J. 1988. Mutation of Ki-*ras* and N-*ras* oncogenes in myelodysplastic syndromes. *Blood* 71: 1707–12

Marchuk, D. A., Saulino, A. M., Tavakkol, R., Swaroop, M., Wallace, M. R., et al. 1991. cDNA Cloning of the type 1 neurofibromatosis gene: complete sequence of the *NF1* gene product. *Genomics* 11: 931–40

Martin, G. A., Viskochil, D., Bollag, G., McCabe, P. C., Crosier, W. J., et al. 1990. The GAP-related domain of the *NF1* gene product interacts with *ras* p21. *Cell* 63: 843–49

Martin, G. A., Yatani, A., Clark, R., Conroy, L., Polakis, P., et al. 1992. GAP domains responsible for *ras* p21-dependent inhibition of muscarinic atrial K⁺ channel currents. *Science* 255: 192–94

McCormick, F. 1989. *ras* GTPase activating protein: signal transmitter and signal terminator. *Cell* 56: 5–8

Menon, A. G., Anderson, K. M., Riccardi, V. M., Chung, R. Y., Whaley, J. M., et al. 1990. Chromosome 17p deletions and p53 gene mutations associated with the formation of malignant neurofibrosarcomas in von Recklinghausen neurofibromatosis. *Proc. Natl. Acad. Sci. USA* 87: 5435–39

Mikol, D., Gulcher, J., Stefansson, K. 1990. The oligodendrocyte-myelin glycoprotein belongs to a distinct family of proteins and contains the HNK-1 carbohydrate. *J. Cell Biol.* 110: 471–80

Mitts, M. R., Bradshaw-Rouse, J., Heideman, W. 1991. Interactions between adenylate cyclase and the yeast GTPase-activating protein *IRA1*. *Mol. Cell. Biol.* 11: 4591–98

Mulvihill, J. J., Parry, D. M., Sherman, J. L., Pikus, A., Kaiser-Kupfer, M. I., et al. 1990. Neurofibromatosis 1 (Recklinghausen disease) and neurofibromatosis 2 (Bilateral acoustic neurofibromatosis). *Ann. Int. Med.* 113: 39–52

Nakagawa, T., Mabry, M., de Bustros, A., Ihle, J. N., Nelkin, B. D., Baylin, S. B. 1987. Introduction of v-Ha-*ras* induces differentiation of cultured human medullary thyroid carcinoma cells. *Proc. Natl. Acad. Sci. USA* 84: 5923–27

Nishi, T., Lee, P. S. Y., Oka, K., Levi, V. A., Tanase, S., et al. 1991. Differential expression of two types of the neurofibromatosis type 1 (*NF1*) gene transcripts related to neuronal differentiation. *Oncogene* 6: 1555–59

Noda, M., Ko, M., Ogura, A., Liu, D. G., Amano, T., et al. 1985. Sarcoma viruses carrying ras oncogenes induce differentiation-associated properties in a neuronal cell line. *Nature* 318: 73–75

O'Connell, P., Leach, R., Cawthon, R., Culver, M., Stevens, J., et al. 1989a. Two *NF1* translocations map within a 600-kilobase segment of 17q11.2. *Science* 344: 1087–88

O'Connell, P., Leach, R., Ledbetter, D., Cawthon, R., Culver, M., et al. 1989b. Fine structure mapping studies of the chromosomal region harboring the genetic defect in neurofibromatosis type 1. *Am. J. Hum. Genet.* 44: 51–57

O'Connell, P., Viskochil, D., Buchberg, A., Fountain, J., Cawthon, R., et al. 1990. The human homologue of murine *evi-2* lies between two translocation breakpoints associated with von Recklinghausen neurofibromatosis. *Genomics* 7: 547–54

Padua, R. A., Carter, G., Hughes, D., Gow, J., Farr, C., et al. 1988. *RAS* mutations in myelodysplasia detected by amplification, oligonucleotide hybridization, and transformation. *Leukemia* 2: 503–10

Ponder, B., Xu, W., Ponder, M., Mathew, C., Smith, B. 1990. *Am. J. Hum. Genet.* (Abstr.) 47(Suppl.): A14

Riccardi, V. M., Eichner, J. E. 1986. *Neurofibromatosis: Phenotype, Natural History, and Pathogenesis.* Baltimore: Johns Hopkins Univ. Press

Ridley, A. J., Paterson, H. F., Noble, M., Land, H. 1988. *ras*-mediated cell cycle arrest is altered by nuclear oncogenes to induce Schwann cell transformation. *EMBO J.* 7: 1635–45

Rubenstein, A. E., Korf, B. R., eds.

1990. *Neurofibromatosis: Handbook for Patients, Families, and Health-Care Professionals.* New York: Thieme. 256 pp.

Schmidt, M. A., Michels, V. V., Dewald, G. W. 1987. Cases of neurofibromatosis with rearrangements of chromosome 17 involving band 17q11.2. *Am. J. Med. Genet.* 28: 771–75

Seizinger, B. R., Rouleau, G. A., Ozelius, L. J., Lane, A. H., Faryniarz, A. G., et al. 1987. Genetic linkage of von Recklinghausen neurofibromatosis to the nerve growth factor receptor gene. *Cell* 49: 589–94

Simon, M. A., Bowtell, D. D. L., Dodson, G. S., Laverty, T. R., Rubin, G. M. 1991. *Ras1* and a putative guanine nucleotide exchange factor perform crucial steps in signaling by the *sevenless* protein tyrosine kinase. *Cell* 67: 701–16

Skuse, G. R. 1990. Technical comments. (Lett.) *Science* 250: 1749

Skuse, G., Kosciolek, B., Rowley, P. 1991. The neurofibroma in von Recklinghausen neurofibromatosis has a unicellular origin. *Am. J. Hum. Genet.* 49: 600–7

Skuse, G. R., Kosciolek, B. A., Rowley, P. T. 1989. Molecular genetic analysis of tumors in von Recklinghausen neurofibromatosis: loss of heterozygosity for chromosome 17. *Genes Chromosomes Cancer* 1: 36–41

Sørensen, S. A., Mulvihill, J. J., Nielsen, A. 1986. Long-term follow-up of von Recklinghausen neurofibromatosis. Survival and malignant neoplasms. *N. Engl. J. Med.* 314: 1010–15

Stacey, D. W., Tsai, M. H., Yu, C. L., Smith, J. K. 1988. Signal transduction. In *Cold Spring Harbor Symp. Quant. Biol.,* LIII: 871–81. New York: Cold Spring Harbor

Stark, M., Assum, G., Krone, W. 1991. A small deletion and an adjacent base exchange in a potential stem-loop region of the neurofibromatosis 1 gene. *Hum. Genet.* 87: 685–87

Suzuki, Y., Suzuki, H., Kayama, T., Yoshimoto, T., Shibahara, S. 1991. Brain tumors predominantly express the neurofibromatosis type 1 gene transcripts containing the 63-base insert in the region coding for GTPase activating protein-related domain. *Biochem. Biophys. Res. Commun.* 181: 955–61

Tong, A. K. F., Fitzpatrick, T. B. 1990. The skin in neurofibromatosis. See Rubenstein & Korf 1990, pp. 88–98

Upadhyaya, M., Cheryson, A., Broadhead, W., Fryer, A., Shaw, D. J., et al. 1990.

A 90-kb DNA deletion associated with neurofibromatosis type 1. *J. Med. Genet.* 27: 738–41

Viskochil, D. H., Buchberg, A. M., Xu, G., Cawthon, R. M., Stevens, J., et al. 1990. Deletions and a translocation interrupt a cloned gene at the neurofibromatosis type 1 locus. *Cell* 62: 187–92

Viskochil, D., Cawthon, R., O'Connell, P., Xu, G., Stevens, J., et al. 1991. The gene encoding the oligodendrocyte-myelin glycoprotein is embedded within the neurofibromatosis type 1 gene. *Mol. Cell. Biol.* 11: 906–12

Wallace, M. R., Anderson, L. B., Saulino, A. M., Gregory, P. E., Glover, T. W., et al. 1991. A de novo Alu insertion results in neurofibromatosis type 1. *Nature* 353: 864–66

Wallace, M. R., Marchuk, D. A., Anderson, L. B., Letcher, R., Odeh, H. M., et al. 1990. Type 1 neurofibromatosis gene: Identification of a large transcript disrupted in three patients. *Science* 249: 182–86

Wang, Y., Boguski, M., Riggs, M., Rodgers, L., Wigler, M. 1991. *Sar1,* a gene from *Schizosaccharomyces pombe* encoding a GAP-like protein that regulates *ras1. Cell Reg.* 2: 453–65

White, R., Nakamura, Y., O'Connell, P., Leppert, M., Lalouel, J.-M., et al. 1987. Tightly linked markers for the neurofibromatosis type 1 gene. *Genomics* 1: 364–67

Xu, G., Lin, B., Tanaka, K., Dunn, D., Wood, D., et al. 1990a. The catalytic domain of the NF1 gene product stimulates *ras* GTPase and complements *IRA* mutants of *S. Cerevisiae. Cell* 63: 835–41

Xu, G., O'Connell, P., Viskochil, D., Cawthon, R., Robertson, M., et al. 1990b. The neurofibromatosis type 1 gene encodes a protein related to GAP. *Cell* 62: 599–608

Yamaguchi-Iwai, Y., Satake, M., Murakami, Y., Sakai, M., Muramatsu, M., Ito, Y. 1990. Differentiation of F9 embryonal carcinoma cells induced by the *c-jun* and activated c-Ha-*ras* oncogenes. *Proc. Natl. Acad. Sci. USA* 87: 8670–74

Yokota, J., Akiyama, T., Fung, Y. K., Benedict, W. F., Namba, Y., et al. 1988. Altered expression of the retinoblastoma (RB) gene in small-cell carcinoma of the lung. *Oncogene* 3: 471–75

Yu, C. L., Tsai, M. H., Stacey, D. W. 1988. Cellular *ras* activity and phospholipid metabolism. *Cell* 52: 63–71

Annu. Rev. Neurosci. 1993. 16:207–22

THE ROLE OF NMDA RECEPTORS IN INFORMATION PROCESSING

N. W. Daw

Department of Ophthalmology, Yale University School of Medicine, New Haven, Connecticut 06510

P. S. G. Stein

Department of Biology, Washington University, St. Louis, Missouri 63130

K. Fox

Department of Physiology, University of Minnesota Medical School, Minneapolis, Minnesota 55455

KEY WORDS: glutamate receptors, amplification, temporal summation, motor rhythms, synaptic transmission

INTRODUCTION

Why have another review on N-methyl-D-aspartate (NMDA) receptors? More than 100 of them have appeared in the last few years, in addition to an excellent book (Watkins & Collingridge 1989). However, most have concentrated on the biophysics, cell biology, and pharmacology of the NMDA receptor and its channel (Collingridge & Lester 1989; Mayer & Westbrook 1987; Thomson 1990), the role of the NMDA receptor in long-term potentiation (LTP) and plasticity (Collingridge & Bliss 1987; Constantine-Paton et al 1990; Rauschecker 1991; Tsumoto 1990), and the role of the NMDA receptor in pathophysiology and cell death (Choi & Rothman 1990; McDonald & Johnston 1990). Only a few (e.g. Dale in Watkins & Collingridge 1989) have discussed the topic of this review— the role of NMDA receptors in normal information processing.

207

0147–006X/93/0301–0207$02.00

The NMDA receptor is activated by glutamate, the excitatory transmitter found in all parts of the central nervous system (CNS). It has several properties that distinguish it from most other receptors. The prime one is that it is both ligand gated and voltage sensitive. As a result, it acts differently from most other receptors. We list its properties and attempt to draw a connection, as far as is possible in 1992, between the properties of the receptor and its channel, and the effect of NMDA agonists and antagonists on information processing in various parts of the CNS.

GENERAL PROPERTIES OF NMDA RECEPTORS

The NMDA Channel Has Voltage-Dependent Properties

The amount of current passed by the NMDA channel is reduced when the cell membrane is hyperpolarized beyond -35 mV, and is quite small at the resting potential of the cell, around -70 mV (Flatman et al 1986; MacDonald & Wotjowicz 1982). In other words, the conductance has a negative slope at membrane voltages more negative than -35 mV, which is caused by the block of the NMDA channel by Mg^{++} (Mayer & Westbrook 1985; Nowak et al 1984). Consequently, in resting slices and isolated cells bathed in physiological concentrations of Mg^{++}, little current flows through the NMDA channel. However, current definitely flows through the channel in live animals in neurons that are usually depolarized from resting potential to near the threshold for firing action potentials (around -55 mV) by incoming excitatory activity.

NMDA Channels Let Calcium into The Cell

Simultaneous intracellular and extracellular recordings of frog motoneurons with ion-sensitive electrodes show that two calcium fluxes can be distinguished: one in response to glutamate that occurs immediately without substantial depolarization, and another in response to depolarization further than -25 mV (Buhrle & Sonnhof 1983). NMDA elicits entry of calcium into the cell (Dingledine 1983), and the calcium that enters the cell in response to glutamate when the membrane is clamped to -60 mV results primarily from NMDA receptors (Mayer et al 1987). How much of the entry of calcium in response to glutamate in a real life situation is caused by NMDA receptors, and how much by voltage-dependent calcium and other glutamate channels, remains to be determined.

NMDA and Non-NMDA Receptors Are Often Found at the Same Synapses

When glutamate pathways in the nervous system are stimulated, both NMDA and non-NMDA contributions are frequently found in the responses. This occurs because both NMDA and non-NMDA receptors are present at the synapses involved. Cells from the spinal cord or hippocampus can be cultured together, and they form synapses across which both NMDA and non-NMDA components are found (Forsythe & Westbrook 1988; Mayer & Westbrook 1984). A similar result is obtained with cells cultured from visual cortex (Jones & Baughman 1988). The point can also be demonstrated by stimulating a cell in a visual cortex slice and recording from a cell that is monosynaptically connected to it (Thomson et al 1989). Seventy percent of synapses on hippocampal cells contain both NMDA and non-NMDA receptors (Bekkers & Stevens 1990), and hot spots for the NMDA effect of glutamate on visual cortex cells in culture are found in the same location as hot spots for the non-NMDA effects of glutamate (Jones & Baughman 1991). In some areas, however, NMDA and non-NMDA receptors may have separate locations (Fox et al 1989; Sillar & Roberts 1991).

The Response to Glutamate at the NMDA Receptor Has a Slow Rise and a Prolonged Effect

Where both NMDA and non-NMDA receptors are found, the synaptic potential has two components: a fast component due to non-NMDA receptors and a slow component due to NMDA receptors (Dale & Roberts 1985; Forsythe & Westbrook 1988). This is reflected in extracellular recordings (Salt & Eaton 1989). The NMDA component of the response has a slow rise time and a prolonged effect that can last 500 msec (Collingridge et al 1988; Forsythe & Westbrook 1988). After the initial activation, the prolonged effect can be shortened by Mg^{++}, which blocks the ion flow through the NMDA channels, but not by aminophosphonovalerate (APV), which antagonizes the binding of glutamate to the NMDA receptor (Hestrin et al 1990; Lester et al 1990). The conclusion is that glutamate activates NMDA channels that stay open for a long period of time. As a result, NMDA and its antagonists tend to affect processes that have a low frequency.

NMDA Agonists Multiply Responses, Whereas Non-NMDA Agonists Add to Them

The manner in which NMDA and non-NMDA receptors contribute to a response can be studied by iontophoresing agonists or antagonists while various levels of response are quantified. This has been done in the visual cortex by measuring the response as a function of the contrast of the stimulus (Fox et al 1990). NMDA increases the slope of the contrast-response curve, whereas quisqualate moves the whole curve upwards.

This result can be partly, but not completely, explained by the voltage-dependence of the NMDA receptor (Fox & Daw 1992).

NMDA Antagonists Alter or Abolish Function in a Wide Variety of Systems

Because glutamate is the main excitatory transmitter in the nervous system, and NMDA receptors are found at most glutamate synapses, one would expect this point to be true. An exhaustive list of all cases in which it has been true would take several hundred references (see Watkins & Collingridge 1989). The remainder of this review discusses a few cases in which particular aspects of the function can be tied into particular properties of the NMDA receptor.

SENSORY INFORMATION PROCESSING

Some sensory pathways use primarily non-NMDA receptors for synaptic transmission, such as inputs to layer IV of the cortex in adult animals (Armstrong-James 1989; Fox et al 1989) and $A\beta$ primary afferent terminations in spinal cord (Dickenson & Sullivan 1987; Haley et al 1990). In these cases, sensory-evoked spike activity is either little affected or completely unaffected by APV treatment. This does not rule out sub-threshold NMDA components, but implies that non-NMDA transmission is dominant. Other sensory pathways employ both non-NMDA and NMDA receptors. For example, in the ventrobasal thalamus, the phasic component of the response to tactile stimulation is mediated by non-NMDA receptors, whereas the long-duration component is due to the NMDA excititory postsynaptic potential (EPSP) (Salt & Eaton 1989, 1991). The temporal characteristics of the NMDA component of the sensory response are presumably important for the function of the system in which they are employed, but this function has not yet been clearly defined in most cases. Three examples are cited below, from the auditory, somatosensory, and visual systems, to give some insight into how NMDA receptors function in a complex sensory system.

NMDA Receptors Amplify Auditory Responses During Echo-Location

In the mustached bat, neurons in the medial geniculate nucleus (MGN) of the thalamus respond to pairs of sounds separated by particular time intervals (Suga et al 1990). Pairs of sounds occur naturally during echo-location. The first sound (pulse) is the initial emission of sound by the bat, and the second (echo) is the reflected sound from the target. Different cells are tuned to different pulse-echo intervals ranging from approximately 0 to 25 ms for different cells, which corresponds to distances of approximately 0–4.5 meters. Thus, target range is encoded among different groups of cells within the MGN.

The basic response of the delay-tuned neurons is to fire a burst of spikes starting at a set latency and lasting between 0 and 30 ms, depending on the strength of the response. Only the initial spike in the burst is solely dependent on non-NMDA receptors. Applying APV iontophoretically to these cells abolishes the later spikes in the train without affecting the first spike of the burst, whereas iontophoresing CNQX only abolishes the first spike in the burst without affecting the later spikes (Butman 1992). Over the range of pulse-echo delays to which these cells respond (they typically have a 2.5–7.5 ms tuning width), the burst of spikes increases in intensity toward the optimum delay and fades to a single spike at pulse-echo intervals at the edge of the response range. The effect of inactivating NMDA-receptors is, therefore, to flatten the tuning curve by reducing responses to all pulse-echo intervals within the tuning range to a single spike (Butman 1992; Suga et al 1990). However, because delay information is encoded by probability of response, the half-width of the tuning curve is only expanded by about 20% when the NMDA component of the response is abolished. The main effect of the NMDA component is, therefore, to amplify the response and convert a probability of firing code into an intensity of firing code.

The exact mechanism by which the tuning curve is produced is unknown at present, but appears to depend on two general NMDA properties: amplification and long-duration EPSPs. The increase in response toward the center of the tuning curve is probably achieved because the NMDA-conductance activated by the initial pulse amplifies the synaptic input generated by the echo. Amplification is possible because of the voltage-dependent properties of the NMDA channel. The large number of spikes encoding the response is probably due to the long duration of the NMDA EPSP.

NMDA Receptors Are Involved in Central Sensitization During Nociception

It has long been known that C-fiber inputs to the spinal cord and brainstem are involved in nociception (Adrian 1931). However, only recently has it been established that glutamate is colocalized with substance-P in their terminals (Battaglia & Rustioni 1988). In addition to the peptidergic, serotonergic, and noradrenergic mechanisms that process tactile nociceptive input in the dorsal horn, glutaminergic transmission also appears to play a role, in part by activating NMDA receptors. Three related phenomena depend on NMDA-receptors: "wind-up" of dorsal horn neurons, the formalin response of dorsal horn neurons, and the modulation of the flexion reflex. All three responses require C-fiber activation for induction, develop slowly over time, are characterized by an amplification of responses to tactile input, and have an enduring quality. All three require NMDA receptors for their induction; in the case of wind-up, induction is also frequency dependent.

Wind-up is characterized by an increased response of cells in laminae IV and V of the dorsal horn after electrical stimulation of primary afferents (Mendell 1966). The stimuli must be of sufficient intensity to recruit C-fibers; just A fiber stimulation will not invoke wind-up. The stimuli must also be delivered at a frequency of at least 0.3 Hz (Dickenson & Sullivan 1990; Mendell 1966). The effect is developed only after several stimuli. On removing the stimuli, the cells continue to produce an after-discharge of spikes that can last for minutes. If the NMDA receptors in the dorsal horn are inactivated by APV before applying stimuli capable of inducing wind-up, only the initial level of response occurs, and the amplification does not develop (Dickenson & Sullivan 1990).

Several results tie this NMDA-dependent property to nociceptive mechanisms. The formalin response, produced by subcutaneous injection of formalin, has a two-phase effect on dorsal horn neurons. The first phase occurs within seconds and consists of a large increase in firing rate that subsides within about ten minutes. The second phase occurs after approximately 25 minutes and can last up to one hour. Only the second phase is related to nociceptor activation (see Dickenson 1990) and blocked by APV (Haley et al 1990). The flexion reflex is produced by pinching the cutaneous surface of the foot and can be increased in vigor by prior activation of C-fibers in the relevant dorsal roots by electrical stimulation (Woolf 1983). This effect has a similar time course to wind-up and, like wind-up, is prevented by systemic NMDA antagonists in a decerebrate animal (Woolf & Thompson 1991).

Two general properties of NMDA receptor action appear to be important in this nociceptive response: the characteristics of response amplification and frequency dependence. In the dorsal horn, the amplification property is necessary for expression of wind-up, whereas frequency dependence is necessary for induction.

NMDA Receptors Amplify Excitatory and Inhibitory Visual Signals

In the visual cortex, NMDA receptors are involved in response amplification over the natural range of visual stimuli. The effect of blocking NMDA receptors is to decrease the response by a constant proportion that is independent of the intensity of the driving input (within the normal response range). This was discovered by varying the input drive to the cell by altering the stimulus contrast, then blocking NMDA receptors iontophoretically with APV. Increasing NMDA-receptor activation increases the response, also by a constant proportion, irrespective of the intensity of the input drive (Fox et al 1990). Quisqualate applied in the same way causes an increase in firing rate by summating with the response, rather than by amplifying it. This fact, together with the local nature of iontophoretic drug application, suggests that the amplification is caused by the intrinsic properties of the NMDA receptors on the cell, rather than a neuronal circuit effect. However, the voltage dependence of the NMDA receptor is not sufficient on its own to account for the effect (Fox & Daw 1992).

Response amplification by NMDA also occurs in layers V and VI of the adult cortex, where the visual response seems to be mediated predominantly by non-NMDA receptors (Fox et al 1989, 1990). An endogenous modulator of these receptors remains to be found, but the prime candidate is the input from the intralaminar nuclei of the thalamus that terminates in layers V and VI (Herkenham 1980) and acts via NMDA receptors (Fox & Armstrong-James 1986). This would provide a mechanism for amplifying the output of the cortex.

Lateral inhibitory mechanisms are prominent in the thalamus. As a result, surround inhibitory receptive fields are more powerful in the lateral geniculate nucleus (LGN) than in the retina. It appears that the NMDA-dependence of the excitatory response seems particularly sensitive to thalamic inhibition. If APV is applied iontophoretically to a LGN cell, it will have an attenuating effect on the center receptive field response. However, the decrease in response caused by APV is far smaller for a large stimulus that encroaches on the inhibitory surround receptive field than for a stimulus confined to the center of the receptive field (Kwon et al 1992). This presumably occurs because NMDA receptors are suppressed when intrathalamic inhibitory mechanisms are activated. Hence, NMDA receptors are inhibited more than non-NMDA receptors by the inhibitory surround. The effect is not due to NMDA receptors being driven proportionally less by a larger stimulus, because APV affects large and small responses by a constant proportion over the normal input range of a neuron, both in cortex (Fox et al 1990) and LGN (Kwon et al 1992).

The NMDA receptors appear to amplify the effect of lateral inhibition in the LGN. Two general properties of NMDA channels make NMDA dependent EPSPs particularly sensitive to inhibition: their voltage sensitivity and the slow rise-time of the conductance increase. The NMDA conductance decreases substantially between -30 and -70 mV and is, therefore, prone to attenuation by $GABA_A$ and $GABA_B$ receptors with reversal potentials near -60 and -90 mV, respectively. The slow rise time of the NMDA conductance means that it is likely to occur at the same time as any disynaptic IPSPs evoked by the same stimulus and be suppressed by them, as the inhibitory postsynaptic potentials (IPSPs) have a similar slow time course (Connors 1992).

CONTRIBUTIONS OF NMDA RECEPTORS TO MOTOR RHYTHMS

NMDA and Non-NMDA Glutamate Agonists Activate Motor Rhythms

Direct application of NMDA to a neural circuit can activate rhythmic bursting motor output resembling that recorded during naturally occurring motor behaviors, e.g. swimming (lamprey, Grillner et al 1991; frog embryo, Roberts et al 1986), stepping (neonatal rat, Smith et al 1988), swallowing (rat, Kessler & Jean 1991), embryonic spontaneous motor output (chick, Barry & O'Donovan 1987), or a sequence of several behaviors, such as wiping, stepping, and jumping (frog, McClellan & Farel 1985). Direct application of non-NMDA glutamate agonists, kainic acid (frog embryo swim, Dale & Roberts 1984; lamprey swim, Brodin et al 1985), quisqualate (rat swallow, Kessler & Jean 1991), or AMPA (lamprey swim, Alford & Grillner 1990), can also evoke rhythmic motor output. Lamprey swimming occurs over a wide frequency range (Brodin et al 1985): NMDA-evoked swim rhythms are in the lower part, and non-NMDA-evoked swim rhythms are in the higher part of this range; activation of either class of receptor can evoke swim rhythms in the middle part of the frequency range. APV blocks NMDA-evoked rhythmic motor output, but not non-NMDA-evoked rhythmic output (Alford & Grillner 1990; Brodin et al 1985; Dale & Roberts 1984; Kessler & Jean 1991). CNQX blocks non-NMDA-evoked rhythmic output, but not NMDA-evoked rhythmic output (Alford & Grillner 1990; Kessler & Jean 1991). Thus, multiple and parallel mechanisms for rhythmicity can exist. For lower frequency swim rhythms in the lamprey, NMDA receptors may play a particularly critical role. In other situations, NMDA receptors may contribute to activation and generation of rhythms, but other mechanisms could also be very important.

Several important characteristics of NMDA receptors in individual cells contribute to rhythm activation and generation. First, temporal summation of NMDA receptor EPSPs can produce a tonic level of excitability in neurons of the circuit generating the motor rhythm (Dale & Roberts 1985). This tonic level of excitability serves as an excitatory drive for these neurons; in addition, it produces phasic rebound excitation following phasic glycinergic inhibition (see model of circuit by Roberts & Tunstall 1990). Thus, rebound excitation and non-NMDA excitation serve as two distinct sources of phasic excitation that contribute to the rhythmicity of the frog embryo swim circuit. Second, individual neurons may generate rhythmic bursting in response to application of NMDA (e.g. Durand 1991; Grillner et al 1991; Hu & Bourque 1992; Tell & Jean 1991). In addition, when sodium action potentials and synaptic interactions among neurons of a circuit are attenuated by tetrodotoxin, individual members of the circuit can display NMDA-evoked rhythmic oscillations (Grillner et al 1991; Sigvardt et al 1985; Tell & Jean 1991; Wallen & Grillner 1987). Specific manipulations, e.g. apamin blockade of calcium-dependent potassium channels or removal of extracellular calcium or magnesium, support the idea that NMDA-evoked cellular rhythms depend upon calcium entry via both NMDA channels and voltage-gated calcium channels followed by opening of calcium-dependent potassium channels (Hu & Bourque 1992; Wallen & Grillner 1987). Neural models that include the voltage dependency and the calcium permeability of NMDA channels reproduce many features of cellular rhythmicity obtained in actual recordings (Brodin et al 1991).

NMDA Receptor Blockade Diminishes Motor Rhythms in Some Preparations

The contribution of NMDA receptors to the activation and generation of motor rhythms has been evaluated by using NMDA receptor antagonists. A variety of results have been obtained. In some preparations, NMDA receptors may either be part of a rhythm-generating circuit, or may contribute to the activation of sensory circuits that activate rhythm-generating circuits, or both. In other preparations, no contribution of NMDA receptors to the activation or generation of rhythmicity has been observed.

NMDA receptor blockade produces various effects in preparations that generate motor rhythms spontaneously. Specifically, blockade does not modify in vitro respiratory rhythms (neonatal rat, Greer et al 1991); alters respiratory rhythms in vivo by producing prolonged inspiration (cat, Foutz et al 1989); lowers the frequency of spontaneous motor activity (chick embryo, Barry & O'Donovan 1987), and abolishes spontaneous low-frequency swim motor rhythms (lamprey, Brodin & Grillner 1985). NMDA receptor blockade also modifies characteristics of sensory-evoked locomotor rhythms (rabbit, Fenaux et al 1991). When a lamprey swim motor rhythm is activated by sensory stimulation, NMDA receptor blockade results in an attenuation of the long-latency, low-frequency portions of the response (Brodin & Grillner 1985; Grillner et al 1991). When glycinergic inhibition in a circuit is diminished, NMDA receptor blockade produces dramatic attenuations of sensory-evoked motor rhythms (frog embryo, Soffe 1989; lamprey, Alford & Williams 1989).

In some of the above experiments, attenuations of excitability may result either from a diminution of NMDA receptor activation in the sensory-processing part and/or in the motor-pattern generating part of the circuit. In other experiments, only a specific portion of the circuit is exposed to the NMDA antagonist. NMDA receptor blockade of spinal cord segments that process rostral scratch cutaneous information results in reductions of sensory-evoked rostral scratch motor output (turtle, Currie & Stein 1992). This effect is seen even when cutaneous afferents are stimulated at low frequency (0.20–0.25 Hz). Attenuation of motor output is associated with diminished activity in spinal neurons that exhibit long after-discharge; activity of these neurons may contribute to the excitability of rostral scratch motor circuits (Currie & Stein 1992). When NMDA receptors are blocked in motor-pattern generating parts of a circuit (and parts of the circuit processing the specific sensory input used to activate the motor pattern are not exposed to NMDA receptor blockade), there is a small diminution of frog embryo swim motor excitability (Dale & Roberts 1984) and no observed diminution of turtle rostral scratch motor excitability (Stein & Schild 1989).

Intersegmental Latencies of Motor Rhythms Can Be Manipulated by NMDA

The motor output in a caudal segment of the spinal cord is delayed from that in a rostral segment, and the difference is called intersegmental latency. When NMDA is applied to a caudal segment in the frog embryo, a decrease in this latency is observed; when NMDA receptors are blocked, an increase is seen (Tunstall & Roberts 1991). Changes in intersegmental latency are also observed in the lamprey when the NMDA concentration applied to rostral spinal segments differs from that applied to caudal spinal segments (Matsushima & Grillner 1992).

Additional work is needed to understand the contributions of NMDA receptors to motor rhythm generation. In some systems, NMDA bath application is sufficient to generate a motor rhythm, and most circuit elements have NMDA receptors; thus, in these systems, NMDA receptors must contribute to motor rhythms. However, NMDA receptors are not always required for rhythmicity; in many systems, motor rhythms are produced following NMDA receptor blockade. Measurements of the contributions of NMDA receptors to those aspects of motor rhythms with time constants near the decay time constant of the NMDA-evoked EPSP will be of special interest in future experiments.

NMDA and Non-NMDA Receptors Affect Different Behaviors of Electric Organ Discharge

Electric fish utilize pulses of electric organ discharge for navigation and communication. Glutamatergic neurotransmission in brain stem circuitry contributes to the electric organ frequency modulations observed during natural behavior. NMDA receptors play a critical role in slow modulations of electric organ frequency that occur during such behaviors as the jamming avoidance response of *Eigenmannia* (Dye et al 1989; Kawasaki & Heiligenberg 1990). In contrast, non-NMDA receptors play a critical role in the more rapid frequency changes of the electric organ observed during certain classes of social behavior, such as "chirping" (Dye et al 1989; Kawasaki & Heiligenberg 1990). The specific modulations of electric organ frequency by either NMDA or non-NMDA receptors are consistent with the temporal characteristics of each of these receptors.

NMDA Receptors Affect Vasomotor Tone and Blood Pressure

Sympathetic preganglionic efferents in specific spinal cord segments control vasomotor tone important for the regulation of blood pressure. NMDA receptor blockade of these spinal segments lowers resting blood pressure by diminishing resting vasomotor output (Bazil & Gordon 1991; Hong & Henry 1992); in addition, blood pressure increases elicited by a variety of stimulation procedures are attenuated by this NMDA receptor blockade (Bazil & Gordon 1991). The particular temporal characteristics of the NMDA receptor may well be suited for the time constants observed during blood pressure modulations.

SUMMARY AND CONCLUSIONS

In this review, we have concentrated on the parallels between the cellular properties of the NMDA receptor and a variety of functional properties within sensory and motor systems. Of course, the NMDA channel exists within the cell in conjunction with a variety of other channels, including non-NMDA channels. Although the NMDA receptor is unique in a cellular sense—it is the only ligand-gated channel that is also voltage dependent and calcium permeable—it is not unique in a functional sense. A cell that has non-NMDA receptors and voltage-sensitive channels will also exhibit nonlinear behavior. Moreover, Buhrle & Sonnhof (1983) demonstrated some time ago that calcium flows into frog motor neurons through more than one type of calcium channel. The contribution to the inflow of calcium from NMDA channels may vary from cell to cell and could easily be a minor proportion of the total.

Many authors have pointed out that the NMDA channel has a low conductance at a resting potential of -70 mV. However, many cells in the nervous system are depolarized from -70 mV by excitatory input. Thus, as pointed out above, NMDA receptors make a contribution to the tonic or spontaneous activity of cells in both visual cortex and spinal cord. In practice, many cells are probably working in a range of membrane potentials where the NMDA channels are always open to some extent. Even in the hippocampal slice where a substantial amount of afferent input is removed, NMDA receptors contribute to spontaneous activity (Sah et al 1989).

Does the NMDA receptor act as a switch? Does it act as an AND gate? The suggestion that it may act as a switch comes from work on LTP in the hippocampus, which is readily produced by high-frequency stimulation and is abolished by APV. However, activation of the NMDA receptor is only the first in a sequence of reactions leading to LTP: In theory, switch-like behavior could also be produced by calcium-buffering systems within dendritic spines, or by enzymatic processes (Lisman 1985; Zador et al 1990). Fox & Daw (1992) have modeled the action of NMDA and non-NMDA receptors that are activated in parallel with each other, and shown that the occurrence of switch-like behavior depends on the relative density of NMDA versus non-NMDA receptors. Switch-like behavior is not seen in the visual cortex, but might be seen in the hippocampus if the relative density of NMDA receptors there was higher than in the visual cortex. Our review did not turn up any specific examples of situations in which switch-like behavior or AND gate behavior has been proved to be caused by NMDA receptors. In fact, quite the opposite: We found that, far from creating a sharp threshold, the voltage dependence of the NMDA receptor generally just leads to an amplification. This area is clearly one for fruitful investigation in the future.

On the positive side, several cellular properties of the NMDA channel—slow rise time, prolonged effect, temporal summation, and amplification of responses due to the voltage dependency—can be related to function. This is particularly true when the function is frequency specific, as in motor rhythms, or develops slowly over time, as in nociception. There are also situations in which NMDA receptors contribute to a low frequency function, and non-NMDA receptors to a high frequency function—jamming avoidance response and "chirping" in the fish electric organ; different frequencies of oscillation in the spinal cord—in systems where both types of receptor are found in parallel with each other. Whereas the role of NMDA receptors in plasticity, LTP, and cell death is now quite well established, investigation of their role in information processing could continue to be fruitful for some time to come.

ACKNOWLEDGMENTS

During the preparation of this review, Nigel Daw was supported by National Institutes of Health (NIH) grants EY00053 and NS29343, Paul Stein by National Science Foundation grant BNS-89-08144, and Kevin Fox by NIH grant NS27759.

Literature Cited

Adrian E. D. 1931. The messages in sensory nerve fibres and their interpretation. *Proc. R. Soc. London Ser. B* 109: 1–18

Alford, S., Grillner, S. 1990. CNQX and DNQX block non-NMDA synaptic transmission but not NMDA-evoked locomotion in lamprey spinal cord. *Brain Res.* 506: 297–302

Alford, S., Williams, T. L. 1989. Endogenous activation of glycine and NMDA receptors in lamprey spinal cord during fictive locomotion. *J. Neurosci.* 9: 2792–2800

Armstrong-James, M. A. 1989. NMDA and non-NMDA excitatory amino acid transmission in construction of receptive fields of rat barrel field neurons. *Soc. Neurosci. Abstr.* 15: 949

Barry, M. J., O'Donovan, M. J. 1987. The effects of excitatory amino acids and their antagonists on the generation of motor activity in the isolated chick spinal cord. *Dev. Brain Res.* 36: 271–76

Battaglia, G., Rustioni, A. 1988. Coexistence of glutamate and substance P in dorsal root ganglion cells of the rat and monkey. *J. Comp. Neurol.* 277: 302–12

Bazil, M. K., Gordon, F. J. 1991. Effect of blockade of spinal NMDA receptors on sympathoexcitation and cardiovascular responses produced by cerebral ischemia. *Brain Res.* 555: 149–52

Bekkers, J. M., Stevens, C. F. 1990. NMDA and non-NMDA receptors are co-localized at individual excitatory synapses in cultured rat hippocampus. *Nature* 341: 230–33

Brodin, L., Grillner, S. 1985. The role of putative excitatory amino acid neurotransmitters in the initiation of locomotion in the lamprey spinal cord. I. The effects of excitatory amino acid antagonists. *Brain Res.* 360: 139–48

Brodin, L., Grillner, S., Rovainen, C. M. 1985. N-methyl-D-aspartate (NMDA), kainate and quisqualate receptors and the generation of fictive locomotion in the lamprey spinal cord. *Brain Res.* 325: 302–6

Brodin, L., Traven, H. G. C., Lansner, A., Wallen, P., Ekeberg, O., Grillner, S. 1991. Computer simulations of N-methyl-D-aspartate receptor-induced membrane properties in a neuron model. *J. Neurophysiol.* 66: 473–84

Burhle, C. P., Sonnhof, U. 1983. The ionic mechanism of the excitatory action of glutamate upon the membranes of motoneurons of the frog. *Pflugers Arch.* 396: 154–62

Butman, J. A. 1992. *Synaptic mechanisms for target ranging in the mustached bat* Pteronotus parnellii. PhD thesis. Washington Univ., St. Lous

Choi, D. W., Rothman S. M. 1990. The role of glutamate neurotoxicity in hypoxic-ischemic neuronal death. *Annu. Rev. Neurosci.* 13: 171–82

Collingridge, G. L., Bliss, T. V. P. 1987. NMDA receptors: their role in long-term potentiation. *Trends Neurosci.* 10: 288–93

Collingridge, G. L., Herron, C. E., Lester, R. A. J. 1988. Synaptic activation of N-methyl-D-aspartate receptors in the Schaffer collateral-commissural pathway of the rat hippocampus. *J. Physiol.* 399: 283–300

Collingridge, G. L., Lester, R. J. 1989. Excitatory amino acid receptors in the vertebrate central nervous system. *Pharmacol. Rev.* 41: 143–210

Connors, B. W. 1992. $GABA_A$ and $GABA_B$-mediated processes in visual cortex. In *Mechanisms of GABA in the Visual System*, ed. R. R. Mize, R. Marc, A. M. Sillito, pp. 335–48. Amsterdam: Elsevier

Constantine-Paton, M., Cline, H. T., Debski, E. 1990. Patterned activity, synaptic convergence, and the NMDA receptor in developing visual pathways. *Annu. Rev. Neurosci.* 13: 129–54

Currie, S. N., Stein, P. S. G. 1992. Glutamate antagonists applied to midbody spinal cord segments reduce the excitability of the fictive rostral scratch reflex in the turtle. *Brain Res.* 581: 91–100

Dale, N., Roberts, A. 1985. Dual-component amino-acid-mediated synaptic potentials: excitatory drive for swimming in *Xenopus* embryos. *J. Physiol.* 363: 35–59

Dale, N., Roberts, A. 1984. Excitatory amino acid receptors in *Xenopus* embryo spinal cord and their role in the activation of swimming. *J. Physiol.* 348: 527–43

Dickenson, A. H. 1990. Recent advances in the physiology and pharmacology of pain: plasticity and its implications for clinical analgesia. *J. Psychopharmacol.* 5: 342–51

Dickenson, A. H., Sullivan, A. F. 1990. Differential effects of excitatory amino acid antagonists on dorsal horn nociceptive neurones in the rat. *Brain Res.* 506: 31–39

Dickenson, A. H., Sullivan, A. F. 1987. Evidence for a role of the NMDA receptor in the frequency dependent potentiation of deep rat dorsal horn nociceptive neurones following C-fibre stimulation. *Neuropharmacology* 26: 1235–38

Dingledine, R. 1983. N-methyl aspartate activates voltage-dependent calcium conductance in rat hippocampal pyramidal cells. *J. Physiol.* 343: 385–405

Durand, J. 1991. NMDA actions on rat abducens motoneurons. *Eur. J. Neurosci.* 3: 621–33

Dye, J., Heiligenberg, W., Keller, C. H., Kawasaki, M. 1989. Different classes of glutamate receptors mediate distinct behaviors in a single brainstem nucleus. *Proc. Natl. Acad. Sci. USA* 86: 8993–97

Fenaux, F., Corio, M., Palisses, R., Viala, D. 1991. Effects of an NMDA-receptor antagonist, MK-801, on central locomotor programming in the rabbit. *Exp. Brain Res.* 86: 393–401

Flatman, J. A., Schwindt, P. C., Crill, W. E. 1986. The induction and modification of voltage-sensitive responses in cat neocortical neurons by N-methyl-D-aspartate. *Brain Res.* 363: 62–77

Forsythe, I. D., Westbrook, G. L. 1988. Slow excitatory postsynaptic currents mediated by N-methyl-D-aspartate receptors on mouse cultured central neurones. *J. Physiol.* 396: 515–33

Foutz, A. S., Champagnat, J., Denavit-Saubie, M. 1989. Involvement of N-methyl-D-aspartate (NMDA) receptors in respiratory rhythmogenesis. *Brain Res.* 500: 199–208

Fox, K., Armstrong-James, M. A. 1986. The role of the anterior intralaminar nuclei and N-methyl-D-aspartate receptors in the generation of spontaneous bursts in rat neocortical neurones. *Exp. Brain Res.* 63: 505–18

Fox, K., Daw, N. W. 1992. A model for the action of NMDA conductances in the visual cortex. *Neural Comput.* 4: 59–83

Fox, K., Sato, H., Daw, N. W. 1990. The effect of varying stimulus intensity on NMDA-receptor activity in cat visual cortex. *J. Neurophysiol.* 64: 1413–28

Fox, K., Sato, H., Daw, N. W. 1989. The location and function of NMDA receptors in cat and kitten visual cortex. *J. Neurosci.* 9: 2443–54

Greer, J. J., Smith, J. C., Feldman, J. L. 1991. Role of excitatory amino acids in the generation and transmission of respiratory drive in neonatal rat. *J. Physiol.* 437: 727–49

Grillner, S., Wallen, P., Brodin, L., Lansner, A. 1991. Neuronal network generating locomotor behavior in lamprey: circuitry, transmitters, membrane properties, and simulation. *Annu. Rev. Neurosci.* 14: 169–99

Haley, J. E., Sullivan, A. F., Dickenson, A. H. 1990. Evidence for spinal N-methyl-D-aspartate receptor involvement in prolonged chemical nociception in the rat. *Brain Res.* 518: 218–26

Herkenham, M. 1980. Laminar organization of thalamic projections to the rat neocortex. *Science* 207: 533–36

Hestrin, S., Sah, P., Nicoll, R. A. 1990. Mechanisms generating the time course of dual component excitatory synaptic currents recorded in hippocampal slices. *Neuron* 5: 247–53

Hong, Y., Henry, J. L. 1992. Glutamate, NMDA and NMDA receptor antagonists: cardiovascular effects of intrathecal administration in the rat. *Brain Res.* 569: 38–45

Hu, B., Bourque, C. W. 1992. NMDA receptor-mediated rhythmic bursting activity in rat supraoptic nucleus neurones in vitro. *J. Physiol.* In press

Jones, K. A., Baughman, R. W. 1991. Both NMDA and non-NMDA receptors are concentrated at synapses on cerebral cortical neurons in culture. *Neuron* 7: 593–603

Jones, K. A., Baughman, R. W. 1988. NMDA and non-NMDA-receptor components of excitatory synaptic potentials recorded from cells in layer V of rat visual cortex. *J. Neurosci.* 8: 3522–34

Kawasaki, M., Heiligenberg, W. 1990. Different classes of glutamate receptors and GABA mediate distinct modulations of a neuronal oscillator, the medullary pacemaker of a gymnotiform electric fish. *J. Neurosci.* 10: 3896–3904

Kessler, J. P., Jean, A. 1991. Evidence that activation of N-methyl-D-aspartate (NMDA) and non-NMDA receptors within the nucleus tractus solitarii triggers swallowing. *Eur. J. Pharmacol.* 201: 59–67

Kwon, Y. H., Nelson, S. B., Toth, L. J., Sur,

M. 1992. Effect of stimulus contrast and size on NMDA receptor activity in cat lateral geniculate nucleus. *J. Neurophysiol.* 68: 182–96

Lester, R. A. J., Clements, J. D., Westbrook, G. L., Jahr, C. E. 1990. Channel kinetics determine the time course of NMDA receptor-mediated synaptic currents. *Nature* 346: 565–67

Lisman, J. E. 1985. A mechanism for memory storage insensitive to molecular turnover: a bistable autophosphorylating kinase. *Proc. Natl. Acad. Sci. USA* 82: 3055–57

MacDonald, J. F., Wojtowicz, J. M. 1982. The effects of L-glutamate and its analogues upon the membrane conductance of central murine neurones in culture. *Can. J. Physiol. Pharmacol.* 60: 282–96

Matsushima, T., Grillner, S. 1992. Neural mechanisms of intersegmental coordination in lamprey: local excitability changes modify the phase coupling along the spinal cord. *J. Neurophysiol.* 67: 373–88

Mayer, M. L., MacDermott, A. B., Westbrook, G. L., Smith, S. J., Barker, J. L. 1987. Agonist- and voltage-gated calcium entry in cultured mouse spinal cord neurons under voltage clamp measured using arsenazo III. *J. Neurosci.* 7: 3230–44

Mayer, M. L., Westbrook, G. L. 1987. The physiology of excitatory amino acids in the vertebrate central nervous system. *Prog. Neurobiol.* 28: 197–276

Mayer, M. L., Westbrook, G. L. 1985. The action of N-methyl-D-aspartic acid on mouse spinal neurones in culture. *J. Physiol.* 361: 65–90

Mayer, M. L., Westbrook, G. L. 1984. Mixed-agonist action of excitatory amino acids on mouse spinal cord neurones under voltage clamp. *J. Physiol.* 354: 29–53

McClellan, A. D., Farel, P. B. 1985. Pharmacological activation of locomotor patterns in larval and adult frog spinal cords. *Brain Res.* 332: 119–30

McDonald, J. W., Johnston, M. V. 1990. Physiological and pathophysiological role of excitatory amino acids during central nervous system development. *Brain Res. Rev.* 15: 41–70

Mendell, L. M. 1966. Physiological properties of unmyelinated fiber projection to the spinal cord. *Exp. Neurol.* 16: 316–32

Nowak, L., Bregetowski, P., Ascher, P., Herbet, A., Prochiantz, A. 1984. Magnesium gates glutamate-activated channels in mouse central neurones. *Nature* 307: 462–65

Rauschecker, J. P. 1991. Mechanisms of visual plasticity: Hebb synapses, NMDA receptors, and beyond. *Physiol. Rev.* 71: 587–615

Roberts, A., Soffe, S. R., Dale, N. 1986. Spinal interneurones and swimming in frog embryos. In *Neurobiology of Vertebrate Locomotion*, ed. S. Grillner, P. S. G. Stein, D. G. Stuart, H. Forssberg, R. M. Herman, pp. 279–306. London: Macmillan

Roberts, A., Tunstall, M. J. 1990. Mutual re-excitation with post-inhibitory rebound: a simulation study on the mechanisms for locomotor rhythm generation in the spinal cord of *Xenopus* embryos. *Eur. J. Neurosci.* 2: 11–23

Sah, P., Hestrin, S., Nicoll, R. A. 1989. Tonic activation of NMDA receptors by ambient glutamate enhances excitability of neurons. *Science* 246: 815–18

Salt, T. E., Eaton, S. A. 1991. Sensory excitatory postsynaptic potentials mediated by NMDA and non-NMDA receptors in the thalamus in vivo. *Eur. J. Neurosci.* 3: 296–300

Salt, T. E., Eaton, S. A. 1989. Function of non-NMDA receptors and NMDA receptors in synaptic responses to natural somatosensory stimulation in the ventrobasal thalamus. *Exp. Brain Res.* 77: 646–52

Sigvardt, K. A., Grillner, S., Wallen, P., Van Dongen, P. A. M. 1985. Activation of NMDA receptors elicits fictive locomotion and bistable membrane properties in the lamprey spinal cord. *Brain Res.* 336: 390–95

Sillar, K. T., Roberts, A. 1991. Segregation of NMDA and non-NMDA receptors at separate synaptic contacts: evidence from spontaneous EPSPs in Xenopus embryo spinal neurons. *Brain Res.* 545: 24–32

Smith, J. C., Feldman, J. L., Schmidt, B. J. 1988. Neural mechanisms generating locomotion studied in mammalian brain stem-spinal cord in vitro. *FASEB J.* 2: 2283–88

Soffe, S. R. 1989. Roles of glycinergic inhibition and N-methyl-D-aspartate receptor mediated excitation in the locomotor rhythmicity of one half of the *Xenopus* embryo central nervous system. *Eur. J. Neurosci.* 1: 561–71

Stein, P. S. G., Schild, C. P. 1989. N-methyl-D-aspartate antagonist applied to the spinal cord hindlimb enlargement reduces the amplitude of flexion reflex in the turtle. *Brain Res.* 479: 379–83

Suga, N., Olsen, J. F., Butman, J. A. 1990. Specialized subsystems for processing biologically important complex sounds: cross correlation analysis for ranging in the bat's brain. *Cold Spring Harbor Symp. Quant. Biol.* 55: 585–97

Tell, F., Jean, A. 1991. Activation of N-methyl-D-aspartate receptors induces endogenous rhythmic bursting activities in nucleus tractus solitarii neurons: an intracellular study on adult rat brainstem slices. *Eur. J. Neurosci.* 3: 1353–65

Thomson, A. M. 1990. Glycine is a coagonist at the NMDA receptor/channel complex. *Prog. Neurobiol.* 35: 53–74

Thomson, A. M., Girdlestone, D., West, D. C. 1989. A local circuit neocortical synapse that operates via both NMDA and non-NMDA receptors. *Br. J. Pharmacol.* 96: 406–8

Tsumoto, T. 1990. Long-term potentiation and depression in the cerebral neocortex. *Jpn. J. Physiol.* 40: 573–93

Tunstall, M. J., Roberts, A. 1991. Longitudinal coordination of motor output during swimming in *Xenopus* embryos. *Proc. R. Soc. London Ser. B* 244: 27–32

Wallen, P., Grillner, S. 1987. N-methyl-D-aspartate receptor-induced, inherent oscillatory activity in neurons active during fictive locomotion in the lamprey. *J. Neurosci.* 7: 2745–55

Watkins, J. C., Collingridge, G. L. 1989. *The NMDA Receptor.* Oxford: Oxford Univ. Press

Woolf, C. J. 1983. Evidence for a central component of post-injury pain hypersensitivity. *Nature* 306: 686–88

Woolf, C. J., Thompson, S. W. 1991. The induction and maintenance of central sensitization is dependent on N-methyl-D-aspartic acid receptor activation; implications for the treatment of post-injury pain hypersensitivity states. *Pain* 44: 293–99

Zador, A. M., Koch, C. K., Brown, T. H. 1990. Biophysical model of a Hebbian synapse. *Neurosci. Abstr.* 16: 492

Annu. Rev. Neurosci. 1993. 16:223–43

PROCESSING OF TEMPORAL INFORMATION IN THE BRAIN

Catherine E. Carr

Department of Zoology, University of Maryland, College Park,
Maryland 20742-4415

KEY WORDS: phase, time, delay, sound localization, owls, bats, electric fish

INTRODUCTION AND SCOPE OF THE REVIEW

Nervous systems can resolve microsecond time differences. These time differences would seem too small for single neurons to resolve, because neural events occur on a millisecond, rather than a microsecond, time scale. Accurate coding of temporal information is widespread, however, and many theories of sensory biology depend upon the detection of signals that are correlated in time. This review summarizes studies of time coding in the central nervous system (CNS) to show how the sensitivity to time differences arises. The first part summarizes behavioral sensitivity to time differences. The second describes specializations for the coding of temporal information in the CNS. The last describes the neural circuits underlying detection of time differences.

The review concentrates on three specialists—weakly electric fish, barn owls, and echolocating bats—because they have well developed abilities to detect time differences. Time coding in other animals is also discussed where possible, although a complete description of the temporal coding of complex stimuli is beyond the scope of this review. Analysis of how animals process temporal information has uncovered many common principles; despite their different neural substrates, time-coding systems implement similar algorithms for the encoding of temporal information. An algorithm refers here to steps and procedures in the processing of a signal (Konishi 1991). In time-coding systems, the timing of the stimulus is coded by phase-locked spikes, timing information is processed in a dedicated pathway in parallel with other stimulus variables, and the time-coding cells have morphological and physiological features suited to their function. Further-

223

0147–006X/93/0301–0223$02.00

more, time-coding circuits show increasing specializations for the extraction of relevant temporal features needed to measure time differences. The circuits that measure time differences in different animals also employ similar algorithms. All circuits that detect time differences depend on some form of delay lines and coincidence detectors. The coincidence detectors are neurons that respond maximally when they receive simultaneous inputs. This can occur when the time difference is compensated for by an equal and opposite delay of the inputs. The coincidence detector circuits use temporal information to compute the new variable of time difference.

BEHAVIOR AND DETECTION OF TEMPORAL INFORMATION

In an influential series of papers, Simmons (1973, 1979) focused attention on the problem of detecting small time differences. He showed that the bat *Eptesicus fuscus* could detect changes as small as 500 ns in the arrival time of jittered sonar echos. There followed several clear demonstrations of sensitivity to small time differences that included discrimination of phase differences between signals on different parts of the body surface in electric fish (Carr et al 1986a; Rose & Heiligenberg 1985) and detection of binaural phase differences for sound localization in owls, bats, and many other mammals (Heffner & Heffner 1992; Masters et al 1985; Moiseff & Konishi 1981). The behavioral studies outlined below identify which features of the stimulus the animal attends to, and describe each animal's ability to detect time differences.

Weakly electric fish generate an electric field around their body by discharging an electric organ. The field is detected by arrays of electroreceptors distributed over the body surface. Phase differences between signals on different parts of the body surface are essential cues for the electrolocation of objects and for communication with conspecifics, and the fish have a well-developed system for encoding and processing phase information (for review, see Heiligenberg 1991). The electric fish *Eigenmannia* produces a quasisinusoidal electric organ discharge in the range of 400–600 Hz. When these fish encounter neighbors with similar discharge frequencies, the electric signals add in the water and produce a beating signal that interferes with each animal's ability to electrolocate, because the small disturbances caused by the presence of a prey item are masked by the large changes in amplitude and phase caused by the interfering signals. Electric fish have developed a response to avoid jamming, whereby if a fish meets a conspecific with a similar frequency, each fish shifts the frequency of its electric organ discharge so as to increase the frequency difference between them (Bullock et al 1972). The fish determines whether its neighbor has a higher or lower frequency than its own by evaluating the fluctuations in amplitude and phase of the beating signal on different parts of its body surface (Heiligenberg et al 1978). Correct performance of the jamming avoidance response, therefore, requires evaluation of differential phase, and this behavior provides an assay for the minimum phase difference that can be perceived by the fish (Carr et al 1986a; Rose & Heiligenberg 1985).

Behavioral experiments have demonstrated that these electric fish detect phase differences in the microsecond range (Carr et al 1986a). The fish is presented with weaker and weaker jamming stimuli that produce smaller and smaller phase differences between different parts of the body surface. Correct performance of jamming avoidance responses to these weak jamming stimuli can be used to show that the fish can detect phase differences as small as 400 ns between different parts of the body surface. Physiological recordings, however, show that the phase-locked responses of afferents are too jittery to permit such fine temporal resolution (see next section). How can the fish, using only the information gathered from any single afferent, reliably perform a correct jamming avoidance response when the maximal phase difference is less than 1 μs? Hyperacuity is a perceptual term to describe the phenomenon whereby sensory thresholds are lower than expected from the properties of individual receptors (see Altes 1989). Rose & Heiligenberg (1985) showed that this temporal hyperacuity must result from the convergence within the CNS of parallel phase-coding channels from sufficiently large areas of the body surface, because the ability to detect the small phase differences diminished when smaller numbers of receptors were stimulated.

The barn owl's ability to detect time differences is acute. These owls can catch mice in total darkness on the basis of auditory cues alone (Konishi 1973; Payne 1971). A sound coming from one side of their body reaches one ear before the other, and the owl translates this interaural time difference into location in azimuth (the horizontal plane). The owl actually derives interaural time differences from the interaural phase differences present in the auditory stimulus. Behavioral experiments show that the owl uses the phase differences between the two ears, rather than stimulus onset time to localize sound, because if sound is presented through earphones, and phase delayed in one ear with respect to the other ear, the owl will turn its head in the direction predicted by the phase difference (Moiseff & Konishi 1981). Barn owls are very accurate (Konishi 1973); Knudsen & Konishi (1979) determined the average error made in free field sound localization by having the owl turn its head toward a sound in the dark and found that the owl's average error in turning its head to a sound was 1.5° (about 1.5 μs interaural time difference).

Most animals use interaural phase differences to localize sound (Fay 1988; Heffner & Heffner 1992). Animals with large heads have much larger time differences available to them; conversely, animals with small heads have to achieve much greater resolution of binaural time differences than a large animal to obtain the same degree of accuracy. For example, barn owls and humans have very similar abilities to localize sound; psychophysical studies have shown that human subjects can localize a frontal tone with an accuracy of about 1° (Mills 1972). The human discrimination task is easier than the barn owl's, because the human head is larger. Sound localization ability is not always well correlated with head size, however. Instead, it is highly correlated with directing the attention of the other senses to the sound source. Animals with narrow fields of best visual acuity require accurate sound localization to direct their gaze, whereas the echolocating mammals, bats, and dolphins use sound localization to direct their biosonar pulses (Heffner & Heffner 1992).

Echolocating microchiropteran bats emit biosonar pulses and listen to the returning echos. They determine target distance by measuring the time delay between the outgoing pulse and the returning echo (Simmons 1973, 1979). A bat can be trained to respond to one of two simultaneously presented stimuli and can detect changes as small as 500 ns in the arrival time of sonar echos when these changes appear as jitter or alterations in arrival time from one echo to the next (Moss & Schitzler 1989; Simmons 1979). Simmons et al's (1990) most recent findings suggest that trained *Eptesicus fuscus* can detect jittered targets with a delay acuity of 10 ns. The psychophysical function relating the bat's performance to the magnitude of the jitter corresponds to a cross-correlation function between the emitted sonar signals and the echos, and the data are consistent with the hypothesis that the bat acts as an ideal sonar receiver that perceives the phase or period of the signals (Moss & Simmons 1992). The bat obtains an acoustic image of the target that is apparently derived from time domain or periodicity information processing by the nervous system (Simmons et al 1990). The neural basis of this nanosecond delay sensitivity is hard to imagine.

ENCODING OF TEMPORAL INFORMATION

Discrimination of small time differences requires accurate transduction and processing of the original stimulus. This section addresses how time coding arises in the periphery, and how this timing information is preserved and improved in the CNS. Furthermore, behavioral hyperacuity is assumed to involve averaging in large neural assembles, in addition to accurate responses from single neurons. Both strategies may be found in systems that have achieved fine temporal resolution.

The Electrosensory System

The electric organ discharge is detected by sensory hair cells, termed electroreceptors. Two kinds of electroreceptors encode the phase or amplitude of the electric signal (Scheich et al 1973; Szabo 1965). Nerve fibers that innervate the phase-coding type of electroreceptor fire one spike on each cycle of the stimulus, phase-locked with little jitter to the zero-crossing of the stimulus. The degree of phase-locking to the stimulus is often quantified by using a measure, termed vector strength (Goldberg & Brown 1969). The distribution of phase angles in the period histogram is measured: If most spikes fall within a small range of phase angles, the vector strength is high (near 1); if spikes are evenly distributed, the vector strength is near 0.

Phase-locking must originate at the level of the receptor, although there are few studies of hair cells that address this issue. Fortunately, in mormyrid electric fish, the Knollenorgan electroreceptor produces a phase-locked spike in response to the electric stimulus (Bennett 1970). These electroreceptors appear to be the only vertebrate hair cells that generate a spike. At the best frequency, about 1.1 kHz, the vector strength is high, about 0.94; it diminishes to 0.2 at 10 kHz (measured from Figure 6 in Hopkins 1986). These frequency-dependent decreases in vector strength have also been found in the auditory system, and they have been interpreted as showing that the neurons ability to phase-lock decreases with frequency (see Weiss & Rose 1988).

In both mormyrid and gymnotiform electric fish, primary afferents convey phase-locked spikes to the lateral line lobe of the medulla, where phase and amplitude coding primary afferents terminate on different cell types (Bell et al 1989; Maler et al 1981; Von der Emde & Bleckmann 1991). Thus, the segregation of phase and amplitude receptors in the skin is reinforced by the central connections formed in the medulla. The separation of phase and amplitude information into two parallel channels is a common feature of all time-coding systems. In electric fish, the two channels not only have separate connections, but also distinct morphology. Phase-coding terminals are specialized for maintenance of phase-locked spikes; the afferents form large club terminals on large spherical cells that have few or no dendrites and thick axons. The spherical cells relay timing information directly to the midbrain torus. These connections are responsible for the fish's ability to detect small time differences, and are discussed below.

The accuracy of phase-coding improves with the progression from receptors to primary afferents to spherical cells to giant cells in the midbrain torus (Figure 1) (Carr et al 1986a). The accuracy of the phase-coding was determined by measuring spikes from phase coders in the medulla and in the midbrain torus. The jitter of these spikes, defined as the standard deviation of the response time to the stimulus, decreased threefold with the progression from medulla to midbrain. The basis for this improvement of accuracy may lie in the convergence from afferents to higher level neurons. Afferents converge upon spherical cells, which converge on giant cells in the torus (Figure 1). Each torus cell receives inputs from between 9 and 16 phase-coding afferents. If giant cells were to average these inputs, they should in theory reduce the jitter by $\sqrt{N}$ of the number of inputs, or 3–4 times. This predicted improvement in jitter is close to the measured decrease in jitter from 30 μs in the medulla to 10 μs in the midbrain. The accuracy of even the best single neurons in these first stations of the time-coding pathway does not match that of the behavior, however (Carr et al 1986a; Rose & Heiligenberg 1985). Electric fish can resolve temporal disparities in the submicrosecond range, a temporal resolution far superior to that observed in the primary electrosensory afferents. This hyperacuity must result from network processing, and is discussed in the next section.

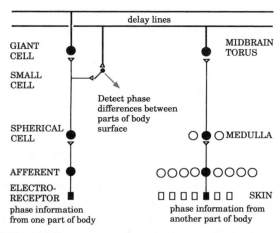

Figure 1 Schematic circuit in the electric fish midbrain for computation of phase differences between signals on any two parts of the body surface. Phase-coding electroreceptors converge on spherical cells in the medulla, which in turn converge in a topographic projection onto giant cell bodies and small cell dendrites in the torus. Giant cells relay the phase-locked signal all over the torus, with their terminals synapsing on the cell bodies of small cells. Small cells can, therefore, compare phase information from different parts of the body surface. Open circles are used to diagram the approximate convergence between different levels. (Modified from Carr et al 1986b and Heiligenberg 1991.)

The Auditory System

The behavioral experiments described in the previous section have shown that most animals use interaural phase differences to localize sound. Evidence of how they do so comes from experiments on the encoding and processing of phase information in the central auditory system. Time-coding systems appear to have changed little during amniote phylogeny, and although barn owls are useful models of temporal coding in the auditory system, studies on time coding in other birds, reptiles, amphibians, and mammals are equally instructive.

Recordings from auditory nerve fibers showed that spikes have a statistical tendency to phase-lock to the waveform of the acoustic stimulus (Kiang et al 1965). Spikes occur most frequently at a particular phase of the tone, although not necessarily in every tonal cycle. Thus, the discharge pattern of a cochlear nerve fiber can encode the phase of a tone with a frequency above 1000 Hz, even though the average discharge rate is low. The general assumption is that the modulating signal at the spike generator in the auditory nerve arises from components of the hair cell receptor potential, via the chemical synapse between the hair cell and the primary afferent. This timing information is degraded for high frequency sounds, presumably because of low pass filter effects in the hair cell (Kidd & Weiss 1990).

Cochlea hair cells encode and transmit both phase and amplitude information to the auditory nerve. Auditory nerve fibers phase-lock to the auditory stimulus and encode amplitude by increases in spike rate. Thus, there can be no predisposition toward coding for either amplitude or phase in the periphery, unlike electric fish. Nevertheless, the same parallel processing of phase and amplitude information that characterizes the electrosensory system is also found in the auditory system. The segregation does not begin at the level of the receptors, but differences in auditory nerve terminals may achieve a similar functional segregation in the CNS. To begin with the avian pattern, auditory nerve afferents enter the brain and then divide into two. One branch ramifies in the dendritic field of the cochlear nucleus angularis, which codes for changes in amplitude, whereas the other branch terminates in the cochlear nucleus magnocellularis, which codes for phase (Takahashi et al 1984). The terminal in the nucleus magnocellularis forms a specialized ending, termed an endbulb of Held (Brawer & Morest 1974). The endbulb synapse conveys the phase-locked discharge of the auditory nerve fibers to their postsynaptic targets in the nucleus magnocellularis. Thus, the synaptic specializations in the auditory nerve accomplish the same goal as the receptor specialization in electric fish. The endbulb is specialized to preserve phase-locked signals. Each endbulb has multiple sites of synaptic contact on the soma to provide the substrate for the preservation of the phase-locked spikes between the auditory nerve fibers and the neurons of the nucleus magnocellularis. The endbulb is a secure and effective connection; physiological measures show that phase-locking is the same or better in the neurons of the nucleus magnocellularis than in the eighth nerve, whereas it is lost in the projection to the amplitude coding nucleus angularis (Sullivan & Konishi 1984).

Phase-locked spikes encode the timing of the stimulus, and the CNS uses this code for the measurement of time disparities. Phase information is preserved and improved, and interaural phase differences are detected, in a circuit composed of the auditory nerve, the cochlear nucleus magnocellularis, and the nucleus laminaris (Figure 2). Many of the features of this circuit may represent specializations for the encoding of timing information. The neurons of the nucleus magnocellularis are morphologically and physiologically specialized for the encoding of temporal information. They have large round cell bodies, a thick axon, and few medium length dendrites (Jhaveri & Morest 1982). In the owl, magnocellular neurons in the high best frequency regions of the nucleus have fewer dendrites than the neurons in the low best frequency region (Carr & Boudreau 1992). Similar changes in dendritic length are found in the nucleus laminaris in the chicken (Smith & Rubel 1979). These reductions in dendritic area might decrease the membrane time constant and improve the speed and accuracy of the phase-locked response to synaptic inputs.

The nucleus magnocellularis relays phase-locked spikes in a bilateral projection to the nucleus laminaris (Figure 2). There is a further improvement in phase-locking between the nucleus magnocellularis and the nucleus laminaris (Carr & Konishi 1990). The observed increase in phase-locking between magnocellular and laminaris neurons may be caused in part by the massive convergence of magnocellular afferents onto laminaris neurons. About 100 inputs from each magnocellular nucleus converge on each laminaris neuron. This convergence may serve two interdependent functions in the owl: It improves the accuracy of the phase-locking to the stimulus, as is the case with converging inputs in electric fish; and, it may produce a cycle-by-cycle representation of the stimulus in the postsynaptic cell and provide a substrate for coincidence detection of phase differences (see next section).

The circuits that encode temporal information in the mammalian auditory system are similar to those in birds, and the projection from auditory nerve fibers to spherical bushy cells in the anteroventral cochlear nucleus is homologous to the projection described in birds and reptiles (Boord 1968; Carr 1992; Ryugo & Fekete 1984; Spzir et al 1990). Mammalian bushy cells are morphologically and physiologically similar to magno-

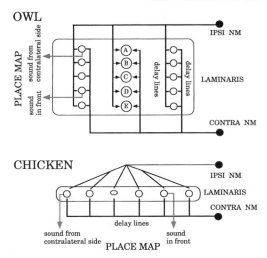

Figure 2 Jeffress model and schematic of brainstem auditory circuits for detection of interaural time differences in the barn owl and the chicken. Owl: Axons from the ipsilateral cochlear nucleus magnocellularis (IPSI NM) divide and enter the nucleus laminaris at several points along the dorsal surface. These axons act as delay lines within laminaris, interdigitating with inputs from the contralateral cochlear nucleus magnocellularis (CONTRA NM). In the center, the owl circuit has been modified to show the principles of the Jeffress model. Each binaural coincidence detector (A–E) fires maximally when inputs from the two sides arrive simultaneously. This can only occur when the interaural phase differences are compensated for by an equal and opposite delay. For example, neuron A fires maximally when sound reaches the contralateral ear first and is delayed by the long path from the contralateral ear so as to arrive simultaneously with the input from the ipsilateral ear. Thus, this array forms a map of interaural time difference in the dorso-ventral dimension of the nucleus. In the owl, sound from the front is mapped towards the ventral surface of the nucleus, and each nucleus appears to contain place maps of the contralateral and part of the ipsilateral hemifield (modified from Konishi 1991).

Chicken: Laminaris cells receive input from the ipsilateral nucleus magnocellularis (IPSI NM) onto their dorsal dendrites and input from the contralateral nucleus magnocellularis (CONTRA NM) onto their ventral dendrites. The ipsilateral inputs arrive simultaneously along the mediolateral extent of the nucleus laminaris, whereas the contralateral axons form delay lines, thus giving off collateral branches along the ventral surface of the nucleus. These delays are hypothesized to form a place map along the mediolateral axis of the nucleus. Delays from the ipsilateral and contralateral sides are approximately equal at the medial portion of the nucleus, and could map sound from the front of the bird. Because contralateral delays are longer than ipsilateral delays in the lateral portion of the nucleus, this region could map sounds that reached the contralateral ear first (modified from Overholt et al 1992).

cellular neurons of birds and reptiles. Bushy cells have large, round somata and thick axons, but also have many more dendrites than magnocellular neurons (Rhode et al 1983; Wu & Oertel 1984). Pioneering in vitro studies of the mouse cochlear nucleus show that bushy cells are well suited to preserve the temporal firing pattern of auditory nerve inputs (Oertel 1985). They fire only one or two spikes in response to electrical stimulation of the auditory nerve and have nonlinear current-voltage relationships around the resting potential (inward rectification). Thus, when bushy cells are depolarized, their input resistance drops, thus causing rapid re-polarization. The effects of excitation are brief and do not summate over time (Oertel 1985; Wu & Oertel 1984). Similar physiological responses may also characterize other phase-coding neurons. Bushy cells project to the medial superior olive in a circuit responsible for encoding and detecting interaural phase differences (see next section).

In addition to the circuit responsible for encoding and detecting inter-aural phase differences, a more mysterious structure has been found in the mammalian auditory system. In echolocating bats and dolphins, the columnar nucleus of the ventral lateral lemniscus (VNLLc) appears to be specialized for encoding temporal information (Covey & Casseday 1991). Its function is currently unknown, but its high degree of development and differentiation in two different mammalian orders suggest that this nucleus plays an important role in echolocation. The cells are innervated by the anteroventral cochlear nucleus and are physiologically suited to temporal coding, with little or no spontaneous activity, broad tuning, and short integration times (Covey & Casseday 1991).

To summarize, time coding is prominent in both the electrosensory and auditory systems. Both systems use similar algorithms with these major features: Phase-locked spikes encode the timing of the stimulus, and the CNS uses this code for the measurement of time disparities. Time and amplitude information is processed in parallel. Phase-coding improves with convergence from the periphery to higher level processing stations. The morphology and physiology of phase-coding cells is suited to their function; they have large cell bodies with few or no dendrites, thick axons, and short duration spikes.

DETECTING TIME DIFFERENCES

Behavioral experiments have shown that animals are capable of great accuracy in detecting time differences. How is this accomplished? The preferred model for detecting temporal disparities depends upon coinci-dence detection and proposes that a neuron that detects time differences responds best to simultaneous inputs. The model was first articulated by Jeffress (1948), with his place theory for detection of interaural time differences. The model circuit is composed of two elements: delay lines and coincidence detectors (Figure 2). The delay lines are created by varying axonal path lengths, and the coincidence detectors are neurons that respond maximally when they receive simultaneous inputs, i.e. when the time difference is exactly compensated for by the delay introduced by the inputs.

The Jeffress model explains not only how interaural time differences are measured, but also how they are encoded. The circuit contains an array of coincidence detectors that receive input from afferent axons that serve as delay lines. Because of its position in the array, each neuron responds only to sound coming from a particular direction; thus, the anatomical place of the neuron encodes the location of the sound (Figure 2). These neurons use the temporal information to compute a new variable, time difference, and transform the time code into a place code. The selectivity of all higher-order auditory neurons to time difference derives from the "labeled-line" output of the place map (Konishi 1986).

Circuits in the fish midbrain, the auditory brainstem of barn owls, cats, and dogs, and the bat thalamus all contain neurons tuned to particular time delays between events. These delay-tuned ensembles employ similar algorithms. The circuits in the auditory brainstem resemble the Jeffress model in all its particulars, whereas the circuits in fish midbrain and bat thalamus differ from the Jeffress model in important respects. The circuit in the fish midbrain does not form a place map; the circuits in the bat thalamus do not contain axonal delay lines.

Detection of Phase Differences in Electric Fish

When *Eigenmannia* is presented with sinusoidally varying electrical fields on two parts of its body, it can distinguish phase differences between these signals smaller than 1 μs. The circuit for detection of these phase differences is as follows: Phase-coding afferents from the medulla project to two cell types of the midbrain torus, synapsing on giant cell bodies and on the dendrites of the small cells (Figure 1). These afferents form local connections that encode the phase of the electric organ discharge from one part of the body surface. The giant cell's axons form horizontal connections that distribute this local phase information to small cells throughout the lamina, so that timing information from one part of the body surface may be compared with any other part. Small cells compare information from one patch of the body surface, through afferent input onto their dendrites, with phase information from any other part of the body surface, through the giant cell input to their cell bodies (Carr et al 1986b). Small cell responses encode either phase advance or phase delay (Heiligenberg & Rose 1985). The biophysical basis of the sensitivity to phase differences is as yet unknown, although a useful computer simulation has been generated (Lytton 1991). The small cell circuit allows the fish to perform all possible comparisons between different parts of the body surface, as required for correct performance of the jamming avoidance response. Therefore, unlike the Jeffress model, the axonal delay lines do not create a place map of the computed stimulus feature, phase difference.

The phase sensitivity of the small cells is not as good as the behavior. Although the convergence of time-coding inputs progressively improves the temporal acuity of single neurons at each level of the time-coding pathway, the smallest temporal disparity detected by the most sensitive neurons in the torus is only 10 μs, a value 20 times higher than that of the behavioral threshold (Carr et al 1986a). If the responses of such neurons are averaged over periods that are much longer than the 300 ms latency of the jamming avoidance response, a sensitivity to 1 μs can be demonstrated (Rose & Heiligenberg 1985). Because the temporal averaging of the electro-physiologist is assumed to be equivalent to spatial averaging or parallel processing of the same signal in the CNS, this suggests that the sensitivity and short latency of the jamming avoidance response requires parallel convergence of large numbers of such toral neurons, which carry temporal information from wide areas of the body surface (Rose & Heiligenberg 1985). This convergence ultimately occurs at the level of single neurons in the prepacemaker nucleus (Kawasaki et al 1988).

To address the question of whether the sensitivity to small temporal disparities even exists at the single neuron level, Kawasaki et al (1988) recorded from neurons in the prepacemaker nucleus. Prepacemaker neurons are located at the top of the sensory hierarchy that determines whether the jamming stimulus is higher or lower in frequency than the fish's own electric organ discharge. Prepacemaker cells modulate their firing with changes in the interfering signal, even with temporal disparities as small as 1 μs. According to Kawasaki et al (1988), it is unlikely that any neuron could discriminate such small temporal disparities in the arrival of two signals with certainty, but small differences in the timing of excitatory and inhibitory inputs could affect firing rates in a probabilistic fashion. Pooling the activity of many such neurons could yield predictable results (Kawasaki et al 1988). Thus, hyperacuity may be achieved by a network of prepacemaker neurons.

Detection of Interaural Phase Differences in Birds and Mammals

The circuits in the auditory brainstem that detect interaural phase differences resemble the Jeffress model. The cochlear nucleus axons act as delay lines, and the laminaris or olivary neurons act as coincidence detectors, to form a circuit that measures and encodes interaural time differences (Figure 2). Physiological studies of the mammalian auditory system by

Goldberg & Brown (1969) and others found neuronal responses consistent with the Jeffress model. Recent results in the barn owl (Carr & Konishi 1990; Sullivan & Konishi 1986), cat (Smith et al 1990; Yin & Chan 1990), and chicken (Overholt et al 1992; Rubel & Parks 1975; Young & Rubel 1983) have described circuits that conform to the model's requirements. There are three major parts to the Jeffress model: delay lines, coincidence detection, and place coding of interaural phase difference. These are discussed in turn.

In the barn owl, magnocellular axons act as delay lines (Carr & Konishi 1988, 1990). They convey the timing or phase of the auditory stimulus in a bilateral projection to the nucleus laminaris, such that axons from the ipsilateral nucleus magnocellularis enter the nucleus laminaris from the dorsal side, whereas axons from the contralateral nucleus magnocellularis enter from the ventral side. Thus, these afferents interdigitate to innervate dorso-ventral arrays of neurons in laminaris in a sequential fashion (Figure 2). For each frequency band, recordings from these interdigitating ipsilateral and contralateral axons show regular changes in delay with depth in the nucleus laminaris (Carr & Konishi 1990). These conduction delays are similar to the 200 μs range of interaural time differences available to the barn owl (Moiseff 1989). The conduction velocities of the delay lines are regulated precisely by the internodal distances within the nucleus laminaris. Normal internode lengths in the CNS are generally greater than 300 μm (Waxman 1975), whereas internode lengths in the delay lines axons in the nucleus laminaris are both short and regular, occurring every 60 μm (Carr & Konishi 1990).

The magnocellular-laminaris circuit in the chicken is similar to the circuit described for the barn owl, except that the nucleus laminaris is not a large nucleus, but a monolayer of bipolar cells oriented in the mediolateral dimension of the brainstem (Jhaveri & Morest 1982; Rubel & Parks 1975; Young & Rubel 1983) (Figure 2). Laminaris cells receive input from the ipsilateral nucleus magnocellularis onto their dorsal dendrites and input from the contralateral nucleus magnocellularis onto their ventral dendrites. The ipsilateral projections splay out to innervate laminaris neurons with approximately equal lengths to each cell, so that ipsilateral inputs arrive simultaneously along the mediolateral extent of the nucleus laminaris. Thus, the ipsilateral axons do not act as delay lines. The contralateral axons act as delay lines; each axon runs along the ventral surface of the nucleus laminaris, thus giving off collateral branches in the nucleus. Estimates of conduction velocity suggest that the contralateral delay line could encode 180 μs of delay (Overholt et al 1992). Patterns of delay lines in the medial superior olive of the cat are very similar to those of the chicken. Axons from the contralateral cochlear nucleus form delay lines across the rostro-caudal axis of the nucleus, whereas the ipsilateral axons form a less organized projection (Smith et al 1990).

The second major requirement of the Jeffress model is that the targets of the delay lines act as coincidence detectors. Goldberg & Brown (1969) showed that olivary neurons acted as coincidence detectors. The neurons of the avian nucleus laminaris and the mammalian medial superior olive phase-lock to both monaural and binaural stimuli and respond maximally when phase-locked spikes from each side arrive simultaneously, i.e. when the difference in the internal conduction delay is nullified by interaural time difference. The number of spikes elicited in response to a favorable interaural time difference is roughly double that elicited by a monaural stimulus; spike counts for unfavorable interaural time differences fall well below monaural response levels. Thus, physiological responses from these coincidence detectors are similar (Carr & Konishi 1990; Goldberg & Brown 1969; Sullivan & Konishi 1984; Yin & Chan 1990). The underlying bio-physical mechanisms of coincidence detection are still unknown. Similarities between olivary cell responses and cross-correlated monaural spike trains have been used to suggest that cross-correlation provides a more general description of coincidence detection (Yin & Chan 1990); cross-correlation models for the computation of interaural phase differences are long standing in the auditory literature (Licklider 1959). The biophysical basis of cross-correlation is unclear, however, and more recent models do not depend upon cross-correlation (Colburn et al 1990; Grün et al 1990).

In one respect, owl laminaris neurons are different to all other cells that detect time differences. The delay line inputs are distributed evenly over the entire somato-dendritic area in the barn owl. In the chicken, laminaris neurons are bipolar, with inputs from the ipsilateral nucleus magno-cellularis restricted to the dorsal dendrites and contralateral inputs to the ventral dendrites of the laminaris neurons (Jhaveri & Morest 1982; Smith & Rubel 1979). In the mammalian medial superior olive, the principal cells are large bipolar ("rabbit ear") cells that segregate inputs from each ear onto the medial and lateral dendritic trees, respectively (Stotler 1953; Warr 1966). Even in the electric fish, the small cells segregate the inputs from the different parts of the body surface onto their soma and dendrite, respectively (Carr et al 1986b). Thus, owl laminaris neurons are the only "comparator" cells that do not segregate the inputs that are to be compared. The functional significance of these morphological differences is unknown.

Jeffress's model is called the place theory, because the array of coincidence detectors encode the computed variable of interaural phase difference by their place in the nucleus (see Figure 2). Each place in the nucleus laminaris or medial superior olive can be characterized by a pair of delays for ipsi- and contralateral axons, and phase-locked spikes arrive at each neuron with delays unique to its place. The monaural delay lines, therefore, form neuronal maps of interaural time difference, or place, that are tapped by coincidence detectors. Evidence for a place map was obtained from the medial superior olive of the cat (Yin & Chan 1990). Cells in the medial superior olive respond to a range of interaural time differences between 0 and 400 μs, thus corresponding to locations in the contralateral sound field. The locations of recordings were marked with lesions; best interaural time differences near 0 μs were located at the anterior pole of the olive, whereas larger time differences were located at more posterior locations. Thus, the medial superior olive appears to contain a map of place in each frequency band. The nucleus laminaris in the chicken may also contain maps of interaural time difference or place organized along the medio-lateral dimension of the nucleus, with best interaural time differences of $0°$ close to the medial edge (Figure 2) (Overholt et al 1992). In the barn owl, the nucleus laminaris contains many maps of interaural time difference or place within each frequency band, repeated along the dorso-ventral dimension of the nucleus (Figure 2). In these maps of interaural time difference, $0°$ is represented near the ventral edge of the nucleus (Sullivan & Konishi 1986). Thus, the barn owl appears to have duplicated the ancestral avian pattern and rotated the place map by $90°$ (Takahashi & Konishi 1988). These duplicated maps may confer increased sensitivity to interaural time differences on this nocturnal predator.

Bats and the Detection of Echo Delays

A bat emits orientation sounds and listens to the returning echos. The delay between the pulse and echo conveys the target distance. For example, a 1 ms echo delay corresponds to a 17.3 cm target distance (Suga 1990). How are these delays encoded and processed in the CNS? At the periphery, neurons respond to both the pulse and the echo, and the time interval between the two grouped discharges is directly related to the echo delay. The medial geniculate body and auditory cortex, however, contain neurons that respond poorly or not at all to the echo alone, but respond strongly when pulse and echo are combined with a particular echo delay. These "combination-sensitive" neurons are called FM-FM cells, and were first found in the auditory cortex of the mustache bat (O'Neill & Suga 1982; Suga et al 1978). Tuning to echo-delay is mapped on the cortical surface to form maps of target range (O'Neill & Suga 1982). To create an FM-FM cell, the bat must differentiate between pulse and echo. Sonar pulses contain four harmonics, and the bat uses the low amplitude first harmonic (FM_1) to mark the pulse and the other harmonics (FM_2, FM_3, FM_4) to mark the echo.

Where are these FM_1-FM_n neurons created? There are no FM_1-FM_n neurons in the inferior colliculus (O'Neill 1985), and delays between pulse- and echo-evoked activity may still be found in the brachium of the inferior colliculus (Kuwabara & Suga 1988). FM_1-FM_n neurons are first found in the medial geniculate body (Olsen & Suga 1991). FM_1 and FM_n channels from the inferior colliculus converge in the medial geniculate body to create FM_1-FM_n combination sensitive neurons (Figure 3). For the FM_1 and FM_n evoked activity to coincide in the medial geniculate body, the FM_1 signal must be delayed with reference to the FM_n. How are the FM_1 delay lines created? The answer is more complex than that from the auditory brainstem because of the time delays involved. The pulse-echo delay axis is from 0.4 to 18 ms, i.e. a range axis of 7–310 cm (Suga 1990). These delays are much larger than the 200 μs interaural time differences available to the barn owl (Carr & Konishi 1990) or 400 μs for the cat (Yin & Chan 1990), and it would be tortuous to create them by varying axonal path lengths. Olsen & Suga (1991) have found evidence for at least two other mechanisms for delaying the FM_1 signal. For best delays below 4 ms, delay tuning depends only on coincidence of pulse and echo excitation.

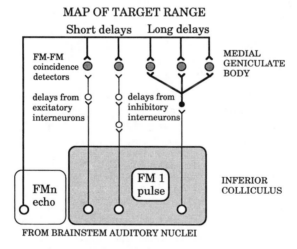

Figure 3 Schematic circuit for creation of delay-tuned FM-FM cells in the medial geniculate body of the thalamus. Auditory information about the pulse (FM_1) and echo (FM_n) travels in different frequency channels from the auditory brainstem to the inferior colliculus, and from there to the medial geniculate body. The neurons tuned to the echo are not delayed in the medial geniculate, whereas neurons tuned to the pulse exhibit both short and long delays. For short delays (0.5–4 ms), the pulse signal is delayed by excitatory relay interneurons. Long delays (4–18 ms) appear to be created by inhibitory interneurons (modified from Suga 1992).

This result is consistent with delay lines created by excitatory relay units (Figure 3). For long best delays (>4 ms), the FM_1 pulse produces inhibition of short latency and variable duration, followed by excitation, consistent with the "inhibitory gate" hypothesis (Sullivan 1982), in which FM_1 delays are created by varying the duration of inhibition. The inhibition may be produced by local circuits within the medial geniculate body (Olsen & Suga 1991).

These inhibitory local circuits in the auditory thalamus have a parallel in the visual thalamus. In the lateral geniculate, inhibition produces the X-lagged cell responses (Mastronade 1987a,b). These geniculate cells differ from their retinal X-afferent inputs and from adjacent nonlagged geniculate X-cells in displaying an early inhibition rather than excitation to the onset of a visual stimulus in their receptive field. The X-lagged response is delayed, because the excitatory input is blocked by inhibition, not because excitation is delayed. Postinhibitory rebound then produces the lagged or phase-shifted response. The response timing differences between lagged and nonlagged cells may be important for calculating direction selectivity in visual cortex (Saul & Humphrey 1990).

CONCLUSIONS

Analysis of the encoding and processing of temporal information has uncovered some common organizational principles. Time-coding systems implement similar algorithms for the encoding and processing of temporal information (Konishi 1991). Despite different neural substrates, CNS time channels share numerous morphological and physiological adaptations to improve the time coding of signal. In the electric sense and audition, a dichotomy exists between coding for the timing or phase of the signal and coding for its amplitude. In electric fish, the dichotomy between phase and amplitude coding begins at the receptor level; in the auditory system, the separation into phase and amplitude coding is derived within the CNS. A similar separation may be found in other sensory systems; visual information is processed in two channels that differ in their spatial and temporal acuity (Livingstone & Hubel 1987; Maunsell & Newsome 1987).

Time-coding systems also implement similar algorithms for the measurement of time differences. These algorithms all depend on delay lines and coincidence detectors in some form. Short delays may be provided by axonal delay lines. Axons can provide functionally significant delays in conduction. Compensatory axonal delay line mechanisms include equalization of path length, differences in conduction velocity, and localized delays, determined by variations in preterminal axon branches (Waxman 1975). Long delays may be introduced by interposed excitatory synapses, whereas the longest delays originate with inhibitory interneurons and regulation of post-inhibitory rebound. Neural circuits that incorporate delays may be widespread in the CNS. Such circuits are also found in many neural models; for example, early models of directional selectivity in the visual system connected two detectors to an AND-NOT gate, one via a delay (reviewed in Marr 1982). The underlying mechanisms of coincidence detection are less well understood than those that generate delays, and the biophysical basis of coincidence detection represents an important challenge. Coincidence detectors transform timing information into a place code for further processing and the formation of perceptions, such as sound location or target range.

ACKNOWLEDGMENTS

I am indebted to M. Konishi and W. Heiligenberg for commenting on drafts of this article. I thank R. A. Eatock, S. Echteler, J. Olsen, and W. E. O'Neill for helpful discussions. N. Suga, R. Hyson, P. Smith, and J. A. Simmons kindly provided unpublished work or preprints. Research in my laboratory has been supported by the National Institutes of Health (DC 00436) and the Sloan Foundation.

Literature Cited

Altes, R. A. 1989. Ubiquity of hyperacuity. *J. Acoust. Soc. Am.* 85: 943–52

Bell, C. C., Zakon, H., Finger, T. E. 1989. Mormyromast electroreceptor organs and their afferent fibers in mormyrid fish: I. Morphology. *J. Comp. Neurol.* 286: 391–409

Bennett, M. V. L. 1970. Comparative physiology: electric organs. *Annu. Rev. Physiol.* 32: 471–531

Boord, R. L. 1968. Ascending projections of the primary cochlear nuclei and nucleus laminaris in the pigeon. *J. Comp. Neurol.* 133: 523–42

Brawer, J. R., Morest, D. K. 1974. Relations between auditory nerve endings and cell types in the cat's anteroventral cochlear nucleus seen with Golgi method and Nomarski optics. *J. Comp. Neurol.* 160: 491–506

Bullock, T. H., Hamstra, R. H., Scheich, H. 1972. The jamming avoidance response of high frequency electric fish. I. General features. *J. Comp. Physiol.* 77: 1–22

Carr, C. E. 1992. The evolution of the central auditory system in reptiles and birds. In *The Evolutionary Biology of Hearing*, ed. D. B. Webster, R. R. Fay, A. N. Popper, pp. 511–44. New York: Springer-Verlag

Carr, C. E., Boudreau, R. E. 1992. Organization of the nucleus magnocellularis and the nucleus laminaris in the barn owl: encoding and measuring interaural time differences. *J. Comp. Neurol.* In press

Carr, C., Heiligenberg, W., Rose, G. 1986a. A time-comparison circuit in the electric fish midbrain. I. Behavior and physiology. *J. Neurosci.* 6: 107–19

Carr, C. E., Konishi, M. 1990. A circuit for detection of interaural time differences in the brainstem of the barn owl. *J. Neurosci.* 10: 3227–46

Carr, C. E., Konishi, M. 1988. Axonal delay lines for time measurement in the owl's brainstem. *Proc. Natl. Acad. Sci. USA* 85: 8311–15

Carr, C. E., Maler, L., Taylor, B. 1986b. A time comparison circuit in the electric fish midbrain. II. Functional morphology. *J. Neurosci.* 6: 1372–83

Colburn, H. S., Han, Y., Culotta, C. P. 1990. Coincidence model of MSO responses. *Hear. Res.* 49: 335–55

Covey, E., Casseday, J. H. 1991. The monaural nuclei of the lateral lemniscus in an echolocating bat: parallel pathways for analyzing temporal features of sound. *J. Neurosci.* 11: 3456–70

Fay, R. R. 1988. *Hearing in Vertebrates: A Psychophysics Databook*. Winnetka, Ill: Hill-Fay Assoc.

Goldberg, J. M., Brown, P. B. 1969. Response of binaural neurons of dog superior olivary complex to dichotic tonal stimuli: Some physiological mechanisms of sound localization. *J. Neurophysiol.* 32: 613–36

Grün, S., Aertsen, A., Wagner, H., Carr, C. 1990. Sound localization in the barn owl: a quantitative model of binaural interaction in the nucleus laminaris. *Soc. Neurosci. Abstr.* 16

Heffner, R. S., Heffner, H. E. 1992. Evolution of sound localization in mammals. In *The Evolutionary Biology of Hearing*, ed. D. B. Webster, R. R. Fay, A. N. Popper, pp. 691–716. New York: Springer-Verlag

Heiligenberg, W. 1991. *Neural Nets in Electric Fish*. Cambridge, Mass: MIT Press

Heiligenberg, W., Baker, C., Matsubara, J. A. 1978. The jamming avoidance response in Eigenmannia revisited: The structure of a neuronal democracy. *J. Comp. Physiol.* 127: 267–86

Heiligenberg, W., Rose, G. 1985. Phase and amplitude computations in the midbrain of an electric fish: intracellular studies of neurons participating in the jamming avoidance response of Eigenmannia. *J. Neurosci.* 5: 515–31

Hopkins, C. D. 1986. Temporal structure of non-propagated electric communication signals. *Brain Behav. Evol.* 31: 43–59

Jeffress, L. A. 1948. A place theory of sound localization. *J. Comp. Physiol. Psychol.* 41: 35–39

Jhaveri, S., Morest, K. 1982. Neuronal architecture in nucleus magnocellularis of the chicken auditory system with observations on nucleus laminaris: A light and electron microscope study. *Neuroscience* 7: 809–36

Kawasaki, M., Rose, G., Heiligenberg, W. 1988. Temporal hyperacuity in single neurons of electric fish. *Nature* 336: 173–76

Kiang, N. Y. S., Watanabe, T., Thomas, E. C., Clark, E. F. 1965. *Discharge Patterns of Single Fibers in the Cat's Auditory Nerve*. Cambridge, Mass: MIT Press

Kidd, R. C., Weiss, T. F. 1990. Mechanisms that degrade timing information in the cochlea. *Hear. Res.* 49: 181–208

Knudsen, E. I., Konishi, M. 1979. Sound localization by the barn owl (Tyto alba). *J. Comp. Physiol.* 133: 1–11

Konishi, M. 1991. Deciphering the brain's codes. *Neural Comput.* 3: 1–18

Konishi, M. 1986. Centrally synthesized maps of sensory space. *Trends Neurosci.* 9: 163–68

Konishi, M. 1973. How the owl tracks its prey. *Am. Sci.* 61: 414–24

Kuwabara, N., Suga, N. 1988. Mechanism for production of "range-tuned" neurons in the mustached bat: delay lines and amplitude selectivity are created by the midbrain auditory nuclei. *Assoc. Res. Otolaryngol. Abstr.* 11: 200

Licklider, J. C. R. 1959. Three auditory theories. In *Psychology: A Study of a Science*, ed. S. Koch, pp. 41–144. New York: McGraw-Hill

Livingstone, M. L., Hubel, D. H. 1987. Psychophysical evidence for separate channels for the perception of form, color, movement and depth. *J. Neurosci.* 7: 3416–68

Lytton, W. W. 1991. Simulations of a phase comparing neuron of the electric fish Eigenmannia. *J. Comp. Physiol.* 169: 117–25

Maler, L., Sas, E., Rogers, J. 1981. The cytology of the posterior lateral line lobe of high frequency weakly electric fish (Gymnotoidei): Dendritic differentiation and synaptic specificity in a simple cortex. *J. Comp. Neurol.* 195: 87–140

Marr, D. 1982. *Vision*. New York: Freeman

Masters, W. M., Moffat, A. J. M., Simmons, J. A. 1985. Sonar tracking of horizontally moving targets by the big brown bat Eptesicus fuscus. *Science* 228: 1331–33

Mastronade, D. N. 1987a. Two classes of single-input X-cells in cat lateral geniculate nucleus. I. Receptive field properties and classification of cells. *J. Neurophysiol.* 57: 357–80

Mastronade, D. N. 1987b. Two classes of single input cell in the lateral geniculate nucleus. II. Retinal inputs and the generation of receptive field properties. *J. Neurophysiol.* 57: 381–413

Maunsell, J. H. R., Newsome, W. T. 1987. Visual processing in monkey extrastriate cortex. *Annu. Rev. Neurosci.* 10: 363–402

Mills, A. W. 1972. Auditory localization. In *Foundations of Modern Auditory Theory*, ed. J. V. Tobias, 2: 303–48. New York: Academic

Moiseff, A. 1989. Binaural disparity cues available to the barn owl for sound localization. *J. Comp. Physiol.* 164: 629–36

Moiseff, A., Konishi, M. 1981. Neuronal and behavioral sensitivity to binaural time differences in the owl. *J. Neurosci.* 1: 40–48

Moss, C. F., Schnitzler, H.-U. 1989. Accuracy of target ranging in echolocating bats: acoustic information processing. *J. Comp. Physiol.* 165: 383–93

Moss, C. F., Simmons, J. A. 1992. Acoustic image representation of a point target in the FM-FM bat, Eptesicus fuscus: Evidence for the perception of echo phase in sonar. *J. Acoust. Soc. Am.* In press

O'Neill, W. E. 1985. Responses to pure tones and linear FM components of the CF-FM biosonar signal by single units in the inferior echolocating of the mustached bat. *J. Comp. Physiol.* 157: 797–815

O'Neill, W. E., Suga, N. 1982. Encoding of target range information and its representation in the auditory cortex of the mustache bat. *J. Neurosci.* 2: 17–31

Oertel, D. 1985. Use of brain slices in the study of the auditory system: Spatial and temporal summation of synaptic inputs in cells in the anteroventral cochlear nucleus of the mouse. *J. Acoust. Soc. Am.* 78: 328–33

Olsen, J. F., Suga, N. 1991. Combination-sensitive neurons in the medial geniculate body of the mustached bat: Encoding of target range information. *J. Neurophysiol.* 65: 1275

Overholt, T., Hyson, R., Rubel, E. W. 1992. A delay-line circuit for coding interaural time differences in the chick brain stem. *J. Neurosci.* 12: 1698–1708

Payne, R. S. 1971. Acoustic localization of prey by barn owls (*Tyto alba*). *J. Exp. Biol.* 54: 535–73

Rhode, W. S., Oertel, D., Smith, P. H. 1983. Physiological response properties of cells labeled intracellularly with horseradish peroxidase in cat ventral cochlear nucleus. *J. Comp. Neurol.* 213: 448–63

Rose, G., Heiligenberg, W. 1985. Temporal hyperacuity in the electric sense of fish. *Nature* 318: 178–80

Rubel, E. W., Parks, T. N. 1975. Organization and development of brainstem auditory nuclei of the chicken: Tonotopic organization of *N. magnocellularis* and *N. laminaris. J. Comp. Neurol.* 164: 411–34

Ryugo, D. K., Fekete, D. M. 1982. Morphology of primary axosomatic endings in the anteroventral cochlear nucleus of the cat: A study of the endbulbs of Held. *J. Comp. Neurol.* 210: 239–57

Saul, A. B., Humphrey, A. L. 1990. Spatial and temporal response properties of lagged and non-lagged cells in cat lateral geniculate nucleus. *J. Neurophysiol.* 64: 206–24

Scheich, H., Bullock, T. H., Hamstra, R. H. 1973. Coding properties of two classes of afferent nerve fibers: High frequency electroreceptors in the electric fish Eigenmannia. *J. Neurophysiol.* 36: 39–60

Simmons, J. A. 1979. Perception of echo phase information in bat sonar. *Science* 204: 1336–38

Simmons, J. A. 1973. The resolution of target range by echolocating bats. *J. Acoust. Soc. Am.* 54: 157–73

Simmons, J. A., Ferragamo, M., Moss, C. F., Stevenson, S. B., Altes, R. A. 1990. Discrimination of jittered sonar echos by the echolocating bat, *Eptesicus fuscus*: The shape of target images in echolocation. *J. Comp. Physiol.* 167: 589–616

Smith, P. H., Joris, P. X., Yin, T. C. T. 1990. Projections of spherical bushy cells to the MSO in the cat: evidence for delay lines. *Neurosci. Abstr.* 16: 723

Smith, Z. D. J., Rubel, E. W. 1979. Organization and development of brainstem auditory nuclei of the chicken: Dendritic gradients in nucleus laminaris. *J. Comp. Neurol.* 186: 213–39

Spzir, M. R., Sento, S., Ryugo, D. K. 1990. The central projections of the cochlear nerve fibers in the alligator lizard. *J. Comp. Neurol.* 295: 530–47

Stotler, W. A. 1953. An experimental study of the cells and connections of the superior olivary complex of the cat. *J. Comp. Neurol.* 98: 401–32

Suga, N. 1992. Parallel-hierarchical processing of complex sounds for specialized auditory function. In *Handbook of Acoustics*, ed. M. J. Crocker. New York: Wiley

Suga, N. 1990. Cortical computational maps for auditory imaging. *Neural Netw.* 3: 3–21

Suga, N., O'Neill, W. E., Manabe, T. 1978. Cortical neurons sensitive to combinations of information bearing elements of biosonar signals in the mustache bat. *Science* 200: 778–81

Sullivan, W. E. 1982. Possible neural mechanisms of target distance coding in auditory system of the echolocating bat *Myotis lucifugis. J. Neurophysiol.* 48: 1033–47

Sullivan, W. E., Konishi, M. 1986. Neural map of interaural phase difference in the owl's brainstem. *Proc. Natl. Acad. Sci. USA* 83: 8400–4

Sullivan, W. E., Konishi, M. 1984. Segregation of stimulus phase and intensity coding in the cochlear nucleus of the barn owl. *J. Neurosci.* 4: 1787–99

Szabo, T. 1965. Sense organs of the lateral line system in some electric fish of the Gymnotidae, Mormyridae and Gynmarchidae. *J. Morphol.* 117: 229–50

Takahashi, T., Konishi, M. 1988. The projections of the cochlear nuclei and nucleus laminaris to the inferior colliculus of the barn owl. *J. Comp. Neurol.* 274: 190–211

Takahashi, T., Moiseff, A., Konishi, M. 1984. Time and intensity cues are processed independently in the auditory system of the owl. *J. Neurosci.* 4: 1781–86

Von der Emde, G., Bleckmann, H. 1991. Extreme phase sensitivity of afferents which innervate mormyromast electroreceptors. *Naturwissenschaften* 78: 131–33

Warr, W. B. 1966. Fiber degeneration following lesions in the anterior ventral cochlear nucleus of the cat. *Exp. Neurol.* 14: 453–74

Waxman, S. G. 1975. Integrative properties and design principles of axons. *Int. Rev. Neurobiol.* 18: 1–40

Weiss, T. F., Rose, C. 1988. A comparison of synchronization filters in different auditory receptor organs. *Hear. Res.* 33: 175–80

Wu, S. H., Oertel, D. 1984. Intracellular injection with horseradish peroxidase of physiologically characterized stellate and bushy cells in slices of mouse anteroventral cochlear nucleus. *J. Neurosci.* 4: 1577–88

Yin, T. C. T., Chan, J. C. K. 1990. Interaural time sensitivity in medial superior olive of cat. *J. Neurophysiol.* 64: 465–88

Young, S. R., Rubel, E. W. 1983. Frequency-specific projections of individual neurons in chick brainstem auditory nuclei. *J. Neurosci.* 7: 1373–78

Annu. Rev. Neurosci. 1993. 16:245–63

INFERIOR TEMPORAL CORTEX: Where Visual Perception Meets Memory

Yasushi Miyashita

Department of Physiology, University of Tokyo, School of Medicine, Hongo, Tokyo 113, Japan

KEY WORDS: long-term memory, association, imagery, retrieval, primate

INTRODUCTION

The neural processes that lead to visual perception and memory, thus subserving the identification of objects, have been assigned to the multisynaptic occipito-temporo-limbic projection that interconnects the striate, prestriate, inferior temporal, and hippocampal cortices (Gross 1972; Maunsell & Newsome 1987; Mishkin et al 1983). Recent advances in our understanding of the organization of the extrastriate cortex revealed a mosaic of more than 25 visual areas beyond the striate cortex (Felleman & Van Essen 1991; Van Essen 1985; Zeki & Shipp 1988). The physical properties of a visual object (such as its size, color, texture, and shape) are analyzed in the multiple subdivisions of the prestriate-posterior temporal complex. The anterior part of inferior temporal cortex has been hypothesized not only to synthesize the analyzed attributes into a unique configuration, but also to work as the storehouse for central representations of the objects (Mishkin 1982; Weiskrantz & Saunders 1984). The supporting evidence for this hypothesis has mainly been given by neuropsychological approaches (Milner 1990; Squire & Zola-Morgan 1991; Weiskrantz 1990). Recently, neuronal correlates of the visual long-term memory were found: The temporal-lobe neurons could reflect learned associative relations among stimuli (Miyashita 1988; Sakai & Miyashita 1991). The findings on this associational mechanism, together with that on the feature selectivity

245

0147–006X/93/0301–0245$02.00

of single cells and functional architectures in the temporal cortex, provide new insight into how the memory-related capacity of this cortex leads to other perceptual properties (Miyashita et al 1991b; Stryker 1991), such as the ability to recognize an object despite different views that result from movement of the observer or the object (Hasselmo et al 1989; Perrett et al 1985). The experimental results reviewed here support the hypothesis that the inferior temporal cortex is a brain region in which visual perception meets memory and imagery.

Because of space limitation, I am unable to cover such important topics as selective attention and related extraretinal input into temporal cortical neurons (reviewed by LaBerge & Brown 1989; Posner & Petersen 1990; Wise & Desimone 1988), although they are inseparable from visual perception. I also restrict my discussion to the results from primates, unless otherwise mentioned, because of the extensive physiological data available for the relevant cognitive functions.

FUNCTIONAL ASPECTS FROM ANATOMICAL AND BEHAVIORAL EVIDENCE

Early evidence that the temporal cortex has a crucial role in memory and perception of objects came largely from anatomical and behavioral experiments (reviewed by Felleman & Van Essen 1991; Milner 1990; Mishkin 1990; Squire & Zola-Morgan 1991).

Neural Connections of Inferior Temporal Cortex

The inferior temporal cortex receives visual input from areas V4, V4t, DP, VOT in the prestriate cortex; areas TF and TH on the parahippocampal gyrus; area TG at the temporal pole; and areas FST at the fundus and STP at the dorsal bank of the superior temporal sulcus, as well as massive connections within the cortex (Felleman & Van Essen 1991; Rockland & Pandya 1979; Shiwa 1987). Among them, V4 was considered to provide a major source of retinal information. However, inferior temporal neurons could clearly respond to visual stimuli after lesioning of V4, although their response was less discriminative and weaker than normal (Desimone et al 1990). Relative contributions from other input have not yet been quantitatively examined. The inferior temporal cortex, in turn, projects to the limbic systems, such as the amygdaloid nuclei (Amaral & Price 1984; Turner et al 1980), and to the hippocampus via the entorhinal cortex (Insausti et al 1987; Van Hoesen & Pandya 1975), as well as to areas 46 and 8a in the frontal cortex (Goldman-Rakic 1988; Seltzer & Pandya

1989a).[1] The anatomical position of inferior temporal cortex is thus referred to as the final link from visual cortices to limbic systems and frontal lobe.

Hypotheses Based on Neuropsychological Evidence

The inferior temporal cortex participates in both visual perception and memory in humans and monkeys. Patients with right anterior temporal lobectomy are mildly impaired in perceptual tasks in which the normal redundancy of the stimuli has been reduced, as in Mooney's Closure Face Test (Milner 1990). They are also markedly impaired in recognition memory for complex visual patterns (such as faces and irregular abstract designs), which cannot be easily coded verbally (Kimura 1963; Milner 1968). These deficits are unrelated to the extent of hippocampal removal in temporal lobectomy (Burke & Nolan 1988; Milner 1990). Thus, the temporal neocortex is itself critically involved in such representational memory. This result closely parallels the well-established findings in monkeys that bilateral excision of the anterior inferior temporal cortex produces a severe and lasting deficit in visual object recognition (Mishkin 1982). Visual pattern discrimination is also impaired in the anterior temporal lesion, but the more posterior, temporo-occipital lesions produce the most severe perceptual deficits in the monkey (Cowey & Gross 1970; Iwai & Mishkin 1969).

Two aspects further characterize the role of this cortex. First, the anterior portion of the inferior temporal cortex was hypothesized to be involved in storing the "prototype" of a visual object (Weiskrantz 1990; Weiskrantz & Saunders 1984). Earlier research showed impairments in size constancy in monkeys with lesions of inferior temporal cortex (Humphrey & Weiskrantz 1969; Ungerleider et al 1977). More recently lesions of this cortex impaired the discrimination of objects transformed by size, orientation, or shadow configuration after learning in untransformed

---

[1] Inferior temporal cortex has been subdivided mainly along the anteroposterior axis, the dorsoventral axis, or a mixture of both (reviewed by Felleman & Van Essen 1991). The distinction between posterior and anterior subdivisions is useful functionally (see below). However, the cytoarchitectonic subdivisions in the Macaque monkey vary from author to author; for example, in the banks of the superior temporal sulcus, Seltzer & Pandya (1978, 1989b) distinguished several subregions, but other authors' schemes differ from this (Felleman & Van Essen 1991). Areas between the rhinal sulcus and the anterior middle temporal sulcus are also classified differently by Van Hoesen & Pandya (1975), Turner et al (1980), Pandya & Yeterian (1985) and Suzuki & Amaral (1990). In this article, I refer to Felleman & Van Essen (1991) unless otherwise mentioned. However, further modifications of the subdivisions are likely necessary, according to the progress in functional characterization (see below).

mode (Weiskrantz & Saunders 1984). The concept of "object-centered representation" is a counterpart in the computational approach (Marr & Nishihara 1978). Single-unit recording data support the hypothesis, as we see in later sections.

Second, there is now compelling evidence that the temporal neocortex contributes to the visual recognition process differently than the limbic cortex does (reviewed by Miyashita et al 1991b; Squire & Zola-Morgan 1991). The distinction can best be interpreted in terms of the hypothesis that memory has at least two components. One is the recently acquired, labile memory that can be readily disrupted by head injury, as retrograde amnesia demonstrates clinically (Russell 1971). The other is the remote, fully consolidated memory (Squire et al 1975). Drug applications selectively depress or facilitate the labile component (McGaugh 1989).

Bilateral damage to the medial temporal region, which includes the hippocampus, amygdala, and adjacent cortex (Milner et al 1968), or only hippocampal CA1 field (Zola-Morgan et al 1986), accompanied a limited, short-span retrograde amnesia in humans. In monkeys, effects of hippocampal lesions (including the entorhinal and parahippocampal cortex) were tested on the retention of 100-object discrimination problems (Zola-Morgan & Squire 1990). The monkeys were severely impaired at remembering recently learned objects, but they remembered objects learned long ago as well as normal monkeys did. These data suggest that the labile component of the long-term memory is localized in the hippocampus and/or adjacent cortex, or is strongly influenced by these structures. Presumably, a later and more consolidated stage of long-term memory is represented in anterior temporal cortex (Miyashita et al 1991b). I discuss recent evidence for the hypothetical memory function of temporal cortex in later sections.

STIMULUS-CODING OF SINGLE CELLS

To know how objects are represented in the neural network of inferior temporal cortex, we must first examine the stimulus-coding properties of single neurons. Early evidence that single inferior temporal neurons respond to visual stimuli came largely from experiments by Gross and colleagues (Gross 1972). Several subsequent studies focused on perceptual, rather than memory, functions and revealed many interesting features of single cells, especially the selective responses to complex objects, such as hands (Desimone et al 1984; Gross et al 1972) and faces (Bruce et al 1981; Perrett et al 1982, 1985; Rolls & Baylis 1986). Here, I summarize progress in the last few years.

Generality of Complex Feature-Coding Cells

Although the presence of cells that selectively respond to the sight of faces and hands was well documented, evidence for other classes of selectivity was meager. Miyashita & Chang (1988) found anterior temporal cortical cells that responded highly selectively to computer-generated fractal pictures (Miyashita et al 1991a) in monkeys who were trained to perform a delayed matching-to-sample task. When the responses to a large set (usually 100) of fractal stimuli were measured, individual neurons commonly responded well to only a small fraction of the stimuli. Different cells responded to different selections from among the stimuli. This result agrees with the findings that some cells were selectively activated by Fourier descriptors (Gochin et al 1991; Schwartz et al 1983).

A different approach was taken by Tanaka et al (1991), who used anesthetized and paralyzed monkey preparation to examine responses of a single cell with an extensive set of stimuli, including real objects and bars and disks. They then carefully determined the stimulus features necessary for maximal activation of the cell. They found that 53% of the cells in the anterior inferior temporal cortex required more complex features than simple bars or disks. The critical features for maximal activation varied from a star- or T-like shape to a complicated combination of shape, color, and texture. Only 12% of tested cells could be maximally activated by the simple bars or disks.

In an earlier study, Desimone et al (1984) stated, "Many IT cells responded equally to nearly every stimulus tested, and most of the stimulus-selective cells gave at least a small response to virtually every stimulus tested, especially visually complex stimuli." Recent results favor the view that the perceptual alphabets represented in single inferior temporal neurons are more selective to certain critical features, although the discrepancy with the previous views may mostly result from a difference in emphasis.

Invariance with Size, Orientation, and Retinal Position

Response invariance across retinal position within the receptive field has been reported since the early discovery that inferior temporal cells have a large receptive field (Gross 1972). Effects of stimulus size, retinal position, or contrast upon single-cell responses were quantitatively described with Fourier descriptor stimuli (Schwartz et al 1983). For two-thirds of the tested cells, the optimum frequency of the Fourier descriptor and the tuning curve along the frequency remained similar over the range of size, although the absolute levels of the response varied. Similar invariance was obtained over retinal translation and constant reversal. Tanaka et al (1991) reported the variety of tolerance to size change and the sharp selectivity for the stimulus orientation. These results were mainly obtained in the anterior part of middle temporal gyrus. When the neurons that were related to a delayed matching-to-sample task were tested with the fractal patterns in the anterior part of inferior temporal gyrus, a higher degree of invariance was found (Miyashita & Chang 1988). Not only the relative stimulus selectivity, but also the absolute levels of the response, remained the same in the majority of tested cells, when sample pictures were manipulated by size reduction, rotation by 90° in a clockwise direction, and transformation of color stimuli into monochrome. Similarly high degrees of response invariance were also found for the face neurons in the superior temporal sulcus (Perrett et al 1985, 1987; Rolls & Baylis 1986). These results support the hypothesis that the anterior portion of inferior temporal cortex is involved in storing the prototype of a visual object (Weiskrantz 1990; Weiskrantz & Saunders 1984).

Functional Architecture

In contrast to striate cortex, columnar organization has not been demon-strated in inferior temporal cortex. However, recent results suggest its occurrence. First, a pair of neurons that were recorded simultaneously on a single electrode (within about 100 μm) had more related feature selectivity than those recorded on different electrodes (Fujita et al 1990; Gochin et al 1991; Higuchi et al 1991). The cells recorded on a single electrode shared selectivity to geometrical shapes or a memory-related property (Higuchi et al 1991). Cross-correlation analysis also showed that occurrence of shared input was more frequent for neuron pairs recorded on the same electrode (Gochin et al 1991). Second, recording a succession of cells along closely spaced microelectrode penetrations revealed a compact cluster of cells that responded selectively to the fractal patterns, running across cortical layers (Higuchi et al 1991). Fujita et al (1991) showed that when electrodes were advanced nearly at a right angle to the cortical surface, they obtained neurons with related stimulus selectivity over a distance of 0.6–1.4 mm, whereas they did so only over 0.2–0.5 mm when electrodes were advanced at an angle of 45°. In the latter type of penetrations, there were two or three separate clusters of neurons that showed similar selectivity, with the gap between adjacent clusters being 0.4–1 mm. All of these results support the conclusion that anterior inferior temporal cortex consists of modules in which neurons with related selectivity cluster across cortical layers.

In the cat visual cortex, oscillatory firing patterns in the frequency range of 40–60 Hz can synchronize across orientation columns with a spatial separation of up to 7 mm (Gray et al 1989). The synchronization of feature-detecting neurons may serve as a mechanism for the binding of different features of an object. This feature-binding mechanism, especially the possible synchronization across cell clusters described above, might be relevant to inferior temporal cortex. Gawne et al (1991) used spectral analysis of local field potentials. With stationary Walsh patterns as the stimuli, the frequency of the peak between 20 and 74 Hz was not significantly related to the pattern of the stimulus at the majority of recording sites, and the pattern accounted for only $8.1 \pm 2.3\%$ of the variance of the frequency. Young et al (1992) used Tanaka's method to activate inferior temporal neurons in anesthetized monkeys effectively, but multi-unit activity autocorrelograms showed no oscillations. On the other hand, Nakamura et al (1991) found neurons with oscillatory activity coupled to colored photographs, especially to familiar objects, in the ventral part of temporal cortex in the monkey performing a visual discrimination task. However, the most common oscillation frequencies were 5–6 Hz. Although the role, or even the existence, of oscillatory activity in temporal cortex is controversial, it remains a challenge how activities in spatially separate modules with related stimulus selectivity correlate themselves.

Regional Differences of Coded Features

The inferior temporal cortex occupies a large area of the primate brain (7.7% of neocortex) (Felleman & Van Essen 1991), and the cortex is probably divided into several subareas with different response selectivity. The region in the fundus of the superior temporal sulcus (area TPO) is well documented to contain a high density of face-selective cells (Bruce et al 1981; Perrett et al 1982, 1985, 1987). Face cells are also found in the ventral bank of the sulcus (TEa and TEm), but few are observed in TE2-TE1 (Baylis et al 1987; Perrett 1987). The cells selectively responsive to complex fractal patterns were found more in the anterior part of inferior temporal gyrus than in the middle temporal gyrus (Miyashita et al 1991b). Baylis et al (1987) related neural response properties in temporal cortex to cyto- and myelo-architectural subdivisions (Seltzer & Pandya 1978). According to Baylis et al's hierarchical cluster analysis, TE3-TEm had little similarity with TE2-TE1. However, a potential qualification to this distinction is that the percentage of selectively tuned neurons in these areas is too low (Figure 6, Baylis et al 1987), compared with the results by Tanaka et al (1991). Tanaka et al (1991) found that in the posterior one-third or one-fourth of inferior temporal cortex (corresponding to TEO of Iwai & Yukie 1987; or PIT of Felleman & Van Essen 1991), most cells could be maximally activated by bars or disk, whereas in the anterior two-thirds or three-quarters (TE$_d$ of Iwai & Yukie 1987; AIT of Felleman &

Van Essen 1991), most cells required more complex features. They failed to find differential distributions in the latter area of cells with different properties. Further insight into the functional subdivisions of this cortex is essential to understanding the detailed computational structures of visual perception.

NEURAL CODE OF ASSOCIATIVE LONG-TERM MEMORY

None of the neurophysiological experiments reviewed above tested whether the selectivity of these neural responses is innately determined or developed during the early critical period (as in some neurons in the striate cortex), or acquired through learning during adulthood. A well-designed experimental strategy was necessary to identify neurons serving visual long-term memory.

One effective strategy was to have monkeys memorize an artificial associative relation among pictures and then to examine whether picture-selective activities of temporal cortical neurons reflect stimulus-stimulus association imposed during learning (Miyashita et al 1991b). If the artificial associative relation among pictures tends to affect the stimulus-selectivity of neurons, the neural selectivity is acquired through learning and represents a neuronal correlate of the associative long-term memory of pictures. So far, two series of such experiments have been successful. First, a visual delayed matching-to-sample task was used, and the effect of a fixed-order presentation of the patterns during the training session was examined (Miyashita 1988). Second, a visual pair-association learning task (Murray et al 1988) was used, and the effect of association was directly examined (Sakai & Miyashita 1991) (Figure 1). Lesion studies have demonstrated that monkeys with bilateral removal of the medial temporal region could learn neither the delayed matching-to-sample task (Mishkin 1982; Squire & Zola-Morgan 1991; Zola-Morgan et al 1989) nor the pair-association task (Murray et al 1988). The type of memory these tasks employed would, therefore, correspond to one that relies on the integrity of these structures.

Long-Term Memory Neurons in a Matching-to-Sample Task

In a trial of the visual delayed matching-to-sample task, sample and match stimuli were successively presented on a video monitor, each for 0.2 sec at a 16 sec delay interval. A set of 97 color patterns was generated by a fractal algorithm with a 32-bit seed of random numbers (Miyashita et al 1991a); the set (learned stimuli) was repeatedly used during a training session in a fixed sequence according to an arbitrary attached number (serial position

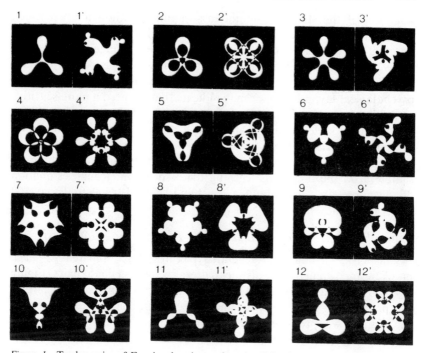

Figure 1 Twelve pairs of Fourier descriptors for stimuli in the pair-association learning task (Sakai & Miyashita 1991).

number, SPN). While extracellular discharges of a neuron were recorded, a sample stimulus was selected not only from the 97 learned patterns, but also from a new set of 97 patterns (new stimuli). Different sets of new stimuli were created for each neuron by using the same algorithm but a different seed. If the consecutively presented patterns tended to be associated together, and if the association was fixed in the choice of effective patterns for a cell, the effective patterns would be correlated along the SPN, in spite of a random presentation of the stimuli during the unit-recording session. The experimental results supported this idea (Miyashita 1988). The effective responses to the learned stimuli indeed cluster along the SPNs, and the clustering was not caused by an artifact in the testing procedure, because the responses simultaneously obtained from the new stimuli were not clustered. Therefore, these neurons are a good candidate for the visual associative long-term memory stores, or at least reflect long-term storage.

It is interesting to ask how many times the monkey should see and/or memorize the fractal pattern that eventually develops the neural representation of the stimulus-stimulus associations. Although a single-cell activity was held by the microelectrode for several hours in this experiment, selectivity to new patterns exhibited no systematic changes (Miyashita et al 1992, unpublished). This may support the view that the cortical memory represents a relatively stable component (see the section on Neuropsychological Hypothesis), or may indicate that we need a more sensitive method to detect the effects of association in the early learning phase. With a longer time scale, no direct answer is available from the experimental procedure described above, because two factors were confounded in the training: One factor was learning the rule of the delayed matching-to-sample task; the other was learning the association. Rough estimation of the upper limit was about a few hundred trials (500 trials/day/100 patterns × 30 days).

Long-Term Memory Neurons in a Pair-Association Task

In this task, 24 computer-generated pictures were prepared for each monkey, and geometrically distinct patterns were sorted into pairs (Figure 1, Sakai & Miyashita 1991). The combination of the paired associates is not predictable, unless they are memorized beforehand. In each trial, a cue stimulus was presented on a video monitor for 1 s. After the 4 s delay period, a choice of two stimuli, the paired associate of the cue and one from a different pair, was shown. Monkeys obtained fruit juice as a reward for correctly touching the paired associate within 1.2 s.

Picture-selective neural responses during the cue period were found in the anterior inferior temporal cortex. Many neurons responded reproducibly to only a few pictures. The cell might have responded to geometrically similar patterns, but in many cases, the strongest and second-strongest responses were ascribed to a particular pair that had no apparent geometrical similarity. Some other cells showed broader tuning and responded to more than three pictures. Nevertheless, paired pictures were among the most effective stimuli for these cells. This type of cell was called a "pair-coding neuron," which manifests selective cue responses to both pictures of the paired associates. These responsive cells tended to be located near one another (1–2 mm in width).

The associational property was analyzed by calculating two coupling indices for each neuron as follows:

$$CI_p = \frac{1}{N_p} \sum_i \sum_{\substack{j \\ i<j, j=i'}} \frac{(x_i-b)(x_j-b)}{(x_{\text{best}}-b)(x_{\text{2nd-best}}-b)} * 100,$$

with $j = i'$ for paired associates,

$$CI_r = \frac{1}{N_r} \sum_i \sum_{\substack{j \\ i<j, j\neq i'}} \frac{(x_i-b)(x_j-b)}{(x_{\text{best}}-b)(x_{\text{2nd-best}}-b)} * 100,$$

with $j \neq i'$ for random combinations,

where x_i denotes a mean discharge rate during the cue period for the ith picture (the ith and i'th pictures belong to a pair), b is a spontaneous discharge rate, x_{best} and $x_{\text{2nd-best}}$ are mean discharge rates for the best and second-best cue-optimal pictures in each cell, and Np and Nr are the total number of combinations for two cases.

One coupling index (denoted as CI_p) measures correlated neural responses to paired associates, whereas the other coupling index (CI_r) estimates responses to other random combinations among 24 pictures. The latter index CI_r serves as an experimental control for untrained association between two pictures. For each cell, a pair index (PI) was defined as equal ($CI_p - CI_r$). The analysis of frequency distribution of PI values demonstrated that the paired associates elicited significantly correlated responses. It was therefore hypothesized that the selectivity of these neurons was acquired through learning of the pair-association task. The cellular mechanisms for these phenomena are not yet clear, but a possible basis for the memory coding lies in the change of synaptic connections through repetitive learning (see below).

Associational Mechanism and Perception

The visual recognition of three-dimensional objects on the basis of their shape poses many difficult problems (Marr & Nishihara 1978), some of which were proposed as being solved in inferior temporal cortex (see the section on Neuropsychological Evidence). Here, I only address the problem of the object view relative to the viewer. The visual image of an object varies not only with the distance, but also with the angle at which it is observed. In the former situation, only the size, not the shape, of the object image changes. The computational problem for this situation is relatively simple, and the invariance of the neural response for this requirement has already been discussed. However, in the latter situation, the two-dimensional shapes of the object themselves change. How can our visual system recognize an object under such a situation? If the central representation of objects is encoded with a viewer-centered coordinate, the visual system would require the storage of a separate appearance for different viewing angles as separate conjunctions of trigger features, obviously a highly inefficient representation. Another proposal was a construction of object-centered three-dimensional representation (Marr & Nishihara 1978). In the investigation of face-selective neurons, some neurons responded differentially to the faces of different individuls, irrespective of the viewing angle (Hasselmo et al 1989; Perrett et al 1985). Some other neurons that responded only when a head underwent rotatory movements were activated by a particular movement independent of the orientation of the moving head (Hasselmo et al 1989). These data indicate existence of the object-centered encoding in temporal cortex. But how can a neuron acquire such selectivity, in spite of dramatic changes in the appearance of the object?

The experiments on visual long-term memory neurons established the idea that the temporal-lobe neurons can reflect learning to associate stimuli on the basis of temporal contiguity, as discussed above. In our experience of the visual world, the different views of an object are nearly always presented in succession, transformed by the movement of the object or the observer. If the temporal cortex contains a general associational mechanism as described, then these views would automatically become associated. This obviates the need for complicated computational processes for performing geometrical transformations or for hierarchically associating conjunctions of trigger features. This hypothesis might be supported by the fact that the local clusters of face neurons contain both view-angle-dependent neurons and independent neurons; thus, the latter neurons emerge locally (Perrett et al 1985). "Pair-coding" neurons are also likely to emerge locally (Sakai & Miyashita 1991). Other supporting evidence came from psychophysics and computational theory. A recent computational theory provided powerful associational algorithms that can learn from a limited number of two-dimensional views of an object and that can interpolate to decide whether new different views are of the same scene (Poggio & Edelman 1990). This proposal was compared with one that needs alignment of internal three-dimensional object models for viewpoint normalization (Ullman 1989). A psychophysical experiment involving computer-generated three-dimensional objects revealed anisotropic generalization to novel views, which supports the two-dimensional view interpolation mechanism (Bulthoff & Edelman 1992). Careful comparison of physiological, psychophysical, and computational data may provide further support for or evidence against the hypothesis.

NEURAL MECHANISMS UNDERLYING RETRIEVAL PROCESS

Electric stimulation of the temporal lobe induces visual experiential response and hallucinations in humans (Penfield & Perot 1963). Although many researchers confirmed evoked complex hallucinations, the original Penfield hypothesis that an anatomical site was related to a specific, repeatable imagery, and thus to a specific "memory record," has been controversial (Halgren et al 1978). No other clues to the memory retrieval process have been reported.

In the pair-association task, Sakai & Miyashita (1991) found another type of neuron with picture-selective activities during the delay period (Figure 2). One picture elicited the strongest response during the cue period from a single neuron. In the trial in which the paired associate of this cue-optimal picture was used as a cue, the same cell exhibited the highest tonic activity during the delay period, in contrast to a weak response during the cue period. This delay activity gradually increased, until the choice of stimuli appeared. Furthermore, the paired associate of the second-best cue-optimal picture still elicited a sustained activity during the delay

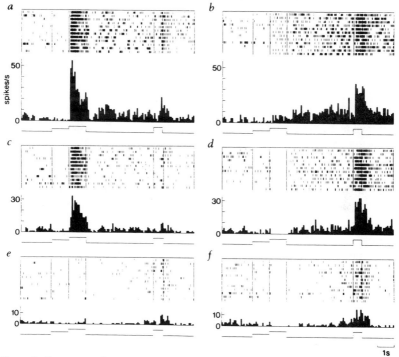

Figure 2   Responses of a pair recall neuron that exhibited picture-selective activity during the delay period, presumably reflecting retrieval of the paired associate. (*a*) Trials for cue 12 (see Figure 1) that elicited the strongest cue responses. (*b*) Trials for cue 12′. (*c*) Trials for cue 1 that resulted in the second-strongest cue response. (*d*) Trials for cue 1′. Note the sustained delay activity and the inhibitory cue response. (*e*) Trials for cue 3 that elicited no response. (*f*) Trials for cue 3′ (Sakai & Miyashita 1991).

period. Other pictures evoked weak or no response. The delay activities were confined to a few cue stimuli in the set. This type of cell was called a "pair-recall neuron," in which the paired associate of a cue-optimal picture elicited the highest delay activity. The delay activity of the pair-recall neurons does not represent mere sensory after-discharge, because it is stronger than the cue response. Significant augmentation of discharge rates was observed for the highest delay activity, whereas delay activity elicited by a cue-optimal picture itself was significantly reduced during the delay period.

Anticipatory neural activities that precede the initiation of movements and increase during the preparatory period have been reported in the primate frontal cortex (Bruce & Goldberg 1985; Funahashi et al 1989). In the pair-association task, the increasing delay activity of pair-recall neurons is not related to motor response, because the monkey could not predict which position should be touched. As noted above, this delay activity is not only picture-selective, but also closely coupled with the paired associate that is not actually seen but retrieved. The neural mechanism for the retrieval process remains to be identified, but it may well involve the pair-recall neurons.

CONSOLIDATION

Retrograde amnesia was thought to be indicative of the existence of con-solidation process (see the section on Neuropsychological Evidence), by which certain evanescent information obtains an enduring representation in long-term memory (McGaugh 1989; Milner 1968, 1990; Squire & Zola-Morgan 1991). Although little is known on the neural basis of the process, two issues deserve discussion here. First is the nature of the signal for association. Structural changes in synapses from which long-lasting memory engrams must result (McGaugh 1989) may be regulated by a molecular machinery that is shared with the process of growth and develop-ment (Kandel 1993). There is compelling evidence that temporal cor-relations in the action potential patterns of young visual synapses deter-mine their relative positions within local regions of neuropile and their convergence onto the same sets of post-synaptic cells (Constantine-Paton et al 1990). This can be shown by the experimentally induced reorgan-ization of inputs produced when two eyes innervate a single tectum in amphibians. The mechanism is strikingly similar to that of the ocular dominance column segregation in mammalian visual cortex (Stryker & Harris 1986). Because the primate temporal cortical neurons can learn to associate stimuli on the basis of temporal contiguity, as discussed above, I propose the possibility that such an associational mechanism based on temporal contiguity may be a general property for consolidation of structural changes in synapses, thus encompassing both memory and development.

Second, consolidation was most simply assumed as a transformation or fixation of short-term memory into long-term memory, which, in visual memory of objects, is represented in the stimulus-selectivity of inferior temporal neurons. The neural basis of short-term memory has been characterized only partially. Evidence has suggested that short-term memory required on-going patterns of activity (Fuster & Jervey 1982; Gnadt & Andersen 1988; Miyashita & Chang 1988). Convergence of elevated activities from different items in the short-term store may lead to temporally correlated synaptic activation and resultant associative structural changes in synapses. On the other hand, another line of evidence suggested that the formation of short-term memory consisted of the modification of synaptic weights, such that familiar, expected, or recently seen stimuli cause the least activation of the cell (Baylis & Rolls 1987; Miller et al 1991; Rolls et al 1989). It will be of critical importance to examine the functional roles of the feedback afferents from the hippocampus to inferior temporal cortex via entorhinal cortex (Insausti et al 1987) in integrating these fragmentary observations into an understanding of the neural proceses of consolidation.

CONCLUSIONS

This review covered new findings on the role of inferior temporal cortex, especially in associative long-term memory and in the integration of perceptual and memory capacities. Three neural bases are crucial to an understanding of this role. First is to know how objects are encoded in single cells. The experimental results indicate that the anterior inferior temporal cortex consists of modules in which neurons with related selectivity cluster across cortical layers. The single-cell selectivity in modules is more sharply tuned than previously believed. The question of how activities in different, but related, modules can be correlated remains open, despite increasing interest and experimentation on the possible contribution of 40–60 Hz oscillatory activity to feature binding. Another challenging question is whether microstimulation or selective pharmacological manipulation of these modules can affect monkey behavior, as successfully demonstrated in area MT (Salzman et al 1990).

Second, the experiments that revealed neuronal correlates of visual long-term memory established the idea that the temporal-lobe neurons can reflect learning to associate stimuli on the basis of temporal contiguity. This idea has many implications. Computational theories had raised the question of how object-centered representation could be constructed from images at different viewing angles. The associational mechanism obviates the need for complicated computational processes. A psychophysical experiment revealed anisotropic generalization to novel views, and thus supported the two-dimensional view interpolation mechanism.

Third is the role of limbic afferents. Cortical memory is likely to be consolidated over time under the influence of the afferents from medial temporal region. Investigation of the time course of this process in single cortical neurons could be the first target of research. The signal content of the limbic afferents onto cortical memory neurons may also be an experimentally accessible question. The nature of the cortico-limbic interaction will probably constitute a major experimental and conceptual challenge in this field in the near future.

Finally, we have now obtained at the single-cell level a first clue to memory retrieval mechanisms. From where and how the retrieval-related activity in the pair-recall neurons originates is unknown, but would be experimentally accessible. The inferior temporal cortex should be a strategic place to understand this intriguing topic in the next several years.

ACKNOWLEDGMENT

This work was supported by a grant from the Japanese Ministry of Education, Science and Culture (02102008), a grant from the Japanese Ministry of Welfare, and a grant from the Human Frontier Science Program.

Literature Cited

Amaral, D. G., Price, J. I. 1984. Amygdalo-cortical projections in the monkey (*Macaca facicularis*). *J. Comp. Neurol.* 230: 465–96

Baylis, G. C., Rolls, E. T. 1987. Responses of neurons in the inferior temporal cortex in short-term and serial recognition memory tasks. *Exp. Brain Res.* 65: 614–22

Baylis, G. C., Rolls, E. T., Leonard, C. M. 1987. Functional subdivisions of the temporal lobe neocortex. *J. Neurosci.* 7: 330–42

Bruce, C. J., Desimone, R., Gross, C. G. 1981. Visual properties of neurons in a polysensory area in superior temporal sulcus of the macaque. *J. Neurophysiol.* 46: 369–84

Bruce, C. J., Goldberg, M. E. 1985. Primate frontal eye fields. I. Single neurons discharging before saccades. *J. Neurophysiol.* 53: 606–35

Bülthoff, H. H., Edelman, S. 1992. Psychophysical support for a two-dimensional view interpolation theory of object recognition. *Proc. Natl. Acad. Sci. USA* 89: 60–64

Burke, T., Nolan, J. R. M. 1988. Material-specific memory deficit after unilateral temporal neocorticectomy. *Soc. Neurosci. Abstr.* 14: 1289

Constantine-Paton, M., Cline, H. T., Debski, E. 1990. Patterned activity, synaptic convergence, and the NMDA receptor in developing visual pathways. *Annu. Rev. Neurosci.* 13: 129–54

Cowey, A., Gross, C. G. 1970. Effects of foveal prestriate and inferotemporal lesions on visual discrimination by rhesus monkeys. *Exp. Brain Res.* 11: 128–44

Desimone, R., Albright, T. D., Gross, C. G., Bruce, C. J. 1984. Stimulus-selective properties of inferior temporal neurons in the macaque. *J. Neurosci.* 4: 2051–62

Desimone, R., Li, L., Lehky, S., Ungerleider, L. G., Mishkin, M. 1990. Effects of V4 lesions on visual discrimination performance and on responses of neurons in inferior temporal cortex. *Soc. Neurosci. Abstr.* 16: 621

Felleman, D. J., Van Essen, D. C. 1991. Distributed hierarchical processing in the primate cerebral cortex. *Cereb. Cortex* 1: 1–47

Fujita, I., Cheng, K., Tanaka, K. 1991. Neurons with related stimulus selectivity are clustered in the Macaque inferotemporal cortex. *Soc. Neurosci. Abstr.* 17: 1283

Fujita, I., Cheng, K., Tanaka, K. 1990. Stimulus selectivity of inferotemporal cortex neurons: Simultaneous recording from adjacent cells. *Soc. Neurosci. Abstr.* 16: 1220

Funahashi, S., Bruce, C. J., Goldman-Rakic, P. S. 1989. Mnemonic coding of visual space in the monkey's dorsolateral prefrontal cortex. *J. Neurophysiol.* 61: 331–49

Fuster, J. M., Jervey, J. P. 1982. Neuronal firing in the inferotemporal cortex of the monkey in a visual memory task. *J. Neurosci.* 2: 361–75

Gawne, T. J., Eskandar, E. N., Richmond, B. J., Optican, L. M. 1991. Oscillations in the responses of neurons in inferior temporal cortex are not driven by stationary visual stimuli. *Soc. Neurosci. Abstr.* 17: 443

Gnadt, J. W., Andersen, R. A. 1988. Memory-related motor planning activity in posterior parietal cortex of macaque. *Exp. Brain Res.* 70: 216–20

Gochin, P. M., Miller, E. X., Gross, C. G., Gerstein, G. L. 1991. Functional interactions among neurons in inferior temporal cortex of the awake macaque. *Exp. Brain Res.* 84: 505–16

Goldman-Rakic, P. S. 1988. Topography of cognition: parallel distributed networks in primate association cortex. *Annu. Rev. Neurosci.* 11: 137–56

Gray, C. M., Konig, P., Engel, A. K., Singer, W. 1989. Oscillatory responses in cat visual cortex exhibit intercolumnar synchronization which reflects global stimulus properties. *Nature* 338: 334–37

Gross, C. G. 1972. Visual functions of inferotemporal cortex. In *Handbook of Sensory Physiology*, ed. R. Jung, 8/3B: 451–82. Berlin: Springer-Verlag

Gross, C. G., Rocha-Miranda, C. E., Bender, D. B. 1972. Visual properties of neurons in inferotemporal cortex of the Macaque. *J. Neurophysiol.* 35: 96–111

Halgren, E., Walter, R. D., Cherlow, D. G., Crandall, P. H. 1978. Mental phenomena evoked by electrical stimulation of the human hippocampal formation and amygdala. *Brain* 101: 83–117

Hasselmo, M. E., Rolls, E. T., Baylis, G. C.,

Nalwa, V. 1989. Object-centered encoding by face-selective neurons in the cortex in the superior temporal sulcus of the monkey. *Exp. Brain Res.* 75: 417–29

Hawkins, R. D., Kandel, E. R., Siegelbaum, S. A. 1993. Learning to modulate transmitter release: Themes and variations in synaptic plasticity. *Annu. Rev. Neurosci.* 16: 425–65

Higuchi, S., Sakai, K., Miyashita, Y. 1991. Functional architecture of pictorial memory neurons in the primate anteroventral temporal cortex. *Neurosci. Res.* 14: S61

Humphrey, N. K., Weiskrantz, L. 1969. Size constancy in monkeys with inferotemporal lesions. *Q. J. Exp. Psychol.* 21: 225–38

Insausti, R., Amaral, D. G., Cowan, W. M. 1987. The entorhinal cortex of the monkey: II. Cortical afferents. *J. Comp. Neurol.* 264: 356–95

Iwai, E., Mishkin, M. 1969. Further evidence on the locus of the visual area in the temporal lobe of the monkey. *Exp. Neurol.* 25: 585–94

Iwai, E., Yukie, M. 1987. Amygdalofugal and amygdalopetal connections with modality-specific visual cortical areas in macaques (*Macaca fuscata, M. mulatta, and M. facicularis*). *J. Comp. Neurol.* 261: 362–87

Kimura, D. 1963. Right temporal-lobe damage. *Arch. Neurol.* 8: 264–71

LaBerge, D., Brown, V. 1989. Theory of attentional operations in shape identification. *Psychol. Rev.* 96: 101–24

Marr, D., Nishihara, H. K. 1978. Representation and recognition of the spatial organization of three dimensional structure. *Proc. R. Soc. London Ser. B* 200: 269–94

Maunsell, J. H. R., Newsome, W. T. 1987. Visual processing in monkey extrastriate cortex. *Annu. Rev. Neurosci.* 10: 363–401

McGaugh, J. L. 1989. Involvement of hormonal and neuromodulatory systems in the regulation of memory storage. *Annu. Rev. Neurosci.* 12: 255–87

Miller, E. K., Li, L., Desimone, R. 1991. A neural mechanism for working and recognition memory in inferior temporal cortex. *Science* 254: 1377–79

Milner, B. 1990. Right temporal-lobe contribution to visual perception and visual memory. In *Vision, Memory and the Temporal Lobe*, ed. E. Iwai, M. Mishkin, pp. 43–53. New York: Elsevier

Milner, B. 1968. Visual recognition and recall after right temporal-lobe excision in man. *Neuropsychologia* 6: 191–209

Milner, B., Corkin, S., Teuber, H. L. 1968. Further analysis of the hippocampal amnesic syndrome: 14-year follow-up study of H.M. *Neuropsychologia* 6: 215–34

Mishkin, M. 1990. Vision, memory, and the temporal lobe: Summary and perspective. In *Vision, Memory and the Temporal Lobe*, ed. E. Iwai, M. Mishkin, pp. 427–36. New York: Elsevier

Mishkin, M. 1982. A memory system in the monkey. *Philos. Trans. R. Soc. London Ser. B* 298: 85–95

Mishkin, M., Ungerleider, L. G., Macko, K. A. 1983. Object vision and spacial vision: Two cortical pathways. *Trends Neurosci.* 6: 414–17

Miyashita, Y. 1988. Neuronal correlate of visual associative long-term memory in the primate temporal cortex. *Nature* 335: 817–20

Miyashita, Y., Chang, H. S. 1988. Neuronal correlate of pictorial short-term memory in the primate temporal cortex. *Nature* 331: 68–70

Miyashita, Y., Higuchi, S., Sakai, K., Masui, N. 1991a. Generation of fractal patterns for probing the visual memory. *Neurosci. Res.* 12: 307–11

Miyashita, Y., Masui, N., Higuchi, S. 1991b. Primal long-term memory in the primate temporal cortex: Linkage between visual perception and memory. In *Representation of Vision*, ed. A. Gorea, pp. 141–52. Cambridge: Cambridge Univ. Press

Miyashita, Y., Rolls, E. T., Cahusac, P. M. B., Niki, H., Feigenbaum, J. D. 1989. Activity of hippocampal neurons in the monkey related to a stimulus-response association task. *J. Neurophysiol.* 61: 669–78

Murray, E. A., Gaffan, D., Mishkin, M. 1988. Role of the amygdala and hippocampus in visual-visual associative memory in rhesus monkeys. *Soc. Neurosci. Abstr.* 14: 2

Nakamura, K., Mikami, A., Kubota, K. 1991. Unique oscillatory activity related to visual processing in the temporal pole of monkeys. *Neurosci. Res.* 12: 293–99

Pandya, D. N., Yeterian, E. H. 1985. Architecture and connections of cortical association area. In *Cerebral Cortex*, ed. A. Peters, E. G. Jones, 4: 3–61. New York: Plenum

Penfield, W. P., Perot, P. 1963. The brain's record of auditory and visual experience. A final summary and discussion. *Brain* 86: 595–696

Perrett, D. I., Mistlin, A. J., Chitty, A. J. 1987. Visual neurones responsive to faces. *Trends Neurosci.* 10: 358–64

Perrett, D. I., Rolls, E. T., Caan, W. 1982. Visual neurones responsive to faces in the monkey temporal cortex. *Exp. Brain Res.* 47: 329–42

Perrett, D. I., Smith, P. A., Potter, D. D., Mistlin, A. J., Head, A. S., et al. 1985. Visual cells in the temporal cortex sensitive to face view and gaze direction. *Proc. R. Soc. London Ser. B* 223: 293–317

Poggio, T., Edelman, S. 1990. A network that learns to recognize three-dimensional objects. *Nature* 343: 263–66

Posner, M. I., Petersen, S. E. 1990. The attention system of the human brain. *Annu. Rev. Neurosci.* 13: 25–42

Rockland, K. S., Pandya, D. N. 1979. Laminar origins and terminations of cortical connections of the occipital lobe in the rhesus monkey. *Brain Res.* 179: 3–20

Rolls, E. T., Baylis, G. C. 1986. Size and contrast have only small effects on the responses to faces of neurons in the cortex of the superior temporal sulcus of the monkey. *Exp. Brain Res.* 65: 38–48

Rolls, E. T., Baylis, G. C., Hasselmo, M. E., Nalwa, V. 1989. The effect of learning on the face selective neurones in the cortex in the superior temporal sulcus of the monkey. *Exp. Brain Res.* 76: 153–64

Russel, W. R. 1971. *The Traumatic Amnesia*. London: Oxford Univ. Press

Sakai, K., Miyashita, Y. 1991. Neural organization for the long-term memory of paired associates. *Nature* 354: 152–55

Salzman, C. D., Britten, K. H., Newsome, W. T. 1990. Cortical microstimulation influences perceptual judgments of motion direction. *Nature* 346: 174–77

Schwartz, E. L., Desinome, R., Albreight, T. D., Gross, C. G. 1983. Shape recognition and inferior temporal neurons. *Proc. Natl. Acad. Sci. USA* 80: 5776–78

Seltzer, B., Pandya, D. N. 1989a. Frontal lobe connections of the superior temporal sulcus in the rhesus monkey. *J. Comp. Neurol.* 281: 97–113

Seltzer, B., Pandya, D. N. 1989b. Intrinsic connections and architectonics of the superior temporal sulcus in the rhesus monkey. *J. Comp. Neurol.* 290: 451–71

Seltzer, B., Pandya, D. N. 1978. Afferent cortical connections and architectonics of the superior temporal sulcus and surrounding cortex in the rhesus monkey. *Brain Res.* 149: 1–24

Shiwa, T. 1987. Corticocortical projections to the monkey temporal lobe with particular reference to the visual processing. *Arch. Ital. Biol.* 125: 139–54

Squire, L. R., Slater, P. C., Chace, P. M. 1975. Retrograde amnesia: Temporal gradient in very long term following electroconvulsive therapy. *Science* 187: 77–79

Squire, L. R., Zola-Morgan, S. 1991. The medial temporal lobe memory system. *Science* 253: 1380–86

Stryker, M. P. 1991. Temporal associations. *Nature* 354: 108

Stryker, M. P., Harris, W. A. 1986. Binocular impulse blockade prevents the formation of ocular dominance columns in cat visual cortex. *J. Neurosci.* 6: 2117–33

Suzuki, W. A., Amaral, D. G. 1990. Cortical inputs to the CA1 field of the monkey hippocampus originate from the perirhinal and parahippocampus cortex but not from area TE. *Neurosci. Lett.* 115: 43–48

Tanaka, K., Saito, H., Fukada, Y., Moriya, M. 1991. Coding visual images of objects in the inferotemporal cortex of the Macaque monkey. *J. Neurophysiol.* 66: 170–89

Turner, B. H., Mishkin, M., Knapp, M. 1980. Organization of the amygdalopetal projections from modality-specific cortical association areas in the monkey. *J. Comp. Neurol.* 191: 515–43

Ullman, S. 1989. Aligning pictorial descriptions: an approach to object recognition. *Cognition* 32: 193–254

Ungerleider, L. G., Ganz, L., Pribram, K. H. 1977. Size constancy in rhesus monkeys: Effects of pulvinar, prestriate, and inferotemporal lesions. *Exp. Brain Res.* 27: 251–69

Van Essen, D. C. 1985. Functional organization of primate visual cortex. In *Cerebral Cortex*, ed. A. Peters, E. G. Jones, 3: 259–329. New York: Plenum

Van Hoesen, G. W., Pandya, D. N. 1975. Some connections of the entorhinal (area 28) and perirhinal (area 35) cortices of the rhesus monkey. I. Temporal lobe afferents. *Brain Res.* 95: 1–24

Weiskrantz, L. 1990. Visual prototypes, memory, and the inferotemporal cortex. In *Vision, Memory and the Temporal Lobe*, ed. E. Iwai, M. Mishkin, pp. 13–28. New York: Elsevier

Weiskrantz, L., Saunders, R. C. 1984. Impairments of visual object transforms in monkeys. *Brain* 107: 1033–72

Wise, S. P., Desimone, R. 1988. Behavioral neurophysiology: insight into seeing and grasping. *Science* 242: 736–41

Young, M. P., Tanaka, K., Yamane, S. 1992. On oscillating neuronal responses in the visual cortex of monkey. *J. Neurophysiol.* 67: 1464–74

Zeki, S., Shipp, S. 1988. The functional logic of cortical connections. *Nature* 335: 311–17

Zola-Morgan, S., Squire, L. R. 1990. The primate hippocampal formation: Evidence for a time-limited role in memory storage. *Science* 250: 299–90

Zola-Morgan, S., Squire, L. R., Amaral, D. G. 1986. Human amnesia and the medial temporal region: Enduring memory impairment following a bilateral lesion limited to field CA1 of the hippocampus. *J. Neurosci.* 6: 2950–67

Zola-Morgan, S., Squire, L. R., Amaral, D. G., Suzuki, W. A. 1989. Lesions of perirhinal and parahippocampal cortex that spare the amygdala and hippocampal formation produce severe memory impairment. *J. Neurosci.* 9: 4355–70

Annu. Rev. Neurosci. 1993. 16:265–97

COMMON PRINCIPLES OF MOTOR CONTROL IN VERTEBRATES AND INVERTEBRATES

K. G. Pearson

Department of Physiology, University of Alberta, Edmonton, Canada T6G 2H7

KEY WORDS: comparative neurobiology, pattern generation, movement

INTRODUCTION

Comparative neurobiology has proven to be a powerful discipline for elucidating the principles that describe the evolution, development, and functioning of the nervous system (Arbas et al 1991; Cohen & Strumwasser 1985; Heiligenberg 1991). Studies on invertebrates have contributed substantially to establishing many of these principles. Some noteworthy examples are the mechanisms that establish the diversity of potassium channels (Jan & Jan 1990; MacKinnon 1991), the ionic mechanisms that generate action potentials (Hodgkin & Huxley 1952), and the cellular and molecular events that underlie learning and memory (Carew & Sahley 1986; Hawkins & Kandel 1990). The analysis of more complex integrative processes has also benefited from studies of invertebrate nervous systems. One example is the influence that the finding of lateral inhibition in the eye of *Limulus* (Hartline et al 1956) had on our understanding of sensory processing in more complex systems. Another is the profound effect that the analysis of rhythmic motor systems in invertebrates has had on investigations on rhythmic pattern generating networks in vertebrates (Getting 1988; Grillner 1981; Grillner & Wallen 1985). More recently, the utility of analyzing the action of neuromodulators on the functioning of neuronal

265

0147–006X/93/0301–0265$02.00

circuits in invertebrates has been recognized (Harris-Warrick 1988; Harris-Warrick & Marder 1991; Kravitz 1988).

This article reviews some of the functional principles that are common to vertebrate and invertebrate motor systems. The parallels between vertebrate and invertebrate motor systems have already been described in numerous reviews (Dekin 1991; Getting 1986, 1988; Kennedy & Davis 1977; Pearson 1976; Pearson & Duysens 1976; Wiersma 1967), but recent advances in our understanding of motor systems in both groups of animals have drawn the parallels even closer. This review primarily focuses on those principles that govern the generation of motor patterns for fairly automatic movements, such as locomotion, respiration, and escape behavior. Where appropriate, comparisons are made with mechanisms that regulate voluntary limb movements in primates. The review is organized into four major sections: motor programing, command systems, afferent regulation, and neuromodulation. I have not attempted to assess the generality of all principles that have been proposed to explain the functioning of invertebrate and vertebrate motor systems or to review exhaustively all studies related to a specific principle. The examples herein are the ones that best illustrate the principle under discussion, and I trust that the principles discussed are sufficient to demonstrate the many similarities in the functional organization of vertebrate and invertebrate motor systems.

MOTOR PROGRAMING

One of the most powerful concepts in the field of motor control is that of motor programing. Marsden et al (1984) have defined a motor program as "a set of muscle commands which are structured before a movement begins and which can be sent to the muscle with the correct timing so that the entire sequence is carried out in the absence of peripheral feedback." The evidence for motor programing comes primarily from observations on movements and motor patterns following deafferentation. Primates, including humans, can perform many voluntary motor tasks after the abolition of afferent input (Ghez et al 1990; Rothwell et al 1982; Sanes et al 1985); more automatic movements, such as walking, swimming, and breathing, can be evoked following deafferentation in other vertebrates and in invertebrates (Delcomyn 1980; Grillner 1981). In no instance, however, has it been found that the movements and the associated motor patterns are identical to those in normal animals. Thus, it is generally acknowledged that afferent feedback modifies the motor programs themselves and/or the output from these programs (see section on Afferent Regulation of Pattern-Generating Networks).

Central Pattern Generation

By far, the most extensive analysis of motor programing has been carried out on rhythmic motor systems. In virtually all systems that have been studied (more than 50 in vertebrates and invertebrates), rhythmic motor patterns can be generated in the absence of afferent input (Delcomyn 1980). The neuronal networks responsible for the central generation of these rhythmic patterns are termed "central pattern generators" (CPGs). The principle of central pattern generation has been the foundation of research on rhythmic motor systems in both vertebrates and invertebrates for more than two decades.

The organization and functioning of CPGs have been reviewed extensively in recent years (Feldman 1986; Getting 1986, 1988; Grillner & Wallen 1985; Grillner et al 1991; Selverston & Moulins 1985); some reviews have specifically compared central pattern generation in vertebrate and invertebrate motor systems (Dekin 1991; Getting 1986, 1988). One of the main conclusions from the analysis of many CPGs is that the neuronal networks are very complex, often exceeding that minimally required to generate the output motor pattern (Dumont & Robertson 1986; Getting 1989). This complexity has made it difficult to formulate generalizations about the organization and functioning of CPGs. Nevertheless, some remarkable superficial similarities have been observed in some circuits, as shown in Figure 1 for the respiratory system in mammals and the swimming system in the mollusc *Tritonia*. This figure also illustrates the circuits for the swimming system in the lamprey and the heartbeat system in the leech. Note the common occurrence of reciprocal inhibition in all four networks, which corresponds to the fact that in most CPGs the majority of interconnections between neurons are inhibitory.

Another conclusion from the analysis of different CPGs is that a wide variety of cellular and synaptic mechanisms are involved in establishing the characteristics of CPGs, but only a subset of these mechanisms function in any particular CPG. For example, in the stomatogastric system of decapod crustacea, the capacity of neurons to burst endogenously or to generate plateau potentials is important for rhythm generation, whereas no neurons have these properties in the swimming system of *Tritonia*. In *Tritonia*, rhythm generation depends largely on time-dependent changes in the strength and sign of the synaptic coupling between neurons. Thus, the general notion today is that CPGs can be assembled in many different ways by using different combinations of basic cellular and synaptic processes in the neurons that make up the CPG network. For detailed discussion of this point, see Getting (1986, 1989) and Dekin (1991).

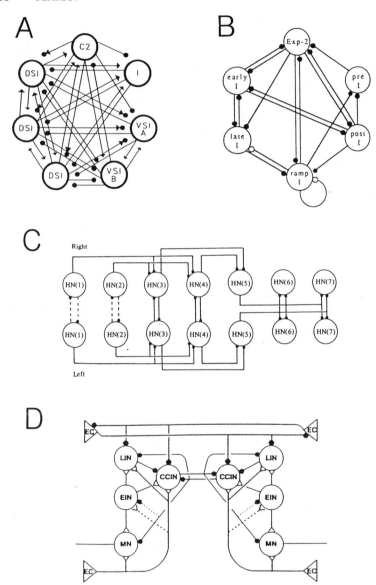

Figure 1 Examples of pattern-generating networks in vertebrate and invertebrate motor systems showing similarities in the network organization. (*A*) Swimming system of *Tritonia* (from Getting 1989). (*B*) Respiratory system of the cat (from Richter 1992). (*C*) Heartbeat system of the leech (from Calabrese 1979). (*D*) Swimming system of the lamprey (from Grillner et al 1991). In all panels, inhibitory connections are indicated by filled circles. Excitatory synapses are indicated by filled triangles in *A*, by open circles in *B*, and open triangles in *C*.

Multifunctional Neuronal Circuits

As most muscles participate in more than one motor task, the issue of how interneuronal systems are organized to generate different motor patterns in the same, or overlapping, set of muscles is raised. One possibility is that separate sets of interneurons produce the motor pattern for each task. This is indeed the case for bifunctional muscles in the locust that control leg movements during walking and wing movements during flight (Ramirez & Pearson 1988) and for two forms of swimmeret beating in the crayfish (Heitler 1985). However, this type of organization appears not to be common, because in most systems that have been examined, there is evidence for the sharing of interneuronal circuitry. Recordings from neurons in numerous pattern-generating networks have shown that many interneurons are active during more than one behavior, with a pattern appropriate for the ongoing behavior: walking, scratching, and posture in cats (Berkinblit et al 1978; Gelfand et al 1988); different forms of respiration during sleep and wakefulness in cats (Lydic & Orem 1979); pyloric and gastric mill rhythms in the stomatogastric system of crabs (Weimann et al 1991); different forms of bending in the leech (Lockery & Kristan 1990); withdrawal and swimming in *Tritonia* (Getting & Dekin 1985); forward and backward scaphognathite beating in crabs (Simmers & Bush 1983); and jumping and kicking in the locust (Gynther & Pearson 1989). There are also good reasons for believing that interneuronal pattern-generating circuitry is shared for many other tasks in both groups of animals. Some likely cases are rhythmic hatching and stepping movements of the legs of chicks (Bekoff 1986; Bekoff et al 1987); different forms of scratching in the turtle (Mortin & Stein 1989); and forward and backward walking in crayfish (Ayers & Davis 1977). Thus, another principle in vertebrate and invertebrate motor systems is that pattern-generating networks can be reorganized to function in more than one motor task.

The basic question now is how does this occur? There are three different sources of input that could act to configure pattern-generating networks: afferent signals from the periphery, signals from other regions of the central nervous system (CNS), and the presence or absence of neuromodulatory substances. A good example of the afferent input configuring a pattern-generating network has been described in the chick (Bekoff et al 1987). Analysis of the motor patterns in intact and deafferented chicks has demonstrated that the motor patterns for hatching and stepping converge toward a common pattern following deafferentation. The differences that remain most likely result from differences in signals from innervated cervical segments, because the posture of the animal is different for the different behaviors (Bekoff et al 1987, 1989). Similarly, different signals from three separate regions of the carapace probably configure a central network for three different forms of scratch in the turtle, with blending of the patterns when sites close to region boundaries are stimulated (Stein et al 1986). An example in invertebrates, in which afferent input configures a pattern-generating network, is found in the escape system of *Tritonia*. Here, afferent depolarization of one set of premotor interneurons is the first in a series of steps that reconfigures a pattern-generating circuit from a state in which it generates a pattern for withdrawal to a state in which it generated the pattern for swimming (Getting & Dekin 1985).

Evidence that central command signals reorganize a pattern-generating network has been obtained in the ventilatory system of crabs (DiCaprio 1990) and the feeding system of gastropod molluscs (McClellan 1983; Susswein & Byrne 1988). In the former system, stimulation of single neurons descending from the brain can reverse the direction of ventilation (possibly by tonically depolarizing critical elements in the pattern-generating network); in the latter, the motor patterns for ingestion and regurgitation can be produced by stimulation of different neurons in the cerebral ganglion. Similarly, spinal pattern-generating networks in vertebrates can function in different modes, depending on the nature of the central command signals. This is best seen in the cat, where the characteristics of posture and stepping depend in the stimulation site within the brain stem and the combination of descending pathways activated (Gelfand et al 1988; Mori 1987).

To explain the flexibility of pattern-generating networks, Grillner (1981, 1985) has proposed that CPGs can be divided into several smaller functional units, which he termed "unit burst generators." The basic idea of this scheme is that the coupling between unit burst generators can be altered, thus allowing various motor patterns to be generated in the same neuronal network. This is analogous to the idea that different patterns of intersegmental coordination in walking and swimming animals can be explained by alterations in the sign of coupling between CPGs that regulate the movements of single limbs or single body segments (Grillner 1981). Support for this scheme has come from the analysis for forward and backward walking in humans and lobsters, in which distinct phase shifts in the timing of activity in different muscles have been noted (Ayers & Davis 1977; Thorstensson 1986), and from studies on the motor patterns associated with three forms of scratching in the turtle (Mortin & Stein 1989; Stein et al 1986). Recent observations on the cat, however, have not been consistent with this proposal (Buford & Smith 1990). The same motor synergy is used for both forward and backward walking in this animal, i.e. no muscle active in the swing phase for forward walking switches to become active in the stance phase for backward walking, and vice versa.

Distinct changes in the amplitude and durations of bursts of activity in many muscles do occur, and there are large differences in posture for the two behaviors. These observations indicate that reconfiguration of pattern-generating networks for different behaviors can be more complex than simply altering the coupling between discrete functional units in the network. This would be consistent with the known mechanisms of re-organization of pattern-generating networks in some invertebrate motor systems, such as in the stomatogastric ganglion of decapod crustacea (Harris-Warrick & Marder 1991) and local bending of the body in leeches (Lockery & Kristan 1990).

In summary, many pattern-generating networks in vertebrates and invertebrates can be configured to generate different motor patterns for different tasks. In most systems, the mechanisms responsible for con-figuring a pattern-generating circuit for a specific task are unknown, but they obviously depend on the nature of afferent, central, and/or neuro-modulatory signals. These signals may tonically bias certain elements within the network, as is the case in the respiratory system of crabs (DiCaprio 1990), or they may alter the cellular and synaptic properties of neurons in the network, as in the stomatogastric ganglia of decapod crustacea (Harris-Warrick & Marder 1991).

COMMAND SYSTEMS

Motor systems in both vertebrates and invertebrates are hierarchically organized into several functional levels. For example, information about spatial coordinates and the kinematics of voluntary limb movements are represented in populations of neurons in the cortex (Georgopoulos 1991; Kalaska 1991), and this information is transformed into the appropriate spatiotemporal pattern of activity in motoneurons by networks lower in the hierarchy, perhaps by networks located mainly in the spinal cord (Bizzi et al 1991). Similarly, the pattern-generating networks that produce rhythmic movements, and the networks that regulate motoneuronal activity for postural adjustments and escape responses, are controlled by higher level command systems that do not specify all the details about how the muscles must be activated to produce the movement (Larimer 1988). One of the main indications for the existence of functionally distinct command systems is that electrical stimulation of discrete sites in the CNSs of vertebrates, and of individual neurons in invertebrates, can evoke coordinated motor acts (see Table 1 for studies that have reported this phenomenon). A major endeavor over the past 20 years has been to establish whether the artificially stimulated neurons actually participate in initiating motor acts in behaving animals. There is now ample evidence

Table 1 Vertebrate and invertebrate motor systems in which stimulation of either localized region in the CNS or single interneurons evokes coordinated motor activity

	Stimulation of localized region in the CNS		
	Behavior	Animal	Reference
Vertebrates			
	walking	cat	Shik et al 1966
		rat	Skinner & Garcia-Rill 1984
		chick	Jacobson & Hollyday 1982
		bird	Steeves et al 1987
	flight	bird	Steeves et al 1987
	swimming	stingray	Droge & Leonard 1983
		turtle	Lennard & Stein 1977
	chewing	rabbit	Dellow & Lund 1971
		guinea pig	Chandler & Goldberg 1984
		cat	Nakamura & Kubo 1978
	scratching	cat	Berkinblit et al 1978
	posture	cat	Mori 1987
Invertebrates			
	walking	locust	Kien 1983
	singing	cricket	Huber 1965
		grasshopper	Hedwig 1986

	Stimulation of single interneurons		
	Behavior	Animal	Reference
Vertebrates			
	escape	fish	Eaton & Hackett 1984
Invertebrates			
	swimmeret	crayfish	Wiersma & Ikeda 1964
	swimming	leech	Weeks & Kristan 1978
		Tritonia	Getting & Dekin 1985
		Clione	Arshavsky et al 1985
	escape	crayfish	Krasne & Wine 1984
	jump	fly	Wyman et al 1984
	locomotion	*Aplysia*	Fredman & Jahan-Parwar 1983
	feeding	*Aplysia*	Rosen et al 1991
		Pleurobranchea	Gillette et al 1978
		Lymnaea	McCrohan & Kyriakides 1989
		Limax	Delaney & Gelperin 1990
	flight	locust	Pearson et al 1985
	singing	cricket	Bentley 1977
	posture	crayfish	Evoy & Kennedy 1967

that this is the case (Gelfand et al 1988; Larimer 1988). The objective of much contemporary research on command systems is aimed at relating the patterns of activity in the neurons of these systems to the characteristics of the motor acts that they control (Camhi & Levy 1989; Georgopoulos

1991; Larimer 1988; Miller et al 1991), and establishing the neuronal mechanisms for transforming command signals into coordinated patterns of motor activity (Bizzi et al 1991; Brodfuehrer & Friesen 1986a,b; Ritzmann & Pollack 1990).

Command Neurons

A concept that has played a prominent role in the analysis of invertebrate motor systems is that behaviors are generated by the selective activation of single neurons known as "command neurons," with each command neuron producing either a different behavior or a specific part of a motor act. This concept arose from the finding that electrical stimulation of single interneurons in the ventral nerve cord of crayfish could produce rhythmic beating of the swimmerets (Wiersma & Ikeda 1964) and subsequent demonstrations of single interneurons initiating coordinated motor acts in many other systems (see Table 1). Several important issues arose from these early studies (see critique by Kupfermann & Weiss 1978), one of which was the criteria for classifying a neuron as a command neuron. Kupfermann & Weiss (1978) proposed that command neurons should be both necessary and sufficient for evoking the motor act. Today, it is clear that only a few neurons have both these properties. Many neurons can evoke motor acts, but only in very specialized systems has it been demonstrated that an individual neuron is necessary for eliciting a specific motor act. One is in teleost fish, where activation of single Mauthner cells produces the initial part of fast turning responses (Eaton & Hackett 1984). Another system is the tail-flip escape system of crayfish, where single spikes in medial or lateral giant interneurons initiate a coordinated tail flip that propels the animal backward or forward and upward, respectively (Krasne & Wine 1984). Both these systems are similar in that they appear to have evolved to maximize the speed of the response at the expense of regulating subsequent features of the escape behavior, such as the direction of swimming. Interestingly, in both systems the command neurons are activated in parallel with other command elements that establish these features.

Although there are few examples of neurons that are both necessary and sufficient for a behavior, the idea that individual neurons have specific roles in the production of a behavior is still regarded as likely (Rosen et al 1991). The strongest evidence for this position comes from studies on the feeding system of gastropod molluscs (Delaney & Gelperin 1990; McClellan 1983; McCrohan & Kyriakides 1989; Susswein & Byrne 1988). Individual neurons have been identified in the cerebral ganglion, which, when stimulated, evoke a characteristic motor pattern in motoneurons that innervate the buccal mass. In some instances, it has been possible to relate the evoked motor pattern to a specific part of the feeding behavior, e.g. swallowing or regurgitation (McClellan 1983; Susswein & Byrne 1988). In general, however, the behavioral significance of the motor pattern has not been established. Moreover, there has often been uncertainty about the number of cerebral neurons activated, because of excitatory coupling between command-like neurons (Susswein & Byrne 1988), and some of these neurons are not always active during the spontaneous expression of the motor pattern (Davis & Kovac 1981; Susswein & Byrne 1988). Thus, at present, the question of whether different aspects of feeding behavior are under the control of a few command-like neurons remains open. This issue has also not been resolved in any other invertebrate motor system.

The idea that closely related behaviors, or different parts of a single behavior, are controlled by separate command systems is also considered likely in vertebrate motor systems. For example, stimulation of different sites in the brain stem of intact and decerebrate cats can evoke coordinated increases and decreases in postural tone (Mori 1987), and separate descending pathways are involved in eliciting walking and scratching in the cat (Gelfand et al 1988).

Population Coding in Command Systems

A principle now firmly established in vertebrate motor systems is that information representing kinematic features of a motor act is coded in the spatiotemporal pattern of activity in large populations of neurons. Individual neurons in the population are active during a wide range of movement, and the command for any particular movement is some sort of weighted average of activity in the active population. So far, the clearest examples of this type of population coding are in the motor cortex of monkeys, where a vector sum of activity in many neurons represents the direction of arm movement (Georgopoulos 1991), and in the superior colliculus of monkeys and cats, where the spatial location of the center of a population of neurons establishes the amplitude and direction of saccadic eye movements (Sparks 1991). A similar mapping is considered likely in the system that commands head movements in the owl in response to auditory stimuli (Masino & Knudsen 1990).

Recent investigations have also demonstrated that population coding is used in many command systems in invertebrates. One clear example has come from investigations on the abdominal positioning system of the crayfish (Jellies & Larimer 1985, 1986; Larimer 1988). By recording from single interneurons in the ventral nerve cord during abdominal positioning and electrically stimulating the same interneurons in quiescent animals, it has been found that many interneurons can elicit positioning behavior (estimated between 40 and 100) and that a large fraction of these neurons are active during naturally evoked positioning. The contribution of individual interneurons to the natural behavior is small or negligible, and no interneuron is solely responsible for any specific feature of the behavior. Larimer (1988) has suggested that the term "command element" be used to describe neurons that can produce a motor act when stimulated, but are not necessary for the natural production of that particular act. So far, the relationship between the overall pattern of activity in the command elements of the positioning system and the features of the behavior, such as amplitude and intersegmental coordination, have not been established, but progressive changes in these features are likely produced by a shifting subset of the population of command elements.

An even more striking parallel with vertebrate motor systems has been found in the system that controls the orientation of insects in response to a wind stimulus (Camhi & Levy 1989; Ritzmann & Pollack 1990). Cockroaches and crickets escape from predators by turning and running in a direction more-or-less opposite to the direction of the predator's approach (Camhi & Tom 1978). This is a complex motor act that involves activation of muscles in all six legs, with the exact pattern of activation depending on the direction of the orientating response. The command signal for eliciting the behavior is mediated via a population of giant interneurons in the ventral nerve cord. These giant interneurons are excited by afferents that arise from receptors in cercal appendages at the end of the abdomen. Recent studies in cockroaches (Camhi & Levy 1989) and crickets (Miller et al 1991) have concluded that the direction of the wind stimulus is coded in the overall activity pattern of the population of giant interneurons. Individual interneurons respond to wind stimuli over a broad range of directions (greater than $90°$), yet the direction of the orienting response, in the cockroach at least, is sensitive to changes in wind direction of less than $20°$. Behavioral studies have not yet been done on the cricket, but theoretical calculations have indicated that the population of broadly tuned giant neurons could discriminate changes in wind direction as small as $7°$ (Theumissen & Miller 1991). The interesting problem now is to determine how the information coded in the population of giant interneurons is transformed into the appropriate pattern of activity in muscles of the six legs. This is exactly analogous to the problem of determining how the population of corticospinal neurons of primates that code the direction of an arm movement activate the proximal muscles of the arm to produce the movement in the correct direction (Georgopoulos 1991). In both systems, we know that the command is largely via interneuronal systems, but the nature of the transformations that occur in these interneuronal systems is almost completely unknown.

AFFERENT REGULATION OF PATTERN-GENERATING NETWORKS

The function of afferent feedback in regulating voluntary and rhythmic movements has been debated for almost a century. An early study by Mott & Sherrington (1895) on unilaterally deafferented monkeys indicated that afferent feedback from the arms was essential for the production of voluntary arm movements. However, later studies in both humans and monkeys showed that under the right conditions, voluntary movements can be made in deafferented limbs [see Sanes et al (1985) for review of early literature]. Recent studies on humans have consistently shown that purposeful arm movements can be made in the absence of sensory feedback, but the control of distal musculature is severely impaired (Rothwell et al 1982; Sanes et al 1985), and there are errors related to the biomechanical properties of the arm (Ghez et al 1990). An analysis of these movement errors has led to the proposal that a major function of afferent feedback in the production of voluntary movement is to update an internal model (motor program) with information about the mechanical state of the limb before the movement begins. The actual movement is then executed largely in an open-loop, feed-forward manner (Ghez et al 1990, 1991).

There has also been a wide range of views on the function of afferent feedback in the production of more automatic, particularly rhythmic, movements (see reviews by Delcomyn 1980; Grillner 1985; Rossignol et al 1988). Today, it is generally acknowledged that afferent feedback has several roles in regulating the production of motor patterns for these types of movement. Some of the most important are establishing details of the temporal order of motor activity; controlling transitions from one phase of a movement to another; and reinforcing ongoing motor activity.

In addition, the regulatory action of afferent feedback is influenced by the pattern-generating networks and/or the command signals that activate these networks (Rossignol et al 1988; Sillar 1989, 1991). Thus, the action of an afferent signal can depend on the state of the pattern generator and/or the motor task.

Also, apart from externally regulating central pattern-generating networks, sensory receptors can be elements within the networks themselves. In the swimming system of the lamprey, for example, mechanoreceptors in the cord (edge-cells) are intimately involved in the generation of the swimming rhythm (Grillner et al 1991); in the flight system of the locust, the wing stretch receptors fulfill all the conventional criteria for elements in a rhythm-generating network (discharge rhythmically during the behavior, change the rhythm when ablated, and reset the rhythm when stimulated) (Pearson et al 1985). Also, nonspiking stretch receptors in the legs of crayfish are probably elements in the rhythm-generating system for walking, as these receptors can reset and entrain the locomotor rhythm when stimulated (Sillar et al 1986). The functional utility of having sensory receptors as integral elements of rhythm-generating networks has not been established, but one likely possibility is that it provides mechanisms for the automatic regulation of frequency in the event of altered external conditions. (See Grillner et al 1991 and Pearson & Ramirez 1990 for a discussion of this idea for the swimming system of the lamprey and the flight system of the locust, respectively.)

Temporal Ordering of Motor Activity

Although voluntary, episodic, and rhythmic movements can all be evoked following deafferentation, the movements tend to be imprecise, and details of the motor pattern are abnormal. This is particularly true for finger and hand movements in primates (Sanes et al 1985) and for behaviors in which there is substantial interaction with the external environment. For example, the locomotor-like pattern produced in immobilized spinal cats has been described by Grillner & Zangger (1979) as being capable of producing ". . . a bad caricature of walking." Similarly, deafferentation produces substantial changes in the locomotor patterns in insects (Pearson 1985) and crustacea (Sillar et al 1987). In insects, these changes are so marked that they have led to questioning the relevance of the concept of central-pattern generation for this behavior (Bässler 1987, 1988; Pearson 1985).

Although it is generally acknowledged that deafferentation degrades the motor pattern for many behaviors, only a few studies have quantitatively assessed the changes produced by deafferentation: mammalian respiration (reviewed by Feldman 1986), chick stepping and hatching (Bekoff et al 1987, 1989), fast paw shaking in cats (Koshland & Smith 1989), and flight in the locust (Pearson & Wolf 1989). Thus, in most motor systems, it is difficult to specify exactly which features of the motor pattern depend on afferent input. Hence, our knowledge about the mechanisms by which afferent signals modify the central pattern-generating network is generally very limited. The only systems in which these mechanisms have been established in any detail are the respiratory system of mammals (Feldman 1986) and the flight system of the locust (Pearson & Ramirez 1990; Wolf & Pearson 1988).

A long-standing issue in the analysis of the role of afferent feedback is the extent to which burst activity in different groups of motoneurons is reflexly generated by phasic afferent signals from proprioceptors (Delcomyn 1980). Surprisingly, there are very few situations in which this has been demonstrated. (Here, I am only considering reflex responses that might be evoked by feedback from proprioceptors, not from extero-ceptors.) One is the generation of extensor activity in the abdominal muscles of a crayfish following an escape flexion (Reichert et al 1981). Another is reflex initiation of the elevator activity in the flight system of the locust (Pearson & Wolf 1988), but here the characteristics of the elevator bursts once initiated are largely established by intrinsic plateau-potential properties of elevator interneurons (Ramirez & Pearson 1991a,b). In the walking system of the cat, extensor-related activity in the bifunc-tional semitendinosus muscle (hip extensor/knee flexor) could be reflexly initiated by phasic input from group Ia afferents when the muscle is lengthened by knee extension and hip flexion at the end to the swing phase (Lundberg 1980; Smith 1986). The connections and discharge patterns of the group Ia afferents are appropriate for this function (Lundberg 1980; Prochazka et al 1989), and extensor-related burst activity is generally absent in immobilized spinal preparations [the only cases they have been reported is following the administration of 4-aminopyridine in DOPA or Clouidine treated animals (Dubuc et al 1986; Pearson & Rossignol 1991)].

The reason that temporal details of motor patterns are established by afferent feedback is to ensure that the motor pattern is appropriate for the biomechanical state of the motor apparatus. That is, for effective movements to be produced, the motor output must be timed exactly to the ongoing positions, movements, and forces in peripheral structures. The sensory signals do not necessarily have to act within negative feedback loops. Indeed, it now appears likely that sensory influences are commonly mediated in an open-loop, feed-forward manner, as has been demonstrated in the regulation of speech movements in humans (Abbs & Gracco 1984) and in the flight system of the locust (Pearson & Ramirez 1990).

Afferent Regulation of Phase Transitions

In many vertebrate and invertebrate motor systems, the transition from one phase of a movement to another is triggered by a phasic afferent signal (Table 2). The Hering-Breuer reflex in the mammalian respiratory system is a good example of this principle. Here, feedback from pulmonary stretch receptors activated during lung inflation terminates the inspiratory phase (Feldman 1986). Another clear example is found in the walking system of the cat, in which near the end of stance a sensory signal switches the motor program from stance to swing (Grillner & Rossignol 1978). This regulatory event is the mechanism that allows decerebrate, spinal, and bipedally walking intact cats to adapt their rate of stepping to the speed of the treadmill upon which they are walking. Two events appear to be involved in regulating the stance to swing transition: extension at the hip (Andersson & Grillner 1983; Grillner & Rossignol 1978) and unloading of leg extensor

Table 2 Rhythmic motor systems in which phase-transitions are controlled by afferent feedback

	Behavior	Animal	Reference
Vertebrates			
	walking	cat	Grillner & Rossignol 1978
			Duysens & Pearson 1980
			Shimamura et al 1984
	scratching	cat	Kuhta & Smith 1990
	respiration	cat	Feldman 1986
	chewing	mammals	Lund & Olsson 1983
	swimming	dogfish	Grillner & Wallen 1982
Invertebrates			
	walking	cockroach	Pearson 1972
			Wong & Pearson 1976
		stick insect	Bässler 1987, 1988
		crayfish	Skorupski & Sillar 1986
			Sillar et al 1986
	flight	locust	Wolf & Pearson 1988
	uropod beating	crab	Paul 1976
	feeding	*Helisoma*	Kater & Rowell 1973

muscles (Conway et al 1987; Duysens & Pearson 1980). The afferents that signal hip extension have not yet been identified, but it is likely that afferents arising from muscles acting at the hip are involved (Andersson & Grillner 1983). The signal that indicates unloading of leg extensors most likely arises from the Golgi tendon organs (group Ib afferents) (Conway et al 1987; Duysens & Pearson 1980; Pearson et al 1992). One hypothesis is that group Ib input during early stance inhibits the generation of flexor activity and only when group Ib input wanes near the end of stance does it release the flexor burst generating system from inhibition, thus enabling the production of the swing phase.

Many of the observations on the afferent control of the stance to swing transition in the cat are closely paralleled by findings in the walking systems of crustacea and insects. In the crayfish, input from nonspiking stretch receptors that signal the extent of remotion of a leg causes the transition from remotion to promotion (Sillar et al 1986); in the stick insect, input from the femoral chordotonal organ near the end of stance can trigger the transition from stance to swing (Bässler 1986, 1987). In cockroaches, there is also evidence that during slow walking, unloading of a leg near the end of stance (signalled by a decrease in activity of cuticular strain detectors, the campaniform sensilla) is a necessary condition for the initiation of the swing phase (Pearson 1972; Zill 1985).

Other invertebrate systems in which afferent signals control phase transitions are the flight system of the locust, in which phasic signals from hair plates at the base of the wings triggers the transition from wing depression to wing elevation (Pearson & Wolf 1988); the walking system of the cockroach, in which the transition from swing to stance is triggered by a phasic signal from a hair plate at the trochanter joint (Wong & Pearson 1976); and the feeding system of the snail *Helisoma*, in which the transition from retraction to protraction is produced by the decline of an inhibitory signal from the system that generates motor activity for protraction (Kater & Rowell 1973). In the feeding system of mammals, the transition from jaw closing to opening may, under some conditions, be triggered by input from periodontal pressure receptors (Lund & Enumono 1988).

The functional significance of regulating phase transitions by afferent feedback has not been fully established. One likely reason is that it limits the terminated movement to a range for effective function, such as limiting the extent of lung inflation during breathing in mammals (Feldman 1986) and the amplitude of wing depression during flight in the locust (Wolf & Pearson 1988). Another reason may be to ensure that a certain phase of a movement is not initiated until a defined biomechanical state of the system has been achieved. This is the likely function in the walking systems of mammals, insects, and crustacea, where afferent feedback functions to prevent the initiation of the swing phase until the leg is unloaded as a result of the animal's weight being carried by the other legs (Pearson & Duyens 1976).

Reinforcing Action of Afferent Feedback

A principle applicable to many vertebrate and invertebrate motor systems is that centrally generated motor activity can be reinforced by afferent feedback (Table 3). In the walking system of the crayfish, for example, two receptors reinforce the activity of motoneurons active during the stance phase. The first is a nonspiking stretch receptor (the T-fiber of the thoracic-coxal muscle receptor organ) that reinforces activity in leg re-motor motoneurons (Siller et al 1986; Skorupski & Sillar 1986). The second is a chordotonal organ at the thoracic-coxal joint that reinforce leg depressor motoneurons (El Manira et al 1991). A particularly interesting aspect of both these reinforcing reflexes is that they only occur when the motor pattern for walking is being expressed; otherwise, the reflex is resistive. This type of reflex reversal from resistive to reinforcing also occurs in the walking system of the stick insect (Bässler 1986). In the forelegs of these animals, a chordotonal organ in the femur is stretched because of flexion of the tibia during stance. In active animals, input from this chordotonal organ reinforces activity in flexor motoneurons (positive feedback); in passive animals, it inhibits these motoneurons.

Table 3 Motor systems in which ongoing motor activity is reinforced by afferent feedback

	Behavior	Animal	Reference
Vertebrates			
	walking	human	Dietz et al 1979
		cat	Andersson & Grillner 1983
			Akazawa et al 1982
	chewing	human	Lamarre & Lund 1975
		rabbit	Lavigne et al 1987
	voluntary movement	human	Burke et al 1978
			Marsden et al 1976
	respiration	human	Newson Davis & Sears 1970
		cat	Sears 1973
Invertebrates			
	walking	cockroach	Pearson 1972
		stick insect	Bässler 1986
		crayfish	Skorupski & Sillar 1986
			El Manira et al 1991
	swimmeret	lobster	Davis 1969
	uropod beating	crab	Paul 1976
	claw closing	crayfish	Wilson & Davis 1965
	jump/kick	locust	Burrows & Pflüger 1988

Reinforcement of stance-related extensor activity also occurs in the walking system of both cats and humans: The magnitude of activity in the ankle extensor muscles can be increased by increasing the length or the load carried by the muscles (Akazawa et al 1982; Dietz et al 1979). Currently, it is thought that this reinforcing reflex is due to feedback from muscle spindle afferents. This reflex may also function during normal walking, because the activity in group Ia afferents increases when the ankle extensors are lengthening at the beginning of the stance phase (Prochazka et al 1989). Another group of afferents that might contribute to the reflex activation of extensor muscles during stance is the group Ib afferent from Golgi tendon organs. These receptors are also active during the stance phase (Prochazka et al 1989). Recent studies in spinal and decerebrate animals have shown that their action on leg extensor motoneurons is switched from inhibition to excitation by procedures that elicit locomotor activity (administration of L-DOPA in spinal cats and stimulation of the brain stem locomotor region in decerebrate cats) (Conway et al 1987; Gossard et al 1990; H. Hultborn 1991, personal communication). This reflex reversal in the walking system of the cat is exactly analogous to the reflex reversal that occurs in the walking systems of stick insects and crayfish.

Reinforcing reflexes have also been described in the feeding system of *Helisoma* (Kater & Rowell 1973) and in the masticatory system of mammals (Lavigne et al 1987; Morimoto et al 1989). In both these systems, motoneurons that supply the muscles active during the period of grinding and biting of food receive the excitatory input from peripheral receptors (receptors in the buccal mass of the snail; periodontal presso-receptors and muscle spindles in mammals).

One function for reinforcing reflexes appears to be load compensation. In all the systems described above, the reinforcing action of peripheral feedback is directed toward motoneurons that are active during the power-stroke, i.e. those that supply muscles actively working against external loads (the body's weight in walking systems and food in feeding systems). Any increase in load results in reflex reinforcement of activity in the load bearing muscles, thus counteracting the increased load.

Task-Dependent and Phase-Dependent Modulation of Reflexes

Reflexes function to adapt posture and movement to changes in the external environment. This requires that reflexes are appropriate for the behavioral state of the animal, including the exact configuration of the body with respect to the external world. It is not surprising, therefore, that the analysis of motor systems in both vertebrates and invertebrates has revealed that the strength of reflex responses depends on the task that the animal is performing and the state of the motor system at different times during the execution of a specific task. A good example of task-dependent modulation of reflex gain is in humans, where the strength of the stretch reflex in leg extensor muscles is high during standing, low during walking, and even lower during running (Capaday & Stein 1986; Edamura et al 1991). The functional significance of the high gain during standing is presumably to give greater stability in the event of unexpected external perturbations. A qualitatively similar phenomenon occurs in the walking system of crustacea, where the gain of resistance reflexes is high while an animal is standing (thus yielding high postural stability), but very low or even reversed when the animal is walking (Barnes 1977; Head & Bush 1991; Skorupski & Sillar 1986). The reason for decreasing the gain of these resistance reflexes is to prevent them from impeding the leg movements during walking. Another example of reflex gain being modulated according to the behavior is in the stick insect, where increasing the gain of the resistance reflex from the femoral chordotonal organ caused the animal to initiate rocking movements, a behavior important for the survival of these animals (Bässler 1983).

An animal's ability to adapt the characteristics of reflexes to the context of a task is often referred to as "motor set" (Matthews 1991; Prochazka 1989). Prochazka (1989) has provided an excellent review of the history of this concept and discussed its relevance to numerous vertebrate and invertebrate motor systems.

Reflexes must be appropriate for the ongoing movement, as well as for the specific task. In rhythmic motor systems, there are now many examples in which the characteristics of a reflex depend on the phase of the ongoing rhythmic activity, termed "phase-dependent modulation" [see reviews by Sillar (1989, 1991) and Rossignol et al (1988) for extensive descriptions of this phenomenon in vertebrates]. One of the best characterized is the modulation of cutaneous reflexes during walking in cats (Forssberg 1979). Stimulation of the dorsum of a paw during the swing phase leads to a marked increase in the amplitude of flexor activity with the functional consequence of elevating the foot to clear the stimulating object. When delivered during that stance phase, the same stimulus causes excitation of extensor motoneurons. This type of reflex reversal also occurs in humans. Electrical stimulation of the tibial nerve during swing yields excitation of the ankle flexor muscle, anterior tibialis, but inhibition when delivered during the stance phase (Duysens et al 1990). In the walking system of the locust, exactly the same phenomenon has been observed (Wolf 1992). By recording intracellularly from motoneurons in a walking locust, Wolf (1992) found that stimulation of the afferents from exteroceptors in the tarsus gave excitation of flexor motoneurons during the swing phase and inhibition of the same motoneurons during the stance phase.

The strength of reflexes can also vary during the course of one phase of a movement. In an active stick insect, for example, the reflex from the femoral chordotonal organ changes from excitation of flexor motoneurons in the early part of the stance phase to inhibition of the same motoneurons at the end of the stance phase (Bässler 1986). In the walking system of the cat, the gain of the stretch reflex in the ankle extensor muscles progressively increases throughout the stance phase (Akazawa et al 1982); in humans, the gain of the stretch reflex in knee extensor muscles is high at the beginning of stance and diminishes as the muscles lengthen during the first half of the stance phase (Dietz et al 1990).

Not only are there functional similarities in the modulation of reflexes in vertebrates and invertebrates, but there are also some striking similarities in the underlying cellular mechanisms. The most obvious is presynaptic modulation of transmission from primary afferents. In crayfish (El Manira et al 1991) and cats (Gossard et al 1989; Gossard & Rossignol 1990), the presynaptic inhibitory input to the terminals of sensory afferents is strongly modulated during locomotor activity. In the crayfish, this modulation strongly influences the strength of transmission from primary afferents to motoneurons (El Manira et al 1991). In both these animals, the strength of the primary afferent depolarization can often be sufficient to excite the afferents antidromically (Dubuc et al 1988; El Manira et al 1991). The functional significance of this phenomenon has not yet been established, but if it occurs in behaving animals it could contribute to blocking of orthodromic signals.

Central regulation of the excitability of interneurons in reflex pathways is also an important mechanism underlying modulation of reflexes in numerous vertebrate and invertebrate motor systems (Büschges & Schmitz 1991; Laurent & Burrows 1989; Moschovakis et al 1991; Nagayama & Hisada 1987; Sillar 1991). A common feature in both groups of animals is the coexistence of parallel pathways that can be differentially modulated. For example, di- and trisynaptic cutaneous reflex pathways in the cat can be differentially regulated during locomotor activity (Moschovakis et al 1991), and the characteristics of reflexes in the uropods of crayfish are determined by central modulation of nonspiking interneurons in at least two separate pathways (Nagayama & Hisada 1987). In the walking system of the stick insect, parallel pathways with antagonistic actions in leg motoneurons have been identified (Büschges 1990; Büschges & Schmitz 1991). Differential modulation of these pathways is considered the basis for the reversal of reflexes from the femoral chordotonal organ when an animal begins to walk (Büschges & Schmitz 1991). Differential modulation of interneurons in parallel antagonistic reflex pathways is also thought to be the basis for phase-dependent reflex reversals in the walking system of cats and humans (Duysens et al 1990; Rossignol et al 1988).

NEUROMODULATION OF PATTERN-GENERATING NETWORKS

One of the major recent advances in the analysis of motor pattern generation is the recognition that the functional properties of neuronal circuits can be modified by the action of neuromodulators (mainly amines and peptides). This enables the circuits to adapt to behavioral needs, as well establish the correct configuration of a neuronal circuit for a specific behavior. From the comparative analysis of the action of neuromodulators on motor systems in invertebrates and vertebrates, three clear functional principles have emerged:

1. Neuromodulators can tonically facilitate or depress ongoing motor acts.
2. Neuromodulators can initiate motor activity and/or prime neuronal circuits to respond more effectively to input from command systems.
3. Neuromodulators can modify the cellular and synaptic properties of neurons within a network, thus enabling the same network to generate different motor patterns for different behaviors.

The function and mechanisms of action of neuromodulators have been reviewed extensively over the past few years (Dickinson 1989; Harris-Warrick 1988; Harris-Warrick & Marder 1991; Katz & Harris-Warrick 1990; Kravitz 1988, 1990; Marder 1991; Sigvardt 1989).

Tonic Modulation of Motor Activity

A common observation in many vertebrate and invertebrate motor systems is that neuromodulators can alter the characteristics of an ongoing motor act, such as the amplitude and speed of the movements and/or the repetition rate of the behavior. Harris-Warrick (1988) lists more than 20 motor systems in vertebrates and invertebrates in which this phenomenon has been observed. Virtually every motor behavior in all animals appears to be subject to modulation. Moreover, in some cases numerous naturally occurring compounds have a modulatory influence on individual motor systems. For example, at least ten different endogenous modulatory substances influence the mammalian respiratory system, whereas at least eight influence the stomatogastric system of decapod crustacea. Despite the ubiquitous occurrence of neuromodulation, the normal function of specific neuromodulators has only been established in a relatively small number of cases. Establishing a functional role requires knowledge about the natural conditions under which neurons releasing neuromodulatory substances are activated, and the behavioral effect of selectively inactivating these neurons or specifically blocking the action of the released modulatory substance. A good example of this is found in the analysis of behavioral arousal associated with feeding in *Aplysia* (Kupfermann & Weiss 1981). Activity of large serotonergic neurons in the cerebral ganglia progressively augments the rate and amplitude of feeding movements by enhancing the strength of muscle contractions and facilitating the output of the CPG for biting. These neurons are normally activated at the onset of feeding, and their ablation drastically reduces the potentiation of feeding movements (Kupfermann & Weiss 1982; Rosen et al 1989).

Activation and Priming of Motor Circuits

In addition to modulating ongoing activity in motor circuits tonically, neuromodulators can also initiate motor activity. Some well known examples are the initiation of swimming in the leech by serotonin (Willard 1981), the initiation of flight in the locust by octopamine (Sombati & Hoyle 1984), and the initiation of locomotion in cats by drugs that either mimic the action of amine or release amines (Grillner 1986). [See Harris-Warrick (1988) for complete listing of motor systems activated by the administration of a neuromodulatory substance.] An issue that has arisen from these findings is whether the amines normally function to activate the pattern-generating networks or whether they are released in conjunction with activity in faster acting command systems to establish the appropriate constellation of cellular and synaptic properties within the network (Harris-Warrick 1988; Sigvardt 1989). Currently, the data favor the latter situation. In the cat, for example, pharmacological depletion of amines does not prevent the initiation of locomotor activity (Steeves et al 1980). The situation in the swimming system of the leech is more complex, because depletion of serotonin in embryos prevents normal swimming, but immersing the animals in water that contains serotonin restores swimming (Glover & Kramer 1982). The fact that this recovery is produced by the nonspecific administration of serotonin suggests that serotonin's normal action is modulatory and does not provide the command signal for swimming. Consistent with this conclusion is that a nonserotonergic, swim-initiating pathway has been identified that acts in parallel with a pathway that activates serotonergic neurons (Brodfuehrer & Friesen 1986b). The importance of the nonserotonergic pathway in initiating swimming follows from the fact that swimming can be prevented when the activity in single interneurons in this pathway is reduced, even though large numbers of serotonergic neurons remain active. Finally, in the flight system of the locust, octopamine also appears to modulate the properties of interneurons in the rhythm-generating network, as it induces plateau-potential properties in flight interneurons at much lower concentrations than required to initiate rhythmic motor activity (Ramirez & Pearson 1991a). Plateau-potential properties are normally induced in flight interneurons at the onset of flight (Ramirez & Pearson 1991b).

Despite these data, the problem of clearly distinguishing between activational and modulatory functions of inputs to pattern-generating networks may prove difficult, particularly in those cases where the presence of neuromodulators is necessary for the generation of motor activity. An additional complication is that conventional transmitters used in activating pattern-generating networks may have modulator-like actions. This is the case in the swimming system of the lamprey, where glutamate, acting via n-methyl-d-aspartate (NMDA) receptors, induces endogenous rhythmicity in some interneurons (Wallen & Grillner 1987).

Functional Reorganization of Motor Networks

The principle that neuromodulators can act to reorganize the functional properties of neuronal networks has come largely from recent studies on the stomatogastric system of decapod crustacea (Harris-Warrick & Marder 1991; Katz & Harris-Warrick 1990) and from studies on slices of tissue from the brains of vertebrates (Llinas 1988; McCormick 1989; Steriade & Llinas 1988). The stomatogastric system of crustaceans consists of four small ganglia that contain neurons controlling movements of the foregut. This system generates several motor patterns, some of which (pyloric, gastric mill, and cardiac sac) are generated in separate sets of neurons. Neuromodulators can reorganize the functioning of these networks in three different ways. The first, and simplest, is that the modulator can act to alter the functioning of a single network and so cause it to produce a different motor pattern for a different behavior. A good example is the action of the peptide proctolin on the gastric mill system in the spiny lobster (Heizel 1988). This system produces two different patterns of tooth movement: squeezing and cut-and-grind. When proctolin is injected into the hemolymph at low concentrations, it evokes the squeezing pattern; at higher concentrations, it switches the movement pattern from squeezing to cut-and-grind. This switching in the functional characteristics of the gastric-mill system also occurs when proctolin is applied to the isolated stomatogastric system, and its basis is the induction of plateau-potential properties in some gastric mill neurons (Heinzel & Selverston 1988). The second action of neuromodulators on the stomatogastric system is to switch neurons from one functional circuit to another, thus allowing a neuron that is normally involved in the production of one motor pattern to participate in the generation of another. In the crab, for example, stimulation of serotonergic sensory neurons causes some gastric-mill neurons to discharge in-phase with the pyloric rhythm (Katz & Harris-Warrick 1991). The third, and most dramatic, action of neuromodulators is to merge the functioning of separate pattern-generating networks completely, so that the composite network generates an entirely new motor pattern unrelated to the patterns generated by the individual networks (Dickinson et al 1990; Meyrand et al 1991).

The extent to which these actions of neuromodulators occurs in other motor systems is difficult to assess. This is largely because only a few motor circuits have been sufficiently well defined that these phenomena could be recognized. Nevertheless, the findings that serotonin and octopamine can produce different postural responses in lobsters (Livingstone et al 1980), that transmitter antagonists can switch the swimming motor patten in the lamprey from alternating to synchronous (Cohen & Harris-Warrick 1984), and that serotonin has powerful actions on the swimming systems of the lamprey and leech (Harris-Warrick & Cohen 1985; Willard 1981), suggest that these phenomena may be quite general. This is reinforced by in vitro studies on vertebrate brain slices, which have demonstrated that naturally occurring neuromodulators can alter the functioning of neuronal circuits in the brain stem, thalamus, and cortex (Llinas 1988). An interesting example, because of its functional relevance, is the modulatory action of histamine on thalamic relay neurons in the cat and guinea pig (McCormick

& Williamson 1991). In the absence of histamine, lateral geniculate relay neurons generate slow oscillatory activity; otherwise, they discharge tonically. This switch in functional properties is considered to be involved in the production of different forebrain rhythms associated with slow-wave sleep and arousal, respectively.

CONCLUSIONS

A general conclusion of this review is that recent progress has emphasized the similarity of concepts now guiding research on motor control in vertebrates and invertebrates. Until quite recently, invertebrate motor behavior was usually regarded as a set of relatively stereotyped behaviors, each governed by a specialized pattern-generating network. This view has changed significantly over the past decade, with the greater emphasis on the experimental analysis of closely related behaviors, on the effects of neuromodulator on pattern generating networks, and on adaptive changes in motor programing. The neuronal networks that regulate invertebrate behaviors are extremely flexible in their functioning. This changed perspective regarding the functioning of invertebrate motor systems corresponds to the general view that vertebrate motor behavior is flexible and governed by highly adaptive neural networks.

The similarities in the functional organization of vertebrate and invertebrate motor systems means that similar problems must be solved in both groups of animals. One is to establish how information represented in populations of command elements is transformed into a spatiotemporal pattern of activity in different motoneuronal pools (Bizzi et al 1991; Georgopoulos 1991; Larimer 1988; Orlovsky 1991). Another is to understand how afferent information is integrated into central pattern-generating networks. Despite the enormous amount of information on the organization of interneurons in reflex pathways in the spinal cord of cats (Baldessira et al 1981; McCrea 1986) and in the thoracic ganglia of insects and crustacea (Burrows 1985; Laurent & Burrows 1989; Nagayama & Hisada 1987), we have very little information about how these interneurons function during normal motor behavior. Finally, we know that neuromodulatory substances have powerful actions on neuronal circuits in both vertebrates and invertebrates, but we have little knowledge about how the different neuromodulators are normally utilized in regulating behavior. Under which circumstances are descending noradrenergic and serotonergic pathways activated to control stepping in mammals? What is the functional significance of the numerous neuromodulators known to exist in some motor systems? Can neuromodulators reorganize neuronal circuits in vertebrate motor systems in the same manner as has been demonstrated in the stomatogastric ganglion of crustacea?

Although I have limited the discussion in this article to the principles governing the immediate production of motor patterns, common functional principles will probably emerge as other aspects of motor function are explored. Indeed, there are already indications for parallels in the mechanisms of sensorimotor transformations in the visuomotor systems of flies and mammals (Hengstenberg 1991), and we know that pattern-generating networks in both vertebrates and invertebrates can be reorganized to recover functionally following damage to the nervous system or ablation of peripheral receptors (Büschges & Pearson 1991; Goldberger & Murray 1980, 1988; Jeannerod 1988; Schildberger & Huber 1988; Vardi & Camhi 1982). There is even a debate about whether the foraging behavior of bees is guided by internal cognitive maps (Gould 1986; Wehner & Menzel 1990).

Given these possibilities for parallels in higher motor functions, and the now clear similarities at lower levels, it is important that we reduce the barriers that separate research on vertebrate and invertebrate motor systems (such as separate sessions at scientific meetings and conference discussions restricted to one group of animals). These barriers imply that there is something fundamentally different in the functioning of the motor systems in these two groups of animals, which as we have seen is not the case. But, more importantly, they tend to isolate researchers from ideas derived from studies on groups of animals other than those under investigation. Ultimately, we want to know how systems of nerve cells function to produce the rich variety of behavior that occurs in every animal. Knowledge drawn from many sources will be necessary for this endeavor to be successful.

Acknowledgments

I thank Tessa Gordon, Victoria Lawson, and Gordon Hiebert for their helpful comments on an early draft of this review. Supported by a program grant from the Medical Research Council of Canada.

Literature Cited

Abbs, J. H., Gracco, V. L. 1984. Control of complex motor gestures: orofacial muscle responses to load perturbations of lips during speech. *J. Neurophysiol.* 51: 705–23

Akazawa, K., Aldridge, J. W., Steeves, J. D., Stein, R. B. 1982. Modulation of stretch reflexes during locomotion in the mesencephalic cat. *J. Physiol.* 329: 553–67

Andersson, O., Grillner, S. 1983. Peripheral control of the cat's step cycle. II. Entrainment of the central pattern generators for locomotion by sinusoidal hip movements

290   PEARSON

during fictive locomotion. *Acta Physiol. Scand.* 118: 229–39

Arbas, E. A., Meinertzhagen, I. A., Shaw, S. R. 1991. Evolution of nervous systems. *Annu. Rev. Neurosci.* 14: 9–38

Arshavsky, Y. I., Beloozerova, I. N., Orlovsky, G. N., Panchin, Y. V., Pavlova, G. A. 1985. Control of locomotion in the marine mollusc Clione limacina. IV. Role of type 12 interneurons. *Exp. Brain Res.* 58: 285–93

Ayers, J. L., Davis, W. J. 1977. Neuronal control of locomotion in the lobster *Homarus americanaus.* I. Motor programs for forward and backward walking. *J. Comp. Physiol. A* 115: 1–27

Baldissera, F., Hultborn, H., Illert, M. 1981. Integration in spinal neuronal systems. In *Handbook of Physiology—The Nervous System*, ed. J. M. Brookhart, V. B. Mountcastle, V. B. Brooks, S. R. Geiger, pp. 509–95. Bethesda, Md: Am. Physiol. Soc.

Barnes, W. J. P. 1977. Proprioceptive influences on the motor output during walking in the crayfish. *J. Physiol.* 73: 543–64

Bässler, U. 1988. Functional principles of pattern generation for walking movements of stick insect forelegs: the role of the femoral chordotonal organ afferences. *J. Exp. Biol.* 136: 125–47

Bässler, U. 1987. Timing and shaping influences on the motor output for walking in the stick insect. *Biol. Cybern.* 55: 397–401

Bässler, U. 1986. Afferent control of walking movements in the stick insect *Cuniculina impigra.* II. Reflex reversal and the release of the swing phase in the restrained foreleg. *J. Comp. Physiol. A* 158: 351–62

Bässler, U. 1983. *Neural Basis of Elementary Behavior in Stick Insects.* Berlin: Springer-Verlag

Bekoff, A. 1986. Ontogeny of chicken motor behaviours: evidence for multi-use limb pattern generating circuitry. In *Neurobiology of Vertebrate Locomotion*, ed. S. Grillner, P. S. G. Stein, D. G. Stuart, H. Forssberg, R. M. Herman, pp. 433–53. London: Macmillan

Bekoff, A., Kauer, J. A., Fulstone, A., Summers, T. R. 1989. Neural control of limb coordination. II. Hatching and walking motor output patterns in the absence of input from the brain. *Exp. Brain Res.* 74: 609–17

Bekoff, A., Nusbaum, M. P., Sabichi, A. L., Clifford, M. 1987. Neural control of limb coordination. I. Comparison of hatching and walking motor output patterns in normal and deafferented chicks. *J. Neurosci.* 7: 2320–30

Bentley, D. 1977. Control of cricket song

patterns by descending interneurons. *J. Comp. Physiol. A* 116: 19–38

Berkinblit, M. B., Deliagina, T. G., Feldman, A. G., Gelfand, I. M., Orlovsky, G. N. 1978. Generation of scratching. II. Nonregular regimes of generation. *J. Neurophysiol.* 41: 1058–69

Bizzi, E., Mussa-Ivaldi, F. A., Giszter, S. 1991. Computations underlying the execution of movement: a biological perspective. *Science* 253: 287–91

Brodfuehrer, P. D., Friesen, W. O. 1986a. From stimulation to undulation: a neuronal pathway for the control of swimming in the leech. *Science* 234: 1002–5

Brodfuehrer, P. D., Friesen, W. O. 1986b. Initiation of swimming activity by trigger neurons in the leech subesophageal ganglion. *J. Comp. Physiol. A* 159: 503–10

Buford, J. A., Smith, J. L. 1990. Adaptive control of backward quadriped walking. II. Hindlimb muscle synergies. *J. Neurophysiol.* 64: 756–66

Burke, D., Hagbarth, K. E., Lofstedt, L. 1978. Muscle spindle responses in man to changes in load during accurate position maintenance. *J. Physiol.* 276: 159–64

Burrows, M. 1985. Nonspiking and spiking local interneurons in the locust. In *Model Neural Networks and Behavior*, ed. A. I. Selverston, pp. 109–25. New York: Plenum

Burrows, M., Pflüger, H. J. 1988. Positive feedback loops from proprioceptors involved in leg movements of the locust. *J. Comp. Physiol. A* 163: 425–40

Büschges, A. 1990. Nonspiking pathways in a joint-control loop of the stick insect *Carausius morosus. J. Exp. Biol.* 151: 133–60

Büschges, A., Pearson, K. G. 1991. Adaptive modifications in the flight system of the locust after the removal of wing proprioceptors. *J. Exp. Biol.* 157: 313–33

Büschges, A., Schmitz, J. 1991. Nonspiking pathways antagonize the resistance reflex in the thoraco-coxal joint of stick insects. *J. Neurobiol.* 22: 224–37

Calabrese, R. L. 1979. The roles of endogenous membrane properties and synaptic interaction in generating the heartbeat rhythm of the leech, *Hirudo medicinalis. J. Exp. Biol.* 82: 163–76

Camhi, J. M., Levy, A. 1989. The code for stimulus direction in a cell assembly in the cockroach. *J. Comp. Physiol. A* 165: 83–97

Camhi, J. M., Tom, W. 1978. The escape behavior of the cockroach *Periplaneta americana.* I. Turning response to wind puffs. *J. Comp. Physiol. A* 128: 193–201

Capaday, C., Stein, R. B. 1986. Amplitude

modulation of the soleus H-reflex in the human during walking and standing. *J. Neurosci.* 6: 1308–13

Carew, T. J., Sahley, C. L. 1986. Invertebrate learning and memory: from behavior to molecules. *Annu. Rev. Neurosci.* 9: 435–88

Chandler, S. H., Goldberg, L. J. 1984. Differentiation of the neural pathways mediating cortically induced and dopaminergic activation of the central pattern generator (CPG) for rhythmical jaw movements in the anesthetized guinea pig. *Brain Res.* 323: 297–301

Cohen, A. H., Harris-Warrick, R. M. 1984. Strychnine eliminates alternating motor output during fictive locomotion in the lamprey. *Brain Res.* 293: 164–67

Cohen, M. J., Strumwasser, F. 1985. *Comparative Neurobiology: Modes of Communication in the Nervous System.* New York: Wiley

Conway, B. A., Hultborn, H., Kiehn, O. 1987. Proprioceptive input resets central locomotor rhythm in the spinal cat. *Exp. Brain Res.* 68: 643–56

Davis, W. J. 1969. Reflex organization in the swimmeret system of the lobster. I. Intrasegmental reflexes. *J. Exp. Biol.* 51: 547–63

Davis, W. J., Kovac, M. P. 1981. The command neuron and the organization of movement. *Trends Neurosci.* 4: 73–76

Dekin, M. S. 1991. Comparative neurobiology of invertebrate motor systems: Implications for the control of breathing in mammals. In *Development Neurobiology of the Control of Breathing*, ed. G. G. Haddad, J. Farber, pp. 111–54. New York: Dekker

Delaney, K., Gelperin, A. 1990. Cerebral interneurons controlling fictive feeding in *Limax maximus*. II. Initiation and modulation of fictive feeding. *J. Comp. Physiol. A* 166: 311–26

Delcomyn, F. 1980. Neural basis of rhythmic behavior in animals. *Science* 210: 492–98

Dellow, P. G., Lund, J. P. 1971. Evidence for central timing of rhythmical mastication. *J. Physiol.* 215: 1–13

DiCaprio, R. 1990. An interneurone mediating motor programme switching in the ventilatory system of the crab. *J. Exp. Biol.* 154: 517–36

Dickinson, P. 1989. Modulation of simple motor patterns. *Semin. Neurosci.* 1: 15–24

Dickinson, P. S., Mecsas, C., Marder, E. 1990. Neuropeptide fusion of two motor-pattern generator circuits. *Nature* 344: 155–58

Dietz, V., Faist, M., Pierrot-Desielligny, E. 1990. Amplitude modulation of the quadriceps H-reflex in the human during the early stance phase of gait. *Exp. Brain Res.* 79: 221–24

Dietz, V., Schmidtbleicher, H. R., Noth, J. 1979. Neuronal mechanisms of human locomotion. *J. Neurophysiol.* 42: 1212–23

Droge, M. H., Leonard, R. B. 1983. Swimming rhythm in decerebrated, paralyzed stingrays: normal and abnormal coupling. *J. Neurophysiol.* 50: 178–91

Dubuc, R., Cabelguen, J.-M., Rossignol, S. 1988. Rhythmic fluctuations of dorsal root potentials and antidromic discharges of primary afferent during fictive locomotion in the cat. *J. Neurophysiol.* 60: 2014–36

Dubuc, R., Rossignol, S., Lamarre, Y. 1986. The effects of 4-aminopyridine on the spinal cord: rhythmic discharges recorded from the peripheral nerves. *Brain Res.* 369: 243–59

Dumont, J. P. C., Robertson, R. M. 1986. Neuronal circuits: an evolutionary perspective. *Science* 233: 849–53

Duysens, J., Trippel, M., Horstmann, G. A., Dietz, V. 1990. Gating and reversal of reflexes in ankle muscles during human walking. *Exp. Brain Res.* 82: 351–58

Duysens, J. D., Pearson, K. G. 1980. Inhibition of flexor burst generation by loading ankle extensor muscles in walking cats. *Brain Res.* 187: 321–32

Eaton, R. C., Hackett, J. T. 1984. The role of the Mauthner cell in fast-starts involving escape in teleost fishes. In *Neural Mechanisms of Startle Behavior*, ed. R. C. Eaton, pp. 213–66. New York: Plenum

Edamura, M., Yang, J. F., Stein, R. B. 1991. Factors that determine the magnitude and time course of the human H-reflex in locomotion. *J. Neurosci.* 11: 420–29

El Manira, A., DiCaprio, R. A., Cattaert, D., Clarac, F. 1991. Monosynaptic interjoint reflexes and their central modulation during fictive locomotion in crayfish. *Eur. J. Neurosci.* 3: 1219–31

Evoy, W. H., Kennedy, D. 1967. The central nervous organization underlying control of antagonistic muscles in the crayfish. I. Types of command fibers. *J. Exp. Zool.* 165: 223–38

Feldman, J. L. 1986. Neurophysiology of respiration in mammals. In *Handbook of Physiology, Section 1: The Nervous System*, ed. F. E. Bloom, 4: 463–524. Bethesda, Md: Am. Physiol. Soc.

Forssberg, H. 1979. Stumbling corrective reaction: a phase dependent compensatory reaction during locomotion. *J. Neurophysiol.* 42: 936–53

Fredman, S. M., Jahan-Parwar, B. 1983. Command neurons for locomotion in Aplysia. *J. Neurophysiol.* 49: 1092–1117

Gelfand, I. M., Orlovsky, G. N., Shik, M. L. 1988. Locomotion and scratching in tetrapods. In *Neural Control of Rhythmic Movements in Vertebrates*, ed. A. H. Cohen, S. Rossignol, S. Grillner, pp. 167–99. New York: Wiley

Georgopoulos, A. P. 1991. Higher order motor control. *Annu. Rev. Neurosci.* 14: 361–78

Getting, P. A. 1989. Emerging principles governing the operation of neural circuits. *Annu. Rev. Neurosci.* 12: 185–204

Getting, P. A. 1988. Comparative analysis of invertebrate central pattern generators. In *Neural Control of Rhythmic Movements in Vertebrates*, ed. A. H. Cohen, S. Rossignol, S. Grillner, pp. 101–28. New York: Wiley

Getting, P. A. 1986. Understanding central pattern generators: insights gained from the study of invertebrate systems. In *Neurobiology of Vertebrate Locomotion*, ed. S. Grillner, P. S. G. Stein, D. G. Stuart, H. Forssberg, R. M. Herman, pp. 231–44. Hong Kong: Macmillan

Getting, P. A., Dekin, M. S. 1985. A model system for integration within rhythmic motor systems. In *Model Neural Networks and Behavior*, ed. A. I. Selverston, pp. 3–20. New York: Plenum

Ghez, C., Gordon, J., Ghilardi, M. F., Christakos, C. N., Cooper, S. E. 1990. Roles of proprioceptive input in the programming of arm trajectories. *Cold Spring Harbor Symp. Quant. Biol.* 55: 837–47

Ghez, C., Hening, W., Gordon, J. 1991. Organization of voluntary movement. *Curr. Opin. Neurobiol.* 1: 664–71

Gillette, R., Kovac, M. P., Davis, W. J. 1978. Command neurons in *Pleurobranchaea* receive synaptic feedback from the motor network they excite. *Science* 199: 798–801

Glover, J. C., Kramer, A. P. 1982. Serotonin analog selectively ablates identified neurons in the leech embryo. *Science* 216: 317–19

Goldberger, M. E., Murray, M. 1988. Patterns of sprouting and implications for recovery of function. In *Functional Recovery in Neurological Disease*, ed. S. G. Waxman, pp. 361–85. New York: Raven

Goldberger, M. E., Murray, M. 1980. Locomotor recovery after deafferentation of one side of the cat's trunk. *Exp. Neurol.* 67: 103–17

Gossard, J. P., Cabelguen, J. M., Rossignol, S. 1989. Intra-axonal recordings of cutaneous primary afferents during fictive locomotion in the cat. *J. Neurophysiol.* 62: 1177–88

Gossard, J. P., Hultborn, H., Barajon, I., Kiehn, O., Conway, B. 1990. Phasic modulation of EPSPs in extensor motoneurones evoked by Ib input during fictive

locomotion in the spinal cat. *Neurosci. Abstr.* 16: 890

Gossard, J. P., Rossignol, S. 1990. Phase-dependent modulation of dorsal root potentials evoked by peripheral nerve stimulation during fictive locomotion in the cat. *Brain Res.* 537: 1–13

Gould, J. L. 1986. The locale map of honey bees: do insects have cognitive maps. *Science* 232: 861–63

Grillner, S. 1986. The effect of L-DOPA on the spinal cord—relation to locomotion and the half-center hypothesis. In *Neurobiology of Vertebrate Locomotion*, ed. S. Grillner, P. S. G. Stein, D. G. Stuart, H. Forssberg, R. M. Herman, pp. 269–77. London: Macmillan

Grillner, S. 1985. Neurobiological bases of rhythmic motor acts in vertebrates. *Science* 228: 143–49

Grillner, S. 1981. Control of locomotion in bipeds, tetrapods and fish. In *Handbook of Physiology, Section 1: The Nervous System, Motor Control*, ed. V. B. Brooks, 2: 1179–1236. Bethesda, Md: Am. Physiol. Soc.

Grillner, S., Rossignol, S. 1978. On the initiation of the swing phase of locomotion in chronic spinal cats. *Brain Res.* 146: 269–77

Grillner, S., Wallen, P. 1985. Central pattern generators for locomotion, with special reference to vertebrates. *Annu. Rev. Neurosci.* 8: 233–61

Grillner, S., Wallen, P. 1982. On peripheral control mechanisms acting on the central pattern generators for swimming in the dogfish. *J. Exp. Biol.* 98: 1–22

Grillner, S., Wallen, P., Brodin, L., Lansner, A. 1991. Neuronal network generating locomotor behavior in lamprey. *Annu. Rev. Neurosci.* 14: 169–200

Grillner, S., Wallen, P., Dale, N., Brodin, L., Buchanan, J., Hill, R. 1987. Transmitters, membrane properties and network circuitry in the control of locomotion in lamprey. *Trends Neurosci.* 10: 34–42

Grillner, S., Zangger, P. 1979. On the central generation of locomotion in the low spinal cat. *Exp. Brain Res.* 34: 241–61

Gynther, I. C., Pearson, K. G. 1989. An evaluation of the role of identified interneurons in triggering kicks and jumps in the locust. *J. Neurophysiol.* 61: 45–57

Harris-Warrick, R. M. 1988. Chemical modulation of central pattern generators. In *Neural Control of Rhythmic Movements in Vertebrates*, ed. A. H. Cohen, S. Rossignol, S. Grillner, pp. 285–332. New York: Wiley

Harris-Warrick, R. M., Cohen, A. H. 1985. Serotonin modulates the central pattern generator for locomotion in the isolated lamprey spinal cord. *J. Exp. Biol.* 116: 27–46

Harris-Warrick, R. M., Marder, E. 1991. Modulation of neural networks for behavior. *Annu. Rev. Neurosci.* 14: 39–58

Hartline, H. K., Wagner, H. C., Ratliff, F. 1956. Inhibition in the eye of *Limulus*. *J. Gen. Physiol.* 39: 651–73

Hawkins, R. D., Kandel, E. R. 1990. Hippocampal LTP and synaptic plasticity in *Aplysia*: possible relationship of associative cellular mechanisms. *Semin. Neurosci.* 2: 391–402

Head, S. I., Bush, B. M. H. 1991. Proprioceptive reflex interactions with central motor rhythms in the isolated thoracic ganglion of the shore crab. *J. Comp. Physiol. A* 168: 445–59

Hedwig, B. 1986. On the role in stridulation of plurisegmental interneurons of the acridid grasshopper *Omocestus viridulus*. I. Anatomy and physiology of descending cephalothoracic interneurons. *J. Comp. Physiol. A* 158: 413–27

Heiligenberg, W. 1991. The neural basis of behavior: a neuroethological view. *Annu. Rev. Neurosci.* 14: 247–68

Heinzel, H. G. 1988. Gastric mill activity in the lobster. II. Proctolin and octopamine initiate and modulate chewing. *J. Neurophysiol.* 59: 551–65

Heinzel, H. G., Selverston, A. I. 1988. Gastric mill activity in the lobster. III. Effects of proctolin on the isolated central pattern generator. *J. Neurophysiol.* 59: 566–85

Heitler, W. J. 1985. Motor programme switching in the crayfish swimmeret system. *J. Exp. Biol.* 114: 521–50

Hengstenberg, R. 1991. Gaze control in the blowfly *Calliphora*: a multisensory, two stage integration process. *Semin. Neurosci.* 3: 19–30

Hodgkin, A. L., Huxley, A. F. 1952. A quantitative description of membrane current and in application to conduction and excitation in nerve. *J. Physiol.* 117: 500–44

Huber, F. 1965. Brain controlled behavior in Orthopterans. In *The Physiology of the Insect Central Nervous System*, ed. J. E. Treherne, J. W. L. Beament, pp. 233–46. New York: Academic

Jacobson, R. D., Hollyday, M. 1982. Electrically evoked walking and fictive locomotion in the chick. *J. Neurophysiol.* 48: 257–70

Jan, L. Y., Jan, Y. N. 1990. How might the diversity of potassium channels be generated? *Trends Neurosci.* 13: 415–19

Jeannerod, M. 1988. Behavioral recovery: a staged process. In *Recovery of Function in the Nervous System*, ed. F. Cohadon, J. Lobo Antunes, pp. 45–61. Padova: Liviana

Jellies, J., Larimer, J. L. 1986. Activity of crayfish abdominal-positioning interneurones during spontaneous and sensory-evoked movements. *J. Exp. Biol.* 120: 173–88

Jellies, J., Larimer, J. L. 1985. Synaptic interactions between neurons involved in the production of abdominal posture in crayfish. *J. Comp. Physiol. A* 156: 861–73

Kalaska, J. 1991. Reaching movements to visual targets: neuronal representations of sensori-motor transformations. *Semin. Neurosci.* 3: 67–80

Kater, S. B., Rowell, C. H. F. 1973. Integration of sensory and centrally programmed components in generation of cyclical feeding activity in *Helisoma trivolvis*. *J. Neurophysiol.* 36: 142–55

Katz, P. S., Harris-Warrick, R. M. 1990. Actions of identified neuromodulatory neurons in a simple motor system. *Trends Neurosci.* 13: 367–72

Kennedy, D., Davis, W. J. 1977. Organization of invertebrate motor systems. In *Handbook of Physiology, Section I, Part 2*, ed. S. R. Geiger, E. R. Kandel, J. M. Brookhart, V. B. Mountcastle, 1: 1023–87. Bethesda, Md: Am. Physiol. Soc.

Kien, J. 1983. The initiation and maintenance of walking in the locust: an alternative to the command concept. *Proc. R. Soc. London* 219: 137–74

Koshland, G. F., Smith, J. L. 1989. Mutable and immutable features of paw-shake responses after hindlimb deafferentation in the cat. *J. Neurophysiol.* 62: 162–73

Krasne, F. B., Wine, J. J. 1984. The production of crayfish tailflip escape responses. In *Neural Mechanisms of Startle Behavior*, ed. R. C. Eaton, pp. 179–212. New York: Plenum

Kravitz, E. A. 1990. Hormonal control of behavior: amines as gain-setting elements that bias behavioral output in lobsters. *Am. Zool.* 30: 595–608

Kravitz, E. A. 1988. Hormonal control of behavior: amines and the biasing of behavioral output in lobsters. *Science* 241: 1775–80

Kuhta, P. C., Smith, J. L. 1990. Scratch responses in normal cats: hindlimb kinematics and muscle synergies. *J. Neurophysiol.* 64: 1653–67

Kupfermann, I., Weiss, K. R. 1982. Activity in an identified serotonergic neuron in free moving *Aplysia* correlates with behavioral arousal. *Brain Res.* 241: 334–37

Kupfermann, I., Weiss, K. R. 1981. The role of serotonin in arousal of feeding behavior in *Aplysia*. In *Serotonin Neurotransmission and Behavior*, ed. B. L. Jacobs, A. Gelperin, pp. 255–87. Cambridge, Mass: MIT Press

Kupfermann, I., Weiss, K. R. 1978. The command neuron concept. *Behav. Brain Sci.* 1: 3–10

Lamarre, Y., Lund, J. P. 1975. Load compensation in human masseter muscles. *J. Physiol.* 253: 21–35

Larimer, J. L. 1988. The command hypothesis: a new view using an old example. *Trends Neurosci.* 11: 506–10

Laurent, G., Burrows, M. 1989. Intersegmental interneurons can control the gain of reflexes in adjacent segments of the locust by their action on nonspiking local interneurons. *J. Neurosci.* 9: 3030–39

Lavigne, G., Kim, J. S., Valiquette, C., Lund, J. P. 1987. Evidence that periodontal presso-receptors provide positive feedback to jaw closing muscles during mastication. *J. Neurophysiol.* 58: 342–58

Lennard, P. R., Stein, P. S. G. 1977. Swimming movements elicited by stimulation of turtle spinal cord. I. Low-spinal and intact preparations. *J. Neurophysiol.* 40: 768–78

Livingstone, M. S., Harris-Warrick, R. M., Kravitz, E. A. 1980. Serotonin and octopamine produce opposite postures in lobsters. *Science* 208: 76–79

Llinas, R. R. 1988. The intrinsic electrophysiological properties of mammalian neurons: insights into central nervous system function. *Science* 242: 1654–64

Lockery, S. R., Kristan, W. B. 1990. Distributed processing of sensory information in the leech. II. Identification of interneurons contributing to the local bending reflex. *J. Neurosci.* 10: 1816–29

Lund, J. P., Enomoto, S. 1988. The generation of mastication by the mammalian central nervous system. In *Neural Control of Rhythmic Movements in Vertebrates*, ed. A. Cohen, S. Rossignol, S. Grillner, pp. 41–72. New York: Wiley

Lund, J. P., Olsson, K. A. 1983. The importance of reflexes and their control during jaw movement. *Trends Neurosci.* 6: 458–63

Lundberg, A. 1980. Half-centres revisited. In *Regulatory Functions of the CNS. Motion and Organization Principles*, ed. J. Szentagothia, M. Palkovits, J. Jamori, pp. 155–67. London: Pergamon

Lydic, R., Orem, J. 1979. Respiratory neurons of the pneumotaxic center during sleep and wakefulness. *Neurosci. Lett.* 15: 187–92

MacKinnon, R. 1991. New insights into the structure and function of potassium channels. *Curr. Opin. Neurobiol.* 1: 14–19

Marsden, C. D., Merton, P. A., Morton, H. B. 1976. Servo action in the human thumb. *J. Physiol.* 257: 1–44

Marsden, C. D., Rothwell, J. C., Day, B. L. 1984. The use of peripheral feedback in the control of movement. *Trends Neurosci.* 7: 253–58

Masino, T., Knudsen, E. I. 1990. Horizontal and vertical components of head movement are controlled by distinct neural circuits in the barn owl. *Nature* 345: 434–37

Matthews, P. B. C. 1991. The human stretch reflex and the motor cortex. *Trends Neurosci.* 14: 87–90

McClellan, A. D. 1983. Higher order neurons in buccal ganglia of *Pleurobranchaea* elicit vomiting motor activity. *J. Neurophysiol.* 50: 658–70

McCormick, D. A. 1989. Cholinergic and noradrenergic modulation of thalamocortical processing. *Trends Neurosci.* 125: 215–21

McCormick, D. A., Williamson, A. 1991. Modulation of neuronal firing mod in cat and guinea pig LGNd by histamine: possible cellular mechanisms of histaminergic control of arousal. *J. Neurosci.* 11: 3188–99

McCrea, D. 1986. Spinal cord circuitry and motor reflexes. In *Exercise and Sport Science Reviews*, ed. K. B. Pandolf, 14: 105–41. New York: Macmillan

McCrohan, C. R., Kyriakides, M. A. 1989. Cerebral interneurones controlling feeding motor output in the snail *Lymnaea stagnalis*. *J. Exp. Biol.* 147: 361–74

Meyrand, P., Simmers, J., Moulins, M. 1991. Construction of a pattern generating circuit with neurons of different networks. *Nature* 351: 60–63

Miller, J. P., Jacobs, G. A., Theunissen, F. E. 1991. Representation of sensory information in the cricket cercal sensory system. I. Response properties of the primary interneurons. *J. Neurophysiol.* 66: 1680–89

Mori, S. 1987. Integration of posture and locomotion in acute decerebrate cats and in awake, freely moving cats. *Prog. Neurobiol.* 28: 161–95

Morimoto, T., Inoue, T., Maduad, Y., Nagashima, T. 1989. Sensory components facilitating jaw-closing muscle activities in the rabbit. *Exp. Brain Res.* 76: 424–40

Mortin, L. I., Stein, P. S. G. 1989. Spinal cord segments containing key elements of the central pattern generators for three forms of scratch reflex in the turtle. *J. Neurosci.* 9: 2285–96

Moschovakis, A. K., Sholomenko, G. N., Burke, R. E. 1991. Differential control of short latency cutaneous excitation in cat FDL motoneurons during fictive locomotion. *Exp. Brain Res.* 83: 489–501

Mott, F. W., Sherrington, C. S. 1895. Experiments upon the influence of sensory nerves upon movement and nutrition of the limbs. *Proc. R. Soc. London* 57: 481

Nagayama, T., Hisada, M. 1987. Opposing parallel connections through crayfish local nonspiking interneurons. *J. Comp. Neurol.* 257: 347–58

Nakamura, Y., Kubo, Y. 1978. Masticatory rhythm in intracellular potential of trigeminal motoneurons induced by stimulation of orbital cortex and amygdala in cats. *Brain Res.* 148: 504–9

Newson Davis, J. N., Sears, T. A. 1970. The proprioceptive reflex control of the intercostal muscles during their voluntary activation. *J. Physiol.* 209: 711–38

Orlovsky, G. N. 1991. Gravistatic postural control in simpler systems. *Curr. Opin. Neurobiol.* 1: 621–27

Paul, D. H. 1976. Role of proprioceptive feedback from nonspiking mechanosensory cells in the sand crab, *Emerita analoga*. *J. Exp. Biol.* 65: 243–58

Pearson, K. G. 1985. Are there central pattern generators for walking and flight in insects? In *Feedback and Motor Control in Invertebrates and Vertebrates*, ed. W. J. P. Barnes, M. Gladden, pp. 307–16. London: Croom Helm

Pearson, K. G. 1976. The control of walking. *Sci. Am.* 235(6): 72–86

Pearson, K. G. 1972. Central programming and reflex control of walking in the cockroach. *J. Exp. Biol.* 56: 321–30

Pearson, K. G., Duysens, J. D. 1976. Function of segmental reflexes in control of stepping in cockroaches and cats. In *Neural Control of Locomotion*, ed. R. M. Herman, S. Grillner, P. S. G. Stein, D. G. Stuart, pp. 519–38. New York: Plenum

Pearson, K. G., Ramirez, J. M. 1990. Influences of input from the forewing stretch receptors on motoneurones in flying locusts. *J. Exp. Biol.* 151: 317–40

Pearson, K. G., Ramirez, J. M., Jiang, W. 1992. Entrainment of the locomotor rhythm in spinal cats by phasic input from group Ib afferents. *Can. J. Physiol. Pharmacol.* In press

Pearson, K. G., Reye, D. N., Parsons, D. W., Bicker, G. 1985. Flight initiating interneurons in the locust. *J. Neurophysiol.* 53: 910–25

Pearson, K. G., Rossignol, S. 1991. Fictive motor patterns in chronic spinal cats. *J. Neurophysiol.* 66: 1874–87

Pearson, K. G., Wolf, H. 1989. Timing of forewing elevator activity during flight in the locust. *J. Comp. Physiol. A* 165: 217–27

Pearson, K. G., Wolf, H. 1988. Connections of hindwing tegulae with flight neurones in the locust, *Locusta migratoria*. *J. Exp. Biol.* 135: 381–409

Prochazka, A. 1989. Sensorimotor gain control: a basic strategy of motor systems? *Prog. Neurobiol.* 33: 281–307

Prochazka, A., Trend, P., Hulliger, M., Vincent, S. 1989. Ensemble proprioceptive activity in the cat step cycle: towards a representative look-up chart. *Prog. Brain Res.* 80: 61–74

Ramirez, J. M., Pearson, K. G. 1991a. Plateau-potentials contribute to the generation of rhythmic depolarizations in locust flight interneurons. *Neurosci. Abstr.* 17: 1580

Ramirez, J. M., Pearson, K. G. 1991b. Octopaminergic modulation of interneurons in the flight system of the locust. *J. Neurophysiol.* 66: 1522–37

Ramirez, J. M., Pearson, K. G. 1988. Generation of motor patterns for walking and flight in motoneurons supplying bifunctional muscles in the locust. *J. Neurobiol.* 19: 257–82

Reichert, H., Wine, J. J., Hagiwara, G. 1981. Crayfish escape behavior: neurobehavioral analysis of phasic extension reveals dual systems for motor control. *J. Comp. Physiol. A* 142: 281–94

Richter, D. R. 1992. Central control of respiratory movements. In *Central Control of the Autonomic Nervous System*, ed. D. Jordan. Chur, Switzerland: Harwood Academic. In press

Ritzmann, R. E., Pollack, A. J. 1990. Parallel motor pathways from thoracic interneurons of the ventral giant interneuron system of the cockroach, *Periplaneta americana*. *J. Neurobiol.* 21: 1219–35

Rosen, S. C., Teyke, T., Miller, M. W., Weiss, K. R., Kupfermann, I. 1991. Identification and characterization of cerebral-to-buccal interneurons implicated in the control of motor programs associated with feeding in *Aplysia*. *J. Neurosci.* 11: 3630–55

Rosen, S. C., Weiss, K. R., Goldstein, R. S., Kupfermann, I. 1989. The role of a modulatory neuron in feeding and satiation in *Aplysia*: effects of lesioning of the serotonergic metacerebral cells. *J. Neurosci.* 9: 1562–78

Rossignol, S., Lund, J. P., Drew, T. 1988. The role of sensory inputs in regulating pattern of rhythmical movements in higher vertebrates. In *Neural Control of Rhythmic Movements in Vertebrates*, ed. A. Cohen, S. Rossignol, S. Grillner, pp. 201–83. New York: Wiley

Rothwell, J. C., Traub, M. M., Day, B. L., Obeso, J. H., Thomas, P. K., Marsden, C. D. 1982. Manual motor performance in a deafferented man. *Brain* 105: 515–42

Sanes, J. N., Mauritz, K. H., Dalakas, M. C., Evarts, E. V. 1985. Motor control in humans with large-fiber sensory neuropathy. *Hum. Neurobiol.* 4: 101–14

Schildberger, K., Huber, F. 1988. Post-lesion

plasticity in the auditory system of the cricket. In *Post-Lesion Neural Plasticity*, ed. H. Flohr, pp. 565–75. Berlin: Springer-Verlag

Sears, T. A. 1973. Servo control of the intercostal muscles. In *New Developments in EMG and Clinical Neurophysiology*, ed. J. E. Desmedt, pp. 404–17. Basel: Karger

Selverston, A., Moulins, M. 1985. Oscillatory neural networks. *Annu. Rev. Physiol.* 47: 29–48

Shik, M. L., Severin, F. V., Orlovsky, G. N. 1966. Control of walking and running by means of electrical stimulation of the midbrain. *Biophysics* 11: 756–65

Shimamura, M., Kogure, I., Fuwa, T. 1984. Role of joint afferents in relation to the initiation of forelimb stepping in thalamic cats. *Brain Res.* 297: 225–35

Sigvardt, K. A. 1989. Modulation of properties of neurons underlying rhythmic movements in vertebrates. *Semin. Neurosci.* 1: 55–66

Sillar, K. T. 1991. Spinal pattern generation and sensory gating mechanisms. *Curr. Opin. Neurobiol.* 1: 583–89

Sillar, K. T. 1989. Synaptic modulation of cutaneous pathways in the vertebrate spinal cord. *Semin. Neurosci.* 1: 45–54

Sillar, K. T., Clarac, F., Bush, B. M. H. 1987. Intersegmental coordination of central neural oscillators for rhythmic movements of the walking legs of crayfish, *Pacifastacus leniusculus. J. Exp. Biol.* 131: 245–64

Sillar, K. T., Skorupski, P., Elson, R. C., Bush, B. M. H. 1986. Two identified afferent neurones entrain a central locomotor rhythm generator. *Nature* 323: 440–43

Simmers, A. J., Bush, B. M. H. 1983. Motor programme switching in the ventilatory system of *Carcinus maenas*: The neuronal basis of bimodal scaphognathite beating. *J. Exp. Biol.* 104: 163–81

Skinner, R. D., Garcia-Rill, E. 1984. The mesencephalic locomotor region (MLR) in the rat. *Brain Res.* 323: 385–89

Skorupski, P., Sillar, K. T. 1986. Phase-dependent reversal of reflexes mediated by the thoracocoxal muscle receptor organ in the crayfish, *pacifastacus leniusculus. J. Neurophysiol.* 55: 689–95

Smith, J. L. 1986. Hindlimb locomotion of the spinal cat: synergistic patterns, limb dynamics and novel blends. In *Neurobiology of Vertebrate Locomotion*, ed. S. Grillner, P. S. G. Stein, D. G. Stuart, H. Forssberg, R. M. Herman, pp. 185–200. Hong Kong: Macmillan

Sombati, S., Hoyle, G. 1984. Generation of specific behaviors in a locust by local release into neuropil of the natural modulator octopamine. *J. Neurobiol.* 15: 481–506

Sparks, D. L. 1991. Sensori-motor integration in the primate superior colliculus. *Semin. Neurosci.* 3: 39–50

Steeves, J. D., Schmidt, B. J., Skovgaard, B. J., Jordan, L. M. 1980. The effect of noradrenaline and 5-hydroxytryptamine depletion on locomotion in the cat. *Brain Res.* 185: 349–62

Steeves, J. D., Sholomenko, G. N., Webster, D. N. S. 1987. Stimulation of the pontomedullary reticular formation initiates locomotion in decerebrate birds. *Brain Res.* 401: 205–12

Stein, P. S. G., Mortin, L. I., Robertson, G. A. 1986. The forms of a task and their blends. In *Neurobiology of Vertebrate Locomotion*, ed. S. Grillner, P. S. G. Stein, D. G. Stuart, H. Forssberg, R. M. Herman, pp. 201–16. Hong Kong: Macmillan

Steriade, M., Llinas, R. R. 1988. The functional states of the thalamus and the associated neuronal interplay. *Physiol. Rev.* 68: 649–742

Susswein, A. J., Byrne, J. H. 1988. Identification and characterization of neurons initiating patterned neural activity in the buccal ganglia of *Aplysia. J. Neurosci.* 8: 2049–61

Theunissen, F. E., Miller, J. P. 1991. Representation of sensory information in the cricket cercal system. II. Information theoretic calculation of system accuracy and optimal tuning-curve widths of four primary interneurons. *J. Neurophysiol.* 66: 1690–1703

Thorstensson, A. 1986. How is the normal locomotor program modified to produce backward walking? *Exp. Brain Res.* 61: 664–68

Vardi, N., Camhi, J. M. 1982. Functional recovery from lesions in the escape system of the cockroach. Behavioral recovery. *J. Comp. Physiol. A* 146: 291–98

Wallen, P., Grillner, S. 1987. N-methyl-D-aspartate receptor-induced, inherent oscillatory activity in neurons active during fictive locomotion in the lamprey. *J. Neurosci.* 7: 2745–55

Weeks, J. C., Kristan, W. B. 1978. Initiation, maintenance and modulation of swimming in medicinal leech by activity of a single neurone. *J. Exp. Biol.* 77: 71–88

Wehner, R., Menzel, R. 1990. Do insects have cognitive maps? *Annu. Rev. Neurosci.* 13: 403–14

Weimann, J. M., Meyrand, P., Marder, E. 1991. Neurons that form multiple pattern generators: identification and multiple activity patterns of gastric/pyloric neurons in the crab stomatogastric system. *J. Neurophysiol.* 65: 111–22

Wiersma, C. A. G. 1967. *Invertebrate Nervous Systems. Their Significance for Mammalian Physiology.* Chicago: Univ. Chicago Press

Wiersma, C. A. G., Ikeda, K. 1964. Interneurons commanding swimmeret movements in the crayfish *Procambarus clarki* (Girard). *Comp. Biochem. Physiol.* 12: 509–25

Willard, A. L. 1981. Effects of serotonin on the generation of the motor program for swimming in the medicinal leech. *J. Neurosci.* 1: 936–44

Wilson, D. M., Davis, W. J. 1965. Nerve impulse patterns and reflex control in the motor system of the crayfish claw. *J. Exp. Biol.* 43: 193–210

Wolf, H. 1992. Reflex modulation in locusts walking on a treadmill—intracellular recordings from motoneurons. *J. Comp. Physiol. A* 170: 443–62

Wolf, H., Pearson, K. G. 1988. Proprioceptive input patterns elevator activity in the locust flight system. *J. Neurophysiol.* 59: 1831–53

Wong, R. K. S., Pearson, K. G. 1976. Properties of the trochanteral hair plate and its function in the control of walking in the cockroach. *J. Exp. Biol.* 64: 233–49

Wyman, R. J., Thomas, J. B., Salkoff, L., King, D. G. 1984. The *Drosophila* giant fiber system. In *Neural Mechanisms of Startle Behavior*, ed. R. C. Eaton, pp. 133–62. New York: Plenum

Zill, S. N. 1985. Proprioceptive feedback and the control of cockroach walking. In *Feedback and Motor Control in Invertebrates and Vertebrates*, ed. W. J. P. Barnes, M. H. Gladden, pp. 187–208. London: Croom Helm

Annu. Rev. Neurosci. 1993. 16:299–321

RECENT ADVANCES IN THE MOLECULAR BIOLOGY OF DOPAMINE RECEPTORS

Jay A. Gingrich and Marc G. Caron

Departments of Cell Biology and Medicine, Howard Hughes Medical Institute Laboratories, Duke University Medical Center, Durham, North Carolina 27710

KEY WORDS: catecholamines, G protein-coupled receptors, signal transduction, basal ganglia, limbic system

INTRODUCTION

The catecholamine dopamine plays a role in the functioning of numerous vertebrate and invertebrate organisms. In mammals, dopamine affects a diverse array of processes, such as motor control, cognition, emotion, neuroendocrine regulation, positive reinforcement, and cardiovascular regulation. Moreover, dopamine responsive systems have been implicated in several pathologic conditions, such as schizophrenia, tardive dyskinesia, Parkinson's disease, Tourette syndrome, hyperprolactinemia, and possibly Huntington's chorea. The quest to find new drugs to combat the symptoms of these conditions has led to the discovery of numerous synthetic agonists and antagonists for dopamine receptors. At the cellular and biochemical level, dopamine also exhibits a wide range of effects, including stimulation and inhibition of adenylyl cyclase (Stoof & Kebabian 1984), stimulation of potassium channels (Sasaki & Sato 1987), alterations of phosphotidyl-inositol (Mahan et al 1990) and arachidonic acid metabolism (Kanterman et al 1991; Piomelli et al 1991), changes in neuronal firing rates (Walters et al 1987), and alterations of gene expression (Gerfen et al 1990). Dopamine mediates these diverse effects through interactions with specific receptor proteins, and one of the challenges has been to identify these receptors and elucidate their properties. By characterizing the structure, function,

299

0147–006X/93/0301–0299$02.00

and cellular localization of these receptors, it should be possible to understand better the wide range of actions that dopamine exerts and to design more effective and selective drugs to ameliorate the symptoms of diseases in which dopamine plays a role.

For the last 10–15 years, it was widely accepted that dopamine acted through only two types of receptors, the so-called D_1 and D_2 dopamine receptors (Kebabian & Calne 1979). In this schema based on pharmacological and biochemical data, the D_1 receptor was responsible for the stimulation of adenylyl cyclase (AC) and had low affinity for butyrophenone neuroleptics, whereas the D_2 receptors had no effect on or mediated the inhibition of AC and possessed high affinity for butyrophenones and substituted benzamides (Stoof & Kebabian 1984). Although this model has significant conceptual utility, recent work in several disciplines has indicated that modification of the two-receptor hypothesis is necessary (reviewed in Andersen et al 1990). In particular, the application of recombinant DNA technology to receptor biology has rapidly expanded our knowledge about dopamine receptors.

The isolation and sequencing of the genes and cDNAs that encode the receptors for dopamine have produced a wealth of structural information about these receptor proteins and have provided the tools to localize precisely the cells that express specific receptor genes; express the receptors in heterologous cell systems and characterize their pharmacology and second-messenger coupling; isolate the promoters and regulatory elements that control receptor gene expression; evaluate the possible genetic linkage of receptor genes to specific disorders; produce mutant receptor proteins to study function; and overexpress the receptor protein to produce antibodies. This article reviews the recent advances made in the cloning of dopamine receptors and in the characterization of their genes, and summarizes recent advances in the areas of protein structure, pharmacology, second-messenger coupling, and anatomical localization.

MOLECULAR CLONING AND GENE STRUCTURE

Molecular Cloning

Currently, two D_1-like and three D_2-like receptor genes have been identified by molecular cloning. The first dopamine receptor cloned was the D_2 receptor reported by Bunzow and colleagues (1988). This receptor was isolated by low stringency screening of a rat brain cDNA library using the hamster β_2 adrenergic receptor cDNA as a probe. Two additional members of this family, the rat D_3 and D_4 receptors, have been cloned by low stringency hybridization using probes derived from the D_2 receptor (Sokoloff et al 1990; Van Tol et al 1991). The first functional D_1-like receptor (referred to herein as the D_{1A} dopamine receptor) was cloned simultaneously in several laboratories (Dearry et al 1990; Monsma et al 1990; Sunahara et al 1990; Zhou et al 1990) by using low stringency screening of libraries or the polymerase chain reaction (PCR) based on sequences of the D_2 dopamine receptor (Bunzow et al 1988). The second member of the D_1-like receptor family was cloned by using similar strategies based on sequences of the cloned D_{1A} receptor. This second D_1-like receptor clone was isolated nearly simultaneously by several groups and has been referred to as the D_5, the D_{1B}, or the $D_{1\beta}$ receptor (Grandy et al 1991a; Sunahara et al 1991; Tiberi et al 1991; Weinshank et al 1991). Because of the high degree of sequence conservation between the D_1 receptor originally cloned and the second member of this family, we favor the nomenclature D_{1A} and D_{1B}.

Although it is now clear that the human D_5 and the rat D_{1B} are species homologues of the same receptor, the data were not so clear initially. The D_5 receptor was cloned from a human, and the D_{1B} from a rat genomic library. The degree of sequence homology [95% identity in the transmembrane segments (TMS), and 85% overall] was in the observed range of homology seen for species homologues. However, in situ hybridization experiments performed in rat brain gave vastly different distributions for both receptors (which are discussed in a later section).

Gene Structure and Chromosomal Localization

The D_{1A} and the D_5/D_{1B} receptor genes are both intronless, a feature shared with a large subset of other G protein-coupled receptors (Dohlman et al 1987). The gene for the D_{1A} receptor was initially localized to the q34–35 locus of human chromosome 5 (Sunahara et al 1990) and later more precisely to 5q35.1 (Grandy et al 1991b). This localization of the D_{1A} receptor is interesting, given the several other catecholamine receptor genes localized to this area of the human genome (Yang-Feng et al 1990). The D_5/D_{1B} receptor gene originally mapped by cytogenetic analysis to the distal region of the short arm of chromosome 4 (4p16.3) (Tiberi et al 1991) has been more precisely localized to region 4p15.1–16.1 by using somatic cell hybrids of chromosome 4 (J. F. Gusella et al 1991, unpublished observations). This region is outside the presumed Huntington's chorea gene locus (Gusella 1989).

An interesting development in the cloning of D_1-like receptors from the genome has been the identification of pseudogenes of the human D_5 receptor (Grandy et al 1991a; Jarvie et al 1991; Weinshank et al 1991). There appears to be two such pseudogenes, and they are 98% identical at the nucleic acids level with each other and 95% identical with the human D_5 receptor. They are judged to be pseudogenes because of the presence of numerous in-frame stop codons that would produce truncated proteins and, presumably, nonfunctional receptors. This hypothesis was borne out by expression studies (Grandy et al 1991a). Although these genes do not produce functional receptors, they appear to be transcribed with an expression pattern similar to the D_5/D_{1B} receptor (Weinshank et al 1991). The presence of these pseudogenes in the human genome could account for some of the anomalous hybridization signals observed during chromosomal localization studies of the D_5/D_{1B} receptors (Tiberi et al 1991; Grandy et al 1991a).

Although numerous G protein-coupled receptors described to date have proven to be intronless within their coding regions, an exception to this rule has been the gene for rhodopsin (Nathans & Hogness 1983). The receptors of the D_2-like dopamine receptor family can now be added to the category of intron-containing G protein-coupled receptors. The D_2 receptor coding region is interrupted by five to six introns (Dal Toso et al 1989; Giros et al 1989; Grandy et al 1989a; Monsma et al 1989; Selbie et al 1989), the D_3 receptor by five (Sokoloff et al 1990), and the D_4 receptor by four (Van Tol et al 1991). The placement of most of the introns within the coding regions of the D_2-like receptors is similar (see Figure 1). Thus, in each of the three receptors, introns are found within the second TMS, in the third cytoplasmic loop, and at the beginning of the sixth TMS. However, the D_3 receptor lacks the fourth intron of the D_2, and the D_4 receptor lacks the third and the fourth introns. The third intron of the D_4 receptor has an unusual intron/exon junction site that lacks the normal donor acceptor sequences, thus making the exact placement of the splice site imprecise (Van Tol et al 1991).

The presence of introns within the coding region of the D_2-like receptors allows for the possibility of alternate splicing of the exons. Such splice variations have been found in other receptor systems, most notably the receptors that form ion permeable channels, where such splice variants form functionally different receptor proteins (Monyer et al 1991). Indeed, splice variants of the D_2 and D_3 receptors have also been described. Two isoforms of the D_2 receptor are generated by alternate splicing of the 87 base pair exon between introns 4 and 5, which differ by 29 amino acids in the intracellular loop IL3 (Table 1). Although the IL3 is thought to be important in the function of G protein recognition and coupling (Kobilka 1992), thus far no functional differences have been reliably attributed to the different D_2 receptor splice variants. However, some regional and developmental differences in the expression of the two forms have been reported (see section on Localization of Receptor mRNA and Protein). The IL3s of the D_3 and D_4 receptors are interrupted by only single introns; no splice variants are then possible in this region of these receptors, unlike

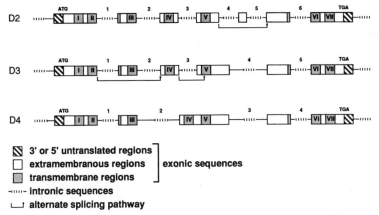

Figure 1 Intron-exon structure of the D_2-like dopamine receptor genes within their coding sequences. This schematic representation of the intron-exon structure of the D_2-like receptor genes indicates only the relative location of introns and the approximate size of the exons. The box-like structures represent the exonic sequences of each receptor, and the lines interrupted by vertical hatching represent the intronic sequences (whose sizes are not represented). The introns are numbered by arabic numerals. Within the exons, the shaded areas numbered with roman numerals indicate the regions coding for putative TMSs, the unshaded areas represent regions lying outside the membrane. The "ATG" and "TGA" sequences symbolize the beginning and end of the translated regions. The diagonally hatched regions represent the presence of untranslated sequences, which are found on the exons that flank the coding region and do not necessarily indicate their relative size. However, in the case of the D_2 and D_3 receptors, the 5' untranslated region is interrupted by introns at positions -32 to -33 relative to the ATG. The arrows indicate pathways of alternate splicing, which have been demonstrated for the D_2 and D_3 receptors, whereby known exons can be lost or alternative splice sites can be employed. In the D_2 receptor, this alternate splicing produces either the so-called long or short forms. In the case of the D_3 receptor, the loss of the second exon (which codes for part of the second and all of the third TMS) leads to a severely truncated receptor (because of a frame shift). The use of an alternate splice donor site in the fifth TMS produces a product with a long open reading frame, which is not apparently functional.

the D_2 receptor. Thus, it is somewhat surprising to find that the human D_3 receptor lacks a 138 nucleotide (46 amino acid) segment of the IL3, which is present in the rat D_3 (Giros et al 1990). Interestingly, this 46 amino acid segment is approximately in the same location as the alternatively spliced exon in the D_2 dopamine receptor gene. This finding could suggest that the human D_3 receptor has an additional intron (in a similar position as the fourth intron of the D_2 receptor) not found in the rat, which might allow the analogous segment to be spliced from the mRNA. Unusual splice variants of the rat D_3 receptor not involving the third IL have been detected by using reverse transcription PCR (Giros et al 1991).

Table 1   Structure, characteristics, and coupling of the cloned dopamine receptor subtypes

Subtype		Species	Coupling	Size (aa's)	Exons	Introns	Chromosome
D₁	**D₁ₐ**	[Human, Rat]	↑ AC	446	1	0	5q35.1
L I K E	**D₅/D₁ʙ**	Rat, Human	↑ AC	475, 477	1	0	4p15.1-16.1
D₂	**D₂ₛ**, **D₂ʟ**	[Human, Rat, Bovine, Xenopus]	↓ AC, ↑ K⁺	415, 444	6, 7	5, 6	11q22-23
L I K E	**D₃**	Rat, Human	?	446, 400	6, ?	5, ?	3p13.3
	D₄	Human	?	387	5	4	11p

This table indicates the various subtypes of dopamine receptors (and the two isoforms of the D_2 receptor). The "species" column indicates the species from which these various receptor subtypes have been cloned. The second messenger coupling for each receptor subtype is indicated in the "coupling" column. The abbreviation AC refers to adenylyl cyclase, with upward or downward arrows indicating either stimulation or inhibition of this enzyme. The abbreviation K^+ indicates the stimulatory effect of D_2 receptors on potassium channel activity. The coupling of the D_3 and D_4 receptors is not known. The deduced size of the receptor protein in amino acid (aa) residues is indicated in the next column. The D_3 receptor differs markedly in size between the rat (446) and the human (400) homologues. The structure of the genes for these receptors is summarized by the exon and intron columns, where the number of exons and introns for the coding region of each receptor gene is indicated. When known, the location of the gene for each receptor in the human genome is indicted in the "chromosome" column. References for information in this table can be found in the text.

Two of these variants have been characterized and appear to be formed by using alternate splice acceptor sites, which then lead to frameshifts in the normal D_3 open reading frame. One of these splice variants was transfected into CHO cells, but expression of the mRNA for these clones failed to produce a protein that could bind dopaminergic ligands (Giros et al 1991). No alternative splice variants of the D_4 receptor have yet been reported.

The human D_2 dopamine receptor gene was localized to the q22–23 locus of chromosome 11 (Grandy et al 1989b) and has been associated with a Taq I restriction fragment length polymorphism (RFLP). This RFLP has been claimed to be associated with alcoholism and other psychiatric conditions (Comings et al 1991), but this finding has proven somewhat controversial (Bolos et al 1990; Gelernter et al 1991a). This RFLP is not linked with markers for schizophrenia (Moises et al 1991), manic depression (Holmes et al 1991), or Tourette syndrome (Gelernter et al 1991b). The gene for the human D_3 dopamine receptor has been localized to the q13.3 locus on chromosome 3 (Giros et al 1991), the same chromosome on which the gene for rhodopsin is found. The gene for the human D_4 dopamine receptor has been localized to the short arm of chromosome 11, the same chromosome as for the D_2 receptor gene (Van Tol et al 1991).

STRUCTURAL FEATURES OF THE RECEPTOR PROTEINS

Transmembrane Homology and Determinants of Binding

All of the receptors for dopamine share relatively high homology within their TMS. The high degree of homology that these receptors share is shown graphically in Figure 2, where darkened residues are those identically conserved between the D_1-like (Figure 2A) and the D_2-like subtypes (Figure 2B). As is evident from Figure 2A, the D_{1A} and the D_5/D_{1B} receptors share a very high homology in their TMS (78% identity). However, the degree of conservation among D_2-like receptors is less. With the D_2 receptor, the D_3 receptor shares a 75% identity and the D_4 receptor shares a 53% identity in the TMS. All of the dopamine receptor subtypes share several conserved residues within their TMS, which are thought to be the minimal requirements for catecholamine binding: the two serine residues in the putative fifth TMS thought to be involved in recognition of the two hydroxyl groups of catecholamines and the aspartic acid residue in the third TMS, which is thought to act as a counterion for the amine moiety in biogenic amines (Strader et al 1987, 1989).

Homology Outside the TMS and G Protein Coupling

It is widely believed that the presumed intracellular domains of seven TMS receptors are responsible for the specificity and coupling of these receptors to guanine nucleotide binding proteins (G proteins), especially those regions that lie closest to TMS segments (reviewed in Dohlman et al 1991). Comparison of the sequences of the D_{1A} and D_5/D_{1B} receptors reveals that these domains have the highest degree of homology for regions outside the TMS. This is consistent with the observed ability of each of these receptors to couple to the same signaling pathway. The D_2 and D_3 receptors also share fairly high homology in these regions, but the D_4 receptor is less homologous. The second-messenger pathway(s) of the D_4 receptor

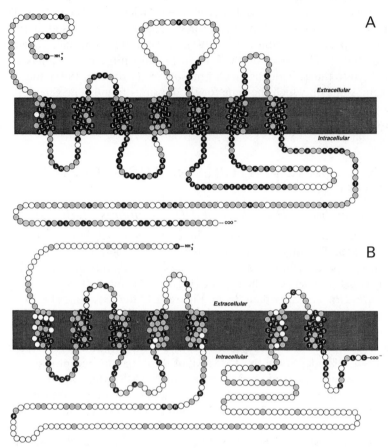

Figure 2 Amino acid identity between dopamine receptor subtypes. Shown here are the hypothetical TM organization of the D_1-like (panel *A*) and the D_2-like (panel *B*) dopamine receptors. Each circle represents an amino acid residue. Blackened circles indicate amino acid residues that are conserved between all three receptor subtypes, shaded circles indicate residues that are identical in two of the three receptors, and unshaded circles indicate residues that are not conserved between any of the three receptor types. In many instances, gaps were necessarily inserted to preserve the alignment of these receptors. (*A*) The structure of the human D_{1A} dopamine receptor subtype is depicted (Dearry et al 1990; Sunahara et al 1990; Zhou et al 1990). Residues that are conserved in this receptor with the D_{1B} (Tiberi et al 1991) and D_5 (Jarvie et al 1991; Sunahara et al 1991) receptor are shaded according to the description above. (*B*) The structure of the rat D_2 receptor (with insert) is depicted in this panel (Bunzow et al 1988) and is compared with the rat D_3 (Sokoloff et al 1990) and the human D_4 (Van Tol et al 1991) dopamine receptors.

is not yet known. As mentioned earlier, the IL3 of the D_2 subtype can differ by presence or absence of a 29 amino acid insert in the middle of this large loop. In the D_3 receptor, this domain includes an extra 46 amino acids in the rat receptor that are not present in the human D_3 receptor. The inability to detect differences in G protein coupling between the different splice variants of the D_2 receptor underscores the fact that the middle portion of these loops does not appear to play a major role in G protein coupling (Einhorn et al 1990).

Potential Sites for Glycosylation, Palmitoylation, and Disulfide Bonding

These receptors also share several potential sites of post-translational modification, which are summarized in Figure 3. Each subtype possesses consensus sites for N-linked glycosylation. The D_{1A} and D_5/D_{1B} possess two such sites: one in the amino terminal domain and another in the second extracellular loop (EL2). The D_2 and D_3 receptors might be multiply glycosylated, with four potential sites available on each receptor in their putative extracellular domains. The D_4 receptor, in contrast, possesses only one potential site for N-linked glycosylation. Also found in the presumed extracellular domains of each of the cloned dopamine receptors are two cysteine residues in EL1 and EL2, which are believed to be disulfide bonded in other receptors (Fraser 1989; Dohlman et al 1990). Likewise, each receptor possesses the cysteine residue in the proximal region of the carboxy tail, which is palmitoylated in rhodopsin and the β_2 adrenergic receptor (O'Dowd et al 1989; Ovchinnikov et al 1988). In the D_1-like receptors, this cysteine residue is found at the beginning of a relatively long carboxyl terminus; for the D_2-like receptors, the relatively short carboxyl terminus ends with this cysteine residue.

Potential Sites for Regulatory Phosphorylation

Both the D_{1A} receptor and the D_2 receptor desensitize in response to agonist exposure (Balmforth et al 1991; Bates et al 1991). Although phosphorylation of G protein-coupled receptors is believed to be involved in desensitization, the roles that protein kinase A, protein kinase C, or receptor kinases (e.g. the family of rhodopsin/β-adrenergic receptor-like kinases) might play in the desensitization process of these receptors have not been determined. Each of these receptors has consensus substrate sites for these different regulatory protein kinases (see Figure 3). Interestingly, the number of these sites and their placement is different among the various subtypes, which suggests that these receptors might differ in their patterns of regulation. Certainly, further work will be required to determine which

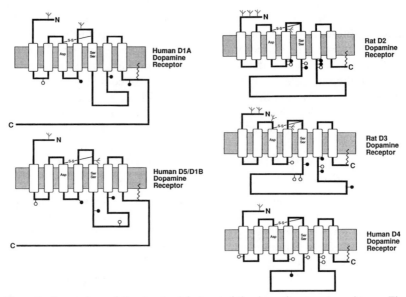

Figure 3 Comparison of the structural features of the dopamine receptor subtypes. The proposed TM topology is illustrated schematically. The shaded horizontal boxes represent the plasma membrane. The vertical boxes represent putative TM domains of these receptors. The dark lines connecting the vertical boxes represent the presumed extramembranous domains of these receptors. Branch-like structures represent the presence of N-linked glycosylation consensus sites (N × S/T) in the receptor sequence. The (S-S) symbol joining the first and second extracellular domains indicates that the two cysteine residues conserved in the β_2 adrenergic receptor (Dohlman et al 1990; Fraser et al 1989), which are believed to be disulfide linked, are present in these sequences. The "Asp" and two "Ser" symbols indicate the presence of the conserved residues that are thought to be the structural determinants of catecholamine recognition in the β_2 adrenergic receptor (Strader et al 1987, 1989). The jagged line near the proximal region of the carboxyl terminus indicates the presence of a cysteine residue that is palmitoylated in rhodopsin (Ovchinnikov et al 1988) and the β_2 adrenergic receptor (O'Dowd et al 1989). Darkened circles represent the presence of PKA consensus sites (R/K R/K × S/T), and open circles represent the presence of PKC consensus sites (S/T × R/K) in the sequence within the carboxy terminal tail of the D_1-like receptors and the third intracellular loop.

of these sites is physiologically relevant, and what effect their phosphorylation has on receptor function.

PHARMACOLOGICAL PROPERTIES

D_1-Like Receptors

As mentioned previously, the D_1-like receptors share very high homology within their TM domains, the regions that are thought to form the ligand-binding site. Thus, it is not surprising that these receptors exhibit very similar ligand binding properties (see Table 2). Each of these receptors exhibits the classical ligand-binding characteristics of D_1 receptors: high affinity for benzazepines (SCH 23390, SKF 38393) and low affinity for butyrophenones (spiperone, haloperidol) and substituted benzamides (sulpiride). An interesting difference between the D_1-like receptors is their relative affinity for dopamine. The human D_5 receptor was reported to have an affinity tenfold higher for dopamine than the D_{1A} receptor, and Tiberi et al (1991) found a threefold higher affinity with the rat D_5/D_{1B} receptor. The only antagonist tested that shows notably different affinities for these three receptors is spiperone (10–20-fold less potent at D_5/D_{1B} than D_{1A}) (see Table 2).

D_2-Like Receptors

The pharmacological properties of the D_2-like receptors differ more than has been observed for the D_1-like receptors. As seen in Table 2, affinities for several antagonists and agonists vary one to two orders of magnitude between subtypes. Each of these receptors, however, has the hallmark ligand-binding characteristics of D_2-like receptors; each possesses high affinity for such butyrophenone compounds as spiperone (0.03–0.6 nM K_D) and haloperidol (0.45–9.8 nM K_D), and low affinity for such benzazepines as SKF 38393 (1800–9560 nM K_D). Of the three D_2-like receptors, the D_4 receptor has the most distinctive pharmacological properties. In general, the D_4 receptor displays lower affinities for most dopaminergic antagonists. For example, the D_2 and D_3 receptors have relatively high affinity for such benzamides as raclopride, whereas the D_4 receptor has surprisingly low affinity for this compound (297 nM K_D, 30–165-fold lower affinity than the D_2 and D_3). However, the D_4 receptor exhibits relatively high affinity for the atypical neuroleptic clozapine (9 nM K_D, 6–20-fold higher than the D_3 and D_2 receptors, respectively). This has led to speculation that the D_4 receptor might be the relevant target of this atypical neuroleptic (Van Tol et al 1991).

SECOND-MESSENGER COUPLING

D_1-Like Receptors

A prediction from the strong structural homology found among the D_1-like receptors in their putative intracellular and extracellular regions was that each of these receptors would couple to the same second-messenger system (Table 1). Indeed, as was originally found for the D_1 receptor in tissues, the D_{1A} and D_5/D_{1B} receptors couple to the stimulation of AC (Dearry et al 1990; Grandy et al 1991a; Monsma et al 1990; Sunahara et

Table 2 Summary of the pharmacological properties of the dopamine receptor subtypes

| | Ligand pharmacology of the D_1-like dopamine receptors | | |
	D_{1A}	D_{1B} rat K_D in nM	D_5 human
Antagonists			
(+)Butaclamol	0.9–3.0	6	5.5*–27
Clozapine	141	343	250–336
SCH 23390	0.11–0.66	0.11	0.3–0.54
SCH 23388	15–41	10	35
cis-Flupenthixol	1.6–5.6	7	6.3–7
Haloperidol	27–203	35	48–151
Spiperone	220	2600	4500
(−)Sulpiride	20,400	21,210	11,000–29,000
Ketanserin	190	368	330*–2500
Agonists			
SKF 38393	87–150	100	43–100
Apomorphine	210–680	240	152–363
Dopamine	2300–12,000	3900	228–235
Fenoldopam	17–28	11	15
NPA	1540–1816	1050	1130

| | Ligand pharmacology of the D_2-like dopamine receptors | | |
	D_2	D_3 K_D in nM	D_4
Antagonists			
(+)Butaclamol	0.8	N.D.	40
Clozapine	56–138	180	9
Haloperidol	0.45–1.0	9.8	5.1
Spiperone	0.03–0.07	0.6	0.08
Raclopride	1.8–10.5	3.5	237
(−)Sulpiride	4.8–46	25	52
Pimozide	2.4	3.7	43
Agonists			
SKF 38393	9560	5000	1800
Apomorphine	24	20	4.1
Bromocryptine	5.3–12.6	7.4	340
Dopamine (high)	2.8–474	25	28
Dopamine (low)	1705–2490	27	450
Quinpirole	576	5.1	46

The K_D values (in nM) of several ligands are compared between the various subtypes of dopamine receptor. In the case of the D_{1A} and the D_2 receptors, often several K_D values for a compound can be found in the literature. In these cases, the range of values that have been reported are given. N.D. indicates that no K_D value has been reported for this compound at this receptor. When appropriate, high and low affinity states for agonists are listed. K_D values for the human D_5 receptor are taken from Sunahara et al (1991) and Weinshank et al (1991). Asterisks indicate that these values are from our own work expressing the human D_5 in COS-7 cells (K. R. Jarvie et al, unpublished).

al 1990, 1991; Tiberi et al 1991; Zhou et al 1990). In addition, it appears that this coupling is very specific. Unlike other receptor systems, such as the 5-HT1A, α_2 adrenergic or the m5 muscarinic receptors, each of which can couple to both AC and phospholipase C (PLC), the D_{1A} and D_5/D_{1B} receptors do not detectably couple to stimulation of the PLC pathway (Dearry et al 1990; Tiberi et al 1991).

D_2-Like Receptors

It has now been extensively demonstrated that both the short and long forms of the cloned D_2 receptor couple to both the inhibition of AC (Albert et al 1990; Dal Toso et al 1989; Neve et al 1989; Vallar et al 1990) and the activation of K^+ channels (Einhorn et al 1990). This observation is consistent with previous observations in physiological systems (Lacey et al 1987; Sasaki & Sato 1987; Stoof & Kebabian 1984). The coupling of the cloned D_2 receptor to these pathways is pertussis toxin-sensitive (Bates et al 1991; Neve et al 1989). The long form of the D_2 receptor has also been reported to potentiate the release of arachidonic acid from CHO cells stimulated with the calcium ionophore, A23187 (Kanterman et al 1991; Piomelli et al 1991). The mechanism of this potentiation appears to be cAMP-independent, not blockable by pertussis toxin, but involving activation of protein kinase C (PKC). Thus far, no differences in coupling efficiency have been observed between the D_2 long and short forms (Einhorn et al 1990).

The second-messenger coupling of the D_3 and D_4 receptors has not been well characterized. Initial attempts to assess the coupling pathway of the D_3 in CHO cells were negative. In these cells, no coupling or high affinity agonist binding was demonstrable (Sokoloff et al 1990).

LOCALIZATION OF RECEPTOR mRNA AND PROTEIN

D_1-Like Receptors

One of the more salient differences between the two D_1 receptor subtypes is their pattern of expression in the central nervous system (CNS). The D_{1A} receptor mRNA (4.2 kb, human and rat) appears to be more abundant and more widely distributed than the D_5/D_{1B} subtype (Fremeau et al 1991; Sunahara et al 1991; Tiberi et al 1991; Weiner et al 1991). The highest levels of D_{1A} mRNA are found in areas of the brain traditionally associated with dopaminergic function, such as the caudate-putamen, the nucleus accumbens, and the olfactory tubercles. In addition, D_{1A} mRNA was also found in the cerebral cortex, limbic system, hypothalamus, and thalamus.

Importantly, no mRNA signal was detected in substantia nigra (SN) pars reticulata, entopeduncular nucleus, globus pallidus, and subthalamic nucleus—areas that possess intermediate to high levels of D_1 receptor binding sites. Moreover, no D_{1A} mRNA signal was detected in the ventral tegmental area (VTA) and the SN compacta, which are the major dopaminergic nuclei in the CNS (Fremeau et al 1991). No message for the D_{1A} receptor was detectable in the kidney, heart, liver, lung, and skeletal muscle (Dearry et al 1990). The only tissue outside the CNS found to express the D_{1A} receptor is the parathyroid gland (Sunahara et al 1990).

In contrast to the widespread distribution of the D_{1A} receptor subtype, the mRNA for the D_5/D_{1B} receptor (3.0–3.7 kb in rat and human) has a remarkably limited distribution in the rat CNS, being confined largely to limbic structures (Tiberi et al 1991). There is no appreciable signal for the D_5/D_{1B} receptor message in striatum, nucleus accumbens, and olfactory tubercle, the primary structures that express the D_{1A} receptor. Instead, the D_5/D_{1B} mRNA is largely limited to two sets of nuclei: the lateral mammillary nuclei and the parafascicular nuclei of the thalamus and several layers of the hippocampus. Originally, we proposed (Tiberi et al 1991) that D_{1B} mRNA was found in the anterior pretectal nuclei; however, more detailed examination of in situ hybridization results reveals those nuclei to be the two adjacent thalamic nuclei. Only weak signals for this receptor mRNA were detectable in the striatum and cerebellum by using the sensitive technique of the PCR to amplify message from various brain regions. Using this same technique, no signal was found in peripheral tissues, such as kidney, heart, lung, and liver.

The distribution of the D_5/D_{1B} receptor is interesting from several perspectives. First, the three areas that express this message do not contain appreciable amounts of D_1 receptor binding sites (Boyson et al 1986; Dawson et al 1985, 1986). This finding suggests that the D_5/D_{1B} receptor protein that is being synthesized in these regions may be translocated axonally to another brain region. This pattern would imply a largely presynaptic localization of the D_5/D_{1B} receptor in these neurons. The second interesting point is that the D_{1A} and the D_5/D_{1B} receptor messages overlap very little, if at all. There do not appear to be any regions in which these two receptor types are coexpressed, except possibly some regions of the hippocampus. However, the D_5/D_{1B} receptor message does overlap with the expression of the D_2 receptor mRNA in the lateral mammillary nuclei (Weiner et al 1991), where D_2 binding sites are also found (Dubois et al 1986).

As was mentioned earlier, some initial confusion was generated by the report (Sunahara et al 1991) that the D_5 receptor (as assessed by using oligonucleotide in situ hybridization and northern blot analysis) was equally localized in striatum, hippocampus, frontal cortex, and hypothalamus, as well as olfactory tubercle, olfactory bulb, and nucleus accumbens, which was a pattern similar to the D_{1A} receptor and completely different from that described for the D_{1B} receptor in rat. This controversy has now been settled in favor of the distribution reported for the rat D_{1B} receptor (Tiberi et al 1991). In situ hybridization using cRNA probes derived from the human D_5 receptor on human brain seems to confirm the limbic distribution found in rat (J. H. Meador-Woodruff and S. J. Watson 1992, personal communication).

D_2-Like Receptors

The distribution of the D_2 dopamine receptor mRNA (~ 2.9 kb in both long and short forms) has now been well characterized by several groups by using either traditional northern blot analysis (Bunzow et al 1988; Giros et al 1989; Monsma et al 1989), PCR amplification of mRNA (Giros et al 1989; Rao et al 1990), or in situ hybridization (Mansour et al 1990; Meador-Woodruff et al 1989; Mengod et al 1989; Monsma et al 1990; Najleranim et al 1989; Weiner & Brann 1989; Weiner et al 1991). The D_2 receptor is expressed most prominently in brain tissues, such as the caudate-putamen, olfactory tubercle, and nucleus accumbens. The mRNA for the D_2 receptor is also found in the SN and in the VTA, the nuclei that give rise to the major dopaminergic tracts of the brain (A9 and A10 nuclei, respectively). This finding indicates that the D_2 receptor is one of the main dopamine receptors that directly control the activity of dopamine-containing neurons. The D_2 receptor is also found outside the CNS in the anterior and neurointermediate lobes of the pituitary gland, which indicates that the D_2 receptor is one of the primary dopamine receptors that regulates hormone release. Interestingly, no D_2 receptor message has been found in the kidney, a tissue known to be responsive to D_2 dopaminergic agents.

In general, there appears to be little tissue specificity to the expression of the long and short forms of the D_2 receptor isoforms. The larger form of the D_2 appears to be the more abundant in every tissue except brain stem (Autelitano et al 1989; Montmayeur et al 1991). Otherwise, these studies confirm that each isoform of the D_2 receptor is present in approximately the same ratio in each tissue where either form is expressed. If any trend exists, it is that in low-expressing tissues, the ratio between the long and short forms tends to be lower (Neve et al 1991). However, differences in the pattern of expression of the two isoforms has been found in development (Guennoun & Bloch 1991; Mack et al 1991).

One of the more striking aspects of the D_3 receptor expression pattern is its tendency to be restricted to structures innervated by the mesolimbic dopaminergic neurons, such as olfactory tubercle, islands of Calleja, and nucleus accumbens, which lead to the speculation that the D_3 receptor might be a primary target of neuroleptic medications, because the anti-psychotic activity of these compounds is thought to occur at these structures. The D_3 mRNA (8.3 kb in rat) was found most abundantly in olfactory tubercle, hypothalamus, and nucleus accumbens. Much weaker signals were observed in basal ganglia tissues, such as caudate-putamen. Significant signals were found in the VTA and the SN, which indicates that, like the D_2 receptor, the D_3 receptor may also act as a dopaminergic autoreceptor. Whether individual neurons coexpress both the D_2 and the D_3 receptor subtypes is not yet known. However, even in areas of highest expression, the D_3 receptor is expressed at levels approximately one order of magnitude lower than the D_2 dopamine receptor. The D_3 receptor is not found in the pituitary or the kidney.

The expression pattern of the D_4 dopamine receptor mRNA (5.3 kb in human, monkey, and rat) has not been well characterized thus far. Northern blot analysis demonstrates that this receptor is expressed most abundantly in medulla, amygdala, midbrain, frontal cortex, and striatum. Lower levels of expression are found in olfactory tubercle and hippocampus. As for the D_3 receptor, the D_4 receptor represents a very rare D_2-like receptor subtype, which is one to two orders of magnitude less abundant than the D_2 dopamine receptor. Table 3 summarizes the specific localization of these various receptor subtypes, as revealed mainly by in situ hybridization and northern blot analysis.

SIGNIFICANCE OF DOPAMINE RECEPTOR HETEROGENEITY

The recent discovery of multiple dopamine receptor subtypes begs the question regarding their possible physiological significance. Although such teleological questions are difficult to answer, several possibilities can be imagined. First, different receptors that couple to different second-messenger pathways allow for a diversity of responses to a single substance. Thus, the existence of D_1-like and D_2-like receptors allows dopamine to either stimulate or inhibit AC in the cell. An additional level of complexity is added to this system, because a single receptor, such as D_2, can activate multiple signal transduction pathways, such as stimulating potassium channels, stimulating arachidonic acid metabolism (PLA2), and inhibiting AC. Whether all of these pathways are elicited under physiological conditions in the same cell by D_2 receptors is not known. Second, different affinities for dopamine would allow different receptors to be activated at low or high levels of dopamine release. Thus far, the different subtypes

Table 3 Summary of the distribution of the mRNA for the various dopamine receptor subtypes

	Subtype	mRNA (kB)	High	Moderate	None
D₁ LIKE	D_{1A}	4.2	Caudate putamen Nucleus accumbens Olfactory tubercle	Cortex Amygdala Hypothalamus Parathyroid gland	Ventral tegmental area Substantia nigra Kidney
	D_5/D_{1B}	3.0-3.7	Lateral mammillary nucleus parafascicular thalamic nucleus Hippocampus		Ventral tegmental area Substantia nigra Kidney
D₂ LIKE	D_{2S} / D_{2L}	2.9	Caudate putamen Olfactory tubercle Nucleus accumbens Pituitary (NIL > Ant.)	Substantia nigra Ventral tegmental area Retina	Kidney
	D_3	8.3	Olfactory tubercle Nucleus accumbens	Hypothalamus Caudate putamen Substantia nigra Septum	Kidney Pituitary
	D_4	5.3	Frontal cortex Medulla Amygdala Midbrain	Caudate putamen	Olfactory tubercle Hypothalamus

For each receptor subtype the size of the mRNA is indicated in kilobases (kb). A range of values indicates that different message sizes are reported in different species or tissues, presumably because of differences in use of polyadenylation signals. The distribution of the mRNA for each receptor is divided arbitrarily into regions that express either high or moderate levels of message. Also included are areas that are interesting because they do not express any message for a given receptor subtype. The information summarized here is taken from data reporting in situ hybridization studies, northern blot analysis, and in some instances reverse transcription PCR. The indication of high and moderate are relative to a given subtype. In the case of the D_5 receptor, the mRNA is expressed approximately at one-tenth the level of the D_{1A} receptor mRNA. This is also the case for the D_3 and D_4 receptors that are expressed at only 1–10% of the level of the D_2 dopamine receptor mRNA.

seem to exhibit different affinities for dopamine. Between the D_{1A} and the D_5/D_{1B} receptors, there is as much as a tenfold difference in their affinity for dopamine and as much as a 150-fold difference between the D_2-like receptors. Third, different subtypes might differ in terms of their regulatory properties. As has been discussed, each dopamine receptor subtype possesses a different array of potential regulatory phosphorylation sites (Figure 3). Differences in the number and location of various sites for regulatory phosphorylation might allow for differences in rate and extent of desensitization. Fourth, different subtypes may also allow for more variation and flexibility in expression during development (see above for D_2 isoforms) or maturation of the brain. Each receptor subtype is presumably under the control of unique promoter elements that differ in terms of their tissue specific expression, developmental programing, and responsiveness to transacting factors, such as steroid hormone receptors, cAMP responsive element binding proteins, and enhancer binding proteins. Lastly, the D_1-like and D_2-like receptors appear to exist both pre- and postsynaptically. It has not yet been determined whether these receptor subtypes have structural features that target them toward the synapse or cell bodies. Certainly, this is not an exhaustive list of possibilities, but determining the functional significance of receptor subtypes will remain a challenge for the foreseeable future.

CONCLUSIONS AND FUTURE DIRECTIONS

Other Subtypes Remaining to Be Cloned?

Thus far, five different genes for functional dopamine receptors have been isolated and characterized, which does not include the mRNA variants that arise from alternate splicing of exons in the D_2 receptor family or the pseudogenes found in the D_1 receptor family. This number already greatly exceeds the original number of receptors predicted by the two-receptor model, but possibly still does not represent the full extent of this family. For example, there have now been several reports of a dopamine receptor with D_1-like pharmacology that can stimulate PLC (Felder et al 1989; Mahan et al 1989; Missale et al 1988). Because none of the D_1-like receptors that have been cloned thus far can stimulate phosphatidylinositol metabolism, there is likely a PLC-coupled D_1 receptor that remains to be cloned and characterized. However, caution needs to be exercised with this possibility, because several receptors coupled to stimulation of AC (TSH and LH receptors) have now been shown, albeit with low efficiency, to couple to the PLC pathway (Gudermann et al 1992; Van Sande et al 1990) and vice versa (Nakajima et al 1992). In addition, the D_1 receptors found in peripheral tissues, such as kidney, appear to have a different ligand-binding pharmacology than their CNS counterparts (reviewed in Andersen et al 1990). Thus far, a message for the cloned D_1-like receptors has not been detected in the kidney, which suggests that yet another subtype might exist in this tissue. Again, however, the recent cloning of a D_{1A} receptor homologue from an opossum kidney cell line (Nash et al 1991) and a D_{1B} receptor from a rat kidney library (Monsma et al 1991) raises doubts as to whether distinct D_1 receptor genes are expressed in the kidney. Similarly, evidence for additional D_2-like receptors in the periphery has also been reported (reviewed in Andersen et al 1990). Thus, the continuing identification of new dopamine receptor subtypes and the cloning of their genes or cDNAs represents a near term challenge of pharmacologists, physiologists, and molecular biologists working in this area.

Further Characterization of Cloned Dopamine Receptors

As stated in the introduction, the cloning of the gene for a receptor protein provides valuable information about the sequence and structure of the gene and the protein for which it codes. However, the availability of the genes for these various dopamine receptors also provides several new approaches for further characterizing these receptors both in vivo and in vitro. Thus, if the identification and cloning of new dopamine receptor subtypes constitutes the short-term goal in this field; the full characterization of these receptors with the tools that cloning provides remains the long-term challenge. In this review, we have described the initial results that have been obtained from these tools, such as the characterization of receptors expressed in heterologous cells, and the expression patterns of these receptors determined by using probes derived from nucleic acid sequences. However, the second wave of such studies is now beginning to make its way into the literature and includes probing of the determinants of ligand binding and G protein coupling through the use of site directed mutagenesis (Neve et al 1991) and receptor chimera (England et al 1991), production of antibodies to characterize the localization of the receptor proteins in a more definite fashion than is possible with ligand auto-radiography (Chazot et al 1991; David et al 1991; Farooqui et al 1991), characterization of the promotor and regulatory elements that control gene expression, and linkage analysis of the receptor genes to various inheritable pathologies.

This is an exciting time for those who are interested in the physiology of dopaminergic systems. Certainly, this new wealth of detailed information should allow new insights into the physiology and functioning of the dopamine systems, as well as what possible role these receptors might have in the pathophysiology or in the symptomatology of the various diseases for which dopaminergic drugs are used as a treatment. Given the new tools at our disposal and the information that they can provide, the hope to expand our basic knowledge, as well as make inroads into the design of new, more efficacious and more specific treatments for dopamine-related diseases, seems very bright indeed.

ACKNOWLEDGMENTS

The authors thank S. J. Watson Jr. and J. H. Meador-Woodruff and M. Tiberi and K. Jarvie for sharing their data before publication; M. Bates, K. Jarvie, M. Tiberi, S. Mestikawy, and M. C. Miquel for their critical reading of this review; M. Bates for his contributions to several of the figures and tables; and D. Staples for her excellent secretarial skill in revising this manuscript. J. A. Gingrich is a recipient of a grant from the Medical Scientist Training Program 5T32GM-07171. Work from this laboratory was supported in part by National Institutes of Health grants NS19576, MH44211, and MH40159.

Literature Cited

Albert, P. R., Neve, K. A., Bunzow, J. R., Civelli, O. 1990. Coupling of a cloned rat dopamine D_2 receptor to inhibition of adenylyl cyclase and prolactin secretion. *J. Biol. Chem.* 265: 2098–2104

Andersen, P. H., Gingrich, J. A., Bates, M. D., Dearry, A., Falardeau, P., et al. 1990. Dopamine receptor subtypes: beyond the D_1/D_2 classification. *Trends Pharmacol. Sci.* 11: 231–36

Autelitano, D. J., Synder, L., Sealfon, S. C., Roberts, J. L. 1989. Dopamine D_2-receptor messenger RNA is differentially regulated by dopaminergic agents in the rat anterior and neurointermediate pituitary. *Mol. Cell. Endocrinol.* 67: 101–5

Balmforth, A. J., Warburton, P., Ball, S. G. 1991. Homologous desensitization of the D1 dopamine receptor. *J. Neurochem.* 55: 2111–16

Bates, M. D., Senogles, S. E., Bunzow, J. R., Liggett, S. B., Civelli, O., Caron, M. G. 1991. Regulation of responsiveness at D_2 dopamine receptors by receptor desensitization and adenylyl cyclase sensitization. *Mol. Pharmacol.* 39: 55–63

Bolos, A. M., Dean, M., Lucas-Derse, S., Ransburg, M., Brown, G. L., Goldman, D. 1990. Population and pedigree studies reveal a lack of association between the dopamine D2 receptor gene and alcoholism. *J. Am. Med. Assoc.* 264: 3156–60

Boyson, S. J., McGonigle, P., Molinoff, P. B. 1986. Quantitative autoradiographic localization of the D_1 and D_2 subtypes of dopamine receptors in rat brain. *J. Neurosci.* 6: 3177–88

Bunzow, J. R., Van Tol, H. H. M., Grandy, D., Albert, P., Salon, J., et al. 1988. Cloning and expression of a rat D_2 dopamine receptor cDNA. *Nature* 336: 783–87

Chazot, P. L., Wilkins, M., Strange, P. G. 1991. Site-specific antibodies as probes of the structure and function of the brain D2 dopamine receptor. *Biochem. Soc. Trans.* 19: 143S

Comings, D. E., Comings, B. G., Muhleman, D., Dietz, G., Shahbahrami, B., et al 1991. The dopamine D_2 receptor locus as a modifying gene in neuropsychiatric disorders. *J. Am. Med. Assoc.* 266: 1793–1800

Dal Toso, R., Sommer, B., Ewert, M., Herb,

A., Pritchett, D. B., et al. 1989. The dopamine D_2 receptor: Two molecular forms generated by alternative splicing *EMBO J.* 8: 4025–34

David, C., Ewert, M., Seeburg, P. H., Fuchs, S. 1991. Antipeptide antibodies differentiate between long and short isoforms of the D_2 dopamine receptor. *Biochem. Biophys. Res. Commun.* 179: 824–29

Dawson, T. M., Gehlert, D. R., McCabe, R. T., Barnett, A., Wamsley, J. K. 1986. D_1 Dopamine receptors in the rat brain: A quantitative autoradiographic analysis. *J. Neurosci.* 6: 2352–65

Dawson, T. M., Gehlert, D. R., Yamamura, H. I., Barnett, A., Wamsley, J. K. 1985. D1 dopamine receptors in the rat brain: Autoradiographic localization using [³H] SCH 23390. *Eur. J. Pharmacol.* 108: 323–25

Dearry, A., Gingrich, J. A., Falardeau, P., Fremeau, R. T. Jr., Bates, M. D., Caron, M. G. 1990. Molecular cloning and expression of the gene for a human D_1 dopamine receptor. *Nature* 347: 72–76

Dohlman, H. G., Caron, M. G., Lefkowitz, R. J. 1987. A family of receptors coupled to guanine nucleotide regulatory proteins. *Biochemistry* 26: 2657–64

Dohlman, H. G., Caron, M. G., DeBlasi, A., Frielle, T., Lefkowitz, R. J. 1990. A role of extracellular disulfide bonded cysteines in the ligand binding function of the β_2 adrenergic receptor. *Biochemistry* 29: 2335–42

Dohlman, H. G., Thorner, J., Caron, M. G., Lefkowitz, R. J. 1991. Model systems for the study of seven-transmembrane-segment receptors. *Annu. Rev. Biochem.* 60: 653–88

Dubois, A., Savasta, M., Curet, O., Scatton, B. 1986. Autoradiographic distribution of the D_1 agonist [3H]SKF 38393, in the rat brain and spinal cord. Comparison with the distribution of D_2 dopamine receptors. *Neuroscience* 19: 125–37

Einhorn, L. C., Falardeau, P., Caron, M. G., Civelli, O., Oxford, G. S. 1990. Both isoforms of the D_2 dopamine receptor couple to a G protein activated K^+ channel when expressed in G4u cells. *Soc. Neurosci. Abstr.* 16: 382

England, B. P., Ackerman, M. S., Barrett,

R. W. 1991. A chimeric D2 dopamine/m1 muscarinic receptor with D_2 binding specificity mobilizes intracellular calcium in response to dopamine. *FEBS Lett.* 279: 87–90

Farooqui, S. M., Brock, J. W., Hamdi, A., Prasad, C. 1991. Antibodies against synthetic peptides predicted from the nucleotide sequence of D_2 receptor recognize native dopamine receptor protein in rat striatum. *J. Neurochem.* 57: 1363–69

Felder, C. C., Jose, P. A., Axelrod, J. 1989. The dopamine-1 agonist, SKF 82526, stimulates phospholipase-C activity independent of adenylate cyclase. *J. Pharmacol. Exp. Ther.* 248: 171–75

Fraser, C. 1989. Site-directed mutagenesis of β-adrenergic receptors. *J. Biol. Chem.* 264: 9266–70

Fremeau, R. J. Jr., Duncan G. E., Fornaretto, M.-G., Dearry, A., Gingrich, J. A., et al. 1991. Localization of D_1 dopamine receptor mRNA in brain supports a role in cognitive, affective and neuroendocrine aspects of dopaminergic transmission. *Proc. Natl. Acad. Sci. USA* 88: 3772–76

Gelernter, J., Pakstis, A. J., Pauls, D. L., Kurlan, R., Gancher, S. T., et al. 1991a. No association between an allele at the D_2 dopamine receptor gene (DRD2) and alcoholism. *J. Am. Med. Assoc.* 266: 1801–7

Gelernter, J., Pakstis, A. J., Pauls, D. L., Kurlan, R., Gancher, S. T., et al. 1991b. Gilles de la Tourette syndrome is not linked to D_2-dopamine receptor. *Arch. Gen. Psychiatry* 47: 1073–77

Gerfen, C. R., Engber, T. M., Mahan, L. C., Susel, Z., Chase, T. N., et al. 1990. D_1 and D_2 dopamine receptor-regulated gene expression of striatonigral and striatopallidal neurons. *Science* 250: 1429–32

Giros, B., Martres, M. P., Pilon, C., Sokoloff, P., Schwartz, J. C. 1991. Shorter variants of the D_3 dopamine receptor produced through various patterns of alternative splicing. *Biochem. Biophys. Res. Commun.* 176: 1584–92

Giros, B., Martres, M. P., Sokoloff, P., Schwartz, J. C. 1990. cDNA cloning of the human dopaminergic D_3 receptor and chromosome identification. *C. R. Acad. Sci. Paris* t311, Ser. 3: 501–8

Giros, B., Sokoloff, P., Martres, M. P., Riou, J. F., Emorine, L. J., Schwartz, J. C. 1989. Alternative splicing directs the expression of the two D_2 dopamine receptor isoforms. *Nature* 347: 923–26

Grandy, D. K., Litt, M., Allen, J. R., Bunzow, J. R., Machioni, M. A., et al. 1989a. The human dopamine D_2 receptor gene is located on chromosome 11 at q22–23 and identifies a Taq 1 RFLP. *Am. J. Hum. Genet.* 45: 778–85

Grandy, D. K., Zhang, Y., Bouvier, C., Zhou, Q.-Y., Johnson, R. A., et al. 1989b. Cloning of the cDNA and a gene for a human D_2 dopamine receptor. *Proc. Natl. Acad. Sci. USA* 86: 9762–66

Grandy, D. K., Zhang, Y., Bouvier, C., Zhou, Q.-Y., Johnson, R. A., et al. 1991a. Multiple human D_5 dopamine receptor genes: A functional receptor and two pseudogenes. *Proc. Natl. Acad. Sci. USA* 88: 9175–79

Grandy, D. K., Zhou, Q.-Y., Allen, L., Litt, R., Magenis, R. E., et al. 1991b. A human D_1 dopamine receptor gene is located on chromosome 5 at q35.1 and identifies an Eco RI RFLP. *Am. J. Hum. Gen.* 47: 828–34

Gudermann, T., Birnbaumer, M., Birnbaumer, L. 1992. Evidence for dual coupling of the murine luteinizing hormone receptor to adenylyl cyclase and phosphoinositide breakdown and Ca^{++} mobilization. *J. Biol. Chem.* 267: 4479–88

Guennoun, R., Bloch, B. 1991. D_2 dopamine receptor gene expression in the rat striatum during ontogeny in an in situ hybridization study. *Dev. Brain Res.* 60: 79–87

Gusella, J. F. 1989. Location cloning strategy for characterizing genetic defects in Huntington's disease and Alzheimer's disease. *FASEB J.* 3: 2036–41

Holmes, D., Brynjolfsson, J., Brett, P., Curtis, D., Petursson, H., et al. 1991. No evidence for a susceptibility locus predisposing to manic depression in the region of the dopamine (D_2) receptor gene. *Br. J. Psychiatry* 158: 635–41

Jarvie, K. R., Silvia, C., Fremeau, R. T. Jr., Gingrich, J. A., Caron, M. G. 1991. Cloning and characterization of a novel human D_1 dopamine receptor subtype and its pseudogene. *Soc. Neurosci. Abstr.* 36: 2

Kanterman, R. Y., Mahan, L. C., Briley, E. M., Monsma, F. J., Sibley, D. R., et al. 1991. Transfected D_2 dopamine receptors mediate the potentiation of arachidonic acid release in Chinese hamster ovary cells. *Mol. Pharmacol.* 39: 364–69

Kebabian, J. W., Calne, D. B. 1979. Multiple receptors for dopamine. *Nature* 227: 93–96

Kobilka, B. K. 1992. Adrenergic receptors as models for G protein-coupled receptors. *Annu. Rev. Neurosci.* 15: 87–114

Lacey, M. G., Mercuri, M. B., North, R. A. 1987. Dopamine acts on D_2 receptors to increase potassium conductance in neurons of the rat substantia nigra zona compacta. *J. Physiol.* 392: 397–416

Mack, K. J., O'Malley, K. L., Todd, R. D. 1991. Differential expression of dopaminergic D2 receptor messenger RNAs during development. *Dev. Brain Res.* 59: 249–51

Mahan, L. C., Burch, R. M., Monsona, F. J. Jr., Sibley, D. R. 1990. Expression of striatae D_1 dopamine receptors coupled to inositae phosphate production and Ca^{2+} mobilization in *Xenopus* oocytes. *Proc. Natl. Acad. Sci. USA* 87: 2196–2200

Mansour, A., Meador-Woodruff, J. H., Bunzow, J. R., Civelli, O., Akil, H., Watson, S. 1990. Localization of the dopamine D_2 receptor mRNA and D_1 and D_2 receptor binding in the rat and pituitary. An in situ hybridization receptor autoradiographic analysis. *J. Neurosci.* 10: 2587–2600

Meador-Woodruff, J. H., Mansour, A., Bunzow, J. R., Van Tol, H. H. M., Watson, S. J. Jr., Civelli, O. 1989. Distribution of D_2 dopamine receptor mRNA in rat brain. *Proc. Natl. Acad. Sci. USA* 86: 7625–28

Mengod, G., Matrinez-Mir, M. I., Vilaro, M. T., Palacios, J. M. 1989. Localization of the mRNA for the dopamine D_2 receptor in the rat brain by in situ hybridization histochemistry. *Proc. Natl. Acad. Sci. USA* 86: 8560–64

Missale, C., Castelletti, L., Memo, M., Carruba, M. O., Spano, P. F. 1988. Identification and characterization of postsynaptic D_1- and D_2-dopamine receptors in the cardiovascular system. *J. Cardiovasc. Pharmacol.* 11: 643–50

Moises, H. W., Gelernter, J., Giuffra, L. A., Zarcone, V., Wetterberg, L., et al. 1991. No linkage between D_2 dopamine receptor gene region and schizophrenia. *Arch. Gen. Psychiatry* 48: 643–47

Monsma, F. J. Jr., Mahan, L. C., McVittie, L. D., Gerfen, C. R., Sibley, D. R. 1990. Molecular cloning and expression of a D_1 dopamine receptor linked to adenylyl cyclase activation. *Proc. Natl. Acad. Sci. USA* 87: 6723–27

Monsma, F. J. Jr., McVittie, L. D., Gerfen, C. R., Mahan, L. C., Sibley, D. R. 1989. Multiple D_2 dopamine receptors produced by alternative RNA splicing. *Nature* 342: 926–29

Monsma, F. J. Jr., Shen, Y., Gerfen, C. R., Mahan, L. C., Jose, P. A., et al. 1991. Molecular cloning of a novel D_1 dopamine receptor from rat kidney. *Soc. Neurosci. Abstr.* 17: 85

Montmayeur, J. P., Bausero, P., Amlaiky, N., Maroteaux, L., Hen, R., Borelii, E. 1991. Differential expression of the mouse D_2 dopamine receptor isoforms. *FEBS Lett.* 278: 239–43

Monyer, H., Seeburg, P. H., Wisden, W. 1991. Glutamate-operated channels: Developmentally early and mature forms

arise by alternative splicing. *Neuron* 6: 799–810

Najleranim, A., Barton, A. J., Harrison, P. J., Heffernan, J., Pearson, R. C. 1989. Messenger RNA encoding the D_2 dopaminergic receptor detected by in situ hybridization histochemistry in rat brain. *FEBS Lett.* 255: 335–39

Nakajima, Y., Tsuchida, K., Negishi, M., Ito, S., Naganishi, S. 1992. Direct linkage of three tachykinin receptors to stimulation of both phosphatidylinositol hydrolysis and cyclic AMP cascades in transfected Chinese hamster ovary cells. *J. Biol. Chem.* 267: 2437–42

Nash, S. R., Fornaretto, M. G., Bates, M. D., Raymond, J. R., Caron, M. G. 1991. Cloning and sequence of a dopamine receptor from opossum kidney (OK) cells that is expressed both in kidney and brain. *Soc. Neurosci. Abstr.* 17: 1090

Nathans, J., Hogness, D. S. 1983. Isolation, sequence analysis, and intron-exon arrangement of the gene encoding bovine rhodopsin. *Cell* 34: 807–14

Neve, K. A., Cox, B. A., Henningsen, R. A., Spanoyannis, A., Neve, R. L. 1991. Pivotal role for aspartate-80 in the regulation of dopamine D_2 receptor affinity for drugs and inhibition of adenylyl cyclase. *Mol. Pharmacol.* 39: 733–39

Neve, K. A., Henningsen, R. A., Bunzow, J. F., Civelli, O. 1989. Functional expression of a rat dopamine D_2 receptor cDNA expressed in a mammalian cell line. *Mol. Pharmacol.* 36: 446–51

Neve, K. A., Neve, R. L., Fidel, S., Janowsky, A., Higgens, G. A. 1991. Increased abundance of alternatively spliced forms of the D_2 dopamine receptor mRNA after denervation. *Proc. Natl. Acad. Sci. USA* 88: 2802–6

O'Dowd, B. F., Hnatowich, M., Caron, M. G., Lefkowitz, R. J., Bouvier, M. 1989. Palmitoylation of the human β_2-adrenergic receptor: Mutation of CYS 341 in the carboxyl tail leads to an uncoupled, non-palmitoylated form of the receptor. *J. Biol. Chem.* 264: 7564–69

Ovchinnikov, Y., Abdulaev, N., Bogachuk, A. 1988. Two adjacent cysteine residues in the C-terminal cytoplasmic fragment of bovine rhodopsin are palmitylated. *FEBS Lett.* 230: 1–5

Piomelli, D., Pilon, C., Giros, B., Sokoloff, P., Martres, M. P., Schwartz, J. C. 1991. Dopamine activation of the arachidonic acid cascade as a basis for D_1/D_2 receptor synergism. *Nature* 353: 164–67

Rao, D. D., McKelvy, J., Kebabian, J., MacKenzie, R. G. 1990. Two forms of the rat D_2 dopamine receptor as revealed by the polymerase chain reaction. *FEBS Lett.* 263: 18–22

Sasaki, K., Sato, M. 1987. A single GTP binding protein regulates K^+ channels coupled with dopamine, histamine, and acetylcholine. *Nature* 325: 259–62

Selbie, L. A., Hayes, G., Shine, J. 1989. The major dopamine D_2 receptor: Molecular analysis of the human D2A subtype. *DNA* 8: 683–89

Sokoloff, P., Giros, B., Martres, M. P., Bouthenet, M. L., Schwartz, J. C. 1990. Molecular cloning and characterization of a novel dopamine receptor (D_3) as target for neuroleptics. *Nature* 347: 146–51

Stoof, J. C., Kebabian, J. W. 1984. Two dopamine receptors: Biochemistry, physiology, and pharmacology. *Life Sci.* 35: 2281–96

Strader, C. D., Candelore, M. R., Hill, W. S., Sigal, I. S., Dixon, R. A. F. 1989. Identification of two serine residues involved in agonist activation of the β-adrenergic receptor. *J. Biol. Chem.* 264: 13572–78

Strader, C. D., Sigal, F. S., Register, R. B., Candelore, M. R., Rands, E., Dixon, R. A. F. 1987. Identification of residues required for ligand binding to the β-adrenergic receptor. *Proc. Natl. Acad. Sci. USA* 84: 4384–88

Sunahara, R. K., Guan, H. C., O'Dowd, B. F., Seeman, P., Laurier, L. G., et al. 1991. Cloning of the gene for a human dopamine D_5 receptor with higher affinity for dopamine than D_1. *Nature* 350: 614–19

Sunahara, R. K., Niznik, H. B., Weiner, D. M., Stormann, T. M., Brann, M. R., et al. 1990. Human dopamine D_1 receptor encoded by an intronless gene on chromosome 5. *Nature* 347: 80–83

Tiberi, M., Jarvie, K. R., Silvia, C., Falardeau, P., Gingrich, J. A., et al. 1991. Cloning, molecular characterization and chromosomal assignment of a gene encoding a novel D_1 dopamine receptor subtype: Differential expression pattern in rat brain compared to the D_{1A} receptor. *Proc. Natl. Acad. Sci. USA* 88: 8111

Vallar, L., Muca, C., Magni, M., Albert, P.,

et al. 1990. Differential coupling of dopaminergic D_2 receptors expressed in different cell types. Stimulation of phosphatidylinositol 4,5-biphosphate hydrolysis in LtK-fibroblasts, hyperpolarization, and cytosolic-free Ca^{2+} concentration decrease in CH4Cl. *J. Biol. Chem.* 265: 10320–26

Van Sande, J., Raspe, E., Perret, J., Lejeune, C., Manbout, C., et al. 1990. Thyrotropin activates both the cyclic AMP and the PIP_2 cascades in CHO cells expressing the human cDNA of TSH receptor. *Mol. Cell. Endocrinol.* 74: R1–R6

Van Tol, H. H. M., Bunzow, J. R., Guan, H. C., Sunahara, R. K., Seeman, P., et al. 1991. Cloning of the gene for a human dopamine D_4 receptor with high affinity for the antipsychotic clozapine. *Nature* 350: 610–14

Walters, J. R., Bergstrom, D. A., Carlson, J. H., Chase, T. N., Brann, A. R. 1987. D_1 dopamine receptor activation required for postsynaptic expression of D_2 agonist effects. *Science* 236: 719–22

Weiner, D. M., Brann, M. R. 1989. The distribution of a dopamine D_2 receptor mRNA in rat brain. *FEBS Lett.* 253: 207–13

Weiner, D. M., Levey, A. I., Sunahara, R. K., Niznik, H. B., O'Dowd, B. F., et al. 1991. D_1 and D_2 dopamine receptor mRNA in rat brain. *Proc. Natl. Acad. Sci. USA* 88: 1859–63

Weinshank, R. L., Adham, N., Macchi, M., Olsen, M. A., Brancheck, T. A., Hartig, P. R. 1991. Molecular cloning and characterization of high affinity dopamine receptor ($D_{1\beta}$) and its pseudogene. *J. Biol. Chem.* 266: 22427–35

Yang-Feng, T. L., Xue, F., Zhong, W., Cotechia, S., Frielle, T., et al. 1990. Chromosomal organization of the adrenergic receptor genes. *Proc. Natl. Acad. Sci. USA* 87: 1516–20

Zhou, Q.-Y., Grandy, D. K., Thambi, L., Kushner, J. A., Van Tol, H. H. M., Cone, R., et al. 1990. Cloning and expression of human and rat D_1 dopamine receptors. *Nature* 347: 76–79

Annu. Rev. Neurosci. 1993. 16:323–45

MOLECULAR BASIS OF NEURAL-SPECIFIC GENE EXPRESSION

Gail Mandel and David McKinnon

Department of Neurobiology and Behavior, State University of New York at Stony Brook, Stony Brook, New York 11794

KEY WORDS: transcription factor, promoter, enhancer, transgenic mouse, neuron

INTRODUCTION

Our understanding of how eukaryotic genes are regulated has improved dramatically over the past decade. This knowledge has been acquired, in part, by exposing the molecular interactions that underlie tissue-specific gene expression, including the very early events involved in cell fate determination. For many tissues, such as the immune system, liver, muscle, and pancreas, DNA elements and their cognate transcription factors that mediate both tissue-specific and developmental regulation of particular genes have been identified. Identification of the DNA elements and transcription factors responsible for expression in the nervous system has been less forthcoming. There are, however, several compelling reasons to study neural-specific gene expression.

First, the mammalian brain is a very heterogeneous and complex organ within which numerous closely related, but distinct, neurons exist in a fantastically elaborated mosaic of cells. This complexity provides an unusual challenge to the specificity of the genetic regulatory apparatus, and some distinct molecular mechanisms may have developed to cope with this complexity. Second, changes in gene expression presumably underlie some aspects of the plasticity of the nervous system, such as learning and memory (Black et al 1987; Goelet et al 1986). These changes are of considerable general interest to neuroscientists, and progress in under-

0147–006X/93/0301–0323$02.00

standing neural gene expression has led to some insights into how this plasticity is achieved (Morgan & Curran 1991). Finally, the development of well-characterized neural-specific promotor/enhancers will provide important tools to manipulate the development and physiology of neurons in vivo through misexpression of molecules by using transgenic animals (Landel et al 1990). These tools can be useful to neuroscientists working on various problems in the nervous system that are not necessarily directly related to gene expression, such as the production of specific neuronal cell lines (Mellon et al 1990) or lineage analysis (Borrelli et al 1989).

This review is largely limited to work done on vertebrates and, more specifically, mammals. Most of the relevant studies have been performed with mammalian genes, because of the ability to analyze foreign gene expression in transgenic mice, a very attractive system for studying vertebrate gene expression. Two other important experimental systems, *Drosophila* and *Caenorhabditis elegans*, offer advantages in terms of speed of analysis and a large battery of genetic mutants. They are considered here only when results are directly relevant to the mammalian studies.

IDENTIFICATION OF DNA ELEMENTS AND TRANSCRIPTION FACTORS INVOLVED IN NEURAL-SPECIFIC EXPRESSION

By analogy with other tissues (Johnson & McKnight 1989; Struhl 1991), neural-specific expression probably occurs primarily at the transcriptional level, through interactions between *cis*-acting DNA elements and transcription factors located in the nervous system. Three approaches for identifying potential transcription factors in the mammalian nervous system have been used. In one approach, novel transcription factors have been isolated by structural similarity to previously identified transcription factors, by using oligonucleotides that specify small conserved amino acid regions as probes in the polymerase chain reaction (PCR). In a second, reciprocal approach, consensus binding sites for previously identified transcription factors, which are not necessarily neural specific, are used to isolate binding proteins in brain. In the third approach, no advantage is taken of previously characterized transcription factors. Instead, novel transcription factors are identified by virtue of their ability to bind DNA elements that mediate neuronal expression of specific genes.

To date, the largest number of mammalian transcription factors implicated in neural-specific expression has been identified by using the first two approaches. However, a disadvantage of isolating brain factors solely by relatedness to other factors is that the identity of the natural target genes often remains unknown. This makes the precise role of the neural transcription factor very difficult to determine. The primary merit of the third approach, which uses functional DNA elements from known neuron-specific genes, is that it is easier to evaluate the contribution of a specific protein in mediating neural specificity. The disadvantage is that the prior identification and characterization of small, defined sequences in the genes are required. The results from experiments that represent the three different approaches are discussed below.

Identification of Novel Transcription Factors by Similarity to Other Known Factors

Transcription factors have been classified on the basis of the primary structures of their DNA-binding domains (for general reviews see Harrison 1991; He & Rosenfeld 1991). One of the largest classes is the "helix-turn-helix" (HTH) proteins. Most of the HTH transcription factors identified in the mammalian nervous system are POU domain proteins, an acronym for the mammalian *Pit1*, *Oct-1/Oct-2*, and *C. elegans Unc 86* proteins (Table 1). These proteins are representatives of this class of HTH proteins, and all share a 150-amino acid domain called the POU domain (Herr et al 1988). This domain is important for site-specific DNA binding and for protein-protein interactions between POU domain proteins and other transcription factors (Rosenfeld 1991). There are two reasons for thinking that the POU domain proteins are important determinants of mammalian neural-specific gene expression: First, POU domain proteins in the invertebrates *Drosophila* and *C. elegans* influence neuronal cell fate (Campos-Ortega & Jan 1991; Chalfie et al 1981; Finney et al 1988). By analogy, a

Table 1 Transcription factors found in the mammalian nervous system

Transcription factor	Tissue-specificity	Brain region
Helix-turn-helix proteins		
Oct-1	ubiquitous	hypothelamus; cerebellum
Oct-2	B-cell lymphocytes; brain	hypothalamus; cerebellum
Pit-1	neural tube; pituitary	N/A
Brn-1	brain	cortex; cerebellum
Brn-2	brain	cortex; cerebellum
Brn-3	brain	sensory ganglia; cerebellum
Tst-1/SCIP	brain, glia, testis	cortex; cerebellum
Unclassified		
Beta	brain	N/A

The data for the HTH proteins is adapted from Table I in He et al (1989). Brain regions indicate location where mRNA coding for an HTH protein is enriched compared with some other regions. The Beta factor was identified in Korner et al (1989).

similar role can be postulated for these proteins in mammals. Second, mRNAs encoding the mammalian POU domain proteins have a restricted distribution in the nervous system, which suggests specialized roles in different populations of neurons (He et al 1989).

The mammalian POU domain proteins *Pit-1*, *Oct-1*, and *Oct-2* share structural similarity to the homeodomain of *Drosophila* genes, a 60-amino acid segment that binds to specific DNA sequences and mediates transcriptional regulation during morphogenesis. Although none of the mammalian transcriptional regulators is exclusively associated with the nervous system, they have all been detected, at some level, in nervous tissue (Table 1). *Oct-1* is ubiquitous, and *Oct-2* is present in both brain and spleen. *Pit-1*, present in restricted regions in the developing nervous system, is only present in the pituitary in the adult (He et al 1989). On the assumption that structurally related POU transcription factors might be important regulators of brain development, novel factors were sought and four new mammalian factors were identified. The brain-specific POU domain proteins, *Brn-1*, *2*, and *3*, were obtained with the PCR reaction by using degenerate oligonucleotides that correspond to conserved amino acids in the POU domain proteins mentioned above (He et al 1989). In addition, *Tst-1*/SCIP cDNA, present in both brain and testis, was isolated. The four new genes encode proteins that are roughly 80% identical to each other in the POU domain.

Transcripts encoding most of the POU domain proteins are present, at similar levels, very early in rat nervous system development. For example, with the exception of *Brn-3*, mRNAs for the other POU domain proteins are readily detected by in situ hybridization histochemistry in the neural plate, as early as embryonic day 10 (He et al 1989). With development, the transcripts become localized differentially in subpopulations of neurons. For example, by embryonic day 16, regions of the hindbrain, brainstem, and spinal cord are enriched for *Brn-1* and *2* transcripts, whereas mRNAs for *Tst-1*/SCIP, *Oct-1*, *Oct-2*, and *Brn-3* are relatively low (He et al 1991). In the adult, *Brn-1* and *Brn-2* exhibit fairly widespread distributions in the cortex, brainstem, and cerebellum. *Brn-3*, *Tst-1*/SCIP, *Oct-1*, and *Oct-2* transcripts are much more restricted in their distribution. For example, in the nervous system, *Tst-1* is mainly present in certain subpopulations of neurons in thalamus and brainstem, in Purkinje cells in the cerebellum, and transiently in myelinating glia. *Oct-1* and *Oct-2* are present at very low levels everywhere, except in the granule cells of the cerebellum and a couple of discrete loci in the hypothalamus. Of particular interest is the finding that *Brn-3* transcripts are found at very high levels in sensory ganglia, which suggests a specific regulatory function for this protein in this set of neurons. Because *Brn-3* shows high structural similarity to the product of the *C. elegans unc 86* gene, which is required for the commitment of sensory neuron differentiation (Chalfie et al 1981; Finney et al 1988), a similar role has been postulated for *Brn-3*, which is found in rat sensory neurons (He et al 1991).

Despite the intriguing anatomical and temporal regulation of the neural POU-domain genes, very little data suggest which neural target genes they act upon. Consequently, the precise roles for these factors in mediating neural-specific expression cannot yet be determined. *Tst-1*/SCIP negatively regulates the rat myelin Po gene, a gene specifying an integral protein of myelin (Monuki et al 1990). In these experiments, a Po fusion gene was constructed that contains a portion of the 5′ flanking region of the Po gene genetically fused to a reporter gene. The chimeric gene was introduced into cultured Schwann cells, where it was expressed at moderately high levels. When a cDNA encoding *Tst-1*/SCIP was cotransfected into Schwann cells along with the Po fusion gene, expression of the reporter gene, driven by Po regulatory sequences, decreased. *Tst-1*/SCIP could repress expression of the Po fusion gene either by directly interacting with a site in the Po gene, or by interfering with the binding of another transcription factor. This mechanistic distinction will have to await further characterization of the factors that regulate the Po gene.

The strongest data demonstrating molecular interactions between a POU homeodomain protein and a neural-specific gene do not come from mammals, but from work done on the *Drosophila* dopa decarboxylase gene (*Ddc*). The *Ddc* gene codes for the terminal enzyme required in the biosynthetic pathways of both dopamine and serotonin. The enzyme is found in approximately 150 dopaminergic and serotonergic neurons in the central nervous system, and at very low levels in glia (Beall & Hirsch 1987). Neuron-specific expression of the *Ddc* gene requires both promoter and distal enhancer elements (Johnson et al 1989). A *Drosophila* POU domain protein (*Cf1-a*), with structural homology to the mammalian *Brn-1/2* proteins and to the *C. elegans unc 86* POU protein, has been isolated. *Cf1-a* binds specifically to an upstream DNA element in the *Ddc* gene (Johnson & Hirsch 1990). Because *unc 86* functions in the specification of dopaminergic and serotonergic neurons in *C. elegans*, *Cf1-a* is suggested to subserve a similar function in *Drosophila* dopaminergic neurons (Johnson & Hirsch 1990).

Although no direct evidence points to a transcriptional regulatory role for *Cf1-a*, the above studies strongly suggest that *Cf1-a* binding activity is required for neural-specific expression of the *Ddc* gene. Recently, Treacy et al (1991) isolated and characterized another member of the POU homeodomain family, I-POU, which lacks a DNA binding domain. I-POU dimerizes with *Cf1-a* and inhibits expression of the *Ddc* gene under some conditions (Treacy et al 1991). The lack of *Ddc* expression in non-dopaminergic neurons could be caused by repression mediated through I-POU (Treacy et al 1991). However, apart from the observation that I-POU is indeed expressed in neurons, the I-POU-*Cf1-a* interaction has no demonstrated biological significance in vivo. The genetic strategies available for analyzing gene function in *Drosophila* will undoubtedly facilitate progress in this area.

The invertebrate systems have provided a framework for conceptualizing the types of transcription factors that may mediate some aspects of mammalian neural-specific expression. One other nonmammalian system warrants mentioning. The neurotrophic virus, Herpes Simplex Virus type I (HSVI), has the interesting biological property that it is replication-deficient in neurons, particularly in sensory neurons (for review, see Roizman & Jenkins 1986). In neurons, the HSVI genome initially assumes a latent state, but occasionally reactivates subsequent to infection to produce the lesions associated with this disease. The regulation of latency has received considerable attention by viral geneticists and neurobiologists alike. Of particular interest is a provocative report that neurons infected with HSV may contain a repressor protein that prevents neural-specific expression of HSV genes involved in viral DNA replication (Kemp et al 1990). An octamer-like sequence, present in an HSVI immediate early gene (IE3) involved in DNA replication, appears to be the target of the repressor activity. Because octamer-binding proteins are found in the nervous system, and at least one of them (*Tst-1*/SCIP) exhibits repressor activity under some conditions, a related factor might be exploited by HSV to regulate viral replication in neuronal and nonneuronal cells differentially. In particular, the POU domain protein *Brn-3*, which is present in relatively high levels in sensory neurons, might be predicted to bind specifically to the octamer-like sequence in the herpes IE3 gene. If this is the case, negative regulation may turn out to play a much greater role in regulating neuronal physiology than currently appreciated.

In addition to the POU domain proteins described above, another class of transcription factors has been identified in the mammalian nervous system (Johnson et al 1990). In this case, mRNA prepared from a sympathoadrenal progenitor cell line was used as a source of template DNA for the PCR. Oligonucleotides specifying amino acids encoded by the *Drosophila achaete-scute* gene complex were used as amplifying primers. This complex of genes is involved in neuronal determination in *Drosophila* (Ghysen & Dambly-Chaudiere 1988). The cloned cDNA sequences, termed MASH-1 and MASH-2, predicted protein domains of approximately 80% amino acid identity to the corresponding regions in *Drosophila achaete-scute* proteins (Johnson et al 1990). MASH-1 mRNA was transiently detected during mammalian embryogenesis in neuronal progenitor cells (embryonic day 10–20), but was not found in nonneural cells. MASH-2 mRNA was present in much lower amounts in neuronal progenitor cells. Although the MASH proteins are likely to be transcriptional regulators of mammalian neural genes, the targets have not yet been identified (Anderson 1993). Therefore, precise biological roles for these novel proteins remain unknown.

The studies described above all take advantage of structural similarities among classes of known transcription factors to identify similar factors in the nervous system. A different approach, which exploits the binding sites for known classes of transcription factors to identify novel transcription factors in brain, has also been pursued. For example, a unique DNA-binding activity that is present in brain, called Beta, has recently been described (Korner et al 1989). This protein(s) binds to a DNA sequence that is the binding site for the B-lymphocyte transcription factor NFkB, and was identified in protein extracts prepared from mammalian hippocampus by gel retardation assays. A palindrome with likeness to the NFkB binding site is present in the enkephalin gene (Korner et al 1989), and a sequence with some similarity to the Beta binding site has also been noted in the upstream regulatory region of the mouse Purkinje cell protein-2 gene (*Pcp-2*) (Vandaele et al 1991).

The studies described above lead to intriguing suggestions, but leave several questions unanswered. For example, evidence indicates that at least two neural proteins can regulate gene activity in cell culture assays (He et al 1991; Korner et al 1989), but what are the in vivo roles of these proteins, and what are their natural target genes? An answer to the former question may have to await the isolation of mouse or invertebrate mutants in which the transcription factors have been inactivated. One obvious approach in mammals takes advantage of recent technical advances in the use of homologous recombination to produce targeted mutations in the mouse genome (Mansour et al 1988), a procedure known as "gene knock-out." Using this approach, the gene encoding the putative transcription factor could be inactivated in embryonic stem cells in culture by replacing the endogenous gene with a mutated counterpart. The stem cells are then used to generate mice in which the transcription factor is now functionally deleted.

The identification of mammalian target genes for specific transcription factors may depend on prior identification of the corresponding invertebrate genes, where the smaller genome size and techniques for direct chromosome localization of genes provide great advantages. One approach has been described that could be used to find genomic target sites for mammalian transcription factors by PCR amplification of DNA sequences that are preferentially bound by an identified factor (Mavrothalassitis et al 1990). Following isolation of the binding site, however, the genes to which the regulatory sequences are attached must still be cloned.

Identification of DNA Sequences Involved in Neural-Specific Expression of Mammalian Genes

Tissue-specific expression of eukaryotic genes is conceptualized as the result of a complex set of DNA-protein and protein-protein interactions (reviewed in He & Rosenfeld 1991; Johnson & McKnight 1989; Struhl 1991). Both positively and negatively acting DNA elements appear to be involved, with the balance determining the absolute level of transcription in a given cell type. In general, much more is known about the molecular basis of transcriptional activation than repression. The mechanisms underlying negative regulation are still enigmatic (but see Levine & Manley 1989). Recently, the functional DNA elements that regulate a few mammalian neural-specific genes have been identified. Using in vitro and in vivo assays of promoter activity, examples of both positive and negative transcriptional regulation of neural-specific genes exist. The experiments are described in the following sections.

IN VITRO STUDIES OF NEURAL-SPECIFIC GENE EXPRESSION

GAP-43 and synapsin I are examples of genes whose neural-specific expression has been studied by the introduction of fusion genes into tissue culture cells. In these studies, and in other studies discussed below, the putative regulatory region of the test gene is separated from the native gene and genetically fused to a reporter gene. The bacterial genes encoding either chloramphenicol acetyltransferase (CAT) (Gorman et al 1982) or β-galactosidase (β-gal) are commonly used as the reporter genes. After introduction into tissue culture cells by standard techniques, transcriptional activity of the fusion gene is measured by determining the amount of reporter activity in the transfected cells.

The GAP-43 gene encodes a growth cone protein involved in neurite outgrowth. In vivo, this gene is expressed exclusively in neural cells, and its expression is most robust in immature neurons (Skene 1989). Nedivi et al (1992) have isolated and characterized a 386 basepair (bp) upstream fragment of the rat GAP-43 gene that is sufficient to drive reporter activity in primary cultures of rat cortical neurons. The same construct cannot direct reporter activity in a variety of nonneural cell lines (Nedivi et al 1992). Two distinct regions, which suppress transcription when assayed individually on the GAP-43 promoter, have been described in the GAP-

43 gene. However, when they are both present, as they are in the native gene, the suppression is dramatically reduced. The mechanism underlying the effects of these sequences is currently unknown. The precise nucleotides responsible for GAP-43 expression, or the transcription factors that bind to them, have not yet been identified within this fragment.

The synapsin I gene encodes a phosphoprotein localized to the synaptic terminals of neurons and may be involved in the modulation of neuro-transmitter release (reviewed in De Camilli et al 1991). Thus, like the GAP-43 gene, synapsin I should be expressed in all neurons. The 5' flanking region of the rat and human genes have been cloned and analyzed for the elements that confer neural-specificity (Sauerwald et al 1990; Thiel et al 1991). A 335 bp regulatory fragment of the rat gene, which contains 5' flanking and untranslated sequences, is sufficient to confer reporter expression in neuroblastoma, but not fibroblast cells. This region contains a C/G-rich stretch of nucleotides, with some likeness to the neurofilament and nerve growth factor receptor genes (Sauerwald et al 1990). However, studies with the human synapsin gene have raised questions about the biological significance of this consensus sequence (Thiel et al 1991). Distal sequences in the 5' flanking region of both the rat and human synapsin genes enhance promoter activity in neural cells. Identification of the puta-tive transcriptional activators will have to await further dissection of the functional elements.

The first example showing a role for negative regulation in mediating neural specificity comes from studies on the mammalian voltage-dependent sodium channel. This ion channel, responsible for the generation of the action potential in several different excitable tissues, is encoded by a large multigene family. In mammals, three different sodium channel genes, termed types I, II, and III (Kayano et al 1988; Noda et al 1986), are expressed almost exclusively in the vertebrate nervous system. The expression patterns for the three genes are anatomically and develop-mentally distinct, with type II having the widest distribution in brain (reviewed in Mandel 1992). The type II sodium channel gene has been studied in some detail with respect to the molecular basis of its cell speci-ficity. The studies have been done by introducing chimeric genes containing the type II 5' flanking region, fused to a reporter gene, into cells in tissue culture. The cloned 5' flanking region (containing 1051 bp of type II 5' flanking and untranslated region) is active in neural and neural-endocrine cell lines that contain functional voltage-dependent sodium channels and express the type II sodium channel gene (Maue et al 1990). The type II fusion gene is also active in primary cultures of rat superior cervical ganglion (D. Leib and G. Mandel, unpublished). By contrast, the same type II-reporter fusion gene is at least 50-fold less active in cultured nonneuronal cells, which express either no sodium channels, or another member of the sodium channel gene family.

Deletional analysis of the type II 5′ flanking region has revealed that its neural specificity can be attributed largely to the presence of several negatively acting elements located distal to the promoter (Kraner et al 1992; Maue et al 1990). Removal of these sequences from the type II minigene derepresses type II promoter activity approximately 100-fold in cell lines that do not express the endogenous type II sodium channel gene and that do not ordinarily express the intact type II-reporter fusion gene (Kraner et al 1992; Maue et al 1990). One of these negatively acting elements, responsible for most of the cell-specific repressor activity, has been characterized in some detail. A 28 bp element has been identified that, in the absence of other sequences, can completely repress transcription from the type II promoter, as well as other promoters, in nonneuronal cells or in neuronal cell types that do not express the type II gene (Kraner et al 1992). The negatively acting element does not share sequence identity with any other known negatively acting elements or silencers and likely represents a unique element.

Negative regulation has also been described for the SCG10 gene, a member of a multigene family that is expressed in the mammalian central and peripheral nervous systems and encodes a 22 kDa protein found in growth cones (Stein et al 1988). SCG10 is expressed in sympathoadrenal progenitor cells and sensory neurons in the embryo, and in brain, sympathetic ganglia, and adrenal medulla in the adult. Like the type II sodium channel, SCG10 mRNA is also present in PC12 cells. Deletional analysis of the cloned 5′ flanking region of the SCG10 gene indicates that a negatively acting element, contained within a 1.6 kilobase fragment, suppresses promoter activity in cells that do not normally express this gene (Mori et al 1990). The element has the properties of a silencer (Brand et al 1985) and is functional on heterologous promoters. The proximal promoter region of the SCG10 gene, when fused to a reporter gene and introduced into different cell lines, is nearly as active in Hela cells as in PC12 cells (Mori et al 1990). With the addition of a DNA fragment containing the silencer activity, however, reporter gene expression is restricted to neuronal cells. Like the type II sodium channel, the effect of the SCG10 silencer is striking: Its removal results in at least a 50-fold increase in promoter activity in the nonneuronal cell type.

The silencer elements described in the neural-specific genes are functionally similar to those described in other genes, such as the insulin and immunoglobulin κ genes (Cordle et al 1991; Pierce et al 1991). However, the DNA sequences of the neural gene silencer elements are distinct from the elements in these other mammalian genes. The transcription factors that suppress transcription of the type II sodium channel and SCG10 genes have not yet been identified. However, gel shift mobility assays and DNAase I footprinting experiments of the type II gene have indicated the presence of specific binding activities in nonneuronal cell types and in subpopulations of neuronal cells that do not express the endogenous type II gene (Kraner et al 1992). This last observation is important, as it suggests that negative regulation is also utilized by neuronal cells to restrict expression of certain genes to subpopulations of neurons in the vertebrate nervous system. As more repressor proteins become isolated and characterized, it will become possible to determine whether silencing is mediated by different molecular mechanisms in neuronal and nonneuronal genes.

The above work has largely focused on the role of cell-specific enhancers in modulating promoter activity of neural-specific genes. There are also hints, from in vitro transcription assays, that specific factors are required for basal promoter activity of some neural genes. In these assays, nuclear extracts prepared from different tissues are used as the source of RNA polymerase and companion transcription factors required for RNA synthesis. The template for the in vitro reaction is a cloned promoter fragment of the test gene. Messenger RNA synthesized in the in vitro reaction is analyzed both quantitatively and by the position of the transcriptional start site, and compared with the results from in vivo transcriptional assays. For example, in vivo, the murine tissue plasminogen activator gene, (*t-pa*), is expressed at high levels in brain, but at relatively low levels in kidney and liver (Rickles & Strickland 1988). Recent studies indicate that the tissue-specific differences are transcriptional and that a cloned *t-pa* promoter fragment is accurately transcribed in an in vitro transcription system that uses brain, but not kidney or liver nuclear extracts. A small GC-rich sequence, similar to the consensus binding site for the ubiquitous *Sp-1* transcription factor, is required for correctly initiated transcription of the *t-pa* promoter (Pecorino et al 1991). An *Sp-1*-like factor, termed BGC, has been identified in the brain extracts. BGC is immunologically distinct from the ubiquitous *Sp-1* factor, but binds the *t-pa* GC sequence in DNase I footprinting and mobility gel shift assays (Pecorino et al 1991). This study suggests that BGC may be involved in expression of a variety of neural-specific genes that contain similar GC-rich sequences.

Promoter activity of nervous system genes may be mediated by a specialized form of another transcription factor, TFIID. TFIID is a classical TATA box-binding factor that forms transcriptional complexes with RNA polymerase II. Tamura et al (1990) have shown that in vitro transcription from the core promoter of the mouse myelin basic protein gene is mediated by a TFIID-like activity enriched in brain, but not liver, extracts. They note that the mouse neurofilament gene promoter, like the myelin basic protein promoter, is also more efficiently transcribed in vitro by the brain-specific TFIID activity, whereas α-1-antitrypsin, a nonneuronal gene, is transcribed preferentially by TFIID purified from liver cells (Tamura et al 1990). A slight twist on this theme is found from studies of the rat brain creatine kinase gene. A TFIID-like activity in brain, called TARP, forms RNA polymerase complexes on a consensus TATA box located at nucleotides -60 upstream from the main in vivo transcriptional start site (Mitchell & Benfield 1990). The binding of TARP is postulated to block initiation at this "biologically irrelevant" site in brain. As these perhaps brain-specific transcription factors become better characterized biochemically, it will be interesting to assess their potential for interacting with the enhancer-binding proteins identified in other studies.

Transient expression assays have provided a convenient means of identifying some of the *cis*-acting DNA elements and transcription factors involved in neural-specific expression. However, only tentative conclusions about endogenous gene regulation can be drawn with this approach. For example, in transient transfection assays, only a limited portion of the gene is usually assayed, and the exogenous DNA is introduced into the cells in abnormally high amounts. Although may minigenes are appropriately regulated, there are also examples in which minigene expression does not mimic expression of the endogenous gene. Therefore, it is desirable, whenever possible, to test hypotheses regarding tissue-specific expression in vivo, usually by the production of transgenic mice that express cloned gene fragments. Recently, transgenic mice have been produced and analyzed for several neural-specific genes (Table 2). These studies are described in the following section.

IN VIVO STUDIES OF NEURAL-SPECIFIC EXPRESSION

Transgenic mice containing new genetic material are produced by injection of the transgene DNA into the pronuclei of fertilized mouse eggs (Hogan et al 1986). The injected eggs are then implanted into a pseudopregnant female, and a fraction of the pups will contain the foreign DNA stably integrated into the chromosome. The typical structure of the newly incorporated transgene is a head-to-tail array of the injected DNA that contains multiple copies of the transgene at a single point of integration in the chromosome (Brinster et al 1985; Lacy et al 1983). The primary advantage of this technique is that expression of the transgene can be tested in the complete range of fully differentiated cell types found in the mouse. The primary disadvantage is that the structure and chromosomal localization of the transgene is not well controlled. Consequently, considerable quan-

Table 2   Regulatory sequences tested in transgenic mice

Gene	Species	Reference
Neuron-specific genes		
Neurofilament (*NF-L*)	human	Julien et al (1987)
SCG10	rat	Wuenschell et al (1990)
Oxytocin	rat	Young et al (1990)
Neuron-specific enolase	rat	Forss-Petter et al (1990)
L7/*Pcp-2*	mouse	Oberdick et al (1990)
		Vandaele et al (1991)
Dopamine β-hydroxylase	human	Mercer et al (1991)
Tyrosine hydroxylase	human	Nagatsu et al (1991)
Amyloid precursor protein	human	Wirak et al (1991)
Neuron-expressed genes		
Gonadotropin-releasing hormone	mouse/rat	Mason et al (1986)
		Mellon et al (1990)
Thy-1	mouse/human	Gordon et al (1987)
Nerve growth factor receptor	human	Patil et al (1990)
Platelet-derived growth factor B-chain	human	Sasahara et al (1991)

titative and qualitative variability can be observed between different lines of animals injected with the same DNA construct. Transgenic mice have become a standard technique to identify the DNA elements responsible for tissue-specific gene expression (Palmiter & Brinster 1986).

Studies Using Intact Neural Genes

The first neuron-specific gene expressed in transgenic mice was the human neurofilament (*NF-L*) gene (Julien et al 1987). A 21.5 kb fragment of the human *NF-L* gene, which contained 14 kb of 5′ flanking sequence and all of the exons encoding the human protein, was introduced into the germ line of transgenic mice. Mice that had incorporated this genomic fragment expressed the human *NF-L* protein in neurons and their processes, but the protein could not be detected in other cell types. The human *NF-L* protein was expressed at a level of 10–30% of the endogenous gene, and the only qualitative difference compared with the mouse gene was that the human gene was expressed later during development. The observation that large (several kb or more) fragments of genomic DNA contain all the *cis*-regulatory sequences necessary for appropriate tissue-specific expression in vivo is a common one and has been observed with other cell-specific promoters (Readhead et al 1987; Swift et al 1984). Other neural genes that have been tested in this way include the human tyrosine hydroxylase gene (Nagatsu et al 1991) and the mouse gonadotropin-releasing hormone gene

(Mason et al 1986). Although these experiments can provide important information about genetic diseases (Mason et al 1986; Readhead et al 1987), they are only minimally informative about the *cis*-acting elements required for neuron-specific expression in vivo, because the cloned DNA fragments are so large.

Studies Using Specific Elements of Neural Genes Attached to a Reporter Gene

Some transgenic experiments using 5′ flanking sequences of neural-specific genes attached to reporter genes have been performed. Regulatory sequences of the SCG10 gene, described above, have been tested in transgenic mice (Wuenschell et al 1990). A construct containing 3.5 kb of 5′ flanking sequence could direct expression of a CAT reporter gene in a tissue specific manner that paralleled expression of the endogenous gene. The pattern of developmental activation of the transgene was similar to the endogenous SCG10 gene. Results from previous in vitro studies had suggested that this 3.5 kb of DNA contained a putative repressor element (Mori et al 1990). To test what effect elimination of the repressor sequences would have, a 0.55 kb construct without these sequences was also tested in vivo. Mice containing this construct still expressed CAT activity in the brain, but now a significant level of CAT expression was found in other tissues, including liver, spleen, and lung, which suggested a loss of tissue specificity in vivo. Expression in nonneuronal tissues was not as high as in the brain, which suggested that brain-specific enhancer elements were also contained within the 0.55 kb construct and could contribute to tissue specificity in vivo. The extent of expression of the 0.55 kb construct in nonneuronal tissues was quite variable between different founder lines, possibly because of differences in the chromosome location of the transgene, as has been suggested for other transgenes (Lacy et al 1983). Wuenschell et al (1990) suggest that the silencer acts by causing the chromatin of the transgene to take up an inactive chromatin structure in nonneuronal cells, thus making it unavailable for transcription. Support for this hypothesis was obtained by using DNAase I sensitivity (HSS) assays (Vandenbergh et al 1989). Two tissue specific, DNAase I hypersensitive sites are found in the endogenous SCG10 gene, close to the transcription start sites. Analysis of these HSS sites in the 0.55 kb transgene suggested that both sites were unprotected in DNA isolated from the brain and liver of transgenic mice, which is consistent with the unregulated expression of this transgene. As expected, the 3.5 kb tissue-specific transgene was unprotected at the two HSS sites in DNA isolated from brain, but was protected in DNA isolated from liver (Vandenbergh et al 1989).

Expression of the products of many neural-specific genes, such as enzymes for the synthesis of neurotransmitters, are restricted to clearly defined subsets of neurons. To test the accuracy with which the regulatory elements of a transgene can distinguish between neuronal subtypes a reporter gene that can give histological resolution of its expression must be used. In a comprehensive study on the human dopamine β-hydroxylase (DBH) gene, a 5.8 kb of 5' flanking sequence of the gene was used to direct expression of the *E. coli lacZ* gene in transgenic mice (Mercer et al 1991). The product of the *lacZ* gene, the β-galactosidase enzyme, was detected in mouse brains by using enzyme histochemistry. The endogenous DBH gene has a complex pattern of expression in subsets of both peripheral and central neurons. In adult transgenic mice, expression of the transgene was almost always restricted to neurons or adrenal chromaffin cells. Most sites normally associated with DBH expression, including post-ganglionic sympathetic neurons and neurons in the locus ceruleus expressed the transgene. In addition, however, expression of the transgene was observed in many ectopic sites that do not normally express the DBH gene in the adult animal. In some cases, the transgene was expressed in tyrosine hydroxylase (TH) positive, DBH negative cells. Because the TH enzyme is immediately upstream from the DBH enzyme in the catecholamine synthesis pathway, Mercer et al (1991) suggested that TH^+, DBH^- neurons, and noradrenergic neurons express a closely related set of transcriptional activators and/or suppressors. In this case, failure of the transgene to differentiate between the two cell types may be caused by a relatively minor failure on the part of the transgene to mimic the endogenous DBH gene. However, only a subset of the TH-immunoreactive cells expressed the transgene, which suggests that this explanation is incomplete (Mercer et al 1991).

In addition to expression of the transgene in DBH^+ and TH^+ neurons, expression also occurred in a range of other neurons that, at least in some cases, were not closely related to noradrenergic neurons (Mercer et al 1991). Some of this ectopic expression might have been due to a failure to turn off expression of the transgene in the adult animal, because the transgene was very widely expressed during development (Kapur et al 1991). A broad range of neuronal precursors transiently express the DBH or TH genes during development, and part of the loss of specificity in the adult animal may result from failure to turn off the transgene in cells that normally express the DBH or TH genes during early development. In addition, the transgene was expressed in regions of the brain for which no explanation was readily apparent. Interpretation of these results was complicated by the use of human regulatory sequences, which may vary subtly in function from the murine elements.

Overall, the pattern of β-galactosidase staining suggested that the human

DBH transgene could direct expression to most brain regions in which the endogenous DBH gene was expressed. However, fine control was lost, particularly of repression of the transgene in nonappropriate neuronal cell types. Similar results have been reported for other brain-specific genes, such as the human tyrosine hydroxylase gene (Nagatsu 1991). Unfortunately, no equivalent murine gene has been tested. These experiments emphasize the importance of negative regulation in the fine control of neuron-specific expression. The large number of closely related cell types in the brain may provide a very stringent test of the extent to which a transgene construct can mimic the negative regulation of the endogenous gene. Chromatin structure probably plays an important part in negative regulation of transcription (Grunstein 1990; Weintraub 1985); thus, the unusual structure and location of the transgene may act to limit the resolution with which transgenes can reproduce the pattern of expression of endogenous genes.

Regulatory squences from the L7/*Pcp-2* gene have been tested in transgenic mice in two different laboratories by using the *lac Z* gene as a reporter (Oberdick et al 1990; Vandaele et al 1991). Expression of the endogenous L7/*Pcp-2* gene is restricted to cerebellar Purkinje cells and retinal bipolar neurons. In one study (Oberdick et al 1990), the transgene included 8 kb of genomic DNA that encompassed 4 kb of 5' flanking region, as well as all the coding exons and 2 kb of 3' flanking sequence. The *lac Z* gene was inserted into the final exon. In this experiment, *lac Z* expression was observed in Purkinje and bipolar cells, as expected, with a low level of ectopic expression in neurons of the interpeduncular nucleus. The level of expression of the *lac Z* gene in Purkinje cells was very variable. Mosaic expression of the native L7/*Pcp-2* gene was not observed. In the second study (Vandaele et al 1991), two constructs were tested, using either 3.5 or 0.4 kb of 5' flanking sequence, plus the first intron and two exons of the L7/*Pcp-2* gene linked to the *lac Z* gene. Expression of the large construct was restricted to Purkinje cells and neurons in the superior colliculus. For the smaller construct, in two of three lines, expression occurred at numerous ectopic sites in the brain, but was always restricted to neurons. For both constructs, expression in Purkinje cells was usually mosaic, and no expression appeared in bipolar cells. These results suggest that *cis*-acting elements contained within the smaller construct can direct expression of the transgene specifically to neurons; upstream elements, found in the larger construct, restrict expression of the transgene to subsets of neurons. One interpretation of these results is that regions distal to the transcription start site act to restrict a proximal, neuron-specific enhancer by some form of negative regulation. Comparison of the two studies suggests that the elements necessary to direct expression of the transgene to retinal bipolar cells might be found in regions 3′ to the transcription start site.

Many of the neural genes that have been studied, such as GAP-43 and neuron-specific enolase (Forss-Petter et al 1990), encode proteins that are ubiquitous in the nervous system. Other genes, such as the L7/*Pcp-2* gene, exhibit a much more restricted distribution. Knowledge of the controlling elements of some of the latter genes may be useful for targeting foreign proteins, either mutant or wild type, to specific anatomical regions within the mammalian nervous system. Currently, the combinations of DNA elements required for targeted expression are poorly understood. No clear structural identities have been noted among the cloned regulatory sequences of different neural-specific genes. A short consensus sequence present in several genes expressed in rat PC12 cells has been noted (Table 3) and expanded to include other representatives with some similarity to this sequence (Vandaele et al 1991). However, this consensus sequence does not appear to be sufficient to confer neural-specific expression of either the GAP-43 (Nedivi et al 1992) or the sodium channel genes (Maue et al 1990).

Ectopic Expression of Transgenes in Neurons

Neurons seem to be unusually susceptible to ectopic expression of trans-genes (Russo et al 1988; Swanson et al 1985), and even promoters not normally expressed in the brain can exhibit expression in a restricted subset of neurons (Kondoh et al 1987). One partial explanation for this phenomenon is simply that the number of phenotypically distinct cell types in the brain is very large compared with most other organ systems. The

Table 3 Sequence similarities among neural-specific genes expressed in PC12 cells

−60	−34	
A C T T G T G A C C A G G A G A T G G A G C T G T C G		rat type II sodium channel
A C C A G G A G A G G G A		rat peripherin
	G G A G C T G T C G	rat peripherin
C C A G G	T G G A G C C G C A G	mouse neurofilament
C C A G G A G A T		rat GAP-43

Adapted from Maue et al (1990). The numbers above the type II sequence refer to the positions upstream from the transcriptional start site. The bold, underlined nucleotides represent differences between the type II gene and all of the others. The peripherin sequences are from Thompson & Ziff (1989), between nucleotides −163 and −154 and −72 to −62, respectively. The neurofilament sequence lies between nucleotides −73 and −62 (Lewis & Cowan 1986). The GAP-43 sequence lies between nucleotides −279 and −286 (Nevidi et al 1992).

accuracy of transgene expression is due to multiple *cis*-acting elements that control transcription, and the combination of these elements can be thought of as a unique cellular address. Any ambiguity in this address will presumably result in ectopic expression. Expression at ectopic sites in the brain may occur more frequently than in other organs, simply because of the much greater probability of inadvertently hitting a neuronal address.

Ectopic expression of transgenes in neurons is interesting, because it suggests both relationships between neurons that are generally thought to be unrelated (Swanson et al 1985), as well as previously unrevealed subsets of neurons within what were thought to be relatively homogeneous sets of neurons (Kondoh et al 1987; Mercer et al 1991). There has been a systematic study of the contribution of different promoters and reporter genes to ectopic expression in the CNS in which several different promoters were linked to different reporter genes, and expression in brain was analyzed by histological techniques (Russo et al 1988). The results were interpreted in terms of a model in which multiple *cis*-acting elements were assumed to occur in both the promoter and reporter regions of the transgene that could bind to transcriptional activators. The conclusion was that both the promoter sequences and the reporter gene can provide enhancer-like elements that act in combination to produce ectopic expression. There are alternative possibilities for at least some of the observations, such as the reporter gene interfering with negative regulation by disrupting the pattern of nucleosome binding, or the disruption of higher order chromatin structures around the promoter by the abnormal juxtapositioning of the promoter with a foreign piece of DNA (Palmiter et al 1991).

One complication with the use of transgenic mice was clearly evident in the analysis of the human DBH transgene expression in the superior cervical ganglia (Mercer et al 1991). In these ganglia, greater than 95% of the neurons are DBH positive, but the percentage of cells that expressed the DBH-*lacZ* transgene was quite variable between animals, ranging from 10 to 95% staining. Individual neurons appeared to express the gene in an all-or-none fashion. Mosaicism was also observed in different brain nuclei, but was not apparent in embryonic animals (Kapur et al 1991). Cellular mosaicism has been observed with several other hemizygous transgenes (McGowan et al 1989). Variability in expression of the transgene by individual cells appears to be correlated with changes in the methylation pattern of the transgene (McGowan et al 1989). Although not an insurmountable problem for studies of regulatory elements in vivo, mosaicism could significantly affect experiments in which a transgene was used to direct expression of a heterologous gene to alter neuronal cellular physiology.

CONCLUSIONS AND FUTURE DIRECTIONS

Given that more genes are specifically expressed in the nervous system than in any other organ system, the coordination of gene expression in the nervous system will presumably require the activities of a correspondingly large number of *cis*-active elements and transcription factors (He & Rosenfeld 1991). Thus, despite the potential for combinatorial interactions between factors (Struhl 1991), which will obviously act to limit the absolute number of factors required, a very large number of *cis*-acting elements and cognate transcription factors remain to be described. The elucidation of the molecular basis for neural-specific gene expression clearly cannot move forward without a more detailed characterization of the factors that bind to and regulate specific genes both in vitro and in vivo.

One question that can be asked is whether any unique molecular mechanisms control neural-specific gene expression. Based on the limited number of genes that have been studied to date, the answer appears to be no. Families of transcription factors that regulate genes in other tissues have now been found in nervous tissue. And, like other tissue-specific genes, neural-specific genes exploit the activities of both positively and negatively acting elements for their regulation. However, evidence is emerging for some quantitative differences between the nervous system and other tissues. In particular, the majority of neural-specific genes that have been studied to date appear to use negative regulation as a major mechanism to provide the fine tuning required for specific expression in subsets of neurons. This is in contrast to the bulk of studies on transcription in other tissues, in which regulation through suppressor-like elements appears to be important for only a minority of genes. Assuming that these preliminary observations hold true, then there are at least two explanations why negative regulation might have become particularly important in the regulation of neural genes. First, to a much greater extent than other tissues, the phenotypic complexity of cells in the nervous system creates a novel situation, with thousands of distinct cell-types within a "single" tissue. This complexity compounds the problem of executing neural-specific expression. In addition to mechanisms specifying "brain-specific" gene expression, as opposed to "liver" or "muscle" expression, mechanisms must also be in place to restrict expression to a particular subset of neurons. Although this situation is obviously not unique in principle (the immune system for example, contains a relatively large number of phenotypically distinct cell types), it is clearly quantitatively more complex. Because a single factor can potentially act as either an activator or a repressor of transcription (Levine & Manley 1989), one way to increase the number of distinct programs of gene expression, while limiting the absolute number of transcription factors that are required for those programs, is to exploit multiple modes of transcription factor activity. Negative regulation may have become increasingly important in the nervous system in response to pressure to make more efficient use of preexisting transcription factors.

A second reason why negative regulation may be particularly important in the nervous system is that inappropriate expression of some neural-specific proteins might be more deleterious to the organism than equivalent "mistakes" in other tissue systems. The central role of the nervous system in the adaptation of the organism to its environment, combined with the large number of neural-specific genes, could make the nervous system unusually susceptible to functional disarray if there was too much leakage of gene expression. Negative regulation might act to add an additional level of repression to some genes, on top of the general transcriptional repression produced by normal chromatin structure. Molecular mechanisms that act through *cis*-acting elements to decrease the probability of inappropriate gene expression, by some as yet poorly defined silencing action, could limit ectopic gene expression by increasing the thermodynamic barrier to random or inappropriate gene expression. Silencing elements might have proliferated in the regulatory regions of neural genes, as an adaptation to counteract the effects of an unusually diverse array of transcriptional activators expressed in the nervous system. This may also have provided more flexibility to modify patterns of gene expression in rapidly evolving regions, such as the forebrain, by limiting the pleiotropic effects caused by changes in transcription factor expression in what must be a very intricate and interwoven network of transcriptional regulatory mechanisms.

ACKNOWLEDGMENTS

This work was supported by National Institutes of Health grant NS22518 to Gail Mandel and NS29755 to David McKinnon. The authors are indebted to Drs. Jane Dixon, Richard H. Goodman, and Simon Halegoua for insightful comments and helpful discussions.

Literature Cited

Anderson, D. J. 1993. Molecular control of cell fate in the neural crest: The sympathoadrenal lineage. *Annu. Rev. Neurosci.* 16: 129–58

Beall, C. J., Hirsh, J. 1987. Regulation of the *Drosophila* dopa decarboxylase gene in neuronal and glial cells. *Genes Dev.* 1: 510–20

Black, I. B., Adler, J. E., Dreyfus, C. F.,

Friedman, W. F., LaGamma, E. F., Roach, A. H. 1987. Biochemistry of information storage in the nervous system. *Science* 236: 1263–68

Borrelli, E., Heyman, R. A., Arias, C., Sawchenko, P. E., Evans, R. M. 1989. Transgenic mice with inducible dwarfism. *Nature* 339: 538–41

Brand, A. H., Breeden, L., Abraham, J.,

Sternglanz, R., Nasmyth, K. 1985. Characterization of a "silencer" in yeast: a DNA sequence with properties opposite to those of a transcriptional enhancer. *Cell* 41: 41–48

Brinster, R. L., Chen, H. Y., Trumbauer, M. E., Yagle, M. K., Palmiter, R. D. 1985. Factors affecting the efficiency of introducing foreign DNA into mice by microinjecting eggs. *Proc. Natl. Acad. Sci. USA* 82: 4438–42

Campos-Ortega, J. A., Jan, Y. N. 1991. Genetic and molecular bases of neurogenesis in *Drosophila* melanogaster. *Annu. Rev. Neurosci.* 14: 399–420

Chalfie, M., Horvitz, H. R., Sulston, J. E. 1981. Mutations that lead to reiterations in the cell lineages of *C. elegans. Cell* 24: 59–69

Cordle, S. R., Whelan, J., Henderson, E., Masuoka, H., Weil, P. A., Stein, R. 1991. Insulin expression in non-expressing cells appears to be regulated by multiple distinct negative-acting control elements. *Mol. Cell. Biol.* 11: 2881–86

De Camilli, P., Benfenati, P., Valtorta, F., Greengard, P. 1991. The synapsins. *Annu. Rev. Cell Biol.* 6: 433–60

Finney, M., Ruvkun, G., Horvitz, H. R. 1988. The *C. elegans* cell lineage and differentiation gene *unc-86* encodes a protein containing a homeodomain and extended sequence similarity to mammalian transcription factors. *Cell* 55: 757–69

Forss-Petter, S., Danielson, P. E., Catsicas, S., Battenberg, E., Price, J., et al. 1990. Transgenic mice expressing β-galactosidase in mature neurons under neuron-specific enolase promoter control. *Neuron* 5: 187–97

Ghysen, A., Dambly-Chaudiere, C. 1988. From DNA to form: the *achaete-scute* complex. *Genes Dev.* 2: 495–501

Goelet, P., Castellucci, V. F., Schacher, S., Kandel, E. R. 1986. The long and the short of long-term memory—a molecular framework. *Nature* 322: 419–22

Gordon, J. W., Chesa, P. G., Nishimura, H., Rettig, W. J., Maccari, J. E., et al. 1987. Regulation of Thy-1 gene expression in transgenic mice. *Cell* 50: 445–52

Gorman, C. M., Moffat, L. F., Howard, B. H. 1982. Recombinant genomes which express chloramphenicol acetyltransferase in mammalian cells. *Mol. Cell. Biol.* 2: 1044–51

Grunstein, M. 1990. Histone function in transcription. *Annu. Rev. Cell Biol.* 6: 643–78

Harrison, S. C. 1991. A structural taxonomy of DNA-binding domains. *Nature* 353: 715–19

He, X., Gerrero, R., Simmons, D. M., Park, R. E., Lin, C. R., et al. 1991. *Tst*-1, a member of the POU domain gene family, binds the promoter of the gene encoding the cell surface adhesion molecule Po. *Mol. Cell. Biol.* 11: 1739–44

He, X., Rosenfeld, M. G. 1991. Mechanisms of complex transcriptional regulation: implications for brain development. *Neuron* 11: 183–96

He, X., Treacy, M. N., Simmons, D. M., Ingraham, H. A., Swanson, L. W., Rosenfeld, M. G. 1989. Expression of a large family of POU-domain regulatory genes in mammalian brain development. *Nature* 340: 35–42

Herr, W., Sturm, R. A., Clerc, R. G., Corcoran, L. M., Baltimore, D., et al. 1988. The POU domain: A large conserved region in the mammalian *pit-1, oct-1, oct-2*, and *Caenorhabditis elegans unc-86* gene products. *Genes Dev.* 2: 1513–16

Hogan, B., Costantini, F., Lacy, E. 1986. *Manipulating the Mouse Embryo*. Cold Spring Harbor, NY: Cold Spring Harbor Lab. 332 pp.

Johnson, J. E., Birren, S. J., Anderson, D. J. 1990. Two rat homologues of *Drosophila achaete-scute* specifically expressed in neuronal precursors. *Nature* 346: 858–61

Johnson, P. F., McKnight, S. L. 1989. Eukaryotic transcriptional regulatory proteins. *Annu. Rev. Biochem.* 58: 799–839

Johnson, W. A., Hirsh, J. 1990. Binding of a *Drosophila* POU-domain protein to a sequence element regulating gene expression in specific dopaminergic neurons. *Nature* 343: 467–70

Johnson, W. A., McCormick, C. A., Bray, S. J., Hirsh, J. 1989. A neuron-specific enhancer of the *Drosophila* dopa decarboxylase gene. *Genes Dev.* 3: 676–86

Julien, J. P., Tretjakoff, I., Beaudet, L., Peterson, A. 1987. Expression and assembly of a human neurofilament protein in transgenic mice provide a novel neuronal marking system. *Genes Dev.* 1: 1085–95

Kapur, R. P., Hoyle, G. W., Mercer, E. H., Brinster, R. L., Palmiter, R. D. 1991. Some neuronal cell populations express human dopamine beta-hydroxylase-lacZ transgenes transiently during embryonic development. *Neuron* 7: 717–27

Kayano, T., Noda, M., Flockerzi, V., Takahashi, H., Numa, S. 1988. Primary structure of rat brain sodium channel III deduced from the cDNA sequence. *FEBS Lett.* 228: 187–94

Kemp, L. M., Dent, C. L., Latchman, D. S. 1990. Octamer motif mediates transcriptional repression of HSV immediate-early genes and octamer-containing cellular promoters in neuronal cells. *Neuron* 4: 215–22

Kondoh, H., Katoh, K., Takahashi, Y., Fujisawa, H., Yokoyama, M., et al. 1987. Specific expression of the chicken delta-crystallin gene in the lens and the pyramidal neurons of the piriform cortex in transgenic mice. *Dev. Biol.* 120: 177–85

Korner, M., Rattner, A., Mauxion, F., Sen, R., Citri, Y. 1989. A brain-specific transcription activator. *Neuron* 3: 563–72

Kraner, S. D., Chong, J. A., Tsay, H.-J., Mandel, G. 1992. Silencing the type II sodium channel gene: a model for neural-specific gene regulation. *Neuron* 9: 37–44

Lacy, E., Roberts, S., Evans, E. P., Burtenshaw, M. D., Costantini, F. D. 1983. A foreign β-globin gene in transgenic mice: integration at abnormal chromosomal positions and expression in inappropriate tissues. *Cell* 34: 343–58

Landel, C. P., Chen, S., Evans, G. A. 1990. Reverse genetics using transgenic mice. *Annu. Rev. Physiol.* 52: 841–51

Levine, M., Manley, J. L. 1989. Transcriptional repression of eukaryotic promoters. *Cell* 59: 405–8

Lewis, S. A., Cowan, N. J. 1986. Anomalous placement of introns in a member of the intermediate filament multigene family: an evolutionary conundrum. *Mol. Cell. Biol.* 6: 1529–34

Mandel, G. 1992. Tissue-specific expression of the voltage-sensitive sodium channel. *J. Membrane Biol.* 125: 1–13

Monuki, E. S., Kuhn, R., Weinmaster, G., Trapp, B. D., Lemke, G. 1990. Expression and activity of the POU transcription factor SCIP. *Science* 249: 1300–3

Mansour, S. L., Thomas, K. R., Capecchi, M. R. 1988. Disruption of the proto-oncogene int-2 in mouse embryo-derived stem cells: a general strategy for targetting mutations to non-selectable genes. *Nature* 336: 348–36

Mason, A. J., Pitss, S. L., Nikokics, K., Szonyi, E., Wilcox, J. N., et al. 1986. The hypogonadal mouse: reproductive functions restored by gene therapy. *Science* 234: 1372–78

Maue, R. A., Kraner, S. D., Goodman, R. H., Mandel, G. 1990. Neuron-specific expression of the rat brain type II sodium channel gene is directed by upstream regulatory elements. *Neuron* 4: 223–31

Mavrothalassitis, G., Beal, G., Papas, T. S. 1990. Defining target sequences of DNA-binding proteins by random selection and PCR: determination of the GCN4 binding sequence repertoire. *DNA Cell. Biol.* 9: 783–88

McGowan, R., Campbell, R., Peterson, A., Sapienza, C. 1989. Cellular mosaicism in

the methylation and expression of hemizygous loci in the mouse. *Genes Dev.* 3: 1669–76

Mellon, P. L., Windle, J. J., Goldsmith, P. C., Padula, C. A., Roberts, J. L., Weiner, R. I. 1990. Immortalization of hypothalamic GnRH neurons by genetically targeted tumorigenesis. *Neuron* 5: 1–10

Mercer, E. H., Hoyle, G. W., Kapur, R. P., Brinster, R. L., Palmiter, R. D. 1991. The dopamine beta-hydroxylase gene promoter directs expression of E. coli lacZ to sympathetic and other neurons in adult transgenic mice. *Neuron* 7: 703–16

Mitchell, M. T., Benfield, P. A. 1990. Two different RNA polymerase II initiation complexes can assemble on the rat brain creatine kinase promoter. *J. Biol. Chem.* 265: 8259–67

Morgan, J. I., Curran, T. 1991. Stimulus-transcription coupling in the nervous system: involvement of the inducible proto-oncogenes *fos* and *jun*. *Annu. Rev. Neurosci.* 14: 421–51

Mori, N., Stein, R., Sigmund, O., Anderson, D. J. 1990. A cell type-preferred silencer element that controls the neural-specific expression of the SCG10 gene. *Neuron* 4: 583–94

Nagatsu, I., Yamada, K., Karasawa, N., Sakai, M., Takeuchi, T., et al. 1991. Expression in brain sensory neurons of the transgene in transgenic mice carrying human tyrosine hydroxylase gene. *Neurosci. Lett.* 127: 91–95

Nedivi, E., Basi, G. S., Akey, I. V., Skene, J. H. P. 1992. A neural-specific GAP-43 core promoter located between unusual DNA elements that interact to regulate its activity. *J. Neurosci.* In press

Noda, M., Ikeda, T., Kayano, T., Suzuki, H., Takeshima, H., et al. 1986. Existence of distinct sodium channel messenger RNAs in rat brain. *Nature* 320: 188–92

Oberdick, J., Smeyne, R. J., Mann, J. R., Zackson, S., Morgan, J. I. 1990. A promoter that drives transgene expression in cerebellar purkinje and retinal bipolar neurons. *Science* 248: 223–26

Palmiter, R. D., Brinster, R. L. 1986. Germline transformation of mice. *Annu. Rev. Genet.* 20: 465–99

Palmiter, R. D., Sandgren, E. P., Avarbock, M. R., Allen, J. M., Brinster, R. L. 1991. Heterologous introns can enhance expression of transgenes in mice. *Proc. Natl. Acad. Sci. USA* 88: 478–82

Patil, N., Lacy, E., Chao, M. V. 1990. Specific neuronal expression of human NGF receptors in the basal forebrain and cerebellum of transgenic mice. *Neuron* 2: 437–47

Pecorino, L. T., Darrow, A. L., Strickland,

S. 1991. In vitro analysis of the tissue plasminogen activator promoter reveals a GC box-binding activity present in murine brain but undetectable in kidney and liver. *Mol. Cell. Biol.* 11: 3139–47

Pierce, J. W., Gifford, A. M., Baltimore, D. 1991. Silencing of the expression of the immunoglobulin kappa gene in non-B cells. *Mol. Cell. Biol.* 11: 1431–37

Readhead, C., Popko, B., Takahashi, N., Shine, H. D., Sidman, R. L., Hood, L. 1987. Expression of a myelin basic protein gene in transgenic shiverer mice: Correction of the dysmyelinating phenotype. *Cell* 48: 703–12

Rickles, R. J., Strickland, S. 1988. Tissue plasminogen activator mRNA in murine tissues. *FEBS Lett.* 229: 100–6

Roizman, B., Jenkins, F. J. 1986. The biologic and molecular properties of Herpesviruses: a summary. In *Herpes and Papilloma Viruses*, ed. G. DePalo, pp. 1–13. New York: Raven

Rosenfeld, M. G. 1991. POU-domain transcription factors: pou-er-ful developmental regulators. *Genes Dev.* 5: 897–907

Russo, A. F., Crenshaw, E. B., Lira, S. A., Simmon, D. M., Swanson, L. W., Rosenfeld, M. G. 1988. Neuronal expression of chimeric genes in transgenic mice. *Neuron* 1: 311–20

Sasahara, M., Fires, J. W. U., Raines, E. W., Gown, A. M., Westrum, L. E., et al. 1991. PDGF B-chain in neurons of the central nervous system, posterior pituitary, and in a transgenic model. *Cell* 64: 217–27

Sauerwald, A., Hoesche, C., Oshwald, R., Kilimann, M. W. 1990. The 5′ flanking region of the synapsin I gene. A G+C rich, TATA and CAAT-less, phylogenetically conserved sequence with cell type-specific promoter function. *J. Biol. Chem.* 265: 14932–37

Skene, J. H. P. 1989. Axonal growth-associated proteins. *Annu. Rev. Neurosci.* 12: 127–56

Stein, R., Mori, N., Matthew, K., Lo, L.-C., Anderson, D. J. 1988. The NGF-inducible SCG10 mRNA encodes a novel membrane-bound protein present in growth cones and abundant in developing neurons. *Neuron* 1: 463–76

Struhl, K. 1991. Mechanisms for diversity in gene expression patterns. *Neuron* 7: 177–81

Swanson, L. W., Simmons, D. M., Arriza, J., Hammer, R., Brinster, R., et al. 1985. Novel developmental specificity in the nervous system of transgenic animals expressing growth hormone fusion genes. *Nature* 317: 363–66

Swift, G. H., Hammer, R. E., MacDonald, R. J., Brinster, R. L. 1984. Tissue-specific expression of the rat pancreatic elastase I gene in transgenic mice. *Cell* 38: 639–46

Tamura, T., Sumita, K., Hirose, S., Mikoshiba, K. 1990. Core promoter of the mouse myelin basic protein gene governs brain-specific transcription in vitro. *EMBO J.* 9: 3101–8

Thiel, G., Greengard, P., Sudhof, T. C. 1991. Characterization of tissue-specific transcription by the human synapsin I gene promoter. *Proc. Natl. Acad. Sci. USA* 88: 3431–35

Thompson, M. A., Ziff, E. B. 1989. Structure of the gene encoding peripherin, an NGF-regulated neuronal-specific type III intermediate filament protein. *Neuron* 2: 1043–53

Treacy, M. N., He, X., Rosenfeld, M. G. 1991. I-POU: a POU domain protein that inhibits neuron-specific gene activation. *Nature* 350: 577–84

Vandaele, S., Nordquist, D. T., Feddersen, R. M., Tretjakoff, I., Peterson, A. C., Orr, H. T. 1991. Purkinje cell protein-2 regulatory regions and transgene expression in cerebellar compartments. *Genes Dev.* 5: 1136–48

Vandenbergh, D. J., Wuenschell, C. W., Mori, N., Anderson, D. J. 1989. Chromatin structure as a molecular marker of cell lineage and developmental potential in neural crest-derived chromaffin cells. *Neuron* 3: 507–18

Weintraub, H. 1985. Assembly and propagation of repressed and derepressed chromosomal states. *Cell* 42: 705–11

Wirak, D. O., Bayney, R., Kundel, C. A., Lee, A., Scangos, G. A., et al. 1991. Regulatory region of human amyloid precursor protein (APP) gene promotes neuron-specific gene expression in the CNS of transgenic mice. *EMBO J.* 10: 289–96

Wuenschell, C. W., Mori, N., Anderson, D. J. 1990. Analysis of SCG10 gene expression in transgenic mice reveals that neural specificity is achieved through selective derepression. *Neuron* 4: 595–602

Young, W. S. III, Reynolds, K., Shepard, E. A., Gainer, H., Castel, M. 1990. Cell-specific expression of the rat oxytocin gene in transgenic mice. *J. Neuroendocrinol.* 2: 917–25

Annu. Rev. Neurosci. 1993. 16:347–68

REGULATION OF ION CHANNEL DISTRIBUTION AT SYNAPSES

Stanley C. Froehner

Department of Physiology, University of North Carolina, Chapel Hill, North Carolina 27599-7545

KEY WORDS: neurotransmitter receptors, receptor clustering, postsynaptic cytoskeleton, extracellular matrix

INTRODUCTION

Synapses are regions of highly specialized contact between two cells. The clustered distribution of both ligand-gated and voltage-activated ion channels is one of the best known examples of molecular specialization in synaptic membranes. At the neuromuscular junction, ion channel clustering is responsible for rapid signal transmission and ensures that the postsynaptic response initiates a propagated action potential. At some excitatory synapses between neurons, potentiation of synaptic strength depends, in part, on coclustering of transmitter receptors. Thus, regulation of the ion channel distribution is not only important in signal transmission, but possibly also in synaptic plasticity.

The mechanisms by which ion channels are anchored at synaptic sites, and the way in which this process is regulated by the nerve, are best understood at the neuromuscular junction. This review focuses on the neuromuscular junction, with discussions of both pre- and postsynaptic elements. Studies of the mechanisms of clustering in the central nervous system (CNS) are less well advanced. Therefore, three of the best characterized examples are considered. Although ion channels are clustered in other parts of the neuron (e.g. nodes of Ranvier, axon hillock), this review is restricted to ion channel distribution at the synapse.

THE NEUROMUSCULAR JUNCTION

In 1943, Kuffler first demonstrated that the synaptic region of skeletal muscle is highly specialized for electrical excitability and sensitivity to

347

0147–006X/93/0301–0347$02.00

acetylcholine. Since then, the neuromuscular junction has been the subject of intense investigation of the mechanisms governing ion channel localization at synapses. Most of this work has focused on the nicotinic acetylcholine receptor (AChR).

Nicotinic Acetylcholine Receptors

AChRs are restricted to the crests of the postjunctional folds, where they are present at 8–10,000 receptors per square micron, a site density at least 1000-fold higher than in the extrasynaptic membrane (Fertuck & Salpeter 1974; Matthews-Bellinger & Salpeter 1978). AChR clustering is the first detectable differentiation of the postsynaptic membrane during neuromuscular synaptogenesis; it occurs within hours of the first contact between nerve and muscle (Steinbach 1981). Maintenance of AChR localization may also be important in synaptic plasticity. During neuromuscular development, multiply innervated endplates are pruned until only a single innervating axon remains. Loss of AChR clusters from postsynaptic sites may be the event that marks a nerve terminal for retraction (Rich & Lichtman 1989). The wealth of information on the structure of muscle AChR, the ability to purify large amounts of postsynaptic membrane from Torpedo electric organ, and the availability of myotube culture systems that express AChR clusters have made this the system of choice in which to study the mechanisms that anchor transmitter receptors beneath nerve terminals and their regulation during synaptogenesis.

Muscle AChR is a pentameric structure composed of homologous subunits, each of which spans the membrane several times (for a recent review, see Galzi et al 1991). AChRs are inherently capable of lateral motion in the membrane, a characteristic that must be overcome to anchor receptors at high concentrations at the synapse. Abundant evidence indicates that the nerve provides the signal to initiate AChR clustering at nascent synapses (Anderson & Cohen 1977; Role et al 1985; Steinbach 1981). Although, in principle, receptor-receptor interactions triggered by a critical concentration in the membrane could be sufficient for AChR clustering, evidence from electric field studies strongly suggests that other molecules are necessary (Stollberg & Fraser 1990a,b).

ACHR CLUSTERING FACTORS   In developing muscle in vivo (Steinbach 1981) and in cultured myotubes (Anderson & Cohen 1977; Role et al 1985), nerve contact initiates AChR clustering. This fact has prompted many laboratories to search for neurally derived factors that stimulate AChR clustering.

Agrin   The basal lamina surrounding each skeletal muscle fiber survives damage-induced degeneration of the muscle fiber and provides a framework for subsequent regeneration of the myotube from satellite cells (Marshall et al 1977; Sanes et al 1978). The synaptic basal lamina is biochemically distinct from that surrounding the rest of the muscle fiber and contains molecules capable of mediating reinnervation by the nerve and initiating differentiation of the pre- and postsynaptic membranes. In regenerating frog muscle, surviving basal lamina directs AChR clustering at the site of the original synaptic junction (Burden et al 1979; McMahan & Slater 1984). This observation has led to the purification, characterization, and cloning of a basal lamina protein called agrin, which stimulates AChR clustering in cultured myotubes.

Agrin solubilized from a basal lamina fraction of Torpedo electric organ causes clustering primarily of preexisting surface AChR without affecting the number of receptors or their metabolism (Wallace 1988). In addition, several other postsynaptic molecules, including the AChR-associated 43K protein, acetylcholine esterase, butyrylcholinesterase, and heparan sulfate proteoglycan, also become organized into clusters in agrin-treated myotubes (Wallace 1989). Monoclonal antibodies that block agrin activity immunoprecipitate proteins of M_r 150,000, 135,000, 95,000, and 70,000 from Torpedo extracts (Nitkin et al 1987). Agrin activity is associated with the 150 kD and 95 kD proteins. All of these polypeptides are probably proteolytic fragments of a larger protein.

Agrin also mediates nerve-induced AChR clustering (Reist et al 1992). Polyclonal antibodies raised against ray agrin recognize chick, but not rat, agrin. AChR clustering on chick myotubes induced by coculture with chick motor neurons was inhibited > 90% by these anti-agrin antibodies. When rat motor neurons were grown with chick myotubes, the antibodies had no effect. Thus, the active agrin molecules are derived from the motor neurons. These findings, and the observation that agrin is found in the cell bodies of motor neurons (Magill-Solc & McMahan 1988) and is transported along their axons (Magill-Solc & McMahan 1990), support the hypothesis that AChR clustering in muscle is mediated by agrin synthesized in and released from motor neurons (McMahan 1990). The muscle also synthesizes and secretes agrin, which becomes part of the synaptic basal lamina (Fallon & Gelfman 1989). Because muscle-derived agrin accumulates at AChR clusters induced by exogenously applied agrin or by coculture with spinal cord neurons, it may also play a role in the clustering process.

The mechanism by which agrin causes AChR clustering is not known. It does not appear to act by direct crosslinking of AChR, however, because agrin-induced clustering is dependent on metabolic energy (Wallace 1988). Furthermore, estimates of the stoichiometry of receptor molecules clustered per molecule of bound agrin are high: At least 100 AChR molecules can be caused to aggregate by one agrin molecule (Nitkin et al 1987). Thus, agrin likely acts catalytically, probably by binding to its own cell surface receptor and activating an intracellular signaling pathway, which culminates in AChR clustering.

Elucidation of the putative intracellular signaling pathway stimulated by agrin is being actively pursued. Agrin requires extracellular calcium for its activity (Wallace 1988), as does the formation of AChR clusters in nerve-muscle cocultures (Henderson et al 1984). Calcium appears to act extracellularly, possibly at the binding step. Because the action of agrin cannot be mimicked by calcium ionophores, increased intracellular calcium is probably not a critical event. Of the many drugs and treatments tested for effects on agrin-induced clustering, only inhibitors of energy metabolism, increased proton concentration, polyanions, and a protein kinase C-activating phorbol ester were effective (Wallace 1988, 1990). The phorbol ester blocks clustering at concentrations similar to that required for protein kinase C activation and augments AChR phosphorylation and dispersal of AChR clusters on myotubes (Ross et al 1988). Heparin and heparan sulfate block nerve-induced AChR clustering at developing neuromuscular junctions (Hirano & Kidokoro 1989) and inhibit the action of agrin (Wallace 1990).

Agrin stimulates phosphorylation of AChRs in chick muscle cultures (Wallace et al 1991). The β, γ, and δ AChR subunits show increased levels of phosphorylation after agrin treatment. Agrin-stimulated phosphorylation of γ and δ subunits was blocked by H-7, an inhibitor of protein serine kinases. However, H-7 had no effect on AChR clustering or on β subunit phosphorylation. Agrin-stimulated phosphorylation of the β subunit occurs on tyrosine residues with a time course that precedes receptor clustering and is blocked by treatments that inhibit agrin-mediated AChR clustering. These results suggest a model in which AChR tyrosine phosphorylation on the β subunit, an early consequence of agrin stimulation, promotes interaction with the postsynaptic cytoskeleton, which then leads to clustering. Experiments on the neuromuscular junction, however, suggest that this model is oversimplified.

Phosphotyrosine revealed by antibody staining is present in the rat diaphragm postsynaptic membrane in a distribution coextensive with AChR (Qu et al 1990). At least some of this phosphotyrosine resides in the AChR β and δ subunits. Postsynaptic tyrosine phosphorylation and AChR clustering are separable events, however (Qu et al 1990). In developing muscle, junctional antiphosphotyrosine is first detected at postnatal day 4, approximately ten days after AChR clustering occurs. Furthermore, denervation causes a marked decrease in antiphosphotyrosine staining, even though AChRs remain clustered. Thus, tyrosine phosphorylation does not appear to be an obligatory step in the earliest stages of AChR clustering.

Full-length cDNA clones encoding rat and chick agrin have recently been isolated (Rupp et al 1991; Tsim et al 1992) by using a partial clone for ray agrin homologue (Smith et al 1992). The deduced amino acid sequences predict rat and chick proteins of approximately 200,000 molecular weight, a size consistent with that of a protein recognized by antibodies made to fusion proteins encoded by the rat clone (Rupp et al 1991). Both agrins contain several conserved structural features, including nine protease inhibitor-like domains near the amino terminus of the protein, a 119 amino acid stretch showing significant homology with various forms of laminin, and four EGF repeats (based on the spacing between six cysteine residues) in the carboxy terminal region. Rat agrin also contains two tripeptide sequences (leu-arg-glu) found in other synaptic basal lamina proteins and thought to be a motor neuron attachment site (Hunter et al 1989a). These homologies with other proteins suggest that agrin may have functions in synapse formation, in addition to AChR clustering.

Recombinant agrins are active in clustering AChR (Campanelli et al 1991; Tsim et al 1992). CHO and COS cells transfected with full-length cDNA encoding rat agrin express a protein of 200–220 kDa that associates with the surface of the cells, possibly in the extracellular matrix. When transfected cells are cocultured with rat muscle fibers, clusters of AChR were often seen at cell contact sites (Campanelli et al 1991). Contact with nontransfected cells or with cells expressing a control protein or a truncated form of agrin did not produce receptor clustering. A cDNA clone encoding the carboxy-terminal half of chick agrin, which lacks the protease inhibitor-like domains and the laminin homologous region, induces AChR clusters that are indistinguishable from those formed by authentic Torpedo agrin (Tsim et al 1992). Some forms of chick agrin derived by alternative splicing of the product of the agrin gene are inactive in AChR clustering (Ruegg et al 1992). These agrin-related proteins are identical in primary structure to agrin, except for the absence of two short stretches of 4 and 11 residues, and may be components of basal lamina at nonsynaptic sites in various tissues (Godfrey 1991).

Identification of the "agrin receptor" is central to understanding how agrin acts. This work is in the early stages, but initial studies appear promising (Nastuk et al 1991). Agrin binding to myotubes is evenly distributed when cells are maintained at 4°C; after several hours at 37°C, however, agrin binding sites redistribute into clusters and become co-localized with AChR. This binding is closely associated with the plasma membrane, is trypsin-sensitive and calcium-dependent, and occurs with an estimated affinity of 5×10^{-10} to 10^{-11} M.

Basic fibroblast growth factor   Recent results suggest a role for basic fibroblast growth factor (bFGF) in postsynaptic induction (Peng et al 1991). Beads coated with bFGF induce AChR clustering at the site of bead-muscle contact to *Xenopus* myotomal muscle cells, whereas beads coated with other proteins or bath-applied bFGF are ineffective (Peng et al 1991). bFGF bead-induced clustering is blocked by tyrphostin, an inhibitor of tyrosine phosphorylation. An intriguing possibility is that agrin and bFGF activate a common pathway, involving tyrosine phosphorylation, that culminates in AChR clustering. Adsorption of such factors as bFGF and local presentation to muscle may also explain the ability of positively charged beads to initiate receptor clustering (Peng & Cheng 1982).

Soluble brain factors   Several laboratories have identified AChR clustering activity in soluble extracts of embryonic brain (Olek et al 1986; Salpeter et al 1982). Aggregation of AChR by embryonic pig brain extract is calcium, energy, and temperature dependent (with a sharp maximum near 37°C). As with agrin (Wallace 1988) and nerve-induced clusters (Kidokoro et al 1980; Role et al 1985; Steinbach 1981), clustering begins with the formation of small microclusters, which appear to coalesce into larger aggregates. One of the active components in brain extract is ascorbic acid (Knaack & Podleski 1985), which may act indirectly on clustering by increasing the number of AChRs (Horovitz et al 1989; Knaack & Podleski 1985) or through increased basal lamina deposition by stimulation of collagen synthesis (Kalcheim et al 1982). Whether additional protein factors are related or identical to agrin awaits purification of the brain proteins responsible for this activity. Other characterized factors, such as acetylcholine receptor-inducing activity (Usdin & Fischbach 1986) and calcitonin gene-related peptide (New & Mudge 1986), enhance AChR levels, but have little direct effect on receptor distribution.

THE EXTRACELLULAR MATRIX   One model of AChR cluster formation is based on the hypothesis that adhesive interactions, presumably involving extracellular matrix (ECM) components, drive the organization of cytoskeletal elements, which in turn cause AChR to become organized (Bloch & Pumplin 1988). Association of ECM with synaptic membrane at early times in neuromuscular synapse formation (Kullberg et al 1977), the presence of clustering molecules in the basal lamina, and a role for ECM in developmental events in the nervous system (Sanes 1983) have all drawn attention to the involvement of the ECM in AChR clustering.

The synaptic ECM contains, in addition to agrin, several well-characterized surface molecules, including a synapse-specific laminin, collagen, heparan sulfate proteoglycan (HSPG), and fibronectin (Anderson & Fambrough 1983; Daniels et al 1984; Hunter et al 1989b; Sanes 1982;

Sanes & Hall 1979). Many of these proteins are also present at clusters of AChR on cultured myotubes (Daniels et al 1984; Silberstein et al 1982), and some, including laminin, HSPG, type IV collagen, and fibronectin, are especially enriched in AChR-rich domains (Dmytrenko et al 1990). A role for collagen in cluster maintenance can be inferred from studies demonstrating that collagenase treatment partially disrupts AChR clusters (Bloch et al 1986).

Direct evidence for the involvement of a muscle ECM protein in AChR clustering comes from studies of a genetic variant. Mouse muscle C2 myotubes spontaneously form clusters of AChRs (Silberstein et al 1982). A genetic variant of C2 cells, the S27 line, is defective in proteoglycan synthesis (Gordon & Hall 1989) and fails to cluster AChRs spontaneously (Gordon et al 1992). Transfected cells expressing recombinant rat agrin overcome the clustering defect when cocultured with S27 cells (Ferns et al 1992). However, only two of four tested forms of agrin, generated by alternative RNA splicing of short sequences near the C-terminal end, are active on S27 cells, whereas all four forms cause AChR clustering on normal C2 cells. The proteoglycan may act as a low affinity receptor that concentrates agrin at the myotube surface, possibly with selectivity for specific forms.

THE POSTSYNAPTIC CYTOSKELETON   Many synapses exhibit dense accumulations of subsynaptic cytoskeletal elements (Heuser & Reese 1977). At the neuromuscular junction, this molecular specialization appears to play a central role in anchoring AChRs in the postsynaptic membrane (see Froehner 1991 for a recent review). Several proteins of this complex appear to be especially important.

The 43K protein   A protein of 43,000 molecular weight, commonly called the 43K protein, was first identified in postsynaptic membranes purified from Torpedo electric organ (Neubig et al 1979, Porter & Froehner 1983; Sobel et al 1977). A potential role for the 43K protein in AChR clustering was inferred from the observation that its removal from the membrane by alkaline extraction (Neubig et al 1979) was accompanied by an increase in lateral and rotational mobility of the AChR (Barrantes et al 1980; Bloch & Froehner 1986; Cartaud et al 1981; Lo et al 1980). The 43K protein is present in approximately equimolar amounts with AChR (LaRochelle & Froehner 1986) and follows a distribution at the synapse indistinguishable from the receptor, both in Torpedo electrocytes (Bridgman et al 1989; Sealock et al 1984) and at the mammalian neuromuscular junction (Bloch & Froehner 1986; Flucher & Daniels 1989; Froehner et al 1981). Crosslinking studies have implicated the β subunit of AChR as the site of interaction with the 43K protein (Burden et al 1983). Recent high-resolution analyses of Torpedo membranes place the 43K protein directly beneath the channel of the AChR (Toyoshima & Unwin 1988) or, alternatively, in a bridge position that could potentially link two adjacent receptors (Mitra et al 1989). The 43K protein is most abundant in tissues that express muscle type AChR (skeletal muscle and electric organ) (LaRochelle & Froehner 1986), but is also found in much lower quantities in some other tissues (Musil et al 1989) and in myoblasts that lack AChR (Frail et al 1989).

A direct demonstration of a role for the 43K protein in AChR clustering comes from coexpression studies. *Xenopus* oocytes injected with synthetic RNA, which encodes the subunits of mouse muscle AChR, express surface receptors in a uniform, diffuse distribution. In contrast, oocytes coinjected with AChR RNA and RNA, which encodes the 43K protein, have surface AChR organized into clusters (Froehner et al 1990). Similar results have also been obtained in quail fibroblasts (Phillips et al 1991a) and in COS cells (Brennan et al 1992). In all cases, the 43K protein is intimately associated with the AChR clusters. A notable difference in these systems is the size of the clusters. Fibroblasts expressing 43K protein and AChR have relatively large clusters of receptors, similar in size to those on cultured myotubes. In contrast, the clusters induced on oocytes and COS cells are much smaller, 1–2 μ in diameter. The basis for this difference is unknown, but suggests regulation of cluster size by proteins or activities that are present in fibroblasts and skeletal muscle, but absent in oocytes. 43K protein-induced clustering of AChR in oocytes and COS cells occurs without any detectable alteration in the single channel properties of the receptor (Brennan et al 1992).

An important feature of the 43K protein is its ability to form clusters in the absence of AChR. In all three cell types, 43K protein expressed alone was organized into clusters on the cytoplasmic side of the membrane. This ability to cluster may arise either from self-association of the 43K protein or by its interaction with other proteins endogenous to the cell membrane.

The mechanism by which 43K protein clusters and interacts with AChR is under active investigation. The amino acid sequences of the 43K protein from three species reveal several conserved structural features that are potentially important for AChR clustering (Baldwin et al 1988; Carr et al 1987; Frail et al 1987, 1988; Froehner 1989). The amino terminal glycine residue is myristoylated (Carr et al 1989; Musil et al 1988). Elimination of this modification by site-directed mutagenesis reduces the number of clusters formed in fibroblasts, but not the ability of 43K protein to interact with AChR (Phillips et al 1991b). A cysteine-rich region near the carboxy terminus conforms to the structure of two zinc fingers (Froehner 1991). A serine residue just downstream of the putative zinc finger may be the site of endogenous phosphorylation (Hill et al 1991). Results with deletion mutants of the 43K protein are consistent with a model in which the amino and carboxy-terminal regions are necessary for membrane association and clustering of the 43K protein, whereas the central region contains sites of interaction with AChR (Phillips et al 1991b).

Although it is now clear that the 43K protein can cluster AChR, the exact role that this activity plays in synapse formation is uncertain. One model (Froehner et al 1990) proposes that the small AChR clusters formed in oocytes are analogous to those that form immediately after nerve contact in vivo (Bevan & Steinbach 1977) or in vitro (Role et al 1985). The appearance of 43K protein at receptor clusters as soon as they are detectable (Burden 1985; Peng & Froehner 1985) is consistent with this model. In contrast, the finding that early AChR clusters in developing electrocytes do not have 43K protein associated with them (Kordeli et al 1989) suggests a model in which the 43K protein stabilizes immature clusters at a later stage in synaptogenesis [but see LaRochelle et al (1990) for an alternative interpretation of this result]. This view is supported by results demonstrating extracellular matrix-induced clustering of AChR in transfected L-cells that lack 43K protein (Hartman et al 1991). A careful study of the developmental appearance of 43K protein in such species as chick or rat, in which the formation of the neuromuscular junction is well characterized, may help resolve this issue.

Other proteins, in addition to the 43K protein, may well be needed for AChR clustering. Immunohistochemical studies have identified several other proteins that are concentrated at the neuromuscular synapse and, thus, may be involved in anchoring AChR. Several of these proteins are discussed below in brief. [For a more detailed discussion, see Froehner (1991) and references therein.]

The spectrin/dystrophin family and associated proteins   Several members of the spectrin/dystrophin family of proteins, characterized by an elongated central rod structure with globular domains at the ends that bind actin and other proteins, are concentrated in the neuromuscular post-synaptic membrane. A unique muscle isoform of β spectrin is one of these proteins (Bloch & Morrow 1989). Clusters on cultured rat myotubes contain alternating subdomains of AChR-rich and AChR-poor membrane. β spectrin is found in the AChR-rich domains, along with the 43K protein (Bloch & Froehner 1986). A non-muscle isoform of actin is present in the postsynaptic membrane and in AChR domains of clusters (Bloch 1986; Hall et al 1981). Removal of β spectrin and actin from the membrane is accompanied by the loss of this domain structure, which implies a role for these proteins in AChR organization (Bloch 1986).

Dystrophin is a 400 kDa cytoskeletal protein whose absence or reduced expression is the primary cause of Duchenne and Becker muscular dystrophies (Hoffman et al 1987). Although dystrophin is found throughout the extrasynaptic sarcolemma, it is also enriched at the postsynaptic membrane of Torpedo electrocytes and mammalian skeletal muscle (Chang et al 1989; Jasmin et al 1990; Sealock et al 1991; Yeadon et al 1991). Recent studies have shown, however, that dystrophin is probably concentrated in the depths of the postjunctional folds (Sealock et al 1991). This finding, along with the observation that AChR clustering appears normal in human Duchenne muscle (Sakakibara et al 1977) and in the dystrophin-minus mdx mouse muscle (Sealock et al 1991), militate against a critical role for dystrophin in AChR clustering. Another member of this family, the dystrophin-related protein (DRP), which is encoded by a separate gene, is more highly concentrated at the endplate and may be situated more favorably for anchoring AChR at this site (Ohlendieck et al 1991).

A protein of M_r 58,000 (58K protein), also originally discovered in electric tissue (Froehner et al 1987), forms independent complexes with Torpedo dystrophin and a protein of M_r 87,000 (87K protein) (Butler et al 1992). Like dystrophin, the muscle homologues of 58K and 87K proteins are found throughout the sarcolemma and are concentrated at muscle endplates (Carr et al 1989; Froehner et al 1987). In addition, the 58K protein is a component of AChR-rich domains of cultured myotubes (Bloch et al 1991). In immunofluorescence experiments on mdx muscle, staining for the 58K protein is much reduced on the sarcolemma, but remains strong at the neuromuscular junction (Butler et al 1992). Retention of the 58K protein at synaptic sites is presumably caused by its interaction with another member of the dystrophin family, possibly DRP or β-spectrin.

Sodium Channels

Electrophysiological (Beam et al 1985; Caldwell et al 1986) and immuno-histochemical (Angelides 1986; Haimovich et al 1987) studies have clearly demonstrated that voltage-activated sodium channels are concentrated in the neuromuscular postsynaptic membrane. This presumably affords a large margin of safety for the generation of action potentials in response to local synaptic depolarization via AChR activation. In contrast to AChR, however, postsynaptic sodium channels are localized primarily in the depths of the junctional folds (Flucher & Daniels 1989). Recent auto-radiographic analysis with radiolabeled α-scorpion toxin has confirmed this distribution (Boudier et al 1992). Sodium channel site densities were $5000/\mu^2$ in the perisynaptic region, whereas the values were about tenfold lower for extrasynaptic membrane.

In contrast to AChR, sodium channels become clustered late in rat neuromuscular development (Lupa et al 1992). Concentration of sodium channels at the synapse first appears during the second week after birth and is not complete until several months later. Furthermore, both embryonic and adult forms of the channel are clustered.

The cytoskeleton appears to play an important role in anchoring sodium channels at selected membrane domains. In brain, sodium channels may be anchored to the spectrin-based cytoskeleton through ankyrin (Davis & Bennett 1984; Srinivasan et al 1988). Ankyrin interacts directly with sodium channels in an equimolar stoichiometry, as demonstrated by binding and copurification studies (Srinivasan et al 1988). A specialized form of erythrocyte ankyrin is localized at nodes of Ranvier, sites of high sodium channel density (Kordeli & Bennett 1991). Colocalization of sodium channels and ankyrin in the neuromuscular postsynaptic membrane (Flucher & Daniels 1989) suggests that similar mechanisms may be relevant at this synapse.

The signals that direct sodium channel concentration at the synapse are not known. Agrin does not appear to play a role in this process (Lupa & Caldwell 1991). In axons, initial segregation of sodium channels depends on interactions with active Schwann cells (Joe & Angelides 1992). The clustered sodium channels then appear to serve as templates for subsequent association of ankyrin and spectrin. This result, along with several other differences listed above, clearly differentiates between the mechanisms of AChR and sodium channel clustering and their regulation.

Calcium Channels

Entry of calcium through voltage-activated calcium channels is the primary activator of neurotransmitter release at the presynaptic terminal (Augustine et al 1987). Exocytosis of synaptic vesicles occurs at specialized regions of the nerve terminal, the active zones. Active zones presumably possess the protein machinery needed for docking of vesicles and the calcium-induced fusion of vesicles with the terminal membrane. In 1981, Pumplin et al speculated that intramembranous particles identified in freeze fracture replicas at these sites represented calcium channels (or aggregations of calcium channels) based on a comparison of their site density with the number of channels expected from electrophysiological measurements. In patients with Lambert-Eaton syndrome, an autoimmune neuromuscular disease in which motor nerve terminals release an insufficient amount of acetylcholine (Vincent et al 1989), there is a paucity and disorder of these particles. The demonstration that LES serum contains anticalcium channel activity (Kim & Neher 1988) lends further credence to the idea that these membrane structures are calcium channels.

This hypothesis has received strong support from localization studies.

Neurons express several types of voltage-activated calcium channels that can be distinguished primarily by their channel properties, voltage dependency, and pharmacological properties (Bean 1989). Although some controversy still exists, N-type calcium channels appear to be largely responsible for ACh release at the neuromuscular junction. N-type channels at the frog neuromuscular junction are inhibited by ω-conotoxin (Kerr & Yoshikami 1984), a peptide toxin isolated from the marine snail *Conus geographus*. Binding sites for fluorescent derivatives of ω-conotoxin in the nerve terminal are nonuniform. Clusters of binding sites, presumably N-type calcium channels, occur in a distribution expected of active zones (Cohen et al 1991; Robitaille et al 1990). Furthermore, double label experiments show that clusters of ω-conotoxin sites are positioned directly across the synaptic gap from postsynaptic AChR clusters (Cohen et al 1991; Robitaille et al 1990) (see Figure 1). In contrast, antibodies to the di-

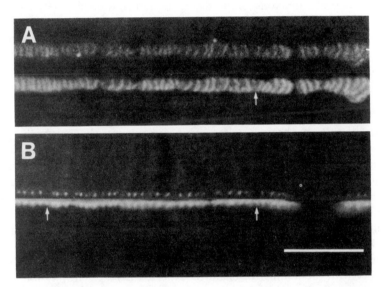

Figure 1 Codistribution of presynaptic calcium channels and postsynaptic acetylcholine receptors at the frog neuromuscular junction. Each figure is a photographic composite showing rhodamine-conjugated ω-conotoxin straining (*top*) and fluoroescein-conjugated α-bungarotoxin staining (*bottom*). (*A*) En face view displaying bands of calcium channels and AChR. (*B*) Side view displaying dots of each channel type. In both cases, the straining patterns for calcium channels and AChR are closely aligned. Arrows mark a few regions of mismatch. Bar in *B*, 10μ. See Cohen et al (1991) for experimental details. Reprinted with permission from *The Journal of Neuroscience.*

hydropyridine-sensitive, L-type channel, do not give detectable staining of presynaptic terminals in brain (Ahlijanian et al 1990).

Alignment of channels across the synaptic space requires communication between the pre- and postsynaptic cells during synapse formation. The nature of molecules involved are unknown, but they appear to reside in the basal lamina. Reinnervation of damaged muscle by regenerating axons occurs even in the absence of the myotube (Marshall et al 1977). When growing axons contact the old synaptic basal lamina left behind by degenerating muscle fibers, presynaptic differentiation is initiated (Sanes et al 1978). Thus, molecular cues remaining in the synaptic basal lamina are sufficient to induce the formation of active zones. Whether calcium channel clustering (and other aspects of presynaptic differentiation) can be induced by agrin, or by some other component of the basal lamina, is worthy of investigation.

This precise arrangement of presynaptic calcium channels and postsynaptic AChR exemplifies the importance of regulation of ion channel distribution in synaptic function. Alignment of these two channel types at the neuromuscular junction ensures that the sites of transmitter release are optimally positioned to activate the largest number of postsynaptic receptors. In the CNS, synaptic integration could presumably be affected by this arrangement: Synapses with highly coordinated alignments of presynaptic calcium channels with postsynaptic transmitter receptors could potentially dominate over those synapses in which the spatial arrangements are less precisely matched.

NEURONAL SYNAPSES

Although CNS synapses have not received the attention directed at the neuromuscular junction, several examples provide strong evidence that ion channel localization is not merely a peculiarity of the stringent requirements for rapid transmission at the neuromuscular junction, but is also important for neuronal synaptic transmission. The possibility of modulation of synaptic strength by regulation of ion channel density may be especially important in the CNS.

Neuronal Nicotinic Receptors

Rapid cholinergic transmission occurs between neurons of the sympathetic and parasympathetic ganglia. Ultrastructural studies of frog sympathetic neurons demonstrated that binding sites for α-bungarotoxin (which blocks cholinergic transmission in this ganglion) and for antibodies to Torpedo AChR were restricted to postsynaptic sites (Marshall 1981). Staining occurred in 0.2–0.5μ patches that were precisely aligned with active zones in the presynaptic terminal. Postsynaptic concentration of nicotinic receptors has also been seen in chick ciliary neurons (Jacob et al 1984). This specialization is formed concomitantly with a large increase in AChR number, soon after synapse formation begins (Jacob 1991). In the frog cardiac ganglion, most regions of anti-AChR staining were localized to postsynaptic sites; however, approximately 20% of clusters were extrasynaptic (Sargent & Pang 1989). These may represent sites recently vacated by innervating neurons. Denervation was accompanied by a small increase in the number of clusters, but a fourfold reduction in their size (Sargent & Pang 1988). Clusters also become more randomly distributed. These results indicate that regulation of AChR clustering by innervation differs in neurons and skeletal muscle.

Although neuronal AChR are clustered at synapses, their site density is lower than at the neuromuscular junction. Studies on chick ciliary ganglion with radiolabeled ^{125}I-neuronal bungarotoxin, which binds to selected types of neuronal AChR, indicate that there are approximately 600 toxin binding sites/μ^2 of postsynaptic membrane surface, or at least 20-fold lower than at the neuromuscular junction (Loring & Zigmond 1987).

Little is known of the mechanisms that anchor AChR at synapses in neurons. Attempts to identify the 43K protein at ganglion synaptic sites by using antibodies against the muscle protein have failed. Either the 43K protein is not present at these cholinergic synapses or a neuronal form of the 43K protein is sufficiently different such that the antibodies used do not recognize it. A cloning approach, analogous to that which has been so successful in identifying neuronal AChR, may prove equally useful in identifying neuronal equivalents of this and other neuromuscular proteins. Agrin has not been detected at CNS synapses, which have much less extracellular matrix than the neuromuscular junction. The expression of agrin in the CNS, however, leaves open the possibility that this protein may be important in AChR organization at neuronal synapses.

Glycine Receptors

Receptors for glycine are members of the same ligand-gated ion channel family as nicotinic AChR. They gate chloride ions and mediate inhibitory synaptic transmission in the spinal cord and the brain. Studies, particularly those by Heinrich Betz and colleagues, have shown that glycine receptors are pentamers composed of two transmembrane subunits (α and β) (Langosch et al 1990). In neurons of the ventral horn of the rat spinal cord, glycine receptors are organized into clusters (Triller et al 1985). Ultrastructural studies revealed that these clusters were confined to the postsynaptic membrane directly across the synapse from active zones.

The glycine receptor copurifies with a peripheral membrane protein of

M_r 93,000, named gephyrin (Langosch et al 1990; Schmitt et al 1987). Gephyrin is found largely at postsynaptic sites of glycinergic inhibitory synapses on the Mauthner cell of the goldfish brainstem (Seitanidou et al 1988). Copurification, gel overlay, and coassembly studies indicate that gephyrin binds polymerized tubulin (Kirsch et al 1991). The binding is high affinity (2.5 nM), shows significant cooperativity, and approaches a stoichiometry of ~ 1 gephyrin:4 tubulin at saturation. Thus, gephyrin may anchor glycine receptors by binding to subsynaptic microtubules. With the recent isolation of cDNA clones encoding gephyrin (Prior et al 1992), coexpression studies similar to those done with the 43K protein and nicotinic receptors should be imminent.

Glutamate Receptors

The central role that glutamate receptors play in long-term potentiation (LTP) has prompted intensive investigation of this large family of ligand-gated ion channels. Current models of LTP predict that two functionally and pharmacologically distinct classes of glutamate receptors, N-methyl-D-aspartate (NMDA) and kainate/quisqualate (non-NMDA), are colocalized at the same postsynaptic site in dendritic spines (reviewed in Kennedy 1989). Antibodies are not yet available to test this idea directly, but functional studies strongly support it. Measurements of miniature synaptic currents in cultured hippocampal neurons showed that about 70% of excitatory synapses express both types of glutamate receptor (Bekkers & Stevens 1989). Jones & Baughman (1991) combined iontophoretic mapping of the distribution of NMDA and non-NMDA receptors along dendrites of cultured neocortical neurons with subsequent immunohistochemical localization of synapses (with antibody to synaptic vesicles). NMDA and non-NMDA responses were colozalized at "hot spots" that coincided with synaptic connections. Subtype-specific antibodies will be very important in determining how this colocalization is achieved. Regulation of the density and relative locations of these two receptor types at an individual synapse may be a mechanism by which synaptic plasticity is modulated.

SUMMARY AND CONCLUSIONS

A model of neuromuscular synaptogenesis predicts that factors released by the nerve, such as agrin, determine the time and site of AChR clustering. These factors are thought to bind to specific receptors on the muscle fiber and activate intracellular signaling pathways, which act ultimately on the AChR, on the postsynaptic cytoskeleton, or on proteins, such as the 43K protein, that link AChR to the cytoskeleton. This model is not yet entirely proven, but provides a clear framework for further studies. Clustering of sodium channels at the neuromuscular postsynaptic membrane may also involve interaction with the cytoskeleton, but the signaling molecules are not yet known. AChR clustering and sodium channel clustering are separable events, both in the molecules involved and in developmental timing.

Ion channel clustering in the CNS is less well understood, but may have features in common with the neuromuscular model. For example, glycine receptors are associated with gephyrin, a protein that mediates linkage to synaptic microtubules, thus raising obvious analogies to muscle AChR and the 43K protein. It is unknown, however, if AChR clustering in neurons is regulated by mechanisms in common with muscle.

Coclustering of ion channels at synapses, both across the synaptic gap and within the same postsynaptic site, is likely to be of major importance to synaptic modulation and integration. Alignment of postsynaptic receptors with presynaptic calcium channels is critical for rapid transmission, but is achieved by mechanisms not yet understood. According to current models, the codistribution of two types of glutamate receptors in the postsynaptic membrane on dendritic spines is important for initiation of LTP. It is reasonable to expect that the basic mechanisms currently being elucidated at the neuromuscular junction will lead to greater understanding of how the CNS uses ion channel distribution to modulate synaptic activity.

ACKNOWLEDGEMENTS

Research in the author's laboratory is supported by grants from the National Institutes of Health (NS14871 and NS27504) and from the Muscular Dystrophy Association. I thank Monroe Cohen for providing the photographs for Figure 1, Robert Sealock for comments on the manuscript, and colleagues for providing papers prior to publication.

Literature Cited

Ahlijanian, M. K., Westenbroek, R. E., Catterall, W. A. 1990. Subunit structure and localization of dihydropyridine-sensitive calcium channels in mammalian brain, spinal cord, and retina. *Neuron* 4: 819–32

Anderson, M. J., Cohen, M. W. 1977. Nerve-induced and spontaneous redistribution of acetylcholine receptors on cultured muscle cells. *J. Physiol.* 268: 757–73

Anderson, M. J., Fambrough, D. M. 1983. Aggregates of acetylcholine receptors are associated with plaques of basal lamina heparan sulfate proteoglycan on the surface of skeletal muscle fibers. *J. Cell Biol.* 97: 1396–1411

Angelides, K. J. 1986. Fluorescently labelled

Na⁺ channels are localized and immobilized to synapses of innervated muscle fibers. *Nature* 321: 63–66

Augustine, G. J., Charlton, M. P., Smith, S. S. 1987. Calcium action in synaptic transmitter release. *Annu. Rev. Neurosci.* 10: 633–93

Baldwin, T. J., Theriot, J. A., Yoshihara, C. M., Burden, S. J. 1988. Regulation of the transcript encoding the 43 kilodalton subsynaptic protein during development and after denervation. *Development* 104: 557–64

Barrantes, F. J., Neugebauer, D.-C., Zingsheim, H. P. 1980. Peptide extraction by alkaline treatment is accompanied by rearrangement of the membrane-bound acetylcholine receptor from Torpedo marmorata. *FEBS Lett.* 112: 73–78

Beam, K. G., Caldwell, J. H., Campbell, J. T. 1985. Na channels in skeletal muscle concentrated near the neuromuscular junction. *Nature* 313: 588–90

Bean, B. P. 1989. Classes of calcium channels in vertebrate cells. *Annu. Rev. Physiol.* 51: 367–84

Bekkers, J. M., Stevens, C. F. 1989. NMDA and non-NMDA receptors are co-localized at individual excitatory synapses in cultured rat hippocampus. *Nature* 341: 230–33

Bevan, S., Steinbach, J. H. 1977. The distribution of alpha-bungarotoxin binding sites on mammalian muscle developing in vivo. *J. Physiol.* 267: 195–213

Bloch, R. J. 1986. Actin at receptor-rich domains of isolated acetylcholine receptor clusters. *J. Cell Biol.* 102: 1447–58

Bloch, R. J., Froehner, S. C. 1986. The relationship of the postsynaptic 43K protein to acetylcholine receptors in receptor clusters isolated from cultured rat myotubes. *J. Cell Biol.* 104: 645–54

Bloch, R. J., Morrow, J. S. 1989. An unusual beta-spectrin associated with clustered acetylcholine receptors. *J. Cell Biol.* 108: 481–94

Bloch, R. J., Pumplin, D. W. 1988. Molecular events in synaptogenesis: nerve-muscle adhesion and postsynaptic differentiation. *Am. J. Physiol.* 254: C345–64

Bloch, R. J., Resneck, W. G., O'Neill, A., Strong, J., Pumplin, D. W. 1991. Cytoplasmic components of acetylcholine receptor clusters of cultured rat myotubes: the 58 kD protein. *J. Cell Biol.* 115: 435–46

Bloch, R. J., Steinbach, J. H., Merlie, J. P., Heinemann, S. 1986. Collagenase digestion alters the organization and turnover of junctional acetylcholine receptors. *Neurosci. Lett.* 66: 113–19

Boudier, J.-L., Le Treut, T., Jover, E. 1992. Autoradiographic localization of voltage-dependent sodium channels on the mouse neuromuscular junction using ^{125}I-α-scorpion toxin. II. Sodium channel distribution on postsynaptic membranes. *J. Neurosci.* 12: 454–66

Brennan, C., Scotland, P. B., Froehner, S. C., Henderson, L. P. 1992. Functional properties of acetylcholine receptors coexpressed with the 43K protein in heterologous cell systems. *Dev. Biol.* 149: 100–11

Bridgman, P. C., Carr, C., Pedersen, S. E., Cohen, J. B. 1989. Visualization of the cytoplasmic surface of Torpedo postsynaptic membrane by freeze-etch and immunoelectron microscopy. *J. Cell Biol.* 105: 1829–46

Burden, S. J. 1985. The subsynaptic 43 kDa protein is concentrated at developing nerve-muscle synapses in vitro. *Proc. Natl. Acad. Sci. USA* 82: 8270–73

Burden, S. J., DePalma, R. L., Gottesman, G. S. 1983. Crosslinking of proteins in acetylcholine receptor-rich membranes: association between the beta-subunit and the 43 kD subsynaptic protein. *Cell* 35: 687–92

Burden, S. J., Sargent, P. B., McMahan, U. J. 1979. Acetylcholine receptors in regenerating muscle accumulate at original synaptic sites in the absence of nerve. *J. Cell Biol.* 82: 412–25

Butler, M. H., Douville, K., Murnane, A. A., Kramarcy, N. R., Cohen, J. B., et al. 1992. Association of the M_r 58,000 postsynaptic protein of electric tissue with Torpedo dystrophin and the M_r 87,000 postsynaptic protein. *J. Biol. Chem.* 267: 6213–18

Caldwell, J. H., Campbell, D. T., Beam, K. G. 1986. Na channel distribution in vertebrate skeletal muscle *J. Gen. Physiol.* 87: 907–32

Campanelli, J. T., Hoch, W., Rupp, F., Kreiner, T., Scheller, R. H. 1991. Agrin mediates cell contact-induced acetylcholine receptor clustering. *Cell* 67: 909–16

Carr, C., Fischbach, G. D., Cohen, J. B. 1989. A novel M_r 87,000 protein associated with acetylcholine receptors in Torpedo electric organ and vertebrate skeletal muscle. *J. Cell Biol.* 109: 1753–64

Carr, C., McCourt, D., Cohen, J. B. 1987. The 43 kDa protein of Torpedo nicotinic postsynaptic membranes: purification and determination of primary sequence. *Biochemistry* 26: 7090–7102

Carr, C., Tyler, A. N., Cohen, J. B. 1989. Myristic acid is the NH_2-terminal blocking group of the 43-kDa protein of Torpedo nicotinic postsynaptic membranes. *FEBS Lett.* 243: 65–69

Cartaud, J., Sobel, A., Rousselet, A., Devaux, P. F., Changeux, J. P. 1981. Consequences of alkaline treatment for the ultrastructure of the acetylcholine-receptor rich membranes from Torpedo marmorata electric organ. *J. Cell Biol.* 90: 418–26

Chang, H. W., Bock, E., Bonilla, E. 1989. Dystrophin in electric organ of Torpedo californica homologous to that in human muscle. *J. Biol. Chem.* 264: 20831–34

Cohen, M. W., Jones, O. T., Angelides, K. J. 1991. Distribution of Ca^{++} channels on frog motor nerve terminals revealed by fluorescent ω-conotoxin. *J. Neurosci.* 11: 1032–39

Daniels, M. P., Vigny, M., Sonderegger, P., Bauer, H. C., Vogel, Z. 1984. Association of laminin and other basement membrane components with regions of high acetylcholine receptor density on cultured myotubes. *Int. J. Dev. Neurosci.* 2: 87–99

Davis, J. Q., Bennett, V. 1984. Brain ankyrin. *J. Biol. Chem.* 13550–59

Dmytrenko, G. M., Scher, M. G., Poiana, G., Baetscher, M., Bloch, R. J. 1990. Extracellular glycoproteins at acetylcholine receptor clusters of rat myotubes are organized into domains. *Exp. Cell Res.* 189: 41–50

Fallon, J. R., Gelfman, C. E. 1989. Agrin-related molecules are concentrated at acetylcholine receptor clusters in normal and aneural developing muscle. *J. Cell Biol.* 108: 1527–35

Ferns, M., Hoch, W., Campanelli, J. T., Rupp, F., Hall, Z. W., Scheller, R. H. 1992. RNA splicing regulates agrin-mediated acetylcholine receptor clustering activity on cultured myotubes. *Neuron* 8: 1079–86

Fertuck, H. C., Salpeter, M. M. 1974. Localization of acetylcholine receptor by ^{125}I-labeled alpha bungarotoxin binding at mouse motor endplates. *Proc. Natl. Acad. Sci. USA* 71: 1376–78

Flucher, B. E., Daniels, M. P. 1989. Distribution of sodium channels and ankyrin in the neuromuscular junction is complementary to that of acetylcholine receptors and the 43 kD protein. *Neuron* 3: 163–75

Frail, D. E., McLaughlin, L. L., Mudd, J., Merlie, J. P. 1988. Identification of the mouse muscle 43,000-dalton acetylcholine receptor-associated protein (RAPsyn) by cDNA cloning. *J. Biol. Chem.* 263: 15602–7

Frail, D. E., Mudd, J., Shah, V., Carr, C., Cohen, J. B., Merlie, J. P. 1987. cDNAs for the postsynaptic M_r 43,000 protein of Torpedo electric organ encode two proteins with different carboxy termini. *Proc. Natl. Acad. Sci. USA* 84: 6302–6

Frail, D. E., Musil, S., Buonanno, A., Merlie, J. P. 1989. Expression of RAPsyn (43K Protein) and nicotinic acetylcholine receptor genes is not coordinately regulated in mouse muscle. *Neuron* 2: 1077–86

Froehner, S. C. 1991. The submembrane machinery for nicotinic acetylcholine receptor clustering. *J. Cell Biol.* 114: 1–7

Froehner, S. C. 1989. Expression of RNA transcripts for the postsynaptic 43 kDa protein in innervated and denervated rat skeletal muscle. *FEBS Lett.* 249: 229–33

Froehner, S. C., Gulbrandsen, V., Hyman, C., Jeng, A. Y., Neubig, R. R., Cohen, J. B. 1981. Immunofluorescence localization at the mammalian neuromuscular junction of the Mr 43,000 protein of Torpedo postsynaptic membrane. *Proc. Natl. Acad. Sci. USA* 78: 5230–34

Froehner, S. C., Luetje, C. W., Scotland, P. B., Patrick, J. 1990. The postsynaptic 43K protein clusters muscle nicotinic acetylcholine receptors in Xenopus oocytes. *Neuron* 5: 403–10

Froehner, S. C., Murnane, A. A., Tobler, M., Peng, H. B., Sealock, R. 1987. A postsynaptic M_r 58,000 (58K) protein concentrated at acetylcholine receptor-rich sites in Torpedo electroplaques and skeletal muscle. *J. Cell Biol.* 104: 1633–46

Galzi, J.-L., Revah, F., Beiss, A., Changeux, J.-P. 1991. Functional architecture of the nicotinic acetylcholine receptor: from electric organ to brain. *Annu. Rev. Pharmacol.* 31: 37–72

Godfrey, E. W. 1991. Comparison of agrin-like proteins from the extracellular matrix of chicken kidney and muscle with neural agrin, a synapse organizing protein. *Exp. Cell Res.* 195: 99–109

Gordon, H., Hall, Z. W. 1989. Glycosaminoglycan variants in the C2 muscle cell line. *Dev. Biol.* 135: 1–11

Gordon, H., Lupa, M., Bowen, D., Hall, Z. W. 1992. A muscle cell variant defective in glycosaminoglycan biosynthesis forms nerve-induced but not spontaneous clusters of the acetylcholine receptor and the 43K protein. *J. Neurosci.* In press

Haimovich, B., Schotland, D. L., Fieles, W. E., Barchi, R. L. 1987. Localization of sodium channel subtypes in adult rat skeletal muscle using channel-specific monoclonal antibodies. *J. Neurosci.* 7: 2957–66

Hall, Z. W., Lubit, B. W., Schwartz, J. H. 1981. Cytoplasmic actin in postsynaptic structures at the neuromuscular junction. *J. Cell Biol.* 90: 789–92

Hartman, D. S., Millar, N. S., Claudio, T. 1991. Extracellular synaptic factors induce clustering of acetylcholine receptors stably expressed in fibroblasts. *J. Cell Biol.* 115: 165–77

Henderson, L. P., Smith, M. A., Spitzer, N. C. 1984. The absence of calcium blocks impulse-evoked release of acetylcholine but not de novo formation of functional neuromuscular synaptic contacts in culture. *J. Neurosci.* 4: 3140–50

Heuser, J. E., Reese, T. S. 1977. Structure of the synapse. In *Handbook of Physiology*, ed. J. M. Brookhart, V. B. Mountcastle. Bethesda, Md: Am. Physiol. Soc.

Hill, J. A., Nghiem, H.-O., Changeux, J.-P. 1991. Serine-specific phosphorylation of nicotinic receptor associated 43K protein. *Biochemistry* 30: 5579–85

Hirano, Y., Kidokoro, Y. 1989. Heparin and

heparan sulfate partially inhibit induction of acetylcholine receptor accumulation by nerve in Xenopus culture. *J. Neurosci.* 9: 1555–61

Hoffman, E. P., Brown, R. H., Kunkel, L. M. 1987. Dystrophin: The protein product of the Duchenne muscular dystrophy locus. *Cell* 51: 919–28

Horovitz, O., Knaack, D., Podleski, T. R., Salpeter, M. M. 1989. Acetylcholine receptor α-subunit mRNA is increased by ascorbic acid in cloned L5 muscle cells: northern blot analysis and in situ hybridization. *J. Cell Biol.* 108: 1823–32

Hunter, D. D., Porter, B. E., Bulock, J. W., Adams, S. P., Merlie, J. P., Sanes, J. R. 1989a. Primary sequence of a motor neuron-selective adhesive site in the synaptic basal lamina protein s-laminin. *Cell* 59: 905–13

Hunter, D. D., Shah, V., Merlie, J. P., Sanes, J. R. 1989b. A laminin-like adhesive protein concentrated in the synaptic cleft of the neuromuscular junction. *Nature* 338: 229–34

Jacob, M. H. 1991. Acetylcholine receptor expression in developing chick ciliary ganglion neurons. *J. Neurosci.* 11: 1701–12

Jacob, M. H., Berg, D. K., Lindstrom, J. M. 1984. Shared antigenic determinants between *Electrophorus* acetylcholine receptor and a synaptic component on chicken ciliary ganglion neurons. *Proc. Natl. Acad. Sci. USA* 81: 3223–27

Jasmin, B. J., Cartaud, A., Ludosky, M. A., Changeux, J.-P., Cartaud, J. 1990. Asymmetric distribution of dystrophin in developing and adult Torpedo marmorata electrocyte: evidence for its association with the acetylcholine receptor-rich membrane. *Proc. Natl. Acad. Sci. USA* 87: 3938–41

Joe, E.-H., Angelides, K. 1992. Clustering of voltage-dependent sodium channels on axons depends on schwann cell contact. *Nature* 356: 333–35

Jones, K. A., Baughman, R. W. 1991. Both NMDA and non-NMDA subtypes of glutamate receptors are concentrated at synapses on cerebral cortical neurons in culture. *Neuron* 7: 593–603

Kalcheim, C., Vogel, Z., Duskin, D. 1982. Embryonic brain extract induces collagen biosynthesis in cultured muscle cells: involvement in acetylcholine receptor aggregation. *Proc. Natl. Acad. Sci. USA* 79: 3077–81

Kennedy, M. B. 1989. Regulation of synaptic transmission in the central nervous system: long-term potentiation. *Cell* 59: 777–87

Kerr, L. M., Yoshikami, D. 1984. A venom peptide with a novel presynaptic blocking action. *Nature* 308: 282–84

Kidokoro, Y., Anderson, M. J., Gruener, R. 1980. Changes in synaptic potential properties during acetylcholine receptor accumulation and neurospecific interactions in Xenopus nerve-muscle cell cultures. *Dev. Biol.* 78: 464–83

Kim, Y., Neher, E. 1988. IgG from patients with Lambert-Eaton syndrome blocks voltage-dependent calcium channels. *Science* 239: 405–8

Kirsch, J., Langosch, D., Prior, P., Littauer, U. Z., Schmitt, B., Betz, H. 1991. The 93-kDa glycine receptor-associated protein binds to tubulin. *J. Biol. Chem.* 266: 22242–45

Knaack, D., Podleski, T. R. 1985. Ascorbic acid mediates acetylcholine receptor increase induced by brain extract in L5 myogenic cells. *Proc. Natl. Acad. Sci. USA* 82: 575–79

Kordeli, E., Bennett, V. 1991. Distinct ankyrin isoforms at neuron cell bodies and nodes of Ranvier resolved using ankyrin R-deficient mice. *J. Cell Biol.* 114: 1243–60

Kordeli, E., Cartaud, J., Nghiem, H.-O., Devillers-Thiery, A., Changeux, J.-P. 1989. Asynchronous assembly of the acetylcholine receptor and of the 43 kD protein in the postsynaptic membrane of developing Torpedo marmorata electrocyte. *J. Cell Biol.* 108: 127–39

Kuffler, S. W. 1943. Specific excitability of the end-plane region in normal and denervated muscle. *J. Neurophysiol.* 6: 99–110

Kullberg, R. W., Lentz, T. L., Cohen, M. W. 1977. Development of the myotomal neuromuscular junction in Xenopus laevis: an electrophysiological and fine-structure study. *Dev. Biol.* 60: 101–29

Langosch, D., Becker, C.-M., Betz, H. 1990. The inhibitory glycine receptor: a ligand-gated chloride channel of the central nervous system. *Eur. J. Biochem.* 194: 1–8

LaRochelle, W. J., Froehner, S. C. 1986. Determination of the tissue distributions and relative concentrations of the postsynaptic 43 kDa protein and the acetylcholine receptor in Torpedo. *J. Biol. Chem.* 261: 5270–74

LaRochelle, W. J., Witzemann, V., Fiedler, W., Froehner, S. C. 1990. Developmental expression of the 43K and 58K postsynaptic membrane proteins and nicotinic acetylcholine receptors in Torpedo electrocytes. *J. Neurosci.* 10: 3460–67

Lo, M. M. S., Garland, P. B., Lamprecht, J., Barnard, E. A. 1980. Rotational mobility of the membrane-bound acetylcholine receptor of Torpedo electric organ measured by phosphorescence depolarization. *FEBS Lett.* 111: 407–12

Loring, R. H., Zigmond, R. E. 1987. Ultrastructural distribution of ^{125}I-toxin F binding sites on chick ciliary neurons: synaptic localization of a toxin that blocks ganglionic nicotinic receptors. *J. Neurosci.* 7: 2153–62

Lupa, M. T., Caldwell, J. H. 1991. Effect of agrin on distribution of acetylcholine receptors and sodium channels on adult skeletal muscle fibers in culture. *J. Cell Biol.* 115: 765–78

Lupa, M. T., Krezemien, D. M., Schaller, K. L., Caldwell, J. H. 1992. Aggregation of sodium channels during development and maturation of the neuromuscular junction. Submitted

Magill-Solc, C., McMahan, U. J. 1990. Synthesis and transport of agrin-like molecules in motor neurons. *J. Exp. Biol.* 153: 1–10

Magill-Solc, C., McMahan, U. J. 1988. Motor neurons contain agrin-like molecules. *J. Cell Biol.* 107: 1825–33

Marshall, L. M. 1981. Synaptic localization of alpha-bungarotoxin binding which blocks nicotinic transmission at frog sympathetic neurons. *Proc. Natl. Acad. Sci. USA* 78: 1948–52

Marshall, L. M., Sanes, J. R., McMahan, U. J. 1977. Reinnervation of original synaptic sites on muscle fiber basement membrane after disruption of the muscle cells. *Proc. Natl. Acad. Sci. USA* 74: 3073–77

Matthews-Bellinger, J., Salpeter, M. M. 1978. Distribution of acetylcholine receptors at frog neuromuscular junctions with a discussion of some physiological implications. *J. Physiol.* 279: 197–213

McMahan, U. J. 1990. The agrin hypothesis. *Cold Spring Harbor Symp. Quant. Biol.* 55: 407–18

McMahan, U. J., Slater, C. R. 1984. Influence of basal lamina on the accumulation of acetylcholine receptors at synaptic sites in regenerating muscle. *J. Cell Biol.* 98: 1453–73

Mitra, A. A. K., McCarthy, M. P., Stroud, R. M. 1989. Three-dimensional structure of the nicotinic acetylcholine receptor, and location of the major associated 43-kD cytoskeletal protein, determined at 22 Å by low dose electron microscopy and x-ray diffraction to 12.5 Å. *J. Cell Biol.* 109: 755–74

Musil, L., Frail, D. E., Merlie, J. P. 1989. The mammalian 43 kD acetylcholine receptor-associated protein (RAPsyn) is expressed in some nonmuscle cells. *J. Cell Biol.* 108: 1833–40

Musil, L. S., Carr, C., Cohen, J. B., Merlie, J. P. 1988. Acetylcholine receptor-associated 43K protein contains covalently-bound myristate. *J. Cell Biol.* 107: 1113–21

Nastuk, M. A., Lieth, E., Ma, J., Cardasis,

C. A., Moynihan, E. B., et al. 1991. The putative agrin receptor binds ligand in a calcium-dependent manner and aggregates during agrin-induced acetylcholine receptor clustering. *Neuron* 7: 807–18

Neubig, R. R., Krodel, E. K., Boyd, N. D., Cohen, J. B. 1979. Acetylcholine and local anesthetic binding to Torpedo nicotinic post-synaptic membranes after removal of non-receptor peptides. *Proc. Natl. Acad. Sci. USA* 76: 690–94

New, H. V., Mudge, A. W. 1986. Calcitonin gene-related peptide regulates muscle acetylcholine receptor synthesis. *Nature* 323: 809–11

Nitkin, R. M., Smith, M. A., Magill, C., Fallon, J. R., Yao, Y. M., et al. 1987. Identification of agrin, a synaptic organizing protein from Torpedo electric organ. *J. Cell Biol.* 105: 2471–78

Ohlendieck, K., Ervasti, J. M., Matsumura, K., Kahl, S. D., Leveille, C. J., Campbell, K. P. 1991. Dystrophin-related protein is localized to neuromuscular junctions of adult skeletal muscle. *Neuron* 7: 499–508

Olek, A. J., Ling, A., Daniels, M. P. 1986. Development of ultrastructural specializations during the formation of acetylcholine receptor aggregates on cultured myotubes. *J. Neurosci.* 6: 487–97

Peng, H. B., Baker, L. P., Chen, Q. 1991. Induction of synaptic development in cultured muscle cells by basic fibroblast growth factor. *Neuron* 6: 237–46

Peng, H. B., Cheng, P.-C. 1982. Formation of postsynaptic specializations induced by latex beads in cultured muscle cells. *J. Neurosci.* 2: 1760–77

Peng, H. B., Froehner, S. C. 1985. Association of the postsynaptic 43K protein with newly formed acetylcholine receptor clusters in cultured muscle cells. *J. Cell Biol.* 100: 1698–1705

Phillips, W. D., Kopta, C., Blount, P., Gardner, P. D., Steinbach, J. H., Merlie, J. P. 1991a. ACh receptor-rich domains organized in fibroblasts by recombinant 43-kilodalton protein. *Science* 251: 568–70

Phillips, W. D., Maimone, M. M., Merlie, J. P. 1991b. Mutagenesis of the 43 kD postsynaptic protein defines domains involved in plasma membrane targeting and AChR clustering. *J. Cell Biol.* 115: 1713–23

Porter, S., Froehner, S. C. 1983. Characterization and localization of the M_r 43,000 proteins associated with acetylcholine receptor-rich membranes. *J. Biol. Chem.* 258: 10034–40

Prior, P., Schmitt, B., Grenningloh, G., Pribilla, I., Multhaup, G., et al. 1992. Primary structure and alternative splice variants of gephyrin, a putative glycine receptor-tubulin linker protein. *Neuron* 8: 1161–70

Pumplin, D. W., Reese, T. S., Llinas, R. 1981. Are the presynaptic membrane particles the calcium channels? *Proc. Natl. Acad. Sci. USA* 78: 7210–13

Qu, Z., Moritz, E., Huganir, R. L. 1990. Regulation of tyrosine phosphorylation of the nicotinic acetylcholine receptor at the rat neuromuscular junction. *Neuron* 4: 367–78

Reist, N., Werle, M. J., McMahan, U. J. 1992. Agrin released by motor neurons induces the aggregation of AChRs at neuromuscular junctions. *Neuron* 8: 865–68

Rich, M. M., Lichtman, J. W. 1989. In vivo visualization of pre- and postsynaptic changes during synapse elimination in reinnervated mouse muscle. *J. Neurosci.* 9: 1781–1805

Robitaille, R., Adler, E. M., Charlton, M. P. 1990. Strategic location of calcium channels at transmitter release sites of frog neuromuscular synapses. *Neuron* 5: 773–79

Role, L. W., Matossian, V. R., O'Brien, R. J., Fischbach, G. D. 1985. On the mechanism of acetylcholine receptor accumulation at newly formed synapses on chick myotubes. *J. Neurosci.* 5: 2197–2204

Ross, A., Rapuano, M., Prives, J. 1988. Induction of phosphorylation and cell surface redistribution of acetylcholine receptors by phorbol ester and carbamylcholine in cultured chick muscle cells. *J. Cell Biol.* 107: 1139–45

Ruegg, M. A., Tsim, K. W. K., Horton, S. E., Kroger, S., Escher, G., et al. 1992. The agrin gene codes for a family of basal lamina proteins that differ in function and distribution. *Neuron* 8: 691–99

Rupp, F., Payan, D. G., Magill-Solc, C., Cowan, D. M., Scheller, R. H. 1991. Structure and expression of a rat agrin. *Neuron* 6: 811–23

Sakakibara, H., Engel, A. G., Lambert, E. H. 1977. Duchenne dystrophy: ultrastructural localization of the acetylcholine receptor and intracellular microelectrode studies of neuromuscular transmission. *Neurology* 27: 741–45

Salpeter, M. M., Spanton, S., Holley, K., Podleski, T. R. 1982. Brain extract causes acetylcholine receptor redistribution which mimics some early events at developing neuromuscular junctions. *J. Cell Biol.* 93: 417

Sanes, J. R. 1983. Roles of extracellular matrix in neural development. *Annu. Rev. Physiol.* 45: 581–600

Sanes, J. R. 1982. Laminin, fibronectin, and collagen in synaptic and extrasynaptic portions of muscle fiber basement membrane. *J. Cell Biol.* 93: 442–51

Sanes, J. R., Hall, Z. W. 1979. Antibodies that bind specifically to synaptic sites on muscle fiber basal lamina. *J. Cell Biol.* 83: 357–70

Sanes, J. R., Marshall, L. M., McMahan, U. J. 1978. Reinnervation of muscle fiber basal lamina after removal of myofibers. Differentiation of regenerating axons at original synaptic sites. *J. Cell Biol.* 78: 176–98

Sargent, P. B., Pang, D. 1989. Acetylcholine receptor-like molecules are found in both synaptic and extrasynaptic clusters on the surface of neurons in the frog cardiac ganglion. *J. Neurosci.* 9: 1062–72

Sargent, P. B., Pang, D. Z. 1988. Denervation alters the size, number and distribution of clusters of acetylcholine receptor-like molecules on frog cardiac ganglion neurons. *Neuron* 1: 877–86

Schmitt, B., Knaus, P., Becker, C.-M., Betz, H. 1987. The M_r 93,000 polypeptide of the postsynaptic glycine receptor is a peripheral membrane protein. *Biochemistry* 26: 805–11

Sealock, R., Wray, B. E., Froehner, S. C. 1984. Ultrastructural localization of the M_r 43,000 protein and the acetylcholine receptor in Torpedo postsynaptic membranes using monoclonal antibodies. *J. Cell Biol.* 98: 2239–44

Sealock, R., Butler, M. H., Kramarcy, N. R., Gao, K.-X., Murnane, A. A., et al. 1991. Localization of dystrophin relative to acetylcholine receptor domains in electric tissue and adult and cultured skeletal muscle. *J. Cell Biol.* 113: 1133–44

Seitanidou, T., Triller, A., Korn, H. 1988. Distribution of glycine receptors on the membrane of a central neuron: an immunoelectron microscopy study. *J. Neurosci.* 8: 4319–33

Silberstein, L., Inestrosa, N. C., Hall, Z. W. 1982. Aneural muscle cell cultures make synaptic basal lamina components. *Nature* 295: 143–45

Smith, M. A., Magill-Solc, C., Rupp, F., Yao, Y.-M. M., Schilling, J. W., et al. 1992. Isolation and characterization of a cDNA that encodes an agrin homolog in the marine ray. *J. Mol. Cell. Neurosci.* In press

Sobel, A., Weber, M., Changeux, J.-P. 1977. Large scale purification of the acetylcholine receptor protein in its membrane-bound and detergent-extracted forms from Torpedo marmorata electric organ. *Eur. J. Biochem.* 80: 215–44

Srinivasan, Y., Elmer, L., Davis, J., Bennett, V., Angelides, K. 1988. Ankyrin and spectrin associate with voltage-dependent sodium channels in brain. *Nature* 333: 177–80

Steinbach, J. H. 1981. Developmental changes in acetylcholine receptor aggregates at rat skeletal neuromuscular junctions. *Dev. Biol.* 84: 267–76

Stollberg, J., Fraser, S. E. 1990a. Acetylcholine receptor clustering is triggered by a change in the density of a nonreceptor molecule. *J. Cell Biol.* 111: 2029–39

Stollberg, J., Fraser, S. E. 1990b. Local accumulation of acetylcholine receptors is neither necessary nor sufficient to induce cluster formation. *J. Neurosci.* 10: 247–55

Toyoshima, C., Unwin, N. 1988. Ion channel of acetylcholine receptor reconstructed from images of postsynaptic membranes. *Nature* 336: 247–50

Triller, A., Cluzeaud, F., Pfeiffer, F., Betz, H., Korn, H. 1985. Distribution of glycine receptors at central synapses: an immunoelectron microscopy study. *J. Cell Biol.* 101: 683–88

Tsim, K. W. K., Ruegg, M. A., Escher, G., Kroger, S., McMahan, U. J. 1992. cDNA that encodes active agrin. *Neuron* 8: 677–89

Usdin, T. B., Fishbach, G. D. 1986. Purification and characterization of a polypeptide from chick brain that promotes the accumulation of acetylcholine receptors in chick myotubes. *J. Cell Biol.* 103: 493–507

Vincent, A., Long, A. B., Newsome-Davis, J. 1989. Autoimmunity to the calcium channel underlies the Lambert-Eaton myasthenic syndrome, paraneoplastic disorder. *Trends Neurosci.* 12: 496–502

Wallace, B. G. 1990. Inhibition of agrin-induced acetylchloine receptor aggregation by heparin, heparan sulfate and other polyanions. *J. Neurosci.* 10: 3576–82

Wallace, B. G. 1989. Agrin-induced specializations contain cytoplasmic, membrane, and extracellular matrix-associated components of the postsynaptic apparatus. *J. Neurosci.* 9: 1294–1302

Wallace, B. G. 1988. Regulation of agrin-induced acetylcholine receptor aggregation by calcium and phorbol ester. *J. Cell Biol.* 107: 267–78

Wallace, B. G., Zhican, Q., Huganir, R. L. 1991. Agrin induces phosphorylation of the nicotinic acetylchloine receptor. *Neuron* 6: 869–78

Yeadon, J. E., Lin, H., Dyer, S. M., Burden, S. J. 1991. Dystrophin is a component of the subsynaptic membrane. *J. Cell Biol.* 115: 1069–76

Annu. Rev. Neurosci. 1993. 16:369–402

HOW PARALLEL ARE THE PRIMATE VISUAL PATHWAYS?

W. H. Merigan[1] *and J. H. R. Maunsell*[2,3]

Center for Visual Science and Departments of [1]Ophthalmology and [2]Physiology, University of Rochester, Rochester, New York 14642

KEY WORDS: vision, cortex, parallel pathways, LGN, macaque monkey

INTRODUCTION

The visual system, like all sensory systems, contains parallel pathways (see Stone 1983). Recently, much emphasis has been placed on the relationship between two subcortical and two cortical pathways. It has been suggested that the cortical and subcortical pathways are continuous, so that distinct channels of information that arise in the retina remain segregated up to the highest levels of visual cortex. According to this view, the visual system comprises two largely independent subsystems that mediate different classes of visual behaviors. In this paper, we evaluate this proposal, which has far-reaching implications for our understanding of the functional organization of the visual system.

The subcortical projection from the retina to cerebral cortex is strongly dominated by the two pathways (M and P pathways) that are relayed by the magnocellular and parvocellular subdivisions of the lateral geniculate nucleus (LGN) (see Shapley & Perry 1986). The importance of these pathways is demonstrated by the fact that they include about 90% of the axons that leave the retinas (Silveira & Perry 1991) and that little vision survives when both pathways are destroyed (Schiller et al 1990a). The P and M pathways maintain their sharp anatomical segregation through the termination of the LGN projection in layer 4C of V1 (striate cortex).

The complex network of connections in primate extrastriate visual cor-

---

[3] Present address: Division of Neuroscience, Baylor College of Medicine, Houston, Texas 77030.

0147–006X/93/0301–0369$02.00

tex has also been described as dominated by two pathways. One pathway includes areas in parietal cortex and is thought to be important for assessing spatial relationships and object motion; the other includes visual areas in temporal cortex and is thought to be more involved in visual identification of colors, patterns, or objects (Ungerleider & Mishkin 1982). Differences between the parietal and temporal pathways can be seen in lesion-induced deficits, in neuronal response properties, and in anatomical connections (see Desimone & Ungerleider 1989; DeYoe et al 1990; Mishkin et al 1983; Ungerleider & Mishkin 1982; Van Essen & Maunsell 1983).

Cumulative anatomical, physiological, and behavioral evidence has suggested a relationship between these subcortical and cortical pathways, eventually leading to explicit proposals of a direct correspondence between them (Livingstone & Hubel 1987; Maunsell 1987). The major components of the parallel pathways involved in this proposal and their interconnections are shown schematically in Figure 1. It has been suggested that the contributions of the M and P pathways remain largely segregated in visual cortex, and each connects to one of the cortical pathways, with the M pathway and the parietal pathway forming one subsystem, and the P pathway and the temporal pathway forming the other. It has further been proposed that several specific visual functions, such as motion, stereopsis, and figure/ground discrimination, could each be attributed to a specific subsystem.

The notion of parallel visual subsystems has been broadly disseminated and popularized (e.g. Kandel et al 1991; Livingstone 1988), and has quickly become widely accepted, owing in part to its great explanatory power and its appealing simplicity. The idea is consistent with an extensive collection of observations, which has been reviewed in detail many times (Desimone & Ungerleider 1989; DeYoe & Van Essen 1988; Felleman & Van Essen 1991; Goodale & Milner 1992; Kaas & Garraghty 1991; Livingstone & Hubel 1987, 1988; Martin 1988; Maunsell 1987; Maunsell & Newsome 1987; Van Essen et al 1992; Zrenner et al 1990). However, a growing number of reports has called the idea of visual subsystems into question. The cortical pathways show appreciable anatomical cross-talk (Felleman & Van Essen 1991; Van Essen et al 1992), some of which is illustrated in Figure 1. Neurophysiological studies have demonstrated functional intermixing (Malpeli et al 1981; Nealey et al 1991), and behavioral studies have contradicted some of the proposed functional segregation (Schiller et al 1990a). Consequently, those outside the immediate field have found it increasingly difficult to know whether it is appropriate to consider the visual system as made up of subsystems.

This question cannot be answered adequately with a simple yes or no. A careful assessment of the available evidence suggests that the con-

Parietal
Pathway

Temporal
Pathway

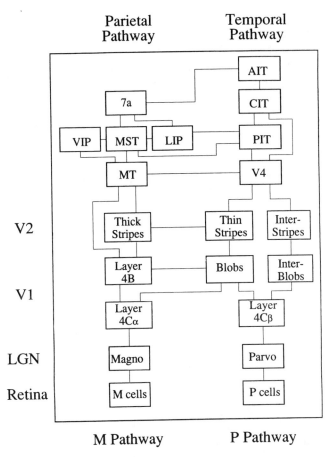

M Pathway   P Pathway

Figure 1 Parallel pathways in the primate visual system. The visual system is shown in schematic form from the retinal ganglion cells (*bottom*) to the higher levels of visual cerebral cortex (*top*). The components of the magnocellular and parietal pathways have been grouped to the left; those of the parvocellular and temporal pathways have been grouped to the right. Lines show established connections between the illustrated components. As in other summaries of visual pathways, many cortical areas and connections have been omitted. Abbreviations: AIT, anterior inferotemporal area; CIT, central inferotemporal area; LIP, lateral intraparietal area; Magno, magnocellular layers of the LGN; MST, medial superior temporal area; MT, middle temporal area; Parvo, parvocellular layers of the LGN; PIT, posterior inferotemporal area; VIP, ventral intraparietal area.

tributions of the M and P subcortical pathways retain some degree of segregation in visual cortex, but that their separation is far from complete and may not justify viewing the components as subsystems. This review elucidates the degree of parallel organization in the visual system through a critical evaluation of the anatomical, physiological, and behavioral evidence for and against parallel visual subsystems. Because the independence of the proposed visual subsystems is contingent on the independence of the M pathway from the P pathway and of the parietal pathway from the temporal pathway, we first consider the degree of segregation of the M and P pathways up to the level of layer 4C in V1. Segregation between the parietal and temporal pathways is then addressed, starting from the level of area V4 and the middle temporal visual area (MT). Finally we address the question of the linkage between the subcortical and cortical pathways, specifically the extent to which they can be considered to correspond. We restrict our discussion to the primate visual system, principally that of the macaque monkey.

THE SEGREGATION OF THE M AND P PATHWAYS

Neurons in the P and M pathways have been extensively studied (see Derrington & Lennie 1984; Ingling & Martinez-Uriegas 1983; Lee et al 1989; Purpura et al 1988; Shapley & Perry 1986). We focus here on observations that bear on the independence of the pathways.

Anatomical Evidence

A striking feature of the P and M pathways is the apparently strict anatomical independence they maintain. Several studies have looked for anatomical evidence of mixing between P and M pathways in the lateral geniculate (Conley & Fitzpatrick 1989; Michael 1988), but have found axonal arbors and dendritic fields in parvocellular and magnocellular layers to be separated. Likewise, the projections from parvocellular layers terminate primarily in V1 layers 4A and 4Cβ, whereas those from magnocellular geniculate terminate in layer 4Cα (Fitzpatrick et al 1985). Both subdivisions of the LGN have connections with layer 6 of VI; however, the cortical projections to the magnocellular and parvocellular layers appear to arise from separate subdivisions within that layer (Fitzpatrick & Einstein 1989; Lund & Boothe 1975). Thus, the anatomical segregation of the P and M pathways appears to be essentially complete.

Physiological Evidence

The physiological responses of cells in the P and M pathways are virtually identical on some dimensions and profoundly different on others. The most striking difference is sensitivity to color. P pathway neurons show color opponency of either the red/green or blue/yellow type, which means that they respond to color change regardless of the relative luminance of the colors (Derrington & Lennie 1984). M neurons, on the other hand, are considered insensitive to color, because they have virtually no response to color alternation when the luminance of the color is balanced. However, even when luminances are balanced (i.e. at isoluminance), M cells show a nonselective response to color transitions (a nonsigned, frequency-doubled response) (Lee et al 1989; Schiller & Colby 1983) that might be used for detecting temporal change.

Cells in the P and M pathways also differ greatly in the time course of their response to visual stimuli. In response to step changes in illumination, P cells show a more tonic response than M cells (Purpura et al 1990; Schiller & Malpeli 1978). P and M pathway neurons are also different in conduction velocity, with the stouter M pathway cells conducting impulses more rapidly (Gouras 1969; Kaplan & Shapley 1982; Schiller & Malpeli 1978). It is not clear if this difference has functional significance, given that total transmission times between retina and visual cortex differ by only a few milliseconds (Lennie et al 1990; Sherman et al 1984). Another physiological difference between P and M cells is their sensitivity to stimulus contrast. The contrast sensitivity of M cells is typically many times that of P cells. P cells rarely respond well to luminance contrasts below 10%, whereas M cells often respond to stimuli with contrasts as low as 2% (Purpura et al 1988; Sclar et al 1990; Shapley et al 1981). This difference in sensitivity might stem from M cells having larger receptive fields (de Monasterio & Gouras 1975), which would indeed result in greater sensitivity (Lennie et al 1990). However, recent studies do not show a marked difference between P and M cells in receptive field center size (Blakemore & Vital-Durand 1986; Crook et al 1988; Derrington & Lennie 1984), a finding that is certainly surprising given the large difference in the size of the dendritic fields of P and M retinal ganglion cells (Perry & Cowey 1985; Rodieck et al 1985) and their great difference in contrast sensitivity.

Along other stimulus dimensions, including spatial response, temporal response, and luminance response, neurons in the P and M pathways have different, but largely overlapping, ranges of sensitivity. M pathway cells are often reported to be responsive to higher temporal and lower spatial frequencies than P cells (Derrington & Lennie 1984; Hicks et al 1983). However, this difference is small, perhaps 15% in peak temporal frequency, cutoff temporal frequency, and peak spatial frequency, and there is a substantial overlap along all of these dimensions between the two classes of cells (Blanckensee 1980; Sherman et al 1984). There also appears to be a large overlap in the range of luminances over which neurons in P and M pathways respond. M pathway cells responded to somewhat lower luminance levels (Purpura et al 1990), but neurons in both pathways respond at rod mediated (scotopic) light levels (Virsu & Lee 1983; Wiesel & Hubel 1966). The possibility that the M pathway may dominate vision under scotopic conditions remains controversial (see Purpura et al 1990).

Some response properties of the P and M pathways show little difference. The most prominent is spatial resolution. As with other properties described above, spatial resolution in both the P and M pathways varies with eccentricity from the fovea and differs in temporal and nasal portions of the visual field. But, at a given eccentricity, neurons in both pathways have virtually the same spatial resolution (Crook et al 1988). This is surprising, because early studies described an approximately threefold difference in the size of receptive fields centers in P and M cells (deMonasterio & Gouras 1975), and spatial resolution is inversely related to receptive field center size among cells that show linear spatial summation, which includes virtually all P and M cells (Shapley et al 1981). However, more recent measurements and calculations have shown that there is little difference in the size of receptive field centers in the two pathways (Blakemore & Vital-Durand 1986; Crook et al 1988; Derrington & Lennie 1984).

Thus, P and M cells in retina and LGN are qualitatively different on only a few dimensions—color opponency, time course of response, and contrast gain. Along some other important dimensions, such as luminance, temporal frequency and spatial frequency (Shapley & Lennie 1985), they each cover a wide range of values, with differences in their mean response, but a large overlap in effective stimuli. Moreover, they appear to show almost no difference in spatial resolution or receptive field center size at any given eccentricity. Thus, although anatomy indicates clear segregation between P and M pathways, physiological properties reveal both differences and similarities. In the following section, we consider behavioral evidence that addresses the functional significance of the above physiological properties.

Behavioral Evidence from Lesions of M or P Pathways

In the past few years, techniques to create selective, localized lesions in the P and M pathways have been developed. These include the use of excitotoxins, such as ibotenic acid, that damage cell bodies, but spare fibers of passage (Schwarcz et al 1979), and acrylamide, which selectively lesions the P pathway (Lynch et al 1992). When combined with careful behavioral measures of visual capabilities, these approaches provide valuable insights into the functional specialization of the P and M pathways. The results of these lesion studies can easily be related, a posteriori, to the known anatomy and physiology of P and M pathway cells. However, we believe that these lesion results could not be fully predicted from current anatomical and physiological knowledge, given the large variety of predictions that would be consistent with this knowledge.

EFFECTS OF M PATHWAY LESIONS   M pathway lesions cause a large decrease in luminance contrast sensitivity for stimuli of higher temporal frequency and lower spatial frequency (Merigan et al 1991a). Figure 2B, which shows sensitivity to temporal frequency measured with a low spatial frequency, illustrates this loss. Loss is restricted to stimuli that include both high temporal and low spatial frequencies. There is no loss of temporal frequency sensitivity at high spatial frequencies, nor loss of sensitivity to high spatial frequency at low temporal frequencies. (This explains why no loss in low spatial frequency sensitivity is seen in Figure 2A, which plots sensitivities measured with a low temporal frequency). The reduction of sensitivity to high temporal and low spatial frequencies results in reduced visibility of rapidly moving or rapidly flickering stimuli. This result is consistent with the greater sensitivity of M pathway neurons at high temporal and low spatial frequencies described above. M lesions cause almost no change in flicker resolution for high contrast stimuli (Merigan

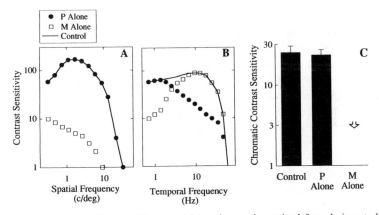

Figure 2   Contrast sensitivity of the P and M pathways determined from lesion studies. (*A*) Spatial contrast sensitivity for stationary gratings (zero temporal frequency). Contrast sensitivity is the inverse of the lowest stimulus contrast that can be detected. The solid line shows sensitivity of the intact monkey; filled circles show the contribution of the P pathway (after M lesions); and open squares the contribution of the M pathway (after P lesions). (*B*) Temporal contrast sensitivity measured at a low spatial frequency. Conventions are the same as in *A*. Labels in panels *A* and *B* should be taken as only relative, because these functions vary in spatial and temporal frequency with eccentricity. (*C*) Color contrast sensitivity in control, as well as after M or P pathway lesions. The arrow indicates the highest chromatic contrast that could be measured. Animals with P pathway lesions could not detect the target as this level. Adapted from Merigan et al (1991a,b).

& Maunsell 1990), although Schiller et al (1990a,b) have reported that such lesions reduce flicker resolution thresholds to about 9 Hz. The difference may be due to lower effctive contrast of the test stimuli in the experiments of Schiller et al (1990a,b). This explanation is illustrated in Figure 2B: The P and M pathways differ little in temporal resolution (highest temporal frequency that can be seen) at high contrasts (small y values), but differ greatly at lower contrasts (large y values). Additional measures indicate that M lesions cause no changes in either visual acuity or color contrast sensitivity (Merigan et al 1991a,b), which suggests little M pathway involvement in these functions. Effects of M pathway lesions on stereo vision and motion perception (Merigan et al 1991a; Schiller et al 1990a) are discussed below in the analysis of possible relationships of P and M to the cortical pathways.

EFFECTS OF P PATHWAY LESIONS   P pathway lesions cause complementary effects to those of M lesions, thus reducing luminance contrast sensitivity for stimuli of higher spatial and lower temporal frequency content (Merigan et al 1991a,b; Merigan & Eskin 1986). Figure 2A, which plots sensitivities measured with a low temporal frequency, shows this effect. Thus, although physiological recordings show that individual P cells do not respond well to contrasts of less than 10% (Shapley et al 1981), these lesion studies indicate that the P pathway can mediate the detection of contrasts as low as 0.5% at low temporal frequencies (Merigan & Eskin 1986), which suggests a large contribution of spatial and probability summation to this behavioral response (Watson 1992). The role of the P pathway in detecting high spatial frequencies is also reflected in the approximately fourfold decrease in visual acuity that follows P lesions (Merigan et al 1991b). These behavioral findings match the superior low temporal frequency response of P cells (Purpura et al 1990), but are clearly not determined by the physiological spatial resolution of P and M cells, which are nearly identical (see above). The superior acuity mediated by the P pathway appears to be caused by the greater sampling density of retinal P ganglion cells (Merigan & Katz 1990), which follows from the approximately 8:1 ratio of P to M retinal ganglion cells (Perry et al 1984; Silveira & Perry 1991).

Perhaps the most dramatic effect of P pathway lesions is an apparently complete loss of color vision (Figure 2C) (Merigan 1989; Merigan et al 1991b; Schiller et al 1990a). This result is consistent with the color opponency of P cells and suggests that the residual frequency-doubled response of M cells at isoluminance (Lee et al 1989; Schiller & Colby 1983) is not sufficient to mediate even the most primitive aspects of color vision, such as detection of chromatic gratings. It is worth noting, in light of the higher luminance contrast sensitivity of the P pathway than of the physiological response of individual P cells, that the color sensitivity mediated by the P pathway is also substantially higher than that of individual P pathway neurons (Derrington & Lennie 1984; Merigan 1989). The effects of P pathway lesions on pattern, texture, and stereoscopic vision (Schiller et al 1990a,b), and on form discrimination (Lynch et al 1992) are discussed later in the context of the relationships of the subcortical and cortical pathways.

The results of these lesion studies show clear differences in the contributions of the P and M pathways to vision. For the most part, these distinctions correspond well with the response properties of their neurons. For example, color vision appears to be subserved by the P pathway. However, these results also point to the risks of predicting behavioral contributions based on physiology. Although M pathway neurons have much higher contrast sensitivity than P cells, M lesions do not affect behavioral contrast sensitivity at low temporal frequencies (Figure 2A). The relatively poor contrast sensitivity of the P pathway appears to be overcome by the greater number of P cells. Likewise, the higher spatial resolution of the P pathway appears to reflect the higher sampling density of the more numerous P cells (Merigan et al 1991b), as P and M pathway cells do not differ in their physiological spatial resolution.

In summary, P and M pathways are anatomically and functionally distinct, but their basic specializations appear to be for low-level properties, such as spatial and temporal frequency. In the next section, we consider specialization of the cortical pathways and subsequently ask whether there are special links between the subcortical and cortical pathways.

THE SEGREGATION OF THE PARIETAL AND TEMPORAL PATHWAYS

Lesions of the Parietal or Temporal Pathways

Although functionally independent visual pathways have been discussed for many years in the context of parallel relationships between cortical and subcortical structures (Schneider 1967; Trevarthen 1968), Ungerleider & Mishkin (1982) proposed that distinct pathways exist within visual cortex itself. The behavioral data relevant to this proposal depended on differences in lesion-induced deficits that follow temporal and parietal lesions. A long-standing clinical literature had shown that lesions in human parietal cortex can cause extreme hemifield neglect and a disruption of visuomotor orientation. When the site of the lesion is temporal cortex, patients frequently have difficulty with form discrimination (agnosia) and problems with visual memory. These syndromes were so distinct that neurologists concluded that the affected regions of the brain were specialized for spatial representation and object recognition, respectively (see Grüsser & Landis 1991).

The behavioral data from monkeys depended on testing animals with selective lesions of temporal or parietal cortex. Pohl (1973) reported one of the very few studies that involved a direct comparison of the effects of parietal and temporal lesion. He tested parietal and temporal lesions by using two tasks in which the animal had to locate which of two food wells contained a reward. In the "landmark" task, an object was placed next to the loaded food well. In the "object discrimination" task, an object was placed near each well, and the animal learned which one marked the loaded well. Groups of monkeys with temporal or parietal lesions were tested on both tasks. Animals with either lesion learned to do both tasks. They differed only in the rate at which they learned the tasks or, in some cases, the rate at which they relearned the tasks after the landmark was switched to the empty well, or the reward-contingent object was switched in the object discrimination task. With the landmark task, there was no difference between parietal and temporal lesions in initial learning. However, monkeys with parietal cortex lesions were slower at learning reversals than were those with temporal lesions. On the object discrimination task, monkeys with temporal lobe lesions made more errors during initial learning, and this difference persisted over several reversals. In a separate group of animals, Pohl tested post-lesion relearning, rather than initial learning, with temporal or parietal lesions after pretraining on a more difficult landmark task. On this task, there was also a clear difference between groups, with the parietal lesioned monkeys showing many more errors.

Other studies that have compared parietal and temporal lesions have not replicated these effects. Ungerleider & Brody (1977) tested acquisition of the landmark task after parietal or temporal lobe lesions and found a greater disruption during reversal learning in monkeys with temporal lesions, a finding at odds with Pohl (1973). Other groups found no deficit in landmark performance after posterior parietal lesions (Petrides & Iversen 1979; Ridley & Ettlinger 1975). These discrepancies might be explained by differences in the methods of testing or in precise placement of the lesions, but they indicate some vagaries of the behavioral observations.

Numerous studies have subsequently tested temporal and parietal lesions with a landmark task (often testing relearning) or an object discrimination task (often testing initial learning). Most of these studies have focused on questions specific to one or the other pathway, such as which regions of parietal cortex are most critical for landmark task performance (e.g. Mishkin et al 1982) and have not attempted further dissociation of temporal and parietal lesion effects. Collectively, the results of these studies give the impression of dissociable effects of parietal and temporal lesions (see Mishkin & Ungerleider 1982); however, the different methods of measurement and analysis make comparisons between studies difficult to interpret.

An additional cause for concern in evaluating comparisons of cortical pathway lesions is that posterior parietal lesions often damage the optic radiations, as indicated by degeneration in the LGN. The visual field defects resulting from such collateral damage complicate the interpretation of these extrastriate lesions. Note that visual field scotomas following medial striate cortex lesions can produce larger effects on the landmark than on the object discrimination task (Mishkin 1966; Ungerleider & Mishkin 1982). Thus, a comparison of parietal and temporal lesions on these two tasks without the added complexity of visual field loss would be most helpful.

Overall, there are few studies that have attempted a double dissociation of the effects of temporal and parietal cortex lesions. Because questions remain regarding reliability of differential effects, the available studies provide only weak support for the segregation of parietal and temporal pathways. Other experiments have studied parietal or temporal lesions separately by using behaviors specifically chosen for the pathway being studied. These studies were generally not designed to distinguish cortical pathways and they provide little evidence to support or refute the notion of segregated pathways, because they typically have not compared the effects of different lesions on the same behaviors.

Included in this category are studies that have examined motion perception and eye movements following parietal pathway lesions made in areas MT, MST, or parietal cortex. Lesions of MT produce severe deficits in motion perception (Newsome & Paré 1988) and eye movements (Dürsteler & Wurtz 1988; Newsome et al 1985). However, if the lesions are not large, these results are transitory, with virtually complete recovery within days. Larger lesions, involving portions of MT and MST (Newsone & Paré 1988; Yamasaki & Wurtz 1991), produce more permanent disruptions of motion perception and both pursuit and saccadic eye movements. Complete, bilateral MT-MST lesions (Pasternak et al 1991) cause persistent disruptions of several aspects of motion perception, including speed thresholds, direction thresholds, and directionally noisy global motion.

Lesions higher in the parietal pathway, in posterior parietal cortex, cause a diffuse syndrome, which is more severe the larger the lesion (Lynch 1980). Unlike humans with parietal lesions, who show profound neglect of the field contralateral to the lesion, monkeys show a relative neglect, termed extinction, when two stimuli are presented simultaneously in ipsilateral and contralateral fields. Eye movements are affected by unilateral lesions, with reduced slow phase OKN and increases in saccadic latency. Profound effects on pursuit eye movements are seen after bilateral lesions. Finally, there is some evidence of spatial disorientation, including disrupted maze performance and difficulty navigating in a familiar room (Sugishita et al 1978).

The function of the temporal pathway has been tested with lesions placed either lower in the pathway, in cortical area V4, or in inferotemporal cortex. Numerous studies have examined the effects of inferotemporal cortex lesions and found that such lesions greatly increase the number of errors made during initial learning of visual discriminations, as well as in recall of previously learned discriminations. Such effects are found for tests along single dimensions (e.g. size, color, luminance), as well as for complex (e.g. shape) discriminations. However, these effects disappear with a variety of manipulations. None are seen in young animals, when easy discriminations are used, for previously overlearned discriminations, or when errors are punished with shock (Gross 1973). Lesions of area V4 show small to moderate deficits in rather complex functions. Desimone et al (1990) found modest deficits in orientation, direction, color, and texture discriminations. Schiller & Lee (1991) reported an exaggeration of the normal finding that stimuli are more detectable from background stimuli if they are distinguished by possessing, rather than lacking, a salient feature (Treisman 1988). Heywood & Cowey (1987) tested color and shape discrimination after making complete bilateral lesions of V4. The relearning of both types of discrimination was disrupted, and tests of color discrimination suggested a small, but permanent, deficit.

In general, it appears widely accepted among investigators that parietal and temporal lesions produce different deficits, although there is little direct evidence bearing on this question. Most studies have not directly compared temporal versus parietal pathways, so we are forced to rely heavily on results from studies that have examined one or the other pathway. This approach is problematic for the purposes of this review, because the effects of parietal or temporal lesions are frequently small and/or transitory, which suggests that the tests used were not well matched to the role of the pathway, that the visual system involves more distributed processing than could be detected with such local lesions (DeYoe & Van Essen 1988), or that other areas rapidly assume functions interrupted by lesions (Newsome & Paré 1988). There is not, at present, sufficient evidence to permit a choice among these alternatives.

Physiology of the Parietal and Temporal Pathways

The cortical visual areas that make up the parietal and temporal pathways are distinguished by the response properties of their neurons. Neurons in the temporal pathway are relatively sensitive to color and form, whereas those in the parietal pathway are more sensitive to the movement of visual stimuli (see Desimone et al 1985; Desimone & Ungerleider 1989; DeYoe & Van Essen 1988; Maunsell 1987; Maunsell & Newsome 1987; Van Essen & Maunsell 1983). Differences in the emphasis on central versus peripheral parts of the visual field also exist (see Ungerleider & Mishkin 1982).

Some of the most striking response selectivities are seen in neurons in the highest levels of the parietal and temporal pathways, in inferior parietal and inferotemporal cortex. Many neurons in inferotemporal cortex are selective for colors, or complex patterns or shapes (occasionally faces or hands) (see Desimone 1991). Surveys of selectivity in inferotemporal cortex find that about one half to two thirds of visually responsive units show obvious selectivity of this sort (Desimone et al 1984; Tanaka et al 1991), although most of these units respond, to some extent, to any visual stimulus. In contrast, neurons in the later stages of the parietal pathway are selective for complex types of motion, such as expansion or rotation (Duffy & Wurtz 1991; Motter & Mountcastle 1981; Saito et al 1986; Sakata et al 1985; Snowden et al 1991; Tanaka et al 1986).

Unfortunately, it is difficult to compare the properties of inferotemporal and parietal neurons precisely based on the available data. As with the behavioral data, few studies have directly compared the response properties in the two regions. There has never been a detailed study of the selectivity of parietal neurons for stimulus form or color, nor have inferotemporal neurons been extensively tested for properties typically studied in parietal areas, such as direction selectivity. One neuronal property that has been examined in both regions is the effect of spatially directed attention in behaving animals, which usually has an opposite effect in the two regions. Spatial attention enhances responses of parietal neurons, but suppresses the responses of inferotemporal neurons (Bushnell et al 1981; Richmond et al 1983). The dearth of single-unit recordings that directly compare the higher levels of the parietal and temporal pathways can be taken as testimony to the confidence of physiologists about the differences between these regions of visual cortex. Nevertheless, the distinctions between parietal and temporal cortex remain documented primarily by incidental observations.

The direct comparisons that have been made between the parietal and temporal pathways mostly concern MT and V4. Figure 3 summarizes the selectivity of neurons in these areas for four of the best studied stimulus dimensions. There is a dramatic distinction in emphasis on color and direction. Numerous studies have shown that whereas MT lacks clear color selectivity (Maunsell & Van Essen 1983b; Movshon et al 1991; Saito et al 1989; Zeki 1974), many neurons in V4 are robustly color selective

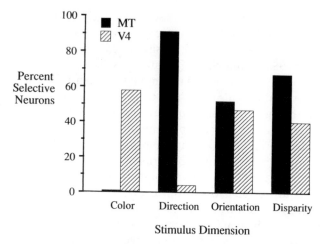

Stimulus Dimension

Figure 3 Stimulus selectivities in MT and V4. Proportions of neurons selective for four stimulus dimensions are plotted for both areas. The data are taken from numerous studies from several laboratories and include both quantitative and subjective assessments of selectivity. In particular, the differences between color and direction selectivity are based entirely on qualitative assessments. Some values are based on data from a single study. Adapted from Felleman & Van Essen (1987).

(Schein & Desimone 1990; Van Essen & Zeki 1978; Zeki 1973). A prevalence of color selectivity persists to the highest levels of the temporal pathway (Desimone et al 1984; Gross et al 1972; Komatsu et al 1992). Direction selectivity, which is also often considered to distinguish neurons in the cortical pathways, is abundant in MT (Albright 1984; Maunsell & Van Essen 1983b; Zeki 1974) and conspicuous in areas in the posterior parietal cortex (see Andersen 1987; Wurtz et al 1990), but rarely found in V4 (Van Essen & Zeki 1978; Zeki 1978). Other stimulus selectivities, such as orientation and disparity sensitivity, do not clearly distinguish the parietal and temporal pathways, even at the level of MT and V4. Selectivity for orientation, speed, binocular disparity, and contrast sensitivity all appear to be comparable in both areas (Cheng et al 1991; see Felleman & Van Essen 1987).

Functional distinctions between the parietal and temporal pathways have recently been demonstrated in humans by using positron-emission tomography (PET). Differential activation of parietal and temporal cortex occurs when subjects view moving or color stimuli (Zeki et al 1991) and when they perform a face-matching or a spatial vision task (Haxby et al 1991). Corbetta and colleagues (1991) compared cortical activation when subjects were required to attend to different aspects of a single stimulus. Subjects viewing moving colored bars were directed to attend to the speed, the shape, or the color of those bars. Attention to speed activated sites in the inferior parietal lobule and the superior temporal sulcus, whereas attention to color or shape activated sites that were closer to the inferior surface of the brain, although the separation of activity was not as clear as in the studies that used different stimuli.

In summary, the PET studies and single-unit recordings support the idea of a parallel division between the parietal and temporal pathways. These differences appear to fall along the lines of identification and recognition versus localization and spatial relationships, although our understanding of the division is far from complete. Also, the extent to which the neuronal properties of the two cortical pathways overlap remains to be established. Until more data become available, it will not be possible to describe physiological differences between the parietal and temporal pathways with precision and certainty.

Anatomy of the Parietal and Temporal Pathways

The distinction between the parietal and temporal pathways is also supported by segregated cortico-cortical connections. For example, the outputs of V4 are directed primarily to inferotemporal cortex, whereas those of MT lead for the most part to parietal cortex (Desimone et al 1980; Maunsell & Van Essen 1983a). The strongest anatomical evidence for parallel organization comes from studies of the extent of overlap in the areas projecting to parietal and temporal cortex that have been done in the same animal. Morel & Bullier (1990) and Baizer and colleagues (1991) injected different neuroanatomical tracers into inferior parietal (in and near LIP and VIP) and inferotemporal cortex (in CIT). Both studies found that although the injections labeled large portions of visual cortex, there was very little overlap in the distributions of the two labels.

This result may seem inconsistent with the numerous cross-connections that have been demonstrated between areas in the two cortical pathways. Some of these cross-connections can be seen in Figure 1, but many more exist. Figure 1, like almost all diagrams of the cortical pathways, shows a highly selected subset of areas and connections. The known connections between areas in macaque cerebral cortex now number over 300, and the complete set provides no clear impression of two parallel pathways (see Figure 4 in Felleman & Van Essen 1991). Indeed, far less segregation would probably be observed in the studies by Morel & Bullier and Baizer et al, had the injections been made at earlier or later levels in the pathways. For example, MT and V4 are reciprocally connected (Maunsell & Van Essen 1983a; Ungerleider & Desimone 1986). However, cross-connections at early and late levels do not preclude anatomical segregation over substantial portions of the parietal and temporal pathways.

The absence of obvious parallel pathways in the complete set of cortical connections is perhaps not surprising. Although quantitative estimates of the number of axons in projections are difficult to obtain, it is a common observation that connections differ greatly in their strength and consistency (e.g. Tanaka et al 1990; see Van Essen 1985). These anatomical differences are presumably reflected in their relative influence on response properties in the recipient areas. "Wiring diagrams" that give equal weight to all pathways may obscure functional relationships. The connections between the two pathways might be weaker than those within pathways, although this question has never been addressed experimentally. Also, some connections might be primarily modulatory and have little influence on stimulus selectivities. For example, there is little identifiable evidence of the connection between MT and V4 in the response properties of their neurons.

In summary, the evidence from behavioral, physiological, and anatomical experiments makes a clear case for the distinctiveness of the parietal and temporal pathways. However, the strength of the observations varies greatly, and the segregation between the cortical pathways is not as striking as that between the M and P pathways. Very few studies have directly compared the parietal and temporal pathways. The anatomical experiments that compared inputs to parietal and inferotemporal cortex are a notable exception and provide some of the best support for distinct cortical pathways. Corroborating evidence is provided by differences in neuronal reponse properties, although direct comparisons of the pathways are lacking in most of these studies. Studies that have compared parietal and temporal lesions show differential effects on spatial and recognition tasks, but these effects are, for the most part, disappointingly weak and primarily involve differences in rates of learning. Because large regions of visual cortex are relatively unexplored, this picture of two parallel cortical pathways may still change.

LINKS BETWEEN THE CORTICAL AND SUBCORTICAL PATHWAYS

The issue of visual subsystems concerns the degree to which the subcortical pathways map onto the cortical pathways in a one-to-one fashion. In this section, we consider first the anatomical and physiological data that address whether contributions from the M and P pathways remain segregated in visual cortex up to the level of MT and V4. We then consider data that compare the neuronal properties and behavioral contributions of the cortical and subcortical pathways.

The Pathways from V1 to MT and V4

ANATOMY   One of the most important contributions to bridging the gap between the subcortical and cortical pathways was made in 1978 by Margaret Wong-Riley, who demonstrated cytochrome oxidase-rich blobs in the superficial layers of striate cortex (see Hendrickson 1985). The blobs and interblobs in V1 (Horton & Hubel 1981; Humphrey & Hendrickson 1980) and the thin stripes, thick stripes, and interstripes in V2 (Livingstone & Hubel 1982; Tootell et al 1983) create subdivisions that could support parallel segregation within these areas. Until the patterns of cytochrome oxidase staining were discovered in V1 and V2, their cytology had been thought to be uniform, and there was little evidence that they contained separate routes for different classes of visual information.

Studies of the connections of the cytochrome oxidase compartments suggest that they make up parallel routes through early visual cortex and that these routes connect to the parietal and temporal pathways in a selective way (Figure 1). Retrograde tracer studies demonstrated that MT receives V2 projections mainly from the thick stripes, whereas projections to V4 arise from both the thin stripes and the interstripes (DeYoe & Van Essen 1985; Shipp & Zeki 1985). The thin stripes and interstripes in V2 in turn interconnect with the blobs and interblobs in V1 (Livingstone & Hubel 1984a). The thick stripes receive input from layer 4B (at least in squirrel monkey) (Livingstone & Hubel 1987), which also sends a direct projection to MT (Lund et al 1976; Maunsell & Van Essen 1983a). Further anatomical evidence for differences between the cytochrome oxidase compartments is provided by the antibody Cat-301, which selectively labels layer 4B, the V2 thick stripes and MT (DeYoe et al 1990).

However, the separation between the compartments is not complete. Although many of the connections of the cytochrome oxidase compartments in V1 and V2 appear to maintain their independence (Livingstone & Hubel 1984b), other relevant connections do not respect their borders. Local circuit neurons make extensive cross-connections within V1 (Yoshioka & Lund 1990). In V2, direct connections exist between the thin and thick stripes (Livingstone & Hubel 1984a), and both are labeled when the pulvinar is injected (Livingstone & Hubel 1982). In owl monkey, thick and thin stripes are also labeled following injection of the dorsomedial visual area (Krubitzer & Kaas 1990). All three stripes receive feedback projections from area V4 (Zeki & Shipp 1989). These cortical and subcortical connections (as well as others not yet identified) could thwart the apparently sharp segregation laid out in other connections of the cytochrome oxidase compartments.

The idea that the M and P pathways correspond to the parietal and temporal pathways rests largely on the cytochrome oxidase compartments in V1 and V2 acting as conduits, thus making selective connections with the subcortical pathways in V1 and selective connections with the parietal and temporal pathways in MT and V4. Although the anatomical evidence for selective connections with MT and V4 is strong (DeYoe & Van Essen 1985; Shipp & Zeki 1985), the connections in V1 appear to maintain only a partial segregation between the contributions of the M and P pathways. Layer 4B receives direct input from $4C\alpha$ (M pathway), but not $4C\beta$ (P pathway) (Lund & Booth 1975; Lund et al 1979), consistent with M pathway dominance in the route that leads to MT and the parietal pathway. On the other hand, although $4C\beta$ projects to the blobs and interblobs in the superficial layers, the blobs in superficial layers also receive major inputs from the M pathway by way of layers 4B and $4C\alpha$ (Blasdel et al 1985; Fitzpatrick et al 1985; Lachica et al 1992; see also Lund 1988).

The anatomical data suggest a partial mixing of M and P pathway contributions in V1, but they are not conclusive about the degree of functional segregation that exists in the early stages of visual cortex. Collectively, interlaminar connections in V1 and the cross-connections between the cytochrome oxidase compartments in V1 and V2 could inter-mix M and P contributions completely before they reach the level of the MT and V4. Alternatively, they could have no appreciable effect, and connections that appear to mix the different pathways might provide only modulatory inputs. Other observations confirm the suggestion that both the M and P channels make substantial contributions to response properties in the superficial layers in V1 and to later stages in the temporal pathway.

PHYSIOLOGY   Segregation between the cytochrome oxidase compartments and their relationships to the M and P pathways have been examined by comparing response properties in the different cytochrome oxidase compartments with each other and with those seen in the M and P pathways. Such comparisons have revealed clear differences between the cytochrome oxidase compartments, consistent with a segregation of streams of infor-mation. However, the physiological segregation is not complete, and its relationship to the M and P pathways is not clear.

Segregation Between the Cytochrome Oxidase Compartments   Observa-tions on the distribution of direction selectivity suggest segregation of function between the cytochrome oxidase compartments. Direction selec-tivity is concentrated in layer 4B in V1 (Dow 1974; Hawken et al 1988; Hubel & Livingstone 1990) and in the thick stripes in V2 (De Yoe & Van Essen 1985; Levitt et al 1990). Moreover, both structures project to MT, which is distinguished from V4 by abundant direction selectivity (Maunsell

& Van Essen 1983b; Zeki 1978). Layer 4B and the V2 thick stripes also appear to be relatively enriched in neurons that are sensitive to binocular disparity (DeYoe & Van Essen 1985; Hubel & Livingstone 1987; Poggio 1984; Ts'o et al 1991).

The blobs contain neurons that are selective for color and relatively unselective for orientation, whereas the converse holds in the interblobs (Blasdel 1992a,b; Landisman et al 1991; Livingstone & Hubel 1987; Tootell et al 1988b; Ts'o et al 1990; Ts'o & Gilbert 1988). Corresponding properties are found in the thin stripes and interstripes in V2 (Hubel & Livingstone 1987; Ts'o et al 1990). Color-sensitive neurons are far less common in layer 4B (Dow 1974; Dow & Gouras 1973; Hubel & Livingstone 1990; Poggio et al 1975; Tootell et al 1988b).

However, the segregation of response properties is not complete. The data of Livingstone & Hubel (1987) suggest that about two thirds of neurons in the blobs are color sensitive, compared with one third in the interblobs. And, Lennie and colleagues (1990) found little evidence for differences in color sensitivity between the blobs and interblobs. Conflicting results have been reported with respect to the segregation of response properties in V2. Hubel & Livingstone (1987) described a remarkably clear-cut division of neurons into three groups: disparity-tuned neurons in thick stripes, color-selective in thin stripes, and end-stopped in interstripes. Other studies have reported only a tendency for thin stripe neurons to be color selective and for thick stripe neurons to be disparity tuned, with neurons in all stripes showing similar selectivities for spatial and temporal frequency (DeYoe & Van Essen 1985; Levitt et al 1990; Ts'o et al 1991). These other studies also found appreciable numbers of color-selective neurons in the interstripes and no sharp division between color-selective and disparity-tuned cells. Some of the discrepancy may arise from difficulties in distinguishing thin and thick stripes anatomically (Crawford & Chodosh 1990). Because the analyses in the latter studies were more thorough and based on objective response measurements and computer-controlled stimuli, it is likely that they provide a more accurate picture and that the segregation of response properties in V2 is far from complete.

Association with the M and P Pathways   The differential distribution of response properties among the subdivisions of V1 and V2 is consistent with the idea that they segregate classes of visual information. Some studies have directly compared response properties in the M and P pathways with those in the V1 and V2 subdivisions to assess the relationship between them. Although, in some cases, correlations have been put forth as evidence for maintained segregation, such comparisons prove difficult to interpret. An example of the difficulties can be seen in the case of contrast sensitivity. Neurons in V1 layer 4B and the V2 thick stripes have high contrast sensitivity (Blasdel & Fitzpatrick 1984; Hawken & Parker 1984; Hubel & Livingstone 1990; Tootell et al 1988a), which matches that in the M pathway. Yet, high contrast sensitivity may not require M pathway contributions. Although P pathway neurons have far less contrast sensitivity, they are not blind to low contrasts. Cortical processing that summed inputs from many relatively insensitive parvocellular neurons could produce a response as sensitive as any in the M pathway (Watson 1992). Summation of this sort has been demonstrated by the existence of cortical neurons that have greater contrast sensitivity than neurons in either subdivision of the LGN (Sclar et al 1990). This consideration leaves most observations about cortical response properties inconclusive regarding contributions from the M and P pathways, whether the observations suggest segregation or mixing. For example, contrast sensitivity comparable to that in the M pathway also exists in the V1 blobs and interblobs (Hawken et al 1988; Hubel & Livingstone 1990; Tootell et al 1988a). This sensitivity might reflect either M pathway contributions provided by the connections with layers 4B and 4Cα (Lachica et al 1992) or summation of P pathway inputs. Because the properties of the M and P pathways overlap so extensively, and summation of responses can increase sensitivities, it is difficult to reach firm conclusions about the presence of M and P pathway contributions from correlations of response properties.

One might hope that color sensitivity, which is perhaps the most distinguishing property of the P pathway, could provide a conclusive test of segregation. This approach is also of limited use. Demonstrating that a cortical neuron possesses a particular response property says little about segregation of the M and P pathways in cortex. Although the color sensitivity in the blobs clearly suggests input from the P pathway, it provides little insight into whether M pathway input is also present. Demonstrating segregation, which depends on showing an absence of one pathway's contribution, is more difficult than demonstrating a contribution. Many forms of processing can reduce sensitivities that exist at earlier levels, and such reductions might be expected as a consequence of cortical processing directed at elaborating more complex representations. Ultimately, the correspondence between response properties can be taken as corroborative, but not conclusive, evidence.

The contributions of the M and P pathways to cortical neurons can be assessed more directly by inactivating individual layers in the LGN (Malpeli & Schiller 1979). The presence or absence of changes in cortical visual responses following block of an LGN layer can be used to infer the contribution of the blocked pathway to the response. Malpeli et al (1981) used this approach to explore M and P contributions to cortical neurons.

They found that the P and M pathways mix within V1, with about 40% of neurons affected when either LGN subdivision was blocked. Subsequent work has shown that substantial M pathway contributions exist in both the blobs and interblobs in the superficial layers (Nealey et al 1991). This observation of M pathway contributions to both the blobs and interblobs differs from the anatomical data that show that layer 4Cα sends axons primarily to the blobs (Lachica et al 1992). The difference might depend on the influence of lateral connections within the superficial layers (Ts'o et al 1986). Also, the sharp distinction that is normally drawn between blobs and interblobs may be artificial (see below).

Although none of the anatomical or physiological data allow precise statements about the degree of segregation of contributions from the M and P pathways in the early stages of cortical processing, collectively they suggest that an incomplete segregation exists. The anatomical connections of layer 4B and the V2 thick stripes point to dominance by the M pathway. However, the existence of some color-sensitive neurons in layer 4B suggests some P pathway contribution. In the other cytochrome oxidase subdivisions, anatomical connections and prevalent color sensitivity suggest P pathway contributions, but selective LGN inactivations and anatomical connections with layers 4Cα and 4B all suggest that the M pathway makes a substantial, and possibly equal, contribution.

Correspondence Between the Cortical and Subcortical Pathways

If the apparent intermixing in V1 and V2 progresses at later stages of processing, the contributions of the M and P pathways might be completely mixed in the first few levels of cortical processing. Were this the case, physiological differences between the parietal and temporal pathways would reflect not differential M and P pathway contributions, but rather differences in the way that a combined signal was processed. It is, therefore, important to consider additional data that compare the subcortical and cortical pathways.

PHYSIOLOGICAL EVIDENCE   Although correlations of response properties are inconclusive for the reasons stated above, the rough correspondence that exists between certain response properties in the M pathway and those in layer 4B and the V2 thick stripes persists in the parietal pathway. Such properties as transient responses and a lack of color specificity in parietal cortical areas led to suggestions of M pathway dominance before segregated pathways had been identified in V1 and V2 (Maunsell & Van Essen 1983b; Motter & Mountcastle 1981). Conversely, the presence of color-selective neurons in the temporal pathways suggest contributions from the P pathways.

Unfortunately, correspondence between response properties in the cortical and subcortical pathways has rarely been demonstrated directly. Perhaps the best basis for a comparison between pathways exists for color. Neurons in the parietal pathway frequently lack color sensitivity (Maunsell & Van Essen 1983b; Robinson et al 1978; Zeki 1974, 1978). The few quantitative studies that have been performed for color sensitivity in the parietal pathway have focused on MT. Most neurons in MT are relatively unresponsive to isoluminant color borders, although many neurons demonstrate some capacity for distinguishing colors (Charles & Logothetis 1989; Dobkins & Albright 1990; Movshon et al 1991; Saito et al 1989). The decline in responsivity at isoluminance in most MT neurons is similar to the behavior of cells in the M pathway (Logothetis et al 1990; Schiller & Colby 1983). However, some neurons in MT do not show such a sharp decline. The residual color response of these cells almost certainly depends on contributions from the P pathway. As described above, poor color sensitivity could result from mixing P pathway inputs that had different color sensitivities.

Selective inactivation of the M or P pathway at the level of the LGN affects responses in the parietal and temporal pathways in ways that fulfill expectations, based on the pattern of partial segregation seen in V1 and V2. The responses of most neurons in V4 are reduced when either the M pathway or the P pathway is blocked (Ferrera et al 1991). In contrast, selective LGN inactivation suggests that the parietal pathway is largely dominated by the M pathway. Blocking the magnocellular layers of the LGN usually eliminates responses in MT and always reduces responses markedly (Maunsell et al 1990). Responses of some MT neurons are also reduced when parvocellular layers are inactivated, but the effects of P pathway block are weaker and less frequently observed.

Overall, the available anatomical and physiological data suggest that the relationship between the M and P pathways and the parietal and temporal pathways is asymmetric. The M pathway seems to dominate the parietal pathway, although some P pathway contributions are found. On the other hand, both the M and P pathways contribute appreciably to the temporal pathway. The segregation of the P and M pathways in extrastriate cortex appears to consist mainly of a partial exclusion of P contributions from the parietal pathway.

The evidence for substantial M pathway contributions to the temporal pathway raises the question of whether this input might be confined to one of the two routes that lead from V1 to V4. If so, one component of the temporal pathway might be dominated by the P pathway. The anatomical routes stemming from the blobs and interblobs remain distinguishable at least to the level of V4, and probably beyond (DeYoe & Sisola 1991; Felleman & McClendon 1991; Zeki & Shipp 1989). They may, therefore, comprise two parallel divisions within the temporal pathway, perhaps serving different visual functions (DeYoe & Van Essen 1988). Because the V1 anatomy suggests that the M pathway contribution goes predominantly to the blobs (Lachica et al 1992), the interblobs might receive primarily P pathway contributions. However, the effects of selective M pathway inactivation show that M and P pathway contributions converge on individual neurons in V1 (Malpeli et al 1981), and that the M contributions are found in the interblobs (Nealey et al 1991). Furthermore, it remains possible that the blobs and interblobs are not components of independent pathways. They may, instead, belong to a single system in which properties vary continuously from blob-like to interblob-like, analogous to the way that neurons in the superficial layers of striate cortex vary between vertical-preferring and horizontal-preferring (see Silverman et al 1989). The borders of the blobs are diffuse and are not marked by interruptions of axonal or dendritic arborizations (Hübener & Bolz 1991; Malach 1991), and there is little compelling evidence for an abrupt physiological transition at the borders of blobs (Born & Tootell 1991). Whatever the actual relationship of the blobs and interblobs, at present there is little indication that either is strongly dominated by P pathway contributions, and the M and P pathways probably mix in the temporal pathway.

BEHAVIORAL EVIDENCE   Two types of behavioral studies bear on the relationship between subcortical and cortical pathways. The first involves attempting to create stimuli that effectively isolate one subsystem. The second involves comparing the effects of lesions of P or M pathways with those following lesions of parietal or temporal cortex.

Selective stimuli   Numerous studies have attempted to stimulate the P pathway alone by using isoluminant chromatic stimuli (Livingstone & Hubel 1987; Ramachandran & Gregory 1978), or random dot stereograms or texture defined stimuli (Cavanagh & Mathers 1989). These studies are not especially useful for exploring the contributions of the P and M pathways. Although the use of isoluminant chromatic stimuli only poorly stimulates neurons in the M pathway (see Cavanagh & Anstis 1991), the P pathway is not fully functional with isoluminance stimuli (Ingling & Grigsby 1990; Merigan et al 1991b). Any observed failures might be due to the lower spatial resolution of chromatic vision (Mullen 1985), the lower effective contrast delivered by chromatic stimuli (Smith & Pokorny 1975), or other limitations of chromatic mechanisms. Thus, the loss of many visual capabilities at isoluminance (Livingstone & Hubel 1987) suggests only that the affected functions are normally mediated by the response of either M or P pathway neurons, or both, to luminance contrast. Likewise, the attempt to stimulate the P pathway selectively by using texture or random dot stimuli (Cavanagh & Mathers 1989) will succeed only if the perception of such stimuli requires the higher acuity of the P pathway (Merigan et al 1991b). This approach is also of questionable value, because there is currently no strong evidence that either chromatic or texture and stereo patterns can selectively stimulate the temporal pathway (Heywood & Cowey 1987; Movshon et al 1991).

Lesions   Before considering lesion studies, it is important to stress that our conclusions are limited by shortcomings of the available data. The major deficiency is that, in most cases, the effects of temporal versus parietal pathway lesions have not been compared using the same visual capacities, although this is not the case for lesions of the M and P pathways. It is also difficult to compare M lesions with parietal pathway lesions or P lesions with temporal pathway lesions for the same reason. Thus, it is difficult to reach strong conclusions about the similarity, or lack thereof, of the effects of lesions of P and M or temporal and parietal pathways. Despite these limitations, the available evidence clearly suggests little relation of the subcortical to the cortical streams.

As described above, lesions of the P pathway disrupt color vision, acuity, and contrast sensitivity for stimuli of low temporal and high spatial frequencies. There is no indication that lesions of the temporal cortical pathway cause any similar effects, although these particular capabilities have not all been tested after temporal pathway lesions. The severe loss of color vision in some human patients with cortical lesions (e.g. Mollon et al 1980) appears as profound as that caused by P lesions, although it differs in that acuity is spared. However, lesions of cortical area V4 in the monkey produce only subtle disruptions of color discrimination (Heywood & Cowey 1987), which suggests that this portion of the temporal pathway is not critical to color vision. Visual acuity also appears not to be reliably reduced by V4 lesions (W. H. Merigan and J. Maunsell, unpublished), and thresholds for acuity, flicker, and orientation discrimination are not affected by IT lesions (Gross 1973).

The most characteristic effects of temporal pathway lesions are disruptions of shape discrimination, which have been seen after V4 (Heywood & Cowey 1987) or inferotemporal cortex lesions (Gross 1972), as well as alterations of visual memory that result from lesions of area TE of inferotemporal cortex (Phillips et al 1988). One might expect comparable effects following P lesions, if the only input to the temporal stream came from the P pathway. Those few studies that have examined shape discrimination after P lesions (Lynch et al 1992; Schiller et al 1990a,b) have found no obvious impairment beyond that which would be expected from the loss of spatial resolution. No studies have examined effects of P lesions on visual memory.

Lesions of the M pathway result in decreased contrast sensitivity for stimuli containing high temporal and low spatial frequencies (see above). No comparable findings have resulted from lesions in the parietal pathway. No alteration in contrast sensitivity for stationary gratings was found by Newsome & Paré (1988) after lesions of area MT (although the temporal frequency content of these stimuli were probably closer to those detected by the P than the M pathway). More recently, Merigan et al (1991c) found no effect of combined MT/MST lesions on the detection of gratings drifting at velocities that included the range preferentially detected by the M pathway (Merigan et al 1991a). Studies of lesions elsewhere in the parietal pathway, such as posterior parietal cortex (Andersen 1987), have not tested effects on contrast sensitivity.

The most characteristic permanent effects of lesions in the parietal pathway include disruption of some aspects of motion perception (Merigan et al 1991c) and subtle changes in eye movements (Dürsteler & Wurtz 1988; Yamasaki & Wurtz 1991) after lesions of areas MT and MST, as well as rather profound changes in eye movements, and evidence of spatial disorientation (Lynch 1980) after parietal lesions. The effects of M pathway lesions described above are very different from these findings.

Lesions of the subcortical M pathway have not borne out the expectation that stereopsis should be dominated by the M pathway (Livingstone & Hubel 1987). This expectation was based on performance at isoluminance and the prevalence of disparity selectivity in layer 4B of V1 and in the thick stripes of V2. Schiller and colleagues (1990a,b) found that stereopsis mediated by high spatial frequency dot patterns was disrupted by P lesions, an effect that may simply reflect reduced visual acuity. M pathway lesions did not disrupt stereopsis.

The question of whether M pathway lesions affect motion perception is more complicated. Formally, motion perception survives M lesions, because direction discrimination at threshold, as well as speed discrimination, is possible in the absence of the M pathway (Merigan et al 1991a). Thus, if a stimulus is visible to a monkey with an M lesion, it can discriminate its direction of motion. We believe that the recent report that M pathway lesions can disrupt motion perception (Schiller et al 1990a,b) may reflect the monkeys' failure to detect the test stimulus after M lesions. Thus, motion perception may be indirectly altered by M lesions, because fast-moving stimuli that may be important to motion perception are difficult to see after an M lesion. This analysis may provide some insight into the impressive anatomical and physiological evidence (above) of a special relationship between the M and the parietal visual pathways. It suggests that the M pathway is not specialized for motion perception, but is specialized for the transmission of middle and high velocity stimuli that are important to some functions of the parietal visual stream. However, neither stereopsis nor motion perception findings support the idea of separate subsystems for specific visual behaviors.

In summary, there appears to be little relationship between the effects of P or M pathway lesions and lesions of temporal or parietal cortex. The lesion effects that have been reported suggest that the subcortical pathways are specialized for the transmission of low-level stimulus properties to the cortical pathways and that the cortical streams themselves may be specialized for more sophisticated visual analysis.

CONCLUDING COMMENTS

Collectively, the available data provide a strong basis for a division between the M and P pathways and between the parietal and temporal pathways, but they suggest that the original notion of parallel visual subsystems that extend from the retina to higher visual cortex must be extensively modified. The mapping between the subcortical and cortical pathways is not simply one to one. Many lines of evidence suggest that the parietal pathway in cortex depends largely on M pathway contributions, but not to the exclusion of contributions from the P pathway. Anatomical, physiological, and behavioral evidence all point to the temporal pathway receiving major contributions from both subcortical pathways. Thus, we are left with an asymmetric organization that is only partially consistent with parallel subsystems.

Why should a partial relationship exist between the cortical and subcortical pathways? We are far from a complete understanding, but insights can be gained by comparing the lines along which each pair of pathways segregates. The P and M pathways are specialized for low-level stimulus features. The P system provides greater spatial resolution, selectivity for color, and the ability to respond to slowly changing or slowly moving stimuli. The M pathway is not selective for color, but achieves much higher sensitivity to moderate or rapidly moving stimuli. Lesion studies suggest that the most fundamental specialization of these two pathways may be the ability to transmit different regions of the "window of visibility" (Watson & Ahumada 1985), i.e. the range of temporal and spatial frequencies that can be seen. In this respect, the P and M pathways resemble specialized detectors that sense different, but overlapping, portions of visible spatial and temporal frequencies. Color vision, which also sharply distinguishes the pathways, may play a less fundamental role, given that similar divisions of spatial and temporal frequencies are seen in species that lack color vision (Stone 1983). Color sensitivity may be a more recently evolved property that has become associated with P pathway, because the spatial or temporal frequencies in that pathway are better suited to color analysis. This perspective on the differences between the P and M pathway sets constraints on the types of possible relationships between subcortical and cortical pathways. In particular, it may help account for the strikingly asymmetric dominance of the parietal pathway by M input. The functions of the parietal pathway (about which we still know little) may almost exclusively depend on the moderate to high velocities transmitted by the M pathway, whereas functions of temporal cortex may require more of the full range of visible spatio-temporal contrasts.

Many readers may wonder how our conclusions can differ so much from the view of parallel subsystems that has reached such prominence. The explanation lies partly in overly enthusiastic acceptance of the notion of parallel subsystems. Few ideas in neuroscience have achieved anything approaching the acceptance that this proposal has received in the six years since it first appeared in print. Two factors have fueled its rapid acceptance. First, it promised to simplify greatly the vast collection of data on visual system organization, thus providing a rationale for the physiological differences between the M and P pathways and between the parietal and temporal pathways. However, simple ideas are not always robust. A relevant example is that the simple and long-standing model of orientation hypercolumns in V1 required fundamental revision following the discovery of cytochrome oxidase blobs. Second, the proposal of parallel subsystems was consistent with an impressively large collection of observations. Although none of those observations is conclusive, their number would make a strong case if each supported parallel subsystems independently. Unfortunately, they do not. For subsystems to exist from the retina to the highest levels of visual cortex segregation must be maintained at every level in the system. If V1 completely intermixed P and M pathway contributions, subsystems of this sort would not exist, regardless of observations that suggest segregation in earlier and later stages. If more than a few of the observations supporting segregation are proven wrong, the case for parallel organization quickly disintegrates.

We expect that the question of parallel pathways will continue to generate intense interest, and it is likely that our understanding will be refined in coming years. Whatever consensus emerges in the future, clearly the simple description that has held sway in recent years is, at best, a rough approximation of the truth.

ACKNOWLEDGMENTS

Preparation of this grant was supported by National Institutes of Health (NIH) EY05911, Office of Naval Research N00014-90-J-1070, and a McKnight Development Award to J. Maunsell, EY08898 to W. H. Merigan and NIH Center Grant EY01319 to the Center for Visual Science. We are grateful to John Assad, Ruth Anne Eatock, Peter Lennie, Tara Nealey, William Newsome, and Tatiana Pasternak for valuable comments on the manuscript.

Literature Cited

Albright, T. D. 1984. Direction and orientation selectivity of neurons in visual area MT of the macaque. *J. Neurosci.* 52: 1106–30

Andersen, R. A. 1987. Inferior parietal lobule function in spatial perception and visuomotor integration. In *Handbook of Physiology Section 1: The Nervous System*, ed. V. B. Mountcastle, F. Plum, S. R. Geiger, pp. 483–518. Bethesda, Md: Am. Physiol. Soc.

Baizer, J. S., Ungerleider, L. G., Desimone, R. 1991. Organization of visual inputs to the inferior temporal and posterior parietal cortex in macaques. *J. Neurosci.* 11: 168–90

Blakemore, C., Vital-Durand, F. 1986. Organization and development of the monkey's lateral geniculate nucleus. *J. Physiol.* 380: 453–91

Blanckensee, H. T. 1980. *Spatio-temporal Properties of Cells in Monkey Lateral Geniculate Nucleus.* Ann Arbor, Mich: Univ. Microfilms Int.

Blasdel, G. G. 1992a. Differential imaging of ocular dominance and orientation selectivity in monkey striate cortex. *J. Neurosci.* 12: 3115–38

Blasdel, G. G. 1992b. Orientation selectivity, preference, and continuity in monkey striate cortex. *J. Neurosci.* 12: 3139–61

Blasdel, G. G., Fitzpatrick, D. 1984. Physiological organization of layer 4 in macaque striate cortex. *J. Neurosci.* 4: 880–95

Blasdel, G. G., Lund, J. S., Fitzpatrick, D. 1985. Intrinsic connections of macaque striate cortex: Axonal projections of cells outside lamina 4C. *J. Neurosci.* 5: 3350–69

Born, R. T., Tootell, R. B. H. 1991. Spatial frequency tuning of single units in macaque supragranular striate cortex. *Proc. Natl. Acad. Sci. USA* 88: 7066–70

Bushnell, M. C., Goldberg, M. E., Robinson, D. L. 1981. Behavioral enhancement of visual responses in monkey cerebral cortex: I. Modulation in posterior parietal cortex related to selective visual attention. *J. Neurophysiol.* 46: 755–72

Cavanagh, P., Anstis, S. 1991. The contribution of color to motion in normal and color-deficient observers. *Vis. Res.* 31: 2109–48

Cavanagh, P., Mathers, G. 1989. Motion: the long and short of it. *Spatial Vis.* 4: 103–29

Charles, E. R., Logothetis, M. K. 1989. The responses of middle temporal (MT) neurons to isoluminant colors. *Invest. Ophthalmol. Vis. Sci.* 30: 427

Cheng, K., Saleem, K. S., Tanaka, K. 1991. Neuronal selectivity for stimulus speed and contrast in the prestriate visual cortical areas V4 and MT of the macaque monkey. *Soc. Neurosci. Abstr.* 17: 441

Conley, M., Fitzpatrick, D. 1989. Morphology of retinogeniculate axons in the macaque. *Vis. Neurosci.* 2: 287–96

Corbetta, M., Miezin, F. M., Dobmeyer, S., Shulman, G. L., Petersen, S. E. 1991. Selective and divided attention during visual discriminations of shape, color, and speed: Functional anatomy by positron emission tomography. *J. Neurosci.* 11: 2382–2402

Crawford, M. L. J., Chodosh, J. 1990. Cytochrome oxidase patterns in V2 cortex of macaque. *Invest. Ophthalmol. Vis. Sci.* 31: 89

Crook, J. M., Lange-Malecki, B., Lee, B. B., Valberg, A. 1988. Visual resolution of macaque retinal ganglion cells. *J. Physiol.* 396: 205–24

deMonasterio, F. M., Gouras, P. 1975. Functional properties of ganglion cells of the rhesus monkey retina. *J. Physiol.* 251: 167–95

Derrington, A. M., Krauskopf, J., Lennie, P. 1984. Chromatic mechanisms in the lateral geniculate nucleus of macaque. *J. Physiol.* 357: 241–65

Derrington, A. M., Lennie, P. 1984. Spatial

and temporal contrast sensitivities of neurons in lateral geniculate nucleus of macaque. *J. Physiol.* 357: 219–40

Desimone, R. 1991. Face-selective cells in the temporal cortex of monkeys. *J. Cogn. Neurosci.* 3: 1–8

Desimone, R., Albright, T. D., Gross, C. G., Bruce, C. 1984. Stimulus-selective properties of inferior temporal neurons in the macaque. *J. Neurosci.* 4: 2051–62

Desimone, R., Fleming, J., Gross, C. G. 1980. Prestriate afferents to inferior temporal cortex: An HRP study. *Brain Res.* 184: 41–55

Desimone, R., Li, L., Lehky, S., Ungerleider, L. G., Mishkin, M. 1990. Effects of V4 lesions on visual discrimination performance and on responses of neurons in inferior temporal cortex. *Soc. Neurosci. Abstr.* 16: 621

Desimone, R., Schein, S. J., Moran, J. Ungerleider, L. G. 1985. Contour, color and shape analysis beyond the striate cortex. *Vis. Res.* 25: 441–52

Desimone, R., Ungerleider, L. G. 1989. Neural mechanisms of visual processing in monkeys. In *Handbook of Neuropsychology*, ed. F. Boller, J. Grafman, 2: 267–99. New York: Elsevier

DeYoe, E. G., Hockfield, S., Garren, H., Van Essen, D. C. 1990. Antibody labeling of functional subdivisions in visual cortex: Cat-301 immunoreactivity in striate and extrastriate cortex of the macaque monkey. *Vis. Neurosci.* 5: 67–81

DeYoe, E. G., Sisola, L. C. 1991. Distinct pathways link anatomical subdivisions of V4 with V2 and temporal cortex. *Soc. Neurosci. Abstr.* 17: 1282

DeYoe, E. G., Van Essen, D. C. 1988. Concurrent processing streams in monkey visual cortex. *Trends Neurosci.* 11: 219–26

DeYoe, E. G., Van Essen, D. C. 1985. Segregation of efferent connections and receptive field properties in visual area V2 of the macaque. *Nature* 317: 58–61

Dobkins, K. R., Albright, T. D. 1990. Color facilitates motion correspondence in visual area MT. *Soc. Neurosci. Abstr.* 16: 1220

Dow, B. M. 1974. Functional classes of cells and their laminar distribution in monkey visual cortex. *J. Neurophysiol.* 37: 927–46

Dow, B. M., Gouras, P. 1973. Color and spatial specificity of single units in rhesus monkey foveal striate cortex. *J. Neurophysiol.* 36: 79–99

Duffy, C. J., Wurtz, R. H. 1991. Sensitivity of MST neurons to optic flow stimuli. I. A continuum of response selectivity to large-field stimuli. *J. Neurophysiol.* 65: 1329–45

Dürsteler, M. R., Wurtz, R. H. 1988. Pursuit and optokinetic deficits following chemi-cal lesions of cortical areas MT and MST. *J. Neurophysiol.* 60: 940–65

Felleman, D. J., McClendon, E. 1991. Modular connections between area V4 and temporal lobe area PITv in macaque monkeys. *Soc. Neurosci. Abstr.* 17: 1282

Felleman, D. J., Van Essen, D. C. 1991. Distributed hierarchical processing in the primate cerebral cortex. *Cereb. Cortex* 1: 1–47

Felleman, D. J., Van Essen, D. C. 1987. Receptive field properties of neurons in area V3 of macaque monkey extrastriate cortex. *J. Neurophysiol.* 57: 889–920

Ferrera, V. P., Nealey, T. A., Maunsell, J. H. R. 1991. Magnocellular and parvocellular contributions to macaque area V4. *Invest. Ophthalmol. Vis. Sci.* 32: 1117

Fitzpatrick, D., Einstein, G. 1989. Laminar distribution and morphology of area 17 neurons projecting to the lateral geniculate nucleus in the macaque. *Soc. Neurosci. Abstr.* 15: 1398

Fitzpatrick, D., Lund, J. S., Blasdel, G. G. 1985. Intrinsic connections of macaque striate cortex: Afferent and efferent connections of lamina 4C. *J. Neurosci.* 5: 3329–49

Goodale, M. A., Milner, A. D. 1992. Separate visual pathways for perception and action. *Trends Neurosci.* 15: 20–25

Gouras, P. 1969. Antidromic responses of orthodromically identified ganglion cells in monkey retina. *J. Physiol.* 204: 407–19

Gross, C. G. 1973. Inferotemporal cortex and vision. In *Progress in Physiological Psychology*, ed. E. Stellar, J. M. Sprague, 5: 77–124. New York: Academic

Gross, C. G. 1972. Visual functions of inferotemporal cortex. *Handbook of Sensory Physiology*, ed. R. Jung, 7/3b: 451–81. Cent. Vis. Inf. Berlin: Springer Verlag

Gross, C. G., Rocha-Miranda, C. E., Bender, D. B. 1972. Visual properties of neurons in inferotemporal cortex of the macaque. *J. Neurophysiol.* 35: 96–111

Grüsser, O. J. Landis, T. 1991. *Visual Agnosias*, v12 of Vision and Visual Dysfunction, ed. J. R. Cronly-Dillon. Boca Raton, Fla: CRC

Hawken, M. J., Parker, A. J. 1984. Contrast sensitivity and orientation selectivity in lamina IV of the striate cortex of old world monkeys. *Exp. Brain Res.* 54: 367–72

Hawken, M. J., Parker, A. J., Lund, J. S. 1988. Laminar organization and contrast sensitivity of direction-selective cells in the striate cortex of the old world monkey. *J. Neurosci.* 8: 3541–48

Haxby, J. V., Grady, C. L., Horwitz, B., Ungerleider, L. G., Mishkin, M., et al. 1991. Dissociation of object and spatial vision processing pathways in human extrastriate cortex. *Proc. Natl. Acad. Sci. USA* 88: 1621–25

Hendrickson, A. E. 1985. Dots, stripes and columns in monkey visual cortex. *Trends Neurosci.* 8: 406–10

Heywood, C. A., Cowey, A. 1987. On the role of cortical area V4 in the discrimination of hue and pattern in macaque monkeys. *J. Neurosci.* 7: 2601–17

Hicks, T. P., Lee, B. B., Vidyasagar, T. R. 1983. The responses of cells in the macaque lateral geniculate nucleus to sinusoidal gratings. *J. Physiol.* 337: 183–200

Horton, J. C., Hubel, D. H. 1981. Regular patchy distribution of cytochrome oxidase staining in primary visual cortex of macaque monkey. *Nature* 292: 762–64

Hubel, D. H., Livingstone, M. S. 1990. Color and contrast sensitivity in the lateral geniculate body and primary visual cortex of the macaque monkey. *J. Neurosci.* 10: 2223–37

Hubel, D. H., Livingstone, M. S. 1987. Segregation of form, color, and stereopsis in primate area 18. *J. Neurosci.* 11: 3378–3415

Hübener, M., Bolz, J. 1991. Cell morphology and blob pattern in monkey striate cortex. *Soc. Neurosci. Abstr.* 17: 117

Humphrey, A. L., Hendrickson, A. E. 1980. Radial zones of high metabolic activity in squirrel monkey striate cortex. *Soc. Neurosci. Abstr.* 6: 315

Ingling, C. R., Grigsby, S. S. 1990. Perceptual correlates of magnocellular and parvocellular channels: seeing form and depth in afterimages. *Vision Res.* 30: 823–28

Ingling, C. R. Martinez-Uriegas, E. 1983. The relationship between spectral sensitivity and spatial sensitivity for the primate r-g X-cell channel. *Vis. Res.* 23: 1495–1500

Kaas, J. H., Garraghty, P. E. 1991. Hierarchical, parallel, and serial arrangements of sensory cortical areas: connection patterns and functional aspects. *Curr. Biol.* 1: 248–51

Kandel, E. R., Schwartz, J. H., Jessel, T. M., eds. 1991. *Principles of Neural Science.* New York: Elsevier. 1135 pp.

Kaplan, E., Shapley, R. M. 1982. X and Y cells in the lateral geniculate nucleus of macaque monkeys. *J. Physiol.* 330: 125–43

Komatsu, H., Ideura, Y., Kaji, H., Yamane, S. 1992. Color selectivity of neurons in the inferior temporal cortex of the awake macaque monkey. *J. Neurosci.* 12: 408–24

Krubitzer, L., Kaas, J. 1990. Convergence of processing channels in the extrastriate cortex of monkeys. *Vis. Neurosci.* 5: 609–13

Lachica, E. A., Beck, P. D., Casagrande, V. A. 1992. Parallel pathways in macaque striate cortex: Anatomically defined columns in layer III. *Proc. Natl. Acad. Sci. USA* 89: 3566–70

Landisman, C. E., Grinvald, A., Ts'o, D. Y. 1991. Optical imaging reveals preferential labeling of cytochrome oxidase-rich regions in response to color stimuli in areas V1 and V2 of macaque monkey. *Soc. Neurosci. Abstr.* 17: 1089

Lee, B. B., Martin, P. R., Valberg, A. 1989. Sensitivity of macaque ganglion cells to luminance and chromatic flicker. *J. Physiol.* 414: 223–43

Lennie, P., Krauskopf, J., Sclar, G. 1990. Chromatic mechanisms in striate cortex of macaque. *J. Neurosci.* 10: 649–69

Levitt, J. B., Kiper, D. C., Movshon, J. A. 1990. Distribution of neuronal response properties in macaque V2. *Soc. Neurosci. Abstr.* 16: 293

Livingstone, M. S. 1988, Art, illusion and the visual system. *Sci. Am.* 256: 78–85

Livingstone, M., Hubel, D. H. 1988. Segregation of form, color, movement, and depth: anatomy, physiology, and perception, *Science* 240: 740–49

Livingstone, M., Hubel, D. H. 1987. Connections between layer 4B of area 17 and thick cytochrome oxidase stripes of area 18 in the squirrel monkey. *J. Neurosci.* 7: 3371–77

Livingstone, M., Hubel, D. H. 1984a. Anatomy and physiology of a color system in the primate visual cortex. *J. Neurosci.* 4: 309–56

Livingstone, M., Hubel, D. H. 1984b. Specificity of intrinsic connections in primate primary visual cortex. *J. Neurosci.* 4: 2830–35

Livingstone, M., Hubel, D. H. 1982. Thalamic inputs to cytochrome oxidase-rich regions in monkey visual cortex. *Proc. Natl. Acad. Sci. USA* 79: 6098–6101

Logothetis, N. K., Schiller, P. H., Charles, E. R., Hurlbert, A. C. 1990 Perceptual deficits and the role of color-opponent and broad-band channels in vision. *Science* 247: 214–17

Lund, J. S. 1988. Anatomical organization of macaque monkey striate visual cortex. *Annu. Rev. Neurosci.* 11: 253–88

Lund, J. S., Boothe, R. G. 1975. Interlaminar connections and pyramidal neuron organisation in the visual cortex, area 17, of the macaque monkey. *J. Comp. Neurol.* 159: 305–34

Lund, J. S., Henry, G. H., MacQueen, C. L., Harvey, A. R. 1979. Anatomical organization of the primary visual cortex (area 17) of the cat: A comparison with area 17 of the macaque monkey. *J. Comp. Neurol.* 184: 599–618

Lund, J. S., Lund, R. D., Hendrickson, A. E., Brunt, A. H., Fuchs, A. F. 1976. The origin of efferent pathways from the primary visual cortex, area 17 the macaque monkey as shown by retrograde transport of horseradish peroxidase. *J. Comp. Neurol.* 164: 287–304

Lynch, J. C. 1980. The functional organization of posterior parietal association cortex. *Behav. Brain Sci.* 3: 485–534

Lynch, J. J., Silveira, L. C. L., Perry, V. H., Merigan, W. H. 1992. Visual effects of damage to P ganglion cells in macaques. *Vis. Neurosci.* In press

Malach, R. 1991. Relationship of biocytin labeled neuronal processes to the cytochrome oxidase (CO) rich blobs in monkey striate cortex. *Soc. Neurosci. Abstr.* 17: 117

Malpeli, J. G., Schiller, P. H. 1979. A method of reversible inactivation of small regions of brain tissue. *J. Neurosci. Methods* 1: 143–51

Malpeli, J. G., Schiller, P. H., Colby, C. L. 1981. Response properties of single cells in monkey striate cortex during reversible inactivation of individual lateral genicular laminae. *J. Neurophysiol.* 46: 1102–19

Martin, K. A. C. 1988. From enzymes to visual perception: a bridge too far? *Trends Neurosci.* 11: 380–87

Maunsell, J. H. R. 1987. Physiological evidence for two visual subsystems. In *Matters of Intelligence*, ed. L. Vaina, pp. 59–87. Dordrecht, Holland: Reidel

Maunsell, J. H. R., Nealey, T. A., DePriest, D. D. 1990. Magnocellular and parvocellular contributions to responses in the middle temporal visual area (MT) of the macaque monkey. *J. Neurosci.* 10: 3323–34

Maunsell, J. H. R., Newsome, W. T. 1987. Visual processing in monkey extrastriate cortex. *Annu. Rev. Neurosci.* 10: 363–401

Maunsell, J. H. R., Van Essen, D. C. 1983a. Anatomical connections of the middle temporal visual area in the macaque monkey and their relationship to a hierarchy of cortical areas. *J. Neurosci.* 3: 2563–86

Maunsell, J. H. R., Van Essen, D. C. 1983b. Functional properties of neurons in the middle temporal visual area of the macaque monkey. I. Selectivity for stimulus direction, speed and orientations. *J. Neurophysiol.* 49: 1148–67

Merigan, W. H. 1991. P and M pathway specialization in the macaque. In *From Pigments to Perception*. ed. A. Valberg, B. B. Lee. New York: Plenum

Merigan, W. H. 1989. Chromatic and achro-

matic vision of macaques: role of the P pathway. *J. Neurosci.* 9: 776–83

Merigan, W. H., Byrne, C., Maunsell, J. H. R. 1991a. Does primate motion perception depend on the magnocellular pathway? *J. Neurosci.* 11: 3422–29

Merigan, W. H., Eskin, T. A. 1986. Spatiotemporal vision of macaques with severe loss of Pb retinal ganglion cells. *Vis. Res.* 26: 1751–61

Merigan, W. H., Katz, L. M. 1990. Spatial resolution across the macaque retina. *Vis. Res.* 30: 985–91

Merigan, W. H., Katz, L. M., Maunsell, J. H. R. 1991b. The effects of parvocellular lateral geniculate lesions on the acuity and contrast sensitivity of macaque monkeys. *J. Neurosci.* 11: 994–1101

Merigan, W. H., Maunsell, J. H. R. 1990. Macaque vision after magnocellular lateral geniculate lesions. *Vis. Neurosci.* 5: 347–52

Merigan, W. H., Pasternak, T., Polashenski, W., Maunsell, J. H. R. 1991c. Permanent deficits in speed discrimination after MT/MST lesions in a macaque monkey. *Invest. Ophthalmol. Vis. Sci.* (Suppl.) 32: 824

Michael, C. R. 1988. Retinal afferent arborization patterns, dendritic field orientations, and the segregation of function in the lateral geniculate nucleus of the monkey. *Proc. Natl. Acad. Sci. USA* 85: 4914–18

Mishkin, M. 1966. Visual mechanisms beyond the striate cortex. In *Frontiers in Physiological Psychology*, ed. R. Russell. New York: Academic

Mishkin, M., Lewis, M. E., Ungerleider, L. G. 1982. Equivalence of parieto-preoccipital subareas of visuospatial ability in monkeys. *Behav. Brain Res.* 6: 41–55

Mishkin, M., Ungerleider, L. G. 1982. Contribution of striate inputs to the visuospatial functions of parieto-preoccipital cortex in monkeys. *Behav. Brain Res.* 6: 57–77

Mishkin, M., Ungerleider, L. G., Macko, K. A. 1983. Object vision and spatial vision: Two cortical pathways. *Trends Neurosci.* 6: 414–17

Mollon, J. D., Newcombe, F., Polden, P. G., Ratcliff, G. 1980. On the presence of three cone mechanisms in a case of total achromatopsia. In *Colour Vision Deficiencies*, ed. G. Verriest, 5: 130–35. Bristol: Hilger

Morel, A., Bullier, J. 1990. Anatomical segregation of two cortical visual pathways in the macaque monkey. *Vis. Neurosci.* 4: 555–58

Motter, B. C., Mountcastle, V. B. 1981. The functional properties of the light-sensitive neurons of the posterior parietal cortex studies in waking monkeys: Foveal sparing and opponent vector organization. *J. Neurosci.* 1: 3–26

Movshon, J. A., Kiper, D., Beusmans, J., Gegenfurtner, K., Zaidi, Q., Carandini, M. 1991. Chromatic properties of neurons in macaque MT. *Soc. Neurosci. Abstr.* 17: 524

Mullen, K. T. 1985. The contrast sensitivity of human color vision to red-green and blue-yellow chromatic gratings. *J. Physiol.* 359: 381–400

Nealey, T. A., Ferrera, V. P., Maunsell, J. H. R. 1991. Magnocellular and parvocellular contributions to the ventral extrastriate cortical processing stream. *Soc. Neurosci. Abstr.* 17: 525

Newsome, W. T., Paré, E. B. 1988. A selective impairment of motion perception following lesions of the middle temporal visual area (MT). *J. Neurosci.* 8: 2201–11

Newsome, W. T., Wurtz, R. H., Dürsteler, M. R., Mikami, A. 1985. Deficits in visual motion processing following ibotenic acid lesions of the middle temporal visual area of the macaque monkey. *J. Neurosci.* 5: 825–40

Pasternak, T., Maunsell, J. H. R., Polashenski, W., Merigan, W. H. 1991. Deficits in global motion perception after MT/MST lesions in a macaque. *Invest. Ophthalmol. Vis. Sci.* (Suppl.) 32: 824

Perry, V. H., Cowey, A. 1985. The ganglion cell and cone distributions in the monkey's retina: Implications for central magnification factors. *Vis. Res.* 25: 1795–1810

Perry, V. H., Oehler, R., Cowey, A. 1984. Retinal ganglion cells which project to the dorsal lateral geniculate nucleus in the macaque monkey. *Neuroscience* 12: 1101–23

Petrides, M., Iversen, S. D. 1979. Restricted posterior parietal lesions in the rhesus monkey and performance on visuospatial tasks. *Brain Res.* 161: 63–77

Phillips, R. R., Malamut, B. L., Bachevalier, J., Mishkin, M. 1988. Dissociation of the effects of inferior temporal and limbic lesions on object discrimination learning with 24-h intertrial intervals. *Behav. Brain Res.* 27: 99–107

Poggio, G. F. 1984. Processing of stereoscopic information in primate visual cortex. In *Dynamic Aspects of Neocortical Function*, ed. G. M. Edelman, W. E. Gall, W. M. Cowan, pp. 613–35. New York: Wiley

Poggio, G. F., Baker, F. H., Mansfield, R. J. W., Sillito, A., Grigg, P. 1975. Spatial and chromatic properties of neurons subserving foveal and parafoveal vision in rhesus monkey. *Brain Res.* 100: 25–59

Pohl, W. 1973. Dissociation of spatial discrimination deficits following frontal and parietal lesions in monkeys. *J. Comp. Physiol. Psychol.* 82: 227–39

Purpura, K., Kaplan, E., Shapley, R. M. 1988. Background light and the contrast gain of primate P and M retinal ganglion cells. *Proc. Natl. Acad. Sci. USA* 85: 4534–37

Purpura, K., Tranchina, D., Kaplan, E., Shapley, R. M. 1990. Light adaptation in the primate retina: analysis of changes in gain and dynamics of monkey retinal ganglion cells. *Vis. Neurosci.* 4: 75–93

Ramachandran, V. S., Gregory, R. L. 1978. Does color provide an input to human motion perception? *Nature* 275: 55–56

Richmond, B. J., Wurtz, R. H., Sato, T. 1983. Visual responses of interior temporal neurons in the awake rhesus monkey. *J. Neurophysiol.* 50: 1415–32

Ridley, R. M., Ettlinger, G. 1975. Tactile and visuo-spatial discrimination performance in the monkey: the effects of total and partial posterior parietal removals. *Neuropsychologia* 13: 191–206

Robinson, D. L., Goldberg, M. E., Stanton, G. B. 1978. Parietal association cortex in the primate: Sensory mechanisms and behavioral modulations. *J. Neurophysiol.* 41: 910–32

Rodieck, R. W., Binmoeller, K. F., Dineen, J. D. 1985. Parasol and midget ganglion cells of the human retina. *J. Comp. Neurol.* 233: 115–32

Saito, H., Tanaka, K., Isono, H., Yasuda, M., Mikami, A. 1989. Directionally selective response of cells in the middle temporal area (MT) of the macaque monkey to the movement of equiluminous opponent color stimuli. *Exp. Brain Res.* 75: 1–14

Saito, H., Yukio, M., Tanaka, K., Hikosaka, K., Fukada, Y., Iwai, E. 1986. Integration of direction signals of image motion in the superior temporal sulcus of the macaque monkey. *J. Neurosci.* 6: 145–57

Sakata, H., Shibutani, H., Kawano, K., Harrington, T. 1985. Neuronal mechanisms of space vision in the parietal association cortex of the monkey. *Vis. Res.* 25: 453–64

Schein, S., Desimone, R. 1990. Spectral properties of V4 neurons in the macaque. *J. Neurosci.* 10: 3369–89

Schiller, P. H., Colby, C. L. 1983. The responses of single cells in the lateral geniculate nucleus of the rhesus monkey to color and luminance contrast. *Vis. Res.* 23: 1631–41

Schiller, P. H., Lee, K. 1991. The role of primate extrastriate area V4 in vision. *Science* 251: 1251–53

Schiller, P. H., Logothetis, N. K., Charles, E. R. 1990a. Role of the color-opponent and broad-band channels in vision. *Vis. Neurosci.* 5: 321–46

Schiller, P. H., Logothetis, N. K., Charles, E. R. 1990b. Functions of the colour-opponent and broad-band channels of the visual system. *Nature* 343: 68–70

Schiller, P. H., Malpeli, J. G. 1978. Functional specificity of lateral geniculate nucleus laminae of the rhesus monkey. *J. Neurophysiol.* 41: 788–97

Schneider, G. E. 1967. Contrasting visuomotor functions of tectum and cortex in the golden hamster. *Psychol. Forschung* 31: 52–62

Schwarcz, R., Hokfelt, T., Fuxe, K., Jonsson, G., Goldstein, M., Terenius, L. 1979. Ibotenic acid-induced neuronal degeneration: a morphological and neurochemical study. *Exp. Brain Res.* 37: 199–216

Sclar, G., Maunsell, J. H. R., Lennie, P. 1990. Coding of image contrast in central visual pathways of the macaque monkey. *Vis. Res.* 30: 1–10

Shapley, R., Kaplan, E., Soodak, R. 1981. Spatial summation and contrast sensitivity of X and Y cells in the lateral geniculate nucleus of the macaque. *Nature* 292: 543–5

Shapley, R., Lennie, P. 1985. Spatial frequency analysis in the visual system. *Annu. Rev. Neurosci.* 8: 547–83

Shapley, R., Perry, V. H. 1986. Cat and monkey retinal ganglion cells and their visual functional roles. *Trends Neurosci.* 9: 229–35

Sherman, S. M., Schumer, R. A., Movshon, J. A. 1984. Functional cell classes in the macaque's LGN. *Soc. Neurosci. Abstr.* 10: 296

Shipp, S., Zeki, S. 1985. Segregation of pathways leading from area V2 to areas V4 and V5 of macaque monkey visual cortex. *Nature* 315: 322–25

Silveira, L. C. L., Perry, V. H. 1991. The topography of magnocellular projecting ganglion cells (M ganglion cells) in the primate retina. *Neuroscience* 40: 217–37

Silverman, M. S., Grosof, D. H., deValois, R. L., Elfar, S. D. 1989. Spatial-frequency organization in primate striate cortex. *Proc. Natl. Acad. Sci. USA* 86: 711–15

Smith, V. C. Pokorny, J. 1975. Spectral sensitivity of the foveal cone photopigments between 400 and 500 nm. *Vis. Res.* 15: 161–71

Snowden, R. J., Treue, S., Erickson, R. G., Andersen, R. A. 1991. The response of area MT and V1 neurons to transparent motion. *J. Neurosci.* 11: 2768–85

Stone, J. 1983. *Parallel Processing in the Visual System.* New York: Plenum. 438 pp

Sugishita, M., Ettlinger, G., Ridley, R. M. 1978. Disturbance of cage finding in the monkey. *Cortex* 14: 431–38

Tanaka, K., Hikosaka, K., Saito, J.-A., Yukie, M., Fukada, Y., Iwai, E. 1986. Analysis of local and wide-field movements in the superior temporal visual areas of the macaque monkey. *J. Neurosci.* 6: 134–44

Tanaka, K., Saito, H.-A., Fukada, Y., Moriya, M. 1991 Coding visual images of objects in the inferotemporal cortex of the macaque monkey. *J. Neurophysiol.* 66: 170–89

Tanaka, M., Lindsley, E., Lausmann, S., Creutzfeldt, O. D. 1990. Afferent connections of the prelunate visual association cortex (areas V4 and DP). *Anat. Embryol.* 181: 19–30

Tootell, R. B. H., Hamilton, S. L., Switkes, E. 1988a. Functional anatomy of macaque striate cortex: IV. Contrast and magnoparvo streams. *J. Neurosci.* 8: 1594–1609

Tootell, R. B. H., Silverman, M. S., DeValois, R. L., Jacobs, G. H. 1983. Functional organization of the second cortical visual area in primates. *Science* 220: 737–30

Tootell, R. B. H., Silverman, M. S., Hamilton, S. L., DeValois, R. L., Switkes, E. 1988b. Functional anatomy of macaque striate cortex: III. Color. *J. Neurosci.* 8: 1569–93

Trevarthan, C. B. 1968. Two mechanisms of vision in primates. *Psychol. Forschung* 31: 229–337

Treisman, A. 1988. Features and objects: the fourteenth Bartlett Memorial lecture. *Q. J. Exp. Psychol.* 40: 201–37

Ts'o, D. Y., Frostig, R. D., Lieke, E. E., Grinvald, A. 1990. Functional organization of primate visual cortex revealed by high resolution optical imaging. *Science* 249: 417–20

Ts'o, D. Y., Gilbert, C. D. 1988. The organization of chromatic and spatial interactions in the primate striate cortex. *J. Neurosci.* 8: 1712–27

Ts'o, D. Y., Gilbert, C. G., Wiesel, T. N. 1991. Orientation selectivity of and interactions between color and disparity subcompartments in area V2 of macaque monkey. *Soc. Neurosci. Abstr.* 17: 1089

Ts'o, D. Y., Gilbert, C. D., Wiesel, T. N. 1986. Relationships between horizontal interactions and functional architecture in cat striate cortex as revealed by cross-correlation analysis. *J. Neurosci.* 6: 1160–70

Ungerleider, L. G., Brody, B. A. 1977. Extrapersonal spatial orientation: the role of posterior parietal, anterior frontal and inferotemporal cortex. *Exp. Neurol.* 56: 265–80

Ungerleider, L. G., Desimone, R. 1986. Cortical connections of visual area MT in the macaque. *J. Comp. Neurol.* 248: 190–222

Ungerleider, L. G., Mishkin, M. 1982. Two cortical visual systems. In *The Analysis of Visual Behavior*, ed. D. J. Ingle, R. J. W. Mansfield, M. S. Goodale, pp. 549–86. Cambridge, Mass: MIT Press

Van Essen, D. C. 1985. Functional organization of primate visual cortex. In *Cerebral Cortex*, ed. E. G. Jones, A. Peters, 3: 259–329. New York: Plenum

Van Essen, D. C., Anderson, C. H., Felleman, D. J. 1992. Information processing in the primate visual system: An integrated systems perspective. *Science* 255: 419–23

Van Essen, D. C., Maunsell, J. H. R. 1983. Hierarchical organization and functional streams in the visual cortex. *Trends Neurosci.* 6: 370–75

Van Essen, D. C., Zeki, S. M. 1978. The topographic organization of rhesus monkey prestriate cortex. *J. Physiol.* 277: 193–226

Virsu, V., Lee, B. B. 1983. Light adaptation in cells of macaque lateral geniculate nucleus and its relation to human light adaptation. *J. Neurophysiol.* 50: 864–78

Watson, A. B. 1992. Transfer of contrast sensitivity in linear visual networks. *Vis. Neurosci.* 8: 65–76

Watson, A. B., Ahumada, A. J. 1985. A model of human visual motion sensing. *J. Opt. Soc. Am. A* 2: 322–42

Wiesel, T. N., Hubel, D. H. 1966. Spatial and chromatic interactions in the lateral geniculate body of the rhesus monkey. *J. Neurophysiol.* 29: 1115–56

Wurtz, R. H., Yamasaki, D. S., Duffy, C. J., Roy, J.-P. 1990. Functional specialization for visual motion processing in primate cerebral cortex. *Cold Spring Harbor Symp. Quant. Biol.* 55: 717–27

Yamasaki, D. S., Wurtz, R. H. 1991. Recovery of function after lesions in the superior temporal sulcus in the monkey. *J. Neurophysiol.* 66: 651–73

Yoshioka, T., Lund, J. S. 1990. Substrates for interaction of visual channels within area V1 of monkey visual cortex. *Soc. Neurosci. Abstr.* 16: 707

Zeki, S. M. 1978. Uniformity and diversity of structure and function in rhesus monkey prestriate cortex. *J. Physiol.* 277: 273–90

Zeki, S. M. 1974. Functional organization of a visual area in the posterior bank of the superior temporal sulcus of the rhesus monkey. *J. Physiol.* 236: 549–73

Zeki, S. M. 1973. Colour coding in rhesus monkey prestriate cortex. *Brain Res.* 53: 422–27

Zeki, S., Shipp, S. 1989 Modular connections between areas V2 and V4 of macaque monkey visual cortex. *Eur. J. Neurosci.* 1: 494–506

Zeki, S., Watson, J. D. G., Lueck, C. J., Friston, K. J., Kennard, C., Frackowiak, R. S. J. 1991. A direct demonstration of functional specialization in human visual cortex. *J. Neurosci.* 11: 641–9

Zrenner, E., Zbramov, I., Akita, M., Cowey, A., Livingstone, M., Valberg, A. 1990. Color perception: Retina to cortex. In *Visual Perception: The Neurophysiological Foundations*, ed. L. Spillmann, J. S. Werner, pp. 163–204. New York: Academic

Annu. Rev. Neurosci. 1993. 16:403–43
Copyright © 1993 by Annual Reviews Inc. All rights reserved

THE DIVERSITY OF NEURONAL NICOTINIC ACETYLCHOLINE RECEPTORS

Peter B. Sargent

Departments of Stomatology and Physiology and the Neuroscience Graduate Program, University of California, San Francisco, California 94143-0512

KEY WORDS: calcium, presynaptic, heterologous expression, transsynaptic regulation, channel gating

The transmitter acetylcholine (ACh) acts on two different classes of receptors: nicotinic and muscarinic. Nicotinic receptors (nAChRs) are directly gated or ionotropic receptors and are members of a supergene family that also includes glycine, $GABA_A$, and $5\text{-}HT_3$ receptors. Muscarinic receptors are metabotropic receptors and belong to a different supergene family. In this review, I concentrate on nAChRs expressed by neurons, which are distinct from skeletal muscle nAChRs. Several recent reviews have treated both neuronal and muscle nAChRs (Claudio 1989; Galzi et al 1991; Lukas & Bencherif 1992; Schmidt 1988; Schuetze & Role 1987), whereas others have focused on neuronal nAChRs (Deneris et al 1991; Lindstrom et al 1987; Luetje et al 1990a; Role 1992).

Over the past several years, the application of molecular, immunological, and physiological techniques has greatly expanded our knowledge of neuronal nAChRs. By applying tools and insight gained from the study of muscle nAChRs, neuronal nAChRs have been purified and their subunits cloned. In addition, whole cell and patch clamp recordings have given us precise information about the function of nAChR ion channels. These studies have demonstrated that neuronal nAChRs are highly diverse: there are several different nAChR subunits, multiple forms of purified nAChRs, and many physiologically distinct classes of nAChR channels. The principal challenge over the next several years will be to bring together molec-

403

0147–006X/93/0301–0403$02.00

ular, biochemical, and physiological observations to define the molecular basis of this diversity and to understand its functional significance.

PHYSIOLOGICAL PROPERTIES OF NEURONAL nAChRS

Nicotinic AChRs are present on autonomic neurons and adrenal chromaffin cells in the peripheral nervous system and on many neurons in the central nervous system (CNS). In nearly all instances, these nAChRs mediate "fast" inward currents, as do their skeletal muscle counterparts. Many of the properties of these neuronal nAChR channels, such as their ion selectivity and gating properties, resemble those of muscle nAChRs. However, neuronal AChRs are clearly distinct from muscle nAChRs and are themselves diverse.

Ion Selectivity

Most neuronal nAChR channels, like muscle nAChR channels, are cation-specific, but do not distinguish readily among cations (e.g. Fieber & Adams 1991; Mulle & Changeux 1990; Nutter & Adams 1991). Neuronal nAChRs have a significant permeability to Ca^{2+}; for example, $P_{Ca}:P_{Na}$ for rat parasympathetic cardiac neurons is 0.7 (Adams & Nutter 1992), and $P_{Ca}:P_{Na}$ for rat PC12 cells is approximately 2.5 (Sands & Barish 1991; see also Vernino et al 1992). By contrast, Decker & Dani (1990) calculated a $P_{Ca}:P_{Na}$ of 0.2 in skeletal muscle. Neuronal nAChR channels expressed heterologously in oocytes also have calcium permeabilities greater than muscle and, in some instances, greater than N-methyl-D-aspartate (NMDA) channels (Séguéla et al 1992; Vernino et al 1992). Calcium entry through neuronal nAChR channels is sufficient to activate calcium-dependent chloride (Mulle et al 1992a; Séguéla et al 1992; Vernino et al 1992) and potassium (Fuchs & Murrow 1992b) conductances and may activate second messenger systems.

Channel Properties

Open-channel conductances of neuronal nAChRs vary from less than 10 picosiemens (pS) to more than 50 pS. Comparison of channel conductance measurements for different preparations is problematic, because they vary as a function of recording conditions, especially the concentrations of divalent cations (Ifune & Steinbach 1991; Mathie et al 1987; Neuhaus & Cachelin 1990). However, under similar recording conditions, clear differences do exist among different neuronal nAChRs (e.g. Mulle et al 1991). The best evidence for functional heterogeneity of neuronal nAChR channels arises from instances in which multiple classes of open-channel

NEURONAL NICOTINIC AChRS   405

conductances are observed in single cells. Two populations of open-channel conductances are present in rat sympathetic neurons (Derkach et al 1987) and chicken ciliary ganglion neurons (Margiotta & Gurantz 1989), and three to four populations are present in rat PC12 cells (Bormann & Matthaei 1983; Ifune & Steinbach 1990), rat cardiac parasympathetic neurons (Adams & Nutter 1992), bovine chromaffin cells (Cull-Candy et al 1988), and chicken sympathetic neurons (Moss et al 1989). Whether distinct conductance states result from nAChRs having different combinations of subunits and/or from nAChR subunits having different post-translational modifications is not known.

Neuronal nAChR channels, like muscle nAChR channels, display bursting behavior. Burst durations of neuronal nAChRs vary from preparation to preparation, and even among different channel types on individual cells. Several nicotinic channels on autonomic neurons show burst lengths with time constants in the range of 5–10 ms (e.g. Kuba et al 1989; Mathie et al 1987; Moss et al 1989; Schofield et al 1985). These times are longer than those recorded from muscle fibers. The functional consequence of longer burst durations is that more current flows through each activated nAChR channel during synaptic transmission.

Agonists and Antagonists

Neuronal nAChRs are distinct from muscle nAChRs in agonist potency. For example, suberyldicholine is generally more effective than ACh on muscle nAChRs, whereas ACh is more effective than suberyldicholine on neuronal nAChRs (e.g. Lukas 1989). Neuronal nAChRs themselves are quite distinct with regard to agonist potency. Mulle et al (1991) found that the order of potency for neurons from the rat interpeduncular nucleus was cytisine > ACh > nicotine; for nAChRs on neurons from the medial habenula, it was nicotine > cytisine > ACh. These differences may be explained by differences in nAChR subunit composition (Luetje & Patrick 1991; discussed below).

Antagonists have also been useful in distinguishing muscle nAChRs from neuronal ones and in distinguishing among different neuronal nAChRs. Paton & Zaimis (1949) demonstrated that decamethonium (C10) was more effective than hexamethonium (C6) in blocking muscle nAChRs, whereas C6 was more effective in autonomic ganglia. This led to the terms "C10 receptor" (muscle) and "C6 receptor" (neuronal). Currently, several antagonists distinguish between neuronal and muscle nAChRs (Lukas 1990).

Snake toxins have proven useful for differentiating between neuronal and muscle nAChRs (reviewed in Chiappinelli 1991; Loring & Zigmond 1988). The first snake toxin to be used widely in studies on nAChRs was α-bungarotoxin (α-Bgt), which generally binds with high affinity and specificity to muscle nAChRs, but not to nAChRs that underlie synaptic transmission in autonomic ganglia. Early reports that synaptic transmission in autonomic ganglia is blocked by α-Bgt were due to the presence of contaminating toxins, one of which has been purified and found to block ganglionic nAChRs at nanomolar concentrations (Chiappinelli 1983; Loring et al 1984; Ravdin & Berg 1979). This toxin is referred to as bungarotoxin 3.1 (Ravdin & Berg 1979), κ-bungarotoxin (Chiappinelli 1983), or toxin F (Loring et al 1984); in this review, I adopt Lindstrom et al's (1987) suggestion that the toxin be named neuronal-bungarotoxin (n-Bgt). [1] In chicken and frog autonomic ganglia, n-Bgt at 5–100 nM blocks nicotinic synaptic transmission and responses of autonomic neurons to nicotinic agonists (Chiappinelli 1983; Loring et al 1984; Ravdin & Berg 1979; Sargent et al 1991). Rat sympathetic ganglion neurons and bovine chromaffin cells show lower affinity for n-Bgt than those in chicken or frog (Higgins & Berg 1987; Nooney et al 1992; Sah et al 1987). For all ganglionic nAChRs, however, n-Bgt is more potent than α-Bgt, which is generally ineffective at blocking ganglionic nAChRs at concentration as high as 1 μM (but see Marshall 1981). Muscle nAChRs, by contrast, have considerably greater affinity for α-Bgt than for n-Bgt.

Many central nAChRs resemble ganglionic nAChRs in their sensitivity to n-Bgt (Calabresi et al 1989; Lipton et al 1987; Loring et al 1989; Schulz & Zigmond 1989; Vidal & Changeux 1989; Wong & Gallagher 1991; Zhang & Feltz 1990). However, nicotinic responses of neurons in the chicken lateral spiriform nucleus are apparently insensitive to 0.5–1.0 μM n-Bgt (Sorenson & Chiappinelli 1990), as are those of neurons in the interpeduncular nucleus and medial habenular nucleus of rats (Mulle et al 1991). The nAChRs underlying the responses of these neurons are pharmacologically distinct from ganglionic nAChRs.

In a few instances, α-Bgt does block central nicotinic responses. In the rat hippocampus, for example, nicotinic responses are virtually eliminated by 20 nM n-Bgt or 300 nM α-Bgt (Alkondon & Albuquerque 1991). These nAChRs thus represent a novel class of neuronal nAChRs, because they are blocked nearly irreversibly by relatively low concentrations of both α-Bgt and n-Bgt. nAChRs with similar properties have been characterized

---

[1] Neuronal-bungarotoxin does not block some neuronal nAChRs (Mulle et al 1991) and it does block muscle-like nAChRs (Luetje et al 1990b), albeit at considerably higher concentrations than needed for neuronal nAChRs. Moreover, still other toxins are present in the venom of *Bungarus multicinctus* that block ganglionic nAChRs (reviewed in Chiappinelli 1991). Thus, the term "neuronal-bungarotoxin" should not be interpreted to mean that there is only one such neuronal-specific toxin or even that the toxin is specific for and effective on all neuronal nAChRs.

in hair cells of the chicken cochlea (Fuchs & Murrow 1992a) and in insects (e.g. Pinnock et al 1988).

Rectification

Nicotinic responses of central and peripheral neurons are distinct from those of muscle in displaying pronounced inward rectification. Inward rectification of whole cell currents occurs independently of current polarity and depends primarily upon membrane potential (Mathie et al 1990; Yawo 1989). Rectification can, in principle, result from a voltage-dependent change in gating and/or ion permeation through single channels. In neurons, as in muscle, ion permeation through single channels displays some inward rectification, and this behavior is dependent upon internal Mg^{2+} (Ifune & Steinbach 1990; Neuhaus & Cachelin 1990; Sands & Barish 1992). With no Mg^{2+} (or Ca^{2+}) on either side of the membrane, single-channel I-V curves are linear, but whole cell currents continue to show inward rectification (Ifune & Steinbach 1990; Mathie et al 1990; Neuhaus & Cachelin 1990). This rectification is due only in small part to a voltage dependence of burst duration, which is weak for neuronal nAChRs (Mathie et al 1990). In rat sympathetic neurons, rectification of whole cell currents results primarily from a reduced probability of channel opening; at $+50$ mV, channels open less than one-tenth as often as at -50 mV (Mathie et al 1990). In rat PC12 cells, rectification is primarily caused by channels closing more rapidly at positive potentials (Ifune & Steinbach 1992; see also Sands & Barish 1992).

STRUCTURE OF NEURONAL nAChRS

Patrick & Stallcup (1977a,b) reported that nicotinic responses elicited from rat PC12 cells were blocked by an antiserum to electric organ nAChR and were thus homologous to muscle nAChRs. This homology suggested to several investigators that cDNA probes for muscle nAChR genes might be useful, under conditions of low stringency, in identifying genes that encode neuronal nAChR subunits.

Neuronal nAChRs Are Encoded by a Family of Related Genes

Recombinant DNA technology has resulted in the identification of nine to ten genes in rat and chicken neural tissue that are homologous to muscle nAChR genes (Table 1; putative nAChR genes from *Drosophila*, cockroach, goldfish, and human species are also listed). These genes encode peptides with four hydrophobic, putative transmembrane domains (M1–M4) in the approximate positions of their muscle counterparts (the struc-

Table 1 Properties of cloned and sequenced neuronal nAChR genes

Species	Subunit	Probe	Mature peptide M_r / # amino acids	Consensus N-linked glycosylation sites[a]	Cysteines[a]	Homology (subunit/%) [conspecific, unless otherwise noted]	References
Rat[b]	α2	chicken α2	55,500 / 484	29, 79, 185	133, 147, 197, 198	α1/48	Wada et al 1988
	α3	mouse α1	54,800 / 474	24, 141	128, 142, 192, 193	α1/52; α2/58	Boulter et al 1986
	α4-1[c]	rat α3, mouse α1	67,100 / 600	24, 141	128, 142, 192, 193	α1/53; α2/68; α3/59	Boulter et al 1987; Goldman et al 1987
	α5	rat β3	48,800 / 424	112, 140, 186	127, 141, 191, 192	α1–α4/44–55	Boulter et al 1990
	α6	PCR	53,300 / 463	24, 141	66, 128, 142, 192, 193	α1, α2, α4, α5/41, 49, 45; α3/59; β2–β4/43–47	Lamar et al 1990; J. Patrick et al, unpublished results
	α7	PCR (primers based on chicken α7, α8)	54,200 / 480	24, 68, 111	116, 128, 142, 200, 201	α2–α5, β2–β4/31–37	Séguéla et al 1992
	β2	rat α3	54,300 / 475	26, 141	128, 142	β1/45; α1–α5/44–50	Deneris et al 1988
	β3	rat α3	50,200 / 434	26, 141	128, 142	β1, β2/40, 44; α1/45 α2–α4/50–56; α5/68	Deneris et al 1989
	β4	rat β2	53,300 / 475	15, 72, 117, 145	132, 146	β1/43; β2/64; β3/44	Duvoisin et al 1989; Boulter et al 1990
	non-α2[d]	universal oligonucleotide	56,100 / 499	41, 98, 143, 171	6, 26, 101, 158, 172	β1/43; α2–α5/46–52 may be an alternatively spliced β4	Isenberg & Meyer 1989
Chicken[e]	α2	chicken α1, γ	58,100 / 505	31, 81	135, 149, 199, 200	α1/50 rat α2/77	Nef et al 1988
	α3	chicken α2	54,800 / 474	24, 141	128, 142, 192, 193	α1/53; α2/65 rat α3/83	Nef et al 1988; Couturier et al 1990b
	α4	chicken α2	68,400 / 599	29, 79	133, 147, 197, 198	α1/51; α2/80; α3/66 rat α4/73	Nef et al 1988
	α5	linkage to α3[f]	49,000 / 425	26, 140, 186	127, 141, 191, 192	α1/45; α2–α4/50–55 rat α5/84	Couturier et al 1990b
	α6	chicken β4	54,100 / 463	23, 140	127, 141, 191, 192	β4/48 rat α6/80	M. Ballivet et al, unpublished results
	α7	chicken α3[g]	54,600 / 480[b]	24, 68, 111[h]	116, 128, 142, 190, 191[h]	α1–α6, β2–β4/35–42 rat α7/87	Schoepfer et al 1990; Couturier et al 1990a
	α8	chicken α7	55,200 / 481	24, 111	116, 128, 142, 190, 191	α7/82	Schoepfer et al 1990
	β2 (nα, nα1)	chicken γ	54,000 / 473	26, 143	130, 144	α1–α4/44–55 rat β2/85	Schoepfer et al 1988; Nef et al 1988
	β3 (nα2)	chicken α5	50,100 / 435	28, 143	130, 144	α5/67 rat β3/82	M. Ballivet et al, unpublished results
	β4 (nα3)	linkage to α3[f]	53,500 / 467	26, 115, 143	130, 144	β2/70; α1–α5/44–52 rat β4/74	Couturier et al 1990b

Drosophila	ALS	chicken α2	62,000 / 547	24, 212	128, 142, 201, 202	mouse α1/36; chicken α2/44	Bossy et al 1988
	SAD [Dα2]	ALS, ARD consensus sequence	61,000 / 535	24, 213	128, 142, 202, 203	ALS/56; ARD/39 chicken α1, α2/35, 41	Sawruk et al 1990a; Jonas et al 1990
	ARD	Torpedo γ	57,300 / 497	24	128, 142	rat α3/46	Hermans-Borgmeyer et al 1986
	SBD	ALS, ARD consensus sequence	57,300 / 493	24	128, 142	ALS/56; SAD/54; ARD/41	Sawruk et al 1990b
Cockroach	αL1	chicken β2	60,600 / 534	24, 212	128, 142, 201, 202	Drosophila ALS/57	Marshall et al 1990
Goldfish	GFα3	Torpedo α1, rat α1, rat α4 (mixed)	58,100 / 489	24, 141	128, 142, 192, 193	rat α3/89; chicken α3/93	Cauley et al 1990; Hieber et al 1990a
	GFβ2	Torpedo α1, rat α1, rat α4 (mixed)	? / 464[i]	26, 143	130, 144	rat β2/84	Hieber et al 1990b
	GFn-α2	Torpedo α1, rat α1, rat α4 (mixed)	53,200 / 434	25, 140	127, 141	rat β2/64; rat β3/82; rat β4/55; rat α2–α4/50–55	Cauley et al 1989
	GFn-α3	Torpedo α1, rat α1, rat α4 (mixed)	53,800 / 438	26, 113, 141, 180	128, 142	rat β2/62; rat β3/85; rat β4/57; GFnα–2,88	Cauley et al 1990
Human	α3	rat α3	54,100 / 474	24, 68, 141	128, 142, 192, 193	rat α3/93	Fornasari et al 1990
	α5	rat α5	51,000 / 446	133, 161, 207	148, 162, 212, 213	rat α5/87	Chini et al 1992
	β2	chicken β2	54,700 / 477	26, 143	130, 144	rat β2/95	Anand & Lindstrom 1990
	β4	rat β2	51,400 / 456	46, 77, 91	128, 142	rat β4/77	F. Clementi et al, unpublished results

[a] Consensus N-linked glycosylation sites and cysteines are given only for the N-terminal extracellular domain (amino acids 1–ca. 210).

[b] The sequence for the rat α6 gene has not been published. A rat α8 gene has not yet been identified.

[c] Two cDNAs were isolated for the α4 gene that were virtually identical over the region of overlap and that are likely to arise from alternative splicing (Goldman et al 1987).

[d] The rat nonα-2 gene differs from the rat β4 gene in having a 69 base insertion, as well as two smaller insertions, a deletion, and an inversion (Deneris et al 1989).

[e] The sequences for the chicken α6 and β3 genes have not yet been published.

[f] The chicken α5 and β4 genes were identified by virtue of their linkage to α3: some recombinant α3 phage contained more restriction fragments recognized by α3 probes that could be accounted for by the presence of α3 alone (Couturier et al 1990b). The α3, α5, and β4 genes are also physically linked in the rat and human genome (Boulter et al 1990; Raimondi et al 1992; see also Sawruk et al 1990b).

[g] Couturier et al (1990a) used an α3 probe to identify the α7 gene, whereas Schoepfer et al (1990) used an oligonucleotide based on the N-terminal sequence of an α-BgtAChR purified by Conti-Tronconi et al (1985).

[h] The N-terminal amino acid of the α7 mature protein cannot be unambiguously determined from the amino acid sequence. The amino acid numbering shown is according to the placement by Schoepfer et al (1990). Couturier et al (1990a) predict that the signal peptide has one additional amino acid (and the mature protein one less).

[i] The GFβ2 clone contains an open reading frame encoding 459 amino acids. The N-terminal amino acid deduced from this clone probably corresponds to amino acid number 6 of the mature rat β2 subunit, with which the GFβ2 subunit is highly homologous.

ture of the muscle nAChR is reviewed in Changeux 1990; Karlin 1991; McCarthy et al 1986). The overall amino acid homology between the products of these genes and muscle genes from the same species is 40–55% (Table 1). This homology approaches 100% in the putative transmembrane regions (especially M1–M3) and in selected stretches of the N-terminal extracellular domain, whereas the amino-acid sequence in the putative cytoplasmic segment between M3 and M4 is divergent. All genes code for a protein with two cysteines separated by 13 residues that align with cysteines 128 and 142 of the muscle α subunit. The deduced gene products fall into two classes, based on whether they have adjacent cysteine residues at positions 192 and 193 (muscle α numbering) that are found in all α subunits from muscle and electric organ and that are affinity alkylated by the ACh agonist bromo acetylcholine (BAC) and by the ACh antagonist 4-(N-maleido)benzyltrimethylammonium (MBTA). Seven of the genes code for such a sequence, and these are assigned to the α class; three do not (non-α or β). The muscle α gene is designated $\alpha 1$, and the seven neuronal α subunits are designated $\alpha 2$–$\alpha 8$, more or less in order of discovery. The α subunits themselves fall into two classes, based on subunit homology. $\alpha 7$ and $\alpha 8$, which encode α-Bgt-binding components, have substantial amino acid homology with one another, but less homology with other α subunits than they have with each other (Table 1).

The neuronal non-α subunits are named $n\alpha 1$–$n\alpha 3$ ($n\alpha$ for non-α) in chicken and $\beta 2$–$\beta 4$ in rat. In this review, I use the terms $\beta 2$, $\beta 3$, and $\beta 4$ for both species to facilitate comparison. Rat and chicken genes of the same name are highly homologous ($> 70\%$ amino acid identity). The β designation does not mean that these gene products most closely resemble muscle β genes, for they don't (Table 1). At least in rat, however, the $\beta 2$ and $\beta 4$ gene products can substitute for the muscle $\beta 1$, but not for the $\alpha 1$, γ, or δ gene products, in generating functional, muscle-like nAChRs in *Xenopus* oocytes following injection of mRNAs (Deneris et al 1988; Duvoisin et al 1989). Based on sequence information alone, there is no reason to classify the three β subunits into a single group, for they are as different from each other as they are from α subunits. Based on heterologous expression studies, however, at least $\beta 2$ and $\beta 4$ belong together, because either can encode for a functional nAChR when coexpressed with $\alpha 2$, $\alpha 3$, or $\alpha 4$ (see below).

Several lines of evidence demonstrate that the genes identified, cloned, and sequenced from rat and chicken nervous tissue encode nAChR subunits. First, nonneuronal cells express functional nAChRs when injected with some combinations of these genes. Second, these genes are expressed in autonomic neurons and in many brain nuclei from which nicotinic responses have been obtained; moreover, the distribution of gene expression matches fairly closely the distribution of high-affinity ^{3}H-ACh and ^{3}H-nicotine binding sites. And third, nAChRs immunopurified from brain are encoded by these genes.

Purification of Neuronal nAChRs

CHICKEN BRAIN   Whiting & Lindstrom (1986a) immunopurified nAChRs from chicken brain by affinity chromatography with a monoclonal antibody (mAb 35) to electric organ nAChR. Passage of chicken brain extracts over a mAb 35-Sepharose column and elution resulted in a preparation that bound ^{3}H-nicotine with high affinity (Whiting & Lindstrom 1986b), as is characteristic for brain nAChRs (Marks & Collins 1982; Schwartz et al 1982), and that contained two bands, at 49 and 59 kd, on denaturing gels. An mAb generated to this immunopurified material (mAb 270) immunoprecipitated a third component at 75 kd from chicken brain extracts in addition to the two components immunopurified by mAb 35 (Whiting et al 1987a). Further analysis showed that there are at least two populations of nAChRs in chicken brain: those containing peptides at 49 and 59 kd (immunoprecipitated by mAb 35), and those containing peptides at 49 and 75 kd (immunoprecipitated by mAb 284) (Whiting et al 1987b). mAb 270 recognizes a band migrating at 49 kd, which is present in both forms. mAb 35 recognizes a distinct protein that also runs at 49 kd (Conroy et al 1992). Whiting & Lindstrom (1987a) found that ^{3}H-MBTA labeled both the 59 and 75 kd peptides, which suggests that the ACh binding subunit of the putative neuronal nAChRs is on the large subunit. Purified neuronal nAChR subunits were initially named α and β in order of increasing molecular weight (M_r) on denaturing gels, by analogy with muscle nAChR subunits. However, once the higher M_r subunits were found to be affinity-labeled by ^{3}H-MBTA and to have N-terminal sequences homologous to α subunits characterized by molecular biological techniques (see below), the nomenclature was altered to correspond to the cDNA nomenclature, in which subunits containing the cysteine 192/193 pair are named α.

Whiting and Lindstrom's results on purified nAChR from chicken brain imply that the neuronal nAChR contains only two different types of subunits. This contrasts with skeletal muscle, in which nAChRs contain four distinct subunits. The following evidence indicates that the material immunopurified by Whiting and Lindstrom comprises a neuronal nAChR: it shows high affinity for ^{3}H-nicotine and is labeled with ACh affinity alkylating agents, and an antiserum to it antigenically modulates functional nAChRs on chicken ciliary ganglion neurons (Stollberg et al 1986). Additional evidence that the immunopurified peptides are nAChR subunits has come from matching them with the predicted products of nAChR genes. Thus, the N-terminal sequence of the 75 kd band (Whiting et al 1991a) corresponds to the sequence predicted for the α4 transcript (Nef et al 1988), and the N-terminal sequence of material running at 49 kd that is recognized by mAb 270 (Schoepfer et al 1988) corresponds to the sequence of the β2 subunit (Nef et al 1988; Schoepfer et al 1988). The 59 kd band in immunopurified chicken nAChR may contain both α2 and trace amounts of α3, because different antisera to bacterial fusion proteins that contain unique sequences of α2 and α3 bind to this band on blots (unpublished results cited in Lindstrom et al 1990). Thus, mAb 270-immunopurified nAChRs appear to have relatively large amounts of a form containing α4 and β2 subunits and a form containing α2 and β2 subunits and trace amounts of a form containing α3 and β2 subunits. Recent findings of Conroy et al (1992) indicate that the α5 subunit, which is recognized by mAb 35, can be assembled with α4, thus raising the possibility that some AChRs contain α4, α5, and β2. Whiting & Lindstrom (1986b) found that mAb 270 immunoprecipitates more than 90% of the high-affinity ^{3}H-nicotine binding sites in chicken brain extracts, which suggests that nAChRs containing β2 subunits account for most of the high-affinity ^{3}H-nicotine binding sites in brain.

MAMMALIAN BRAIN   Whiting & Lindstrom (1987b) immunopurified nAChRs from rat brain with anti-chicken nAChR mAb 270 and found it to contain two bands on denaturing gels, with apparent M_rs of 52 and 80 kd (see also Dwork & Desmond 1991; analogous subunits have been immunopurified from bovine brain by Whiting & Lindstrom 1988). The 80 kd subunit was affinity alkylated with ^{3}H-MBTA, which suggests that it contains the ACh binding site (Whiting & Lindstrom 1987a). The N-terminal amino acid sequence of the 80 kd subunit corresponds to the sequence encoded by the α4 transcript (Goldman et al 1987; Whiting et al 1987a), and that of the 52 kd subunit corresponds to the sequence encoded by the β2 transcript (Schoepfer et al 1988; Wada et al 1988). These results indicate that the principal form of rat brain nAChR immunopurified by mAb 270 contains α4 and β2. Immunoprecipitation of detergent-solubilized rat brain extracts with either mAb 270 (specific for a 52 kd band in blots) or mAb 299 (specific for a 80 kd band on blots) removed >90% of the high-affinity ^{3}H-nicotine binding sites from solution. This suggests that an nAChR containing α4 and β2 subunits accounts for a majority of the high-affinity ^{3}H-nicotine binding sites in rat brain. Similar results were noted by Flores et al (1992), who found that antisera to α4 and to β2 fusion proteins each removed >90% of the high-affinity ^{3}H-cytisine binding sites from rat brain extracts. Flores et al (1992) also found that the α4-specific antisera immunoprecipitated virtually all β2-reactive material from extracts, and vice versa. Thus, the α4 and β2 subunits in rat brain must associate principally (but not necessarily exclusively) with one another. The finding that β2-containing nAChRs account for most of the high-affinity agonist binding sites in both chicken and rat brain suggests that central nAChRs lacking β2 may resemble ganglionic nAChRs in having a lower agonist affinity (e.g. Kemp & Morley 1986).

CHICKEN CILIARY GANGLION   Purification of nAChRs from ciliary ganglion extracts with mAb 35-Sepharose yielded a fraction displaying three bands on denaturing gels, at 49, 52, and 60 kd (Halvorsen & Berg 1990). The 60 kd peptide is likely to be an α3 gene product, because it is recognized by α3-specific antibodies (Schoepfer et al 1989; Vernallis et al 1991). The 52 kd peptide appears to be a β4 gene product, because it is recognized by an mAb specific for a β4 fusion protein (Vernallis et al 1991). The 49 kd peptide is probably an α5 gene product, because it binds several mAbs that recognize α5 in chicken brain (Conroy et al 1992; Halvorsen & Berg 1990). These results suggest that mAb 35-immunopurified nAChRs from chicken ciliary ganglion neurons contains α3, α5, and β4 subunits. Whether these all assemble into a single AChR must await further analysis.

α-BGT RECEPTORS   In vertebrate brain, nicotinic receptors that recognize α-Bgt with high affinity (α-BgtAChRs) are distinct from those that do not (nAChRs). For example, α-BgtAChRs do not display high-affinity binding for ^{3}H-nicotine (Wonnacott 1986), whereas nAChRs do (Whiting & Lindstrom 1986b, 1988). α-Bgt-affinity purification of α-BgtAChRs from brain yields fractions containing one to five bands on denaturing gels in the M_r range of 45–70 kd (Conti-Tronconi et al 1985; Gotti et al 1991; Kemp et al 1985; Norman et al 1982; Whiting & Lindstrom 1987b). ^{3}H-MBTA and ^{3}H-BAC alkylate a component with an apparent M_r of about 55 kd, which presumably is an α subunit for α-BgtAChRs (Kemp et al 1985; Norman et al 1982). The components that sometimes copurify with this subunit may represent other subunits that help form an α-BgtAChR hetero-oligomer, or they may simply be associated peptides or proteolytic breakdown products that are not an integral part of the native α-BgtAChR.

Schoepfer et al (1990) demonstrated that the majority of α-BgtAChRs in chicken brain contain α7 and/or α8 subunits, because mAbs to α7 and α8 fusion proteins immunoprecipitate a large fraction of the high-affinity ^{125}I-α-Bgt binding sites (an α7-specific mAb alone immunoprecipitates >90% of α-BgtAChRs). The apparent M_r of the α7 gene product on denaturing gels, 57 kd, corresponds closely with the size of the subunit affinity alkylated by ^{3}H-MBTA and ^{3}H-BAC (Kemp et al 1985; Norman et al 1982). These studies demonstrate that the α7 and α8 gene products represent the principal agonist binding subunits of α-BgtAChRs. Immuno-adsorption of α-BgtAChRs with an mAb to α7 does not precipitate detectable amounts of material recognizable with mAbs to α4 or β2; conversely, mAb 270 (β2) immunoadsorption does not precipitate detectable α7- or α8-like immunoreactivity (Schoepfer et al 1990). This suggests that α7 and α8 subunits do not commonly associate with α4 or β2 subunits to form nAChRs in chicken brain. McLane et al (1990) showed that peptides unique to α5 bind α-Bgt (as do those from α1, α7, and α8), whereas peptides from the corresponding regions of α2–α4 do not. However, whether α5-specific antibodies can immunoprecipitate ^{125}I-α-Bgt binding sites from brain extracts, is not known.

In the invertebrate CNS, in contrast to the vertebrate CNS, nicotinic receptors are often blocked by α-Bgt (e.g. David & Sattelle 1984). α-Bgt-affinity purified material from cockroach CNS displays an overall size similar to electric organ nAChR ($\approx$300 kd), but a single band on denaturing gels (65 kd; Breer et al 1985). Reconstitution of α-Bgt-affinity purified material into planar lipid bilayers produced channels gateable with nicotinic agonists that were blocked by the ACh antagonist d-tubocurarine (α-Bgt apparently was not tested; Hanke & Breer 1986). Marshall et al (1990) cloned and sequenced a cockroach gene (αL1) that is 46% homologous to the rat α3 gene and that contains adjacent cysteines at positions 201 and 202 (Table 1). Injection of αL1 mRNA into *Xenopus* oocytes resulted in the expression of functional channels gated by nicotine and blocked by both α-Bgt and n-Bgt. Thus, αL1 homo-oligomers can form functional nAChR channels that mimic those found on neurons in the cockroach CNS. The αL1 transcript, which has a predicted amino acid M_r of 61 kd (Table 1), may encode the 65 kd component purified by Breer et al (1985). These results suggest that αBgt- and n-Bgt-blockable nAChRs in cockroach CNS can be formed by αL1 homo-oligomers.

In *Drosophila*, which is unrelated to cockroach evolutionarily, molecular biological studies have led to the cloning and sequencing of two α and two non-α genes (Table 1). Antisera to fusion proteins for two of these genes (ALS, ARD), singly or in combination, immunoprecipitate $\approx$25% of the ^{125}I-α-Bgt binding sites in *Drosophila* head membranes and may associate, possibly with other subunits, to form an α-Bgt-sensitive nAChR in *Drosophila* (Schloss et al 1988, 1991).

In sum, molecular and biochemical approaches have identified a family of nAChR subunits that fall into two classes (ACh binding or α subunits, and "structural" or β subunits), based on the presence of adjacent cysteines and on reactivity with ACh affinity alkylating agents. Immunoprecipitation studies suggest that many neuronal nAChRs are composed of both α and β subunits. Although some nAChRs may contain only two distinct subunits (e.g. α4β2), other nAChRs might contain two or more kinds of α subunits (Conroy et al 1992), possibly in addition to β subunits.

HETEROLOGOUS EXPRESSION OF NEURONAL nAChR GENES

Functional Expression of Neuronal nAChR Genes

The most striking confirmation that nAChR-like genes indeed encode for nAChR subunits has come from heterologous expression studies. Boulter et al (1987), Wada et al (1988), and Deneris et al (1989) found that *Xenopus* oocytes expressed functional nAChRs after injection of mRNAs for the rat $\alpha 2$, $\alpha 3$, or $\alpha 4$ gene in combination with either the $\beta 2$ or $\beta 4$ gene. Injection of any one mRNA species alone did not lead to the appearance of functional responses, except $\alpha 4$, which produced small depolarizations in response to high concentrations of ACh in about one third of the trials. The same six combinations of chicken genes are functional when expressed following injection of cDNAs into the oocyte nucleus, with transcription under the control of a heat shock or SV40 promoter (Ballivet et al 1988; Couturier et al 1990b). These results demonstrate that functional nAChRs can be synthesized from specific combinations of α and β subunits and complement the biochemical findings, which suggest that native nAChRs consist of α/β heteromers.

The demonstration that the $\alpha 7$ gene encodes a functional nicotinic channel was first provided by Couturier et al (1990a; see also Bertrand et al 1992; Revah et al 1991; Séguéla et al 1992), who found that voltage-clamped oocytes previously injected with $\alpha 7$ cDNA respond to nicotinic ligands by producing inward currents. These channels are unusual in the sense that β subunit expression is apparently not needed; barring the existence of endogenous β subunits in the oocyte, these nAChRs must consist of $\alpha 7$ homo-oligomers. Currents from $\alpha 7$ nAChRs are blocked by low concentrations of α-Bgt ($IC_{50} = 0.7$ nM) and less potently by n-Bgt (Couturier et al 1990a). Coinjection of $\beta 2$, $\beta 3$, or $\beta 4$ cDNA, along with $\alpha 7$ cDNA, did not result in functional responses with properties any different than those expressed in oocytes injected with $\alpha 7$ cDNA alone. Thus, in oocytes at least, $\alpha 7$ subunits apparently cannot coassemble with known neuronal β subunits to form hetero-oligomers that have functional properties distinct from those of $\alpha 7$ homo-oligomers.

The rat $\beta 3$ gene does not form functional nAChRs when coinjected as mRNA with $\alpha 2$, $\alpha 3$, or $\alpha 4$ genes (Deneris et al 1989). The rat $\alpha 5$ and $\alpha 6$ genes likewise do not participate in the formation of functional nicotinic channels in oocytes when injected in combination with several other α and β genes as mRNAs (Boulter et al 1990; J. Patrick, personal communication). These subunits may form functional nAChRs when coexpressed with undiscovered (or untested) subunits (Conroy et al 1992), or they may be members of another family of ligand-gated channels. Still another possibility is that these gene products require post-translational modifications not made by the oocyte in order to function.

Whiting et al (1991b) stably expressed nAChRs in mouse fibroblasts by transfecting them with chicken $\alpha4$ and $\beta2$ cDNAs. The fibroblasts acquire functional responses that are pharmacologically similar to those in chicken brain, where both $\alpha4$ and $\beta2$ are widely expressed (Morris et al 1990). The availability of host cells that stably express nAChRs and that can be grown in large numbers presents many advantages over oocytes and may become the favored method of expressing nAChRs in the future.

Functional Diversity Among Heterologously Expressed nAChRs

OPEN-CHANNEL CONDUCTANCES   Single-channel properties show considerable diversity among heterologously expressed nAChRs. Papke et al (1989) found unique populations of nicotinic channels, based on their open-channel conductance, following injection of oocytes with $\beta2$ mRNA in combination with $\alpha2$, $\alpha3$, or $\alpha4$ mRNA. Cell-attached patches displayed two populations of open-channel conductances after injection with $\alpha2$ and $\beta2$ mRNA (mean conductances: 34 and 16 pS) and with $\alpha3$ and $\beta2$ mRNA (15 and 5 pS), and usually a single population of 13 pS channels after injection with $\alpha4$ and $\beta2$ mRNA. Because most of these mRNAs do not produce functional channels when expressed on their own, functional nAChRs that appear following injection of pairs of mRNAs are likely to contain both classes of subunits. How can nAChRs that consist of, say, $\alpha3$ and $\beta2$ subunits give rise to two populations of openings? One possibility is that there is some promiscuity in the way these subunits assemble, either in their ratios [e.g. $(\alpha3)_2(\beta2)_2$, $(\alpha3)_3(\beta2)_2$, $(\alpha3)_2(\beta2)_3$, $(\alpha3)_3(\beta2)_3$] or in their arrangement around the central pore (e.g. $\alpha\beta\alpha\beta\beta$ and $\alpha\alpha\beta\beta\beta$). Another possibility is that channels yielding different conductances have the same subunit stoichiometry and arrangement, but differ in their post-translational modification.

The conductance of the nicotinic channel can be influenced by the β subunit, as well as by the α subunit. Thus, injection of mRNAs for $\alpha3$ and $\beta2$ yielded two populations of channels that have mean conductances of 15 and 5 pS, whereas injection of mRNAs for $\alpha3$ and $\beta4$ yielded populations of channels that have mean conductances of 22 and 13 pS when measured under the same conditions (Papke & Heinemann 1991).

The channel conductances recorded by Papke et al (1989) for heterologously expressed nAChRs are generally smaller than those observed for native neuronal nAChRs; the values they measured for $\alpha4\beta2$ channels are smaller than those measured for $\alpha4\beta2$ channels by Charnet et al (1992), who found three classes of channels with mean conductances of 34, 22, and 12 pS. Papke et al (1989) recorded channel conductances at room temperature and in the presence of 1.8 mM Ca^{2+}, whereas Charnet et al (1992) recorded at 13°C in the absence of Ca^{2+}. Increased extracellular Ca^{2+} reduces single-channel conductance through nAChR channels (Mulle et al 1992b; Vernino et al 1992); this reduction may be sufficient to explain the difference in results. However, native nAChR channels often show conductances larger than 20 pS, even in the presence of physiological levels of Ca^{2+}. Whether this discrepancy reflects upon an unphysiological characteristic of the oocyte expression system is difficult to know, because the conductance(s) of native $\alpha4\beta2$ channels is not known. In fact, it has not yet been possible to ascribe any native nicotinic channel property to an nAChR of defined subunit composition. Functional knock out experiments may be the best approach to this problem (Listerud et al 1991).

GATING   The gating properties of heterologously expressed nicotinic channels are dependent upon their subunit composition. Papke et al (1989) found that each class of channels expressed after injection of oocytes with mRNAs from rat $\beta2$ and from $\alpha2$, $\alpha3$, and $\alpha4$ genes could be distinguished by their distribution of open times. In all instances, open-time distribution could be adequately described by the sum of two exponential functions; one describing brief openings (open times of 0.1–0.4 ms) and one describing longer openings, or bursts (open times of 2–7 ms). Channels formed with $\alpha3$ and $\beta4$ subunits show longer brief openings and longer bursts than those formed with $\alpha3$ and $\beta2$ subunits (Papke & Heinemann 1991).

AGONIST POTENCY   Luetje & Patrick (1991) measured the rank order potency for the agonists cytisine, nicotine, ACh, and 1,1-dimethyl-4-phenylpiperazinium on nAChRs formed from rat $\beta2$ or $\beta4$ subunits in combination with $\alpha2$, $\alpha3$, or $\alpha4$ subunits. Each pair-wise injection yielded nAChRs with a unique rank order potency. Surprisingly, the nature of the β subunit had at least as much influence upon potency as the nature of the α subunit. The most striking effect noted was that upon sensitivity to cytisine: nAChRs containing the $\beta4$ subunit were more sensitive to cytisine than to any other agonist tested, whereas nAChRs containing the $\beta2$ subunit were almost completely insensitive to cytisine.

    Gross et al (1991) noted an effect of the α subunit upon agonist affinity in oocytes injected with chicken nAChR cDNAs. Heterologously expressed nAChRs made from $\alpha4$ and $\beta2$ cDNAs had higher affinity for ACh and showed less desensitization than nAChRs made from $\alpha3$ and $\beta2$ cDNAs. Luetje & Patrick (1991) did not observe a pronounced difference in ACh affinity between $\alpha3\beta2$ and $\alpha4\beta2$ nAChRs made from rat transcripts, which suggests that homologous rat and chicken nAChRs may not function similarly. The affinities of heterologously expressed $\alpha3\beta2$ and $\alpha3\beta4$ nAChRs for ACh (Bertrand et al 1990; Gross et al 1991) are higher than affinities measured for native chicken nAChRs (Margiotta et al 1987a), but perhaps only because the expressed and native nAChRs being compared do not have the same subunit composition.

ANTAGONIST POTENCY   The sensitivity of heterologously expressed nAChRs to n-Bgt is dependent upon the nature of the subunit content (Luetje et al 1990b). Incubation of oocytes with 0.1 μM n-Bgt for 30 minutes completely blocks ACh-induced currents elicited following washout of the toxin in oocytes previously injected with rat $\alpha3$ and $\beta2$ mRNAs; under the same conditions, n-Bgt is completely ineffective at blocking function of $\alpha2\beta2$ nAChRs (the sensitivity of $\alpha4\beta2$ nAChRs is intermediate between $\alpha2\beta2$ and $\alpha3\beta2$ nAChRs). Site-directed mutagenesis studies have shown that the amino acid at position 198 of the α subunit (glutamine in $\alpha3$, proline in $\alpha2$) is critically important in determining the difference in n-Bgt sensitivity between $\alpha3\beta2$ and $\alpha2\beta2$ nAChRs (C. Luetje and J. Patrick, personal communication). All three classes of $\beta2$-containing nAChRs were blocked by neosurugatoxin, which is highly effective on all neuronal nAChRs thus far tested (Luetje et al 1990b). Interestingly, the nature of the β subunit again influences the characteristics of expressed nAChRs; ACh-induced currents elicited from $\alpha3\beta2$ nAChRs are blocked by pre-incubation with 0.1 μM n-Bgt, but those elicited from $\alpha3\beta4$ nAChRs are completely insensitive (Duvoisin et al 1989).

In sum, heterologous expression studies have revealed that the subunit composition of nAChRs can affect their open-channel conductance, their gating behavior, and their sensitivity to both agonists and antagonists. It is too early to say whether these differences can be used as diagnostic tools for identifying native nAChRs, as no one has yet been able to identify a population of native channels as arising from a specific combination of subunits in order to compare their properties with those of heterologously expressed channels.

Stoichiometry

Neuronal nAChRs may have an $\alpha:\beta$ stoichiometry of 2:3, because their activation is typically characterized by Hill coefficients of about 1.5 and because their size (≈300 kd) is consistent with a pentameric structure. Anand et al (1991) used ^{35}S-methionine to label metabolically nAChRs expressed following injection of chicken $\alpha4$ and $\beta2$ mRNAs into *Xenopus* oocytes. They found 1.46 times more ^{35}S in β subunits than in α subunits after correction was made for their methionine content. This ratio is very close to the expected value of 1.5 for a stoichiometry of $\alpha_2\beta_3$, which suggests that expressed $\alpha 4\beta 2$ nAChRs are pentamers that contain two α and three β subunits. Comparable results have been obtained by Whiting et al (1991b) for $\alpha 4\beta 2$ nAChRs stably expressed in mouse fibroblasts. Cooper et al (1991) concluded that $\alpha 4\beta 2$ nAChRs have two α and three β subunits based on the number of channel conductance populations observed when oocytes express a mixture of $\alpha 4$, $\beta 2$, and an $\alpha 4$ or $\beta 2$ whose putative M2 transmembrane domain has been altered by site-directed mutagenesis. A stoichiometry of $\alpha_2\beta_3$ for native nAChRs would seem likely based on these findings and by analogy with muscle nAChRs, which are pentamers containing two α subunits.

Looking for Matches Between Native and Expressed Receptors

Do we know enough about the functional properties of heterologously expressed nAChRs to make educated guesses about the identity of native nAChRs present on individual neurons? In the rat interpeduncular nucleus (IPN), for example, at least six nAChR genes are expressed: $\alpha 2$, $\alpha 3$, $\alpha 4$, $\alpha 5$, $\beta 2$, and $\beta 4$ (although not all necessarily in common neurons). The channel conductance of IPN neurons measured by Mulle et al (1991), 35 pS, is very close to that of the largest class of heterologously expressed $\alpha 2\beta 2$ nAChRs (34 pS; Papke et al 1989); therefore, the $\alpha 2\beta 2$ combination may be the principal class of nAChRs on IPN neuronal cell bodies. Mulle et al (1991) found that cytisine was at least as effective as ACh in eliciting responses from IPN neurons. The rank order potency data for agonists in heterologously expressed nAChRs, however, suggest that cytisine is ineffective on $\beta 2$-containing nAChRs, whereas it is most effective on $\beta 4$-containing nAChRs (Luetje & Patrick 1991). nAChRs on IPN neurons may, therefore, be $\alpha 2\beta 4$ or $\alpha 4\beta 4$, the channel conductances of which are not known (the channel conductance of $\alpha 3\beta 4$ nAChRs channels does not match that of native channels in the IPN).

Attempts to match native nAChRs with heterologously expressed nAChRs of defined composition fail in the rat medial habenula and in rat PC12 cells. In the medial habenula, for example, nAChR channel conductance (26 pS) is closest to the larger conductance class of heterologously expressed $\alpha 3\beta 4$ channels (22 pS), but the agonist rank order potency for habenular nAChRs resembles neither $\alpha 3\beta 4$ nAChRs nor any other class of heterologously expressed nAChR (Mulle et al 1991, Luetje & Patrick 1991). The difficulty in finding matches between native and heterologously expressed nAChRs raises an important question. Do oocytes faithfully translate mRNAs and modify and assemble their translation products? Results from experiments on muscle nAChRs suggest that oocytes can translate and assemble subunits for muscle nAChRs accurately (reviewed in Lingle et al 1992). If the oocytes are not at fault, then the failure to match native and expressed nAChRs may mean that additional nAChR genes are yet to be identified or that native nAChRs consist of subunit combinations that have not yet been tested in oocytes. In particular, native nAChRs may consist of more than one α and/or β type. At least two approaches can be envisioned that could reveal the presence of hetero-α- and hetero-β-containing nAChRs. One approach is to inject multiple α and/or β subunit mRNAs into oocytes and search for channels that have properties not seen when only one α and one β are injected. Another approach is to perform sequential immunoprecipitations with subunit-specific antibodies (Conroy et al 1992).

DISTRIBUTION OF NEURONAL nAChRS

Neuronal nAChRs in the central and peripheral nervous systems have been visualized with ligands and with antibodies, and nAChR mRNAs have been detected with cDNA probes. In general, these approaches have yielded similar results. Mapping studies in brain have provided a wealth of information about which nAChR genes are expressed and where, whereas mapping studies in autonomic ganglia, done at considerably higher spatial resolution, have revealed the distribution and density of nAChRs on the neuronal surface and within the cell.

Mapping nAChRs in Brain

LIGANDS   Clarke et al (1985) mapped high-affinity binding sites in rat brain for three nAChR ligands: ^{3}H-ACh, ^{3}H-nicotine, and ^{125}I-α-Bgt. The distribution of binding of ^{3}H-ACh and ^{3}H-nicotine were virtually identical to each other and distinct from that of binding sites for ^{125}I-α-Bgt. Binding of ^{125}I-α-Bgt was high in the cerebral cortex (especially in layers I and IV), hypothalamus, hippocampus, inferior colliculus, and in a few brainstem nuclei (see also Hunt & Schmidt 1978a). Binding of ^{3}H-ACh and ^{3}H-nicotine was high in the interpeduncular nucleus; in several thalamic nuclei, including the medial habenula; and in the superior colliculus. Moderate labeling was noted in the presubiculum, in layers III and IV of the cerebral cortex, in parts of the striatum, and in the dorsal tegmental nuclei. Weak labeling was noted in the cerebellum, and virtually no labeling was seen in the hypothalamus or the hippocampus. London et al (1988) analyzed the increase in glucose utilization following systemic infusion of nicotine and found a somewhat similar pattern of labeling, which suggests that the sites recognized with high affinity by ^{3}H-ACh and ^{3}H-nicotine represent functional nAChRs. Moreover, nicotinic responses can be demonstrated in some of these areas, including the medial habenula and the interpeduncular nucleus.

ANTIBODIES   Immunocytochemical studies on tissue sections of rat brain using anti-nAChR mAbs, some of which are specific for the β2 subunit, have revealed a staining pattern somewhat similar to that seen for ^{3}H-nicotine and ^{3}H-ACh (Deutch et al 1987; Swanson et al 1987). In chicken brain, the distribution of immunostaining noted with anti-α7 and anti-α8 mAbs is different from that seen with an anti-β2 mAb (Britto et al 1992; see also Watson et al 1988). This finding is consistent with the radioligand studies, as both suggest that the distribution of nAChRs recognized by α-Bgt (α-BgtAChRs) is distinct from those having high affinity for agonists (nAChRs).

The subcellular distribution of nAChR immunoreactivity in brain has been largely unexplored. Schroder et al (1989) found that mAb WF 6 bound to synaptic membranes in rat and human cortex; this mAb competes for the α-Bgt binding site on Torpedo nAChR and may recognize an α-BgtAChR in brain. Synaptic membranes are labeled with ^{125}I-α-Bgt or HRP-α-Bgt in rat brain (Hunt & Schmidt 1978b) and in chicken retina (Vogel et al 1977). No study has yet demonstrated that anti-bodies of defined nAChR subunit specificity label synaptic membranes in brain.

GENE EXPRESSION   Neuronal nAChR genes are expressed almost exclusively in neurons, many of which have functional nAChRs. The differences noted among distinct genes in their pattern of expression among brain regions suggest that different nAChR genes are expressed by unique subsets of neurons. Within the CNS, nAChR gene expression has been assessed with a combination of Northern blots, RNAse protection assays, and in situ hybridization assays. In situ hybridization assays have revealed that at least one nAChR gene is expressed in numerous areas within rat brain and that each gene is expressed in a distinct pattern (Séguéla et al 1992; Wada et al 1989). For example, the α2 subunit is expressed at high levels only in parts of the interpeduncular nucleus, whereas the α4 subunit is expressed strongly in a large number of areas, including the substantia nigra pars compacta, the ventral tegmental area, and the medial habenula. The extent of expression of the α3 gene is intermediate between α2 and α4. Among the β subunits, β2 is expressed at detectable levels in almost all parts of the brain, whereas the expression of β3 is more restricted (Deneris et al 1989). Duvoisin et al (1989) reported that β4 is expressed at high levels only in the medial habenula, but β4 is actually expressed more widely (Dineley-Miller & Patrick 1992). The distribution of mRNAs for α2, α3, α4, and β2 genes in chicken brain is generally similar to that in rat for the corresponding genes (Morris et al 1990). The α4 and β2 genes are expressed widely in chicken brain, whereas α3 expression is more limited.

The anatomical distribution of rat brain regions expressing nAChR genes correlates fairly well with the distribution of high-affinity nicotine and ACh binding sites (Clarke et al 1985) and with the distribution of anti-nAChR antibody binding (Swanson et al 1987). A reasonably good correspondence has also been noted between the distribution of the α-BgtAChR gene α7 and the distribution of high-affinity binding sites for ^{125}I-α-Bgt (Clarke et al 1985; Séguéla et al 1992). Differences in the distribution of nAChRs and of their mRNAs would be expected wherever nAChRs accumulate at axon terminals, because the protein would then lie some distance from the mRNA. For example, the superficial layers of the rat superior colliculus have high levels of mAb 270 immunoreactivity but no detectable β2 mRNA (Swanson et al 1987; Wada et al 1989). Even if protein transport from the cell body is discounted, however, a close match between immunocytochemical and hybridization findings may not occur invariably, because mRNA and protein levels need not be tightly correlated (erythrocytes are an extreme example).

When in situ hybridizations are performed with liquid emulsion rather than with films, labeling can be detected at the level of individual cell bodies. In instances where tissue labeling is weak, staining of a small subset of neurons may be strong, such as in the cerebellar cortex, where a small number of Purkinje cells are intensely labeled with a rat α2 probe (Wada et al 1989; see also Matter et al 1990). It would be useful to know whether the same subpopulation of Purkinje cells express α2 at all times, or whether all Purkinje cells express α2 sporadically and transiently.

High-Resolution Mapping of nAChRs in Autonomic Ganglia

Marshall (1981) used an antiserum to electric organ nAChR and immunoperoxidase and electron microscopic techniques in bullfrog sympathetic ganglia to show that neuronal nAChR-like molecules are highly concentrated in the synaptic membrane. Immunoreactivity was present within that portion of the synaptic cleft delineated by the nerve terminal's active zone. Similar results have been found in the chicken ciliary ganglion by Jacob et al (1984) and in the frog cardiac ganglion by Sargent & Pang (1989), who used mAbs to electric organ nAChR.

The density of nAChRs in the synaptic membrane has been measured in the chicken ciliary ganglion by Loring & Zigmond (1987), who used quantitative electron microscopic autoradiography and ^{125}I-n-Bgt, which blocks nAChR function in the ganglion. n-Bgt binds to two sites on the neuronal surface, one of which is also recognized by α-Bgt. When the site shared with α-Bgt site is blocked, ^{125}I-n-Bgt binds selectively to synaptic sites at a density of approximately $600/\mu m^2$. Generally similar findings were made in cultured rat superior cervical ganglion cells (Loring et al 1988), which can form nicotinic synapses with each other under certain culture conditions. At the site exclusively recognized by ^{125}I-n-Bgt, the density of binding sites is approximately $4800/\mu m^2$ in the synaptic membrane and about 1% of this in the extrasynaptic membrane. α-BgtAChRs are present on chicken ciliary ganglion neurons, but, in contrast to nAChRs, are not concentrated at synaptic sites (Jacob et al 1983; Loring et al 1985; Messing & Gonatas 1983).

The estimates of ^{125}I-n-Bgt binding site density in chicken ciliary ganglion neurons is considerably lower than the density of α-Bgt binding sites estimated for the synaptic membrane in adult skeletal muscle (10,000–20,000 sites/μm^2; Fertuck & Salpeter 1976). Given the small size of the nAChR-rich patches on these cells (Jacob & Berg 1988), there may be only a few hundred nAChRs available per active zone to bind transmitter. Consequently, a quantum of transmitter may saturate receptors and yet only produce a small amount of current (Dryer & Chiappinelli 1987).

nAChR FUNCTION

Evidence for Presynaptic nAChRs in Brain

nAChRs in autonomic ganglia are responsible for mediating fast excitation, and it is generally assumed that the same holds true for central nAChRs. Nicotinic responses have been elicited from neuronal cell bodies in several areas of brain, including retina, spinal cord, hippocampus, respiratory nuclei of the brainstem, cerebral cortex, cerebellar cortex, thalamus (medial habenula, medial and lateral geniculate nuclei), interpeduncular nucleus, septal nucleus, substantia nigra, striatum, hypothalamus, and locus coeruleus in rats, and the retina and lateral spiriform nucleus in chickens (see citations in Clarke 1990; also Alkondon & Albuquerque 1991; Cobbett et al 1985; Lipton et al 1987; Mulle & Changeux 1990; Mulle et al 1991; Phelan & Gallagher 1992; Schulz & Zigmond 1989; Sorensen & Chiappinelli 1990; Vidal & Changeux 1989; Wong & Gallagher 1991; Zhang & Feltz 1990). However, definitive evidence that central synaptic transmission is mediated by nAChRs only exists at the motor neuron-Renshaw cell synapse in the spinal cord (reviewed in Nicoll et al 1990). The presence of functional nAChRs on many neuronal cell bodies suggests that nAChRs mediate postsynaptic nicotinic responses widely in the nervous system; attempts to document this function have not yet been successful (Brown et al 1983), but presumably await further characterization of central synaptic pathways.

Although nicotinic responses can be elicited from many different areas of the brain, they are not as prominent as might be expected based on mapping studies. This discrepancy may be explained if many of the nAChRs in brain are located on nerve terminals, from which recordings are rarely made. In the rat IPN, Brown et al (1984) and Mulle et al (1991) characterized one such set of nAChRs found on the terminals of neurons that arise from the medial habenula (Clarke et al 1986). Another population of physiologically characterized presynaptic nAChRs are those in the striatum on terminals of dopaminergic neurons that originate in the substantia nigra pars compacta (Clarke & Pert 1985; Giorguieff-Chesselet et al 1979; Schulz & Zigmond 1989). Radioligand binding studies in the cat visual cortex strongly suggest that presynaptic nAChRs are also found there (Parkinson et al 1988; Prusky et al 1987). Although the function of presynaptic nAChRs is presumably to modulate transmitter release, axo-axonic synapses have not been reported on central terminals bearing nAChRs. If ACh does affect the function of these terminals, it may do so via a paracrine mechanism.

Immunochemical studies have complemented radioligand and physiological ones by suggesting that some nAChRs in brain are presynaptic. Swanson et al (1983) found that several anti-electric organ nAChR mAbs, including mAb 35, labeled the lateral spiriform nucleus (SpL) and specific layers of the chicken optic tectum, which is the principal site of termination of SpL neurons and contains axon terminals bearing nAChR immunoreactivity (Britto et al 1992). The SpL also displays high-affinity ^{3}H-nicotine binding, and most of its neurons have functional nAChRs (Sorenson & Chiappinelli 1990). Lesioning the SpL resulted in loss of immunoreactivity both there and in the ipsilateral tectum (Swanson et al 1983), which suggests that the labeling in the tectum is to a presynaptic nAChR (see also Swanson et al 1987). However, some nAChRs are probably tectal in origin (Matter et al 1990; Morris et al 1990) and may be lost following degeneration of afferents (e.g. Hattori & Fibiger 1982).

Additional evidence for presynaptic central nAChRs has come from an analysis of the retinotectal system of both goldfish and frog. Henley et al (1986b) found that anti-electric organ nAChR mAbs bound to layers of the goldfish optic tectum that receive a projection from the contralateral retina. Following injection of ^{35}S-methionine into the retina, Henley et al (1986a) could immunoprecipitate labeled material from both the eye and the optic tectum. This suggested that nAChRs recognized by mAbs in the goldfish tectum are, at least in part, synthesized in the retina and transported to the tectum (see also Cauley et al 1990). A similar story has emerged from studies in the frog, where nAChR-like immunoreactivity is found in the optic tract and optic tectum and where removal of the contralateral eye results in loss of virtually all staining (Sargent et al 1989). Electron microscopic examination of immunolabeled nAChRs in the tectum revealed that they are present not at synaptic sites, but rather on the nonsynaptic surfaces of vesicle-bearing profiles (Sargent et al 1989). These results suggest that nAChRs in the tectum are found on the terminals of retinal ganglion cell afferents. The principal source of cholinergic terminals in the tectum is the nucleus isthmi (Ricciuti & Gruberg 1985); conceivably, ACh released from these terminals may bind nAChRs on retinal afferents and modify their release properties. However, the nucleus isthmi neurons do not terminate in the same tectal layers as do the retinal afferents and are, therefore, not likely to influence them by direct axo-axonic synapses. Rather, the influence may be paracrine.

α-BgtAChRs

Schoepfer et al (1990) showed that most α-BgtAChRs in chicken brain contain $\alpha7$ subunits, and Couturier et al (1990a) showed that $\alpha7$ encodes a functional nAChR that is blocked by nanomolar concentrations of α-Bgt. In brain, however, nicotinic responses are rarely blocked by α-Bgt (reviewed in Quik & Geertsen 1988). Exceptions to this generalization are in the rat hippocampus (Alkondon & Albuquerque 1991) and, perhaps, in the cerebellum (de la Garza et al 1987). Retinotectal transmission in goldfish and toads was originally thought to be mediated by an α-Bgt-blockable mechanism (reviewed in Oswald & Freeman 1981). More recently, however, Langdon & Freeman (1987) have suggested that retino-tectal transmission in goldfish is mediated by an amino acid. The failure to find numerous α-Bgt-blockable nicotinic responses in brain is puzzling, given the behavior of heterologously expressed $\alpha7$ homo-oligomers. Perhaps these responses are not detected because of some unusual property, such as rapid desensitization. Alternatively, $\alpha7$ subunits may associate with other subunits in vivo to form oligomers whose function is not blocked by α-Bgt.

In the peripheral nervous system, α-BgtAChRs have been best characterized in chicken autonomic ganglia. Chicken ciliary ganglion neurons express $\alpha7$ subunits (Vernallis et al 1991) and possess high-affinity binding sites for α-Bgt; yet, α-Bgt does not detectably affect nAChR function at concentrations that saturate its binding site (Ravdin & Berg 1979). Smith et al (1985) showed that ^{125}I-α-Bgt binds with high affinity to a single class of sites in extracts of cultured chicken ciliary ganglion neurons. These sites are distinct from nAChRs, because they can be regulated independently of ACh sensitivity (Smith et al 1983), and because α-Bgt-Sepharose does not adsorb binding sites for mAb 35, which recognizes nAChRs on these neurons (Smith et al 1985). Moreover, an mAb to an $\alpha 7$ fusion protein immunoprecipitated only ^{125}I-α-Bgt binding sites, whereas an mAb to an $\alpha 3$ fusion protein immunoprecipitated only ^{125}I-mAb 35 binding sites (Vernallis et al 1991). Thus, nAChRs containing $\alpha 3$ subunits are largely distinct from α-BgtAChRs containing $\alpha 7$ subunits. What function do α-BgtAChRs serve in the chicken ciliary ganglion? Recent experiments by Vijayaraghavan et al (1992) indicate that activation of α-BgtAChR may raise intracellular Ca^{2+} levels. Fluo-3 imaging of cultured neurons shows that 50 nM α-Bgt blocks an increase in intracellular Ca^{2+} induced by 1 μM nicotine. The calcium signal is dependent upon the presence of extracellular Ca^{2+} and is blocked by Cd^{2+}, nifedipine, and D600, which suggests that it is caused, or at least triggered, by calcium current through voltage-dependent calcium channels. Although a nonspecific effect of these inhibitors upon α-BgtAChRs cannot be ruled out, the more likely explanation is that nicotine acts upon α-BgtAChRs to produce depolarizing inward currents, which activate voltage-dependent calcium channels. Curiously, however, no α-Bgt blockable nicotinic currents can be demonstrated by intracellular recording (Vijayaraghavan et al 1992). If binding of nicotine to α-BgtAChRs leads to activation of voltage-dependent calcium channels via a depolarization, then nicotine-induced calcium signals should be abolished when neurons are voltage clamped.

In chicken sympathetic ganglia, as in ciliary ganglia, α-Bgt binds with high affinity to the neuronal surface and yet does not block functional nAChRs (e.g. Listerud et al 1991; Smith et al 1985). Cultured chicken sympathetic neurons express $\alpha 7$ mRNA in addition to mRNAs for $\alpha 3$, $\alpha 4$, $\alpha 5$, $\beta 2$, and $\beta 4$ subunits (Listerud et al 1991). Treatment of neurons with $\alpha 7$ antisense oligonucleotide results in a 30% reduction in whole-cell currents in response to saturating doses of ACh (Listerud et al 1991). This result suggests that $\alpha 7$ subunits contribute to functional nAChRs. The fact that nicotinic responses of normal cells are not sensitive to α-Bgt implies that $\alpha 7$-containing nAChRs are not homo-oligomers, because heterologously expressed $\alpha 7$ homo-oligomers are blocked by α-Bgt (Couturier et al 1990a). Rather, the nAChRs may be hetero-oligomers formed with other α and/or β subunits. When neurons are treated with $\alpha 3$ antisense oligonucleotide, whole-cell responses to ACh are reduced by 60%, and the remaining currents are partially sensitive to α-Bgt. This might be explained if knock out of $\alpha 3$ subunits leads to the expression of $\alpha 7$ homo-oligomers. However, the slow rate of desensitization of nAChRs in $\alpha 3$ antisense oligonucleotide-treated sympathetic neurons contrasts with the very rapid desensitization of heterologously expressed $\alpha 7$ homo-oligomers (Couturier et al 1990a). Another possibility is that both $\alpha 3$ and $\alpha 7$ subunits normally combine with β subunits to produce an oligomer that is insensitive to α-Bgt; knock out of $\alpha 3$ subunits might then shift the balance of $\alpha 3$ and $\alpha 7$ subunits that associate with β subunits. The presence of significant numbers of $\alpha 3\alpha 7$-containing "hetero-α" nAChRs in chicken sympathetic neurons would distinguish them from neurons in the ciliary ganglion and brain of chickens, where $\alpha 7$ subunits and $\alpha 3$ subunits are largely or completely restricted to separate oligomers (Schoepfer et al 1990; Vernallis et al 1991).

REGULATION OF NEURONAL nAChRS

Developmental Changes in nAChR Number and Properties

Development of ciliary ganglion neurons in chicken embryos is accompanied by a gradual increase in ACh sensitivity that is at least partly caused by an increase in nAChR density (Engisch & Fischbach 1990; Margiotta & Gurantz 1989). Development of these neurons is also accompanied by a change in ACh affinity and in open-channel duration for the principal channel class (40 pS; Margiotta & Gurantz 1989). Neurons taken from chicken sympathetic ganglia at about the time of innervation have different populations of nAChR channels, based on their mean conductances and open times, than neurons taken from ganglia that have been innervated for about a week (Moss & Role 1992; Moss et al 1989). Two obvious objectives are to understand the molecular basis of the change in nAChR properties and to learn whether innervation causes these changes.

Transsynaptic Regulation of Neuronal nAChRs

TRANSIENT EXPRESSION OF nAChR GENES IN BRAIN   In situ hybridization and Northern blot studies performed in developing brain have suggested that the expression of nAChR genes is increased, sometimes transiently, when neurons receive or make synaptic connections. Neurons in the chicken lateral spiriform nucleus do not express detectable levels of $\alpha 2$ mRNA until embryonic day 11 (E11), when choline acetyltransferase-containing fibers first enter the nucleus (Daubas et al 1990). In the chicken optic lobe, expression of the $\beta 2$ gene, but not the $\alpha 4$ gene, is increased more than tenfold between E6 and E12, after which it falls precipitously (Matter et al 1990). In situ hybridization studies show that $\beta 2$ expression occurs in a rostrocaudal gradient over the same time that retinal afferents invade the tectum. Removal of the eye cup at E2 produces more than a tenfold loss of $\beta 2$ expression compared with controls. This suggests that $\beta 2$ expression by optic lobe neurons is transiently stimulated by arriving retinal afferents. The significance of this transient expression is not clear, however, because retinal ganglion cells in chickens are probably not cholinergic.

Expression of nAChR genes can also be influenced by contact with target tissue. In adult goldfish retina, Hieber et al (1992) found an increase in expression of $\alpha 3$ and three distinct β genes 15 days after optic nerve crush, when retinal ganglion cell axons are reforming connections in the optic tectum. If the optic nerve is repeatedly sectioned, so that retinal ganglion cell terminals do not reform connections, no increase in expression of nAChR genes occurs, although tubulin mRNA expression increases manyfold. This intriguing result suggests that retinal ganglion cells are retrogradely stimulated to express nAChR genes once their terminals have arrived at their target.

INITIAL EXPRESSION OF nAChRS IN AUTONOMIC GANGLIA   Does the initial expression of nAChRs and the sustained increase in nAChR number and density on chicken autonomic neurons depend upon the presence of the preganglionic axon? In the ciliary ganglion, Jacob (1991) found that nAChRs first could be detected by immunocytochemical techniques at about the time preganglionic axon terminals contact ganglion cells (E4.5– 5). Engisch & Fischbach (1992) examined the effects of removal of the accessory oculomotor nucleus (AON: the source of preganglionic neurons in chickens) upon the development of ACh sensitivity in ciliary ganglion neurons. Removal of the AON at E4 did not reduce the sensitivity of surviving neurons taken from E14 or E18 embryos, as compared with neurons in sham-operated embryos. These experiments indicate that the appearance of functional AChRs on ciliary ganglion neurons and the sizable developmental increase in ACh sensitivity is not dependent upon contact by preganglionic terminals, nor upon any diffusible factors released by preganglionic neurons. These results are consistent with the findings of McEachern et al (1989), who found that denervation of post-hatched chickens did not alter sensitivity to ACh or to carbachol, a slowly hydro-lyzable analogue of ACh. However, at first sight, these results appear to conflict with those obtained on chicken sympathetic neurons in culture, where innervation has a significant effect upon ACh and carbachol sensitivity (Role 1988). ACh sensitivity was two to four times higher on innervated sympathetic neurons than on uninnervated ones in cultures that contain a source of preganglionic or somatic motor neurons (Gardette et al 1991; Role 1988). Moreover, neurons grown in medium conditioned by spinal cord explants had elevated levels of ACh sensitivity compared with those grown in unconditioned medium (Role 1988). These results suggest that cholinergic neurons can induce nAChRs on sympathetic neurons and that some of their influence is exerted by diffusible factors. These results may not be in conflict with those of Engisch & Fischbach (1992), because an alternate influence in vivo might promote the expression of nAChRs by ganglion neurons in the absence of preganglionic axons. The cholinergic neurons within the ganglion may themselves supply this influence, although it is not likely to be mediated by direct synaptic contact, because intraganglionic synaptic contacts are rare in deafferented ganglia (Furber et al 1987).

Primary sensory neurons isolated from the newborn rat nodose ganglion are sensitive to ACh, and this sensitivity is regulated by culture conditions. Coculture with satellite cells (glial cells) suppresses both ACh sensitivity and formation of synaptic connections between sensory neurons (Baccaglini & Cooper 1982; Cooper 1984). ACh sensitivity has been noted in other primary sensory neurons, including those from dorsal root ganglia of chickens (Boyd et al 1991) and bullfrogs (Morita & Katayama 1989). nAChRs synthesized by sensory neurons may function as presynaptic nAChRs within the CNS.

Marshall (1985) has shown that innervation in bullfrog sympathetic ganglia can influence nAChR gating properties. Lumbar sympathetic ganglia contain two classes of neurons, B cells and C cells, which are innervated by distinct classes of preganglionic axons. Voltage-clamp studies reveal that the time constant describing the decay of synaptic current, which is determined by nAChR burst kinetics, is about twice as long for C cells as for B cells (10 ms versus 5 ms). Marshall (1985) found that synaptic currents of B cells that were cross-innervated by C preganglionic axons decayed with a time constant of 10 ms, whereas synaptic currents of B cells reinnervated by B preganglionic axons decayed with a time constant of 5 ms. Thus, the properties of the nAChRs are under preganglionic control. Papke & Heinemann (1991) found that the average burst duration can be influenced by the subunit composition of rat nAChRs expressed in oocytes. In the bullfrog sympathetic ganglia, B and C preganglionic axons could induce expression of nAChRs that have different subunit composition.

DENERVATION   In the chicken ciliary ganglion, denervation for ten days leads to a threefold reduction in the number of nAChRs recognized by mAb 35 (Boyd et al 1988; Jacob & Berg 1987). The loss of nAChRs is due to a reduction in the intracellular pool (Jacob & Berg 1988); the number of surface nAChRs is not significantly changed (McEachern et al 1989). Former synaptic sites on denervated ciliary ganglion neurons are identifiable in the electron microscope and were labeled with mAb 35 (Jacob & Berg 1988), which suggests that synaptic nAChR clusters remain behind following loss of the nerve terminal, as they do in adult muscle. The sensitivity of ciliary ganglion neurons to ACh or carbachol was also not altered upon denervation for ten days (McEachern et al 1989), and thus the number of functional nAChRs is unchanged following denervation of these neurons.

The failure of denervation to alter ACh sensitivity of ciliary ganglion neurons is consistent with results on bullfrog sympathetic neurons (Dunn & Marshall 1985), but contrasts with earlier results on amphibian para-sympathetic neurons (Dennis & Sargent 1979; Kuffler et al 1971; Roper 1976). Brown (1969) reported that denervation increased sensitivity of cat sympathetic neurons to ACh, but not carbachol, and suggested that denervation supersensitivity to ACh is caused by a loss of acetyl-cholinesterase (AChE), which declines in mammalian sympathetic ganglia upon denervation (see also Bird & Aghajanian 1975). The role of AChE in denervation supersensitivity in the frog cardiac ganglion was examined by Streichert & Sargent (1992), who showed that denervation for two to three weeks increases ACh sensitivity, but does not alter sensitivity to either carbachol or ACh when applied after inhibition of extracellular AChE. Moreover, denervation did not appear to increase the number of nAChRs on the surface of these neurons, assessed by the binding of ^{125}I-n-Bgt, which blocks nearly all functional nAChRs in the ganglion (Sargent et al 1991). These results indicate that denervation supersensitivity in the frog cardiac ganglion is not caused by an increase in the number of nAChRs, but by a reduction in the effectiveness of AChE.

Denervation of frog cardiac neurons alters the distribution and size of nAChR clusters (Sargent & Pang 1988). On normally innervated neurons, most nAChR clusters are located at synaptic sites, which are concentrated at the base of the cell. nAChR clusters on denervated neurons, however, are one-fourth the size of those on normal neurons and are widely dis-tributed over the cell surface. Thus, synaptic nAChR clusters in the frog cardiac ganglion do not survive short periods of denervation, as they do in mature skeletal muscle and in the chicken ciliary ganglion (Jacob & Berg 1988). Instead, their behavior resembles that of immature neuromuscular junctions, where junctional nAChR clusters disperse following denervation (Kuromi & Kidokoro 1984; Slater 1982).

The failure of denervation to increase the number of nAChRs in auto-nomic ganglia implies that neuronal nAChRs are regulated differently from muscle nAChRs, where denervation increases nAChR number many-fold (Hartzell & Fambrough 1972). However, this comparison may not be fair, because the large increase in nAChR number caused by denervation of myofibers results from expression of nAChR mRNAs by extrasynaptic nuclei, which are absent in neurons. If the behavior of only the synaptic nuclei, which are under the local influence of the motor neuron, could be compared with that of the neuronal nucleus, then the situation in muscle and nerve might be deemed comparable.

AXOTOMY   Axotomy results in a pronounced loss of neuronal nAChRs in the ciliary ganglion of post-hatched chickens (Brenner & Martin 1976; Jacob & Berg 1987, 1988; McEachern et al 1989). Halvorsen et al (1991) identified a factor in eye extracts that, when added to cultures of ciliary ganglion neurons, increases nAChR number. Retrograde transport of such a factor may stimulate ganglion neurons to synthesize nAChRs, and loss of this factor upon axotomy may be one cause for the loss of nAChRs. However, Engisch & Fischbach (1990) found that removal of the eye at an early developmental stage does not markedly alter the subsequent development of ACh sensitivity in surviving ciliary ganglion neurons. In these experiments, the cells never innervate their target and so are not directly damaged by the surgery. The subsequent development of normal ACh sensitivity in these cells suggests that the principal effect of axotomy on ciliary ganglion neurons in post-hatched chickens results from damage and not from the loss of target influences. Alternatively, ciliary ganglion neurons may develop a dependence upon their target only after making contact with it. Yet another possibility is that the neurons that do survive loss of their target are unusual; neurons that depend upon their target, and eventually die in its absence, may fail to acquire normal levels of ACh sensitivity.

Functional and Nonfunctional nAChRs

ACh sensitivity is reduced severalfold by elevated K^+ when chicken ciliary ganglion neurons are cultured in medium containing eye extract, whereas the number of mAb 35 binding sites, which represent nAChRs (Halvorsen & Berg 1987; Jacob et al 1984; Smith et al 1986), is unchanged (Smith et al 1986). A careful examination of functional nAChRs on neurons grown in normal and in high K^+ media revealed them to be indistinguishable with regard to reversal potential, EC_{50}, Hill coefficient, mean open-channel conductance, and distribution of open-channel times (Margiotta et al 1987a). By comparing the whole-cell currents elicited by saturating doses of ACh with the average current passing through a single open channel, Margiotta et al (1987a) estimated that neurons grown under normal conditions have about 2000 functional channels/cell, whereas those grown in elevated K^+ have about 700 functional channels/cell. The number of mAb 35 binding sites is about 100,000/cell under either condition. Apparently, only a small fraction of the nAChRs, identified immunochemically, can be opened by ACh. Ciliary ganglion neurons thus contain two pools of nAChRs on their surface: a large nonfunctional pool and a small functional pool. In a related study, Margiotta et al (1987b) found that incubation of ciliary ganglion neurons for six hours with 8-Br-cAMP and IBMX (a phosphodiesterase inhibitor) increased ACh sensitivity two- to threefold without affecting the number of mAb 35 binding sites. Here again, the increase in sensitivity could not be explained by a difference in the functional properties of nAChRs; rather, 8-Br-cAMP increases the number of functional nAChRs. Margiotta et al (1987b) argue that 8-Br-cAMP acts by converting pre-existing, nonfunctional nAChRs to functional ones, rather than by de novo synthesis of functional nAChRs, because the effect is not blocked by protein synthesis inhibitors. It is tempting to speculate that this pool of pre-existing, nonfunctional nAChRs lies on the cell surface, although an intracellular source cannot be ruled out. Does 8-Br-cAMP act directly upon nAChR channels? Margiotta and colleagues found that an increase in ACh sensitivity could be observed within two to five minutes of initiating a whole-cell recording with cAMP in the recording pipette. Thus, cAMP is presumably acting from the inside the cell, perhaps via a protein phosphorylation step. Vijayaraghavan et al (1990) found that incubation of cultures with 8-Br-cAMP and IBMX indeed leads to the phosphorylation of the $\alpha3$ subunit of mAb 35-immunopurified nAChRs, although it is not yet known whether this phosphorylation is necessary or sufficient to produce an increase in ACh sensitivity.

The possible existence of interconvertible pools of functional and nonfunctional nAChRs presents marvelous possibilities for the cell. Within a matter of seconds or minutes, neurons could alter their sensitivity to ACh manyfold. Interconversion of nAChRs between nonfunctional and functional pools is a potential mechanism underlying changes in synaptic efficacy. Nonfunctional nAChRs present a nightmare for the experimenter, however, because there is no assay for them (yet), and because their very existence can be inferred only after the sort of painstaking analysis performed by Margiotta et al (1987a,b). There is no evidence for the existence of a large pool of nonfunctional nAChRs in muscle (Sine & Steinbach 1986).

CONCLUSIONS

Six years ago, Schuetze & Role (1987) concluded that "Compared to muscle nAChRs, little is known about neuronal nAChRs." Six years later, this statement still rings true. However, we have learned a lot about neuronal nAChRs in the intervening time. Perhaps the most striking finding is the considerable diversity of neuronal nAChRs. This diversity takes several forms: open-channel conductance, gating, and agonist and antagonist potencies. At least one source of this diversity is the multiplicity of nAChR genes. Already, ten vertebrate genes have been identified that encode putative nAChR subunits. In theory, these genes, in various combinations of α and β subunits, could encode some 20 different nAChRs, even discounting the possibility of hetero-α- and hetero-β-containing nAChRs. Given the distribution of these genes in the nervous system and the results of heterologous expression studies, probably only a fraction of possible combinations actually occurs. Nonetheless, with at least seven different functional nAChRs constructed to date in oocytes ($\alpha 7$, $\alpha 2\beta 2$, $\alpha 3\beta 2$, $\alpha 4\beta 2$, $\alpha 2\beta 4$, $\alpha 3\beta 4$, $\alpha 4\beta 4$), and with the possibility that additional nAChR types may be formed by mixing α or β subunits, neuronal nAChRs represent a remarkably diverse family of molecules.

Neuronal nAChRs have several properties that distinguish them from muscle nAChRs and that suggest that they might have novel functions in addition to those associated with mediation of fast, excitatory synaptic transmission. Neuronal nAChR channels have a greater Ca^{2+} permeability than muscle nAChR channels, and calcium entry through neuronal nicotinic channels may lead to the activation of second messenger systems and the modulation of cell function. Another novel feature of neuronal nAChRs, illustrated thus far only for chicken ciliary ganglion neurons, is the presence of apparently interconvertible pools of functional and nonfunctional nAChRs.

Although the explosive phase of research on neuronal nAChRs over the past few years has narrowed the gap of understanding between muscle and neuronal nAChRs, it has also raised many questions. I conclude this review with a partial list of these questions.

1. What are nAChRs doing in the brain? In very few instances is ACh known to mediate fast, excitatory transmission at central synapses. Many nAChRs may be presynaptic, but we still have no understanding of how these nAChRs work.
2. How promiscuous is assembly of nAChR subunits in neurons? Do neurons that synthesize multiple α (or β) subunits assemble them into common oligomers? If so, are all possible combinations assembled (let's hope not)? How do neurons regulate assembly of nAChRs?
3. Are there more nAChR genes yet to be identified in chickens and rats? If so, do the new ones form functional nAChRs by assembling with those previously identified and for which no function has yet been found (e.g. $\alpha 6$, $\beta 3$)?
4. How widespread is the existence of nonfunctional pools of surface nAChRs? Can the efficacy of synaptic transmission be regulated by interconversion of pre-existing nonfunctional and functional nAChRs?
5. Does calcium entry through nAChR channels play an important role in cell regulation?
6. What do α-BgtAChRs do? Do $\alpha 7$ subunits form homo-oligomers in the brain, and are their properties similar to those expressed in oocytes?

7. Finally, why are there so many neuronal nAChR genes? Presumably, different gene products serve different functions and may be directed to different parts of the neuron. The details are likely to interest and occupy neuroscientists for some time to come.

ACKNOWLEDGMENTS

I thank Drs. Darwin Berg, Michele Jacob, Zach Hall, Jon Lindstrom, Jim Patrick, Lorna Role, and Joe Henry Steinbach for insightful comments on drafts of this review and for many enjoyable and fruitful discussions. I also thank Drs. David Adams, Darwin Berg, Jean-Pierre Changeux, Vinny Chiappinelli, Francesco Clementi, Christopher Flores, John Dani, Gerald Fishbach, Joel Gallagher, Dan Goldman, Kenneth Kellar, Henry Lester, Jon Lindstrom, Ron Lukas, Christophe Mulle, Jim Patrick, Guillermo Pilar, Lorna Role, and Joe Henry Steinbach for sharing unpublished results.

Cited results from my own laboratory have been supported by National Institutes of Health grant NS 24207.

Literature Cited

Adams, D. J., Nutter, T. J. 1992. Calcium permeability and modulation of nicotinic acetylcholine receptor-channels in rat parasympathetic neurons. *J. Physiol. (Paris).* In press

Alkondon, M., Albuquerque, E. X. 1991. Initial characterization of the nicotinic acetylcholine receptors in rat hippocampal neurons. *J. Recept. Res.* 11: 1001–21

Anand, R., Conroy, W. G., Schoepfer, R., Whiting, P., Lindstrom, J. 1991. Neuronal nicotinic acetylcholine receptors expressed in *Xenopus* oocytes have a pentameric quarternary structure. *J. Biol. Chem.* 266: 11192–98

Anand, R., Lindstrom, J. 1990. Nucleotide sequence of the human nicotinic acetylcholine receptor β2 subunit gene. *Nucleic Acids Res.* 18: 4272 (Abstr.)

Baccaglini, P., Cooper, E. 1982. Influences on the expression of acetylcholine receptors on rat nodose neurones in cell culture. *J. Physiol.* 324: 441–51

Ballivet, M., Nef, P., Couturier, S., Rungger, D., Bader, C. R., et al. 1988. Electrophysiology of a chick neuronal nicotinic acetylcholine receptor expressed in *Xenopus* oocytes after cDNA injection. *Neuron* 1: 847–52

Bertrand, D., Ballivet, M., Rungger, D. 1990. Activation and blocking of neuronal nicotinic acetylcholine receptor reconstituted in *Xenopus* oocytes. *Proc. Natl. Acad. Sci. USA* 87: 1993–97

Bertrand, D., Devillers-Thiery, A., Revah, F., Galzi, J.-L., Hussy, N., et al. 1992. Unconventional pharmacology of a neuronal nicotinic receptor mutated in the channel domain. *Proc. Natl. Acad. Sci. USA* 89: 1261–65

Bird, S. J., Aghajanian, G. K. 1975. Denervation supersensitivity in the cholinergic septohippocampal pathway: a microiontophoretic study. *Brain Res.* 100: 355–70

Bormann, J., Matthaei, H. 1983. Three types of acetylcholine-induced single channel currents in clonal rat pheochromocytoma cells. *Neurosci. Lett.* 40: 193–97

Bossy, B., Ballivet, M., Spierer, P. 1988. Conservation of neural nicotinic acetylcholine receptors from *Drosophila* to vertebrate central nervous systems. *EMBO J.* 7: 611–18

Boulter, J., Connolly, J., Deneris, E., Goldman, D., Heinemann, S., Patrick, J. 1987. Functional expression of two neuronal nicotinic acetylcholine receptors from cDNA clones identifies a gene family. *Proc. Natl. Acad. Sci. USA* 84: 7763–67

Boulter, J., Evans, K., Goldman, D., Martin, G., Treco, D., et al. 1986. Isolation of a cDNA clone coding for a possible neural nicotinic acetylcholine receptor α-subunit. *Nature* 319: 368–74

Boulter, J., O'Shea-Greenfield, A., Duvoisin, R. M., Connolly, J. G., Wada, E., et al. 1990. α3, α5, and β4: three members of the rat neuronal nicotinic acetylcholine receptor-related gene family form a gene cluster. *J. Biol. Chem.* 265: 4472–82

Boyd, R. T., Jacob, M. H., Couturier, S., Ballivet, M., Berg, D. K. 1988. Expression and regulation of neuronal acetylcholine receptor mRNA in chick ciliary ganglia. *Neuron* 1: 495–502

Boyd, R. T., Jacob, M. H., McEachern, A. E., Caron, S., Berg, D. K. 1991. Nicotinic acetylcholine receptor mRNA in dorsal root ganglion neurons. *J. Neurobiol.* 22: 1–14

Breer, H., Kleene, R., Hinz, G. 1985. Molecular forms and subunit structure of the acetylcholine receptor in the central nervous system of insects. *J. Neurosci.* 5: 3386–92

Brenner, H. R., Martin, A. R. 1976. Reduction in acetylcholine sensitivity of axotomized ciliary ganglion cells. *J. Physiol.* 260: 159–75

Britto, L. R., Keyser, K. T., Lindstrom, J. M., Karten, H. J. 1992. Immunohistochemical localization of α-bungarotoxin binding proteins in the chick brain. *J. Comp. Neurol.* 317: 325–40

Brown, D. A. 1969. Responses of normal and denervated cat superior cervical ganglia to some stimulant compounds. *J. Physiol.* 201: 225–36

Brown, D. A., Docherty, R. J., Halliwell, J. V. 1984. The action of cholinomimetic substances on impulse conduction in the habenulointerpeduncular pathway of the rat in vitro. *J. Physiol.* 353: 101–9

Brown, D. A., Docherty, R. J., Halliwell, J. V. 1983. Chemical transmission in the rat interpeduncular nucleus in vitro. *J. Physiol.* 341: 655–70

Calabresi, P., Lacey, M. G., North, R. A. 1989. Nicotinic excitation of rat ventral tegmental neurones in vitro studies by intracellular recording. *Br. J. Pharmacol.* 98: 135–40

Cauley, K., Agranoff, B. W., Goldman, D. 1990. Multiple nicotinic acetylcholine receptor genes are expressed in goldfish retina. *J. Neurosci.* 10: 670–83

Cauley, K., Agranoff, B. W., Goldman, D. 1989. Identification of a noval nicotinic acetylcholine receptor structural subunit expressed in goldfish retina. *J. Cell Biol.* 108: 637–45

Changeux, J.-P. 1990. Functional architecture and dynamics of the nicotinic acetylcholine receptor: an allosteric ligand-gated ion channel. *Fidia Res. Found. Neurosci. Found. Lect.* 4: 21–168

Charnet, P., Labarca, C., Cohen, B. N., Davidson, N., Lester, H. A., Pilar, G. 1992. Pharmacological and kinetic properties of α4β2 neuronal nicotinic acetylcholine receptors expressed in *Xenopus* oocytes. *J. Physiol.* 450: 375–94

Chiappinelli, V. A. 1991. κ-Neurotoxins and α-neurotoxins: effects on neuronal nicotinic acetylcholine receptors. In *Snake Toxins*, ed. A. L. Harvey, pp. 223–58. New York: Pergamon

Chiappinelli, V. A. 1983. Kappa-bungarotoxin: a probe for the neuronal nicotinic receptor in the avian ciliary ganglion. *Brain Res.* 277: 9–21

Chini, B., Clementi, F., Hukovic, N., Sher, E. 1992. Neuronal type α-bungarotoxin receptors and the α5 nicotinic receptor subunit gene are expressed in neuronal and non-neuronal human cell lines. *Proc. Natl. Acad. Sci. USA* 89: 1572–76

Clarke, P. B. S. 1990. The central pharmacology of nicotine: electrophysiological approaches. In *Nicotine Pharmacology: Molecular, Cellular, and Behavioral Aspects*, ed. S. Wonnacott, M. A. H. Russell, I. P. Stolerman, pp. 158–93. Oxford: Oxford Univ. Press

Clarke, P. B. S., Hamill, G. S., Nadi, N. S., Jacobowitz, D. M., Pert, A. 1986. [3H]-Nicotine- and [125I]-α-bungarotoxin-labeled nicotinic receptors in the interpeduncular nucleus of rats. II. Effects of habenular deafferentation. *J. Comp. Neurol.* 251: 407–13

Clarke, P. B. S., Pert, A. 1985. Autoradiographic evidence for nicotine receptors on nigrostriatal and mesolimbic dopaminergic neurons. *Brain Res.* 348: 355–58

Clarke, P. B. S., Schwartz, R. D., Paul, S. M., Pert, C. B., Pert, A. 1985. Nicotinic binding in rat brain: autoradiographic comparison of [3H]acetylcholine, [3H]nicotine, and [125I]-α-bungarotoxin. *J. Neurosci.* 5: 1307–15

Claudio, T. 1989. Molecular genetics of acetylcholine receptor-channels. In *Molecular Neurobiology*, ed. D. M. Glover, B. D. Hames, pp. 63–142. Oxford: Oxford Univ. Press

Cobbett, P., Mason, W. T., Poulain, D. A. 1985. Intracellular analysis of control of rat supraoptic neurone (SON) activity in vitro by acetylcholine. *J. Physiol.* 371: 216P (Abstr.)

Conroy, W. G., Vernallis, A. B., Berg, D. K. 1992. The α5 gene product assembles with multiple acetylcholine receptor subunits to form distinctive receptor subtypes in brain. *Neuron.* In press

Conti-Tronconi, B. M., Dunn, S. M. J., Barnard, E. A., Dolly, J. O., Lai, F. A., et al. 1985. Brain and muscle nicotinic acetylcholine receptors are different but homologous proteins. *Proc. Natl. Acad. Sci. USA* 82: 5208–12

Cooper, E. 1984. Synapse formation among developing sensory neurones from rat nodose ganglia grown in tissue culture. *J. Physiol.* 351: 263–74

Cooper, E., Couturier, S., Ballivet, M. 1991. Pentameric structure and subunit stoichiometry of a neuronal acetylcholine receptor. *Nature* 350: 235–38

Couturier, S., Bertrand, D., Matter, J.-M., Hernandez, M.-C., Bertrand, S., et al. 1990a. A neuronal nicotinic acetylcholine receptor subunit (α7) is developmentally regulated and forms a homo-oligomeric channel blocked by α-BTX. *Neuron* 5: 847–56

Couturier, S., Erkman, L., Valera, S., Rungger, D., Bertrand, S., et al. 1990b. α5, α3, and non-α3: three clustered avian genes encoding neuronal nicotinic acetylcholine receptor-related subunits. *J. Biol. Chem.* 265: 17560–67

Cull-Candy, S. G., Mathie, A., Powis, D. A. 1988. Acetylcholine receptor channels and their block by clonidine in cultured bovine chromaffin cells. *J. Physiol.* 402: 255–78

Daubas, P., Devillers-Thiery, A., Geoffroy, B., Martinez, S., Bessis, A., Changeux, J.-P. 1990. Differential expression of the neuronal acetylcholine receptor α2 subunit gene during chick brain development. *Neuron* 5: 49–60

David, J. A., Sattelle, D. B. 1984. Actions of cholinergic pharmacological agens on the cell body membrane of the fast coxal depressor motoneurone of the cockroach (*Periplaneta americana*). *J. Exp. Biol.* 108: 119–36

de la Garza, R., McGuire, T. J., Freedman, R., Hoffer, B. J. 1987. Selective antagonism of nicotine actions in the rat cerebellum with α-bungarotoxin. *Neuroscience* 23: 887–91

Decker, E. R., Dani, J. A. 1990. Calcium permeability of the nicotinic acetylcholine receptor: the single-channel calcium influx is significant. *J. Neurosci.* 10: 3413–20

Deneris, E. S., Boulter, J., Swanson, L. W., Patrick, J., Heinemann, S. 1989. β3: a new member of nicotinic acetylcholine receptor gene family is expressed in brain. *J. Biol. Chem.* 264: 6268–72

Deneris, E. S., Connolly, J., Boulter, J., Wada, E., Wada, K., et al. 1988. Primary structure and expression of β2: a novel subunit of neuronal nicotinic acetylcholine receptors. *Neuron* 1: 45–54

Deneris, E. S., Connolly, J., Rogers, S. W., Duvoisin, R. 1991. Pharmacological and functional diversity of neuronal nicotinic acetylcholine receptors. *Trends Pharmacol. Sci.* 12: 34–40

Dennis, M. J., Sargent, P. B. 1979. Loss of extrasynaptic acetylcholine sensitivity upon reinnervation of parasympathetic ganglion cells. *J. Physiol.* 289: 263–75

Derkach, V. A., North, R. A., Selyanko, A. A., Skok, V. I. 1987. Single channels activated by acetylcholine in rat superior cervical ganglion. *J. Physiol.* 388: 141–51

Deutch, A. Y., Holliday, J., Roth, R. H., Chun, L. L. Y., Hawrot, E. 1987. Immunohistochemical localization of a neuronal nicotinic acetylcholine receptor in mammalian brain. *Proc. Natl. Acad. Sci. USA* 84: 8697–8701

Dineley-Miller, K., Patrick, J. 1992. Gene transcripts for the nicotinic acetylcholine receptor subunit, β4, are distributed in multiple areas of the rat central nervous system. *Mol. Brain Res.* In press

Dryer, S. E., Chiappinelli, V. A. 1987. Analysis of quantal content and quantal conductance in two populations of neurons in the avian ciliary ganglion. *Neuroscience* 20: 905–10

Dunn, P. M., Marshall, L. M. 1985. Lack of nicotinic supersensitivity in frog sympathetic neurones following denervation. *J. Physiol.* 363: 211–25

Duvoisin, R. M., Deneris, E. S., Patrick, J., Heinemann, S. 1989. The functional diversity of the neuronal nicotinic acetylcholine receptors is increased by a novel subunit: β4. *Neuron* 3: 487–96

Dwork, A. J., Desmond, J. T. 1991. Purification of a nicotinic acetylcholine receptor from rat brain by affinity chromatography directed at the acetylcholine binding site. *Brain Res.* 552: 119–23

Engisch, K. L., Fischbach, G. D. 1992. The development of ACh- and GABA-activated currents in embryonic chick ciliary ganglion neurons in the absence of innervation in vivo. *J. Neurosci.* 12: 1115–25

Engisch, K. L., Fischbach, G. D. 1990. The development of ACh- and GABA-activated currents in normal and target-deprived embryonic chick ciliary ganglia. *Dev. Biol.* 139: 417–26

Fertuck, H. C., Salpeter, M. M. 1976. Quantitation of junctional and extrajunctional acetylcholine receptors by electron microscopic autoradiography after ^{125}I-α-bungarotoxin binding at mouse neuromuscular junctions. *J. Cell Biol.* 69: 144–58

Fieber, L. A., Adams, D. J. 1991. Acetylcholine-evoked currents in cultured neurones dissociated from rat parasympathetic cardiac ganglia. *J. Physiol.* 434: 215–37

Flores, C. M., Rogers, S. W., Pabreza, L. A.,

Wolfe, B. B., Kellar, K. J. 1992. A subtype of nicotinic cholinergic receptor in rat brain is comprised of α-4 and β-2 subunits and is up-regulated by chronic nicotine treatment. *Mol. Pharm.* 41: 31–37

Fornasari, D., Chini, B., Tarroni, P., Clementi, F. 1990. Molecular cloning of human neuronal nicotinic receptor α3-subunit. *Neurosci. Lett.* 111: 351–56

Fuchs, P. A., Murrow, B. W. 1992a. A novel cholinergic receptor mediates inhibition of chick cochlear hair cells. *Proc. R. Soc. London Ser. B* 248: 35–40

Fuchs, P. A., Murrow, B. W. 1992b. Cholinergic inhibition of short (outer) hair cells of the chick's cochlea. *J. Neurosci.* 12: 800–9

Furber, S., Oppenheim, R. W., Prevette, D. 1987. Naturally-occurring neuron death in the ciliary ganglion of the chick embryo following removal of preganglionic input: evidence for the role of afferents in ganglion cell survival. *J. Neurosci.* 7: 1816–32

Galzi, J.-L., Revah, F., Bessis, A., Changuex, J.-P. 1991. Functional architecture of the nicotinic acetylcholine receptor: from electric organ to brain. *Annu. Rev. Pharmacol.* 31: 37–72

Gardette, R., Listerud, M. D., Brussaard, A. B., Role, L. W. 1991. Developmental changes in transmitter sensitivity and synaptic transmission in embryonic chicken sympathetic neurons innervated in vitro. *Dev. Biol.* 147: 83–95

Giorguieff-Chesselet, M. F., Kemel, M. L., Wandscheer, D., Glowinski, J. 1979. Regulation of dopamine release by presynaptic nicotinic receptors in rat striatal slices: effects of nicotine in a low concentration. *Life Sci.* 25: 1257–62

Goldman, D., Deneris, E., Luyten, W., Kochlar, A., Patrick, J., Heinemann, S. 1987. Members of a nicotinic acetylcholine receptor gene family are expressed in different regions of the mammalian central nervous system. *Cell* 48: 965–73

Gotti, C., Ogando, A. E., Hanke, W., Schlue, R., Moretti, M., Clementi, F. 1991. Purification and characterization of an α-bungarotoxin receptor that forms a functional nicotinic channel. *Proc. Natl. Acad. Sci. USA* 88: 3258–62

Gross, A., Ballivet, M., Rungger, D., Bertrand, D. 1991. Neuronal nicotinic acetylcholine receptors expressed in *Xenopus* oocytes: role of the α subunit in agonist sensitivity and desensitization. *Pflugers Arch.* 419: 545–51

Halvorsen, S. W., Berg, D. K. 1990. Subunit composition of nicotinic acetylcholine receptors from chick ciliary ganglia. *J. Neurosci.* 10: 1711–18

Halvorsen, S. W., Berg, D. K. 1987. Affinity labeling of neuronal acetylcholine receptor subunits with an α-neurotoxin that blocks receptor function. *J. Neurosci.* 7: 2547–55

Halvorsen, S. W., Schmid, H. A., McEachern, A. E., Berg, D. K. 1991. Regulation of acetylcholine receptors on chick ciliary ganglion neurons by components from the synaptic target tissue. *J. Neurosci.* 11: 2177–86

Hanke, W., Breer, H. 1986. Channel properties of an insect neuronal acetylcholine receptor protein reconstituted in planar lipid bilayers. *Nature* 321: 171–74

Hartzell, H. C., Fambrough, D. M. 1972. Acetylcholine receptors: distribution and extrajunctional density in rat diaphragm after denervation correlated with acetylcholine sensitivity. *J. Gen. Physiol.* 60: 248–62

Hattori, T., Fibiger, H. C. 1982. On the use of lesions of afferents to localize neurotransmitter receptor sites in the striatum. *Brain Res.* 238: 245–50

Henley, J. M., Lindstrom, J. M., Oswald, R. E. 1986a. Acetylcholine receptor systhesis in retina and transport to optic tectum in goldfish. *Science* 232: 1627–29

Henley, J. M., Mynlieff, M., Lindstrom, J. M., Oswald, R. E. 1986b. Interaction of monoclonal antibodies to electroplaque acetylcholine receptors with the α-bungarotoxin binding site of goldfish brain. *Brain Res.* 364: 405–8

Hermans-Borgmeyer, I., Zopf, D., Ryseck, R.-P., Hovemann, H., Betz, H., Gundelfinger, E. D. 1986. Primary structure of a developmentally regulated nicotinic acetylcholine receptor from *Drosophila*. *EMBO J.* 5: 1503–8

Hieber, V., Agranoff, B. W., Goldman, D. 1992. Target-dependent regulation of retinal nicotinic acetylcholine receptor and tubulin RNAs during optic nerve regeneration in goldfish. *J. Neurochem.* 58: 1009–15

Hieber, V., Bouchey, J., Agranoff, B. W., Goldman, D. 1990a. Nucleotide and deduced amino acid sequence of the goldfish neural nicotinic acetylcholine receptor α-3 subunit. *Nucleic Acids Res.* 18: 5293 (Abstr.)

Hieber, V., Bouchey, J., Agranoff, B. W., Goldman, D. 1990b. Nucleotide and deduced amino acid sequence of the goldfish neural nicotinic acetylcholine receptor β-2 subunit. *Nucleic Acids Res.* 18: 5307 (Abstr.)

Higgins, L. S., Berg, D. K. 1987. Immunological identification of a nicotinic acetylcholine receptor on bovine chromaffin cells. *J. Neurosci.* 7: 1792–98

Hunt, S. P., Schmidt, J. 1978a. Some obser-

438   SARGENT

vations on the binding patterns of α-bungarotoxin in the central nervous system of the rat. *Brain Res.* 157: 213–32

Hunt, S. P., Schmidt, J. 1978b. The electron microscopic autoradiographic localization of α-bungarotoxin binding sites within the central nervous system of the rat. *Brain Res.* 142: 152–59

Ifune, C. K., Steinbach, J. H. 1992. Inward rectification of acetylcholine-elicited currents in rat pheochromocytoma cells. *J. Physiol.* In press

Ifune, C. K., Steinbach, J. H. 1991. Voltage-dependent block by magnesium of neuronal nicotinic acetylcholine receptor channels in rat pheochromocytoma cells. *J. Physiol.* 443: 683–701

Ifune, C. K., Steinbach, J. H. 1990. Rectification of acetylcholine-elicited currents in PC12 pheochromocytoma cells. *Proc. Natl. Acad. Sci. USA* 87: 4794–98

Isenberg, K. E., Meyer, G. E. 1989. Cloning of a putative neuronal nicotinic acetylcholine receptor subunit. *J. Neurochem.* 52: 988–91

Jacob, M. H. 1991. Acetylcholine receptor expression in developing chick ciliary ganglion neurons. *J. Neurosci.* 11: 1701–12

Jacob, M. H., Berg, D. K. 1988. The distribution of acetylcholine receptors in chick ciliary ganglion neurons following disruption of ganglionic connections. *J. Neurosci.* 8: 3838–49

Jacob, M. H., Berg, D. K. 1987. Effects of preganglionic denervation and postganglionic axotomy on acetylcholine receptors in the chick ciliary ganglion. *J. Cell Biol.* 105: 1847–54

Jacob, M. H., Berg, D. K., Lindstrom, J. M. 1984. Shared antigenic determinants between Electrophorus acetylcholine receptor and a synaptic component on chicken ciliary ganglion neurons. *Proc. Natl. Acad. Sci. USA* 81: 3223–27

Jacob, M. H., Berg, D. K. 1983. The ultrastructural localization of α-bungarotoxin binding sites in relation to synapses on chick ciliary ganglion neurons. *J. Neurosci.* 3: 260–71

Jonas, P., Baumann, A., Merz, B., Gundelfinger, E. D. 1990. Structure and developmental expression of the Dα2 gene encoding a novel acetylcholine receptor protein from *Drosophila melanogaster*. *FEBS Lett.* 269: 264–68

Karlin, A. 1991. Explorations of the nicotinic acetylcholine receptor. *Harvey Lect.* 71: 71–107

Kemp, G., Bentley, L., McNamee, M. G., Morley, B. J. 1985. Purification and characterization of the α-bungarotoxin binding protein from rat brain. *Brain Res.* 347: 274–83

Kemp, G., Morley, B. J. 1986. Ganglionic AChRs and high affinity nicotinic binding sites are not equivalent. *FEBS Lett.* 205: 265–68

Kuba, K., Tanaka, E., Kumamoto, E., Minota, S. 1989. Patch clamp experiments on nicotinic acetylcholine receptor-ion channels in bullfrog sympathetic ganglion cells. *Pflugers Arch.* 414: 105–12

Kuffler, S. W., Dennis, M. J., Harris, A. J. 1971. The development of chemosensitivity in extrasynaptic areas of the neuronal surface after denervation of parasympathetic ganglion cells in the heart of the frog. *Proc. R. Soc. London Ser. B* 177: 555–63

Kuromi, H., Kidokoro, Y. 1984. Denervation disperses acetylcholine receptor clusters at the neuromuscular junction in *Xenopus* cultures. *Dev. Biol.* 104: 421–27

Lamar, E., Miller, K., Patrick, J. 1990. Amplification of genomic sequences identifies a new gene, α6, in the nicotinic acetylcholine receptor gene family. *Soc. Neurosci. Abstr.* 16: 681

Langdon, R. B., Freeman, J. A. 1987. Pharmacology of retinotectal transmission in the goldfish: effects of nicotinic ligands, strychnine, and kynurenic acid. *J. Neurosci.* 7: 760–73

Lindstrom, J., Schoepfer, R., Conroy, W. G., Whiting, P. 1990. Structural and functional heterogeneity of nicotinic receptors. *CIBA Found. Symp.* 152: 23–52

Lindstrom, J., Schoepfer, R., Whiting, P. 1987. Molecular studies of the neuronal nicotinic acetylcholine receptor family. *Mol. Neurobiol.* 1: 281–337

Lingle, C. J., Maconochie, D., Steinbach, J. H. 1992. Activation of skeletal muscle nicotinic acetylcholine receptors. *J. Membrane Biol.* 126: 195–217

Lipton, S. A., Aizenman, E., Loring, R. H. 1987. Neural nicotinic acetylcholine responses in solitary mammalian retinal ganglion cells. *Pflugers Arch.* 410: 37–43

Listerud, M., Brussaard, A. B., Devay, P., Colman, D. R., Role, L. W. 1991. Functional contribution of neuronal AChR subunits by antisense oligonucleotides. *Science* 254: 1518–21

London, E. D., Connolly, R. J., Szikszay, M., Wamsley, J. K., Dam, M. 1988. Effects of nicotine on local cerebral glucose utilization in the rat. *J. Neurosci.* 8: 3920–28

Loring, R. H., Aizenman, E., Lipton, S. A., Zigmond, R. E. 1989. Characterization of nicotinic receptors in chick retina using a snake venom neurotoxin that blocks neuronal nicotinic receptor function. *J. Neurosci.* 9: 2423–31

Loring, R. H., Chiappinelli, V. A., Zigmond,

R. E., Cohen, J. B. 1984. Characterization of a snake venom neurotoxin which blocks nicotinic transmission in the avian ciliary ganglion. *Neuroscience* 11: 989–99

Loring, R. H., Dahm, L. M., Zigmond, R. E. 1985. Localization of α-bungarotoxin binding sites in the ciliary ganglion of the embryonic chick: an autoradiographic study at the light and electron microscopic level. *Neuroscience* 14: 645–60

Loring, R. H., Sah, D. W. Y., Landis, S. C., Zigmond, R. E. 1988. The ultrastructural distribution of putative nicotinic receptors on cultured neurons from the rat superior cervical ganglion. *Neuroscience* 24: 1071–80

Loring, R. H., Zigmond, R. E. 1988. Characterization of neuronal nicotinic receptors by snake venom neurotoxins. *Trends Neurosci.* 11: 73–78

Loring, R. H., Zigmond, R. E. 1987. Ultrastructural distribution of ^{125}I-toxin F binding sites on chick ciliary neurons: synaptic localization of a toxin that blocks ganglionic nicotinic receptors. *J. Neurosci.* 7: 2153–62

Luetje, C. W., Patrick, J. 1991. Both α- and β-subunits contribute to the agonist sensitivity of neuronal nicotinic acetylcholine receptors. *J. Neurosci.* 11: 837–45

Luetje, C. W., Patrick, J., Séguéla, P. 1990a. Nicotine receptors in the mammalian brain. *FASEB J.* 4: 2753–60

Luetje, C. W., Wada, K., Rogers, S., Abramson, S. N., Tsuji, K., Heinemann, S., Patrick, J. 1990b. Neurotoxins distinguish between different neuronal nicotinic acetylcholine receptor subunit combinations. *J. Neurochem.* 55: 632–40

Lukas, R. J. 1990. Heterogeneity of high-affinity nicotinic [^{3}H]-acetylcholine binding sites. *J. Pharmacol. Exp. Ther.* 253: 51–57

Lukas, R. J. 1989. Nicotinic acetylcholine receptor diversity: agonist binding and functional potency. *Prog. Brain Res.* 79: 117–27

Lukas, R. J., Bencherif, M. 1992. Heterogeneity and regulation of nicotinic acetylcholine receptors. *Int. Rev. Neurobiol.* 34: 25–131

Margiotta, J. F., Berg, D. K., Dionne, V. E. 1987a. The properties and regulation of functional acetylcholine receptors on chick ciliary ganglion neurons. *J. Neurosci.* 7: 3612–22

Margiotta, J. F., Berg, D. K., Dionne, V. E. 1987b. Cyclic AMP regulates the proportion of functional acetylcholine receptors on chicken ciliary ganglion neurons. *Proc. Natl. Acad. Sci. USA* 84: 8155–59

Margiotta, J. F., Gurantz, D. 1989. Changes in the number, function, and regulation of nicotinic acetylcholine receptors during

neuronal development. *Dev. Biol.* 135: 326–39

Marks, M. J., Collins, A. C. 1982. Characterization of nicotine binding in mouse brain and comparison with the binding of α-bungarotoxin and quinuclidinyl benzilate. *Mol. Pharmacol.* 22: 554–64

Marshall, J., Buckingham, S. D., Shingai, R., Lunt, G. G., Goosey, M. W., et al. 1990. Sequence and functional expression of a single α subunit of an insect nicotinic acetylcholine receptor. *EMBO J.* 9: 4391–98

Marshall, L. M. 1985. Presynaptic control of synaptic channel kinetics in sympathetic neurones. *Nature* 317: 621–23

Marshall, L. M. 1981. Synaptic localization of α-bungarotoxin binding which blocks nicotinic transmission at frog sympathetic neurons. *Proc. Natl. Acad. Sci. USA* 78: 1948–52

Mathie, A., Colquhoun, D., Cull-Candy, S. G. 1990. Rectification of currents activated by nicotinic acetylcholine receptors in rat sympathetic ganglion neurones. *J. Physiol.* 427: 625–55

Mathie, A., Cull-Candy, S. G., Colquhoun, D. 1987. Single-channel and whole-cell currents evoked by acetylcholine in dissociated sympathetic neurons of the rat. *Proc. R. Soc. London Ser. B* 232: 239–48

Matter, J.-M., Matter-Sadzinski, L., Ballivet, M. 1990. Expression of neuronal nicotinic acetylcholine receptor genes in the developing visual system. *EMBO J.* 9: 1021–26

McCarthy, M. P., Earnest, J. P., Young, E. F., Choe, S., Stroud, R. M. 1986. The molecular biology of the acetylcholine receptor. *Annu. Rev. Neurosci.* 9: 383–413

McEachern, A. E., Jacob, M. H., Berg, D. K. 1989. Differential effects of nerve transection on the ACh and GABA receptors of chick ciliary ganglion neurons. *J. Neurosci.* 9: 3899–3907

McLane, K. E., Wu, X., Conti-Tronconi, B. M. 1990. Identification of a brain acetylcholine receptor α subunit able to bind α-bungarotoxin. *J. Biol. Chem.* 265: 9816–24

Messing, A., Gonatas, N. K. 1983. Extrasynaptic localization of α-bungarotoxin receptors in cultured chick ciliary ganglion neurons. *Brain Res.* 269: 172–76

Morita, K., Katayama, Y. 1989. Bullfrog dorsal root ganglion cells having tetrodotoxin-resistant spikes are endowed with nicotinic receptors. *J. Neurophysiol.* 62: 657–64

Morris, B. J., Hicks, A. A., Wisden, W., Darlison, M. G., Hunt, S. P., Barnard, E. A. 1990. Distinct regional expression of nicotinic acetylcholine receptor genes in chick brain. *Mol. Brain Res.* 7: 305–15

Moss, B. L., Role, L. W. 1992. Enhanced ACh sensitivity is accompanied by changes in ACh receptor channel properties and segregation of ACh receptor subtypes on sympathetic neurons during innervation in vivo. *J. Neurosci.* In press

Moss, B. L., Schuetze, S. M., Role, L. W. 1989. Functional properties and developmental regulation of nicotinic acetylcholine receptors on embryonic chicken sympathetic neurons. *Neuron* 3: 597–607

Mulle, C., Changeux, J.-P. 1990. A novel type of nicotinic receptor in the rat central nervous system characterized by patch-clamp techniques. *J. Neurosci.* 10: 169–75

Mulle, C., Choquet, D., Korn, H., Changeux, J.-P. 1992a. Calcium influx through nicotinic receptor in rat central neurons: its relevance to cellular regulation. *Neuron* 8: 135–43

Mulle, C., Lena, C., Changeux, J.-P. 1992b. Potentiation of nicotinic receptor response by external calcium in rat central neurons. *Neuron* 8: 937–45

Mulle, C., Vidal, C., Benoit, P., Changeux, J.-P. 1991. Existence of different subtypes of nicotinic acetylcholine receptors in the rat habenulo-interpeduncular system. *J. Neurosci.* 11: 2588–97

Nef, P., Oneyser, C., Alliod, C., Couturier, S., Ballivet, M. 1988. Genes expressed in the brain define three distinct neuronal nicotinic acetylcholine receptors. *EMBO J.* 7: 595–601

Neuhaus, R., Cachelin, A. B. 1990. Changes in the conductance of the neuronal nicotinic acetylcholine receptor channel induced by magnesium. *Proc. R. Soc. London Ser. B* 241: 78–84

Nicoll, R. A., Malenka, R. C., Kauer, J. A. 1990. Functional comparison of neurotransmitter subtypes in the mammalian central nervous system. *Physiol. Rev.* 70: 513–65

Nooney, J. M., Lambert, J. J., Chiappinelli, V. A. 1992. The interaction of κ-bungarotoxin with the nicotinic receptor of bovine chromaffin cells. *Brain Res.* 573: 77–82

Norman, R. I., Mehraban, F., Barnard, E. A., Dolly, J. D. 1982. Nicotinic acetylcholine receptor from chick optic lobe. *Proc. Natl. Acad. Sci. USA* 79: 1321–25

Nutter, T. J., Adams, D. J. 1991. The permeability of neuronal nicotinic receptor-channels to monovalent and divalent inorganic cations. *Biophys. J.* 59: 34a (Abstr.)

Oswald, R. E., Freeman, J. A. 1981. α-Bungarotoxin binding and central nervous system nicotinic acetylcholine receptors. *Neuroscience* 6: 1–14

Papke, R. L., Boulter, J., Patrick, J., Heine-
mann, S. 1989. Single-channel currents of rat neuronal nicotinic acetylcholine receptors expressed in *Xenopus* oocytes. *Neuron* 3: 589–96

Papke, R. L., Heinemann, S. F. 1991. The role of the β4-subunit in determining the kinetic properties of rat neuronal nicotinic acetylcholine α3-receptors. *J. Physiol.* 440: 95–112

Parkinson, D., Kratz, K. E., Daw, N. W. 1988. Evidence for a nicotinic component to the actions of acetylcholine in cat visual cortex. *Exp. Brain Res.* 73: 553–68

Paton, W. D. M., Zaimis, E. J. 1949. The pharmacological actions of polymethylene bistrimethylammonium salts. *Br. J. Pharmacol.* 4: 381–400

Patrick, J., Stallcup, B. 1977a. α-Bungarotoxin binding and cholinergic receptor function on a rat sympathetic nerve line. *J. Biol. Chem.* 252: 8629–33

Patrick, J., Stallcup, W. B. 1977b. Immunological distinction between acetylcholine receptor and the α-bungarotoxin-binding component on sympathetic neurons. *Proc. Natl. Acad. Sci. USA* 74: 4689–92

Phelan, K. D., Gallagher, J. P. 1992. Direct muscarinic and nicotinic receptor-mediated excitation of rat medial vestibular nucleus neurons in vitro. *Synapse* 10: 349–58

Pinnock, R. D., Lummis, S. C. R., Chiappinelli, V. A., Sattelle, D. B. 1988. κ-Bungarotoxin blocks an α-bungarotoxin-sensitive nicotinic receptor in the insect central nervous system. *Brain Res.* 458: 45–52

Prusky, G. T., Shaw, C., Cynander, M. S. 1987. Nicotine receptors are located on lateral geniculate terminals in cat visual cortex. *Brain Res.* 412: 131–38

Quik, M., Geertsen, S. 1988. Neuronal nicotinic α-bungarotoxin sites. *Can. J. Physiol. Pharmacol.* 66: 971–79

Raimondi, E., Rubboli, F., Moralli, D., Chini, B., Fornasari, D., et al. 1992. Chromosomal localization and physical linkage of the genes encoding the human α3, α5, and β4 neuronal nicotinic receptor subunits. *Genomics* 12: 849–50

Ravdin, P. M., Berg, D. K. 1979. Inhibition of neuronal acetylcholine sensitivity by α-toxins from Bungarus multicinctus venom. *Proc. Natl. Acad. Sci. USA* 76: 2072–76

Revah, F., Bertrand, D., Galzi, J.-L., Devillers-Thiery, A., Mulle, C., et al. 1991. Mutations in the channel domain alter desensitization of a neuronal nicotinic receptor. *Nature* 353: 846–49

Ricciuti, A. J., Gruberg, E. R. 1985. Nucleus isthmi provides most tectal choline acetyltransferase in the frog Rana pipiens. *Brain Res.* 341: 399–402

Role, L. W. 1992. Diversity in primary structure and function of neuronal nicotinic acetylcholine receptor channels. *Curr. Opin. Neurosci.* 2: 254–62

Role, L. W. 1988. Neural regulation of acetylcholine sensitivity in embryonic sympathetic neurons. *Proc. Natl. Acad. Sci. USA* 85: 2825–29

Roper, S. 1976. The acetylcholine sensitivity of the surface membrane of multiply innervated parasympathetic ganglion cells in the mudpuppy before and after partial denervation. *J. Physiol.* 254: 455–73

Sah, D. W. Y., Loring, R. H., Zigmond, R. E. 1987. Long-term blockade by toxin F of nicotinic synaptic potentials in cultured sympathetic neurons. *Neuroscience* 20: 867–74

Sands, S. B., Barish, M. E. 1992. Neuronal nicotinic acetylcholine receptor currents in phaeochromocytoma (PC12) cells: dual mechanisms of rectification. *J. Physiol.* 447: 467–87

Sands, S. B., Barish, M. E. 1991. Calcium permeability of neuronal nicotinic acetylcholine receptor channels in PC12 cells. *Brain Res.* 560: 38–42

Sargent, P. B., Bryan, G. K., Streichert, L. C., Garrett, E. N. 1991. Denervation does not alter the number of neuronal bungarotoxin binding sites on autonomic neurons in the frog cardiac ganglion. *J. Neurosci.* 11: 3610–23

Sargent, P. B., Pang, D. Z. 1989. Acetylcholine receptor-like molecules are found in both synaptic and extrasynaptic clusters on the surface of neurons in the frog cardiac ganglion. *J. Neurosci.* 9: 1062–72

Sargent, P. B., Pang, D. Z. 1988. Denervation alters the size, number, and distribution of clusters of acetylcholine receptor-like molecules on frog cardiac ganglion neurons. *Neuron* 1: 877–86

Sargent, P. B., Pike, S. H., Nadel, D. B., Lindstrom, J. M. 1989. Nicotinic acetylcholine receptor-like molecules in the retina, retinotectal pathway, and optic tectum of the frog. *J. Neurosci.* 9: 565–73

Sawruk, E., Schloss, P., Betz, H., Schmitt, B. 1990a. Heterogeneity of *Drosophila* nicotinic acetylcholine receptors: SAD, a novel developmentally regulated α-subunit. *EMBO J.* 9: 2671–77

Sawruk, E., Udri, C., Betz, H., Schmitt, B. 1990b. SDB, a novel structural subunit of the Drosophila nicotinic acetylcholine receptor, shares its genomic localization with two α-subunits. *FEBS Lett.* 273: 177–81

Schloss, P., Betz, H., Schroder, C., Gundelfinger, E. D. 1991. Neuronal acetylcholine

receptors in *Drosophila*: antibodies against an α-like and a non-α-subunit recognize the same high-affinity α-bungarotoxin binding complex. *J. Neurochem.* 57: 1556–62

Schloss, P., Hermans-Borgmeyer, I., Betz, H., Gundelfinger, E. D. 1988. Neuronal acetylcholine receptors in *Drosophila*: the ARD protein is a component of a high-affinity α-bungarotoxin binding complex. *EMBO J.* 7: 2889–94

Schmidt, J. 1988. Biochemistry of nicotinic acetylcholine receptors in the vertebrate brain. *Int. Rev. Neurobiol.* 30: 1–38

Schoepfer, R., Conroy, W. G., Whiting, P., Gore, M., Lindstrom, J. 1990. Brain α-bungarotoxin binding protein cDNAs and MAbs reveal subtypes of this branch of the ligand-gated ion channel gene superfamily. *Neuron* 5: 35–48

Schoepfer, R., Halvorsen, S. W., Conroy, W. G., Whiting, P., Lindstrom, J. 1989. Antisera against an acetylcholine receptor α3 fusion protein bind to ganglionic but not brain nicotinic acetylcholine receptors. *FEBS Lett.* 257: 393–99

Schoepfer, R., Whiting, P., Esch, F., Blacher, R., Shimasaki, S., Lindstrom, J. 1988. cDNA clones coding for the structural subunit of a chicken brain nicotinic acetylcholine receptor. *Neuron* 1: 241–48

Schofield, G. G., Weight, F. F., Adler, M. 1985. Single acetylcholine channel currents in sympathetic neurons. *Brain Res.* 342: 200–3

Schroder, H., Zilles, K., Maelicke, A., Hajos, F. 1989. Immunohisto- and cytochemical localization of cortical nicotinic cholinoceptors in rat and man. *Brain Res.* 502: 287–95

Schuetze, S. M., Role, L. W. 1987. Developmental regulation of nicotinic acetylcholine receptors. *Annu. Rev. Neurosci.* 10: 403–57

Schulz, D. W., Zigmond, R. E. 1989. Neuronal bungarotoxin blocks the nicotinic stimulation of endogenous dopamine release from rat striatum. *Neurosci. Lett.* 98: 310–16

Schwartz, R. D., McGee, R., Kellar, K. J. 1982. Nicotinic cholinergic receptors labeled by [³H] acetylcholine in rat brain. *Mol. Pharmacol.* 22: 56–62

Séguéla, P., Wadiche, J., Dineley-Miller, K., Dani, J. A., Patrick, J. W. 1992. Molecular cloning, functional properties and distribution of rat brain α7: a nicotinic cation channel highly permeable to calcium. *J. Neurosci.* In press

Sine, S. M., Steinbach, J. H. 1986. Activation of acetylcholine receptors on clonal mammalian BC3H-1 cells by low concentrations of agonist. *J. Physiol.* 373: 129–62

Slater, C. R. 1982. Neural influence on the

postnatal changes in acetylcholine receptor distribution at nerve-muscle junctions in the mouse. *Dev. Biol.* 94: 23–30

Smith, M. A., Margiotta, J. F., Berg, D. K. 1983. Differential regulation of acetylcholine sensitivity and α-bungarotoxin-binding sites on ciliary ganglion neurons in cell culture. *J. Neurosci.* 3: 2395–2402

Smith, M. A., Margiotta, J. F., Franco, A., Lindstrom, J. M., Berg, D. K. 1986. Cholinergic modulation of an acetylcholine receptor-like antigen on the surface of chick ciliary ganglion neurons in cell culture. *J. Neurosci.* 6: 946–53

Smith, M. A., Stollberg, J., Lindstrom, J. M., Berg, D. K. 1985. Characterization of a component in chick ciliary ganglia that cross-reacts with monoclonal antibodies to muscle and electric organ acetylcholine receptor. *J. Neurosci.* 5: 2726–31

Sorenson, E. M., Chiappinelli, V. A. 1990. Intracellular recording in avian brain of a nicotinic response that is insensitive to κ-bungarotoxin. *Neuron* 5: 307–15

Stollberg, J., Whiting, P. J., Lindstrom, J. M., Berg, D. K. 1986. Functional blockade of neuronal acetylcholine receptors by antisera to a putative receptor from brain. *Brain Res.* 378: 179–82

Streichert, L. C., Sargent, P. B. 1992. The role of acetylcholinesterase in denervation supersensitivity in the frog cardiac ganglion. *J. Physiol.* 445: 249–60

Swanson, L. W., Lindstrom, J., Tzartos, S., Schmued, L. C., O'Leary, D. D. M., Cowan, W. M. 1983. Immunohistochemical localization of monoclonal antibodies to the nicotinic acetylcholine receptor in chick midbrain. *Proc. Natl. Acad. Sci. USA* 80: 4532–36

Swanson, L. W., Simmons, D. M., Whiting, P. J., Lindstrom, J. 1987. Immunohistochemical localization of neuronal nicotinic receptors in the rodent central nervous system. *J. Neurosci.* 7: 3334–42

Vernallis, A. B., Conroy, W. G., Corriveau, R. A., Halvorsen, S. W., Berg, D. K. 1991. AChR gene products in chick ciliary ganglia: transcripts, subunits, and receptor subtypes. *Soc. Neurosci. Abstr.* 17: 12

Vernino, S., Amador, M., Luetje, C. W., Patrick, J., Dani, J. A. 1992. Calcium modulation and high calcium permeability of neuronal nicotinic acetylcholine receptors. *Neuron* 8: 127–34

Vidal, C., Changeux, J.-P. 1989. Pharmacological profile of nicotinic acetylcholine receptors in the rat prefrontal cortex: an electrophysiological study in a slice preparation. *Neuroscience* 29: 261–70

Vijayaraghavan, S., Pugh, P. C., Zhang, Z.-W., Rathouz, M. M., Berg, D. K. 1992. Nicotinic receptors that bind α-bun-

garotoxin on neurons raise intracellular free Ca^{++}. *Neuron* 8: 353–62

Vijayaraghavan, S., Schmid, H. A., Halvorsen, S. W., Berg, D. K. 1990. Cyclic AMP-dependent phosphorylation of a neuronal acetylcholine receptor α-type subunit. *J. Neurosci.* 10: 3255–62

Vogel, Z., Maloney, G. J., Ling, L., Daniels, M. P. 1977. Identification of synaptic acetylcholine receptor sites in retina with peroxidase-labeled-α-bungarotoxin. *Proc. Natl. Acad. Sci. USA* 74: 3268–72

Wada, E., Wada, K., Boulter, J., Deneris, E., Heinemann, S., et al. 1989. Distribution of α2, α3, α4, and β2 neuronal nicotinic receptor subunit mRNAs in the central nervous system: a hybridization histochemical study in the rat. *J. Comp. Neurol.* 284: 314–35

Wada, K., Ballivet, M., Boulter, J., Connolly, J., Wada, E., et al. 1988. Functional expression of a new pharmacological subtype of brain nicotinic acetylcholine receptor. *Science* 240: 330–34

Watson, J. T., Adkins-Regan, E., Whiting, P., Lindstrom, J. M., Podleski, T. R. 1988. Autoradiographic localization of nicotinic acetylcholine receptors in the brain of the zebra finch (*Poephila guttata*). *J. Comp. Neurol.* 274: 255–64

Whiting, P., Esch, F., Shimasaki, S., Lindstrom, J. 1987a. Neuronal nicotinic acetylcholine receptor β-subunit is coded for by the cDNA clone α4. *FEBS Lett.* 219: 459–63

Whiting, P. J., Lindstrom, J. M. 1988. Characterization of bovine and human neuronal nicotinic acetylcholine receptors using monoclonal antibodies. *J. Neurosci.* 8: 3395–3404

Whiting, P., Lindstrom, J. 1987a. Affinity labelling of neuronal acetylcholine receptors localizes the neurotransmitter binding site to their β subunit. *FEBS Lett.* 213: 55–60

Whiting, P. J., Lindstrom, J. 1987b. Purification and characterization of nicotinic acetylcholine receptor from rat brain. *Proc. Natl. Acad. Sci. USA* 84: 595–99

Whiting, P. J., Lindstrom, J. M. 1986a. Purification and characterization of a nicotinic acetylcholine receptor from chick brain. *Biochemistry* 25: 2082–93

Whiting, P., Lindstrom, J. 1986b. Pharmacological properties of immuno-isolated neuronal nicotinic receptors. *J. Neurosci.* 6: 3061–69

Whiting, P. J., Liu, R., Morley, B. J., Lindstrom, J. M. 1987b. Structurally different neuronal nicotinic acetylcholine receptor subtypes purified and characterized using monoclonal antibodies. *J. Neurosci.* 7: 4005–16

Whiting, P. J., Schoepfer, R., Conroy, W. G., Gore, M. J., Keyser, K. T., et al. 1991a. Expression of nicotinic acetylcholine receptor subtypes in brain and retina. *Mol. Brain Res.* 10: 61–70

Whiting, P., Schoepfer, R., Lindstrom, J., Priestley, T. 1991b. Structural and pharmacological characterization of the major brain nicotinic acetylcholine receptor subtype stably expressed in mouse fibroblasts. *Mol. Pharmacol.* 40: 463–72

Wong, L. A., Gallagher, J. P. 1991. Pharmacology of nicotinic receptor-mediated inhibition in rat dorsolateral septal neurones. *J. Physiol.* 436: 325–46

Wonnacott, S. 1986. α-Bungarotoxin binds to low-affinity binding sites in rat brain. *J. Neurochem.* 47: 1706–12

Yawo, H. 1989. Rectification of synaptic and acetylcholine currents in the mouse submandibular ganglion cells. *J. Physiol.* 417: 307–22

Zhang, Z. W., Feltz, P. 1990. Nicotinic acetylcholine receptors in porcine hypophyseal intermediate lobe cells. *J. Physiol.* 422: 83–101

Annu. Rev. Neurosci. 1993. 16:445–70

GLIAL BOUNDARIES IN THE DEVELOPING NERVOUS SYSTEM

Dennis A. Steindler

Department of Anatomy and Neurobiology, College of Medicine,
The University of Tennessee, Memphis, Tennessee 38163

KEY WORDS: astrocytes, recognition molecules, cordones, glial scars

INTRODUCTION

Advances in the biology of glial cells, particularly the development of new
molecular markers and the refinement of in vitro techniques, have now
revealed many intriguing glial-neuronal interactions that occur both during nervous system development and following injury. In recent years,
many excellent reviews have focused on the well-accepted guidance role
of radial glia during neuronal migration (Hatten et al 1990; Rakic 1988,
1990) and numerous other aspects of neuron-glia interactions that influence cellular phenotype and differentiation, circuitry formation, and synaptic activity (e.g. see Abbott 1991; Miller et al 1989; Vernadakis 1988).
This review conceptualizes transient glial boundaries that surround functional groups of neurons, their dendrites, and axons, during neural
development. These boundaries are comprised of unique glial cells and
glycoconjugates (i.e. glycoproteins, glycolipids, and glycosaminoglycans),
which have been referred to as "cordones" (Steindler et al 1989b). Boundaries are widespread during central nervous system (CNS) pattern formation and disappear following synaptic stabilization (Steindler et al
1989a,b, 1990). These specialized cells may express numerous neurite growth-
inhibitory molecules (Schwab 1990; Schwab & Schnell 1991; Snow et al
1990b; Steindler et al 1990), as well as attractant molecules, that ultimately
guide neurites to boundary regions where their neuritic arbors accumulate
and fasciculate (Laywell & Steindler 1991). Astrocyte/recognition molecule
boundaries may represent intermediaries between neurons and neurites

445

0147–006X/93/0301–0445$02.00

(Tolbert & Oland 1989), and between neurons and their extracellular environment.

One system that has been under extensive scrutiny for more than two decades is the somatosensory cerebral cortical barrel field of the rodent (Woolsey 1990; Woolsey & Van der Loos 1970). The barrel field is a distinct geometric representation of the contralateral facial whiskers, with individual vibrissae being represented by a neuronal unit called a "barrel." Each barrel consists of neuron-rich sides that surround a neuron-poor hollow: each of these units is then surrounded by a cell sparse interbarrel septae. The pathway between whiskers and cortical barrels has a patterned representation in each station of the neuraxis, i.e. the brainstem trigeminal nuclei and the thalamic ventrobasal complex known as the barreloid field (Belford & Killackey 1979; O'Brien et al 1987; Van der Loos 1976; Welt and Steindler 1977). This system has been extensively used as a model for analyzing boundary-neurite interactions and characterizing cellular and molecular processes because of its specific and predictable patterning, and because structure and function are represented so clearly (Crossin et al 1989; Jhaveri et al 1991; Schlaggar & O'Leary 1991; Steindler et al 1989a,b). Striosomes within the caudate-putamen of the basal ganglia (Fishell & van der Kooy 1987; Graybiel & Hickey 1982; O'Brien et al 1992; Steindler et al 1988) have also shown putative events and molecules involved in the formation of a functional neural pattern. The patterning of the striosomes and the patch/matrix of the striatum is not as clearly laid out as the barrel field. The topography is not as consistent from animal to animal and does not clearly represent a single specific structure or function; however, striatal boundaries do separate specific groups of neurons and, thus, lend themselves to intense scrutiny even without a precise structure-function corollary.

Extracellular matrix (ECM) molecules involved in neuronal migration and process outgrowth are also studied extensively with regard to patterning and cell migrations (for review, see Reichardt & Tomaselli 1991; Sanes 1989), as well as to surface proteins that comprise the growing class of recognition molecules that may identify distinct classes of cells and their processes (Dodd & Jessel 1988; Perris & Bronner-Fraser 1989; Schachner 1990). This increased attention to ECM molecules is surprising, considering that not long ago, ECM was an unknown element in the CNS. Much of our understanding of predominantly glial-derived ECM associated molecules has occurred within the last ten years. Determining their functions during morphogenesis by using bioassays has only barely begun. Immunocytochemical and in situ hybridization studies of specific ECM molecules have just started to demonstrate how different ECM components may be involved in CNS pattern formation (Brunso-Bechtold et al 1992; Lander 1989; Reichardt & Tomaselli 1991; Sanes 1989; Sheppard et al 1991). Several studies have revealed the extremely complicated actions of these molecules on neuron/neurite attachment and growth through complex adhesive, anti-adhesive, and transmembrane morphoregulatory (Edelman 1992) events. These interactions ultimately affect the position of young neurons and the patterning of their axonal and dendritic arbors within developing CNS structures (Autillo-Touati et al 1988; Chamak et al 1987; Chamak & Prochiantz 1989; Crossin et al 1990; Faissner & Kruse 1990; Lochter et al 1991; Morganti et al 1990; Pesheva et al 1989; Snow et al 1990a). Extracellular matrix molecules provide a unique surface for signals involved in cell-cell and cell-substrate interactions and ultimately convey identity and positional information to migratory cells and their growing processes (Hatten et al 1990; Laywell et al 1992; Laywell & Steindler 1991; McKeon et al 1991; Schachner 1990). New evidence for glial-neuronal signaling includes ionic channel and transmitter receptors that act both similarly and distinctly from those of neurons (Barres et al 1990; Bevan 1990; Kettenmann & Ransom 1988). Although not considered here, ionic transport and other aspects of membrane biophysics certainly play important roles in the interactions between glia and neurons during morphogenesis. The reader is directed to the papers listed above, as well as recent issues of *Seminars in the Neurosciences* (Jessen & Mirsky 1990) and *Annals of the New York Academy of Sciences* (Abbott 1991), which are dedicated to the neurobiology of glia for more complete reviews on the various aspects of glial cell structure, biochemistry, function, and development.

One glial-associated ECM molecule has been studied extensively by several different laboratories. This is the astrocyte-derived, oligomeric, glycoprotein tenascin (also called cytotactin, J1, hexabrachion, myotendinous antigen; see Erickson 1989). This molecule is part of a family of proteins and sugars (e.g. with fibronectin type III and EGF repeats and the L2/HNK-1 carbohydrate epitope; see Edelman 1987; Schachner 1990) involved in the mechanisms of cellular interaction during morphogenesis and wound healing. Although tenascin is not the only molecule involved in cellular interactions, recognition, and positioning, it has given us the most insight into the complex and, at times, extremely confusing events that occur during development and wound healing (e.g. see Crossin et al 1990; Erickson 1989; Laywell et al 1992; Schachner 1990).

A SHORT HISTORY OF GLIAL/GLYCOCONJUGATE BOUNDARY STUDIES

The role of glia and associated molecules during neuronal migration has been well documented (Hatten et al 1990; Rakic 1990). Herrup (1987) was the first to review glial-associated "invisible boundaries." At that time, only sporadic references to neural boundaries could be found in the literature, and no one study seemed to suggest that transient brain boundaries represent a general organizational feature of neural morphogenesis. The first reports of these boundaries used immunocytochemistry for a glial-specific marker (glial fibrillary acidic protein, GFAP) and lectin cytochemistry. These papers describe transient accumulations of astrocyte processes and undetermined glycoconjugates seen around developing CNS structures (Cooper & Steindler 1986a,b). Since then, boundaries have been observed in a variety of CNS areas, including segments, midlines, nuclear divisions, the subcompartmentalization of functional units, and axonal pathways (see Table 1), during development. The similarity between neuromorphogenetic boundaries and boundaries described in epithelial and mesenchymal structures in invertebrates and vertebrates has already been noted (Steindler et al 1989b; Tosney 1991).

Boundary appearance at the junction of nuclei and other forms of compartments, including borders of major nervous system subdivisions, suggests that boundaries may be genetically programmed in accordance with lineage restrictions (Lumsden 1990; Lumsden & Keynes 1989; Stern & Keynes 1988). Recent studies have established the existence of DNA sequences associated with several genes that affect body patterning and segmentation in the *Drosophila* and in the mouse (homeobox genes), and some of these genes are associated with pattern formation. The developmental regulation of boundaries may reflect turnover mechanisms, such as changes in saccharide components (Steindler & Cooper 1987), enzyme degradation of the protein cores, and many other mechanisms (second messengers and other transmembrane signaling events) that might alter boundary molecules or their expression. The sugars of glycoconjugates themselves may likewise be signals for different states of morphogenesis (Steindler & Cooper 1987), as altering saccharidic sequences and glycoconjugate synthesis with β-D-xyloside adversely affects pattern formation in the cultured whole rat embryo (Morriss-Kay & Crutch 1982), including the near abolishment of somites that may depend on glycoconjugate-rich boundaries between the developing segments.

BOUNDARIES AND COMPARTMENTALIZATION DURING DEVELOPMENT IN A VARIETY OF SPECIES

Boundaries are found in all parts of the nervous system (see Table 1), from segmental and longitudinal glial boundaries in the *Drosophila* CNS (Jacobs & Goodman 1989; Jacobs et al 1989; Klämbt et al 1991) to the axon pathway boundaries found in both the central and peripheral nervous

Table 1   Developmental neural boundaries

Segmental Boundaries	*Drosophila*: Stern & Keynes 1988; Jacobs et al 1989 Leech: Thorey & Zipser 1991 Zebrafish: Trevarrow et al 1990 Chick: Lumsden & Keynes 1989; Lawrence 1990; Fraser et al 1990; Layer & Alber 1990; Lumsden 1990; Stern et al 1991 Mouse and rat: Steindler et al 1989b; Suzue et al 1990
Midline Boundaries Tectum Raphe Spinal Cord	*Drosophila*: Klämbt et al 1990 Rodent brainstem raphe: Van Hartesveldt et al 1986; McKanna & Cohen 1989; Mori et al 1990 Rat and hamster spinal cord roof plate and midbrain tectum: Snow et al 1990b
Nuclear Boundaries Inferior Olive Lateral Geniculate	Mouse inferior olive, other brainstem and thalamic nuclei: Steindler & Cooper 1987; Steindler et al 1989a,b, 1990 Hamster hypothalamus: Suarez et al 1987 Tree shrew and ferret lateral geniculate: Hutchins & Casagrande 1988; Brunso-Bechtold et al 1992
Functional Sub-Compartment Boundaries Olfactory Glomeruli Barrel Field 	Olfactory glomeruli in the moth: Tolbert & Oland 1989 Hedgehog: Valverde & Lopez-Mascaraque 1991 Whisker barrels in rodents: Cooper & Steindler 1986a,b; Steindler et al 1989a,b; Crossin et al 1989; Crandall et al 1990; Woolsey 1990; Schlaggar & O'Leary 1991 Neostriatal mosaic in mouse: Steindler et al 1988; O'Brien et al 1992
Axon Pathway Boundaries	*Drosophila*: Jacobs & Goodman 1989 Lizard spinal cord: Bodega et al 1990 Chick peripheral nervous system: Tosney 1991; Oakley & Tosney 1991; Wehrle-Haller et al 1991 Chick central nervous system: Silver & Rutishauser 1984; Silver et al 1987; Chiquet 1989 Rat and cat central nervous system: Schwab & Schnell 1991; Silver et al 1992
Synaptic Boundaries Astrocyte Axon With Bouton Neuron	LaFarga et al 1984; Raisman 1985; Steindler & Cooper 1986; Kosaka & Hama 1986; Fujita et al 1989; Atoji et al 1989

systems of the chick (Oakley & Tosney 1991; Silver et al 1987; Tosney 1991). They apparently serve the common function of surrounding major groups of neurons and their projections and, possibly, act as landmarks or barriers for migratory neurons and their growing neurites. Midline boundaries, as revealed by glia and novel proteins, may prevent neurites from crossing to the wrong side of the brain, thus, preserving laterality

(Brunso-Bechtold & Henkel 1988; McKanna & Cohen 1989; Snow et al 1990b; Thorey & Zipser 1991; Van Hartesveldt et al 1986). Nuclear boundaries (Hutchins & Casagrande 1988, 1990; Steindler & Cooper 1987; Steindler et al 1990; Suarez et al 1987) might keep a functional group of target neurons together, just as axon pathway boundaries would cordon off different fiber tracts within the developing white matter (Bodega et al 1990; Schwab & Schnell 1991). Table 1 is a brief overview of the different types of boundaries found in the nervous system, and the species they have been found in, to date.

Segmentation has been most studied in invertebrates. In addition to the simplicity of these species and amenability to analyses of all the cellular players, many well-characterized genetic mutations exist that affect particular aspects of morphogenesis. Studies in *Drosophila* have demonstrated that there may be genetic programs for cellular and molecular diversity at boundary sites, as seen in the expression of the segmentation gene *fushi tarazu*, which controls cell fate during neurogenesis (Doe et al 1988). Several other mutations in *Drosophila* are candidates for affecting the expression of either glia or recognition molecules during pattern formation, including *patchy*, *notch*, and the *engrailed* and *decapentaplegic* (dpp) genes (see Raftery et al 1991). The latter two provide insights into the expression of an anterior-posterior compartment boundary during development of the primordia of the adult appendages, the imaginal disks. Gelbart and collaborators (Raftery et al 1991) have shown that the dpp gene is expressed at or near the anterior-posterior boundary. Interestingly the product of the dpp gene is ". . . a signaling molecule of the TGF-β family of secreted proteins . . ." (Gelbart 1989), as it is known to regulate the expression of tenascin (Pearson et al 1988). The dpp studies suggest that TGF-β is essential for proximal-distal outgrowth of adult appendages in *Drosophila*; considering tenascin's role in neural boundaries, homologous or similar genes could affect development of compartments in the vertebrate brain.

Lumsden and colleagues (Fraser et al 1990; Lumsden 1990; Lumsden & Keynes 1989) have demonstrated that segmentation exists in vertebrate embryos by focusing their studies on cellular and molecular dispositions in the chick embryo spinal cord, brainstem rhombomeres, and diencelphalic neuromeres. Patterson and coworkers (Suzue et al 1990) have also reported that an antibody, ROCA 1, which binds to glial cells, defines rostro-caudal gradients during development of the rat spinal cord. Most recently, Stern et al (1991) and Lim et al (1991) have shown the existence of periodic lineage restrictions in the developing chick spinal cord that owe their maintenance to adjacent somite mesoderm, as seen by spinal cord segmental clones that no longer observe segmental boundaries when the pattern is surgically disturbed. Another intriguing study in vertebrates was done in the zebrafish. Trevarrow et al (1990) have stained neuronal and glial components of hindbrain segments that resemble rhombomeres. They describe a row of glial fibers ("curtain-like structures") between each center and border region of individual hindbrain segments. These investigators suggest that these curtains could be adhesive or mechanical barriers between the centers and borders and may provide substrate cues for growing axons within the segments. Finally, a segmental boundary, shown by astrocytic molecules, has been described in the mouse at the junction of the brain and spinal cord (Steindler et al 1989b).

Segmentation of the leech nervous system is "... prefigured by myogenic cells at the embryonic midline expressing a muscle-specific matrix protein ..." (Thorey & Zipser 1991). Thus, distinct cellular and molecular interactions underlie compartmentalization of this simple nervous system, and similar events may occur during vertebrate neurogenesis. The midline neural boundaries referred to in Table 1 may literally separate the left and right halves of the developing nervous system in *Drosophila* (Klämbt et al 1989), as well as in rodents (McKanna & Cohen 1989; Mori et al 1990; Van Hartesveldt et al 1986). The cells in a midline boundary in the roof plate of the developing optic tectum and spinal cord have been described as primitive, neuroepithelial cells that may be destined to become boundary glia (Snow et al 1990b). These same cells seem to express molecules not unlike the muscle proteins described by Thorey & Zipser (1991) in the leech, including keratan sulfate proteoglycan, which Snow et al (1990a,b) suggest is a repulsive molecule that could help deter axons from crossing the midline during development of the dorsal spinal cord and optic tectum. Lesions of midline boundaries during development, such as those present in the roof plate of the developing optic tectum (Poston et al 1988; Schneider 1973; Snow et al 1990b), lead to an anomalous crossing of retinal projections across the tectal midline, which suggests that these boundaries may be crucial in the development of laterality in the CNS.

The other types of boundaries summarized in Table 1, including nuclear boundaries around such structures as the brainstem inferior olive (Steindler & Cooper 1987), different thalamic nuclei (Brunso-Bechtold et al 1992; Hutchins & Casagrande 1988, 1990; Steindler et al 1990), and the hypothalamus (Suarez et al 1987), may represent landmarks and somehow deter neurites of these structures from crossing into adjacent, functionally distinct, neuropilar or fiber bundle territories. The subcompartment boundaries are seen in the olfactory system, where they surround individual glomeruli (for review, see Tolbert & Oland 1989, 1990; Valverde & Lopez-Mascaraque 1991), the cortical barrel field in rodents, where the boundaries separate the individual whisker barrels (for review, see Steindler et al 1990); and in the neostriatal mosaic of the early postnatal mouse, where patch/matrix boundaries also are defined by astrocytic glyco-conjugates (O'Brien et al 1992). Subcompartment boundaries are discussed in greatest detail, because of their amenability to perturbation and because the functions of their "systems" are so well known. Synaptic boundaries are also discussed later in the review. Axonal pathway boundaries could be considered white matter versions of the neuropilar boundaries described around nuclei or subcompartments, such as barrels. Silver et al (1987) described the presence of boundary cells and the absence of certain adhesion molecules in an axon-refractory barrier that separates the developing optic and olfactory projections. Chiquet (1989), in discussing Schwab's hypotheses on the functions of oligodendrocyte-derived, myelin-associated inhibitory glycoproteins, such as "NI-35" and "NI-250" (see Schwab & Schnell 1991), notes that "In the spinal cord of newborn rats he found that myelination does not occur uniformly, but tends to form boundaries around prospective fiber tracts in a mosaic manner." By using an x-irradiating procedure developed by Gilmore (1963) that eliminates oligodendrocytes in the developing spinal cord, Schwab showed that, under these conditions, axons no longer obeyed their axonal pathway boundaries and would often take aberrant paths over large areas of the spinal white and gray matter (Schwab & Schnell 1991). Therefore, Chiquet (1989) suggested that axonal pathway boundaries may ". . . serve to separate the territories of different fiber tracts from each other in a complex CNS."

WHAT STUDIES OF BOUNDARIES AND COMPARTMENTS HAVE TAUGHT US SO FAR

Studies of compartment boundaries in insect epithelia have revealed a unique boundary cell type that is either programmed to be unique, or owes its differences to other factors, such as being ". . . at the leading edge of the segment . . . [and confronting] the most posterior cells of the segment ahead . . ." (Blennerhassett & Caveney 1984). Retrovirus and fluorescent dye spread studies (Fraser & Bryant 1985) demonstrate lineage boundaries and segmental restriction. Dye-spread studies during development in the mollusk have revealed a mosaic plan of communication compartments that contain progeny of groups of founder cells that are separated by boundaries (Serras & van den Biggelaar 1987). Giaume et al (1991) have shown that a single gap junction protein (a connexin) is present in cultured mouse astrocytes and affects current flow across coupled astrocytes. As Blennerhassett & Caveney (1984) noted, boundary cells of insect epi-thelium "may modulate the size and range of molecules able to pass from segment to segment at different times in development . . . [and their] . . . reduced junctional communication and characteristic behaviour may allow these cells to remain neutral in their gradient properties and hence serve to establish and maintain discontinuity between the segmental compartments." There might also be a common sugar epitope within many or all boundary molecules. In bioassays, peanut agglutinin binding glycoconjugates seem to have an inhibitory action on neurite extension (Davies et al 1990), thus forming a mechanical-type barrier to neuritic outgrowth. A differential adhesivity, as also revealed by polysialylated NCAM staining in rhombomeres boundaries, suggests that boundaries may ". . . comprise a more tightly coherent and less mobile group of 'border' cells" (Lumsden 1990). The peanut lectin also binds to mesodermal elements, which act as a barrier to growing spinal cord axons, in the developing chick hindlimb (Oakley & Tosney 1991; Tosney 1991), to boundaries of rhombomeres in the chick (Layer & Alber 1990), and to barrel and neostriatal boundaries in the mouse (O'Brien et al 1992; Steindler et al 1990).

Early studies suggested that barrel boundaries preceded the appearance of the adult cellular barrels in the somatosensory cortex, which indicated that the boundaries related to primary events in the formation or differentiation of these functional units (Cooper & Steindler 1986a,b). All evidence showed that map-conveying or "blueprinting" afferents from the thalamus arrived later in the first postnatal week. These initial studies emphasized that the roles for afferent systems in shaping the boundary needed further study. With the advent of new axonal tracing techniques, such as DiI, as well as immunocytochemical studies of transmitters and novel antigens of growing axons (Erzurumlu et al 1990; Rhoades et al 1990), thalamocortical afferents were later found to be organized within barrel patterns at or before (Jhaveri et al 1991) boundary appearance (postnatal day 0–4, or "Stage I–II" of development; Steindler et al 1989b). Thus, the afferents have been determined to have a role of primary importance in the establishment of barrel patterns (Senft & Woolsey 1991).

Because vibrissae-map conveying afferents from the thalamus and other sources, including the serotonergic raphe (Rhoades et al 1990), arrive in the developing layer IV of the cortex around the time that glial/glycoconjugate boundaries appear, they probably interact with each other. Radial glia act as guideposts for the radial ingrowth of cortico-cortical axons during development (Norris & Kalil 1991), which could be a general rule for axonal guidance in the cerebral cortex, because thalamocortical and other afferent axons exhibit a distinct radial organization. Transforming radial glia and migrating young astrocytes, which are most concentrated with barrel boundaries within the first postnatal week (Cooper & Steindler 1986b, 1989; Crandall et al 1990), may get the message to segregate their processes within barrel patterns in the overlying cortical plate via inter-actions with afferents that are growing into the deeper layers of the young cortex. Deafferenting lesion experiments (Cooper & Steindler 1989; Crossin et al 1989; Steindler et al 1990) have demonstrated an obvious effect of afferent input loss on the pattern of barrel boundaries. The arrival of afferents that have begun to sort functionally, possibly based on patterned activity and nearest neighbor associations (Shatz & Stryker 1988; Sur et al 1990), might sculpture the boundary pattern as a result of their release of enzymes or other morphogens. These molecules could then alter the glial/glycoconjugate disposition within the barrel field forming the barrel hollows, thus allowing neurons and their processes to sort within barrels or resort within a nonrestricted zone following whisker injury (for review, see Woolsey 1990). Transplant studies, which use visual and somato-sensory cortex during critical periods in barrel field formation and peanut agglutinin to visualize the boundaries, support this theory by providing evidence that sensory cortical modules are "multipotential" (Schlaggar & O'Leary 1991). These investigators further suggest that the early develop-ing cortex is a tabula rasa that can be molded into any type of cortex by thalamocortical axons (O'Leary 1989; Schlaggar & O'Leary 1991).

At late embryonic and early postnatal times, the ECM molecules in the cortical plate have a rather uniform and dense distribution (Sheppard et al 1991; Steindler et al 1989a,b, 1990). Subcortical afferents may directly affect the expression of boundary molecules by astrocytes, or possibly release enzymes, such as plasminogen activator and metalloproteases, that degrade these molecules. These enzymes are present in and can be released by growth cones (McGuire & Seeds 1990; Pittman 1985). This would contribute to the image of "hollowing out," as seen in the distribution of ECM molecules in the developing barrel field. The degradation of poten-tially inhibitory ECM molecules may allow neuritic exploration and synaptogenesis to occur within the developing synaptic neuropil of barrel hollows. Extracellular matrix molecules might allow axonal growth, limit branching, and encourage fasciculation (Wehrle & Chiquet 1990; Wehrle-Haller et al 1991), thus helping to confine spatially and provide landmarks to prune back the barrel-like arbors of the thalamic projections that ori-ginally defy the barrel plan through their exuberant projections early in development (Senft & Woolsey 1991). The interactions between glia and neurons during pattern formation events may utilize recognition molecules that bind to distinct surface receptors. Whether these extracellular mol-ecules bind members of the integrin family (Hynes 1987; Reichardt & Tomaselli 1991) or other macromolecules (e.g. talin; Horwitz et al 1986) that possess transmembrane domains, the signals that they transduce can pass to the cytoskeleton through second messenger and/or ionic mechanisms (Schachner 1990). These intracellular changes could alter motility and process outgrowth such that glia, neurons, and neurites distribute with particular patterns.

There is evidence, however, that afferents alone are not responsible for target neuronal patterning. In the neurological mutant mouse, *reeler*, the thalamic barreloid field is delineated by completely normal boundaries in the first postnatal week (O'Brien et al 1987). Presumably normal thalamo-cortical projections yield an obviously abnormal cortical barrel field in *reeler*. Although the visual appearance of this field is abnormal, certain aspects of the functional field are still intact (O'Brien et al 1987; Steindler et al 1990; Welt & Steindler 1977). Additional studies are needed to determine if the abnormal cortical cytoarchitecture seen in the adult *reeler* is directly attributable to abnormal glia and glycoconjugate dispositions in the cortex (Steindler et al 1990). Future studies utilizing transgenic mice and gene "knock-out" experiments to assess the consequences of single genes that control the expression of boundary cells and molecules will provide additional information on the roles of boundaries during development and their interactions with afferents. The olfactory glomerulus of the moth has been convincingly compared with the rodent barrel field, including the presence of boundary glia around forming glomerular units and the ingrowth of afferents from a sensory organ (there are also many similarities to the vertebrate olfactory bulb, see Raisman 1985; Valverde & Lopez-Mascaraque 1991). In vivo and in vitro perturbation schemes have determined that the lesioning of any glial or neuronal components result in abnormal glomerular pattern formation (Oland et al 1988; Tolbert & Oland 1989, 1990). These findings emphasize the requirement for all of the glial and neuronal players during the development of normal functional units during neurogenesis. Sensory systems, in particular, would be extremely dependent on peripheral cues for the development of appropriate representations (Cooper & Steindler 1989; Woolsey 1990). This makes sense, because these maps and circuits are attempting to represent the periphery precisely, and the absence or increased representation of peripheral targets would need accompanying alterations in central receptive fields, synaptic territories, and overall map organization.

Motor system structures, such as the neostriatum, seem to be less dependent on afferent cues than sensory systems. Compartmentalization, as seen in the neostriatal mosaic, arises from genetic and intrinsic cell-cell interactions, with local neurons determining a significant amount of their own pattern formation before afferent ingrowth (Fishell & van der Kooy 1987; Krushel et al 1989). The glial/glycoconjugate boundaries around developing patches and matrix do not completely disappear following 6-hydroxydopamine (6-OHDA) lesions of the developing dopaminergic nigrostriatal system (O'Brien et al 1992). Neostriatal neuronal dendrites might then defy their mosaic compartment boundaries as a result of this boundary molecule loss. The idea of "hard" versus "soft" boundaries, which was advanced in these studies, related to the observation that, in the barrel field, the boundaries completely disappeared in a whisker lesion-related deafferented region and that all cellular elements (i.e. neuronal cell bodies, dendritic arbors) redistributed within the perturbated barrel row, as well as within adjacent intact, and sometimes enlarged, barrels (Cooper & Steindler 1989; Crossin et al 1990; Steindler et al 1990). The soft boundary was represented by the partial loss, or fading, of the neostriatal mosaic boundaries following 6-OHDA lesions during development. This softening presumably allows the dendrites to defy their patch/matrix programming in search of afferent input, but there is no obvious reorganization of neuronal cell bodies (O'Brien et al 1992). The theory of soft boundaries is partially supported by the work of Greenough and colleagues, who have shown that 6-OHDA lesions of the noradrenergic nucleus, locus coeruleus, result in dendritic defiance of cellular barrel boundaries (Loeb et al 1987). However, an accompanying down regulation of boundary molecules in this paradigm has not yet been demonstrated.

System-specific molecules may also play a role in boundary-neuron interactions. Pattern formation within limbic or association systems may be quite different from sensory systems, with intrinsic guidelines for development, including the programmed expression of such molecules as LAMP (the Limbic Associated Membrane Protein described by Levitt and coworkers; see Zacco et al 1990), which appear to play a much greater role. Astrocytes cultured from different CNS areas express different neuropeptides and their associated genes (Shinoda et al 1989). Prochiantz and collaborators have described the expression of distinct glycoconjugates by glia in different CNS structures (Autillo-Touati et al 1988; Barbin et al 1988; Chamak et al 1987; Chamak & Prochiantz 1989). Numerous other studies have alluded to astrocyte heterogeneity across structures (Shinoda et al 1989; Wilkin et al 1990), as well as within the same structure, through clonal analyses of glial cell lineages (Miller & Szigeti 1991; Vaysse & Goldman 1990). Also, boundaries are inhabited by a distinct set of astrocytes (Cooper & Steindler 1986b, 1989; Crandall et al 1990). These immature astrocytes, from different areas of the brain, could thus convey region-specific information to ingrowing afferent systems, which would search for clues to help guide them to their appropriate targets. In culture studies, young cells migrate toward higher concentrations of ECM molecules (e.g. Basara et al 1985); thus, different amounts of ECM molecules may result in graded effects on neuron/neurite migration and growth. Small amounts of such molecules could hinder movement or outgrowth, whereas larger amounts might completely restrict these elements. System-specific neuronal molecules have also been described (Hockfield & McKay 1983; Zacco et al 1990) that allow afferent axons to provide information to astrocytes (Grossfeld et al 1988) and furnish a means for two-way communication between glia and neurons.

PATTERN FORMATION BOUNDARIES DISAPPEAR FOLLOWING SYNAPTIC STABILIZATION

Numerous studies support a notion of novel molecules mediating different cell-cell interactions between astrocytes, oligodendrocytes, neurons, and nonneural (e.g. pial or vascular related) cells. Extracellular matrix molecules most likely play a major role in such interactions. In studies of glycoconjugates present in adult CNS neuropil, again using certain lectins with distinct sugar-binding properties in cytochemical studies as described above during development, lectin staining patterns revealed a more discrete boundary plan in which the synapse seemed to be the structure now cordoned off by astrocyte processes and associated glycoconjugates (Steindler & Cooper 1986). We also have seen small amounts of molecules, such as tenascin and the chondroitin sulfate-containing 473 proteoglycan (Faissner 1988), associated with these perisynaptic sites, and they seem to exist at adult nodes of Ranvier (Black & Waxman 1988; ffrench-Constant et al 1986; Rieger et al 1986). The astrocyte processes themselves produce a net-like pattern on the surfaces of neurons where they cordon off synapses (Kosaka & Hama 1986; Lafarga et al 1984; Raisman 1985; Valverde & Lopez-Mascaraque 1991). Lectin-bound glycoconjugates, as well as antibody binding of tenascin and proteoglycans, similarly reveal this perisynaptic net (Atoji et al 1989; Fujita et al 1989; Steindler & Cooper 1986). The synaptic boundaries are more subtle, and their detection requires higher resolution microscopy, compared with developmental boundaries around larger structures (e.g. nuclei or barrels) that can often be seen by the naked eye in tissue sections processed for cytochemistry or immunocytochemistry. These smaller "microscopic" boundaries could be involved in a variety of functions, including the making and breaking of synaptic connections. They might deter neuritic sprouting. They may function in other aspects of synaptic transmission, including the isolation of synaptic complexes that possess different neurochemical transmitters. In such a case, the synaptic boundary may act much like the developmental cordone or larger "macroscopic" boundaries (which separate cells and molecules of adjacent yet different compartments) in isolating the synaptic active zone from other neurochemicals used at nearby synapses.

It should not be surprising to discover that astrocytes and their associated molecules serve a microscopic role in boundary function in the adult, considering their role in macroscopic boundaries of development. There is a consistent theme of astrocytes as "barriers" or "compartmentalizers" in the review by Kimelberg & Norenberg (1989) and in Ferderoff & Vernadakis's series on *Astrocytes* (1986). Astrocyte end-feet at the pial surface, which form the glial limitans, are a sort of boundary around the entire nervous system that helps separate neural elements from mesodermal tissues. They also comprise a major portion of the "blood-brain-barrier" (for review, see Stewart & Coomber 1986), which helps protect the brain from toxins, viruses, and macromolecules. The theory of "scale independence," as associated with pattern formation, fractals, and chaos (Gleick 1987), suggests that similar elements perform similar functions in both small and large scale interactions in nature. Perhaps macroscopic boundaries, such as those around barrels and striosomes, and microscopic boundaries, such as the neuronal surface glial net, are related not only in their structure (e.g. boundary astrocytes and expression of the same ECM molecules) and function of cordoning off specific areas, but also in their ontogeny.

A rather uniform distribution of ECM and adhesion molecules precedes the development of the barrel field (Sheppard et al 1991; Steindler et al 1990). Before synaptogenesis, neurons also appear to be embedded in a dense, homogeneous sea of these same molecules. The glial/glycoconjugate network of boundaries could be sculptured out of this sea by afferent projections at the level of both the developmental boundary and the synaptic boundary. Alternatively, the macroscopic and microscopic boundaries could be a type of caulk that is expressed by glia at the final stages of map and circuit stabilization. Studies by Meshul et al (1987) provide evidence that astrocytes play a significant role in determining synaptic density. The authors have observed increases in synaptic density of presumed Purkinje cell axon collaterals on Purkinje cells in cerebellar cultures following exposure to cytosine arabinoside, which destroys granule cells and arrests development of surviving glia, thus preventing the glia from being sufficient in number and maturity to ensheathe fully the Purkinje cell. They further observed decreases in synaptic number when these cultures were transplanted to glial-rich preparations, in which the Purkinje cells were ensheathed by the astrocytes. Thus, presynaptic axons seem to interact with an astroglial intermediary before contacting a target neuron. This notion is further supported by the observations of Raisman (1985), who looked at the astrocytic arrangement in the olfactory bulb, where afferent projections are continually replaced throughout life. Here, distinct astrocyte types relate to primary olfactory and vomeronasal afferent projections, some of which cordon off olfactory glomeruli and seem to form synaptic nets on neuronal surfaces (Valverde & Lopez-Mascaraque 1991). The astrocytes in this system are strategically positioned in areas where the afferents must grow in order to establish contacts with their target neurons. Raisman (1985), therefore, proposes that these glia might have a causal, or even permissive, role in the ingrowth of olfactory axons into the bulb, and that they might be important for ". . . the unique regenerative capacity of this system." It is not known whether the afferents could interact directly with their target neurons, make synapses, and induce the release of astrocytic adhesion and ECM molecules, which would then seal them in place.

The sealing of functional synapses should be the final event in circuitry stabilization. Glia and their associated molecules are in a position that could successfully deter further neuritic growth. The astrocyte is well positioned to regulate the molecular environment of its associated neuronal set, thus taking up transmitters, helping regulate activity and, attempting to prevent toxicity by sequestering toxins, viruses, and other harmful substances. The synaptic boundaries might also isolate synapses from one another, by keeping apart their different chemical messengers. At the same time, they possess the machinery necessary to affect chemical and electrical transmission, including ion channels that interact with counterparts present in the membranes of their set of neurons and axons (for a review of such events at the node of Ranvier, see Black & Waxman 1988). The astrocyte would now be a different sort of compartmentalizer, which maintains the status quo in normal brain functions, until something happens: trauma.

BRAIN LESIONS IN THE ADULT MAKE BOUNDARIES REAPPEAR

The distinct geometry of the barrel field, with boundary and nonboundary astrocytes seen during development, was used to study potential patterning of wound-related astrocytes in adult injury (Laywell & Steindler 1991). Following small stab lesions that involved portions of the adult somatosensory cortical barrel field, newly born (using tritiated thymidine labeling), GFAP-positive reactive astrocytes surround the wound in no apparent pattern, but newly born GFAP-negative astrocytes seemed to reside mainly within barrel boundaries (often in close association with blood vessels), as well as within the lesion itself. The glial/ECM molecules tenascin and the 473 proteoglycan also seemed to be more concentrated within barrel boundaries away from the wound. These findings suggest that a vestige of the developmental cordone plan might remain, and cordone patterns might be reactivated following lesions in the adult. By using both in situ hybridization and immunocytochemistry, Laywell et al (1992) have shown that tenascin is up-regulated in a set of GFAP-positive cortical astrocytes within the first 72 hours following a stab wound in the adult mouse. Boundary astrocytes are labeled by this same tenascin riboprobe during development (Laywell et al 1992). In the lesioned adult somato-sensory cerebral cortex, the halo of up-regulated astrocytes is extremely small, just surrounding the puncture site. In the cerebellum following stab wounds, the halo of tenascin immunoreactivity is more extensive and permeates many layers of the cerebellum, despite mRNA up-regulation appearing only in the Golgi epithelial cells. Laywell et al (1992) postulate that boundary molecules may diffuse away from the lesion site.

The enhanced expression of ECM molecules (also see McKeon et al 1991) by an apparent subpopulation of astrocytes in adult lesions, appears to be one of the earliest molecular changes of wound-associated neural cells. In vitro studies of tenascin (Crossin et al 1990; Faissner & Kruse 1990; Grierson et al 1990; Lochter et al 1991) suggest that, under certain conditions, this molecule can discourage either neuronal spreading or neuritic growth. In vitro bioassays (Snow et al 1990a) have demonstrated that certain glycoconjugates, including chondroitin sulfate and keratan sulfate-containing proteoglycans, provide an inhibitory substrate to most neuritic growth. Studies by Schwab and collaborators (Schnell & Schwab 1990; Schwab 1990) implicate oligodendrocyte-derived inhibitory macro-molecules in failed regenerative attempts in the CNS white matter. Rudge & Silver (1990) have made an astroglial model "scar in a dish," in which unique morphologies and biochemistries of scar astrocytes that pre-dominantly inhibit neurite growth exist. They suggest that the ". . . inhi-bition is more probably due to the expression of molecules on the surface of the adult scar that either directly inhibit growth cones or inhibit them indirectly by occluding neurite-promoting factors in the extracellular matrix or on the astrocyte surface." The molecules associated with brain wounds, glial scars, and developmental boundaries (for review, see Silver & Steindler 1990) have complex actions and interactions that in either situation might be ". . . saying no to growth cones" (Patterson 1988).

We know a little about components of the so-called glial or astrocytic scar from classic studies, such as those by Ramon y Cajal (1928), Clemente (1955), and Reier & Houle (1988). To understand the potential barrier functions of a glial scar further, we must attempt to understand a key component of this structure—the boundary astrocyte. Geisert & Stewart (1991) have shown that the astrocyte surface is a changing substrate, with older astrocytes having a more deleterious effect on neurite growth than cells taken from neonatal brains. In their studies of compartment boun-daries in insect epithelia, Blennerhassett & Caveney (1984) noted that

"Other distinct behaviours of the [compartmental] border cells include their rapid migration and proliferation during wound healing." The astrocytes that express tenascin might be permanent boundary cells, newly generated astrocytes that recapitulate the differentiation sequences of biochemical changes seen during development, or cells that "de-differentiate" in response to lesion-associated events (the release of growth factors, cytokines, and other molecules from the vasculature or from injured cells; Nieto-Sampedro et al 1988). If the boundary astrocyte observed in adult brain lesions is the neuroepithelial version of Blennerhassett & Caveney's epithelial cell, then the astrocytic retention of developmental boundary programs may allow them to play a crucial role in the healing of brain wounds. This same astrocyte in the injured adult brain may alter the distribution of its perisynaptic boundary processes and its expression of boundary ECM molecules that cordon off synapses. These cellular and molecular changes might be associated with failed regeneration in the CNS. And, the tenascin glycoprotein, alone or in combination with other glycoconjugates including proteoglycans, may influence the regrowth of neurites in adult brain wounds through as yet undefined mechanisms.

CONCLUSION: Both Macroscopic and Microscopic Boundaries May Restrict Neuritic Growth During Development and Following Injury

The different lines of evidence for glial and glycoconjugate boundaries suggest that they are expressed at late stages of pattern formation (Lumsden 1990) or synaptic stabilization. Thus, the macroscopic barrel boundaries, and the microscopic synaptic boundaries, could be a type of caulk or cement that finally seals different neural units in place. Peanut agglutinin binding in boundaries around rhombomeres also appear at late stages in their maturation (Lumsden 1990; Lumsden & Keynes 1989). However, Lumsden and other investigators have noted that the neurons within rhombomere compartments may sort "independent of a unique set of border cells" (Lumsden 1990), with the expression of their own unique surface proteins (e.g. NCAM, L1), thus binding them together and reinforcing their likeness. Therefore, both intrinsic and extrinsic programs could exist for compartmentalization during development, and a boundary that contains molecules that might allure, yet prevent, neurons and neurites from crossing compartments could reinforce and then secure an existing pattern following commitment. But, what brings a pattern or a pre- and postsynaptic interaction to commitment?

Although there is cellular diversity within such units as barrels [lineage studies indicate that the cells of barrels are polyclones (Goldowitz 1987;

Price et al 1991)] there is a commitment to some aspects of unit organization early in embryogenesis that appears to be independent of afferent inputs. Kuljis & Rakic (1990), looking at visual cortex organization following early eye removal, note the presence of certain characteristic features of visual cortex organization (e.g. "hypercolumns") in the absence of retinal influences. In the case of barrel field or synaptic boundaries, the commitment to secure a functional unit (i.e. a barrel or a synapse) seems to depend on the presence of afferents, and deafferentation has different effects on boundaries during development versus after lesions in the adult. The presence of afferents during development and the loss of inputs following adult lesions might contribute to structural and molecular changes in astrocytes, including an altered expression of GFAP and ECM molecules (Laywell et al 1992; Laywell & Steindler 1991). As mentioned above, different types of astrocytes may also facilitate or deter neuritic growth. Reactive astrocytes in general seem to be nonpermissive (Rudge & Silver 1990), but subsets of lesion-associated astrocytes may express different permissive and nonpermissive growth substrates (e.g. proteoglycans, including glial hyaluronate-binding protein and other proteoglycan-related molecules; Gallo et al 1987; Mansour et al 1990). When reactive or immature astrocytes are cultured, they express a variety of glycoconjugates that together may affect neurite growth. Tenascin is but one of these glycoconjugates that, in vitro, is associated with certain inhibitory astrocytes. Whether it is the molecules, or the astrocytes themselves, clearly there is an association between astrotypy and regenerative capabilities of mature CNS neurons. Perhaps the surface of old versus young astrocytes can differentially affect neurite growth; immature astrocytes provide an extremely favorable substrate for neuritic growth, and immature astrocytes transplanted to the adult visual cortex induce map plasticity that is normally seen just during development; see Muller & Best 1989. There may also be diverse effects of astrocytes and their molecules during development, when young neurons search for their adult position and begin to send out their processes.

The boundaries seen around developing neural units most likely affect neuronal position and neurite growth, thus producing the mature image of functionally segregated neurons and their dendritic arbors. Many functional implications for glial cell recognition or boundary molecules exist, and they have been extensively reviewed by aforementioned investigators who are directly testing the actions of these molecules in unique in vitro bioassays. I could not attempt to tie all of their findings together, not just because they are too extensive for this review, but also because the actions of these molecules on neural cells are at best confounding and at worst contradictory. Recognition molecules, such as tenascin, have been described, in different in vitro bioassays, as being adhesive and favoring attachment to culture dish substrates, anti-adhesive, growth-promoting, or growth-inhibitory (for review, see Halfter et al 1989; Lochter et al 1991). Some of these discrepancies may be a result of comparing fibroblasts, neural crest cells, neurons, and glia, or the presence or absence of membrane receptors for this molecule on different cells at different times (Wehrle & Chiquet 1990). Faissner & Kruse (1990), looking at attachment and growth of cerebellar astrocytes and neurons on tenascin versus other ECM molecule substrates, found that tenascin was overall repulsive. There may also be isoforms of ECM molecules, as described by Kaplony et al (1991), who have noted different tenascin domains that may facilitate migration of corneal epithelial cells. Lochter et al (1991) have also suggested that the way in which boundary molecules are presented to neuronal cell bodies and growing dendrites (in bound versus soluble form) can distinctly affect these aspects of neuronal differentiation. These authors showed that this molecule is truly multifunctional, because in substrate-bound form it appears to promote neurite growth, whereas in soluble form it appears to attenuate neurite outgrowth. These investigators also propose that tenascin ". . . contains a neurite outgrowth promoting domain that is indistinguishable from the cell-binding site and presumably not involved in the inhibition of neurite outgrowth or cell spreading." Tenascin is not the only multifunctional ECM molecule, because chondroitin and keratan sulfate proteoglycans, as well as laminin, all exhibit complex interactions with cells and each other in culture assays. Perhaps one of the most intriguing descriptions for a tenascin action is that described by Wehrle & Chiquet (1990), relating to its accumulation along developing peripheral nerves. These investigators suggest that tenascin could have different effects on different parts of the growing axon, the main fiber versus the growth cone. Interactions with other cell adhesion or substrate adhesion molecules could also be involved in this process. Tenascin accumulation along growing axon fibers may promote adhesivity with other axons thus facilitating fasciculation. On the other hand, because they showed that growth cones can locomote on tenascin alone, the molecule at the growing tips of axons could promote growth cone movements and axon extension (Wehrle & Chiquet 1990). The presence or absence of particular ECM and adhesion molecules could even affect different aspects of neuronal differentiation, including the expression and growth of dendrites versus axons (Chamak et al 1987; Chamak & Prochiantz 1989).

Following trauma, glial/glycoconjugate boundaries are up-regulated and form a scar at the wound site. However, if the glial scar limits neurite growth following lesions in the adult brain, why does this occur? The reappearance of developmental boundary molecules may secure intact circuitry components and deter the formation of anomalous synaptic connections that might form amid the established circuitries of the mature brain. These circuitries are far more complicated and protracted compared with those present in the embryonic or early postnatal brain. Abnormal connections can have severe consequences, as seen in certain neurological mutant mice that exhibit severe behavioral disturbances. Thus, the adult glial boundary may limit the losses by sealing off the wound site and reducing the chances for inappropriate connections to form. The problem of guiding sprouting axons to the appropriate target would appear to be a more pressing problem than merely surmounting the wound site (as already has been somewhat accomplished by using molecular interventions, e.g. Schnell & Schwab 1990). Achieving the reexpression of growth and guidance factors possibly made by deafferented neurons in the lesioned brain is still perplexing. Thus, failed regeneration might be caused by glial/glycoconjugate boundary reappearance that owes its expression to a programmed compartmental boundary plan used during normal brain development, and the loss of neurite directional signals needed within axon pathways between a wound and various targets. Thus, in addition to surmounting macroscopic boundaries, such as the scar, regenerating axons would also have to find their way through what are now adult inhibitory pathways to interact with hopefully amicable synaptic boundaries within their appropriate targets in order to reestablish functional circuits.

ACKNOWLEDGMENTS

Thanks to Kristine Harrington, Leslie Tolbert, and Natalie Kaufman for their help with this review. I am grateful for other great collaborators—Tom O'Brien, Eric Laywell, Monte Gates, Nigel Cooper, Andreas Faissner, Melitta Schachner, and Jerry Silver. Some of the work discussed here was funded by National Institutes of Health grant NS 20856 and National Science Foundation grant BNS 8911514.

Literature Cited

Abbott, N. J., ed. 1991. *Glial-Neuronal Interaction. Ann. NY Acad. Sci.*, vol. 633. New York: NY Acad. Sci. 640 pp.

Atoji, U., Hori, Y., Sugimura, M., Suzuki, Y. 1989. Extracellular matrix of the superior olivary nuclei in the dog. *J. Neurocytol.* 18: 599–610

Autillo-Touati, A., Chamak, B., Araud, D., Vuillet, J., Seite, R., Prochiantz, A. 1988. Region-specific neuro-astroglial interactions: ultrastructural study of the in vitro expression of neuronal polarity. *J. Neurosci. Res.* 19: 326–42

Barbin, G., Katz, D. M., Chamak, B., Glowinski, J., Prochiantz, A. 1988. Brain astrocytes express region-specific surface glycoproteins in culture. *Glia* 1: 96–103

Barres, B. A., Chun, L. L. Y., Corey, D. P. 1990. Ion channels in vertebrate glia. *Annu. Rev. Neurosci.* 13: 441–74

Basara, M. L., McCarthy, J. B., Barnes, D. W., Furcht, L. T. 1985. Stimulation of haptotaxis and migration of tumor cells by serum spreading factor. *Cancer Res.* 45: 2487–94

Belford, G. R., Killackey, H. P. 1979. The development of vibrissae representation in subcortical trigeminal centers of the neonatal rat. *J. Comp. Neurol.* 188: 63–74

Bevan, S. 1990. Ion channels and neurotransmitter receptors in glia. *Semin. Neurosci.* 2: 467–81

Black, J. A., Waxman, S. G. 1988. The perinodal astrocyte. *Glia* 1: 169–83

Blennerhassett, M. F., Caveney, S. 1984. Separation of developmental compartments by a cell type with reduced junctional permeability. *Nature* 309: 361–64

Bodega, G., Suárez, I., Rubio, M., Fernández, B. 1990. Distribution and characteristics of the different astroglial cell types in the adult lizard (*Lacerta lepida*) spinal cord. *Anat. Embryol.* 181: 567–75

Brunso-Bechtold, J. K., Agee, D., Sweatt, A. J. 1992. Immunohistochemical evidence for transient expression of fibronectin in the developing dorsal lateral geniculate nucleus of the ferret. *J. Comp. Neurol.* 315: 275–86

Brunso-Bechtold, J. K., Henkel, C. K. 1988. A possible role of glia in the decussation of hind brain auditory fibers in the developing ferret. *Soc. Neurosci. Abstr.* 14: 748

Chamak, B., Fellous, A., Glowinski, J., Prochiantz, A. 1987. MAP2 expression and neuritic outgrowth and branching are coregulated through region-specific neuroastroglial interactions. *J. Neurosci.* 7: 3163–70

Chamak, B., Prochiantz, A. 1989. Influence of extracellular matrix proteins on the expression of neuronal polarity. *Development* 106: 483–91

Chiquet, M. 1989. Neurite growth inhibition by CNS myelin proteins: a mechanism to confine fiber tracts? *Trends Neurosci.* 12: 1–3

Clemente, C. D. 1955. Structural regeneration in the mammalian central nervous system and the role of neuroglia and connective tissue. In *Regeneration in the Central Nervous System*, ed. W. F. Windle, pp. 147–61. Springfield, Ill: Thomas

Cooper, N. G. F., Steindler, D. A. 1989. Critical period-dependent alterations of the transient body image in the rodent cerebral cortex. *Brain Res.* 489: 167–76

Cooper, N. G. F., Steindler, D. A. 1986a. Lectins demarcate the barrel subfield in the somatosensory cortex of the early postnatal mouse. *J. Comp. Neurol.* 249: 157–68

Cooper, N. G. F., Steindler, D. A. 1986b. Monoclonal antibody to glial fibrillary

acidic protein reveals a parcellation of individual barrels in the early postnatal mouse somatosensory cortex. *Brain Res.* 380: 341–48

Crandall, J. E., Misson, J.-P., Butler, D. 1990. The development of radial glia and radial dendrites during barrel formation in mouse somatosensory cortex. *Dev. Brain Res.* 55: 87–94

Crossin, K. L., Hoffman, S., Tan, S.-S., Edelman, G. M. 1989. Cytotactin and its proteoglycan ligand mark structural and functional boundaries in somatosensory cortex of the early postnatal mouse. *Dev. Biol.* 136: 381–92

Crossin, K. L., Prieto, A. L., Hoffman, S., Jones, F. S., Friedlander, D. R. 1990. Expression of adhesion molecules and the establishment of boundaries during embryonic and neural development. *Exp. Neurol.* 109: 6–18

Davies, J. A., Cook, G. M. W., Stern, C. D., Keynes, R. J. 1990. Isolation from chick somites of a glycoprotein fraction causes collapse of dorsal root ganglion growth cones. *Neuron* 2: 11–20

Dodd, J., Jessel, T. M. 1988. Axon guidance and the patterning of neuronal projections in vertebrates. *Science* 242: 692–99

Doe, C. Q., Hiromi, Y., Gehring, W. J., Goodman, C. S. 1988. Expression and function of the segmentation gene *fushi tarazu* during drosophila neurogenesis. *Science* 239: 170–75

Edelman, G. M. 1992. Morphoregulation. *Dev. Dyn.* 193: 2–10

Edelman, G. M. 1987. CAMs and Igs: Cell adhesion and the evolutionary origins of immunity. *Immunol. Rev.* 100: 11–45

Erickson, H. P. 1989. Tenascin: an extracellular matrix protein prominent in specialized embryonic tissues and tumors. *Annu. Rev. Cell Biol.* 5: 71–92

Erzurumlu, R. S., Jhaveri, S., Benowitz, L. I. 1990. Transient patterns of GAP-43 expression during formation of barrels in the rat somatosensory cortex. *J. Comp. Neurol.* 292: 443–56

Faissner, A. 1988. Monoclonal antibody identifies a proteoglycan expressed by a subclass of astrocytes. *Soc. Neurosci. Abstr.* 14: 920

Faissner, A., Kruse, J. 1990. J1/tenascin is a repulsive substrate for central nervous system neurons. *Neuron* 5: 627–37

Fedoroff, S., Vernadakis, A., eds. 1986. *Astrocytes*, vols. 1–3. Orlando, Fla: Academic

Fishell, G., van der Kooy, D. 1987. Pattern formation in the striatum: Developmental changes in the distribution of striatonigral neurons. *J. Neurosci.* 7: 1969–78

Fraser, S. E., Bryant, P. J. 1985. Patterns

of dye coupling in the imaginal disk of *Drosophila melanogaster. Nature* 317: 533–36

Fraser, S., Keynes, R., Lumsden, A. 1990. Segmentation in the chick embryo hindbrain is defined by cell lineage restrictions. *Nature* 344: 431

ffrench-Constant, C., Miller, R. H., Kruse, J., Schachner, M., Raff, M. C. 1986. Molecular specialization of astrocyte processes at nodes of Ranvier in optic nerve. *J. Cell Biol.* 102: 844–52

Fujita, S. C., Tada, Y., Murakami, J., Hayashi, M., Matsumura, M. 1989. Glycosaminoglycan-related epitopes surrounding different subsets of mammalian central neurons. *Neurosci. Res.* 7: 117–30

Geisert, E. E., Stewart, A. M. 1991. Changing interactions between astrocytes and neurons during CNS maturation. *Dev. Biol.* 143: 335–45

Gallo, V., Bertolotto, A., Levi, G. 1987. The proteoglycan chondroitin sulfate is present in a subpopulation of cultured astrocytes and in their precursors. *Dev. Biol.* 123: 282–85

Gelbart, W. M. 1989. The *decapentaplegic* gene: a TGF-β homologue controlling pattern formation in *Drosophila. Development* (Suppl.): 65–74

Giaume, C., Fromaget, C., Aoumari, A. E., Cordier, J., Glowinski, J., Gros, D. 1991. Gap junctions in cultured astrocytes: single-channel currents and characterization of channel-forming protein. *Neuron* 6: 133–43

Gilmore, S. A. 1963. The effects of X-irradiation on the spinal cords of neonatal rats. II. Histological observations. *J. Neuropathol. Exp. Neurol.* 22: 294–301

Gleick, J. 1987. *Chaos.* New York: Penguin. 352 pp.

Goldowitz, D. 1987. Cell partitioning and mixing in the formation of the CNS: Analysis of the cortical somatosensory barrels in chimeric mice. *Dev. Brain Res.* 35: 1–9

Graybiel, A. M., Hickey, T. L. 1982. Chemospecificity of ontogenetic units in the striatum: Demonstration by combining [3H] thymidine neuronography and histochemical staining. *Proc. Natl. Acad. Sci. USA* 79: 198–202

Grierson, J. P., Petroski, R. E., Ling, D. S. F., Geller, H. N. 1990. Astrocyte topography and tenascin/cytotactin expression: correlation with the ability to support neuritic outgrowth. *Dev. Brain Res.* 55: 11–19

Grossfeld, R. M., Klinge, M. A., Lieberman, E. M., Stewart, L. C. 1988. Axon-glia transfer of a protein and a carbohydrate. *Glia* 1: 292–300

Halfter, W., Chiquet-Ehrismann, R., Tucker, R. P. 1989. The effect of tenascin and embryonic basal lamina on the behavior and morphology of neural crest cells in vitro. *Dev. Biol.* 132: 14–25

Hatten, M. E., Fishell, F., Stitt, T. N., Mason, C. A. 1990. Astroglia as a scaffold for develolpment of the CNS. *Semin. Neurosci.* 2: 455–65

Herrup, K. 1987. Glial cells and the formation of invisible boundaries in development (or, peanut barrels in the brain). *Trends Neurosci.* 10: 443–44

Hockfield, S., McKay, R. D. G. 1983. A surface antigen expressed by a subset of neurons in the vertebrate nervous system. *Proc. Natl. Acad. Sci. USA* 80: 5758–61

Horwitz, A., Duggan, K., Buck, C., Beckerle, M. C., Burridge, K. 1986. Interaction of plasma membrane fibronectin receptor with talin—a transmembrane linkage. *Nature* 320: 531–33

Hutchins, J. B., Casagrande, V. A. 1990. Development of the lateral geniculate nucleus: interactions between retinal afferent, cytoarchitectonic, and glial cell process lamination in ferrets and tree shrews. *J. Comp. Neurol.* 298: 113–28

Hutchins, J. B., Casagrande, V. A. 1988. Glial cells develop a laminar pattern before neuronal cells in the lateral geniculate nucleus. *Proc. Natl. Acad. Sci. USA* 85: 8316–20

Hynes, R. 1987. Integrins: A family of cell surface receptors. *Cell* 48: 549–54

Jacobs, J. R., Goodman, C. S. 1989. Embryonic development of axon pathways in the *Drosophila* CNS. I. A glial scaffold appears before the first growth cones. *J. Neurosci.* 9: 2402–11

Jacobs, J. R., Hiromi, Y., Patel, N. H., Goodman, C. S. 1989. Lineage, migration, and morphogenesis of longitudinal glia in the *Drosophila* CNS as revealed by molecular lineage marker. *Neuron* 2: 1625–31

Jessen, K. R., Mirsky, R. 1990. Introduction: Neurobiology of glia. *Semin. Neurosci.* 2: 421

Jhaveri, S., Erzurumlu, R. S., Crossin, K. 1991. Barrel construction in rodent neocortex: role of thalamic afferents versus extracellular matrix molecules. *Proc. Natl. Acad. Sci. USA* 88: 4489–93

Kaplony, A., Zimmermann, D. R., Fischer, R. W., Imhof, B. A., Odermatt, B. F., et al. 1991. Tenascin M_r 220 000 isoform expression correlates with corneal cell migration. *Development* 112: 605–14

Kettenmann, H., Ransom, B. R. 1988. Electrical coupling between astrocytes and between oligodendrocytes in mammalian cell cultures. *Glia* 1: 64–73

Kimelberg, H. K., Norenberg, M. D. 1989. Astrocytes. *Sci. Am.* 260: 66–76

Klämbt, C., Jacobs, J. R., Goodman, C. S. 1991. The midline of the *Drosophila* central nervous system: a model for the genetic analysis of cell fate, cell migration, and growth cone guidance. *Cell* 64: 801–15

Kosaka, T., Hama, K. 1986. Three-dimensional structure of astrocytes in the rat dentate gyrus. *J. Comp. Neurol.* 249: 242–60

Krushel, L. A., Connolly, J. A., van der Kooy, D. 1989. Pattern formation in the mammalian forebrain: patch neurons from the rat striatum selectively reassociate in vitro. *Dev. Brain Res.* 47: 137–42

Kuljis, R. O., Rakic, P. 1990. Hypercolumns in primate visual cortex can develop in the absence of cues from photoreceptors. *Proc. Natl. Acad. Sci. USA* 87: 5303–6

Lafarga, M., Berciano, M. T., Blanco, M. 1984. The perineuronal net in the fastigial nucleus of the rat cerebellum, a Golgi and quantitative study. *Anat. Embryol.* 170: 79–85

Lander, A. D. 1989. Understanding the molecules of neural cell contacts: emerging patterns of structure and function. *Trends Neurosci.* 12: 189–95

Lawrence, P. A. 1990. Compartments in vertebrates? *Nature* 344: 382–83

Layer, P. G., Alber, R. 1990. Patterning of chick brain vesicles as revealed by peanut agglutinin and cholinesterases. *Development* 109: 613–24

Laywell, E. D., Dörries, U., Bartsch, U., Faissner, A., Schachner, M., Steindler, D. A. 1992. Enhanced expression of the developmentally regulated extracellular matrix molecule tenascin following adult brain injury. *Proc. Natl. Acad. Sci. USA* 89: 2634–38

Laywell, E. D., Steindler, D. A. 1991. Boundaries and wounds, glia and glycoconjugates: cellular and molecular analyses of developmental partitions and adult brain lesions. *Ann. NY Acad. Sci.* 633: 122–41

Lim, T.-M., Jaques, K. F., Stern, C. D., Keynes, R. J. 1991. An evaluation of myelomeres and segmentation of the chick embryo spinal cord. *Development* 113: 227–38

Lochter, A., Vaughan, L., Kaplony, A., Prochiantz, A., Schachner, M., Faissner, A. 1991. J1/tenascin in substrate-bound and soluble form displays contrary effects on neurite outgrowth. *J. Cell Biol.* 113: 1159–71

Loeb, E. P., Chang, F. F., Greenough, W. T. 1987. Effects of neonatal 6-hydroxy-dopamine treatment upon morphological organization of the posteromedial barrel subfield in the mouse somatosensory cortex. *Brain Res.* 403: 113–20

Lumsden, A. 1990. The cellular basis of segmentation in the developing hindbrain. *Trends Neurosci.* 13: 329–35

Lumsden, A., Keynes, R. 1989. Segmental patterns of neuronal development in the chick hindbrain. *Nature* 337(2): 424–28

Mansour, H., Asher, R., Dahl, D., Labkovsky, B., Perides, G., Bignami, A. 1990. Permissive and non-permissive reactive astrocytes: immunofluorescence study with antibodies to the glial hyaluronate-binding protein. *J. Neurosci. Res.* 25: 300–11

McKanna, J. A., Cohen, S. 1989. The EGF receptor kinase substrate p35 in the floor plate of the embryonic rat CNS. *Science* 243: 1477–79

McGuire, P. G., Seeds, N. W. 1990. Degradation of underlying extracellular matrix by sensory neurons during neurite outgrowth. *Neuron* 4: 633–42

McKeon, R. J., Schreiber, R. C., Rudge, J. S., Silver, J. 1991. Reduction of neurite outgrowth in a model of glial scarring following CNS injury is correlated with the expression of inhibitory molecules on reactive astrocytes. *J. Neurosci.* 11: 3309–3411

Meshul, C. K., Seil, F. J., Herndon, R. M. 1987. Astrocytes play a role in regulation of synaptic density. *Brain Res.* 402: 139–45

Miller, R. H., ffrench-Constant, C., Raff, M. C. 1989. The macroglial cells of the rat optic nerve. *Annu. Rev. Neurosci.* 12: 517–34

Miller, R. H., Szigeti, V. 1991. Clonal analysis of astrocyte diversity in neonatal rat spinal cord cultures. *Development* 113: 353–62

Morganti, M. C., Taylor, J., Pesheva, P., Schachner, M. 1990. Oligodendrocyte-derived J1-160/180 extracellular matrix glycoproteins are adhesive or repulsive depending on the partner cell type and time of interaction. *Exp. Neurol.* 109: 98–110

Mori, K., Ikeda, J., Hayaishi, O. 1990. Monoclonal antibody R2D5 reveals midsagittal radial glial system in postnatally developing and adult brainstem. *Proc. Natl. Acad. Sci. USA* 87: 5489–93

Morriss-Kay, G. M., Crutch, B. 1982. Culture of rat embryos with; β-D-xyloside: evidence of a role for proteoglycans in neuralation. *Anatomy* 134: 491–506

Muller, C. M., Best, J. 1989. Ocular dominance plasticity in adult cat visual cortex after transplantation of cultured astro-cytes. *Nature* 342: 427–30

Nieto-Sampedro, M., Lim, R., Hicklin, D. J., Cotman, C. W. 1988. Early release of glia maturation factor after rat brain injury. *Neurosci. Lett.* 86: 361–65

Norris, C. R., Kalil, K. 1991. Guidance of callosal axons by radial glia in the develop-ing cerebral cortex. *J. Neurosci.* 11: 3481–92

Oakley, R. A., Tosney, K. A. 1991. Peanut agglutinin and chondroitin-6-sulfate are molecular markers for tissues that act as barriers to axon advance in the avian embryo. *Dev. Biol.* 147: 187–206

O'Brien, T. F., Faissner, A., Schachner, M., Steindler, D. A. 1992. Afferent-boundary interactions in the developing neostriatal mosaic. *Dev. Brain Res.* 65: 259–67

O'Brien, T. F., Steindler, D. A., Cooper, N. G. F. 1987. Abnormal glial and glyco-conjugate dispositions in the somato-sensory cortical barrel field of the early postnatal reeler mutant mouse. *Dev. Brain Res.* 32: 309–17

Oland, L. A., Tolbert, L. P., Mossman, K. L. 1988. Radiation-induced reduction of the glial population during development disrupts the formation of olfactory glo-meruli in an insect. *J. Neurosci.* 8: 353–67

O'Leary, D. D. M. 1989. Do cortical areas emerge from a protocortex? *Trends Neuro-sci.* 12: 400–6

Patterson, P. H. 1988. On the importance of being inhibited, or saying no to growth cones. *Neuron* 1: 263–67

Pearson, C. A., Pearson, D., Shibahara, S., Hofsteenge, J., Chiquet-Ehrismann, R. 1988. Tenascin: cDNA cloning and induc-tion by TGF-beta. *EMBO J.* 7: 2977–81

Perris, R., Bronner-Fraser, M. 1989. Recent advances in defining the role of the extra-cellular matrix in neural crest develop-ment. *Comm. Dev. Neurobiol.* 1: 61–83

Pesheva, P., Spiess, E., Schachner, M. 1989. J1-160 and J1-180 are oligodendrocyte-secreted nonpermissive substrates for cell adhesion. *J. Cell Biol.* 109: 1765–78

Pittman, R. N. 1985. Release of plasminogen activator and a calcium dependent metalloprotease from cultured sym-pathetic and sensory neurons. *Dev. Biol.* 110: 91–101

Poston, M. R., Jhaveri, S., Schneider, G., Silver, J. 1988. Damage of a midline boundary and formation of a tissue bridge allows the misguidance of optic axons across the midline in hamsters. *Soc. Neurosci. Abstr.* 14: 594

Price, J., Williams, B., Moore, R., Read, J., Grove, E. 1991. Analysis of cell lineage in the rat cerebral cortex. *Ann. NY Acad. Sci.* 633: 56–63

Raftery, L. A., Sanicola, M., Blackman, R. K., Gelbart, W. M. 1991. The relation-ship of *decapentaplegic* and *engrailed* expression in *Drosophila* imaginal disks: do these genes mark the anterior-posterior compartment boundary? *Development* 113: 27–33

Raisman, F. 1985. Specialized neuroglial arrangement may explain the capacity of vomeronasal axons to reinnervate central neurons. *Neuroscience* 14: 237–54

Rakic, P. 1990. Principles of neural cell migration. *Experientia* 46: 882–91

Rakic, P. 1988. Specification of cerebral cortical areas. *Science* 241: 170–77

Ramon y Cajal, S. 1928. *Degeneration and Regeneration of the Nervous System.* Lon-don: Oxford Univ. Press. 769 pp.

Reichardt, L. F., Tomaselli, K. J. 1991. Extracellular matrix molecules and their receptors: functions in neural develop-ment. *Annu. Rev. Neurosci.* 14: 531–70

Reier, P. J., Houle, J. D. 1988. The glial scar: its bearing on axonal regeneration and transplantation approaches to CNS repair. In *Advances in Neurology: Func-tional Recovery in Neurological Disease,* ed. S. G. Waxman, pp. 87–138. New York: Raven

Rhoades, R. W., Bennett-Clarke, C. A., Chiata, N. L., White, F. A., Macdonald, G. J., et al. 1990. Development and lesion induced reorganization of the cortical rep-resentation of the rat's body surface as revealed by immunocytochemistry for serotonin. *J. Comp. Neurol.* 293: 190–207

Rieger, F., Daniloff, J. K., Pincon-Raymond, M., Crossin, K. L., Grumet, M., Edelman, G. M. 1986. Neuronal cell adhesion molecules and cytotactin are co-localized at the node of Ranvier. *J. Cell Biol.* 103: 379–91

Rudge, J. S., Silver, J. 1990. Inhibition of neurite outgrowth on astroglial scars in vitro. *J. Neurosci.* 10: 3594–3603

Sanes, J. R. 1989. Extracellular matrix mol-ecules that influence neural development. *Annu. Rev. Neurosci.* 12: 491–516

Schachner, M. 1990. Functional impli-cations of glial cell recognition molecules. *Semin. Neurosci.* 2: 497–507

Schlaggar, B. L., O'Leary, D. D. M. 1991. Potential of visual cortex to develop an array of functional units unique to somatosensory cortex. *Science* 252: 1556–60

Schneider, G. E. 1973. Early lesions of superior colliculus: Factors affecting the formation of abnormal retinal projec-tions. *Brain Behav. Evol.* 8: 73–109

Schnell, L., Schwab, M. E. 1990. Axonal regeneration in the rat spinal cord pro-duced by an antibody against myelin-associated neurite growth inhibitors. *Nature* 343: 269–72

Schwab, M. E. 1990. Myelin-associated inhibitors of neurite growth. *Exp. Neurol.* 109: 2–5

Schwab, M. E., Schnell, L. 1991. Channeling of developing rat corticospinal tract axons by myelin-associated neurite growth inhibitors. *J. Neurosci.* 11709–21

Senft, S. L., Woolsey, T. A. 1991. Growth of thalamic afferents into mouse barrel cortex. *Cereb. Cortex* 1: 308–35

Serras, F., van den Biggelaar, J. A. M. 1987. Is a mosaic embryo also a mosaic of communication compartments. *Dev. Biol.* 120: 132–38

Shatz, C. J., Stryker, M. P. 1988. Prenatal tetrodotoxin infusion blocks segregation of retinogeniculate afferents. *Science* 242: 87–89

Sheppard, A. M., Hamilton, S. K., Pearlman, A. L. 1991. Changes in the distribution of extracellular matrix components accompany early morphogenetic events of mammalian cortical development. *J. Neurosci.* 11: 3928–48

Shinoda, H., Marini, A. M., Cosi, C., Schwartz, J. P. 1989. Brain region and gene specificity of neuropeptide gene expression in cultured astrocytes. *Science* 245: 415–17

Silver, J., Edwards, M., Levitt, P. 1992. Immunocytochemical demonstration of early appearing astroglial structures that form boundaries and pathways along axon tracts in the fetal cat brain. *J. Comp. Neurol.* In press

Silver, J., Poston, M., Rutishauser, U. 1987. Axon pathway boundaries in the developing brain. I. Cellular and molecular determinants that separate the optic and olfactory projections. *J. Neurosci.* 7: 2264–73

Silver, J., Rutishauser, U. 1984. Guidance of optic axons in vivo by a preformed adhesive pathway on neuroepithelial endfeet. *Dev. Biol.* 106: 485–99

Silver, J., Steindler, D. A., eds. 1990. Axonal boundaries and inhibitory mechanisms during neural development and regeneration. *Exp. Neurol.* 109: 1–139

Snow, D. M., Lemmon, V., Carrino, D. A., Caplan, A. I., Silver, J. 1990a. Sulfated proteoglycans in astroglial barriers inhibit neurite outgrowth in vitro. *Exp. Neurol.* 109: 111–30

Snow, D. M., Steindler, D. A., Silver, J. 1990b. Molecular and cellular characterization of the glial roof plate of the spinal cord and optic tectum: a possible role for a proteoglycan in the development of an axon barrier. *Dev. Biol.* 138: 359–76

Steindler, D. A., Cooper, N. G. F. 1987. Glial and glycoconjugate boundaries during postnatal development of the central nervous system. *Dev. Brain Res.* 36: 27–38

Steindler, D. A., Cooper, N. G. F. 1986. Wheat germ agglutinin binding sites in the adult mouse cerebellum: Light and electron microscopic studies. *J. Comp. Neurol.* 249: 170–85

Steindler, D. A., Cooper, N. G. F., Faissner, A., Schachner, M. 1989a. Boundaries defined by adhesion molecules during development of the cerebral cortex: the J1/tenascin glycoprotein in the mouse somatosensory cortical barrel field. *Dev. Biol.* 131: 243–60

Steindler, D. A., Faissner, A., Schachner, M. 1989b. Brain "cordones": transient boundaries of glia and adhesion molecules that define developing functional units. *Comm. Dev. Neurobiol.* 1: 29–60

Steindler, D. A., O'Brien, T. F., Cooper, N. G. F. 1988. Glycoconjugate boundaries during early postnatal development of the neostriatal mosaic. *J. Comp. Neurol.* 267: 357–69

Steindler, D. A., O'Brien, T. F., Laywell, E., Harrington, K., Faissner, A., Shachner, M. 1990. Boundaries during normal and abnormal brain development: in vivo and in vitro studies of glia and glycoconjugates. *Exp. Neurol.* 109: 35–56

Stern, C. D., Jaques, K. F., Lim, T.-M., Fraser, S. E., Keynes, R. F. 1991. Segmental lineage restrictions in the chick embryo spinal cord depend on the adjacent somites. *Development* 113: 239–44

Stern, C. D., Keynes, R. F. 1988. Spatial patterns of homeobox gene expression in the developing mammalian CNS. *Trends Neurosci.* 11: 190–92

Stewart, P. A., Coomber, B. L. 1986. Astrocytes and the blood-brain barrier. In *Astrocytes*, ed. S. Fedoroff, A. Vernadakis, pp. 311–28. Orlando: Academic

Suarez, I., Fernandez, B., Bodega, F., Tranque, P., Olmos, G., Garcia-Segura, L. M. 1987. Postnatal development of glial fibrillary acidic protein immunoreactivity in the hamster arcuate nucleus. *Dev. Brain Res.* 37: 89–95

Sur, M., Pallas, S. L., Roe, A. W. 1990. Cross-modal plasticity in cortical development: differentiation and specification of sensory neocortex. *Trends Neurosci.* 13: 227–33

Suzue, T., Kaprielian, Z., Patterson, P. H. 1990. A monoclonal antibody that defines rostrocaudal gradients in the mammalian nervous system. *Neuron* 5: 421–31

Thorey, I. S., Zipser, B. 1991. The segmentation of the leech nervous system is prefigured by myogenic cells at the embryonic midline expressing a muscle-specific matrix protein. *J. Neurosci.* 11: 1786–99

Tolbert, L. P., Oland, L. A. 1990. Glial cells form boundaries for developing insect olfactory glomeruli. *Exp. Neurol.* 109: 19–28

Tolbert, L. P., Oland, L. A. 1989. A role for glia in the development of organized neuropilar structures. *Trends Neurosci.* 2: 70–75

Tosney, K. W. 1991. Cells and cell-interactions that guide motor axons in the developing chick embryo. *BioEssays* 13: 17–25

Trevarrow, B., Marks, D. L., Kimmel, C. B. 1990. Organization of hindbrain segments in the Zebrafish embryo. *Neuron* 4: 669–79

Valverde, F., Lopez-Mascaraque, L. 1991. Neuroglial arrangements in the olfactory glomeruli of the hedgehog. *J. Comp. Neurol.* 307: 658–74

Van der Loos, H. 1976. Barreloids in mouse somatosensory thalamus. *Neurosci. Lett.* 2: 1–6

Van Hartesveldt, C., Moore, B., Hartman, B. K. 1986. Transient midline raphe glial structure in the developing rat. *J. Comp. Neurol.* 253: 175–84

Vaysse, P. J.-J., Goldman, J. E. 1990. A clonal analysis of glial lineages in neonatal forebrain development in vitro. *Neuron* 5: 227–35

Vernadakis, A. 1988. Neuron-glia interrelations. *Int. Rev. Neurobiol.* 30: 149–224

Wehrle, B., Chiquet, M. 1990. Tenascin is accumulated along developing peripheral nerves and allows neurite outgrowth in vitro. *Development* 110: 401–45

Wehrle-Haller, B., Koch, M., Baumgartner, S., Spring, J., Chiquet, M. 1991. Nerve-dependent and -independent tenascin expression in the developing chick limb bud. *Development* 112: 627–37

Welt, C., Steindler, D. A. 1977. Somatosensory cortical barrels and thalamic barreloids in reeler mutant mice. *Neuroscience* 2: 755–66

Wilkin, G. P., Marriott, D. R., Cholewinski, A. J. 1990. Astrocyte heterogeneity. *Trends Neurosci.* 13: 43–45

Woolsey, T. A. 1990. Peripheral alteration and somatosensory development. In *Development of Sensory Systems in Mammals*, ed. J. Coleman, pp. 461–516. New York: Wiley

Woolsey, T. A., Van der Loos, H. 1970. The structural organization of layer IV in the somatosensory region (SI) of the mouse cerebral cortex: The description of a cortical field composed of discrete cytoarchitectonic units. *Brain Res.* 17: 205–42

Zacco, A., Cooper, V., Chantler, P. D., Fisher-Hyland, S., Horton, H. L., Levitt, P. 1990. Isolation, biochemical characterization and ultrastructural analysis of the limbic system-associated protein (LAMP), a protein expressed by neurons comprising functional neural circuits. *J. Neurosci.* 10: 73–90

Annu. Rev. Neurosci. 1993. 16:471–507

THE CHEMICAL NEUROANATOMY OF SYMPATHETIC GANGLIA

Lars-Gösta Elfvin,[1] *Björn Lindh*[1] *and Tomas Hökfelt*[2]

Departments of [1]Anatomy and [2]Histology and Neurobiology, Karolinska Institutet, S-104 01 Stockholm, Sweden

KEY WORDS: autonomic ganglia, neuropeptides, coexistence, chemical coding, immunohistochemistry

INTRODUCTION

The study of sympathetic ganglia has become increasingly important in the understanding of peripheral vegetative reflex patterns and their development and plasticity. Structural, histochemical, physiological, pharmacological, and biophysical studies have all contributed to a change in the classical concept that the ganglia are basically simple relay stations. The ganglia, which can deliver excitation and inhibition to and within the viscera and control the secretory and vascular systems, clearly possess a considerable degree of integrative activity. And, in many regions, they can function quite independently from central control. Recent work with immunohistochemistry has addressed the cellular localization of various neuropeptides, in addition to the classical transmitters noradrenaline (NA) and acetylcholine (ACh), and the results have been remarkably elucidative. New information is constantly being procured with respect to the synthesis, storage, and release of neuroactive substances in the ganglia and in their afferent and efferent processes. Axonal tracing techniques have been extensively used, either alone or in combination with immunohistochemistry, to clarify the origin and projections of transmitter-identified neurons related to the ganglia. Furthermore, studies of ganglionic transmission that have used patch clamp techniques and pharmacological methods, for example, have contributed to an increased knowledge of receptors, ionic

471

0147–006X/93/0301–0471$02.00

channels, and ionic current characteristics, as well as signal transduction mechanisms.

This review summarizes, with a morphological/histochemical perspective, the field of sympathetic ganglion structure and function, mainly in mammals. We also pinpoint some of the more urgent problems in the field. Some basic aspects of the organization of the ganglia are considered first. Most of the article, however, is devoted to a description of the ganglia as revealed by immunohistochemistry and axonal tracing methods and the functional implications of these findings. Extensive reviews and books dealing with the autonomic nervous system (ANS) have already discussed the sympathetic ganglia elaborately (Björklund et al 1988; Elfvin 1983; Gabella 1974, 1976, 1985; Gibbins 1990; Karczmar et al 1986; Pick 1970; Simmons 1985; Skok 1973). Therefore, we review a relatively restricted number of papers on the structure and function of ganglia. We emphasize the distribution of several peptides, which are present in sympathetic ganglia and coexist with classical transmitters and, thus, chemically code subpopulations of cells within the system.

GENERAL ORGANIZATION

Anatomical Considerations

The ANS, as defined by Langley (1921), consists of the sympathetic, parasympathetic, and enteric nervous systems (see Gilman et al 1990). The efferent division of the sympathetic part of the ANS is structurally organized as a two-neuron pathway from the central nervous system (CNS) to the peripheral effector organs. In the sympathetic ganglia, the information is transmitted from the pre- to the postganglionic neurons, and the ganglia are characterized as paravertebral, prevertebral, and previsceral or terminal by their anatomical localization. The paravertebral ganglia are located on both sides of the vertebrate column, thereby forming the two sympathetic chains or trunks. The ganglia are connected with their respective spinal nerves via white and gray communicating rami. Preganglionic spinal fibers reach the ganglia via the white rami, and postganglionic axons leave through the gray rami to join the spinal nerves and distribute to the effector organs. The prevertebral ganglia consist of the celiac, the superior mesenteric, and the inferior mesenteric ganglia and are located on the abdominal aorta and along the arteries that supply the same tissues. The cell bodies of most sympathetic preganglionic neurons that project to the paravertebral ganglia are in the intermediolateral (IML) cell column in Rexed's lamina VII of the thoracic and upper lumbar spinal cord. In addition, there are preganglionic neurons in nuclei more medially located and laterally extending into the white matter. The spinal preganglionic fibers to the prevertebral ganglia traverse the paravertebral ganglia and reach the prevertebral ganglia via several nerve trunks (splanchnic nerves). The previsceral ganglia are mainly parasympathetic, but small populations of sympathetic previsceral ganglia, located near or within their target organs, also occur. They are not considered here.

The postganglionic sympathetic neurons in mammalian ganglia are multipolar cells (De Castro 1932; Ramón y Cajal 1911), with synapses mainly located on their dendrites (Elfvin 1963, 1971), although axosomatic synaptic junctions also occur. Sympathetic ganglia in lower animals, such as amphibians, may lack dendrites and have the synapses on the soma and the proximal portion of the axon (Pick 1970; Taxi 1976; Watanabe 1983). Each ganglion cell is innervated by spinal preganglionic fibers. In the rat, the fibers vary in number, with many to ganglia that control head and thoracic organs and relatively few to ganglia that control the gut, kidney, and pelvic organs (Strack et al 1988). In addition, there is evidence for the presence of synaptic contacts of axon collaterals from visceral sensory fibers traversing especially prevertebral ganglia (Aldskogius et al 1986; Matthews & Cuello 1982; Quigg et al 1990), and from axons coming from postganglionic cells in neighboring sympathetic ganglia (Bowers & Zigmond 1981; Dalsgaard & Elfvin 1982). One major difference between the para- and prevertebral ganglia is that the prevertebral ganglia also receive an input by fibers from neurons in the gastrointestinal wall, thus forming a peripheral feedback system. So far, we have no clearcut evidence for such a control system in the paravertebral ganglia.

From early physiological studies, the concept had emerged that functionally distinct populations of neurons exist in sympathetic ganglia, which are innervated by preganglionic fibers from different levels of the spinal cord. The physiological work on the superior cervical ganglion (SCG) of the guinea pig by Njå & Purves (1977) also showed that each neuron is innervated by preganglionic fibers from about four spinal segments, but usually one of the segments dominates. Their work also indicated that ganglion cells that innervate a common target organ would be scattered in a random fashion in the ganglia. However, several lines of evidence based on denervation experiments (Jacobowitz & Woodward 1968; Matthews & Raisman 1972), axonal tracing experiments (Bowers & Zigmond 1979; Dail & Barton 1983; Flett & Bell 1991), and immunohistochemistry (Lindh et al 1986a; Macrae et al 1986) suggest that the sympathetic ganglia may be subdivided into regions of neurons that project to specific targets. Immunohistochemical findings strongly indicate that both the principal ganglion cells and the nerve fibers of extrinsic origin, which impinge on the ganglion cells, are chemically coded with regard to their content of messenger molecules (Heym et al 1991; Krukoff et al 1985; Lindh et al 1988b). Moreover, the chemically coded neuron populations in the prevertebral ganglia innervate different targets within the intestinal wall (Costa et al 1986). Therefore, different types of chemically and functionally distinct neuronal pools exist in the ganglia, each pool being influenced by specific afferent inputs or by the demands of the target organ as a whole or tissue and/or cell type in the target organ.

Our knowledge of the exact origin of the nerve fibers in the transmitter- and peptide-containing networks, as well as the detailed projections of the cell bodies in sympathetic ganglia, is incomplete. However, axonal tracing studies alone or in combination with immunohistochemistry and various types of denervation experiments have helped clarify much of the connectivity. With respect to transmitter and peptide content, there are species differences in the patterns of coexistence of these substances, and their functions may vary accordingly. Differences sometimes occur between reports concerned with peptide localization and distribution in the same species. These differences may be explained by using different antisera or different tissue processing and staining.

Classical Transmitters

The main transmitter of the peripheral sympathetic nervous system of mammals is NA (von Euler 1946), and the majority of the principal ganglion cells are noradrenergic, as first shown with the formaldehyde-induced histofluorescence technique (Falck et al 1962; Eränkö & Härkönen 1963; Norberg & Hamberger 1964) and later verified immunohisto-chemically with antiserum against dopamine-β-hydroxylase (Geffen et al 1969), the catecholamine-synthesizing enzyme that converts dopamine to NA. There is also a certain number of cells that are not noradrenergic but cholinergic, as first shown with a histochemical method (Koelle & Friedenwald 1949) by demonstrating the ACh metabolizing enzyme, acetylcholine esterase (AChE), in a population of cells in the cat stellate and lumbosacral ganglia L5-S2 (Holmstedt & Sjöqvist 1959; Sjöqvist 1962), and later by using antisera toward AChE (Lindh et al 1989a). The cholinergic cells, which are mainly present in paravertebral ganglia and amount to 10–15% of the total cell population in stellate and lumbar sympathetic ganglia, also express immunoreactivity toward the ACh-synthesizing enzyme choline acetyltransferase (ChAT) (Buckley et al 1967). In contrast to the paravertebral ganglia, the number of presumable cholinergic cells in prevertebral ganglia is low, probably about 1%. The identification of cholinergic neurons in peripheral tissues, unlike in the CNS (Wainer et al 1984), has been hampered by difficulties in using antibodies to the generally accepted cholinergic marker ChAT. Although a few papers have appeared (Furness et al 1983, 1984; Lindh et al 1986b;

Steele et al 1991; Suzuki et al 1990), the final mapping of peripheral cholinergic neurons remains to be done.

Small Intensely Fluorescent Cells

Small intensely fluorescent (SIF) cells (Eränkö & Härkönen 1963, 1965; Norberg et al 1966), which contain a biogenic amine, are distributed throughout the ANS, but their function is not completely understood. The SIF cells are sometimes referred to as small granule-containing cells, paraganglionic cells, paraneurons, or chromaffin and chromaffin-like cells (Eränkö 1976; Eränkö et al 1980; Taxi 1979). Type I SIF cells (Williams et al 1976) are considered to function as interneurons between pre- and postganglionic sympathetic neurons and usually occur singly or in small groups in paravertebral ganglia. Type II SIF cells (Williams et al 1976) are clustered in paraganglion-like arrangements and considered to be paracrine or endocrine cells; they are mainly located in the prevertebral ganglia. The SIF cells are innervated by spinal preganglionic nerves and possibly by axons that originate from the principal ganglionic neurons (Furness & Sobels 1976). In addition, the cells form specialized contacts with one another, some of which are gap junction-like (Andersson Forsman & Elfvin 1991).

The SIF cells usually contain a catecholamine, dopamine (Libet & Owman 1974; Norberg et al 1966), NA or adrenaline (Elfvin et al 1975), but some cells store serotonin (5-HT) (Verhofstad et al 1981). In addition, immunohistochemical studies have revealed that SIF cells of sympathetic ganglia in several mammalian species, including humans, contain several neuropeptide-like immunoreactivities (LI), such as enkephalin (ENK)-LI (Hervonen et al 1980, 1981; Schultzberg et al 1979); substance P (SP)-LI (Schultzberg et al 1983); neuropeptide Y (NPY)-LI (Lundberg et al 1983; Macrae et al 1986); calcitonin gene-related peptide (CGRP)-LI (Kummer & Heym 1988); methionine ENK-Arg-Gly-Leu-LI bombesin/gastrin-releasing peptide (BOM/GRP)-LI (Helén et al 1984); and neurotensin (NT)-LI (Heym et al 1984). α-neo-endorphin-LI, dynorphin (DYN)-LI, and leucine (leu)-ENK-LI coexist in SIF cells of the SCG of the guinea pig (Kummer et al 1986); in the guinea pig, inferior mesenteric ganglion (IMG) of as many as five peptide-LIs, including that of vasoactive intestinal polypeptide (VIP) (Heym et al 1984) and somatostatin (SOM), were shown in single SIF-cells by Chiba & Masuko (1989). Inagaki et al (1986) have demonstrated the occurrence of human atrial natriuretic polypeptide-LI in SIF cells of the rat. There is also evidence for the presence of gamma-aminobutyric-acid (GABA), the major inhibitory transmitter in the CNS, in SIF cells. Thus, the GABA-synthesizing enzyme, L-glutamate decarboxylase (GAD), the GABA inactivating transaminase, and GABA-LI immunoreactivity have been demonstrated in SIF cells of rat sympathetic ganglia (Häppölä et al 1987a; Kenny & Ariano 1986; Wolff et al 1986). The presence of histamine was shown in SIF cells of the rat SCG (Häppölä et al 1985). The significance of all these neuroactive substances, in many cases colocalized in single SIF cells and usually together with an amine transmitter, remains as yet unclear.

PARAVERTEBRAL GANGLIA

Peptidergic Cell Bodies

Since the discovery of SP (Chang et al 1971; Chang & Leeman 1970; von Euler & Gaddum 1931) in nerve fibers (Hökfelt et al 1977c) and of SOM (Brazeau et al 1973) in noradrenergic ganglion cell bodies (Hökfelt et al 1977a) of sympathetic ganglia, numerous neuropeptides have been identified in sympathetic ganglia, in addition to the classical transmitters (Figure 1).

NORADRENERGIC CELLS   Large numbers of SOM positive cells have been observed in prevertebral ganglia, and randomly scattered SOM-containing noradrenergic neurons have been found in the SCG of guinea pig and rat

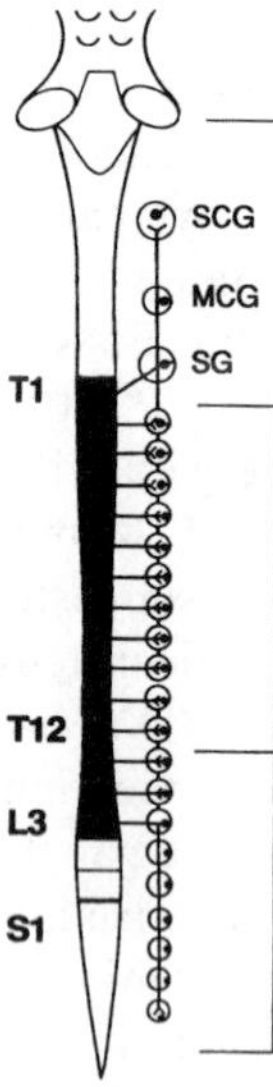

| | CAT | | GUINEA PIG | |
	Noradrenergic	Cholinergic	Noradrenergic	Cholinergic
SCG / MCG / SG	NPY SOM ENK	VIP CGRP SP NT	NPY NPY + DYN DYN SOM ENK	VIP VIP + NPY
	NPY	VIP	NPY	VIP
	NPY NPY + GAL GAL NT	VIP VIP+CGRP+SP NT	NPY NPY + DYN DYN	VIP VIP + NPY

Figure 1   Major subpopulations of peptide-containing postganglionic noradrenergic and cholinergic neurons in paravertebral ganglia of the sympathetic trunk in cat and guinea pig are shown. In the schematic drawing of the spinal cord, the extent of the sympathetic preganglionic nuclei projecting to the ganglia is indicated in black. Abbreviations: SCG, superior cervical ganglion; MCG, middle cervical ganglion; SG, stellate ganglion.

(Hökfelt et al 1977a). In fact, SOM was the first peptide shown to be colocalized with a classical transmitter. Later studies have confirmed and extended the observation of localization of SOM in noradrenergic neurons to other species, including humans (Järvi et al 1987). The opioid peptide ENK (Hughes et al 1975) was demonstrated in the guinea pig and rat SCG (Schultzberg et al 1979). Application of the mitosis inhibitor colchicine increased the number of ENK-positive cells, and colocalization with DBH was observed in the rat SCG. Later, Shimosegawa et al (1985) showed that as many as 55% of the cells in rat lumbar paravertebral ganglia contained immunoreactivity to met-ENK after treatment with colchicine. Immunoreactivity to met-ENK-Arg-Phe was found in neurons of rat SCG (Häppölä et al 1987b). Finally, Hanley et al (1984) have reported that vasopressin-LI occurs in neuronal perikarya of the SCG and other paravertebral ganglia of rat and monkey.

There is strong evidence that a large population of the paravertebral post-ganglionic noradrenergic neurons contains NPY (Gibbins 1991; Lundberg et al 1983, 1984), a 36 amino acid peptide belonging to the pancreatic poly-peptide family (Tatemoto 1982; Tatemoto et al 1982). As many as 90% of the neurons in the lower lumbar ganglia of the guinea pig are immuno-histochemically NPY-positive, according to McLachlan & Llewellyn-Smith (1986). Galanin (GAL)-LI, a 29 amino acid peptide (Tatemoto et al 1983), has been found in a large population of the NPY-containing principal ganglion cells in cat paravertebral ganglia (Kummer 1987; Lindh et al 1989a). There are also GAL-containing cells that lack NPY (Lindh et al 1989a). Moreover, a DYN-like peptide is present in about 90% of the NPY-containing neurons in the SCG of the guinea pig (Gibbins & Morris 1987), as well as in a subgroup of neurons lacking NPY-LI (Gibbins 1989). DYN-LI is not restricted to the cervical sympathetic ganglia of the guinea pig, because single DYN-IR cells have also been demonstrated in lumbar ganglia (Heym et al 1990). Recently, peptide YY(PYY)-LI has been reported to occur in neurons and nerve fibers in the SCG of the rat. Whereas 10–30% of the total number of neurons were immunoreactive for PYY, all PYY-immunoreactive (IR) cells were immunoreactive for NPY (Häppölä et al 1990). In general, these peptides have been described to be distributed in cells with no distinct preferred orientation, and further work is needed to get a more complete picture of the peptidergic cell distributions. Additional studies are also needed to clarify whether there are major differences in expression of neuropeptides in the sympathetic ganglion cells of different mammalian species.

NON-ADRENERGIC CELLS Whereas the above-described peptides are mainly present in noradrenergic neurons, immunoreactivity of several other peptides appears to be predominantly expressed in non-nora-drenergic cells. Thus, cell bodies containing VIP (Said & Mutt 1970) and CGRP (Amara et al 1982; Rosenfeld et al 1983) have been demonstrated in paravertebral ganglia (Hökfelt et al 1977b; Kummer & Heym 1988). The VIP-containing cells are either distributed in small clusters or single and scattered, apparently at random in the ganglia. The CGRP cells are usually single. In the cat stellate and lumbar sympathetic ganglia, the VIP positive cells constitute 10–15% of the neurons (Lundberg et al 1979). In other species, such as the pig, there are more VIP-positive cells in the thoracic ganglia than in the stellate and lumbar ganglia (Hill & Elde 1989). In the cat, many VIP cells are AChE positive and probably correspond to cholinergic cells (Lundberg et al 1979). In some presumably cholinergic neurons, CGRP coexists with VIP, as shown in the cervical, stellate, and lumbar ganglia of the cat (Kummer & Heym 1988; Lindh et al 1987, 1989b). An interesting observation is that VIP and NPY are sometimes colocalized (Heym et al 1990). In general, these cells do not appear to contain tyrosine hydroxylase (TH)-IR and may thus presumably be cholinergic, as is found in cranial parasympathetic ganglia (Suzuki et al 1990). In thoracic ganglia of the pig, however, VIP coexists with moderate DBH-IR in some cells (Hill & Elde 1989).

Considerable effort has been made to study the possible expression of SP (Pernow 1953, 1983) in ganglionic neuronal cell bodies. According to Kessler and coworkers (1981, 1982, 1985), the SP-like material in the rat SCG increases 200-fold after three days in culture. However, neither SP-LI nor SP mRNA have so far been visualized in rat SCG neurons in vivo, except in the studies by Ariano & Tress (1983). A faint SP-immuno-reactivity was described in some neurons of the SCG, middle cervical ganglion, and stellate ganglion of the cat by Kummer & Heym (1988). After prolonged incubation, SP-IR was demonstrated in middle cervical and stellate ganglia of the dog, in about 6% of the cells (Darvesh et al 1987). In situ hybridization studies on lumbar paravertebral ganglia of the cat show that SP mRNA is, in fact, present in a population of CGRP- and VIP-IR cholinergic neurons (Lindh et al 1989b).

Recent findings have provided evidence that neuronal cell bodies containing NT-LI occur in paravertebral ganglia of the cat (B. Lindh et al, in preparation). The majority of these cells are TH-negative, although some examples of colocalization with TH-IR have been encountered. Elution-restaining experiments suggest that the NT-LI is present in the scattered type of the VIP-containing cholinergic neurons, which is also immuno-reactive to CGRP, peptide histidine isoleucine (PHI), and SP (see Lindh et al 1987, 1989b).

Apart from classical neurotransmitters and neuropeptides, some postganglionic neuronal cells contain GABA, and GAD and GABA-trans-aminase have been demonstrated in cell bodies of the rat SCG (Häppölä et al 1987a; Kenny & Ariano 1986). This indicates that GABA may be produced and metabolized in these cells. The functional significance of the GABA-ergic cells requires further studies. Whether the cells contain NA or ACh and/or a neuropeptide is not known.

Peptidergic Fibers

As mentioned above, nerve fibers immunoreactive to neuropeptides occur in varying numbers in paravertebral ganglia. Immunoreactivity to SP (Hökfelt et al 1977c), VIP (Hökfelt et al 1977b), cholecystokinin (CCK) (Larsson & Rehfeld 1979), BOM/GRP (Schultzberg 1983), and CGRP (Gibbins et al 1987; Lee et al 1985) have been demonstrated in the SCG and stellate ganglion of the rat and guinea pig. Furthermore, ENK-IR fibers occur in SCG and stellate ganglion of cat, rat, and guinea pig (Schultzberg et al 1979) and in lumbosacral ganglia of the cat (Lindh 1992b), as well as in human paravertebral ganglia (Hervonen et al 1980, 1981). NPY-, VIP-, SP-, CGRP-, met-ENK-Arg-Phe, leu-ENK- and DYN-IR fibers were described in guinea pig lumbar ganglia (Heym et al 1990). A few DYN-IR fibers have also been demonstrated in the SCG of the guinea pig (Gibbins et al 1987). NT-IR fibers have been demonstrated in the stellate and other sympathetic chain ganglia of the cat (Lundberg et al 1982b), as well as in the guinea pig paravertebral ganglia (Reinecke et al 1983). SOM-IR fibers were seen in the stellate ganglion (Schultzberg & Lindh 1988) and in lumbosacral ganglia (Lindh et al 1992) of the cat.

In the stellate ganglion of the guinea pig, the SP-IR fibers are restricted to the dorsal region of the ganglion (Dalsgaard et al 1983b; Heym et al 1991), with only single fibers distributed in the remaining part of the ganglion. Only the latter fibers also contain CGRP-IR (Heym et al 1991). In the SCG, only a sparse network of SP-IR fibers occurs (Dalsgaard et al 1983b). The ENK-IR fibers of the guinea pig stellate ganglion are also regionally distributed (Lindh et al 1986b), apparently along the ventral portion of the ganglion (Heym et al 1991).

Neurons containing acetylcholine, the coexisting classical transmitter in many autonomic neurons, have been less easily identified in the periphery (see above). However, a dense network of ChAT-positive fibers was found in the SCG and stellate ganglia of the guinea pig (Lindh et al 1986b). Although the ChAT-immunoreactive fibers occurred in all parts of the SCG, they often seemed to have a somewhat regional distribution, as they form more intensely fluorescent networks around certain groups of cells. The ChAT-positive fibers in the SCG disappeared after cutting the cervical sympathetic trunk and were assumed to be of spinal preganglionic origin.

Connectivity and Projections

MOTOR AFFERENTS   The major source of the cholinergic fibers in the ganglia is the spinal preganglionic nuclei. Cell bodies in the IML cell column of the cat spinal cord contain ENK, NT, SOM, and SP, which are sometimes colocalized (Krukoff 1987; Krukoff et al 1985), and NT was observed in IML perikarya of the guinea pig (Reinecke et al 1983). CGRP-IR (Yamamoto et al 1989) and VIP-IR (Baldwin et al 1991) have been demonstrated in preganglionic neurons projecting to the SCG of the rat. These neurons thus represent a likely source for many of the fibers in the ganglia that contain these peptides. In fact, cholinergic preganglionic perikarya, which also contain enkephalin, have been directly demonstrated in the rat (Kondo et al 1985). The distribution of ChAT-positive fibers in the stellate ganglion of the guinea pig was complementary to a network of ENK-positive fibers (Lindh et al 1986b), which suggests that the two fiber groups represent different preganglionic projections. Although most of the peptide-containing preganglionic fibers of spinal origin are probably cholinergic, the possibility cannot be excluded that some of the peptidergic preganglionic neurons lack ACh and perhaps contain another classical transmitter (see Lindh et al 1986b and above).

Recently, the cholinergic innervation of the mouse SCG has been analyzed with immunohistochemistry for ChAT (Kasa et al 1991). The ChAT-positive fibers were also heterogeneously distributed in the mouse SCG with more ChAT-stained fibers and varicosities in the cranial part of the ganglion.

SENSORY AFFERENTS   Retrograde and anterograde axonal tracing with horseradish peroxidase (HRP) or fluorescent dyes has been used in the investigations of the extrinsic sources of the above-described pathways related to the paravertebral ganglia (Baldwin et al 1991; Dalsgaard & Elfvin 1979, 1981; Oldfield & McLachlan 1978; Quigg et al 1990; Strack et al 1988), thus extending earlier results obtained mainly with denervation techniques. In a study with wheat germ agglutinin, or choleratoxin-subunit B-conjugated HRP (Quigg et al 1990), anterogradely transported from upper thoracic dorsal root ganglia, both afferent-sensory fibers and postganglionic cell bodies were labeled in the guinea pig stellate ganglion. This suggested that synaptic contacts occur between axon collaterals of primary afferent fibers traversing the ganglia and postganglionic neurons, thus indicating a transsynaptic transport of the tracer through these contacts. The transport may be related to receptor density or synaptic activity, as Jankowska (1985) has suggested for motoneurons. Trophic molecules, retrogradely transported from the peripheral targets, cannot only regulate the form of the ganglionic neurons, but may influence their preganglionic innervation transsynaptically (Purves et al 1988). Perhaps primary afferents may also be involved in similar transsynaptic regulatory processes at the sympathetic ganglion level. The primary sensory fibers projecting to the stellate ganglion are probably SP and/or CGRP containing and contribute to the SP and CGRP fiber networks in the ganglion, as well as in other paravertebral ganglia. As far as we know, however, no studies using a combination of axonal tracing and immunohistochemistry have been performed to demonstrate the origin of these peptidergic fibers from dorsal root ganglia.

INTRAGANGLIONIC FIBERS   The TH- and DBH-containing nerve fibers in the ganglia may, in part, originate from the noradrenergic postganglionic neurons. Ultrastructural analyses after intracellular application of HRP (Kiraly et al 1989; Kondo et al 1980) have been performed to reveal dendrodendritic and dendrosomatic interconnections between the postganglionic neurons in rat SCG, thus confirming earlier results obtained by three-dimensional reconstructions from electron micrographs of ultrathin serial sections of cat SCG (Elfvin 1963). The noradrenergic neurons may also be the origin for some of the fibers containing NPY, DYN, SOM, and ENK-IR. An intraganglionic origin for the SP, CGRP, and VIP-IR nerves must also be considered, because the cholinergic cell population contains these immunoreactivities and may give rise to intraganglionic collaterals. The SIF cells represent another likely source for some of the peptide- and amine-containing fibers in the ganglia.

CELL POPULATIONS AND THEIR AFFERENT INPUT   There is now evidence that the functional units of cells are also differentiated with respect to extrinsic control. For instance, in the lower lumbar and sacral ganglia of the cat, a chemically coded input to the noradrenergic and cholinergic neurons has been disclosed. Thus, the noradrenergic, NPY-, and GAL-containing cells are preferentially surrounded by NT- and SP-IR fibers, whereas SOM- and ENK-IR fibers are mainly associated with the cholinergic neurons, as revealed by double-staining experiments (Lindh et al 1991). In the cervicothoracic ganglia of the cat, three types of CGRP/VIP-IR neurons were distinguished, based on the peptidergic fiber network surrounding the perikarya (Kummer & Heym 1988): those with SOM-IR fibers, those with DBH- and met-ENK-IR fibers, and those with DBH- and NT-IR fibers. The clustered VIP-IR neurons lacking CGRP-IR were surrounded by fine SOM-IR fibers and coarse CGRP-IR fibers, whereas patches of NT-fibers were observed complementary to these VIP-IR cell groups. In the stellate ganglion of the guinea pig, a further neurochemical and somatotopic organization has been revealed (Heym et al 1991), with an ENK-IR network surrounding NPY/TH-IR cell bodies and VIP-IR neurons innervated by SP-IR fibers. The ENK-IR surrounded NPY/TH-IR-neurons project to the lung, judging from combined retrograde axonal tracing and immunohistochemistry, whereas the TH-IR neurons devoid of ENK-IR innervation project to the hairy skin of forelimb and neck and to skeletal muscle and fat. Further subdivisions of these cell groups will likely be detected in the future.

The difference in peptidergic afferent input to the various populations of neurons in the paravertebral ganglia may reflect a high degree of specificity in central control and is likely to be intimately related to the differences in target innervation. The sympathetic activity in skin nerves has completely different characteristics, as compared with the sympathetic signals to the muscles (Wallin 1990; Wallin & Fagius 1986). Thus, the activity in the muscle nerves is spontaneous and controlled by the baroreceptors, i.e. related to heart and respiration rhythms. In contrast, the skin nerves almost completely lack baroreceptor control, but are influenced by, for instance, temperature and emotional factors.

Various experimental approaches have been used to study the factors that govern the establishment of peptide connections. One example is a recent study in which the lumbosacral paravertebral ganglia of the kitten have been analyzed after sciatic nerve resection and a neuroma formation (Lindh et al 1992). Three to six months after nerve lesion, CGRP-IR had disappeared from the scattered cholinergic cell population, and an almost complete loss of SOM-IR nerve fibers was seen in ganglia ipsilateral to the operation. The ENK-IR fiber network appeared to be unaffected. Another distinct change was the occurrence of a fairly dense CGRP-IR varicose network not present in control animals. The latter fibers have been interpreted to be recurrent sprouts of sensory fibers, probably from the neuroma, which enter the ganglia. The functional implications of this finding are obscure. A coupling between nociceptive and sympathetic fibers after neuroma formation had been suggested earlier (Devor 1983; Jänig & McLachlan 1984). Causalgic pain syndromes are associated with effects of the sympathetic nervous system activity, such as regulation of cutaneous blood flow and perspiration. Relief from such pain can often be accomplished by interruption of the sympathetic nervous supply (Devor 1983), which indicates an activation of C-fiber nociceptors by sympathetic activity. An upregulation of adrenergic $\alpha2$-receptors has recently been observed in sensory nerve terminals after nerve injuries (Sato & Perl 1991).

CELL POPULATIONS AND THEIR EFFERENT PROJECTIONS   The neurons of sympathetic ganglia innervate a large variety of peripheral tissues, and the net effect of the nerve impulses in these neurons depends upon the colocalized peptide(s), other possible neuromodulators, and type of pre- and postsynaptic receptor(s). Each subpopulation acting on a specific target is presumably chemically coded and appears, in most instances, to release more than one neuroactive substance, i.e. the transmission is plurichemical (Furness et al 1989a). Certain recent findings, which are particularly interesting to the transmitter and peptide expression in specific groups of neurons in the paravertebral ganglia, are briefly described here.

Apparently, two noradrenergic GAL-IR populations of neurons exist in paravertebral ganglia of the cat, one with and the other without NPY (Lindh et al 1989a). Many NA neurons, which contain both NPY- and GAL-IR, are present in the lower lumbar ganglia. Most of these neurons innervate blood vessels in the foot pads and probably in skeletal muscle (Figure 2). NPY interacts with NA in vascular control, and because the NPY-effect cannot be blocked by adrenoceptor antagonists, NPY is thought to exert its effect independently of NA and adrenoceptors (Lundberg & Hökfelt 1986). The functional role of GAL in sympathetic neurons remains to be established. Ekblad et al (1985) have demonstrated that GAL inhibits acetylcholine release in the intestine, and Ohhashi & Jacobowitz (1985) showed that it potentiates the electrically, as well as NA, induced

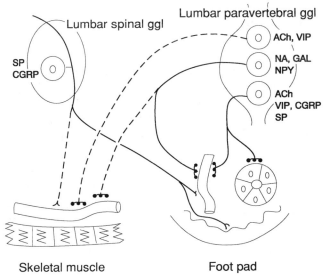

Figure 2 Schematic drawing illustrating the neuronal subpopulations of the L7 paravertebral sympathetic and spinal ganglia and their projections to the foot pad and skeletal muscle of the hind leg in the cat. The noradrenergic NPY/GAL neurons and the cholinergic VIP/CGRP/SP neurons project to the blood vessels of the foot pad. The latter neurons also project to the sweat glands. These pathways have been revealed with a combination of axonal tracing and immunohistochemistry. Sensory fibers containing SP and CGRP also project to blood vessels in the foot pad and are present beneath the epidermis. Presumable pathways, not yet verified with the described technique, are indicated with stippled lines.

contraction of the vas deferens. Recently, McKeon et al (1991) found that GAL acts as an inhibitory transmitter in the mudpuppy cardiac ganglion.

The blood vessels in the foot pad and skeletal muscle are also innervated by cholinergic neurons. In cat ganglia, these cells can be divided into at least two subpopulations: In one, scattered cells containing VIP/PHI-, CGRP-, SP-, and possibly NT-IR; in the other, the neurons are clustered in groups in the peripheral parts of the ganglia with cells immunoreactive to VIP/PHI, but not to CGRP and SP. The scattered neurons in the lumbar paravertebral ganglia of the cat mainly innervate exocrine sweat glands, whereas neurons of the other subgroup innervate blood vessels in the skeletal muscle, presumably precapillary resistance vessels (Landis 1988; Landis & Fredieu 1986; Lindh et al, 1988a, 1989b; Lindh & Hökfelt 1990). ACh, VIP, CGRP, and SP are all vasodilators, and Lundberg et al (1981) had suggested that VIP is responsible for the atropine resistant vasodilation seen after stimulation of cholinergic nerves. When colocalized with SP, CGRP may, at least in part, exert its effect by inhibiting an endopeptidase responsible for inactivation of SP (Le Greves et al 1985, 1989).

An additional CGRP-IR cell population containing NPY-IR, but not NA, has recently been demonstrated in the guinea pig stellate ganglion (Kummer & Heym 1991). Whether this cell group is also cholinergic is not known, but appears likely. The target(s) of the CGRP+/NPY+ cells is also unknown. In the thoracic ganglia T_2–T_5 of the cat, there appears to be very few CGRP-IR perikarya, whereas VIP-IR cells that have no colocalized CGRP and are arranged in clusters occur (Kummer & Heym 1988). A further non-noradrenergic subgroup of neurons in the stellate ganglion of the guinea pig that contains VIP and NPY was recently disclosed to project to smooth muscle of the airways (Bowden & Gibbins 1992).

Blood flow regulation in terms of peptide influences is apparently much more diversified among species than hitherto believed. By using retrograde axonal transport and double-labeling immunofluorescence techniques, Gibbins & Morris (1990) characterized three populations of noradrenergic neurons that innervate blood vessels in the guinea pig on the basis of their specific content of neuropeptides. Neurons innervating large distributing arteries contained NPY-IR. Neurons to smaller cutaneous arteries in addition contained IR to prodynorphin-derived peptides, whereas the smallest arterioles and arteriovenous anastomoses had no NPY-IR, but IR to prodynorphin-derived peptides. Two thirds of the nerve cell bodies in the SCG retrogradely labeled from the eartips contained IR to DYN A (1–8) or DYN A (1–17). The presence of very low levels of IR to DYN A (1–17) in the terminals of the perivascular sympathetic nerves led the authors to suggest that DYN A (1–17) was being processed to DYN A (1–

8) during transport to the terminals. In contrast, DYN A (1–17)-IR is found in the axons of sympathetic neurons, which also contain NPY-IR, that innervate the guinea pig iris (Gibbins & Morris 1987). This finding indicates that the DYN precursor processing is related to the target tissue innervated. The perivascular sympathetic neurons containing pro-dynorphin-derived peptides, but not NPY, appear to be involved in the regulation of thermoregulatory cutaneous vascular circuits.

PREVERTEBRAL GANGLIA

Peptidergic Cell Bodies

Almost all postganglionic neurons in prevertebral sympathetic ganglia are noradrenergic. Many of these neurons also contain peptides. And, on the basis of NA/peptide coexistence, the principal ganglion cells can be divided into subpopulations (Fig. 3). In the celiac-superior mesenteric ganglion complex (CSMG) of the guinea pig, where the chemical anatomy of the neuronal cell bodies has been extensively studied, three major noradrenergic cell populations occupy specific domains of the ganglion (Lindh et al 1986a; Macrae et al 1986). The SOM-positive cells (NA/SOM neurons), which represent 25% of the total cell population (Lindh et al 1986a), are mainly present in the anterior inferior part of the ganglion, the superior mesenteric pole. The NPY-positive cells (NA/NPY neurons) constitute 65% of all ganglion cells (Lindh et al 1986a) and are present bilaterally in the posterior superior parts of the ganglion, i.e. the celiac poles. Thus, the NA/NPY and NA/SOM neurons have a complementary intraganglionic distribution. Approximately 10% of the noradrenergic ganglion cells appear to lack peptide immunoreactivity (NA/-neurons) (Lindh et al 1986a; Macrae et al 1986). In addition, about 1% of the cells in the CSMG, most of which are non-noradrenergic, are VIP/PHI-containing (Hökfelt et al 1977b). Many of the VIP/PHI-positive cells contain NPY-LI and, occasionally, TH-LI. A few cells also contain immunoreactivity to DYN (Lindh et al 1988a; Macrae et al 1986). The non-noradrenergic neurons probably contain ACh as the main transmitter, although a direct demonstration of their cholinergic nature is lacking. SOM- (Hökfelt et al 1977a), NPY- (Lundberg et al 1982a, 1983), and VIP/PHI-positive (Hökfelt et al 1977b) neurons have also been demonstrated in the rat and cat CSMG. In an ongoing study on cat, NT-containing neurons have also been found. The majority of these cells are TH-negative. Some NT-containing cells are also NPY-negative, but some neurons contain both peptides (B. Lindh et al, in preparation). GAD and GABA-transaminase are present in nerve cells of the CSMG (Häppölä et al 1987a), which indicates that some of the cells are GABA-ergic.

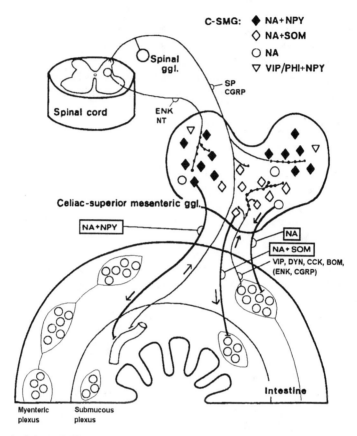

Figure 3 Schematic illustration of the CSMG and its connections with the spinal cord and intestine with special reference to peptides and coexistence systems. In the CSMG, two main populations of ganglion cells are seen characterized by the presence of NA + NPY and NA + SOM, respectively. The former cells are located in the upper lateral parts of the ganglion, whereas the latter occupy its mid-lower portion. A small population of cells contain VIP plus NPY. Some NA cells seem to lack a peptide. The NA + SOM neurons project to the submucous ganglia, the NA + NPY neurons innervate blood vessels in the intestinal wall, and the cell bodies containing only NA project to the myenteric ganglia. Projections from the intestine to the CSMG from the myenteric ganglia contain multiple peptides, including VIP, DYN, CCK, BOM, ENK, and CGRP. They seem to innervate exclusively NA + SOM cell bodies. The fibers from the spinal cord contain peptides, possibly ENK and NT, but these peptides are not colocalized. The spinal ganglion neurons contain SP and CGRP, which sometimes are colocalized.

In the inferior mesenteric ganglion (IMG) of the guinea pig, rat, and cat, principal ganglion cells contain immunoreactivity to SOM (Hökfelt et al 1977a), NPY (Lundberg et al 1982a, 1983), and VIP/PHI (Hökfelt et al 1977b). In the guinea pig IMG, about 60% of the cells are SOM-containing (Hökfelt et al 1977a), whereas the NPY-positive cells represent 20% of the ganglion cell population (McLachlan & Llewellyn-Smith 1986). Also in this ganglion, a small number of cells are VIP-containing (Webber & Heym 1988). Recently, principal ganglion cells in the pig IMG containing IR to CGRP, GAL, ENK, NPY, SOM, and VIP in various combinations with one another and with TH and DBH have been described (Majewski et al 1991).

Peptidergic Fibers

The ganglion cells in the CSMG and IMG of guinea pig, rat, and cat are surrounded by peptide-containing nerve fibers immunoreactive to SP (Hökfelt et al 1977c), VIP/PHI (Hökfelt et al 1977b), ENK (Schultzberg et al 1979), CCK (Larsson & Rehfeld 1979), BOM/GRP (Schultzberg 1983), DYN (Dalsgaard et al 1983c; Vincent 1984), and CGRP (Lee et al 1985). NT fibers have been described in prevertebral ganglia of the cat (Lundberg et al 1982b) and guinea pig (Reinecke et al 1983). Fibers with SOM- and NPY-like IR have also been seen in the prevertebral ganglia. The SOM-containing fibers appear to contain DBH-IR and may correspond to processes, most likely dendrites, of the principal ganglion cells (Lindh et al 1988b). Also the NPY-positive fibers may contain NA (Järvi et al 1986). In addition, GABA-like immunoreactivity has been demonstrated in nerve fibers of prevertebral ganglia in the guinea pig (Hills et al 1988).

The peptidergic nerve fibers in the prevertebral ganglia exhibit different patterns of distribution. The SP- and CGRP-IR fibers appear mostly as single thin fibers with relatively long intervaricose segments and a random general distribution. BOM/GRP-IR and CCK-IR fibers are generally more sparse, compared with the VIP-, ENK-, and DYN-positive fibers. The ENK-positive fibers in the guinea pig CSMG are of two different types: intensely fluorescent fibers arranged in patches around groups of ganglion cells, and a denser network of weakly fluorescent fibers (Schultzberg et al 1979). In the guinea pig CSMG, the VIP-, CCK-, BOM/GRP-, DYN-, and weakly fluorescent ENK-IR fibers are almost exclusively related to the SMG pole of the ganglion surrounding SOM-containing cells (Lindh et al 1986a, 1988b; Macrae et al 1986). Fibers containing neuromedin U-like immunoreactivity surround small groups of nerve cell bodies in the celiac ganglia (Furness et al 1989b). These fibers originate from neurons in the small intestine.

Connectivity and Projections

As a result of axonal tracing and nerve lesion experiments performed on the CSMG and IMG, in combination with immunohistochemistry, the SP-IR and most of the CGRP-IR fibers are suggested to be of sensory origin, whereas the CCK-, BOM/GRP-, and VIP-, as well as a minor part of the CGRP-IR, are mainly derived from neurons in the gut wall (Figure 4) (Dalsgaard et al 1982a, 1983a; Lindh et al 1988b; Matthews & Cuello 1982; Schultzberg & Dalsgaard 1983; Webber & Heym 1988; see also Furness & Costa 1987). There is also evidence that VIP-containing neurons in the guinea pig pelvic ganglia project via the hypogastric nerves to the IMG (Dalsgaard & Elfvin 1982a; Dalsgaard et al 1983a) and contribute to the VIP-containing fiber network. The two opioid peptides ENK and DYN have separate origins. The majority of the ENK-positive fibers are strongly fluorescent, arranged in a patch-like fashion and represent spinal preganglionic neurons (Dalsgaard et al 1982b), whereas the weakly fluorescent fiber network derives from the gut. The major part of the DYN-IR fibers is of enteric origin, and only a minor part originates from dorsal root ganglion cells (Dalsgaard et al 1983c; Lindh et al 1988b), or possibly from postganglionic cells containing DYN (Lindh et al 1988b; Macrae et al 1986). The NT-IR fibers are of spinal preganglionic origin (Lundberg et al 1982b; Reinecke et al 1983). The GABA-IR fibers appear to originate in the gut wall, most likely from GABA-containing neurons in the myenteric plexus (Hills et al 1988).

SENSORY AFFERENTS   Several lines of evidence indicate that primary sensory fibers traversing the prevertebral ganglia give off collateral branches, which form synaptic contacts with dendrites of the CSMG and IMG postganglionic neurons (Aldskogius et al 1986; Baker et al 1980; Costa et al 1986; Elfvin & Dalsgaard 1977; Matthews & Cuello 1982, 1984; Matthews et al 1987). Many of the varicose CGRP-positive fibers in the CSMG also appear to be collaterals of primary sensory nerves from thoracic dorsal root ganglia (Lindh et al 1988b). CGRP-positive nerve fibers and terminals were found that also contained SP-LI (Lindh et al 1988b), and colocalization of SP and CGRP has been demonstrated in dorsal root ganglion cell bodies (Gibbins et al 1987).

INTESTINAL CONNECTIONS   With respect to CCK-, VIP-, BOM/GRP-, DYN-, as well as the weakly fluorescent ENK-network and the minor part of the CGRP-IR fibers, Costa et al (1986) have suggested that they, at least in part, may coexist and thus derive from the same gut neurons. The multipeptidergic input from the gut is closely related to the cholinergic input, thus acting on receptors of the principal ganglion cells (Crowcroft

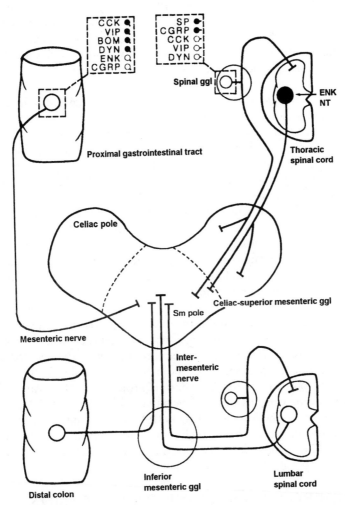

Figure 4 Schematic drawing illustrating the origin of the peptide-containing nerve fibers projecting to the guinea pig CSMG. The peptidergic nerve fibers reach the ganglion via the connecting nerve trunks: 1. The thoracic splanchnic nerves contain axons that originate in sensory neurons in the spinal ganglia and sympathetic preganglionic axons that originate in the thoracic spinal cord. 2. Mesenteric nerves carry axons of enteric neurons with cell bodies located in the myenteric plexus of the proximal gastrointestinal tract. 3. Intermesenteric nerves contain axons from lumbar spinal ganglia and sympathetic preganglionic nerves that originate in the lumbar spinal cord and processes of enteric neurons in the distal colon. The peptides are indicated in the drawing. The main projections to the CSMG for each peptide is indicated with a filled circle.

et al 1971; Crowcroft & Szurszewski 1971). The ganglion cells may thus integrate peripheral afferent input from visceral mechanosensory fibers with synaptic input from central preganglionic sympathetic nerve fibers and participate in a highly specific manner in peripheral reflexes between visceral organs, particularly intestino-intestinal reflexes (Szurszewski 1981).

The noradrenergic neurons of the CSMG and the IMG innervate abdominal and pelvic viscera, including the gastrointestinal tract. Costa, Furness, and collaborators (Costa & Furness 1984; Costa et al 1986; Furness & Costa 1987) have shown that the three different subpopulations of noradrenergic neurons in the CSMG have separate targets in the wall of the guinea pig small intestine. Thus, the NA/NPY neurons innervate intestinal blood vessels involved in the regulation of blood flow. Both NA and NPY are vasoconstrictors (Edvinsson et al 1984; Lundberg & Tatemoto 1982). The NA/SOM neurons project to submucous ganglia and act on secretomotor neurons that influence water and electrolyte secretion. NA/- neurons innervate myenteric ganglia and reduce gastro-intestinal motility by inhibiting ACh release from enteric excitatory neurons and by contracting sphincters.

IMG peptide-containing neurons may also have other targets. Post-ganglionic noradrenergic neurons project to the pelvis via the hypogastric nerves and innervate pelvic organs, such as the lower urogenital tract. There are, however, rather few studies concerned with the detailed mapping of peptidergic projections to the pelvis from the IMG. In a recent study of the pig IMG, GAL-IR neurons projected to the ovary. These GAL-containing neurons were TH-negative (Majewski et al 1991).

With respect to the function of the VIP/PHI-containing principal ganglion neurons in prevertebral ganglia, our knowledge is very limited. As mentioned above, most of them are presumably cholinergic, and they may have an intraganglionic function.

PEPTIDERGIC GANGLIONIC TRANSMISSION

Functional Considerations

The functional role of the complex peptidergic innervation of sympathetic ganglia is not yet clarified in detail, but it has become increasingly evident that the neuropeptides in the nerve fiber networks of the sympathetic ganglia are modulating transmission and related events in the ganglia. The neurons in the ganglia respond by changes in their membrane potential to the actions of neurotransmitters, neuropeptides, and other neuro-modulators that alter the ionic permeability of the cell membrane through the activation of appropriate surface receptors. Nicotinic cholinergic receptors (AChRs) are responsible for a fast excitatory postsynaptic potential (F-EPSP), which triggers the initiation of the postsynaptic spike (Skok 1983). Activation of other receptors are usually followed by slow changes in the membrane potential, which may play a modulatory role (Kobayashi & Tosaka 1983). Three types of slow potentials are known: slow inhibitory (S-IPSP), the nature of which is still controversial (Koketsu 1986); slow excitatory (S-EPSP), which is muscarinic; and late slow excitatory (LS-EPSP), usually regarded as peptidergic (Dun 1983).

The LS-EPSP was first described in principal neurons of the ninth–tenth paravertebral sympathetic ganglia of the bullfrog, and found not to be blocked by cholinergic antagonists (Nishi & Koketsu 1968). Subsequently, LS-EPSP responses in the IMG of the guinea pig were demonstrated (Dun & Karczmar 1979; Konishi et al 1979a; Neild 1978). Further studies have revealed the existence of late slow noncholinergic EPSPs in the celiac ganglion of the guinea pig (Dun & Ma 1984), in the IMG of the rabbit (Simmons & Dun 1985), and in mammalian paravertebral ganglia, such as the SCG of the rabbit (Ashe & Libet 1981).

LHRH-LIKE PEPTIDE   In a search for the possible transmitter involved in generating the LS-EPSP in frog ganglia, a LHRH-like peptide was identified (Jan & Jan 1982; Jan et al 1979, 1980). The substance was found by radioimmunoassay in the bullfrog sympathetic chain and was not identical to mammalian LHRH. The peptide was apparently located in the pre-ganglionic C fibers, as demonstrated by its accumulation proximal to the cut of the spinal nerves, and also identified in fibers that terminate on the cells of C type. Because the B cells also exhibit the LS-EPSP, the transmitter must diffuse large distances from the release site to elicit the response in these cells.

SUBSTANCE P   With respect to the transmitter mediating the LS-EPSP in mammalian ganglia, evidence indicates that the slow long-lasting depo-larization of the postganglionic neurons in the IMG of the guinea pig is evoked by SP (Dun & Karczmar 1979; Konishi et al 1979a), presumably released from the above-mentioned axon collaterals of primary sensory fibers traversing the IMG. These synapses represent about 5% of the total number of synapses with the principal ganglion cells (Matthews et al 1987). Interestingly, early work applying purified SP to the SCG of the cat increased the contraction of the nictitating membrane because of gangli-onic stimulation, presumably through a direct SP-action on the post-ganglionic neurons (Beleslin et al 1960).

In the prevertebral ganglia, the noncholinergic excitatory potentials seem to be brought about mainly by release of putative peptide transmitters from primary afferents, thereby playing a role in viscero-visceral reflexes.

The situation in paravertebral ganglia, however, is less clear and not systematically studied, but there is evidence for a sensory projection to paravertebral ganglia, as mentioned above (Oldfield & McLachlan 1978; Quigg et al 1990).

A concept regarding an additional functional role of the sensory fibers has recently begun to crystallize. Neuropeptide-containing primary sensory afferents that convey information from peripheral tissues to the spinal cord also seem to participate in the efferent regulation of the peripheral tissues they innervate, such as the efferent regulation of inflammatory, immune, and wound-healing responses. This may be a direct effect; however, in this context, the possibility must now be considered that the sensory neuron acts indirectly via the sympathetic nervous system, presumably at the ganglion level and maybe more peripherally. As Mantyh et al (1992) noted, this interaction between the sensory and sympathetic neurons may be significant in maintaining the integrity of a peripheral tissue jointly innervated by the two systems.

NEUROTENSIN   It has recently become increasingly obvious that peptides in preganglionic neurons may also be important for excitatory events, both in prevertebral and paravertebral ganglia. For instance, the integration of incoming signals to the IMG appears to involve not only the peptidergic input from axon collaterals of dorsal root primary afferents, but also noncholinergic and cholinergic inputs from fibers of visceral origin, as well as the cholinergic input from spinal preganglionic fibers. Recent studies on the guinea pig IMG (Stapelfeldt & Szurszewski 1989a,b,c) indicate that the sympathetic spinal preganglionic fibers may contain a peptide (apart from opioids, see below) that is responsible for a slow noncholinergic signal. This substance appears in guinea pig to be NT or a NT-like peptide, which can facilitate release of SP from mechanosensory fibers, as well as exert a direct postsynaptic action (Figure 5).

In paravertebral ganglia, NT has also been suggested to be released from preganglionic fibers. An NT-mediated noncholinergic excitation of ganglion cells was demonstrated in the stellate ganglion of the cat (Bachoo & Polosa 1988; Maher et al 1991). The LS-EPSP modulating the cholinergic potentials may then play several important roles in the information processing in the ganglia. The diversity in putative transmitters, as well as the variability in type of fibers from which they are released, indicates a very high degree of specialization of function with respect to noncholinergic excitatory transmission in the different ganglia.

OPIOID PEPTIDES   An inhibitory function is usually ascribed to the opioid peptides. In the guinea pig IMG, met-ENK, leu-ENK, and (D-Ala^2)-met-enkephalinamide (here referred to as ENK) all depress the F-EPSP elicited

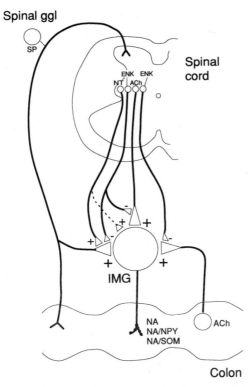

Figure 5 Schematic drawing of the pathways connecting the IMG with the distal colon and the lumbar spinal cord in the guinea pig. Neurons in the colon wall, which are cholinergic, provide an excitatory input to the sympathetic ganglion cells in the IMG. Apart from ACh, these cells contain several neuropeptides. Of these, cholecystokinin and VIP-containing neurons are mechanosensory. Other mechanosensory fibers originate in spinal ganglion cells and contain SP. They provide excitatory synapses to the sympathetic ganglion cells in the IMG. Fibers originating in the preganglionic spinal cord nuclei contain ACh and either ENK or NT. The ACh-containing neurons form an excitatory input to the ganglion cells. The ENK pathway inhibits release of ACh and SP, whereas the NT pathway facilitates release of SP and gives rise to an excitatory potential in some IMG neurons (stippled line). Excitatory neurotransmitters and peptides are indicated as +; inhibitory, as −. Modified from Stapelfeldt & Szurszewski (1989c).

by preganglionic stimulation (Konishi et al 1979b). Because ENK was not effective in blocking iontophoretically applied ACh, a presynaptic action, presumably through axo-axonal contacts, was assumed. Similarly, ENK inhibited SP released presynaptically from the sensory axon collaterals (Jiang et al 1982). Thus, ENK, or a closely related peptide, may be liberated from certain preganglionic fibers to act as a transmitter for presynaptic inhibition (Konishi et al 1979b). However, the actions of ENK and DYN released from fibers originating in the gut (see above) are unknown.

Results supporting a functional role of ENK on transmission in paravertebral ganglia are mainly derived from studies on the cat SCG and stellate ganglion. ENK-IR disappears from the stellate ganglion after prolonged preganglionic, but not postganglionic, stimulation (Bachoo et al 1987), which suggests a release of the peptide from preganglionic nerves. Prosdocimi et al (1986) demonstrated that ENK administered locally on stellate ganglia inhibited the tachycardia elicited by preganglionic nerve stimulation and reduced the compound action potential in the cardiac nerves. Analyses of the cat SCG have shown that met-ENK-Arg-Phe can reduce the release of ACh caused by preganglionic stimulation (Araujo & Collier 1987) and that this effect could be partly antagonized by naloxone. Furthermore, evidence suggests that an opioid peptide with inhibitory action on nicotinic (Zhang et al 1991), as well as noncholinergic transmission (Bachoo et al 1989), is released by preganglionic stimulation. These findings go along with the idea that ENK is an inhibitory modulator, pre-, and/or postsynaptic, in paravertebral ganglia.

VASOACTIVE INTESTINAL POLYPEPTIDE   The effects of VIP on transmission events have also been analyzed in some detail. Intra-arterial injection of VIP to the SCG of the cat facilitated or unmasked the late component of 5-HT-induced postganglionic discharge (Kawatani et al 1987), and VIP modulates the muscarinic responses (Kawatani et al 1985).

In the IMG of the guinea pig, VIP or a VIP-like peptide present in mechanosensory afferent nerves that originate in the colonic wall may be one of the transmitters involved in the slow noncholinergic depolarization of the membrane potential in principal ganglion cells (Love & Szurszewski 1985, 1987). The noncholinergic depolarization was evoked not only by electrical stimulation of the lumbar colonic nerves, but also by radial distension of the colon (Peters & Kreulen 1986). Also, CCK can induce a slow noncholinergic response (Mo & Dun 1986; Schumann & Kreulen 1986). Because VIP and CCK may be colocalized with several other peptides, such as DYN, CGRP, and ENK, in these colonic neurons, and possibly also with GABA (see above), further studies are likely to reveal additional effects on the ganglion cells by stimulation of these colonic mechanoreceptor neurons.

Noncholinergic transmitters or modulators may control ganglion cell functions other than those directly related to synaptic transmission. An acute increase in TH activity occurs in the cat SCG after VIP administration, presumably through a direct action on the preganglionic neurons (Ip et al 1982). The increased TH activity was mediated via a cAMP-dependent phosphorylation of TH involving Ca^{++} channels (Ip et al 1985; Zigmond et al 1991). One likely source of the peptide was VIP-containing preganglionic neurons in the spinal cord projecting to the SCG (Baldwin et al 1991; Sasek & Zigmond 1989). VIP also induces the hydrolysis of inositol-containing phospholipids in the SCG of the rat (Audigier et al 1986; Durroux et al 1987).

OTHER MEDIATORS   Neurokinin A (NKA)-like immunoreactivity has been found in fibers of the IMG of the guinea pig, and this tachykinin causes a slow noncholinergic EPSP similar to that induced by SP (Saria et al 1987). Other mediators may thus be effective in eliciting a LS-EPSP. In the guinea pig celiac ganglion, the noncholinergic response in some cells appears to be mediated by 5-HT; in other cells, an unidentified substance appears to be involved (Dun & Ma 1984).

In summary, these examples of recent studies show that neuron populations associated with sympathetic ganglia produce and presumably release peptides, which may act as modulators and sometimes have a transmitter-like action in sympathetic ganglia. Some peptides may directly influence the signal transduction system, as well as determine the level of transmitter available for release.

Neuropeptide Receptors

Earlier studies had suggested that several neuropeptide receptors might be present in sympathetic ganglia (Audigier et al 1986; Niwa et al 1986). In a recent elegant receptor binding study, Mantyh et al (1992) demonstrated that, in the SCG of the rat and rabbit and the SMG of the rabbit, high levels of binding sites are present for several peptide ligands, such as CCK, GAL, SOM, SP, and VIP. Within each ganglion, binding sites are expressed within discrete populations of ganglion cells for each of these neuropeptides. On the other hand, no binding sites were observed for BOM, CGRP, and NKA. The results further support the assumption that discrete groups of neurons in sympathetic ganglia may be regulated by neuropeptides released from pre- and postganglionic neurons, dorsal root ganglion neurons, and, in case of prevertebral ganglia, enteric neurons.

Subcellular Localization of Neuropeptides

Several intriguing questions are related to synthesis, storage, and release mechanisms of the peptides in the ganglia. Immunocytochemical studies at the electron microscope level, using the peroxidase-antiperoxidase method, have been performed to analyze the localization of peptides in nerve cell bodies and nerve terminals (Kondo 1983). SP (Kondo & Yui 1981; Matthews et al 1982), VIP (Kondo & Yui 1982a), ENK (Kondo & Yui

1982b), GRP (Kondo et al 1983), BOM, CCK (Hamaji et al 1989), SOM (Feher & Burnstock 1991; Leranth et al 1980), NPY (Järvi et al 1986), CGRP (Lee et al 1987), and DYN (Elfvin et al 1989) have all been studied with this method. A common finding in these studies is that the immunoreactive material of the peptides is present in synaptic-like terminals and associated with the large dense core vesicles. SOM (Feher & Burnstock 1991; Leranth et al 1980), NPY (Järvi et al 1986), and VIP (Järvi et al 1989) were also shown in perikarya of the postganglionic neurons, which indicates that at least some of the SOM-, NPY-, and VIP-containing terminals are processes from the principal ganglion cells. The neuropeptides are generally assumed to be synthesized in the cell body, packaged into vesicles, and transported through the cell processes to the terminal endings in the target region. However, in situ hybridization studies have raised the question concerning the possibility that peptides are synthesized in nerve endings, by demonstrating that oxytocin mRNA is present in nerve terminals in the posterior pituitary (Jirikowski et al 1990). Further studies are needed to clarify this interesting possibility with respect to peptides in the ANS. It is also important to find out whether the peptides, when colocalized, are present in the same organelle. If the messengers are sorted into different vesicles, the possibility cannot be excluded that they are transported into different neuronal processes, thus segregating intra- and extraganglionic release sites. Sossin et al (1990) have shown that this possibility exists in *Aplysia*, where two peptides produced in the same neuron were routed into different terminal domains.

In spite of the many studies dealing with the identification of the peptide material at the ultrastructural level, only limited evidence has been presented for release by an exocytotic process of the immunoreactive material in the mammalian ANS (see Klein & Thureson-Klein 1990). On the other hand, the dense core vesicles containing the immunoreactive material are, in most cases, located extrajunctionally, which indicates that the release may not be in association with the synaptic cleft, but at some distance from the active site (see Klein & Thuresson-Klein 1990). The clarification of the release mechanism of the peptides, especially in relation to release of classical transmitters (see Lundberg & Hökfelt 1986), represents one of the more fascinating challenges in future work.

CONCLUDING REMARKS

The introduction of axonal tracing techniques and immunohistochemistry have revolutionized the study of the nervous system. The main purpose of this review has been to summarize the results obtained with these techniques applied to the analyses of the structure and function of sympathetic ganglia. The abundance of different peptides in the CNS is matched by that in the ganglia. However, the relatively simple organization and the easy accessibility of the ganglia, which permit sophisticated experimental manipulations, have led to a better understanding of the connectivity patterns and their relation to transmitter/peptide localization in the ganglia, as compared with the CNS. What initially appeared as a random distribution of peptides in relation to the classical transmitters has emerged as complex, but well organized and functionally meaningful, expression patterns. Thus, one of the major findings has been the discovery of distinct neuronal, partly chemically coded, subpopulations related to specific targets. These subgroups, in turn, seem to be related to discrete, again chemically coded, afferent nerve fiber networks, which constitute local reflex arcs that apparently can operate with little or no central control and a high degree of specificity. This integrative, partly noncentrally dominated, activity may be one important reason for having autonomic ganglia. The exact role of the peptides is still poorly understood, but the frequent finding of colocalization of several peptides in any one neuron suggests that the physiological effects of neuropeptides on the targets may depend upon the different combinations of substances that are released at the synaptic terminals. Hopefully, the recent development of several peptide, and especially nonpeptide peptide antagonists, will further clarify the complicated processes taking place at these terminals. In fact, such compounds, paired with the knowledge of the pathways harboring specific peptides, may in the future form the basis for designing drugs to treat disturbances related to the autonomic nervous system.

ACKNOWLEDGEMENTS

This work was supported by grants from the Swedish Medical Research Council (5189, 2887), Karolinska Institutet, Stiftelsen Lars Hiertas Minne, Swedish Medical Society, Stiftelsen Ruth och Richard Julins Fond, and Konung Gustav Vs och Drottning Victorias Stiftelse. The skillful secretarial assistance of Mrs. Marianne Rapp is gratefully acknowledged.

Dr. Björn Lindh died unexpectedly in the summer of 1991 at the age of 33 in the jogging track from cardiac arrest.

Literature Cited

Aldskogius, H., Elfvin, L.-G., Forsman, C. 1986. Primary sensory afferents in the inferior mesenteric ganglion and related nerves of the guinea pig. An experimental study with anterogradely transported wheat germ agglutinin-horseradish per-oxidase conjugate. *J. Auton. Nerv. Syst.* 15: 179–90

Amara, S. G., Jonas, V., Rosenfeld, M. G., Ong, E. S., Evans, R. M. 1982. Alternative processing in calcitonin gene expression generates mRNAas encoding different

498 ELFVIN, LINDH & HÖKFELT

polypeptide products. *Nature* 198: 240–44

Andersson Forsman, C., Elfvin, L.-G. 1991. The ultrastructure of specialized contacts between small intensely fluorescent (SIF) cells in the inferior mesenteric ganglion of the guinea pig. *J. Submicrosc. Cytol. Pathol.* 23: 313–17

Araujo, D. M., Collier, B. 1987. Effect of endogenous opioid peptides on acetylcholine release from the cat superior cervical ganglion: selective effect of a heptapeptide. *J. Neurosci.* 7: 1698–1704

Ariano, M. A., Tress, E. L. 1983. Co-localization of cyclic GMP in superior cervical ganglion with peptide neurotransmitters. *Brain Res.* 289: 362–65

Ashe, J. H., Libet, B. 1981. Orthodromic production of noncholinergic slow depolarizing response in the superior cervical ganglion of the rabbit. *J. Physiol.* 320: 333–46

Audigier, S., Barberis, C., Jard, S. 1986. Vasoactive intestinal polypeptide increases inositol phospholipid breakdown in the rat superior cervical ganglion. *Brain Res.* 376: 363–67

Bachoo, M., Ciriello, J., Polosa, C. 1987. Effects of preganglionic stimulation on neuropeptide-like immunoreactivity in the stellate ganglion of the cat. *Brain Res.* 400: 377–82

Bachoo, M., Polosa, C. 1988. Cardioacceleration produced by close intraarterial injection of neurotensin into the stellate ganglion of the cat. *Can. J. Physiol. Pharmacol.* 66: 408–12

Bachoo, M., Yip, R., Polosa, C. 1989. Effects of naloxone and prolonged preganglionic stimulation on noncholinergic transmission in the superior cervical ganglion of the cat. *Can. J. Physiol. Pharmacol.* 68: 1093–99

Baker, S. C., Cuello, A. C., Matthews, M. R. 1980. Substance P-containing synapses in a sympathetic ganglion, and their possible origin as collaterals from sensory nerve fibres. *J. Physiol. (London)* 308: 76P–77P

Baldwin, C., Sasek, C. A., Zigmond, R. E. 1991. Evidence that some preganglionic sympathetic neurons in the rat contain vasoactive intestinal peptide or peptide histidine isoleucine amide-like immunoreactivities. *Neuroscience* 40: 175–84

Beleslin, D., Radnarovic, B., Varagic, V. 1960. The effect of substance P on the superior cervical ganglion of the cat. *Br. J. Pharmacol.* 15: 10–13

Björklund, A., Hökfelt, T., Owman, C., eds. 1988. The peripheral nervous system. In *Handbook of Chemical Neuroanatomy*, Vol 6. Amsterdam: Elsevier

Bowden, J. J., Gibbins, J. L. 1992. Vasoactive intestinal peptide and neuropeptide Y coexist in non-noradrenergic sympathetic neurons to guinea pig trachea. *J. Auton. Nerv. Syst.* 38: 1–20

Bowers, C. W., Zigmond, R. E. 1981. Sympathetic neurons in lower cervical ganglia send axons through the superior cervical ganglion. *Neuroscience* 6: 1783–91

Bowers, C. W., Zigmond, R. E. 1979. Localization of neurons in the rat superior cervical ganglion that project into different postganglionic trunks. *J. Comp. Neurol.* 185: 381–92

Brazeau, P., Vale, W., Burgus, R., Ling, N., Butcher, M., et al. 1973. Hypothalamic peptide that inhibits the secretion of immunoreactive pituitary growth hormone. *Science* 179: 77–79

Buckley, G., Consolo, S., Sjöqvist, F. 1967. Cholinacetylase in innervated and denervated sympathetic ganglia and ganglion cells of the cat. *Acta Physiol. Scand.* 71: 348–56

Chang, M. M., Leeman, S. E. 1970. Isolation of a sialogic peptide from bovine hypothalamic tissue and its characterization as substance P. *J. Biol. Chem.* 245: 4784–90

Chang, M. M., Leeman, S. E., Niall, H. D. 1971. Amino acid sequence of substance P. *Nature* 232: 86–87

Chiba, T., Masuko, S. 1989. Coexistence of multiple peptides in small intensely fluorescent (SIF) cells of inferior mesenteric ganglion of the guinea pig. *Cell Tissue Res.* 255: 523–27

Costa, M., Furness, J. B. 1984. Somatostatin is present in a subpopulation of noradrenergic nerve fibers supplying the intestine. *Neuroscience* 13: 911–19

Costa, M., Furness, J. B., Gibbins, I. L. 1986. Chemical coding of enteric neurons. In *Progress in Brain Research*, ed. T. Hökfelt, K. Fuxe, B. Pernow, 68: 217–39. Amsterdam: Elsevier

Crowcroft, P. J., Holman, M. E., Szurszewski, J. H. 1971. Excitatory input from the distal colon to the inferior mesenteric ganglion in the guinea pig. *J. Physiol.* 219: 443–61

Crowcroft, P. J., Szurszewski, J. H. 1971. A study of the inferior mesenteric and pelvic ganglia of guinea pigs with intracellular electrodes. *J. Physiol.* 219: 421–41

Dail, W. G., Barton, S. 1983. Structure and organization of mammalian sympathetic ganglia. In *Autonomic Ganglia*, ed. L.-G. Elfvin, pp. 3–25. Chichester/New York: Wiley

Dalsgaard, C.-J., Elfvin, L.-G. 1982. Structural studies on the connectivity of the inferior mesenteric ganglion of the guinea pig. *J. Auton. Nerv. Syst.* 5: 265–77

Dalsgaard, C.-J., Elfvin, L.-G. 1981. The distribution of sympathetic preganglionic neurons projecting onto the stellate ganglion of the guinea pig. A horseradish peroxidase study. *J. Auton. Nerv. Syst.* 4: 327–37

Dalsgaard, C.-J., Elfvin, L.-G. 1979. Spinal origin of preganglionic fibers projecting onto the superior cervical ganglion and inferior mesenteric ganglion of the guinea pig as demonstrated by the horseradish peroxidase technique. *Brain Res.* 172: 139–43

Dalsgaard, C.-J., Hökfelt, T., Elfvin, L.-G., Skirboll, L., Emson, P. 1982a. Substance P-containing primary sensory neurons projecting to the inferior mesenteric ganglion: evidence from combined retrograde tracing and immunohistochemistry. *Neuroscience* 7: 647–54

Dalsgaard, C.-J., Hökfelt, T., Elfvin, L.-G., Terenius, L. 1982b. Enkephalin-containing sympathetic preganglionic neurons projecting to the inferior mesenteric ganglion: evidence from retrograde tracing and immunohistochemistry. *Neuroscience* 7: 2039–50

Dalsgaard, C.-J., Hökfelt, T., Schultzberg, M., Lundberg, J. M., Terenius, L., et al. 1983a. Origin of peptide-containing fibers in the inferior mesenteric ganglion of the guinea pig: immunohistochemical studies with antisera to substance P, enkephalin, vasoactive intestinal polypeptide, cholecystokinin and bombesin. *Neuroscience* 9: 191–211

Dalsgaard, C.-J., Schultzberg, M., Vincent, S. R., Elfvin, L.-G. 1983b. Substance P-like immunoreactivity in guinea pig sympathetic ganglia. In *Substance P*, ed. P. Skrabanek, D. Powell, pp. 141–2. Dublin: Boole

Dalsgaard, C.-J., Vincent, S. R., Hökfelt, T., Christensson, I., Terenius, L. 1983c. Separate origins for the dynorphin and enkephalin immunoreactive fibers in the inferior mesenteric ganglion of the guinea pig. *J. Comp. Neurol.* 221: 482–89

Darvesh, S., Nance, D. M., Hopkins, D. A., Armour, J. A. 1987. Distribution of neuropeptide-like immunoreactivity in intact and chronically decentralized middle cervical and stellate ganglia of dogs. *J. Auton. Nerv. Syst.* 21: 167–80

De Castro, F. 1932. Sympathetic ganglia normal and pathological. In *Cytology and Cellular Pathology of the Nervous System*, ed. W. Penfield, 1: 319–79. New York: Harper (Hoeber)

Devor, M. 1983. Nerve pathophysiology and mechanisms of pain in causalgia. *J. Auton. Nerv. Syst.* 7: 371–84

Dun, N. J. 1983. Peptide hormones and transmission in sympathetic ganglia. In *Autonomic Ganglia*, ed. L.-G. Elfvin, pp. 345–66. Chichester/New York: Wiley

Dun, N. J., Karczmar, A. G. 1979. Actions of substance P on sympathetic neurons. *Neuropharmacology* 18: 215–18

Dun, N. J., Ma, R. C. 1984. Slow non-cholinergic excitatory potentials in neurones of the guinea-pig coeliac ganglia. *J. Physiol.* 351: 47–60

Durroux, T., Barberis, C., Jard, S. 1987. Vasoactive intestinal polypeptide and carbachol act synergistically to induce the hydrolysis of inositol containing phospholipids in the rat superior cervical ganglion. *Neurosci. Lett.* 75: 211–15

Edvinsson, L., Ekblad, E., Håkanson, R., Wahlstedt, C. 1984. Neuropeptide Y potentiates the effects of various vasoconstrictor agents on rabbit blood vessels. *Br. J. Pharmacol.* 83: 519–25

Ekblad, E., Håkanson, R., Sundler, F., Wahlstedt, C. 1985. Galanin-neuromodulatory and direct contractile effects on smooth muscle preparations. *Br. J. Pharmacol.* 86: 241–46

Elfvin, L.-G., ed. 1983. *Autonomic Ganglia*. Chichester/New York: Wiley

Elfvin, L.-G. 1971. Ultrastructural studies on the inferior mesenteric ganglion of the cat. *J. Ultrastruct. Res.* 37: 411–48

Elfvin, L.-G. 1963. The ultrastructure of the superior cervical sympathetic ganglion of the cat. *J. Ultrastruct. Res.* 8: 441–76

Elfvin, L.-G., Dalsgaard, C.-J. 1977. Retrograde axonal transport of horseradish peroxidase in afferent fibers of the inferior mesenteric ganglion of the guinea pig. Identification of the cells of origin in dorsal root ganglia. *Brain Res.* 126: 149–53

Elfvin, L.-G., Hökfelt, T., Goldstein, M. 1975. Fluorescence microscopical, immunohistochemical and ultrastructural studies on sympathetic ganglia of the guinea pig, with special reference to the SIF cells and their catecholamine content. *J. Ultrastruct. Res.* 51: 377–96

Elfvin, L.-G., Lindh, B., Hökfelt, T., Terenius, L. 1989. An ultrastructural study of dynorphin-immunoreactive nerve fibers and terminals in the celiac-superior mesenteric ganglion of the guinea pig. *Brain Res.* 502: 341–48.

Eränkö, O. 1976. SIF cells. Structure and function of the small intensely fluorescent sympathetic cells. *Fogarty Int. Cent. Proc.* No. 30

Eränkö, O., Härkönen, M. 1965. Monoamine-containing small cells in the superior cervical ganglion of the rat and an organ composed of them. *Acta Physiol. Scand.* 65: 511–12

Eränkö, O., Härkönen, M. 1963. Histochemical demonstration of fluorogenic amines in the cytoplasm of sympathetic ganglion cells of the rat. *Acta Physiol. Scand.* 52: 285–86

Eränkö, O., Soinila, S., Päivärinta, H., eds. 1980. *Histochemistry and Cell Biology of Autonomic Neurons, SIF Cells and Paraneurons.* New York: Raven

Falck, B., Hillarp, N.-Å., Thieme, G., Torp, A. 1962. Fluorescence of catecholamines and related compounds condensed with formaldehyde. *J. Histochem. Cytochem.* 10: 348–54

Fehér, E., Burnstock, G. 1991. Ultrastructure and distribution of somatostatin-like immunoreactive neurons and nerve fibers in the coeliac ganglion of cats. *Cell Tissue Res.* 263: 567–72

Flett, D. L., Bell, C. 1991. Topography of functional subpopulations of neurons in the superior cervical ganglion of the rat. *J. Anat.* 177: 55–66

Furness, J. B., Costa, M. 1987. *The Enteric Nervous System.* Edinburgh: Churchill Livingstone

Furness, J. B., Costa, M., Eckenstein, F. 1983. Neurones localized with antibodies against choline acetyltransferase in the enteric nervous system. *Neurosci. Lett.* 40: 105–9

Furness, J. B., Costa, M., Keast, J. R. 1984. Choline acetyltransferase and peptide immunoreactivity of submucous neurons in the small intestine of the guinea pig. *Cell Tissue Res.* 237: 328–36

Furness, J. B., Morris, J. L., Gibbins, J. L., Costa, M. 1989a. Chemical coding of neurons and plurichemical transmission. *Annu. Rev. Pharmacol. Toxicol.* 29: 289–306

Furness, J. B., Pompolo, S., Murphy, R., Giraud, A. 1989b. Projections of neurons with neuromedin U-like immunoreactivity in the small intestine of the guinea pig. *Cell Tissue Res.* 257: 415–22

Furness, J. B., Sobels, G. 1976. The ultrastructure of paraganglia associated with the inferior mesenteric ganglia in the guinea pig. *Cell Tissue Res.* 171: 123–29

Gabella, G. 1985. Autonomic nervous system. In *The Rat Nervous System,* ed. G. Paxinos, pp. 325–53. Australia: Academic

Gabella, G. 1976. *Structure of the Autonomic Nervous System.* London: Chapman & Hall

Gabella, G. 1974. Ganglia of the autonomic nervous system. In *The Peripheral Nerve,* ed. J. I. Hubbard. New York: Plenum

Geffen L. B., Livett, D. G., Rush, R. A. 1969. Immunohistochemical localization of protein components of catecholamine storage vesicles. *J. Physiol. (London)* 204: 593–605

Gibbins, I. L. 1991. Vasomotor, pilomotor and secretomotor neurons distinguished by size and neuropeptide content in superior cervical ganglia of mice. *J. Auton. Nerv. Syst.* 34: 171–84

Gibbins, I. L. 1990. Peripheral autonomic nervous system in *The Human Nervous System,* ed. G. Paxinos, pp. 93–123. New York: Academic

Gibbins, I. L. 1989. Dynorphin-containing pilomotor neurons in the superior cervical ganglion of guinea pig. *Neurosci. Lett.* 197: 45–50

Gibbins, I. L., Furness, J. B., Costa, M. 1987. Pathway-specific patterns of the co-existence of substance P, calcitonin gene-related peptide, cholecystokinin and dynorphin in neurons in the dorsal root ganglia of the guinea pig. *Cell Tissue Res.* 248: 417–37

Gibbins, I. L., Morris, J. L. 1990. Sympathetic noradrenergic neurons containing dynorphin but not neuropeptide Y innervate small cutaneous blood vessels of guinea pigs. *J. Auton. Nerv. Syst.* 29: 137–50

Gibbins, I. L., Morris, J. L. 1987. Coexistence of neuropeptides in sympathetic, cranial autonomic and sensory neurons innervating the iris of the guinea pig. *J. Auton. Nerv. Syst.* 21: 67–82

Goodman Gilman, A., Rall, T. W., Nies, A. S., Taylor, P., eds. 1990. *The Pharmacological Basis of Therapies.* New York: Pergamon, 8th ed.

Hamaji, M., Kawai, Y., Kawashima, Y., Tokyama, M. 1989. An electron microscopic study on VIP-, Bom- and CCK-like immunoreactive terminals in the celiac-superior mesenteric ganglion of the guinea pig. *Brain Res.* 488: 283–87

Hanley, M. R., Benton, H. P., Lightman, S. L., Todd, K., Bone, E. A., et al. 1984. A vasopressin-like peptide in the mammalian sympathetic nervous system. *Nature* 309: 258–61

Helén, P., Panula, P., Yang, H.-YT., Rapaport, S. I. 1984. Bombesin/Gastrin-releasing peptides (GRP)- and Met^5-Enkephalin Arg^6-Gly^7-Leu^8-like immunoreactivities in small intensely fluorescent (SIF) cells and nerve fibers of rat sympathetic ganglia. *J. Histochem. Cytochem.* 32: 1131–38

Hervonen, A., Linnoila, I., Pickel, V. M., Helen, P., Pelto-Huikko, M., et al. 1981. Localization of (met^5)- and (leu^5)-enkephalin-like immunoreactivity in nerve terminals in human paravertebral sympathetic ganglia. *Neuroscience* 6: 323–30

Hervonen, A., Pelto-Huikko, M., Helén, P., Alho, H. 1980. Electron microscopic localization of enkephalin-like immunoreactivity in axon terminals of human sympathetic ganglia. *Histochemistry* 70: 1–6

Heym, C., Kummer, W., Gleich, A., Asmar, R., Liu, N. 1991. The guinea pig stellate ganglion: neurochemical and somatotopic organization. *J. Auton. Nerv. Syst.* 33: 104–5.

Heym, C., Reinecke, M., Weihe, E., Forssmann, W. G. 1984. Dopamine-β-hydroxylase-, neurotensin-, substance P-, vasoactive intestinal polypeptide-, and enkephalin-immunohistochemistry of paravertebral and prevertebral ganglia in the cat. *Cell Tissue Res.* 235: 411–18

Heym, C., Webber, R., Horn, M., Kummer, W. 1990. Neuronal pathways in the guinea-pig lumbar sympathetic ganglia as revealed by immunohistochemistry. *Histochemistry* 93: 547–57

Hill, E. L., Elde, R. 1989. Vasoactive intestinal peptide distribution and colocalization with dopamine-β-hydroxylase in sympathetic chain ganglia of pig. *J. Auton. Nerv. Syst.* 27: 229–39

Hills, J. M., King, B. F., Mirsky, R., Jessen, K. R. 1988. Immunohistochemical localization and electrophysiological actions of GABA in prevertebral ganglia in guinea pig. *J. Auton. Nerv. Syst.* 22: 129–40.

Holmstedt, B., Sjöqvist, F. 1959. Distribution of acetocolinesterase in various sympathetic ganglia. *Acta Physiol.* 47: 284–96

Hughes, J., Smith, T. W., Kosterlitz, H. W., Fothergill, L. A., Morgan, B. A., Morris, H. R. 1975. Identification of two related pentapeptides from the brain with potent opiate agonist activity. *Nature* 258: 577–79

Häppölä, O., Päivärinta, H., Soinila, S., Wu, J.-Y., Panula, P. 1987a. Localization of L-glutamate decarboxylase and GABA transaminase immunoreactivity in the sympathetic ganglia of the rat. *Neuroscience* 21: 271–81

Häppölä, O., Soinila, S., Päivärinta, H., Panula, P. 1987b. (Met[5])-enkephalin-Arg[6]-Phe[7]- and (Met[5])-enkephalin-Arg[6]-Gly[7]-Leu[8]-immunoreactive nerve fibers and neurons in the superior cervical ganglion of the rat. *Neuroscience* 21: 283–95

Häppölä, O., Soinila, S., Päivärinta, H., Panula, P., Eränkö, O. 1985. Histamine-immunoreactive cells in the superior cervical ganglion and in the coeliac-superior mesenteric ganglion complex of the rat. *Histochemistry* 82: 1–3

Häppölä, O., Wahlestedt, C., Ekman, R., Soinila, S., Panula, P., Håkanson, R. 1990. Peptide YY-like immunoreactivity in sympathetic neurons of the rat. *Neuroscience* 39: 225–30

Hökfelt, T., Elfvin, L.-G., Elde, R., Schultz-

berg, M., Goldstein, M., Luft, R. 1977a. Occurrence of somatostatin-like immunoreactivity in some peripheral sympathetic noradrenergic neurons. *Proc. Natl. Acad. Sci. USA* 74: 3587–91

Hökfelt, T., Elfvin, L.-G., Schultzberg, M., Fuxe, K., Said, S. I., Goldstein, M. 1977b. Immunohistochemical evidence of vasoactive intestinal polypeptide-containing neurons and nerve fibers in sympathetic ganglia. *Neuroscience* 2: 885–96

Hökfelt, T., Elfvin, L.-G., Schultzberg, M., Goldstein, M., Nilsson, G. 1977c. On the occurrence of substance P-containing fibers in sympathetic ganglia: immunohistochemical evidence. *Brain Res.* 132: 29–41

Inagaki, S., Kubota, Y., Kito, S., Kangawa, K., Matsuo, H. 1986. Immunoreactive atrial natriuretic polypeptides in the adrenal medulla and sympathetic ganglia. *Regul. Pept.* 15: 249–60

Ip, N. Y., Baldwin, C., Zigmond, R. E. 1985. Regulation of the concentration of adenosine 3.′5′ cyclic monophosphate and the activity of tyrosine hydroxylase in the rat superior cervical ganglion by three neuropeptides of the secretin family. *J. Neurosci.* 5: 1947–54

Ip, N. Y., Ho, C. K., Zigmond, R. E. 1982. Secretin and vasoactive intestinal peptide acutely increase tyrosine 3-monooxygenase in the rat superior cervical ganglion. *Proc. Natl. Acad. Sci. USA* 79: 7566–69

Jan, L. Y., Jan, Y. N. 1982. Peptidergic transmission in sympathetic ganglia of the frog. *J. Physiol. (London)* 327: 219–46

Jan, Y. N., Jan, L. Y., Kuffler, S. 1980. Further evidence for peptidergic transmission in sympathetic ganglia. *Proc. Natl. Acad. Sci. USA* 77: 5008–12

Jan, Y. N., Jan, L. Y., Kuffler, S. 1979. A peptide as a possible transmitter in sympathetic ganglia of the frog. *Proc. Natl. Acad. Sci. USA* 76: 1501–5

Jacobowitz, D. M., Woodward, J. K. 1968. Adrenergic neurons in the cat superior cervical ganglion and cervical sympathetic trunk. A histochemical study. *J. Pharmacol. Exp. Ther.* 162: 213–26

Jankowska, E. 1985. Further indications of retrograde transneuronal transport of WGA-HRP by synaptic activity. *Brain Res.* 341: 403–8

Jiang, Z. G., Simmons, M. A., Dun, N. J. 1982. Enkephalinergic modulation of noncholinergic transmission in mammalian prevertebral ganglia. *Brain Res.* 235: 185–91

Jirikowski, G. F., Sanna, P. P., Bloom, F. E. 1990. mRNA coding for oxytocin is present in axons of the hypothalamoneurohypophysical tract. *Proc. Natl. Acad. Sci. USA* 87: 7400–4

Jänig, W., McLachlan, E. 1984. On the fate of sympathetic and sensory neurons projecting into a neuroma of the superficial peroneal nerve in the cat. *J. Comp. Neurol.* 255: 301–11

Järvi, R., Helén, P., Hervonen, A., Pelto-Huikko, M. 1989. Vasoactive intestinal peptide (VIP)-like immunoreactivity in the human sympathetic ganglia. *Histochemistry* 90: 347–51

Järvi, R., Helén, P., Pelto-Huikko, M., Hervonen, A. 1986. Neuropeptide Y (NPY)-like immunoreactivity in rat sympathetic neurons and small granule-containing cells. *Neurosci. Lett.* 67: 223–27

Järvi, R., Pelto-Huikko, M., Helén, P., Hervonen, A. 1987. Somatostatin-like immunoreactivity in human sympathetic ganglia. *Cell Tissue Res.* 249: 1–5

Karczmar, A. G., Koketsu, K., Nishi, S., eds. 1986. *Autonomic and Enteric Ganglia. Transmission and its Pharmacology.* New York/London: Plenum

Kasa, P., Dobo, E., Wolff, J. R. 1991. Cholinergic innervation of the mouse superior cervical ganglion: light- and electron-microscopic immunocytochemistry for choline acetyltransferase. *Cell Tissue Res.* 265: 151–58

Kawatani, M., Rutigliano, M. J., De Groat, W. C. 1987. Vasoactive intestinal polypeptide facilitates the late component of the 5-hydroxytryptamine-induced discharge in the cat superior cervical ganglion. *Neurosci. Lett.* 73: 59–64

Kawatani, M., Rutigliano, M. J., De Groat, W. C. 1985. Depolarization and muscarinic excitation induced in a sympathetic ganglion by vasoactive intestinal polypeptide. *Science* 229: 879–81

Kenny, S. L., Ariano, M. A. 1986. The immunofluorescence localization of glutamate decarboxylase in the rat superior cervical ganglion. *J. Auton. Nerv. Syst.* 17: 211–15

Kessler, J. A. 1985. Differential regulation of peptide and catecholamine characters in cultures of sympathetic neurons. *Neuroscience* 15: 827–39

Kessler, J. A., Adler, J., Bohn, M., Black, I. B. 1981. Substance P in sympathetic neurons: regulation by impulse activity. *Science* 214: 335–36

Kessler, J. A., Black, I. B. 1982. Regulation of substance P in adult rat sympathetic ganglia. *Brain Res.* 234: 182–87

Kiraly, M., Favrod, P., Matthews, M. R. 1989. Neuroneuronal interconnections in the rat superior cervical ganglion: Possible anatomical bases for modulatory interactions revealed by intracellular horse-radish peroxidase labelling. *Neuroscience* 33: 617–42

Klein, R. L., Thureson-Klein, Å. K. 1990. Neuropeptide co-storage and exocytosis by neuronal large dense-cored vesicles: how good is the evidence? In *Current Aspects of the Neuroscience*, ed. N. N. Osborne, pp. 219–58. London: Macmillan

Kobayashi, H., Tosaka, T. 1983. Slow synaptic actions in mammalian sympathetic ganglia, with special reference to the possible roles played by cyclic nucleotides. In *Autonomic Ganglia*, ed. L.-G. Elfvin, pp. 281–307. Chichester/New York: Wiley

Koelle, G. B., Friedenwald, J. S. 1949. A histochemical method for localizing cholinesterase activity. *Proc. Soc. Exp. Biol. Med.* 70: 617–22

Koketsu, K. 1986. Inhibitory transmission: Slow inhibitory postsynaptic potential. In *Autonomic and Enteric Ganglia. Transmission and Its Pharmacology*, ed. A. G. Karczmar, K. Koketsu, S. Nishi, pp. 201–23. New York/London: Plenum

Kondo, H. 1983. Ultrastructure of peptidergic neurons in the mammalian sympathetic ganglion—A review. *Biomed. Res.* (Suppl.) 4: 145–50

Kondo, H., Dun, N., Pappas, G. D. 1980. A light and electron microscopic study of the rat superior cervical ganglion cells by intracellular HRP-labeling. *Brain Res.* 197: 193–99

Kondo, H,., Iwanaga, T., Yanaihara, N. 1983. On the occurrence of gastrin releasing peptide (GRP)-like immunoreactive nerve fibers in the celiac ganglion of rats. *Brain Res.* 289: 326–29

Kondo, H., Kuramoto, H., Wainer, B. H., Yanaihara, N. 1985. Evidence for the coexistence of acetylcholine and enkephalin in the sympathetic preganglionic neurons of rats. *Brain Res.* 335: 309–14

Kondo, H., Yui, R. 1982a. An electron microscopic study on VIP-like immunoreactive nerve fibers in the celiac ganglion of guinea pigs. *Brain Res.* 237: 227–31

Kondo, H., Yui, R. 1982b. An electron microscopic study on enkephalin-like immunoreactive nerve fibers in the celiac ganglion of guinea pigs. *Brain Res.* 252: 142–45

Kondo, H., Yui, R. 1981. An electron microscopic study on substance P-like immunoreactive nerve fibers in the celiac ganglion of guinea pigs. *Brain Res.* 222: 134–37

Konishi, S., Tsunoo, A., Otsuka, M. 1979a. Substance P and noncholinergic excitatory synaptic transmission in guinea-pig sympathetic ganglia. *Proc. Jpn. Acad.* 55(B): 525–30

Konishi, S., Tsunoo, A., Otsuka, M. 1979b. Enkephalin presynaptically inhibits cholinergic transmission in sympathetic ganglia. *Nature* 282: 515–16

Krukoff, T. L. 1987. Coexistence of neuropeptides in sympathetic preganglionic neurons of the cat. *Peptides* 8: 109–12

Krukoff, T. L., Ciriello, J., Calaresu, F. R. 1985. Segmental distribution of peptide-like immunoreactivity in cell bodies of the thoracolumbar sympathetic nuclei of the cat. *J. Comp. Neurol.* 240: 90–102

Kummer, W. 1987. Galanin- and neuropeptide Y-like immunoreactivities coexist in paravertebral sympathetic neurons of the cat. *Neurosci. Lett.* 78: 127–31

Kummer, W., Heym, C. 1991. Different types of calcitonin gene-related peptide-immunoreactive neurons in the guinea-pig stellate ganglion as revealed by triple-labelling immunofluorescence. *Neurosci. Lett.* 128: 187–90

Kummer, W., Heym, C. 1988. Neuropeptide distribution in the cervico-thoracic paravertebral ganglia of the cat with particular reference to calcitonin gene-related peptide immunoreactivity. *Cell Tissue Res.* 252: 463–71

Kummer, W., Heym, C., Colombo, M., Lang, R. 1986. Immunohistochemical evidence for extrinsic and intrinsic opioid systems in the guinea pig superior cervical ganglion. *Anat. Embryol.* 174: 401–5

Landis, S. C. 1988. Neurotransmitter plasticity in sympathetic neurons. *Handbook of Chemical Neuroanatomy: The peripheral nervous system*, 6: 65–115. Amsterdam: Elsevier

Landis, S. C., Fredieu, J. R. 1986. Coexistence of calcitonin gene-related peptide and vasoactive intestinal peptide in cholinergic sympathetic innervation of rat sweat glands. *Brain Res.* 377: 177–81

Langley, J. N. 1921. *The Autonomic Nervous System.* Cambridge: Heffer

Larsson, L. I., Rehfeld, J.-F. 1979. Localization and molecular heterogenity of cholecystokinin in the central and peripheral nervous system. *Brain Res.* 165: 201–18

Lee, Y., Hayashi, N., Hillyard, C. J., Girgis, S., MacIntyre, I., et al. 1987. Calcitonin gene-related peptide-like immunoreactivity sensory fibers form synaptic contact with sympathetic neurons in rat celiac ganglion. *Brain Res.* 407: 149–51

Lee, Y., Takami, K., Kawai, Y., Girgis, S., Hillyard, C. J., MacIntyre, I., Emson, P. C. and Tohyyama, M. 1985. Distribution of calcitonin gene-related peptide in the rat peripheral nervous system with reference to its coexistence with substance P. *Neuroscience* 15: 1227–37

Le Greves, P., Nyberg, F., Hökfelt, T., Terenius, L. 1989. Calcitonin gene-related peptide is metabolized by an endopeptidase hydrolyzing substance P. *Regul. Pept.* 25: 277–86

Le Greves, P., Nyberg, F., Terenius, L., Hökfelt, T. 1985. Calcitonin gene-related peptide is a potent inhibitor of substance P degradation. *Eur. J. Pharmacol.* 115: 309–11

Leranth, Cs., Williams, T. H., Jew, J. Y., Arimura, A. 1980. Immuno-electron microscopic identification of somatostatin in cells and axons of sympathetic ganglia in the guinea pig. *Cell Tissue Res.* 212: 83–89

Libet, B., Owman, C. 1974. Concomitant changes in formaldehyde-induced fluorescence of dopamine interneurons and in slow inhibitory postsynaptic potentials of rabbit superior cervical ganglion, induced by stimulation of preganglionic nerve or by a muscarinic agent. *J. Physiol. (London)* 237: 635–62

Lindh, B., Hökfelt, T. 1990. Structural and functional aspects on acetylcholine peptide coexistence in the autonomic nervous system. *Prog. Brain Res.* 84: 175–91

Lindh, B., Haegerstrand, A., Lundberg, J. M., Hökfelt, T., Fahrenkrug, J., et al. 1988a. Substance P-, VIP- and CGRP-like immunoreactivities coexist in a population of cholinergic postganglionic sympathetic nerves innervating sweat glands in the cat. *Acta Physiol. Scand.* 134:: 569–70

Lindh, B., Hökfelt, T., Elfvin, L.-G. 1988b. Distribution and origin of peptide-containing fibers in the celiac superior mesenteric ganglion of the guinea-pig. *Neuroscience* 26: 1037–71

Lindh, B., Hökfelt, T., Elfvin, L.-G., Terenius, L., Fahrenkrug, J., et al. 1986a. Topography of NPY-, somatostatin, and VIP-immunoreactive, neuronal subpopulations in the guinea pig celiac-superior mesenteric ganglion and their projections to the pylorus. *J. Neurosci.* 6: 2371–83

Lindh, B., Staines, W., Hökfelt, T., Terenius, L., Salvaterra, P. M. 1986b. Immunohistochemical demonstration of choline acetyltransferase-immunoreactive preganglionic nerve fibers in guinea pig autonomic ganglia. *Proc. Natl. Acad. Sci. USA* 83: 5316–20

Lindh, B., Lundberg, J. M., Hökfelt, T., Elfvin, L.-G., Fahrenkrug. J., Fischer, J. 1987. Coexistence of CGRP- and VIP-like immunoreactivities in a population of neurons in the cat stellate ganglia. *Acta Physiol. Scand.* 131: 475–76

Lindh, B., Lundberg, J. M., Hökfelt, T.

1989a. NPY-, galanin, VIP/PHI-, CGRP-and substance P-immunoreactive neuronal subpopulations in cat autonomic and sensory ganglia and their projections. *Cell Tissue Res.* 256: 259–73

Lindh, B., Pelto-Huikko, M., Schalling, M., Lundberg, J. M., Hökfelt, T. 1989b. Substance P mRNA is present in a population of CGRP-immunoreactive cholinergic postganglionic sympathetic neurons of the cat: Evidence from combined in situ hybridization and immunohistochemistry. *Neurosci. Lett.* 107: 1–5

Lindh, B., Pelto-Huikko, M., Risling, M., Hökfelt, T. 1991. Cholinergic sympathetic neurons. *J. Auton. Nerv. Syst.* 33: 106–7

Lindh, B., Risling, M., Remahl, S., Hökfelt, T. 1992. Distribution of peptide-containing neurons and nerve fibers in lumbosacral ganglia of the cat: An immunohistochemical study in normal animals and after sciatic nerve resection. *Neuroscience.* Submitted

Love, J. A., Szurszewski, J. H. 1987. The electrophysiological effects of vasoactive intestinal polypeptide in the guinea-pig inferior mesenteric ganglion. *J. Physiol.* 394: 67–84

Love, J. A., Szurszewski, J. H. 1985. Effects of vasoactive intestinal polypeptide on neurons of the guinea pig inferior mesenteric ganglion. *Fed. Proc. Fed. Am. Soc. Exp. Biol.* 44: 1718

Lundberg, J. M., Änggård, A., Emson, P. C., Fahrenkrug, J., Hökfelt, T. 1981. Vasoactive intestinal polypeptide and cholinergic mechanisms in cat nasal mucosa: studies on choline acetyltransferase and release of vasoactive intestinal polypeptide. *Proc. Natl. Acad. Sci. USA* 78: 5255–59

Lundberg, J. M., Hökfelt, T. 1986. Multiple co-existence of peptides and classical transmitters in peripheral autonomic and sensory neurons—functional and pharmacological implications. In *Progress in Brain Research*, ed. T. Hökfelt, K. Fuxe, B. Pernow, 68: 241–62. Amsterdam: Elsevier

Lundberg, J., Hökfelt, T., Änggård, A., Terenius, L., Elde, R., et al. 1982a Organizational principles in the peripheral sympathetic nervous system: Subdivisions by coexisting peptides (somatostatin, avian pancreatic polypeptide and vasoactive intestinal polypeptide-like immunoreactive materials). *Proc. Natl. Acad. Sci. USA* 79: 1303–7

Lundberg, J. M., Hökfelt, T., Schultzberg, M., Uvnäs-Wallensten, K., Köhler, C., Said, S. I. 1979. Occurrence of vasoactive intestinal polypeptide (VIP)-like immunoreactivity in certain cholinergic neurons

of the cat. Evidence from combined immunohistochemistry and acetylcholinesterase staining. *Neuroscience* 4: 539–59

Lundberg, J. M., Rökaeus, Å., Hökfelt, T., Rosell, S., Brown, M., Goldstein, M. 1982b. Neurotensin-like immunoreactivity in the preganglionic sympathetic nerves and in the adrenal medulla of the cat. *Aca Physiol. Scand.* 114: 153–55

Lundberg, J. M., Tatemoto, K. 1982. Pancreatic polypeptide family (APP, BPP, NPY and PYY) in relation to α-adrenoceptor-resistant sympathetic vasoconstriction. *Acta Physiol. Scand.* 116: 393–402

Lundberg, J. M., Terenius, L., Hökfelt, T., Goldstein, M. 1983. High levels of neuropeptide Y in peripheral noradrenergic neurons in various mammals including man. *Neurosci. Lett.* 42: 167–72

Lundberg, J. M., Terenius, L., Hökfelt, T., Tatemoto, K. 1984. Comparative immunohistochemical and biochemical analysis of pancreatic polypeptide-like peptides with special reference to presence of neuropeptide Y in central and peripheral neurons. *J. Neurosci.* 4: 2376–86

Macrae, I. M., Furness, J. B., Costa, M. 1986. Distribution of subgroups of noradrenaline neurons in the coeliac ganglion of the guinea pig. *Cell Tissue Res.* 244: 173–80

Maher, E., Bachoo, M., Cernacek, P., Polosa, C. 1991. Dynamics of neurotensin stores in the stellate ganglion of the cat. *Brain Res.* 562: 258–64

Majewski, M., Kummer, W., Kaleczyc, J., Heym, C. 1991. Peptidergic pathways in the inferior mesenteric ganglion of the pig. *J. Auton. Nerv. Syst.* 33: 109–10

Mantyh, P. W., Catton, M. D., Allen, C. J., Labenski, M. E., Maggio, J. E., Vigna, S. R. 1992. Receptor binding sites for cholecystokinin, galanin, somatostatin, substance P and vasoactive intestinal polypeptide in sympathetic ganglia. *Neuroscience* 46: 739–54

Matthews, M. R., Connaughton, M., Cuello, A. C. 1987. Ultrastructure and distribution of substance P-immunoreactive sensory collaterals in the guinea pig prevertebral sympathetic ganglia. *J. Comp. Neurol.* 258: 28–51

Matthews, M. R., Cuello, A. C. 1984. The origin and possible significance of substance P immunoreactive networks in prevertebral ganglia and related structures in the guinea pig. *Philos. Trans. R. Soc. London* 306: 247–76

Matthews, M. R., Cuello, A. C. 1982. Substance P-immunoreactive peripheral branches of sensory neurons innervate guinea pig sympathetic neurons. *Proc. Natl. Acad. Sci. USA* 79: 1668–72

Matthews, M. R., Raisman, G. 1972. A light and electron microscopic study of the cellular response to axonal injury in the superior cervical ganglion of the rat. *Proc. R. Soc. London Ser. B.* 181: 43–79

McKeon, T. W., Konopka, L. M., Parson, R. L. 1991. Galanin as an inhibitory transmitter in the mudpuppy cardiac ganglion. *J. Auton. Nerv. Syst.* 3: 119–20

McLachlan, E. M. Llewellyn-Smith, I. J. 1986. The immunohistochemical distribution of neuropeptide Y in lumbar pre- and paravertebral sympathetic ganglia of the guinea pig. *J. Auton. Nerv. Syst.* 17: 313–24

Mo, N., Dun, N. J. 1986. Cholecystokinin octapeptide depolarizes guinea pig inferior mesenteric ganglion cells and facilitates nicotinic transmission. *Neurosci. Lett.* 64: 263–68

Neild, T. O. 1978. Slowly developing depolarization of neurones in the guinea pig inferior mesenteric ganglion following repetitive stimulating of the preganglionic nerves. *Brain Res.* 140: 231–39

Nishi, S., Koketsu, K. 1968. Early and late after discharges of amphibian sympathetic ganglion cells. *J. Neurophysiol.* 31: 109–21

Niwa, M., Shigematsu, K., Plunkett L., Saavadra, J. M. 1986. High affinity substance P binding sites in rat sympathetic ganglia. *Fed. Proc.* pp. 694–97

Njå, A., Purves, D. 1977. Specific innervation of guinea pig superior cervical ganglion cells by preganglionic fibres arising from different levels of the spinal cord. *J. Physiol. (London)* 264: 565–83

Norberg, K.-A., Hamberger, B. 1964. The sympathetic adrenergic neuron. Some characteristics revealed by histochemical studies on the intraneuronal distribution of the transmitter. *Acta Physiol. Scand.* (Suppl.) 63: 238: 1–42

Norberg, K.-A., Ritzén, M., Ungerstedt, U. 1966. Histochemical studies on a special catecholamine-containing cell type in sympathetic ganglia. *Acta Physiol. Scand.* 67: 260–70

Ohhashi, T., Jacobowitz, M. 1985. Galanin potentiates electrical stimulation and exogenous norepinephrine-induced contractions in the rat vas deferens. *Regul. Pept.* 12: 163–71

Oldfield, B. J., McLachlan, E. M. 1978. Localization of sensory neurons transversing the stellate ganglion of the cat. *J. Comp. Neurol.* 182: 915–22

Pernow, B. 1983. Substance P. *Pharmacol. Rev.* 35: 85–141

Pernow, B. 1953. Studies on substance P. Purification, occurrence and biological actions. *Acta Physiol. Scand.* 29 (Suppl.) 105: 1–90

Peters, S., Kreulen, D. L. 1986. Fast and slow synaptic potentials produced in a mammalian sympathetic ganglion by colon distension. *Proc. Natl. Acad. Sci. USA* 83: 1941–44

Pick, J. 1970. *The Autonomic Nervous System.* Philadelphia: Lippincott

Prosdocimi, M., Finesso, M., Gorio, A. 1986. Enkephalin modulation of neural transmission in the cat stellate ganglion: pharmacological actions of exogenous opiates. *J. Auton. Nerv. Syst.* 17: 217–30

Purves, D., Snider, W. D., Voyvodic, J. T. 1988. Trophic regulation of nerve cell morphology and innervation in the autonomic nervous system. *Nature* 336: 123–28

Quigg, M., Elfvin, L.-G., Aldskogius, H. 1990. Anterograde transsynaptic transport of WGA-HRP to postganglionic sympathetic cells in the stellate ganglion of the guinea pig. *Brain Res.* 518: 173–78

Ramón y Cajal, S. 1911. *Histologie du Système Nerveux de l'Homme et des Vertébrés.* Madrid: Inst. Ramón y Cajal (2nd ed., 1972)

Reinecke, M., Forssmann, W. G., Thiekötter, G., Triepel, J. 1983. Localization of neurotensin-immunoreactivity in the spinal cord and peripheral nervous system of the guinea pig. *Neurosci. Lett.* 37: 37–42

Rosenfeld, M. G., Mermod, J. J., Amara, S. G., Swanson, L. W., Sawchenko, P. E., et al. 1983. Production of a novel neuropeptide by the calcitonin gene via tissue-specific RNA processing. *Nature* 304: 129–32

Said, S. I., Mutt, V. 1970. Polypeptide with broad biological activity: isolation from small intestine. *Science* 169: 1217–18

Saria, A., Ma, R. C., Dun, N. J., Theodorsson-Norheim, E., Lundberg, J. M. 1987. Neurokinin A in capsaicin-sensitive neurons of the guinea pig inferior mesenteric ganglia: An additional putative mediator for the non-cholinergic excitatory postsynaptic potential. *Neuroscience* 21: 951–58

Sasek, C. S., Zigmond, R. E. 1989. Localization of vasoactive intestinal peptide- and peptide histidine isoleucine amide-like immunoreactivities in the rat superior cervical ganglion and its nerve trunks. *J. Comp. Neurol.* 280: 522–32

Sato, J., Perl, E. R. 1991. Adrenergic excitation of cutaneous pain receptors induced by peripheral nerve injury. *Science* 251: 1608–10

Schultzberg, M. 1983. Bombesin-like immunoreactivity in sympathetic ganglia. *Neuroscience* 8: 363–74

Schultzberg, M., Dalsgaard, C.-J. 1983. Enteric origin of bombesin immunoreactive fibres in the rat coeliac-superior mesenteric ganglion. *Brain Res.* 269: 190–95

Schultzberg, M., Hökfelt, T., Terenius, L., Elfvin, L.-G., Lundberg, J. M., et al. 1979. Enkephalin immunoreactive nerve fibers and cell bodies in sympathetic ganglia of the guinea-pig and rat. *Neuroscience* 4: 249–70

Schultzberg, M., Lindh, B. 1988. Transmitters·and peptides in autonomic ganglia. In *Handbook of Chemical Neuroanatomy: The Peripheral Nervous System*, 6: 297–326

Schultzberg, M., Hökfelt, T., Lundberg, J. M., Dalsgaard, C.-J., Elfvin, L.-G. 1983. Transmitter histochemistry of autonomic ganglia. In *Autonomic Ganglia*, ed. L.-G. Elfvin, pp. 205–33. Chichester, UK: Wiley

Schumann, M. A., Kreulen, D. A. 1986. Action of cholecystokinin octapeptide and CCK-related peptides on neurons in the inferior mesenteric ganglion of guinea pig. *J. Pharmacol. Exp. Ther.* 239: 618–25

Shimosegawa, T., Koizumi, M., Toyota, T., Goto, Y., Kobayashi, S., et al. 1985. Methionine-enkephalin-Arg⁶-Gly⁷-Leu⁸-immunoreactive nerve fibers and cell bodies in lumbar paravertebral ganglia and in the celiac-superior mesenteric ganglion complex of the rat: an immunohistochemical study. *Neurosci. Lett.* 57: 169–74

Simmons, M. A. 1985. The complexity and diversity of synaptic transmission in the prevertebral sympathetic ganglia. *Prog. Neurobiol.* 24: 43–93

Simmons, M. A., Dun, N. J. 1985. Synaptic transmission in the rabbit inferior mesenteric ganglion. *J. Auton. Nerv. Syst.* 14: 335–50

Sjöqvist, F. 1962. *Cholinergic sympathetic ganglion cells*. MD thesis, Univ. Stockholm

Skok, V. I. 1973. *Physiology of Autonomic Ganglia*. Tokyo: Igaku Shoin

Skok, V. I. 1983. Fast synaptic transmission in autonomic ganglia. In *Autonomic Ganglia*, ed. L.-G. Elfvin, pp. 265–279. Chichester/New York: Wiley

Sossin, W. S., Sweet, C. A., Scheller, R. H. 1990. Dale's hypothesis revisited: different neuropeptides derived from a common prohormone are targeted to different processes. *Proc. Natl. Acad. Sci. USA* 87: 4845–48

Stapelfeldt, W. H., Szurszewski, J. H. 1989a. The electrophysiological effects of neurotensin on neurones of guinea-pig pre-vertebral sympathetic ganglia. *J. Physiol.* 411: 301–23

Stapelfeldt, W. H., Szurszewski, J. H. 1989b. Neurotensin facilitates release of substance P in the guinea-pig inferior mesenteric ganglion. *J. Physiol.* 411: 325–45

Stapelfeldt, W. H., Szurszewski, J. H. 1989c. Central neurotensin nerves modulate colo-colonic reflex activity in the guinea pig inferior mesenteric ganglion. *J. Physiol.* 411: 347–65

Steele, P. A., Brookes, S. J. H., Costa, M. 1991. Immunohistochemical identification of cholinergic neurons in the myenteric plexus of guinea-pig small intestine. *Neuroscience* 45: 227–39

Strack, A. M., Sawyer, W. B., Marubio, L. M., Loewy, A. D. 1988. Spinal origin of sympathetic preganglionic neurons in the rat. *Brain Res.* 455: 187–91

Suzuki, N., Hardebo, E., Kåhrström, J., Owman, C. 1990. Neuropeptide Y co-exists with vasoactive intestinal polypeptide and acetylcholine in parasympathetic cerebrovascular nerves originating in the sphenopalatine, otic, and internal carotid ganglia of the rat. *Neuroscience* 36: 507–19

Szurszewski, J. H. 1981. Physiology of mammalian prevertebral ganglia. *Annu. Rev. Physiol.* 43: 53–68

Tatemoto, K. 1982. Neuropeptide Y: complete amino-acid sequence of the brain peptide. *Proc. Natl. Acad. Sci. USA* 79: 5485–89

Tatemoto, K., Carlquist, M., Mutt, V. 1982. Neuropeptide Y: a novel brain peptide with structural similarities to peptide YY and pancreatic polypeptide. *Nature* 296: 659–60

Tatemoto, K., Rökaeus, Å., Jörnvall, H., McDonald, M. J., Mutt, V. 1983. Galanin: A novel biologically active peptide from porcine intestine. *FEBS Lett.* 164: 124–28

Taxi, J. 1976. Morphology of the autonomic nervous system. In *Frog Neurobiology*, ed. R. Llinas, W. Precht, pp. 93–150. Berlin: Springer-Verlag

Taxi, J. 1979. The chromaffin and chromaffin-like cells in the autonomic nervous system. *Int. Rev. Cytol.* 57: 283–343

Verhofstad, A. A. J., Steinbusch, H. W. M., Penke, B., Varga, J., Joosten, H. 1981. Serotonin-immunoreactive cells in the superior cervical ganglion of the rat. Evidence for the existence of separate serotonin- and catecholamine-containing small ganglionic cells. *Brain Res.* 212: 39–49

Vincent, S. R., Dalsgaard, C.-J., Schultzberg, M., Hökfelt, T., Christensson, I., Terenius, L. 1984. Dynorphin-immunoreactive neurons in the autonomic nervous system. *Neuroscience* 11: 973–87

von Euler, U. S. 1946. A specific sympathomimetic ergone in sympathetic nerve fibers (sympathin) and its relation to adrenaline and nor-adrenaline. *Acta Physiol. Scand.* 12: 73–97

von Euler, U. S., Gaddum, J. H. 1931. An unidentified depressor substance in certain tissue extracts. *J. Physiol.* (*London*) 72: 74–87

Wainer, B. H., Levey, A. I., Mufson, E. J., Mesulam, M. M. 1984. Cholinergic systems in mammalian brain identified with antibodies against choline acetyltransferase. *Neurochem. Int.* 6: 163–82

Wallin, B. G. 1990. Neural control of human skin blood flow. *J. Auton. Nerv. Syst.* 30: S185–90

Wallin, B. G., Fagius, J. 1986. The sympathetic nervous system in man—aspects derived from microelectrode recordings. *Trends Neurosci.* 9: 63–67

Watanabe, H. 1983. The organization and fine structure of autonomic ganglia of amphibia. In *Autonomic Ganglia*, ed. L.-G. Elfvin, pp. 183–201. Chichester/New York: Wiley

Webber, R. H., Heym, C. 1988. Immunohistochemistry of biogenic polypeptides in nerve cells and fibers of the guinea pig inferior mesenteric ganglion after perturbations. *Histochemistry* 88: 287–97

Williams, T. H., Chiba, T., Black, A. C., Bhalla, R. C., Jew, J. 1976. Species variation in SIF cells of superior cervical ganglia: are there two functional types? *Fogarty Int. Cent. Proc.* 30: 143–62

Wolff, J. R., Joó, F., Kasa, P., Storm-Mathiesen, J., Toldi, J., Balcar, V. J. 1986. Presence of neurons with GABA-like immunoreactivity in the superior cervical ganglion of the rat. *Neurosci. Lett.* 71: 157–62

Yamamoto, K., Senba, E., Matsunaga, T., Tohyama, M. 1989. Calcitonin gene-related peptide containing sympathetic preganglionic and sensory neurons projecting to the superior cervical ganglion of the rat. *Brain Res.* 487: 158–64

Zigmond, R. E., Baldwin, C., Hyatt-Sachs, H., Rittenhouse, A., Sasek, C., Schwarzschild, M. A. 1991. Regulation of catecholamine biosynthesis in sympathetic neurones by neuropeptides and by depolarization. *J. Auton. Nerv. Syst.* 33: 120–21

Zhang, C., Bachoo, M., Polosa, C. 1991. Naloxone-sensitive inhibition of nicotinic transmission in the superior cervical ganglion of the cat. *Brain Res.* 548: 29–34

Annu. Rev. Neurosci. 1993. 16:509–30

THE PROCESSING OF SINGLE WORDS STUDIED WITH POSITRON EMISSION TOMOGRAPHY

S. E. Petersen and J. A. Fiez

Department of Neurology and Neurological Surgery, Washington University, St. Louis, Missouri 63110

KEY WORDS: language, cognition, neuroimaging, brain blood flow, aphasia

INTRODUCTION

Language, the most profoundly human of all abilities, has been a focus of study in many disciplines. Cognitive psychological studies of normals have been used to describe language-related information processing. The correlation of language-related behavioral deficits with the site of nervous system injury has been used to relate language processing to brain function.

Over the past decade, there has been an explosion in the use of imaging technology, including positron emission tomography (PET), to study the structure and function of the human brain. This article explores how functional imaging with PET has been applied to the study of language, particularly the study of the processing of single words (lexical processing).

To understand how PET is used in the study of lexical processing, one has to have some understanding of the technology of PET, a framework in which to explore the processing of words, and knowledge of the questions that can be addressed by the specific application of PET studies to issues in lexical processing. This paper follows that path with a brief general introduction to PET, an outline of some aspects of lexical processing, and a presentation of selected PET studies that have addressed issues in lexical processing.

509

0147–006X/93/0301–0509$02.00

PET BASICS

Positron emission tomography creates pictures of the distribution of radiation within the central opening of a doughnut-shaped PET scanner. The rationale behind functional imaging is that those parts of the brain that "work harder" (i.e. have high levels of neuronal activity) have higher blood flow or metabolism (Lassen et al 1978), and damaged and/or disconnected areas may have aberrant levels of blood flow or metabolism (Baron et al 1989). Radioactive substances, called tracers, are employed to "image" different physiological processes, such as brain blood flow or metabolism. By comparing images of activity, information relating to localization of certain functions can be obtained. For general reviews of PET technology and methodology, see Raichle (1989) and Stytz & Frieder (1990).

The methods and data analyses used are specific to the application, so those issues are addressed in the sections on specific studies below.

LEXICAL PROCESSING BASICS

In the design and interpretation of PET studies, it is necessary to utilize information from other types of studies on lexical processing. At an intuitive level, there are several different kinds of internal coding that can be done at the level of the single word, including how a word looks (orthographic codes) and sounds (phonological codes) and what it means (semantic codes). These internal codes can be experimentally isolated [for a review of cognitive psychological experimental methodology, see Posner (1978)]. For instance, differences in the internal coding of phonology and articulation can be demonstrated by having subjects perform articulatory and phonological processing tasks concurrently (Shallice et al 1985). Other investigations have focused on orthographic (e.g. Carr & Pollatsek 1985; Glushko 1979), phonological (e.g. Jackson & Morton 1984; Tanenhaus et al 1980), and semantic (e.g. Davidson 1986; Neely 1977) aspects of lexical processing.

Studies of behavioral deficits following damage to particular brain regions have provided the most information on the neural substrates underlying different kinds of lexical coding [see reviews by Caramazza (1988) and Damasio (1992)], although nontomographic blood flow measurements (Lassen et al 1978), event-related potential recording (Kutas & Van Petten 1990), and other methods have also contributed. For instance, specific deficits in reading (e.g. Damasio & Damasio 1983; Henderson 1986), understanding auditory input (e.g. Yaqub et al 1988), and programming speech output (e.g. Geschwind 1979) have been documented following damage to particular brain regions.

STUDIES OF LEXICAL PROCESSING WITH PET

Functional studies of lexical processing with PET fall into two main categories. Although both utilize functional images, their general approaches are quite different.

The first type of study defines brain areas of functional abnormality in conjunction with behavioral studies of language deficits. This use of PET to study abnormal populations is a natural extension of work using classical lesion/behavior.

The second type of study uses activity-based measures to provide images of brain activation during performance of lexical processing tasks, most often in normal subjects. These studies represent a chance to study the function of the normal human brain in vivo.

Imaging of Functional Abnormality

RATIONALE   There is now strong evidence that structural damage is frequently associated with hypometabolism in structurally intact regions distant to the site of damage (Baron 1989; Metter et al 1989; Powers & Raichle 1985). Based upon observations of the rapidity by which these distant effects can develop, the initial hypometabolism is most likely caused by the loss of afferent input to the secondary region. Behavioral deficits similar to those associated with structural damage have been associated with regions of secondary hypometabolism, which leads to the term "metabolic lesions." Despite uncertainty regarding mechanisms responsible for the production of and recovery from metabolic lesions, descriptions of the relationships between such lesions and behavioral deficits should be extremely useful.

METHODS   Under normal conditions at rest, blood flow, oxygen metabolism, and glucose metabolism are tightly correlated (Lassen et al 1978), and all three are useful for indirectly measuring neuronal activity. In such conditions as ischemia or transient activation, this coupling breaks down (Baron et al 1989; Fox & Raichle 1986). For this reason, glucose metabolism, rather than blood flow, has been measured in nearly all PET studies of aphasic patients.

[18]F-labeled fluorodeoxyglucose (FDG), a competitive substrate for glucose, is the tracer typically used to measure glucose metabolism. The accumulation of FDG in the brain is proportional to the amount of glucose utilized. The simplest and most commonly used method requires that the tracer reach a steady- or near steady-state before measuring the accumulation of FDG with the PET scanner (Reivich et al 1979; Sokoloff et al 1977). During this period (about 40 minutes), subjects must continuously

perform the task of interest. The task condition for studies of abnormal populations is most often a "resting" condition, e.g. the patients are told to close their eyes, and their ears may be plugged.

To analyze metabolic images, each individual PET image is usually first divided into standard regions of interest (Herholz et al 1985; Mazziotta et al 1983). Two general approaches have been used to identify which regions are metabolically abnormal. For one approach, the average regional metabolic values of controls are determined for each region; based upon these values and the range of normal variation, a criterion is established to define "abnormal regional metabolism" (e.g. see Kushner et al 1987; Metter et al 1987). Because of the numerous comparisons to be made, and the small size of the abnormal group (in many cases, only an individual is compared), this criterion does not reflect a statistically rigorous value for significantly different regional metabolism, but rather a descriptive measure of regional metabolic normality. Another approach has been to avoid any comparison to normal metabolic regional values and instead determine, in a large patient group, which regional values correlate most closely with behavioral measures of impairment (e.g. see Karbe et al 1989; Metter et al 1989).

AREAS OF INVESTIGATION   The use of PET to extend lesion-behavior analysis to include information about metabolic lesions is still relatively new, but has provided insight into mechanisms of aphasia. Several studies have provided evidence that the distribution of metabolic lesions can help account for inconsistencies in the relationship between structural lesions and behavioral deficits. For example, various types of aphasia have been reported following damage to thalamic and basal ganglia structures (Crosson 1985; Naeser et al 1982). However, aphasia is not always present following subcortical structural damage, which leads many investigators to question the importance of subcortical structures in language processing. Aphasia following subcortical structural damage is apparently correlated with cortical hypometabolism, particularly in left posterior temporoparietal regions (Karbe et al 1989, 1990; Metter et al 1986, 1988).

Other investigations have focused on identifying the structural and metabolic location of lesions associated with specific behavioral deficits. In one report, left temporoparietal hypometabolism, caused by either structural damage or a secondary metabolic effect, was found in all 44 aphasic patients examined (Metter et al 1990). In another report, standard diagnostic aphasia exams were used to classify patients into a variety of subtypes, such as Broca's and Wernicke's aphasia. The location of structural lesions ranged from damage limited to subcortical structures to widespread damage of frontal, temporal, and parietal regions of the left hemisphere. The degree of temporoparietal hypometabolism correlated most strongly with impaired performance on tests of language comprehension. The presence of frontal hypometabolism, caused by either structural or metabolic lesions, reliably discriminated Broca's from Wernicke's aphasics (Metter et al 1989). Other studies have also found correlations between temporal/temporoparietal hypometabolism and impaired language comprehension (Karbe et al 1989, 1990; Tyrrell et al 1990).

Finally, PET provides a unique means to monitor functional changes over time. Several different investigators have now reported the results of sequential PET scans of aphasic patients. Correlations between reductions in secondary hypometabolism and recovery of language functions have been reported. Conversely, patients who do not show significant improvement tend to show persistence of metabolic abnormalities (Kushner et al 1987; Metter et al 1986). [Also see a study by Vallar et al (1988), who address this issue by using a related technique to measure metabolism.]

Mapping Functional Activation

RATIONALE   The basic rationale for using PET activation studies is that the performance of any task places specific information processing demands on the brain. These demands are met through changes in neural activity in various functional areas of the brain (Posner et al 1988). Changes in neuronal activity produce changes in local blood flow (Frostig et al 1990; Raichle 1989), which can be measured with PET.

In the case of studies from our group, the design and interpretation of these experiments is based on a particular view of the localization of function in the brain (see Posner et al 1988). Understanding from integrative neuroscience and cognitive science has encouraged a limited framework for understanding functional localization. The main idea of this framework is that elementary operations, defined on the basis of information processing analyses of task performance, are localized in different regions of the brain. Because many such elementary operations are involved in any cognitive task, a set of distributed functional areas must be orchestrated in the performance of even simple cognitive tasks.

Although these seem like trivial statements, it is what is not said that is important. A functional area of the brain is not a task area; there is no "tennis forehand area" to be discovered. Likewise, no area of the brain is devoted to a very complex function; "attention" or "language" is not localized in a particular Brodmann area or lobe. Any task or "function" utilizes a complex and distributed set of brain areas.

The areas involved in performing a particular task are distributed in different locations in the brain, but the processing involved in task performance is not diffusely distributed among them. Each area makes a specific contribution to the performance of the task, and the contribution is determined by where the area resides within its richly connected parallel, distributed hierarchy [such as that described for the visual system by Felleman & van Essen (1991)]. The difficult but exciting job to which PET activation studies contribute is the identification of specific sets of computations with particular areas of the brain. Given the complexity of this job, no single experiment or image provides compelling evidence at this level of analysis. Information from multiple PET studies and the results from other methodologies must be considered. This section attempts not only to address specific issues in lexical processing, but also provide examples of how converging evidence can aid in the interpretation of specific results.

METHODS   Positron emission tomography activation methods have been extensively described, and a brief description follows.

In designing or understanding PET activation studies, one should realize the performance characteristics of the methodologies for these studies. Currently, the spatial resolution of reconstructed images for activation studies is well above 1 cm, but the accuracy of localizing a single source is on the order of 2–5 mm (see, e.g. Fox et al 1987a; Mintun et al 1989). These spatial capabilities allow for some topographic mapping within large primary and extra-primary sensory and motor areas (e.g. Grafton et al 1991), but most of the study is at the level of the identification of functional areas involved in the performance of particular tasks (e.g. Zeki et al 1991).

Although earlier studies utilized different tracers, including FDG, most activation studies in normals currently measure blood flow by using ^{15}O. ^{15}O has a short half-life (about 2 min) and acquisition time (about 40 s–4 min) compared with other tracers, which allows several scans to be performed in a single session (Herscovitch et al 1983; Raichle et al 1983).

Because several scans can be made in a scan session, different tasks can be performed in each subject. Because within-subject designs are used, comparison images between different task conditions can be created, with the idea of imaging activity changes related to specific elementary operations of a task. Different techniques have been developed to allow the creation of inter- or intrasubject averaged images, or data sets that increase the signal-to-noise ratio (Evans et al 1988; Fox et al 1985, 1988; Friston et al 1991b; Mazziotta et al 1991). Recent reports (Steinmetz & Seitz 1991) have asserted that although such averaging might be appropriate for primary areas, association and higher level cortical areas have much more anatomical variability and, thus, are inappropriately studied with averaging. Fox & Pardo (1991) have shown that although activation in frontal and visual association cortex shows slightly higher variability across subjects than in primary areas, this difference is small and the averaging techniques are still quite effective.

Once created, images are usually searched with computerized routines to localize areas of change in stereotactic coordinates (e.g. Mintun et al 1989). Because these images have been cast into stereotactic coordinates, most often one of the Talairach stereotactic brain atlas spaces (Talairach et al 1967; Talairach & Tournoux 1988), the results of experiments from different groups can be compared with one another in a relatively direct way.

Different types of statistical analyses have been applied to identify areas of significant change. The most common approach is to survey images without a priori assumptions about the locations of significant changes. This presents more difficult statistical problems, because of the numerous spatial locations that are not statistically independent from one another and on which there are small numbers of observations (Fox et al 1988; Friston et al 1991a; Worsley et al 1992). The issue of statistical analysis is an aspect of PET activation studies in which further advances should occur.

FUNCTIONAL MAPPING INVESTIGATIONS   A survey study of lexical processing (Petersen et al 1988, 1989) was designed to examine, at several levels, the processing of single words (see Table 1 and Figures 1–3). This study identified several regions potentially involved in lexical processing and provided a framework for the development and interpretation of subsequent PET investigations of language processing. This study looked at three levels of change: simple presentation of the common English nouns, compared with no presentation; repetition aloud of the nouns, compared with simple presentation; and generation aloud of a verb appropriate to the presented noun, compared with repetition. The active state from one level acted as the control for the next level of analysis. Each of these levels was assessed in one set of scans with visual input and in another set of scans with auditory input.

Table 1   Paradigm design for Petersen et al (1988) study[a]

Control state	Active state	Added component
Fixation point only	Passive words (aud or vis)	Passive sensory processing, automatic word-level coding
Passive words (aud or vis)	Repeat words	Articulatory coding motor programming and output
Repeat words	Generate verbs (uses and actions)	Semantic/syntactic association, selection for action

[a] Each task (except fixation only) was performed in one set of scans with visual presentation of words and in another set of scans with auditory presentation. All subtractions are made within modality of presentation.

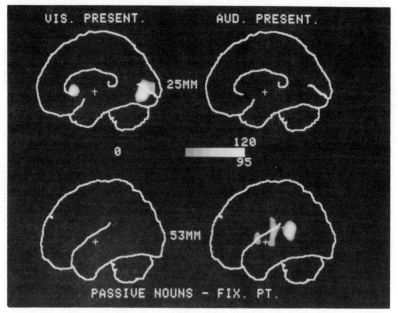

Figure 1   Sagittal slices (*anterior left, posterior right*) through averaged subtraction images during passive presentation of visual (*left slices*) and auditory words with a control condition of simple fixation. The upper slices are taken 25 mm left of midline, and the lower slices are taken 53 mm left of midline. The upper left image shows extrastriate visual cortex activation posteriorly for the visual presentation with no activation in the corresponding slice (*upper right*) for auditory activation. The opposite holds true for the more lateral slices. Several areas are active for auditory presentation (*lower right*) that are not present for visual presentation. The most posterior activation for auditory words is discussed in the text as the temporoparietal region.

Passive presentation of words appeared to activate modality-specific primary and extraprimary sensory-processing areas (Figure 1). When words are presented visually (without any task demands), several areas of extrastriate visual cortex are activated in both hemispheres. The presentation of auditory words activated areas bilaterally along the superior temporal gyrus, as well as a left-lateralized area in the temporoparietal cortex.

When the repeat aloud tasks were compared with the passive tasks, similar areas of activation for auditory and visual presentations were found (as would be expected, because the sensory-specific activation would be subtracted away). For both auditory and visual cues, speech output produced activation in areas that have been implicated in some aspects of motor coding or programming, including primary sensorimotor mouth cortex, the supplementary motor area (SMA), and regions of the cerebel-

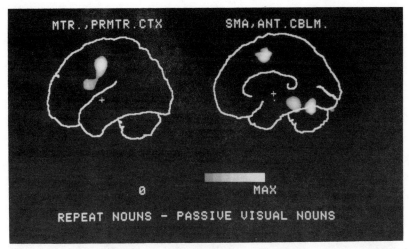

Figure 2   Sagittal slices through averaged substraction images when subjects repeat aloud visually presented words with a control condition of visual presentation. The left image is taken about 40 mm left of midline and shows activation in primary motor (upper activation) and premotor cortex (lower activation). The right image is taken near the midline and shows activation in supplementary motor cortex (SMA, upper activation) and midline cerebellum (lower activation).

lum. Several areas that might be considered lateral premotor regions in Sylvian-opercular cortex and a left-lateralized region on the lateral surface of the frontal cortex at or near inferior area 6 were also activated (Figure 2).

The generate verb task made additional processing demands. Here, the subtraction condition was the repeat aloud condition, so the sensory input was identical in the active and control condition, and the motor output was very similar. Again, the comparison should have subtracted away differences between the auditory and visual input versions of the task, and this was most often the case. For the generate subtraction, two foci in anterior cingulate cortex were activated; several regions of the left anterior inferior prefrontal cortex and the right inferior lateral cerebellum (Figure 3).

We now examine each of the areas discussed above.

Extrastriate visual cortex   Recent studies (Marrett et al 1990; Wise et al 1991b) have also shown extrastriate activation with the presentation of visual words. With this evidence alone, however, the activations in extrastriate cortex could be accounted for by any of several factors. The activations could be due to the simple visual features of the stimuli; processing at some higher level, such as the letter or word levels; or some intermediate between these stages. Subsequent experiments have addressed these questions.

In one experiment, different sets of words and word-like visual stimuli

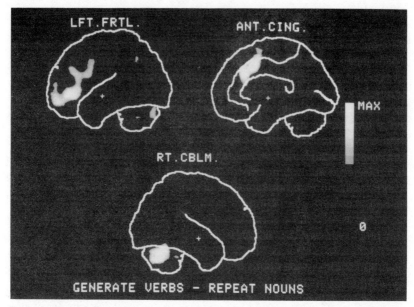

Figure 3 Sagittal slices through averaged subtraction images when subjects generate verbs appropriate to visually presented nouns with a control condition of simple repetition of visually presented nouns. The upper left image, taken about 40 mm left of midline, shows activation in several prefrontal regions. The upper right image is taken near midline and shows activation in the anterior cingulate. The lower image (*anterior right, posterior left*) is taken 25 mm right of midline and shows activation in lateral cerebellum.

were presented in separate scans, without other task demands (Petersen et al 1990). Again, the main control condition was simple fixation on a crosshair. One of the sets consisted of real common nouns to replicate the condition from the original study. In another scan, the stimuli consisted of strings of letter-like forms that were matched for many visual features to actual letters. These stimuli were called "false-font." A third set of stimuli were strings of random consonants, which were called letter-string stimuli. These stimuli were chosen to determine whether the activation was caused by processing at the level of analysis of single letters. The final set of stimuli was a series of letter strings that followed the spelling rules of English, but were not real words, e.g. "tweal" (pseudowords).

All four stimulus types produce extrastriate activation that is laterally placed in the hemispheres (Figure 4). This activation seems most easily interpreted as related to general visual processing at the level of common visual features or characteristics shared by all of the stimulus sets.

Although the lateral activations are produced by presentation of all four types of visual work-like stimuli, left medial extrastriate cortex activation

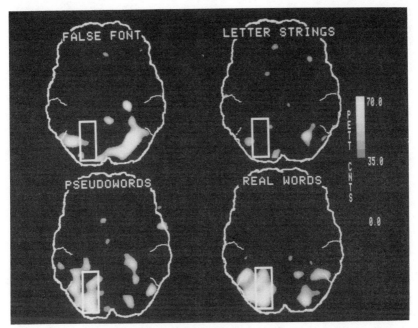

Figure 4 Horizontal slices taken through averaged subtraction images when subjects see sets of word-like stimuli with a control condition of simple fixation (anterior is to the top of the images, and left hemisphere is to the left of the images). The upper-left image represents activation when a series of false font stimuli are presented, the upper right when random letter strings are presented, the lower left when pseudowords are presented, and the lower right when real words are presented for all four images, there is activation in the posterior cortex in lateral extrastriate cortex. Only for the pseudoword and real word conditions is there activation in medial extrastriate cortex (*box*).

is found for real and pseudowords, but not for letter strings and false font stimuli (box on Figure 4). These differences are probably not due to the simple visual characteristics of these stimuli, because these were carefully matched. This area (or these areas) is likely part of a system that makes distinctions based on information about the combinations of letters that English words can regularly assume, a level of analysis called orthographic regularity. This level of distinction is not innate; it must have been learned as the subjects were becoming literate, and, judging by the location of the medial activation, the effects show up early in the visual processing stream. This particular level of distinction, where words and pseudowords are distinguished from other word-like stimuli, is consistent with cognitive psychology experiments about the visual processing of words. There is a well-known effect that the letters inside of words are more efficiently processed (can be seen at lower thresholds or are responded to more quickly) than the letters inside of random letter strings. This word superiority effect extends to pseudowords, such as those presented in our studies (see Carr & Pollatsek 1985).

Temporal areas   Activation near Heschel's gyrus and the middle portion of the superior temporal gyrus has been found for a wider range of auditory stimuli than solely auditorily presented words, e.g. clicks, tones, words, orthographically regular nonwords, and real words played backwards (Lauter et al 1985; Mazziota et al 1982; Wise et al 1991a,b), which suggests a localization to primary and surrounding extraprimary auditory cortex.

Posterior to these areas, it appears as if activation is dependent upon more than aural presentation of a stimulus. Wise et al (1991a) report bilateral activation of a posterior superior temporal gyral near the location of the left-lateralized temporoparietal area found by Petersen et al (1988, 1989) during passive presentation of auditory words. Wise and collaborators used three different task conditions relative to a resting baseline control. For one task, subjects listened to aurally presented nonwords; in the other tasks, subjects made decisions about whether aurally presented pairs of nouns, or nouns and verbs, were appropriately matched. The presentation rate was varied across the tasks and subjects (26–60 words presented per minute). In a fourth condition, subjects were aurally presented with nouns at a rate of 15 per minute and asked to think silently of as many appropriate verbs as they could for each presented noun. Left-lateralized posterior temporal gyral activation was seen for this condition as well, despite a lack of significant activation in Heschel's gyrus and the middle superior temporal gyrus. These posterior temporal/temporoparietal regions do not appear to be activated by simple auditory stimuli, including tones, clicks, or rapidly presented synthetic syllables (Lauter et al 1985; Mazziota et al 1982). Temporoparietal activation has been found when subjects performed rhyme-detection tasks on visually presented words (Petersen et al 1989).

Different interpretations have been placed upon the posterior temporal and temporoparietal regions of activation. Wise et al (1991a) have suggested that the posterior temporal area is related to word comprehension and semantic processing, based largely upon the observation that a left posterior region is active during a task when subjects must silently generate verbs appropriate for aurally presented nouns. Because more anterior areas were not significantly active during the silent verb generation task, and activity in the left posterior temporal region appears uncorrelated with the rate of stimulus presentation, the authors suggest that the activation probably does not represent some form of phonological processing.

Other findings present difficulties for the conclusion that posterior temporal areas are related to word comprehension in a general way. Most troubling is the lack of significant activation observed in a similar verb-generation task in which subjects were asked to say aloud appropriate verbs for visually (in contrast to auditorily) presented nouns (Petersen et al 1988), and a condition in which subjects were asked to say aloud either as many jobs or words beginning with the letter "a" as they could think of in three minutes (in which case no stimuli were presented during the scan, either auditorily or visually) (Frith et al 1991b). For both of these generation tasks, subjects verbally, rather than silently, produced responses, and the control condition consisted of verbal output (read aloud visually presented words and count aloud, respectively).

Wise and collaborators have suggested that the failure to replicate posterior temporal activation may be accounted for on the basis of verbal output. They raise two issues. One possibility is that word repetition may activate the semantic (i.e. posterior superior gyral region) system. Thus, in the subtraction of generate verbs versus read aloud visually presented nouns, semantic activation during the verb-generation task might have been canceled out (Wise et al 1991a,b). Arguing against this interpretation is the lack of any temporal activation during any level of lexical analysis upon visually presented words, e.g. passive visual presentation-fixation point, or repeat visual word-passive visual words (Petersen et al 1988).

Based upon the lack of superior temporal gyral activation during reading aloud of visual words versus passive viewing of visually presented words, Wise and collaborators also suggest that perhaps the process of self-vocalization causes a reduction of activity in temporal regions and, thus, might cancel out any activation produced by semantic analysis/word comprehension during any generation condition in which subjects say aloud their responses (Frith et al 1991b; Wise et al 1991a,b). This explanation is only plausible if the additional semantic processing demands of the generate task (above the repeat task) produces no further neuronal activity.

An alternative explanation for the posterior temporal/temporoparietal results is that the activation is related to auditory or phonological processing that includes activity related to short-term storage. This would explain its ubiquitous presence and modulation in auditory tasks, its presence in a task that requires phonological processing of visual information, and its absence in a semantic association task (the generate a verb task) when the input for the task is presented visually. Such an explanation might also explain why a low presentation rate of auditory words (15 words per minute) might not produce significant activation in early auditory regions, but might produce significant activation in posterior regions when the task requires continued storage of each presented word for the interval between stimuli.

All of this is not to say that some aspects of semantic processing or word comprehension are not present in the posterior temporal lobe. The lesion literature relates damage to posterior cortex with comprehension deficits much more strongly than anterior cortex [see reviews by Damasio (1992) and Geschwind (1979)]. Regions in the temporal lobe, other than the particular temporoparietal region discussed here, might be more directly related to computations of a semantics or word meaning. Some data recently produced in PET studies (Marrett et al 1990; Raichle et al 1991) are consistent with this possibility.

More fundamentally, the comprehension of normal language is not likely to be subsumed in a single enclosed localization. The computation of meaning must occur above the level of the single word in its context within the grammatical structure. Because many lexical items have different meanings dependent on context (such as "play"), several types of coding must be coordinated to compute meaning. Rather than attempt to find the region of the brain related to word comprehension, an experimental analysis of the contribution of different areas to semantic processing should be the goal.

Lateral premotor areas   Simple repetition of visual and auditory words evoke activation in Sylvian-opercular and premotor regions. The left-sided regions surround the traditionally defined Broca's area. Damage to Broca's area in the left hemisphere is often thought to be associated with speech production deficits and agrammatism (Geschwind 1965; Mohr et al 1978). Although the lateral premotor activation could be related to some specifically linguistic functions, both simple tongue and hand movements (Fox et al 1987b) caused similar lateral premotor activation. Mohr et al (1978) have shown that lesions that are confined to classically defined Broca's area produce motor and praxis deficits without specific language involvement. The full-blown syndrome of Broca's aphasia requires much larger lesions.

Similar lateral and opercular activation was also found in a study of selective and divided attention to visual features of color, speed, and shape (Corbetta et al 1991) (Figure 5). In the selective conditions, a subject judged changes in only one of the attributes during the scan; in the divided condition, the subjects monitored for changes in all of the attributes simultaneously. For each of the selective conditions, a lateral or opercular activation was found, but none was seen in the divided condition. The divided condition produced activation in prefrontal and cingulate cortex (see below).

In each of the conditions that produce lateral premotor-Sylvian-opercular activation, efficient response selection can be based on simple or limited

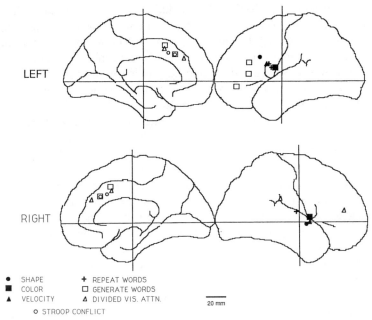

LEFT

RIGHT

● SHAPE   + REPEAT WORDS
■ COLOR   □ GENERATE WORDS
▲ VELOCITY   △ DIVIDED VIS. ATTN.
    ○ STROOP CONFLICT

20 mm

Figure 5 Diagrammatic representation of the location of frontal, prefrontal, and cingulate activations in different tasks. The filled symbols represent activation seen when subjects selectively responded to changes in a single visual attribute (e.g. color) of a visual array in the visual attention experiment referred to in the text. The open triangle refers to the divided condition from the same study. The open circle refers to activation seen in the Stroop conflict paradigm (Pardo et al 1990), and the open square to regions activated in the generate verb condition with visual presentation. There is a cluster of points in the anterior cingulate across a range of conditions and a cluster of points in posterior frontal cortex across several conditions. Tentative explanations for these clusterings are presented in the text.

processing of input. In the word repetition conditions, the relationship between a visual or auditory word, and articulation of that word, is an overlearned association that might be mediated by left-medial extrastriate cortex (for visual words) and parietotemporal cortex (for auditory words). In the selective conditions of visual attention experiments, the information on which the decision is made is limited to a single attribute, perhaps allowing much of the processing to occur in different extrastriate cortical regions.

Anterior cingulate When the demands of a particular task do not have a strong stimulus response association, which allows for the use of posterior processing regions and lateral premotor regions, another set of networks seems to become involved. These include the use of lateral prefrontal regions (see below) as processing centers for complex associations and the use of the anterior cingulate as the response selection, or premotor stage.

The anterior cingulate region activated in the generate a verb task has now been seen in several other conditions (Figure 5). In the conflict condition of the Stroop color naming task, subjects name the color of ink in which a word is printed, rather than say aloud the word itself, e.g. say "green" when the word "red" is presented in green ink (Pardo et al 1990). In a divided attention condition, subjects were asked to monitor simultaneously visual arrays for changes in the color shape or speed of a visual stimulus (Corbetta et al 1991). And, in another condition, subjects were asked to pick an arbitrary direction for movement when given a signal to move (Frith et al 1991a). All of these conditions were also associated with activation of lateral prefrontal regions (see below). Although cingulate activation tends to occur in more difficult conditions, e.g. generating a word versus simple repetition, psychophysical results (Corbetta et al 1991) provide evidence against a general task difficulty interpretation of cingulate function.

Consistent with this view of the cingulate as being related to response selection based on complex stimulus-response associations, are results from other methodologies. Large midline lesions, including SMA and anterior cingulate, often produce akinetic mutism, a syndrome in which spontaneous speech is extremely rare (Barris & Schuman 1953; Masdau et al 1978; Nielsen & Jacobs 1951). Recent single-unit recording work has shown greater activation for many anterior cingulate neurons in monkeys when the animals are performing more complex sensory-motor tasks than in more simple conditions (Shima et al 1991).

Prefrontal activations   There are two issues related to understanding the prefrontal activations in lexical processing experiments. One concerns the general nature of the task involved. The other is more specific and relates to the domain of specific information necessary to perform a task. As discussed above, both prefrontal and anterior cingulate activation appear to be critically dependent upon the nature of the relationship between presented stimuli and expected responses in a manner that cuts across many different types of tasks (Figure 5). For lexical tasks, manipulation of the degree to which a response to a stimulus is overlearned appears to effect the degree of observed prefrontal activation strongly. Thus, when subjects are merely asked to read aloud or repeat presented words, prefrontal activation is not found. When subjects are required to generate a response to a presented noun, but the response is practiced and "automatic," prefrontal activation is not found. However, when a response is not habitual, or overlearned, prefrontal activation is found (Frith et al 1991b; Petersen et al 1988; Raichle et al 1991).

The second issue is the domain of information that is used to perform the tasks in the less automatic conditions. Activation has been seen in left dorsolateral prefrontal cortex across a wide range of conditions. Petersen et al (1988, 1989) found activation in the verb generation subtractions, for both visual and auditory input. Activation was also found in tasks where subjects silently monitored a list of words for members of a semantic category, although it was weak. Further studies with simple visual presentation of words and word-like stimuli (described in the extrastriate visual cortex section) showed that when simple visual activation is subtracted away (by subtracting the false font condition from the pseudoword and real word conditions), there is greater activation in left inferior prefrontal cortex for real words than for pseudowords (Petersen et al 1990). The combination of results led us to suggest that some computation related to semantic processing or association between words gives rise to this activation. The difficulty has been in understanding what this computation might actually be, because as stated earlier, comprehension is much more affected by posterior than frontal lesions.

There is evidence that some aspects of semantic processing might be affected by frontal lesions, particularly in semantic priming tasks (Milberg & Blumstein 1981; Swinney et al 1989). The interpretation of the priming results has been that frontal lesions produce difficulties with access to semantic codes for lexical items. However, neither the PET nor the lesion results can distinguish whether the prefrontal activation represents processing the semantic code itself, or the process of accessing it for a particular task. It seems plausible that the prefrontal activation described is caused by task specific use of semantic information, particularly when the response to a stimulus is not an automatic or overlearned association (such as simple repetition of the seen or heard item).

Goldman-Rakic (1988) has outlined a similar interpretation of prefrontal cortical function. Her work in macaques has implicated areas in anterior prefrontal cortex in monkeys as being involved in higher order transformations or representations of information, particularly when an animal is involved in working memory tasks. An animal with lesions in this region has difficulty withholding a preponent, direct response to the stimulus when asked to hold and transform the information and act on that transformed representation. This description holds for many of the tasks that have produced prefrontal activation in PET.

Goldman-Rakic (1988) also hypothesizes that different subregions of dorsolateral prefrontal cortex may perform similar transformations on different input information. Different prefrontal areas have been associated with each of the tasks that produced cingulate activation. For example, the divided attention condition of the attention to visual features experiment produced activation in right prefrontal cortex.

Cerebellum   The activation in right lateral inferior cerebellum is anatomically distinct from activation found with the repeat words and other motor tasks, which argues for a "cognitive," rather than sensory or motor, computation being related to this activation. The lateralization to the right cerebellar hemisphere is consistent with observations that each cerebellar hemisphere is anatomically and functionally related to the contralateral (in this case left, language-dominant) cerebral hemisphere. A role for the cerebellum in cognition has been advanced [e.g. see reviews by Leiner et al (1991) and Schmahmann (1991)].

This activation motivated a study of a 49-year-old man (RC1) with right cerebellar damage on a variety of tasks involving complex nonmotor processing, including the generate verb task, which produced right cerebellar PET activation (Fiez et al 1992). RC1's performance on standard tests of intelligence and language was excellent. He did, however, have profound deficits in two areas: practice-related learning and detection of errors that manifested itself across several tasks. In the generate verb task, he often produced nonverb associates and could not improve his performance on this and several related habit formation tasks, even though both his short- and long-term memory appeared to be above normal in function. These results suggest that some functions performed by the cerebellum can be generalized beyond a purely motor domain consistent with the earlier PET activation (Thompson 1986).

CONCLUSION

These studies have shown that PET can contribute to our understanding of the brain mechanisms that underlie the processing of single words and, by extension, to other areas of study. However, PET should be viewed as evolutionary, rather than revolutionary. The results are most interpretable when they occur in an environment of information from other types of studies (e.g. the psychological studies of visual word processing coupled to the visual word studies of extrastriate cortex, and the use of lesion-behavior evidence to help interpret PET studies, as in the case of the cerebellar activation in the verb generation experiments). Because of the complexity underlying an understanding of even the reduced arena of the processing of single words, it is best to accept constraints from many different modalities, including the relatively recent contributions of PET activation studies.

ACKNOWLEDGMENTS

The authors wish to thank Paula Jost for her help in preparing the manuscript and Chris Reid for reading an earlier version of the manuscript. This work was supported by the McDonnell Center for Higher Brain Function, and National Institutes of Health grants EY08775 and NS06833. J. A. Fiez received fellowship support from the National Science Foundation and the Mr. and Mrs. Spencer T. Olin Program for Women.

Literature Cited

Baron, J. C. 1989. Depression of energy metabolism in distant brain structures: Studies with positron emission tomography in stroke patients. *Semin. Neurol.* 9: 281–85

Baron, J. C., Frackowiak, R. S. J., Herholz, K., Jones, T., Lammertsma, A. A., et al. 1989. Use of PET Methods for Measurement of Cerebral Energy Metabolism and Hemodynamics in Cerebrovascular Disease. *J. Cereb. Blood Flow Metab.* 9: 723–42

Barris, R. W., Schuman, H. R. 1953. Bilateral anterior cingulate gyrus lesions. *Neurolology* 3: 44–52

Caramazza, A. 1988. Some aspects of language processing revealed through the analysis of acquired aphasia: The lexical system. *Annu. Rev. Neurosci.* 11: 395–421

Carr, T. H., Pollatsek, A. 1985. Recognizing printed words: A look at current models. In *Reading Research*, ed. D. Besner, T. G. Weller, G. E. MacKinnon, pp. 2–73. New York: Academic

Corbetta, M., Miezin, F. M., Dobmeyer, S., Shulman, G. L., Petersen, S. E. 1991. Selective and divided attention during visual discriminations of shape, color, and speed: functional anatomy by positron emission tomography. *J. Neurosci.* 11: 2383–2402

Crosson, B. 1985. Subcortical functions in language: A working model. *Brain Lang.* 25: 257–92

Damasio, A. R. 1992. Aphasia. *N. Engl. J. Med.* 326: 531–39

Damasio, A. R., Damasio, H. 1983. The anatomic basis of pure alexia. *Neurolology* 33: 1573–83

Davidson, B. J. 1986. Activation of semantic and phonological codes during reading. *J. Exp. Psychol. Learn. Mem. Cogn.* 12: 201–7

Evans, A. C., Beil, C., Marrett, S., Thompson, C. J., Hakim, A. 1988. Anatomical-functional correlation using an adjustable MRI-based region of interest atlas with positron emission tomography. *J. Cereb. Blood Flow Metab.* 8: 513–30

Felleman, D. J., Van Essen, D. C. 1991. Distributed hierarchical processing in the primate cerebral cortex. *Cereb. Cortex* 1: 1–47

Fiez, J. A., Petersen, S. E., Cheney, M. K., Raichle, M. E. 1992. Impaired nonmotor learning and error detection associated with cerebellar damage: A single-case study. *Brain* 115: 155–78

Fox, P. T., Miezin, F. M., Allman, J. M., Van Essen, D. C., Raichle, M. E. 1987a. Retinotopic organization of human visual cortex mapped with positron emission tomography. *J. Neurosci.* 7: 913–22

Fox, P. T., Mintun, M. A., Reiman, E. M., Raichle, M. E. 1988. Enhanced detection of focal brain responses using intersubject averaging and change-distribution analysis of subtracted PET images. *J. Cereb. Blood Flow Metab.* 8: 642–53

Fox, P. T., Pardo, J. V. 1991. Does intersubject variability in cortical functional organization increase with neural "distance" from the periphery? In *Exploring Brain Functional Anatomy with Positron Emission Tomography*, ed. P. T. Fox, J. V. Pardo, pp. 125–44. Chichester: Wiley (Ciba Found. Symp. 163)

Fox, P. T., Pardo, J. V., Petersen, S. E., Raichle, M. E. 1987b. Supplementary motor and premotor responses to actual and imagined hand movement with positron emission tomography. *Soc. Neurosci. Abstr.* 13: 1433

Fox, P. T., Perlmutter, J. S., Raichle, M. E. 1985. A stereotactic method of anatomical localization for positron emission tomography. *J. Comput. Assist. Tomogr.* 9: 141–53

Fox, P. T., Raichle, M. E. 1986. Focal physiological uncoupling of cerebral blood flow and oxidative metabolism during somatosensory stimulation in human subjects. *Proc. Natl. Acad. Sci. USA* 83: 1140–44

Friston, K. J., Frith, C. D., Liddle, P. F., Frackowiak, R. S. J. 1991a. Comparing functional (PET) images: The assessment of significant change. *J. Cereb. Blood Flow Metab.* 11: 690–99

Friston, K. J., Frith, C. D., Liddle, P. F., Frackowiak, R. S. J. 1991b. Plastic transformation of PET images. *J. Comput. Assist. Tomogr.* 15: 634–39

Frith, C. D., Friston, K., Liddle, P. F., Frackowiak, R. S. J. 1991a. Willed action and the prefrontal cortex in man: a study with PET. *Proc. R. Soc. London Ser. B* 244: 241–46

Frith, C. D., Friston, K., Liddle, P. F., Frackowiak, R. S. J. 1991b. A PET study of word finding. *Neuropsychologia* 29: 1137–48

Frostig, R. D., Lieke, E. E., Ts'o, D. Y., Grinvald, A. 1990. Cortical functional architecture and local coupling between neuronal activity and the microcirculation revealed by in vivo high-resolution optical imaging of intrinsic signals. *Proc. Natl. Acad. Sci. USA* 87: 6082–86

Geschwind, N. 1979. Specializations of the human brain. *Sci. Am.* 158–68

Geschwind, N. 1965. Disconnection syndromes in animals and man, Part 1. *Brain* 88: 237–94

Glushko, R. J. 1979. The organization and activation of orthographic knowledge in reading aloud. *J. Exp. Psychol. Percept. Perform.* 5: 674–91

Goldman-Rakic, P. S. 1988. Topography of cognition: Parallel distributed networks in primate association cortex. *Annu. Rev. Neurosci.* 11: 137–56

Grafton, S. T., Woods, R. P., Mazziotta, J. C., Phelps, M. E. 1991. Somatotopic mapping of the primary motor cortex in humans: Activation studies with cerebral blood flow and positron emission tomography. *J. Neurophysiol.* 66: 735–43

Henderson, V. W. 1986. Anatomy of posterior pathways in reading: A reassessment. *Brain Lang.* 29: 119–33

Herholz, K., Pawlik, G., Wienhard, K., Heiss, W.-D. 1985. Computer assisted mapping in quantitative analysis of cerebral positron emission tomograms. *J. Comp. Assist. Tomogr.* 9: 154–61

Herscovitch, P., Markham, J., Raichle, M. E. 1983. Brain blood flow measured with intravenous $H_2{}^{15}O$. I. Theory and error analysis. *J. Nucl. Med.* 24: 782–89

Jackson, A., Morton, J. 1984. Facilitation of auditory word recognition. *Mem. Cogn.* 12: 568–74

Karbe, H., Herholz, K., Szelies, B., Pawlik, G., Wienhard, K., et al. 1989. Regional metabolic correlates of Token test results

in cortical and subcortical left hemispheric infarction. *Neurolology* 39: 1083–88

Karbe, H., Szelies, B., Herholz, K., Heiss, W.-D. 1990. Impairment of language is related to left parieto-temporal glucose metabolism in aphasic stroke patients. *J. Neurol.* 237: 19–23

Kushner, M., Reivich, M., Alavi, A., Greenberg, J., Stern, M., et al. 1987. Regional cerebral glucose metabolism in aphemia: A case report. *Brain Lang.* 31: 201–14

Kutas, M., Van Petten, C. 1990. Electrophysiological perspectives on comprehending written language. In *New Trends and Advanced Techniques in Clinical Neurophysiology*, ed. P. M. Rossini, F. Mauguiere, pp. 155–67. New York: Elsevier

Lassen, N. A., Ingvar, D. H., Skinhoj, E. 1978. Brain function and blood flow. *Sci. Am.* 239: 62–71

Lauter, J., Herscovitch, P., Formby, C., Raichle, M. E. 1985. Tonotopic organization in human auditory cortex revealed by positron emission tomography. *Hear. Res.* 20: 199–205

Leiner, H. C., Leiner, A. L., Dow, R. S. 1991. The human cerebrocerebellar system: its computing, cognitive, and language skills. *Behav. Brain Res.* 44: 113–28

Marrett, S., Bub, D., Chertkow, H., Meyer, E., Gum, T., et al. 1990. Functional neuroanatomy of visual single word processing studied with PET/MRI. *Soc. Neurosci. Abstr.* 16: 27

Masdau, J. C., Schoene, W. C., Funkenstein, H. 1978. Aphasia following infarction of the left supplementary motor area. *Neurolology* 28: 1220–23

Mazziotta, J. C., Pelizzari, C. C., Chen, G. T., Brookstein, F. L., Valentino, D. 1991. Region of interest issues: The relationship between structure and function in the brain. *J. Cereb. Blood Flow Metab.* 11: A51–56

Mazziotta, J. C., Phelps, M. E., Carson, R. E., Kuhl, D. E. 1982. Tomographic mapping of human cerebral metabolism: Auditory stimulation. *Neurolology* 32: 921–37

Mazziotta, J. C., Phelps, M., Plummer, D., Schwab, R., Halgren, E. 1983. Optimization and standardization of anatomical data in neurobehavioral investigations using positron computed tomography. *J. Cereb. Blood Flow Metab.* 3: S266–67

Metter, E. J., Hanson, W. R., Jackson, C. A., Kempler, D., van Lancker, D., et al. 1990. Temporoparietal cortex in aphasia. *Arch. Neurol.* 47: 1235–38

Metter, E. J., Jackson, C., Kempler, D., Riege, W. H., Hanson, W. R., et al. 1986. Left hemisphere intracerebral hemorrhages studied by (F-18)-fluorodeoxy-glucose PET. *Neurolology* 36: 1155–62

Metter, E. J., Kempler, D., Jackson, C., Hanson, W. R., Mazziotta, J. C., et al. 1989. Cerebral glucose metabolism in Wernicke's, Broca's, and conduction aphasia. *Arch. Neurol.* 46: 27–34

Metter, E. J., Kempler, D., Jackson, C. A., Hanson, W. R., Riege, W. H., et al. 1987. Cerebellar glucose metabolism in chronic aphasia. *Neurology* 37: 1599–1606

Metter, E. J., Riege, W. H., Hanson, W. R., Jackson, C. A., Kempler, D., et al. 1988. Subcortical structures in aphasia: An analysis based on (F-18)-fluorodeoxy-glucose, positron emission tomography, and computed tomography. *Arch. Neurol.* 45: 1229–34

Milberg, W., Blumstein, S. E. 1981. Lexical decision and aphasia: evidence for semantic processing. *Brain Lang.* 14: 371–85

Mintun, M. A., Fox, P. T., Raichle, M. E. 1989. A highly accurate method of localizing regions of neuronal activation in the human brain with positron emission tomography. *J. Cereb. Blood Flow Metab.* 9: 96–103

Mohr, J. P., Pessin, M. S., Finkelstein, S., Funkenstein, H. H., Duncan, G. W., et al. 1978. Broca aphasia: Pathologic and clinical. *Neurology* 28: 311–24

Naeser, M., Alexander, M. P., Helm-Estabrooks, N., Levine, H. L., Laughlin, S. A., et al. 1982. Aphasia with predominantly subcortical lesion sites. *Arch. Neurol.* 39: 2–14

Neely, J. H. 1977. Semantic priming and retrieval from lexical memory: Roles of inhibitionless spreading activation and limited-capacity attention. *J. Exp. Psychol. Gen.* 106: 226–54

Nielsen, J. M., Jacobs, L. L. 1951. Bilateral lesions of the anterior cingulate gyri. *Bull. Los Angeles Neurol. Soc.* 16: 231–34

Pardo, J. V., Pardo, P. J., Janer, K. W., Raichle, M. E. 1990. The anterior cingular cortex mediates processing selection in the Stroop attentional conflict paradigm. *Proc. Natl. Acad. Sci. USA* 87: 256–59

Petersen, S. E., Fox, P. T., Posner, M. I., Mintun, M., Raichle, M. E. 1989. Positron emission tomographic studies of the processing of single words. *J. Cogn. Neurosci.* 1: 153–70

Petersen, S. E., Fox, P. T., Posner, M. I., Mintin, M., Raichle, M. E. 1988. Positron emission tomographic studies of the cortical anatomy of single-word processing. *Nature* 331: 585–89

Petersen, S. E., Fox, P. T., Snyder, A., Raichle, M. E. 1990. Activation of extrastriate and frontal cortical areas by visual words and work-like stimuli. *Science* 249: 1041–44

Posner, M. I. 1978. *Chronometric Explorations of Mind.* Englewood Heights, NJ: Erlbaum Assoc.

Posner, M. I., Petersen, S. E., Fox, P. T., Raichle, M. E. 1988. Localization of cognitive functions in the human brain. *Science* 240: 1627–31

Powers, W. J., Raichle, M. E. 1985. Positron emission tomography and its application to the study of cerebrovascular disease in man. *Stroke* 16: 361–76

Raichle, M. E. 1989. Developing a functional anatomy of the human brain with positron emission tomography. *Curr. Neurol.* 9: 161–78

Raichle, M. E., Fiez, J., Videen, T. O., Fox, P. T., Pardo, J. V., et al. 1991. Practice-related changes in human brain functional anatomy. *Soc. Neurosci. Abstr.* 17: 21

Raichle, M. E., Martin, W. R. W., Herscovitch, P., Mintun, M. A., Markham, J. 1983. Brain blood flow measured with intravenous $H_2^{15}O$. II. Implementation and validation. *J. Nucl. Med.* 24: 790–98

Reivich, M., Kuhl, D., Wolf, A., Greenberg, J., Phelps, M., et al. 1979. The [^{18}F] fluorodeoxyglucose method for the measurement of local cerebral glucose utilization in man. *Circ. Res.* 44: 127–37

Schmahmann, J. D. 1991. An emerging concept. The cerebellar contribution to higher function. *Arch. Neurol.* 48: 1178–87

Shallice, T., McLeod, P., Lewis, K. 1985. Isolating cognitive modules with the dual task paradigm: Are speech perception and production separate processes? *Q. J. Exp. Psychol.* 37A: 507–32

Shima, K., Aya, K., Mushiake, H., Inase, M., Aizawa, H., et al. 1991. Two movement-related foci in the primate cingulate cortex observed in signal-triggered and self-paced forelimb movements. *J. Neurophysiol.* 65: 188–202

Sokoloff, L., Reivich, M., Kennedy, C., Des Rosiers, M. H., Patlak, C. S., et al. 1977. The [^{14}C] deoxyglucose method for the measurement of local cerebral glucose utilization: Theory, procedure, and normal values in the conscious and anesthetized albino rat. *J. Neurochem.* 28: 897–916

Steinmetz, H., Seitz, R. J. 1991. Functional anatomy of language processing: Neuroimaging and the problem of individual variability. *Neuropsychologia* 29: 1149–61

Stytz, M. R., Frieder, O. 1990. Three-dimensional medical imaging modalities: An overview. *Crit. Rev. Biomed. Eng.* 18: 1–25

Swinney, D., Zurif, E., Nicol, J. 1989. The effects of focal brain damage on sentence processing: An examination of the neurological organization of a mental module. *J. Cogn. Neurosci.* 1: 25–37

Talairach, J., Szikla, G., Tournoux, P. 1967. *Atlas d'Anatomie Stereotaxique du Telencephale.* Paris: Masson

Talairach, J., Tournoux, P. 1988. *Co-Planar Stereotaxic Atlas of the Human Brain.* New York: Thieme

Tanenhaus, M. K., Flanigan, H. P., Seidenberg, M. S. 1980. Orthographic and phonological activation in auditory and visual word recognition. *Mem. Cogn.* 8: 513–20

Thompson, R. F. 1986. The neurobiology of learning and memory. *Science* 223: 941–47

Tyrrell, P. J., Warrington, E. K., Frackowiak, R. S. J., Rossor, M. N. 1990. Heterogeneity in progressive aphasia due to focal cortical atrophy. *Brain* 113: 1321–36

Vallar, G., Perani, D., Cappa, S. F., Messa, C., Lenzi, G. L., et al. 1988. Recovery from aphasia and neglect after subcortical stroke: neuropsychological and cerebral perfusion study. *J. Neurol. Neurosurg. Psychiatry* 51: 1269–76

Wise, R., Chollet, F., Hadar, U., Friston, K., Hoffner, E., et al. 1991a. Distribution of cortical neural networks involved in word comprehension and word retrieval. *Brain* 114: 1803–17

Wise, E., Hadar, U., Howard, D., Patterson, K. 1991b. Language activation studies with positron emission tomography. In *Exploring Brain Functional Anatomy with Positron Emission Tomography*, pp. 218–34. Chichester: Wiley (Ciba Found. Symp. 163)

Worsley, K. J., Evans, A. C., Marrett, S., Neelin, P. 1992. A three-dimensional statistical analysis for CBF activation studies in human brain. *J. Cereb. Blood Flow Metab.* In press

Yaqub, B. A., Gascon, G. G., Alnosha, M., Whitaker, H. 1988. Pure word deafness (acquired verbal auditory agnosia) in an arabic speaking patient. *Brain* 111: 457–66

Zeki, S., Watson, J. D. G., Lueck, C. J., Friston, K. J., Kennard, C., et al. 1991. A direct demonstration of functional specialization in human visual cortex. *J. Neurosci.* 11: 641–49

Annu. Rev. Neurosci. 1993. 16:531–46

MODELING OF NEURAL CIRCUITS: What Have We Learned?

A. I. Selverston

Department of Biology, University of California, San Diego, La Jolla,
California 92093

KEY WORDS: computational neuroscience, cell models, circuit models, neural
networks, biological models

INTRODUCTION

The recent introduction of new, biologically inspired learning algorithms
has led to an upsurge of interest in circuit modeling (Hopfield & Tank
1986; McClellan & Rummelhart 1986; Rummelhart et al 1986). The performance of these algorithms can be extremely interesting and has led to
a great deal of excitement (see Crick 1989) in some circles. When units
representing neurons were connected into simple circuits, they could be
"taught" to store patterns by adjusting the strengths of entries in a connectivity matrix; the circuits can then recall the patterns when challenged.
These new "neural network" models, as they are called, could also solve
some difficult classes of problems better than conventional computer algorithms. The fact that information could be stored as a distributed array of
synaptic weights did seem surprisingly brain-like and suggested applications among diverse disciplines. Several recent reviews and books dealing
with neural network modeling are available (Aleksander & Morton 1990;
Anderson & Rosenfeld 1988; Byrne & Berry 1989; Durbin et al 1989;
Hawkins & Bower 1989; Koch & Segev 1989; MacGregor 1987; Wasserman 1989). Because of their obvious analogy to nervous systems and the
fact that they had widespread applicability to real world problems, neural
network models promised to usher in a new era of multidisciplinary coop-
eration between neurobiologists, computer scientists, engineers, and others
interested in processing information. Implicit in these efforts was the idea
that biologists would learn more about the brain by viewing it as a com-

531

putational engine and that engineers would convert information about real nervous systems into devices with practical applications. But, an additionally important benefit was to cause a reexamination of many older types of models. The new models were much different conceptually than the older models, which generally tried to mimic the biophysics of real cells and circuits.

In this review, the impact of both old and new efforts at circuit modeling is examined in terms of what has actually been learned about real nervous systems. We address only one of many interested constituencies, neuroscientists, and attempt to answer the question of what we have learned, from the viewpoint of understanding how brains compute. No attempt is made to consider neural modeling from a purely theoretical perspective, particularly where efforts are unconstrained by real data or not amenable to experimental verification.

What Has Caused the Recent Upsurge in Interest?

There are many reasons why the "new" circuit modeling has generated so much interest. First, the learning algorithms have widespread applicability to both the neurobiological and psychological communities. They are also clearly relevant to emerging engineering and computer technologies.

Second, the amount of new biological data available has increased enormously since the last upsurge in neural modeling, so there is a lot to model. The many invertebrate circuits that have been more or less completely described provide extensive data upon which models may be based (Selverston & Moulins 1985), and modelers also have available a large amount of more coarse-grained circuit information about the mammalian brain, particularly the visual system (Douglas & Martin 1991; Gilbert 1983).

But perhaps the one factor that has attracted biologists most to modeling has been the confluence of two ideas—the Hebb synapse and the n-methyl d-aspartate (NMDA) receptor. Hebb (1949) suggested that synaptic strength could be altered by the correlated timing of activity in a pre- and postsynaptic neuron. This idea, now called the Hebb rule, became the basis for changing weights in most of the new neural network algorithms. The NMDA receptor is a glutamate receptor subtype thought to be responsible for long-term potentiation (LTP), a form of learning that can be studied at the cellular level in hippocampal slice preparations. Because the NMDA receptor has to be depolarized for LTP to occur (Malinow & Miller 1986), and this would happen if a presynaptic cell fired concurrently with the postsynaptic cell, NMDA receptor properties provide strong physiological underpinning to the idea of Hebbian synapses and a link between neural network modelers and physiologists (Brown et al 1989).

Finally, the availability of relatively cheap work stations and desktop computers played a major role in running the new models. This hardware enabled the development of neural networks to occur at a prodigious rate. These computers also sped up the process of running enough equations to capture the key currents of single cells in more biophysical models at a reasonable computing cost. Powerful graphics allowed the output of the models to be displayed in a way familiar to physiologists, i.e. with time-variant potentials (Wilson & Bower 1989).

About Computational Neuroscience

The term computational neuroscience has grown in popularity as a way of bringing together the different fields interested in modeling. The emphasis is in using biological principles as the basis for developing new algorithms. For years, neuroscientists using computational methods have made fundamental discoveries about the nervous system. The pioneering research of Cole (1968), Hodgkin & Huxley (1952), Moore and coworkers (1966), Rall (1962), and Gerstein & Mandelbrot (1964) has been at both the cellular and systems levels. But, although the problem space can be defined, no current theories adequately explain events above the level of single cells. Computational neuroscience seeks to develop such theories (Churchland et al 1989) at all levels, but also to bind different levels to one another. Computational neuroscience includes theoretical findings not closely tied to experimental data. Many biologically inspired theories at circuit and even higher levels of analysis have been suggested. Have they informed us much about the brain? For the most part, probably not. Either the theories are not sufficiently biologically grounded or they simply cannot be verified empirically. Such theorizing, even when presumably neuron based, may be useful at some point, but is of little help to the contemporary experimentalist.

Cell Models and Circuit Models

Models of fundamental cellular processes are deeply entrenched in neurobiology, indeed the Donnan Equilibrium; the Nernst and Goldman equations, which describe the resting potential; and the Hodgkin-Huxley equations, which describe the action potential, are often the first things learned in an introductory neurophysiology course. Other phenomena, like the statistical interpretation of transmitter release and the cable equations, are routinely used in the study of today's cutting-edge problems (Bekkers & Stevens 1991; Segev et al 1989). Even more abstract cellular models, such as those that consider transmitter-receptor interactions (Zagotta & Aldrich 1990), are invaluable in the analysis of patch-clamp data.

When moving to the level of describing neuronal circuits, however, the value of modeling becomes much less generalizable and, therefore, more controversial. There are historically embedded reasons for this (see Perkel 1988 for a good review). In general, the record was one of over-simplification, overpromise, hyperbole, and, most important, insufficient correspondence to real nervous systems. Each new wave of circuit modeling, which arose at intervals of 20 years or so, eventually lost credibility when it could not solve some particular class of problem. Cellular models, on the other hand, were robust, because they were derived from experimental data and sought to describe only a limited range of phenomena that could be experimentally verified and widely applicable.

Circuit models fall into two classes. In one class, neurons are computational primitives that incorporate the input-output relations of single cells. They sum the weighted "synaptic" inputs from other units and produce some output. The second class uses as much biophysical data as possible in representing neurons and as much anatomical and physiological data as are available in constructing circuits. The first class, often referred to as abstract, can be justified with the argument that models with too much biophysical reality were unnecessarily complicated and too computationally intensive to explain more global phenomena. By radically simplifying the computational process, abstract modeling would be much better suited to deal with the more complex information processing involved in higher brain functioning (Churchland et al 1989). One impetus for this approach was the work of Marr (1982), whose initial level of analysis was top-down and largely independent of neural implementation. That is, before looking at the "wetware" implementation of some neural task, Marr advocated first describing what was being computed and then identifying an algorithm for the computation.

Because in most cases we do not know all the biophysical details of each neuron or the precise way in which they are synaptically arranged, such abstract simplification is not entirely unjustified. Besides, some psychologists and philosophers traditionally take a "black box" approach to the nervous system, which holds that the actual hardware does not matter anyway—it is the software that is important.

Which Simplifications in a Model are Justified?

The sine qua non of any model is that it represents a simplification of the details that are actually present in a cell or system of cells. This question of simplification is crucial in modeling, and the degree to which simplifications can be made depends on both the level of question being asked and the current level of knowledge pertaining to that level. It is also a function of what might be called computational elegance and the hardware available to implement the model. There is usually a trade-off that must be made between computational elegance and biological fidelity. We can illustrate this at different levels. A truly accurate model of the current flow in a single cell must consider the cell's three-dimensional morphology, its membrane electrical properties, the location of receptors, the spike-initiating zones, and the type and distribution of membrane channels. If we try and model this cell as a system of modular, parallel conductance compartments, then most cells will need hundreds, maybe thousands, of compartments to predict accurately the flow of synaptic current in them. This can be done with currently available hardware (Miller 1990; O'Donnell et al 1985), but represents a level of reductionism that is far too complicated to be used in circuit modeling. For most purposes, therefore, single neurons must be simplified, but still be able to capture the essential integrative properties. It would be best if, at a minimum, neurons are described by compartments representing the soma, the dendritic input-output sites, and a spike-initiating zone, plus conductances that enable a simulation of at least some of the more common physiological properties, such as postinhibitory rebound, bursting, plateauing, and adaptation. Most of the more realistic, biophysically based models currently available include provisions for handling most of these parameters.

When neurons are represented as grossly simplified input-output units, they do not begin to capture the biological richness of real neurons, but they may capture the global computation that occurs in the system. It is the heart of a long-standing feud between physiologists, who want all of the details (which makes the model as complicated as the nervous system), and modelers, who want computational tractability (which may have nothing to do with the nervous system). One of the reasons why this current round of circuit modeling may be different from those in the past is that by constructing "hybrid" models, it now may be possible to meet the expectations of both camps.

What Can We Reasonably Expect to Learn from Circuit Modeling?

Neural modeling can help us understand and manipulate complex systems. Because neuronal interactions are nonlinear, the dynamic behavior between just two synaptically connected neurons is not intuitive. Neural modeling attempts to help explain such dynamic processes and, therefore, is not like molecular neurobiology, for example, with which it has been erroneously compared (Anderson & Palca 1988). Molecular neurobiology is acquiring extremely rigorous data, but eventually these data will have to be put into some framework to make them useful in understanding how the brain works. Modeling is necessary for providing such a framework and, hopefully, will be able to link such disparate levels of analysis. Circuit modeling can make suggestions about how to interpret higher-level exper-imental data and it should be able to make predictions about how complex sets of data interact. For experimentalists who are often awash in such data, a model can suggest which information is most important for a particular task. Neurobiologists can ask these specific questions:

1. How does the output of the model compare with biological output? Does it have the same dynamic range and properties under different operating conditions as the biological system?
2. How does the output of the simulated circuit compare with the biological circuit when it is perturbed by current injection into single cells, by lesions of cells or groups of cells, by pharmacological manipulations, or by the application of neural modulators?
3. What are the roles of the individual neurons or classes of neurons in a circuit? Are some neurons more important than others? Which ones are redundant?
4. Which cellular properties are the most important in the functioning of the circuit? Which can be eliminated from the representation of the neurons without serious effects?
5. Can the model generalize to other small circuits or scale up to larger ones easily? Is the model unique only to the system under study?
6. Does the model suggest ways of interpreting the data? Does it make predictions? Does it suggest new experiments?
7. Can the model be used as a tool to adjust parameters or to simplify analysis?

Very "theoretical" models, i.e. those that are not neuronally based, can sometimes suggest experiments. Theoretical models that make reasonable biological assumptions can often produce results consistent with real input-output data. Although interesting in their own right, they are basically irrelevant if a rigorous biophysical interpretation of neural function is the goal.

To gain some appreciation of circuit modeling, we can consider some selected examples of the main types that we have been discussing. First, we will look at some biophysically based, realistic models of both small and large circuits. We can compare these with examples of some more simplified or abstract models, which, although computationally based, capture some of the basic aspects of real neurons. Finally, we will look at some neuron-based hybrid models, which contain both abstract and realistic components.

REALISTIC MODELS

Realistic or biophysically based models seek to incorporate the most important physiologically measurable parameters. The challenge is to simplify the situation to provide a computationally tractable set of equations, while still representing the cellular properties essential to integration. Some classic papers clearly demonstrate the effectiveness of looking at neural circuits in this way. Modeling studies by Rall & Shepherd (1968) of the laminar field potentials in the mammalian olfactory bulb, for example, predicted the existence of reciprocal synaptic connections between granule and mitral cells. These synapses had actually been observed by Reese in an electron microscopic study (Rall et al 1966).

Another example is the synchronization of discharges in the pharmacologically disinhibited hippocampal slice. Traub & Wong (1982) predicted that bursts in a single pyramidal cell could evoke bursts in one other pyramidal cell. This was subsequently confirmed experimentally by dual recordings (Miles & Wong 1983). Their model also predicted that stimulation of a single neuron could, in some cases, evoke synchronous firing in large numbers of cells that had been disinhibited. This also was experimentally verified (Miles & Wong 1983).

Both of these examples relied on interactions between computation and experiment. Such interactions are particularly appropriate for modeling small circuits (Getting 1989; Marder 1992). Even when detailed circuitry is available, the dynamic output of the system cannot be determined intuitively.

Realistic Modeling of Small Circuits

Several examples of modeling small central pattern generating circuits are discussed in a review by Mulloney & Perkel (1988). In an example from their own work, they sought to test the hypothesis that postinhibitory rebound could stabilize alternate bursting in two neurons connected to each other with reciprocal inhibitory synapses (Perkel & Mulloney 1974). The mechanisms involved in the production of alternate bursting by reciprocal inhibition have been investigated for many years. The problem is to explain how neurons make the transition from the firing state to the inhibited state. The interaction between intrinsic cellular properties, such as bursting, and the strength and time course of synaptic interactions can produce many possible firing patterns and provide a cogent example of why theoretical studies are needed. In the Mulloney-Perkel study, neurons were modeled as single compartments with resting potentials, thresholds, and single time constants. In addition, neurons were given a parameter to simulate postinhibitory rebound (PIR). This parameter increased when the neuron was hyperpolarized and decreased when depolarized. As with many cell simulations, no attempt was made to model the ionic basis for this phenomenon, but only to find a mathematical way of describing the phenomenon itself. Synaptic inputs were simulated as voltages added to or subtracted from the resting potential in proportion to their distance from a specified reversal potential.

Despite the model's simplicity, it demonstrated which role PIR might play in alternate bursting. A subsequent analysis of the role of PIR, which used a model containing voltage dependent currents, confirmed the earlier model (Mulloney et al 1981). This analysis also included a study of the role of electrical coupling between the two reciprocal inhibitory neurons. When an oscillating pair contained neurons that had similar but not identical properties, the burst patterns could drift, lock in synchrony, entrain in antiphase, entrain at some intermediate phase, or be suppressed. The behavior depended on the relative strengths of the chemical and electrical coupling, as well as the amount of depression built into the chemical synapses. Trying to establish such relationships experimentally would be extremely difficult. The modeling study suggested how complicated the behavior of even very small neural networks can be.

Another recent study of the relationship between burster cells and electronic coupling in small CPG circuits has shown equally interesting results. In this case, specific experimental data were the impetus for modeling a group of pacemaker neurons in the pyloric rhythm of the lobster stomatogastric ganglion (Kepler et al 1990). Previous work had shown that the frequency of proctolin-modulated bursting in the isolated AB neuron was about 2 Hz (Hooper & Marder 1987). However, the frequency of the pyloric rhythm in proctolin was only 1 Hz when the pyloric circuit was isolated from the rest of the central nervous system (CNS). Hooper & Marder reasoned that, because two PD cells that were electrically coupled to the AB were not affected by proctolin and were not getting any other modulatory input, they acted like passive cells to decrease the AB burst frequency. In the model, the AB neuron was represented as a simple oscillator by using the Fitzhugh (1961) equations. By changing the parameters in these equations and other similar simple models, it is possible to alter both the frequency and the wave form of the bursting neuron. This allows the modeler to simulate the AB cell as having a slowly depolarizing interburst interval with a rapid hyperpolarization after the burst (inward current dominated period) or a rapidly depolarizing interburst interval and a slowly repolarizing plateau phase (outward current dominated period). When the model AB neuron was coupled to a hyperpolarized passive neuron, it behaved differently in each case. When dominated by inward currents, it decreased in frequency when the electrical coupling between the cells was increased. When dominated by outward currents, the AB cell first increased in frequency when the coupling strength was increased, but then decreased. The model made specific predictions about the interactions between bursting and nonbursting neurons that could be understood in biophysical terms. When the passive neuron was hyperpolarized, it added an outward current during the entire cycle. When the cell depolarizes slowly, the additional outward current acts to retard the depolarization and thus decrease the frequency. When the cell hyperpolarizes for most of the cycle period, the current from the passive cell accelerates the repolarization phase so the period increases. As coupling is increased, the plateau phase is truncated to the point that an oscillator that started out with a slow repolarization phase would now have equal duration inward and outward current phases, thus giving the oscillator its minimum period. Further increases in coupling result in further decreases in frequency. In subsequent tissue culture studies, it has been possible to couple a bursting neuron with a passive RC circuit and demonstrate experimentally that the results of the modeling study were correct (Sharp & Marder 1991).

Another very successful attempt at gaining insight into the dynamics of a small CPG has been with the mollusc *Tritonia*. Here, the network generating the escape swim pattern has been worked out in detail (Getting & Dekin 1985) and shown to contain only three major classes of neurons: dorsal swim interneurons (DSI), ventral swim interneurons (VSI), and a cerebral cell (C2). Although the synaptic connections between these neurons are complex, the pattern was thought to be generated by the synaptic interactions alone, i.e. without the necessity of burster neurons, so common in other CPGs. The basis of the oscillation was postulated to be represented by reciprocal inhibition between the DSI and VSI, paralleled by delayed excitation between DSI and C2. How these and other synaptic interactions contributed to the formation of the pattern could not be determined without modeling the circuit. Getting (1989) used an updated version of the Perkel model called MARIO. First, the input/ output relationship for each neuron was defined by specifying passive membrane properties and repetitive firing characteristics. Second, the time course and strength for each synaptic action was determined. Third, the entire network was assembled and given an input equivalent to normal sensory activation. The model could produce the oscillatory pattern with only minor adjustments, if the cells were given small bias currents equivalent to the tonic drive seen in the animal, but not incorporated into the model. This kind of result is often observed for models of dynamic systems if all of the parameters have been properly adjusted, so it is important to see how such simulations respond under different operating conditions. In this case, the model appeared to be quite robust, producing cycle periods similar to those seen in real swims. The simulated swim also displayed an asymmetric increase in cycle period, which compared favorably with the real data (Lennard et al 1980). Other perturbations, such as initiation of the swim pattern in vitro by the tonic depolarization of C2, could also be duplicated by the model. Note that the parameters were not chosen because of the assembled network's ability to reproduce the biological swim pattern. But, it would be interesting to see if similar parameters could be found by using abstract models, like recurrent back propagation, as a systems identification tool. (This may, in fact, be the most useful way in which back prop can be used by an experimentalist.)

What is the value of this particular model in understanding small CPGs in general? Such a simulation strategy can begin to provide insights into how many physiological details need to be incorporated to construct a robust model of a particular system. This can give us an idea of just how detailed other models of this genre need to be. It can begin to explain the roles of individual cells or cell parameters in generating the overall pattern. Unfortunately, it tells us very little about general principles, and such principles may emerge only after we have produced specific models of many systems independently, each with its unique attributes, but based on known biophysical building blocks.

Realistic Modeling of Large Circuits

A different set of assumptions are made to generate realistic models of larger circuits. The properties of individual neurons in the CNS of vertebrates must be extrapolated to large classes, and the synaptic wiring, so precise in invertebrates, must be dealt with statistically. For example, the in vitro hippocampal slice preparation can provide information from single neurons by means of the whole cell patch clamp technique (Kay & Wong 1986; Numann et al 1987; Sah et al 1988). The slice preparation, under appropriate conditions, can display population behaviors similar to various forms of epilepsy, i.e. more than 25% of the cells fire in a defined time period. If certain assumptions are made (for details, see Traub & Miles 1991), a network model can be constructed in which the "neurons" are randomly connected with unitary synaptic interactions, like those made from dual intracellular recordings. The output of the network can be studied under a variety of conditions, such as the strength of the $GABA_A$ receptors. Results of these studies show, for example, that with GABA inhibition blocked, a localized stimulus can produce single synchronized bursts and predict the effects on propagation of partial disinhibition. This result, as well as many others derived from this model, can be tested experimentally and are, therefore, extremely useful in guiding experimental approaches.

The olfactory cortex has been another area of the CNS in which circuit modeling has been useful. In one series of studies, pyramidal neurons were modeled as membranes with three types of synaptically activated currents to account for afferent inputs, feedforward and feedback inhibition, and various kinds of associational inputs (Wilson & Bower 1989). For large-scale simulations, each neuron was represented as a single compartment, but required five compartments to model spatial current flow within the neuron, to account for extracellular field potentials. The neurons were synaptically connected based on physiologically derived data. The model contained input fibers from mitral cells in the olfactory bulb, excitatory association fibers from other cells within the cortex (with an influence that decreased with distance), inhibition from local feedback, and some afferents. Despite the fact that both synaptic weights and intrinsic neuronal connections could only be approximated, the model can accurately mimic physiological data, if the parameters are adjusted properly. The intracellular responses to olfactory tract stimulation, the EEG responses, and wave-like responses traveling across the cortex in response to fictive odors, all look surprisingly real. As with simulations of the hippocampus, removal of inhibition can produce epileptic-like synchronized bursting.

One way of dealing with the huge numbers of neurons in the brain is to select an animal model with fewer and more experimentally accessible cells. The lamprey has been suggested as a good animal model for the study of central pattern generators, because it has far fewer cells than mammals, and some aspects of their circuitry can be elucidated electrophysiologically. Much knowledge has been gained about circuits that produce locomotion in the lamprey nervous system, but this information is difficult to generalize to other neural systems, because of the major anatomical differences between the lamprey and higher vertebrates. The lamprey spinal cord has about 100 segments, and each segment has about 1200 neurons (Rovainen 1974). During fictive locomotion, four main cell types appear to be active: approximately 50 cross-caudal interneurons (CC), 5 inhibitory interneurons (IN), and 50 excitatory interneurons (EIN) per hemisegment, plus 50 lateral interneurons (LIN) in the entire spinal cord (Sigvardt & Williams 1992). A simple circuit composed of three classes of spinal interneuron—the CCs, EINs, and LINs—has been proposed and could produce oscillatory behavior when modeled (Grillner et al 1988). This original model has now been expanded into a larger, more biophysically based model containing both voltage and ligand-gated conductances (Ekeberg et al 1991; Grillner & Matsushima 1991). Each neuron is treated electrically as one soma compartment and a three-compartment dendritic tree. There are voltage-dependent Na, K, and Ca channels simulated by using the Hodgkin-Huxley formalisms, with the addition of a Ca-dependent potassium channel. The excitatory and inhibitory synapses are modeled as increases in chloride or a combined sodium/potassium conductance. By using these parameters, the shapes of the action potentials and the after potentials were adjusted to experimental values. Although the simulations consisted of reduced networks of only 18 neurons, their combined activity resembled the output of the biological network. But, more importantly, the model circuit closely mimicked such experimental results as a general increased level of frequency after excitation, bursting following NMDA receptor activation, and NMDA dose response curves. The model could satisfactorily account for single segment output. When sensory feedback was included in a multisegmental model, entrainment could be observed, with the overall behavior best described as a system of coupled oscillators. By treating the system this way, others have accounted for the segmental delay necessary for the production of swimming in a quantitative manner (Sigvardt & Williams 1992).

COMPUTATION-BASED MODELS

Models based on back propagation have been used to study such high-level phenomena as the visual detection of shape from shading (Lehky & Sejnowski 1988), learning to pronounce English text (Sejnowski & Rosenberg 1987), and response properties of posterior parietal neurons involved in determining the spatial location of external objects (Zipser & Anderson 1988). Lehky & Sejnowski used a three-layer back-prop model to determine the size and principal surface curvature based on the shading in an image. The input layer contained 122 on- and off-center-surround units similar to those described in the literature. The desired output of the system was for 24 output units, each tuned differently to curvature and orientation. The idea was to construct a feature map from the input, so that they would signal the curvature and shape of the input independent of position or direction of illlumination. By using a hidden layer of 27 units, the network was trained, i.e. its weights adjusted, until after 40,000 trials it could accurately recognize the shapes.

No biological significance was attached to the training algorithm, only to the results obtained from the fully trained network. Here, there were some surprises. The network performed well, despite the fact that specific units responding to curvature have never been found experimentally. Further, although the input units had classical center-surround receptors, the hidden units acquired the properties of cortical "simple" cells, i.e. they responded best to bars moving across the receptive field. The output units, on the other hand, took on properties similar to visual cortical complex end-stopped bar cells. The idea that end-stopped bar cells could be reporting on curvature and depth perception has emerged directly from the model and predicts a new role for these cells, which can be experimentally tested.

A different back-prop model has been used to simulate the response properties of posterior parietal neurons involved in integrating the position of an object on the retina with eye position (Zipser & Andersen 1988). Cells in the posterior parietal cortex respond to both of these signals and combine them to form a spatial representation of the external object. Lesion studies in this area of the brain produce deficits in both motor performance and perception. No single neurons appear to code the spatial representation; instead, each neuron responds to different combinations of retinal and eye positions. Their model suggests how information about location in the visual field that is independent of eye position might be extracted from a population of parietal cortical cells. This study was novel in that it used back prop to generate an algorithm from a randomly connected network, rather than specifying a neural model to implement a computational algorithm. The authors even suggest that, despite the fact that back prop is not possible biologically, it can discover the same algorithms used by the brain for visual processing.

HYBRID MODELS

Hybrid models seek to exploit the computational elegance of abstract modeling, while simultaneously building some biological reality into the computation. One example of this approach has been Lockery's work on the local bending reflex of the leech (Lockery et al 1989). Initially treated as a straightforward backward propagating model, a refined model used actual physiological recordings from leech bending motor neurons to determine the input-output function of a three-layered network. To find out how the interneurons between the sensory and motor neurons computed the reflex, a subpopulation of interneurons, which contributed to one type of bend, was identified, and their responses to patterns of sensory input were measured. These data were then used as the training set. The model was made more realistic by adding experimentally derived physiological constraints. For example, the weights could only be positive, i.e. excitatory; the outputs of units that received no net input were zero; and there were ipsilateral inhibitory connections between inhibitory and excitatory motorneurons. After training, the hidden units could receive inputs, surprisingly like those found in the biological interneurons, and the local bending function could be achieved with only those neurons present, which suggests that a further search for interneurons engaged in local bending was probably not necessary.

Further biological realism was incorporated into the leech model by using a recurrent back-propagation algorithm, with neurons having both input resistance and membrane time constants. Neurons could also be linked via electrical synapses, which suggests the possibility of using multicompartmental models to represent neurons. After training, the matches between the model interneurons and real interneurons were as good as, and in some ways better than, the matches obtained with the original back-prop algorithm.

Impact of Circuit Modeling on Neuroscience

Despite the major criticisms, circuit and, implicitly, cellular modeling is gaining acceptance by large numbers of neuroscientists. These techniques can be used both as tools and as explanation; in either case, they provide insights into the dynamics of complex nonlinear systems impossible to achieve in any other way. One must not be misled by fancy computer graphics or models, which promise more than they can deliver. Learning that occurs by changing weights of connections in neural network models is interesting, but may have little relevance to real nervous systems. In any case, these network models have led to a resurgence in interest in statistical and systems approximation techniques, which may turn out to be relevant to higher brain function at some point in the future.

Sigvardt & Williams (1992) have suggested that, in developing a new model, one should ask three questions: Is the model at the appropriate level? Have the number of unknown parameters been kept to a minimum? And, are the assumptions valid biologically? These are probably as clearly stated a set of rules as one could use in this still emerging field. Clearly, models are meant to simplify. If so many parameters are put into the model that it approaches the complexity of the real system, nothing may be gained from modeling.

The real challenge for the future will be for circuit modeling to provide some linkage with the enormous amount of data being generated at the cellular and molecular levels. It will be crucial to know how molecular properties give rise to the biophysical properties of the single cell, and, in turn, how these delimit the neuron's computational range. Local recordings, multielectrode recordings, and recordings from cortical slices will provide a wealth of data that can be incorporated into circuit models. Already, neural mapping algorithms have strongly influenced thinking about cortical mapping and shown effects that were not obvious. We know now that modeling is not a panacea, but simply one more tool with which to study and help explain how nervous systems work. Successes will be incremental and limited. Models of circuits will probably have to be individually crafted to fit each system under study. Nevertheless, as the hyperbole subsides, and as simulation techniques become better and more widely used, biologically constrained circuit models are likely to play a permanent and increasingly important role in brain research.

ACKNOWLEDGMENTS

The author is indebted to Eve Marder and Ron Harris-Warrick for helpful discussions of the manuscript, and is supported by the National Institutes of Health (09322 & 25916) and the O.N.R. (N00014-91-J-1720).

Literature Cited

Aleksander, I., Morton, H. 1990. *An Introduction to Neural Computing*. New York: Chapman & Hall

Anderson, A., Palca, J. 1988. Who knows how the brain works? *Nature* 335: 489–91

Anderson, J. A., Rosenfeld, E. 1988. *Neurocomputing: Foundations of Research*. Cambridge, Mass: MIT Press

Bekkers, J., Stevens, C. 1991. Presynaptic mechanism for long term potentiation in the hippocampus. *Nature* 346: 724–29

Brown, T. H., Ganong, A. H., Kariss, E. W., Keenan, C. L., Kelso, S. R. 1989. Long term potentiation in two synaptic systems of the hippocampal brain slice. In *Neural Models of Plasticity*, ed. J. H. Byrne, W. O. Berry. New York: Academic

Byrne, J. H., Berry, W. O., eds. 1989. *Neural Models of Plasticity*. San Diego: Academic Press

Churchland, P. S., Koch, C., Sejnowski, T. J. 1989. What is computational neuroscience? In *Computational Neuroscience*, ed. E. Schwartz. Cambridg, Mass: MIT Press

Cole, K. S. 1968. *Membranes, Ions and Impulses*. Berkeley, Calif: Univ. Calif. Press

Crick, F. 1989. The recent excitement about neural networks. *Nature* 337: 129–32

Douglas, R., Martin, K. 1991. Opening the grey box. *Trends Neurosci.* 14: 286–93

Durbin, R., Miall, C., Mitchison, G. 1989. *The Computing Neuron*. Wokingham: Addison-Wesley

Ekeberg, O., Wallen, P., Lanser, A., Traven, H., Brodin, L., Grillner, S. 1991. A computer based model for realistic simulations of neural networks. *Biol. Cybern.* 65: 81–90

Fitzhugh, R. 1961. Impulses and physiological state in theoretical models of nerve membrane. *Biophys. J.* 55: 847–81

Gerstein, G., Mandelbrot, B. 1964. Random walk models for the spike activity of a single neuron. *Biophys. J.* 1964: 41–68

Getting, P. A. 1989. Reconstruction of small neuronal networks. In *Methods in Neuronal Modeling*, ed. C. Koch, I. Segev. Cambridge, Mass: MIT Press

Getting, P. A., Dekin, M. S. 1985. Mechanisms of pattern generation underlying swimming in Tritonia. IV. Gating of a central pattern generator. *J. Neurophysiol.* 53: 466–79

Gilbert, C. D. 1983. Microcircuitry of the visual cortex. *Annu. Rev. Neurosci.* 6: 217–47

Grillner, S., Matsushima, T. 1991. The neural network underlying locomotion in lamprey-synaptic and cellular mechanisms. *Nature* 7: 7–15

Hawkins, R. D., Bower, G. H. 1989. *Computational Models of Learning in Simple Neural Systems*. San Diego: Academic

Hebb, D. O. 1949. *The Organization of Behavior: A Neuropsychological Theory*. New York: Wiley

Hodgkin, A. L., Huxley, A. F. 1952. Currents carried by sodium and potassium ions through the membrane of the giant axon of Loligo. *J. Physiol.* 116: 449–72

Hooper, S. L., Marder, E. 1987. Modulation of central pattern generator by the peptide. *J. Neurosci.* 7: 2097–12

Hopfield, J. J., Tank, D. W. 1986. Computing with neural circuits: A model. *Science* 233: 625–33

Kay, A. R., Wong, R. K. S. 1986. Isolation of neurons suitable for patch-clamping from adult mammalian central nervous systems. *J. Neurosci. Methods* 16: 227–38

Kepler, T. B., Marder, E., Abbott, L. F. 1990. The effect of electrical coupling on the frequency of model neuronal oscillators. *Science* 248: 83–85

Koch, C., Segev, I. 1989. *Methods in Neuronal Modeling: From Synapses to Networks*. Cambridge, Mass: MIT Press

Lehky, S., Sejnowski, T. 1988. Network model of shape-from-shading neural function arises from both receptive and projective fields. *Nature* 333: 452–54

Lennard, P. R., Getting, P. A., Hume, R. I. 1980. Central pattern generator mediating swimming in Tritonia. II. Initiation, maintenance and termination. *J. Neurophysiol.* 44: 165–73

Lockery, S. R., Wittenberg, G., Kristan, W. B. Jr, Cottrell, G. 1989. Functions of identified interneurons in the leech elucidated by neural networks using back-propagation. *Nature* 340: 668–71

MacGregor, R. J. 1987. *Neural & Brain Modeling*. San Diego: Academic

Manilow, R., Miller, J. P. 1986. Postsynaptic hyperpolarization during conditioning reversibly blocks induction of long-term potentiation. *Nature* 320: 529–31

Marder, E. 1992. Modulating membrane properties of neurons: role in information processing. In *Exploring Brain Functions: Models in Neuroscience*, ed. J. Altman. Chichester, England: Wiley

Marr, D. 1982. *Vision*. San Francisco: Freeman

McClelland, J. L., Rummelhart, D. E. 1986. A distributed model of human learning and memory. In *Parallel Distributed Processing*, ed. J. L. McClelland, D. E. Rummelhart. Cambridge, Mass: MIT Press

Miles, R., Wong, R. K. S. 1983. Single neurons can initiate synchronized population discharge in the hippocampus. *Nature* 306: 371–73

Miller, J. P. 1990. Computer modelling at the single-neuron level. *Nature* 347: 783–84

Moore, G. P., Perkel, D. H., Segundo, J. P. 1966. Statistical analysis and functional interpretation of neuronal spike data. *Annu. Rev. Physiol.* 28: 493–522

Mulloney, B., Perkel, D. H. 1988. The roles of synthetic models in the study of central pattern generators. In *Neural Control of Rhythmic Movements in Vetebrates*, ed. A. H. Cohen, S. Rossignol, S. Grillner, pp. 415–53. New York: Wiley

Mulloney, B., Perkel, D. H., Budelli, R. W. 1981. Motor pattern production: interaction of electrical and chemical synapses. *Brain Res.* 229: 25–33

Numann, R. E., Wadman, W. J., Wong, R. K. S. 1987. Outward currents of single hippocampal cells obtained from the adult guinea-pig. *J. Physiol.* 393: 331–53

O'Donnell, P., Koch, C., Poggio, T. 1985. Demonstrating the nonlinear interaction between excitation and inhibition in dendritic trees using computer-generated color graphics: a film. *Soc. Neurosci. Abstr.* 11: 141.2

Perkel, D. H. 1988. Logical neurons: the enigmatic legacy of Warren McCulloch. *Trends Neurosci.* 11: 9–12

Perkel, D. H., Mulloney, B. 1974. Motor pattern production in reciprocally inhibitory neurons exhibiting postinhibitory rebound. *Science* 185: 181–83

Rall, W. 1962. Electrophysiology of a dendrite neuron model. *Biophys. J.* 2: 145–67

Rall, W., Shepherd, G. M. 1968. Theoretical reconstruction of field potentials and dendrodendritic synaptic interactions in olfactory bulb. *J. Physiol.* 31: 884–91

Rall, W., Shepherd, G. M., Reese, T. S. 1966. Dendrodendritic synaptic pathway for inhibition in the olfactory bulb. *Exp. Neurol.* 14: 44–56

Rummelhart, D. E., Hinton, G. E., Williams, R. J. 1986. Learning internal representations by error propagation. In *Parallel Distributed Processing*, ed. D. Rummelhart, J. L. McClelland. Cambridge, Mass: MIT Press

Sah, P., Gibb, A. J., Gage, P. W. 1988. The sodium current underlying action potentials in guinea-pig hippocampal CA1 neurons. *J. Gen. Physiol.* 91: 373–98

Segev, I., Fleshman, J. W., Burke, R. E. 1989. Compartmental models of complex neurons. In *Methods in Neuronal Modeling*, ed. C. Koch, I. Segev. Cambridge, Mass: MIT Press

Sejnowski, T. J., Rosenberg, C. R. 1987. Parallel networks that learn to pronounce English text. *Complex Syst.* 1: 145–68

Selverston, A. I., Moulins, M. 1985. Oscillatory neural networks. *Annu. Rev. Physiol.* 47: 29–48

Sharp, A., Marder, E. 1991. Artificially coupled stomatogastric neurons in primary cell culture. *Soc. Neurosci. Abstr.* 17: 50.15

Sigvardt, K., Williams, T. 1992. Models of central pattern generators as oscillators: the lamprey locomotor CPG. *Semin. Neurosci.* 4: In press

Traub, R. D., Miles, R. 1991. *Neuronal Networks of the Hippocampus*. Cambridge: Cambridge Univ. Press

Traub, R. D., Wong, R. K. S. 1982. Cellular mechanism of neuronal synchronization in epilepsy. *Science* 216: 745–47

Wasserman, P. D. 1989. *Neural Computing: Theory & Practice*. New York: Van Nostrand Reinhold

Wilson, M. A., Bower, J. M. 1989. The simulation of large-scale neural networks. In *Methods in Neuronal Modeling*, ed. C. Koch, I. Segev, pp. 291–334. Cambridge, Mass: MIT Press

Zagotta, W. N., Aldrich, R. W. 1990. Voltage-dependent gating of Shaker A-type potassium channels in *Drosophila* muscle. *J. Gen. Physiol.* 95: 29

Zipser, D., Anderson, R. A. 1988. A backpropagation programmed network that simulates response properties of a subset of posterior parietal neurons. *Nature* 331: 679–84

Annu. Rev. Neurosci. 1993. 16: 547–63

NEUROANATOMY OF MEMORY

S. Zola-Morgan and L. R. Squire

Veterans Affairs Medical Center and University of California, San Diego, California 92093-0603

KEY WORDS: amnesia, declarative memory, medial temporal lobe, hippocampus, diencephalon, basal forebrain, frontal lobe

INTRODUCTION

Three important developments have occurred in the area of memory during the past decade. The first was the recognition that there is more than one kind of memory (Cohen 1984; Schacter 1987; Squire 1982; Tulving 1985). Declarative memory (or, explicit memory) affords the capacity for conscious recollections about facts and events. This is the kind of memory that is usually referred to when the terms "memory" or "remembering" are used in ordinary language. Declarative memory can be contrasted with nondeclarative (or implicit) memory, a heterogeneous collection of nonconscious abilities that includes the learning of skills and habits, priming, and some forms of classical conditioning. In these cases, experience cumulates in behavioral change, but without affording access to any memory content. The distinction between declarative and nondeclarative memory is fundamental, because it has turned out that different kinds of memory are supported by different brain systems.

The second important development was the establishment of an animal model of human amnesia in the monkey (Mahut & Moss 1984; Mishkin 1982; Squire & Zola-Morgan 1983). In the 1950s, Scoville & Milner (1957) described the severe amnesia that followed bilateral surgical removal of the medial temporal lobe (patient H.M.). This important case demonstrated that memory is a distinct cerebral function, dissociable from other perceptual and cognitive abilities. Subsequently, surgical lesions of the medial temporal lobe in monkeys, which approximated the damage sustained by patient H.M., were shown to reproduce many features of human memory impairment. In particular, both monkeys and humans were

547

0147–006X/93/0301–0547$02.00

impaired on tasks of declarative memory, but fully intact at skill and habit learning and other tasks of nondeclarative memory. This achievement set the stage for identifying which structures and connections within the medial temporal lobe are important for declarative memory.

The third development was the emergence of new technologies for studying anatomy and function in living subjects. Magnetic resonance imaging (MRI) is beginning to provide detailed information about the anatomy of damage in patients with memory impairment (spatial resolution < 1.0 mm). Studies using positron emission tomography (PET) provide images of regional blood flow or local glucose metabolism in the brains of normal subjects as they perform specific tasks of learning and memory.

This review summarizes recent findings concerning the anatomy of memory. We focus on the brain regions where damage can cause impairment of declarative memory: the medial temporal lobe, the diencephalon, and the basal forebrain (for other reviews, see Damasio 1984; Markowitsch 1988; Markowitsch & Pritzel 1985; Squire 1987).

THE MEDIAL TEMPORAL LOBE

During the last several years, work with monkeys and new information from patients have identified the structures in the medial temporal lobe that are important for declarative memory. These structures are the hippocampus (including the dentate gyrus and subicular complex) and adjacent cortical areas that are anatomically related to the hippocampus, especially the entorhinal, perirhinal, and parahippocampal cortices (Squire & Zola-Morgan 1991). The work with monkeys has depended on several tasks known to be sensitive to human amnesia (Squire et al 1988), including retention of simple object discriminations and the simultaneous learning of multiple pairs of objects (eight-pair concurrent discrimination learning). The most widely used memory task has been trial-unique delayed nonmatching to sample (Mishkin & Delacour 1975). In this test of recognition memory, the monkey first sees a sample object. Then after a delay, the original object and a novel object are presented together, and the monkey must displace the novel object to obtain a food reward. New pairs of objects are used on each trial. A recent report questioned the usefulness of the delayed nonmatching task for studying memory in monkeys (Ringo 1991). Percent correct data were transformed to a discriminability measure (using the d' measure from signal detection theory) to reanalyze the data from several laboratories. Performance appeared to be just as impaired at short retention intervals as at long retention intervals. If true, this finding would raise the possibility that the impairment caused by medial temporal lobe lesions in monkeys is in perception, attention, or some other cognitive function, rather than memory. However, the studies that were surveyed by Ringo (1991) had been designed primarily to assess the severity of impairment, not to compare short and long retention intervals. Studies with delayed nonmatching to sample that have used appropriate designs (Alvarez-Royo et al 1992; Overman et al 1990; Zola-Morgan & Squire 1985b, Figure 5) show clearly that medial temporal lobe lesions impair performance at long delays, but not at short delays. This is true whether the data are analyzed by using percent correct or the d' measure. This finding is consistent with the facts of human amnesia and supports the validity of the delayed nonmatching task for studying recognition memory in monkeys.

The current era of studies in the monkey began with a large medial temporal lobe removal to approximate the damage in amnesic patient H.M. (Mishkin 1978). This lesion has been termed the H^+A^+ lesion (Squire & Zola-Morgan 1988), where H refers to the hippocampus (including the dentate gyrus and the subicular complex); A, the amygdala; and $^+$, the cortical regions adjacent to the hippocampus and the amygdala that are necessarily damaged when either of these structures is removed by using a direct surgical approach (i.e. the perirhinal, entorhinal, and parahippocampal cortices). The H^+A^+ lesion produces severe memory impairment (Mahut et al 1981; Mishkin 1978; Zola-Morgan & Squire 1985a).

Memory is also impaired following a lesion that involves only the posterior portion of the medial temporal lobe (the H^+ lesion), although the impairment is not as severe as with the H^+A^+ lesion (Mishkin 1978; Zola-Morgan et al 1989a). The H^+ lesion involves the hippocampus proper, the dentate gyrus, the subicular complex, the posterior portion of the entorhinal cortex, and the parahippocampal cortex. Recent studies indicate that the more severe memory impairment associated with H^+A^+ lesions, as compared with H^+ lesions, results from cortical damage, not from amygdala damage. An important clue came from a study in which the amygdala was damaged separately (the A lesion), and the cortex adjacent to the amygdala was spared (Zola-Morgan et al 1989b). Monkeys with A lesions performed as well as normal monkeys on four different memory tasks, including delayed nonmatching to sample. In addition, extending the H^+ lesion forward to include the amygdala (the H^+A lesion) did not exacerbate the memory impairment associated with H^+ lesions on any of these tasks. Similar findings have been reported in several studies in the rodent (Squire 1992a, Table 3).

These findings focused attention on the cortex adjacent to the amygdala, i.e. the perirhinal and entorhinal cortices (see also Murray 1992). Neuroanatomical evidence had shown that the perirhinal and the caudally adjacent parahippocampal cortices provide nearly two thirds of the cortical input to the entorhinal cortex (Insausti et al 1987a). Because entorhinal cortex is, in turn, the major source of projections to the hippocampus and dentate gyrus, there was reason to suppose that damage to the perirhinal cortex might affect memory. Moreover, in a behavioral study of monkeys with removals of anteroventral temporal cortex, the most affected animal had a lesion involving perirhinal cortex (Horel et al 1987). When the H^+ lesion was extended forward to include the perirhinal cortex (the H^{++} lesion), memory impairment was greater than after H^+ or H^+A lesions (Zola-Morgan et al 1993). The impairment following H^{++} lesions remained stable for more than one year after surgery. Finally, monkeys with lesions of the perirhinal and parahippocampal cortices, which included damage to projections to the entorhinal cortex from other cortical areas (the PRPH lesion), exhibited long-lasting memory impairment in both the visual and tactual modalities (Suzuki et al 1993; Zola-Morgan et al 1989c).

These findings point to the importance for normal memory function of the hippocampus and adjacent cortical regions, including the perirhinal, parahippocampal, the entorhinal cortices. The perirhinal, entorhinal, and parahippocampal cortices are not simply routes by which information from the neocortex can reach the hippocampus. The fact that the memory deficit is more severe when these cortical regions are damaged (e.g. H^{++} lesion versus H^+A or H^+ lesion) indicates that these regions must also be important for memory function. The implication is that information from neocortex need not reach the hippocampus itself for some memory storage to occur.

The Involvement of the Hippocampus in Memory

Although the hippocampal region has been linked to memory function since patient H.M. was first described (Scoville & Milner 1957), the hippocampus itself has only recently been identified as a critical structure. Neuropathological findings from a patient with permanent circumscribed memory impairment following global ischemia (patient R.B.) revealed a bilateral lesion involving the entire CA_1 field of the hippocampus (Zola-Morgan et al 1986; for a related case, see Victor & Agamanolis 1990). This result suggested that damage to the hippocampus itself is sufficient to produce a clinically significant and long-lasting memory impairment. Additional information has come from high-resolution MRI studies of patients with circumscribed memory impairment, which revealed that the hippocampal formation was reduced in size (Press et al 1989; Squire et al 1990). Finally, studies of regional cerebral blood flow using PET have been carried out with normal subjects while they performed tasks of reading, word completion from three-letter stems, and recall from a recently presented word list using three-letter stems as cues (Squire et al 1992). The largest area of activation in the memory recall task was in the posterior medial temporal lobe in the region of the hippocampus and the parahippocampal gyrus. No activation was detected in the amygdala.

In the monkey, two approaches have been used to assess the role of the hippocampus itself. First, stereotaxic neurosurgery was combined with MRI to improve the accuracy of circumscribed surgical lesions (Alvarez-Royo et al 1991). Monkeys prepared using this technique (the H lesion) were as impaired as monkeys with H^+ lesions on the delayed nonmatching to sample task when delays reached ten minutes (Clower et al 1991). Overall, however, as measured by performance at the shorter delay intervals of the delayed nonmatching to sample task, as well as by performance on two other tasks (retention of object discriminations and eight-pair concurrent discrimination learning), the monkeys with H lesions were less impaired than monkeys with H^+ lesions. This finding is consistent with the idea that the cortex adjacent to the hippocampus makes a contribution to memory, in addition to the contribution made by the hippocampus itself.

The second approach was to establish an animal model of global ischemia in the monkey (Zola-Morgan et al 1992). This procedure consistently produced a highly selective pattern of damage: bilateral loss of CA_1 and CA_2 pyramidal cells of the hippocampus, together with substantial bilateral loss of somatostatin-staining cells in the hilar region of the dentate gyrus (the ISC lesion). Cell loss was greater in the caudal portion of the hippocampus than in the rostral portion. Except for patchy loss of cerebellar Purkinje cells, significant damage was not detected outside the hippocampus. On the delayed nonmatching to sample task, monkeys with ISC lesions were about as impaired as monkeys with H^+ lesions. However, like the H group, the ISC group performed significantly better than the H^+ monkeys on other tasks. Thus, the overall level of memory impairment following ISC lesions was similar to the level associated with H lesions and less than the level associated with H^+ lesions.

These findings from monkeys make several points. They support the long-standing view that the hippocampus is important for memory. Indeed, even incomplete damage to the hippocampus is sufficient to impair memory. Although the original findings from patient R.B. made this same point, it had been difficult to exclude entirely the possibility that some additional neural damage might have occurred in R.B. that was not detected in histological examination. However, because ISC monkeys obtained better memory scores overall than H^+ monkeys, it seems reasonable to think that the ISC animals (and, by extension, patient R.B.) did not have widespread neuropathological damage affecting memory beyond what was detected histologically. Finally, the finding that even partial damage to the hippocampus produced a significant and enduring memory impairment in monkeys contrasts sharply with the absence of impairment following virtually complete lesions of the amygdala. Experiments with rats and monkeys suggest that the amygdala is important for other kinds of memory, including the development of conditioned fear and other forms of affective memory in which the valence of a neutral stimulus is strongly altered by experience (Davis 1986; Gallagher et al 1990; Kesner 1992; LeDoux 1987; McGaugh 1989).

THE DIENCEPHALON

Damage to the midline diencephalic region was first linked to amnesia in humans nearly a century ago (Gudden 1896). Although it is now accepted that medial diencephalic damage is sufficient to cause severe amnesia, the specific structures and connections that must be damaged to cause memory impairment have not yet been identified. The two structures most frequently implicated have been the mammillary nuclei (MN) and the mediodorsal thalamic nucleus (MD) (Markowitsch 1988; Victor et al 1989). The idea that damage to the MN impairs memory originated in the finding that the MN are consistently damaged in alcoholic Korsakoff's syndrome. However, the MN are not the only site of damage. In two thorough studies of postmortem material, in which significant memory impairment was well documented during life (Mair et al 1979; Mayes et al 1988), four patients exhibited marked neuronal loss in the medial MN together with a band of gliosis in the medial thalamus located along the wall of the third ventricle adjacent to the medial magnocellular portion of MD. Neuropathological findings from Korsakoff's syndrome have led to the view that damage to MD itself is critical, either alone (Victor et al 1989) or in combination with MN (Butters 1984).

During the past several years, new data have become available concerning memory loss and medial thalamic damage. One study used computed tomography (CT) to identify the common damage in seven patients with memory impairment following medial thalamic infarctions (von Cramon et al 1985). This analysis identified as the important sites the mammillothalamic tract and the ventral portion of the internal medullary lamina, which forms the ventrolateral boundary of MD. A second radiographic study of two amnesic patients with bilateral thalamic infarctions suggested that the lesions responsible for memory impairment damaged the mammillothalamic tract and the inferior thalamic peduncle at the level of the anterior nucleus (Graff-Radford et al 1990; for an additional single case, see Malamut et al 1992). Both studies concluded that disconnection of both the anterior nucleus and MD from other structures is required to produce severe memory impairment. At the same time, it remains unclear how much memory impairment would occur following damage limited to either nucleus alone or to other medial thalamic nuclei, such as the intralaminar nuclei. Thus, MD has been identified as damaged in several single-case studies of thalamic infarction (cf. Guberman & Stuss 1981; Winocur et al 1984), but additional damage was also present, as would be expected given that the thalamic arteries supply more than one thalamic nucleus.

The idea that amnesia results when several diencephalic nuclei are damaged conjointly is consistent with the radiographic findings from patient N.A. (Squire et al 1989a). This individual developed amnesia, especially for verbal material, following a penetrating stab wound to the brain, and CT scans had initially indicated a lesion in the region of the left mediodorsal nucleus. Subsequently, MRI studies revealed more extensive damage in the left thalamus. In addition, the injury likely damaged the mammillothalamic tract, and the MN appeared to be damaged bilaterally. The thalamic damage involved the internal medullary lamina, the ventral portion of MD, the intralaminar nuclei, and the ventral lateral and ventral anterior nuclei. Mori et al (1986) described a patient with a very similar left thalamic lesion caused by infarction.

In the monkey, circumscribed bilateral MN lesions produced a measurable memory impairment (Aggleton & Mishkin 1985; Zola-Morgan et al 1989a), but one that was mild compared with the impairment associated with lesions of the hippocampal formation or related cortex. More severe memory impairment also occurred following lesions that included the anterior thalamic nucleus, MD, and midline nuclei (Aggleton & Mishkin 1983a). This impairment was greater than when the lesion involved either the anterior or posterior half of the larger lesion (Aggleton & Mishkin 1983b).

Additional information comes from the recent development of an animal model of alcoholic Korsakoff's syndrome in the rat (Mair et al 1988). Rats that were recovered from approximately two weeks of pyrithiamine-induced thiamine deficiency exhibited diencephalic lesions similar to the lesions that occur in Korsakoff's syndrome: bilaterally symmetric lesions in the MN and in the medial thalamus in the area of the internal medullary lamina (Mair et al 1988). Additional studies of this animal model of Korsakoff's syndrome used radio frequency lesions to damage separately the internal medullary lamina, the MN, or the midline nuclei. Rats with radio frequency lesions of the internal medullary lamina were impaired on a spatial alternation task to the same extent as rats with thiamine deficiency. Rats with MN lesions or midline nuclei damage performed normally (Mair & Lacourse 1992; Mair et al 1992).

The recent studies with experimental animals are consistent with the findings from human amnesia in showing the importance of damage within the medial thalamus for producing memory loss, especially damage in the internal medullary lamina. Lesions in the internal medullary lamina would be expected to disconnect or damage several thalamic nuclei, including intralaminar nuclei and MD. Evidence from rats and monkeys suggests that the MD may be an important structure (Aggleton & Mishkin 1983b; Mair et al 1991; Zola-Morgan & Squire 1985b). The separate contributions of MD, the anterior nucleus, and the intralaminar nuclei remain to be explored systematically with well-circumscribed lesions.

THE BASAL FOREBRAIN

Some patients with ruptured aneurysms of the anterior communicating artery exhibit persisting memory impairment, together with personality change. The critical damage reportedly involves the basal forebrain (Alexander & Freedman 1984; Damasio et al 1985a,b; Phillips et al 1987). The basal forebrain is the primary source of cholinergic innervation of cortex. It includes the medial septal nucleus and the diagonal band of Broca, which project to the hippocampal formation mainly through the fornix, and the nucleus basalis, which projects widely to frontal, parietal, and temporal cortices (Mesulam et al 1983). The idea that basal forebrain damage, and damage to cholinergic neurons in particular, can impair memory gained additional support from reports that patients with Alzheimer's disease, who exhibit memory impairment as a prominent early symptom, show decreased activity of choline acetyltransferase (ChAT) in the cortex and hippocampus and markedly reduced cell numbers in the basal forebrain (Coyle et al 1983).

Recent work has raised questions about the relationship between cholinergic dysfunction and memory impairment. For example, some behavioral effects of damage to nucleus basalis in rats are not related to cholinergic dysfunction. Thus, lesions of the nucleus basalis produced by quisqualic acid injections produced less severe behavioral impairment and sometimes no impairment, compared with lesions produced by ibotenic acid, despite the fact that quisqualic acid results in larger decreases in cortical levels of ChAT than did ibotenic acid (Dunnett et al 1987). In macaque monkeys, combined ibotenate lesions of the nucleus basalis, the medial septal nucleus, and the diagonal band, but not separate lesions of these areas, produced significant memory impairment (Aigner et al 1991). However, the performance of the impaired group recovered fully by about six months after surgery. These findings suggest that extensive damage to the basal forebrain, not just nucleus basalis lesions, is necessary for even transient memory impairment to be observed. Squirrel monkeys with basal forebrain lesions exhibited long-lasting behavioral deficits (Irle & Markowitsch 1987). However, the work in rodents just described raises the possibility that cholinergic dysfunction is not responsible for these impairments.

Lesions of the basal forebrain can impair memory, but a range of other behavioral deficits has also been described (for reviews, see Dekker et al 1991; Fibiger 1991; Kesner 1988; Olton & Wenk 1987). In rats, for example, deficits in attention have been reported to be the principal cognitive effect of nucleus basalis lesions (Robbins et al 1989). Although early work suggested that similar cognitive effects occurred following damage to any of the components of the basal forebrain (Hepler et al 1985), it is now clear that medial septal lesions and nucleus basalis lesions produce strikingly different effects. In an important study, Olton et al (1988) compared the performance of rats with ibotenic acid lesions of the medial septal area or nucleus basalis with the performance of rats with surgical lesions of the fornix or frontal cortex. Rats with damage to the medial septum or the fornix exhibited similar deficits on a memory task that required accurate timing of the duration of a tone. Neither group was impaired on a second, divided-attention task in which animals had to time the duration of a tone through a period when an interfering tone was also present. In contrast, damage to either the nucleus basalis or frontal cortex impaired performance on the divided-attention task, but had no effect on the memory task. These results indicate that the components of the basal forebrain are involved in different cognitive functions. Only medial septal damage produced a clear memory impairment, possibly by direct disruptive effects on the hippocampal formation (Buzsaki & Gage 1989; Mizumori et al 1989). Indeed, the strong anatomical connections between the basal forebrain and the medial temporal lobe suggest that the effects of basal forebrain damage in monkeys and humans, including patients with Alzheimer's disease, result from disruption of information processing within the hippocampus and other medial temporal lobe structures (Damasio et al 1985b; Squire 1987). Interestingly, in the case of Alzheimer's disease, neuropathological studies have found prominent pathology in the entorhinal cortex and the subiculum of the hippocampal formation (Hyman et al 1984), as well as in the perforant pathway, the principal source of cortical input to the hippocampus (Hyman et al 1986). These abnormalities effectively disconnect the hippocampus from widespread areas of neocortex and could be sufficient in themselves to account for the memory impairment associated with this disease.

In summary, work with rats, monkeys, and humans indicates that unitary formulations of basal forebrain function are not appropriate. The nucleus basalis appears to be more important in attentional functions than in memory functions. The medial septal area, as well as the diagonal band, can influence memory functions, perhaps by virtue of the strong anatomical connections, including cholinergic connections, to the hippocampal formation.

FROM BRAIN STRUCTURES TO BRAIN SYSTEMS

The identification of critical brain structures in the medial temporal lobe and the midline diencephalon provides only a first step to understanding the neuroanatomy of memory. The connections among these regions, and between these regions and putative sites of long-term memory storage in neocortex, must also be identified. For perceptual processing in neocortex to persist as long-term memory, information from neocortex must reach medial temporal lobe structures (Mishkin 1982; Squire & Zola-Morgan 1991). Projections from neocortex arrive initially in the parahippocampal cortex (area TF/TH) and perirhinal cortex. Further processing then occurs at the next stage, the entorhinal cortex, and in the several stages of the hippocampal formation (dentate gyrus, CA_3, and CA_1). This connectivity provides the hippocampus and related structures with access to ongoing cortical activity at widespread sites throughout the neocortex. Information can then be returned to neocortex via the subiculum and entorhinal cortex.

Information processed in the medial temporal lobe is also routed to critical areas for memory in the diencephalon. Thus, the mammillary nuclei receive a major input from the subiculum of the hippocampal formation through the fornix, and the mammillary nuclei originate a major projection to the anterior nucleus through the mammillothalamic tract. The hippocampal formation also sends direct projections to the anterior nucleus. The mediodorsal nucleus of the thalamus, in addition to its well-described projections from the amygdala, receives a fairly prominent projection from perirhinal cortex. Somewhat weaker projections to the mediodorsal nucleus also originate in the subiculum and in area TF of parahippocampal cortex (Amaral 1987). In addition, the basal forebrain has widespread projections to medial temporal lobe and can potentially modulate its function (Insausti et al 1987b).

One important target of diencephalic and medial temporal lobe structures is the frontal lobe, especially ventromedial frontal cortex. The anterior nucleus and the mediodorsal nucleus project to both ventromedial and dorsolateral frontal cortex. In addition, both the entorhinal cortex and the subiculum send significant projections to ventromedial cortex, especially its medial orbital surface (Insausti et al 1987a; Carmichael & Price 1991). One possibility is that the ventromedial frontal cortex, together with the medial temporal lobe and medial thalamus, constitutes a component of the neural system essential for the formation of long-term memory (Bachevalier & Mishkin 1986). Another possibility is that the medial temporal lobe and medial thalamus work conjointly to establish long-term memory and that the projections to the frontal lobe provide a route by which recollections can be translated into action. The frontal lobes are important in guiding behavior at the time of both information encoding and information retrieval, especially when information must be organized and retained for temporary use in short-term (or working) memory (Fuster 1989; Goldman-Rakic 1987).

Recently, Irle & Markowitsch (1990) reported that squirrel monkeys with conjoint bilateral lesions of five different structures (the hippocampus, the amygdala, the anterior thalamic region, the mediodorsal thalamic nucleus, and the septum) performed better on the delayed nonmatching to sample task than monkeys with lesions limited to one or two of these structures (the hippocampus, the hippocampus plus amygdala, or the anterior and mediodorsal thalamic regions). If true, this finding would be unique and important, because of the implication that massive damage to the declarative memory system is somehow less disruptive than damage to specific structures within the system. However, the data presented are not compelling. First, the study involved only two monkeys in each of the single and double-lesion groups. Second, monkeys in the multiple (fivefold) lesion group were significantly impaired on the delayed nonmatching task. Third, a close reading of the paper indicates that, contrary to the proposal in the paper (p. 86), the comparison between the monkeys in the fivefold lesion group ($n = 4$) and the three groups of monkeys with single or double lesions ($n = 6$; Table 4, p. 89) did not approach statistical significance.

BRAIN SYSTEMS AND MEMORY

Medial temporal lobe structures and the medial thalamus are components of a memory system that is essential for the formation of long-term declarative memory. Memory depends on this system for only a limited period of time after learning. This conclusion rests partly on the finding that remote memory is often fully intact in amnesic patients (Squire et al 1989b) and on the finding of temporally graded retrograde amnesia in prospective studies of monkeys (Zola-Morgan & Squire 1990) and rats (Kim & Fanselow 1992; Winocur 1990) with lesions. Thus, medial temporal lobe and medial thalamic structures are not the repository for permanent memory. This system is required at the time of learning and during a lengthy period thereafter, while a slow-developing, more permanent memory is established elsewhere, presumably in neocortex (Squire 1992a).

Short-term memory is independent of these brain structures and is fully intact in amnesia, whether it is assessed in the conventional manner by verbal tests of digit span or by tests of nonverbal short-term memory, including spatial short-term memory (Cave & Squire 1992). Skills and habits, priming, and some forms of conditioning are also independent of the medial temporal lobe and the medial thalamus. Whereas declarative memory depends on an interaction between the neocortex and these structures, many skills and habits depend on the neocortex and the neostriatum (Packard et al 1989; Wang et al 1990). Perceptual priming likely depends on posterior cortical areas, such as extrastriate cortex in the case of visual priming (Squire et al 1992). Classical conditioning of skeletal musculature depends on essential pathways in the cerebellum (Thompson 1986). Declarative and nondeclarative memory can seem rather similar to each other. For example, an animal can select an object on the delayed non-matching to sample test or can select the same object when it is presented in a task of habit learning [e.g. the 24-hour concurrent discrimination task (Malamut et al 1984)]. However, these are different kinds of learning, the resulting knowledge has different characteristics, and different brain systems are involved (Squire 1992b).

The question naturally arises as to whether damage to the medial temporal lobe or to the medial thalamus produces similar or different kinds of memory impairment. Although the two regions probably make distinct contributions to normal memory, it is also possible that the two regions belong to a larger functional system and that their separate contributions would be difficult to detect with behavioral measures. Although possible differences have been proposed between diencephalic and medial temporal lobe amnesia (Parkin 1984), there is currently little evidence to support such a difference. For example, whereas one early suggestion concerned differences in the rate of forgetting in long-term memory, McKee & Squire (1992) recently demonstrated that amnesic patients with confirmed medial temporal lobe lesions or diencephalic lesions have virtually identical forgetting rates for information within long-term memory.

Another suggestion, influenced by work with rodents, has been that the hippocampus is involved especially in computing and storing information about allocentric space (O'Keefe & Nadel 1978). In our view, however, spatial memory is better understood as a good example of the broader category of (declarative) memory abilities, which includes memory for spatial locations, but also includes memory for word lists, faces, odors, and tactual impressions (Squire & Cave 1991). For amnesic patients with confirmed damage to the hippocampal formation or diencephalon, spatial memory impairment was proportional to the severity of impairment on other measures of declarative memory (Cave & Squire 1991); for additional discussion, see *Hippocampus* 1991, 1: 221–92.

One sense in which one should expect to find functional specialization within the medial temporal lobe system follows from the fact that anatomical connections from different parts of neocortex enter the system at different points. For example, parietal cortex projects to parahippocampal cortex, but not to perirhinal cortex, and inferotemporal cortex projects more strongly to perirhinal cortex than to parahippocampal cortex (Suzuki et al 1991). These anatomical facts provide a way to understand why anterior medial temporal lobe lesions, which damage perirhinal cortex, and posterior medial temporal lobe lesions, which damage parahippocampal cortex, might differentially affect spatial memory (Parkinson et al 1988).

CONCLUSIONS

Cumulative and systematic research with monkeys and rats and related research with humans has identified structures and connections important for declarative memory in the medial temporal lobe and the midline diencephalon. The important structures within the medial temporal lobe are the hippocampus, and adjacent, anatomically related entorhinal, perirhinal, and parahippocampal cortices. The amygdala is not a part of this system. The important structures in the diencephalon appear to be the anterior thalamic nucleus, the mediodorsal nucleus, and connections to and from the medial thalamus within the internal medullary lamina. With respect to the basal forebrain, the nucleus basalis appears to be involved more in attentional functions than in memory functions. Other components of the basal forebrain can influence memory functions by virtue of their anatomical projections to the hippocampal formation.

The declarative memory system is fast, has limited capacity, and has a crucial function beginning at the time of learning in establishing long-term memories. This function involves binding together the multiple areas in neocortex that together subserve perception and short-term memory of whole events. Gradually, the neocortex comes to support long-term memory storage independently of the medial temporal lobe and diencephalon.

ACKNOWLEDGMENTS

This work is supported by the Medical Research Service of the Department of Veterans Affairs, the Office of Naval Research, National Institutes of Health grant NS19063, National Institute of Mental Health grant MH24600, and the McKnight Foundation. We thank D. Amaral, P. Alvarez-Royo, R. Clower, N. Rempel, S. Ramus, and W. Suzuki for their contributions to work summarized here.

Literature Cited

Aggleton, J. P., Mishkin, M. 1985. Mammillary-body lesions and visual recognition in monkeys. *Exp. Brain Res.* 58: 190–97

Aggleton, J. P., Mishkin, M. 1983a. Visual recognition impairment following medial thalamic lesions in monkeys. *Neuropsychologia* 21: 189–97

Aggleton, J. P., Mishkin, M. 1983b. Memory impairments following restricted medial thalamic lesions. *Exp. Brain Res.* 52: 199–209

Aigner, T. G., Mitchell, S. J., Aggleton, J. P., DeLong, M. R., Struble, R. G., et al. 1991. Transient impairment of recognition memory following ibotenic-acid lesions of the basal forebrain in macaques. *Exp. Brain Res.* 86: 18–26

Alexander, M. P., Freedman, M. 1984. Amnesia after anterior communicating artery aneurysm rupture. *Neurology* 34: 752–57

Alvarez-Royo, P., Clower, R. P., Zola-Morgan, S., Squire, L. R. 1991. Stereotaxic lesions of the hippocampus in monkeys: determination of surgical coordinates and analysis of lesions using magnetic resonance imaging. *J. Neurosci. Methods* 38: 223–32

Alvarez-Royo, P., Zola-Morgan, S., Squire, L. R. 1992. Impairment of long-term memory and sparing of short-term memory in monkeys with medial temporal lobe lesions: a response to Ringo. *Behav. Brain Res.* In press

Amaral, D. G. 1987. Memory: Anatomical organization of candidate brain regions. In *Handbook of Physiology: The Nervous System, V. Higher Functions of the Nervous System*, ed. J. M. Brookhart, V. B. Mountcastle, pp. 211–94. Bethesda: Am. Physiol. Soc. 5th ed.

Bachevalier, J., Mishkin, M. 1986. Visual recognition impairment follows ventromedial but not dorsolateral prefrontal lesions in monkeys. *Behav. Brain Res.* 20: 249–61

Butters, N. 1984. Alcoholic Korsakoff's syndrome: an update. *Semin. Neurol.* 4: 226–44

Buzsaki, G., Gage, F. H. 1989. Absence of long-term potentiation in the subcortically deafferented dentate gyrus. *Brain Res.* 484: 94–101

Carmichael, S. T., Price, J. L. 1991. Orbital prefrontal cortex in the monkey: Structurally distinct areas with specific inputs connected in parallel networks. *Soc. Neurosci. Abstr.* 17: 1584

Cave, C. B., Squire, L. R. 1992. Intact verbal and nonverbal short-term memory following damage to the human hippocampus. *Hippocampus* 2: 151–64

Cave, C. B., Squire, L. R. 1991. Equivalent impairment of spatial and nonspatial memory following damage to the human hippocampus. *Hippocampus* 1: 329–40

Clower, R. P., Alvarez-Royo, P., Zola-Morgan, S., Squire, L. R. 1991. Recognition memory impairment in monkeys with selective hippocampal lesions. *Soc. Neurosci. Abstr.* 17: 338

Cohen, N. S. 1984. Preserved learning capacity in amnesia: evidence for multiple memory systems. In *Neuropsychology of Memory*, ed. L. R. Squire, N. Butters, pp. 83–103. New York: Guilford

Coyle, J. T., Price, D. L., DeLong, M. R. 1983. Alzheimer's Disease: A disorder of cortical cholinergic innervation. *Science* 219: 1184–90

Damasio, A. 1984. The anatomical basis of memory disorders. *Semin. Neurol.* 4: 223–25

Damasio, A. R., Eslinger, P. J., Damasio, H., Van Hoesen, G. W., Cornell, S. 1985a. Multimodal amnesic syndrome following bilateral temporal and basal forebrain damage. *Arch. Neurol.* 42: 252–59

Damasio, A. R., Graff-Radford, N. R., Eslinger, P. J., Damasio, H., Kassell, N. 1985b. Amnesia following basal forebrain lesions. *Arch. Neurol.* 42: 263–71

Davis, M. 1986. Pharmacological and anatomical analysis of fear conditioning using the fear-potentiated startle paradigm. *Behav. Neurosci.* 100: 814–24

Dekker, A. J. A. M., Conner, D. J., Thal, L. J. 1991. The role of cholinergic projections from the nucleus basalis in memory. *Neurosci. Biobehav. Rev.* 15: 299–317

Dunnett, S. B., Whishaw, I. Q., Jones, G. H., Bunch, S. T. 1987. Behavioural biochemical and histochemical effects of different neurotoxic amino acids injected into nucleus basalis magnocellularis of rats. *Neuroscience* 22: 441–69

Fibiger, H. C. 1991. Cholinergic mechanisms in learning, memory and dementia: a review of recent evidence. *Trends Neurosci.* 14: 220–23

Fuster, J. M. 1989. *The Prefrontal Cortex.* New York: Raven. 2nd ed.

Gallagher, M., Graham, P. W., Holland, P. 1990. The amygdala central nucleus and appetitive Pavlovian conditioning: Lesions impair one class of conditioned behavior. *J. Neurosci.* 10: 1906–11

Goldman-Rakic, P. S. 1987. Circuitry of primate prefrontal cortex and regulation of behavior by representational memory. In *Handbook of Physiology*, ed. J. M. Brookhart, V. B. Mountcastle, pp. 373–407. Bethesda: Am. Physiol. Soc. 5th ed.

Graff-Radford, N. R., Tranel, D., Van Hoesen, G. W., Brandt, J. 1990. Diencephalic amnesia. *Brain* 113: 1–25

Guberman, A., Stuss, D. 1981. The syndrome of bilateral paramedian thalamic infarction. *Neurology* 33: 540–46

Gudden, H. 1896. Klinische und anatomische Beitrage zur Kenntniss der multiplen Alkoholneuritis nebst Bemerkungen uber die Regenerationsvorgange im peripheren Nervensystem. *Arch. Psychiatr. Nervenkr.* 28: 643–741

Hepler, D. J., Olton, S. D., Wenk, G. L., Coyle, J. T. 1985. Lesions in nucleus basalis magnocellularis and medial septal area of rats produce qualitatively similar memory impairments. *J. Neurosci.* 5: 866–73

Horel, J. A., Pytko-Joiner, E., Voytko, M. L., Salsbury, K. 1987. The performance of visual tasks while segments of the inferotemporal cortex are suppressed by cold. *Behav. Brain Res.* 23: 29–42

Hyman, B. T., Van Hoesen, G. W., Damasio, A. R., Barnes, C. L. 1984. Alzheimer's disease: cell-specific pathology isolates the hippocampal formation. *Science* 225: 1161–70

Hyman, B. T., Van Hoesen, G. W., Kromer, L. J., Damasio, A. R. 1986. Perforant pathway changes and the memory impairment of Alzheimer's disease. *Ann. Neurol.* 20: 472–81

Insausti, R., Amaral, D. G., Cowan, W. M. 1987a. The entorhinal cortex of the monkey: II. Cortical afferents. *J. Comp. Neurol.* 264: 356–95

Insausti, R., Amaral, D. G., Cowan, W. M. 1987b. The entorhinal cortex of the monkey: III. Subcortical afferents. *J. Comp. Neurol.* 264: 396–408

Irle, E., Markowitsch, H. J. 1990. Functional recovery after limbic lesions in monkeys. *Brain Res. Bull.* 25: 79–92

Irle, E., Markowitsch, H. J. 1987. Basal forebrain-lesioned monkeys are severely impaired in tasks of association and recognition memory. *Ann. Neurol.* 22: 735–43

Kesner, R. P. 1992. Learning and memory in rats with an emphasis on the role of the amygdala. In *The Amygdala*, ed. J. Aggleton, pp. 379–400. New York: Wiley

Kesner, R. P. 1988. Reevaluation of the contribution of the basal forebrain cholinergic system to memory. *Neurobiol. Aging* 9: 609–16

Kim, J. J., Fanselow, M. S. 1992. Modality-specific retrograde amnesia of fear. *Science* 256: 675–77

LeDoux, J. 1987. Emotion. In *Handbook of Physiology: The Nervous System, V. Higher Functions of the Nervous System,*

ed. J. M. Brookhart, V. B. Mountcastle, pp. 419–60. Bethesda: Am. Physiol. Soc. 5th ed.

Mahut, H., Moss, M. 1984. Consolidation of memory: The hippocampus revisited. In *Neuropsychology of Memory*, ed. L. R. Squire, N. Butters, pp. 297–315. New York: Guilford

Mahut, H., Moss, M., Zola-Morgan, S. 1981. Retention deficits after combined amygdalo-hippocampal and selective hippocampal resections in the monkey. *Neuropsychologia* 19: 201–25

Mair, R. G., Anderson, C. D., Langlais, P. J., McEntree, W. J. 1988. Behavioral impairments, brain lesions and monoaminergic activity in the rat following a bout of thiamine deficiency. *Behav. Brain Res.* 27: 223–39

Mair, R. G., Knoth, R. L., Rabehenuk, S. A., Langlais, P. J. 1991. Impairment of olfactory, auditory, and spatial serial reversal learning in rats recovered from pyrithiamine induced thiamine deficiency. *Behav. Neurosci.* 105: 360–74

Mair, R. G., Lacourse, D. M. 1992. Radiofrequency lesions of thalamus produce delayed non-matching to sample impairments comparable to pyrithiamine-induced encephalopathy in rats. *Behav. Neurosci.* 106: 634–45

Mair, R. G., Robinson, J. K., Koger, S. M., Fox, G. D., Zhang, Y. P. 1992. Delayed non-matching to sample is impaired by extensive, but not by limited lesions of thalamus in the rat. *Behav. Neurosci.* 106: 646–56

Mair, W. G. P., Warrington, E. K., Weiskrantz, L. 1979. Memory disorder in Korsakoff psychosis. A neuropathological and neuropsychological investigation of two cases. *Brain* 102: 749–83

Malamut, B. L., Graff-Radford, N., Chawluk, J., Grossman, R. I., Guy, R. C. 1992. Memory in a case of bilateral thalamic infarction. *Neurology* 42: 163–69

Malamut, B. L., Saunders, R. C., Mishkin, M. 1984. Monkeys with combined amygdalo-hippocampal lesions succeed in object discrimination learning despite 24-hour intertrial intervals. *Behav. Neurosci.* 98: 759–69

Markowitsch, H. J. 1988. Diencephalic amnesia: a reorientation towards tracts? *Brain Res. Rev.* 13: 351–70

Markowitsch, H. J., Pritzel, M. 1985. The neuropathology of amnesia. *Progr. Neurobiol.* 25: 189–287

Mayes, A. R., Meudell, P. R., Mann, D., Pickering, A. 1988. Location of lesions in Korsakoff's syndrome: Neuropsychological and neuropathological data on two patients. *Cortex* 24: 367–88

McGaugh, J. L. 1989. Involvement of hormonal and neuromodulatory systems in the regulation of memory storage. *Annu. Rev. Neurosci.* 12: 255–87

McKee, R. D., Squire, L. R. 1992. Equivalent forgetting rates in long-term memory for diencephalic and medial temporal lobe amnesia. *J. Neurosci.* 12: 3765–72

Mesulam, M. M., Mufson, E. J., Levey, A. I., Wainer, B. H. 1983. Cholinergic innervation of cortex by the basal forebrain: Cytochemistry and cortical connections of the septal area, diagonal band nuclei, nucleus basalis (substantia innominata), and hypothalamus in the rhesus monkey. *J. Comp. Neurol.* 214: 170–97

Mishkin, M. 1982. A memory system in the monkey. *Philos. Trans. R. Soc. London (Biol.)* 298: 85–92

Mishkin, M. 1978. Memory in monkeys severely impaired by combined but not separate removal of the amygdala and hippocampus. *Nature* 273: 297–98

Mishkin, M., Delacour, J. 1975. An analysis of short-term visual memory in the monkey *J. Exp. Psychol. (Anim. Behav.)* 1: 326–34

Mizumori, S. J. Y., Barnes, C. A., McNaughten, B. C. 1989. Reversible inactivation of the medial septum: selective effects on the spontaneous unit activity of different hippocampal cell types. *Brain Res.* 500: 99–106

Mori, E., Yamadori, A., Mitani, Y. 1986. Left thalamic infarction and disturbance of verbal memory: A clinicoanatomical study with a new method of computed tomographic stereotaxic lesion localization. *Ann. Neurol.* 20: 671–76

Murray, E. A. 1992. Medial temporal lobe structure contributing to recognition memory: the amygdaloid complex versus the rhinal cortex. In *The Amygdala*, ed. J. Aggleton, pp. 453–70. New York: Wiley

O'Keefe, J., Nadel, L. 1978. *The Hippocampus as a Cognitive Map.* London: Oxford Univ. Press

Olton, D. S., Wenk, G. L. 1987. Dementia: animal models of the cognitive impairments produced by degeneration of the basal forebrain cholinergic system. In *Psychopharmacology: The Third Generation of Progress*, ed. H. Y. Meltzer, pp. 941–54. New York: Raven

Olton, D. S., Wenk, G. L., Church, R. M., Meck, W. H. 1988. Attention and the frontal cortex as examined by simultaneous temporal processing. *Neuropsychologia* 26: 307–18

Overman, W. H., Ormsby, G., Mishkin, M. 1990. Picture recognition vs. picture discrimination learning in monkeys with medial temporal removals. *Exp. Brain Res.* 79: 18–24

Packard, M. G., Hirsh, R., White, N. M. 1989. Differential effects of fornix and caudate nucleus lesions on two radial maze tasks: Evidence for multiple memory systems. *J. Neurosci.* 9: 1465–72

Parkin, A. J. 1984. Amnesic syndrome: A lesion-specific disorder? *Cortex* 20: 479–508

Parkinson, J. K., Murray, A. E., Mishkin, M. 1988. A selective mnemonic role for the hippocampus in monkeys: Memory for the location of objects. *J. Neurosci.* 8: 4159–67

Phillips, S., Sangalang, V., Sterns, G. 1987. Basal forebrain infarction: A clinicopathologic correlation. *Arch. Neurol.* 44: 1134–38

Press, G. A., Amaral, D. G., Squire, L. R. 1989. Hippocampal abnormalities in amnesic patients revealed by high-resolution magnetic resonance imaging. *Nature* 341: 54–57

Ringo, J. L. 1991. Memory decays at the same rate in macaques with and without brain lesions when expressed in d' or arcsine terms. *Behav. Brain Res.* 42: 123–34

Robbins, T. W., Everitt, B. J., Ryan, C. N., Marston, H. M., Jones, G. H., Page, K. J. 1989. Comparative effects of quisqualic and ibotenic acid-induced lesions of the substantia innominata and globus pallidus on attentional function in the rat: further implications for the role of the cholinergic neurons of the nucleus basalis in cognitive processes. *Behav. Brain Res.* 35: 221–40

Schacter, D. 1987. Implicit memory: History and current status. *J. Exp. Psychol. Learn. Mem. Cogn.* 13: 501–18

Scoville, W. B., Milner, B. 1957. Loss of recent memory after bilateral hippocampal lesions. *J. Neurol. Neurosurg. Psychiatry* 20: 11–21

Squire, L. R. 1992a. Memory and the hippocampus: A synthesis from findings with rats, monkeys and humans. *Psychol. Rev.* 99: 195–231

Squire, L. R. 1992b. Declarative and nondeclarative memory: Multiple brain systems supporting learning and memory. *J. Cogn. Neurosci.* 4: 232–43

Squire, L. R. 1987. *Memory and Brain.* New York: Oxford Univ. Press

Squire, L. R. 1982. The neuropsychology of human memory. *Annu. Rev. Neurosci.* 5: 241–73

Squire, L. R., Amaral, D. G., Press, G. A. 1990. Magnetic resonance measurements of hippocampal formation and mammillary nuclei distinguish medial temporal lobe and diencephalic amnesia. *J. Neurosci.* 10: 3106–17

Squire, L. R., Amaral, D. G., Zola-Morgan, S., Kritchevsky, M., Press, G. 1989a. Description of brain injury in amnesic patient N.A. based on magnetic resonance imaging. *Exp. Neurol.* 105: 23–35

Squire, L. R., Cave, C. B. 1991. The hippocampus, memory, and space. *Hippocampus* 1: 269–71

Squire, L. R., Haist, F., Shimamura, A. P. 1989b. The neurology of memory: Quantitative assessment of retrograde amnesia in two groups of amnesic patients. *J. Neurosci.* 9: 828–39

Squire, L. R., Zola-Morgan, S. 1991. The medial temporal lobe memory system. *Science* 253: 1380–86

Squire, L. R., Zola-Morgan, S. 1988. Memory: Brain systems and behavior. *Trends Neurosci.* 11: 170–75

Squire, L. R., Zola-Morgan, S. 1983. The neurology of memory: The case for correspondence between the findings for human and non-human primate. In *The Physiological Basis of Memory*, ed. J. A. Deutsch, pp. 199–268. New York: Academic. 2nd ed.

Squire, L. R., Zola-Morgan, S., Chen, K. 1988. Human amnesia and animal models of amnesia: Performance of amnesic patients on tests designed for the monkey. *Behav. Neurosci.* 11: 210–21

Squire, L. R., Ojemann, J. G., Meizin, S. M., Petersen, S. E., Videen, T. O., et al. 1992. Activation of the hippocampus in normal humans: A functional anatomical study of human memory. *Proc. Natl. Acad. Sci. USA* 89: 1837–41

Suzuki, W., Zola-Morgan, S., Squire, L. R., Amaral, D. G. 1993. Lesions of the perirhinal and parahippocampal cortices in the monkey produce long lasting memory impairments in the usual and tactual modalities. *J. Neurosci.* In press

Thompson, R. F. 1986. The neurobiology of learning and memory. *Science* 233: 941–47

Tulving, E. 1985. How many memory systems are there? *Am. Psychol.* 40: 385–98

Victor, M., Adams, R. D., Collins, G. H. 1989. *The Wernicke-Korsakoff Syndrome and Related Neurological Disorders due to Alcoholism and Malnutrition.* Philadelphia: Davis. 2nd ed.

Victor, M., Agamanolis, D. 1990. Amnesia due to lesions confined to the hippocampus: A clinical-pathologic study. *J. Cogn. Neurosci.* 2: 246–57

von Cramon, D. Y., Hebel, N., Schuri, U. 1985. A contribution to the anatomical

basis of thalamic amnesia. *Brain* 108: 993–1008

Wang, J., Aigner, T., Mishkin, M. 1990. Effects of neostriatal lesions on visual habit formation of rhesus monkeys. *Soc. Neurosci. Abstr.* 16: 617

Winocur, G. 1990. Anterograde and retrograde amnesia in rats with dorsal hippocampal or dorsomedial thalamic lesions. *Behav. Brain Res.* 38: 145–54

Winocur, G., Oxbury, S., Roberts, R., Agnetti, V., Davis, D. 1984. Amnesia in a patient with bilateral lesions to the thalamus. *Neuropsychologia* 22: 123–43

Zola-Morgan, S., Squire, L. R. 1985a. Medial temporal lesions on monkeys impair memory in a variety of tasks sensitive to human amnesia. *Behav. Neurosci.* 99: 22–34

Zola-Morgan, S., Squire, L. R. 1985b. Amnesia in monkeys following lesions of the mediodorsal nucleus of the thalamus. *Ann. Neurol.* 17: 558–64

Zola-Morgan, S., Squire, L. R. 1990. The primate hippocampal formation: Evidence for a time-limited role in memory storage. *Science* 250: 288–90

Zola-Morgan, S., Squire, L. R., Amaral, D. G. 1989a. Lesions of the hippocampal formation but not lesions of the fornix or the mammillary nuclei produce long-lasting memory impairment in monkeys. *J. Neurosci.* 9: 897–912

Zola-Morgan, S., Squire, L. R., Amaral, D. G. 1989b. Lesions of the amygdala that spare adjacent cortical regions do not impair memory or exacerbate the impairment following lesions of the hippocampal formation. *J. Neurosci.* 9: 1922–36

Zola-Morgan, S., Squire, L. R., Amaral, D. G., Suzuki, W. 1989c. Lesions of perirhinal and parahippocampal cortex that spare the amygdala and the hippocampal formation produce severe memory impairment. *J. Neurosci.* 9: 4355–70

Zola-Morgan, S., Squire, L. R., Amaral, D. G. 1986. Human amnesia and the medial temporal region: Enduring memory impairment following a bilateral lesion limited to field CA1 of the hippocampus. *J. Neurosci.* 6: 2950–67

Zola-Morgan, S., Squire, L. R., Rempel, N., Clower, R. L. 1992. Enduring memory impairment in monkeys after ischemic damage to the hippocampus. *J. Neurosci.* 12: 2582–96

Zola-Morgan, S., Squire, L. R., Clower, R. P., Rempel, N. L. 1993. Damage to the perirhinal cortex but not the amygdala exacerbates memory impairment following lesions to the hippocampal formation. *J. Neurosci.* 13: 251–65

Annu. Rev. Neurosci. 1993. 16:565–95
Copyright © 1993 by Annual Reviews Inc. All rights reserved

INHIBITORS OF NEURITE GROWTH

M. E. Schwab, J. P. Kapfhammer, C. E. Bandtlow

Brain Research Institute, University of Zurich, CH-8029 Zurich, Switzerland

KEY WORDS: nervous system development, guidance and pathway formation, growth cones, regeneration

INTRODUCTION

An important step in the development of the nervous system is the directed outgrowth of neurites, which brings them into contact with their target cells in the peripheral and central nervous systems. Until recently, most of the studies on the mechanism and regulation of neurite growth have focused exclusively on the positive, neurite growth-promoting effects. Early in this century, careful descriptions of developmental and regenerative processes, along with early tissue culture experiments, suggested the existence of specific neurite growth-promoting factors (Harrison 1910; Ramon y Cajal 1928, 1929). The detection of nerve growth factor (NGF) (Cohen 1960; Levi-Montalcini & Hamburger 1951) led to the classical theory of neurotrophic factors, which is currently being complemented and transformed by the rapid increase in information on families of soluble factors that have neurite outgrowth promoting, survival, and differentiation effects (Thoenen 1991). As a result of the growing awareness of the importance of direct cell-cell and cell-matrix interactions in the early 1970s, membrane-bound and extracellular matrix (ECM) molecules were analyzed for their effects on neurite growth, cell adhesion, and axonal guidance (for reviews, see Edelman & Crossin 1991; Jessell 1988; Reichardt & Tomaselli 1991; Sanes 1989; Takeichi 1991). Again, the focus was on the positive, growth-promoting signals.

Evidence for the existence of negative, neurite growth inhibitory signals has been found only recently. The first evidence was provided by in vitro

565

0147–006X/93/0301–0565$02.00

observations that could not be explained by a lack of trophic factors or growth-promoting substrates. Subsequently, cell biological and biochemical studies led to the isolation of membrane-bound proteins that have inhibitory activity, from various tissue sources, including the developing and adult central nervous system (CNS). Certain neurotransmitters were also shown to lead to growth cone arrest. And, multifunctional molecules that can exert both positive and negative effects have been found, and a variety of observations in the developing central and peripheral nervous systems further suggest that there may be many more such molecules that can be appropriately called genuine inhibitors of neurite growth. Other molecules that primarily affect either the direction of extension or the site of termination of growth cones may be referred to as repulsors or instructive inhibitors of neurite growth.

Adhesive interactions of growth cones with their substrates are a necessary condition for growth. However, the notion that there is a direct relationship between the strength of adhesion exerted by a particular substrate and its growth-promoting property (Letourneau 1975) is too simplistic. Examples exist that the same molecules can be poor adhesive substrates, but nevertheless growth promoting (e.g. tenascin, laminin) (Gundersen 1987; Wehrle & Chiquet 1990). However, none of the known molecules present in embryonic somites or optic tectum or produced by oligodendrocytes are anti-adhesive in the sense that they would prevent growth cone adhesion to a particular substrate or cell. In fact, tight adhesions of growth cone filopodia occur concomitantly with growth arrest (Bandtlow et al 1990; Kapfhammer & Raper 1987a; Vos et al 1991).

In this review, we consider the different systems in which the presence of molecules with inhibitory activities has been demonstrated and the current state of their molecular characterization, as well as the signaling mechanisms through which their inhibitory actions may be mediated.

NEGATIVE SIGNALS FOR GROWTH CONES ASSOCIATED WITH SPECIFIC PROTEINS

Inhibitory Glycoproteins in Chick Embryo Somites

In early embryonic development, neural crest cells migrate out from the dorsal neural tube to form the ganglia of the peripheral nervous system (PNS) (Davies & Lumsden 1990). Somewhat later, motor axons grow out from the spinal cord. Interestingly, crest cells and motor fibers are found exclusively in the anterior part of the forming sclerotomes (Keynes & Stern 1984). Neither N-CAM, N-cadherin, fibronectin, nor laminin are differentially distributed in anterior and posterior somites; however, tenascin has been found to be associated with the immigrating neural crest cell themselves (Stern et al 1989). Interestingly, lectin peanut agglutinin (PNA) staining is sharply localized to the posterior half of each somite (Davies et al 1990; Stern et al 1986). In coculture experiments, cells from posterior positions of somites appear to arrest neurite growth (Stern et al 1986). By using lectin-affinity chromatography, two molecules of molecular weight 48 kD and 55 kD have been identified from posterior somatic tissue (Davies et al 1990). These proteins are membrane bound and have to be solubilized with detergent. In two different bioassays on dorsal root ganglion (DRG) neurites, PNA-affinity column purified material from somites was used, with the following results: If DRGs were plated onto this purified, substrate-bound material, neurite outgrowth was extremely limited, significantly less than on bovine serum albumin, which, by itself, is a rather poor substrate. And, if DRG neurites growing on laminin were confronted with the PNA-binding somite proteins incorporated into liposomes, growth cone collapse occurred. A rabbit antiserum raised against the 48 and 55 kDa components eliminated the growth cone collapsing activity (Davies et al 1990). When chick embryo sections were stained with this antiserum, a selective staining of the posterior half of the sclerotomes was observed, closely comparable to the staining with PNA.

No amino acid sequence data are available yet on these proteins, but interestingly, proteins that cross react with the antibodies are found in gray, but not white, matter of adult chicken brains (Keynes et al 1990). The final identification of this material awaits full biochemical purification and cloning. In summary, these studies strongly suggest that a negative signal, probably associated with 48 and 55 kD PNA-binding membrane glycoproteins, is present in posterior halves of sclerotomes. This signal is present before any contact with neural crest cells or axons and appears to influence cell migration and axon growth in the forming PNS crucially.

A Molecule with an Inhibitory Activity on Neurite Growth Involved in Map Formation in the Retinotectal System

Retinal axons project in a topographic manner onto the optic tectum: The temporal retinal fibers terminate in the anterior part, and the nasal fibers in the posterior part of the tectum. This model system for neuronal pathfinding has been studied extensively in vitro and in vivo (Stirling 1991; Udin & Fawcett 1988). In tissue culture experiments, Bonhoeffer & Huf (1980) demonstrated that embryonic chick retinal fibers extend preferentially on tectal, as compared with retinal, cell monolayers. In similar experiments, retinal axons responded differentially to tectal cells from different positions along the anterior-posterior axis of the tectum. Temporal axons preferentially extend on cells from more anterior tectal regions, which is in agreement with the normal distribution of these fibers in vivo

(Bonhoeffer & Huf 1982). This finding is consistent with the assumption of a gradient of a molecule present on the tectum and recognized by temporal retinal axons. Interestingly, nasal axons show no preference for monolayers from any part of the tectum.

These findings have been confirmed in a new assay that allowed the use of membrane preparations, instead of cell monolayers. In this new assay, membranes are adsorbed to a filter and can be arranged as narrow alternating stripes. In this way, any preference of retinal axons toward membranes derived from different parts of the tectum can be readily detected (Walter et al 1987b). Using this assay, temporal retinal axons again were shown to have a strong preference for anterior, rather than posterior, tectal membranes. This preference is lost after heat treatment of the posterior tectal membranes but interestingly, not after treatment of the anterior tectal membranes (Walter et al 1987a). This suggests that the observed preference is not the result of an attraction of the temporal fibers to the anterior tectal membranes, but rather their avoidance of posterior membranes presumably due to the presence of a heat-sensitive inhibitory activity. Similar inhibitory activities could be detected in tectal membrane preparations from fish and mouse (Godement & Bonhoeffer 1989; Vielmetter & Steurmer 1989).

In the growth cone collapse assay developed by Raper & Kapfhammer (1990, see below), a reconstituted extract of posterior tectal membranes induces the collapse of temporal, but not nasal, retinal growth cones (Cox et al 1990). Biochemical studies show that an activity that causes growth cone collapse and avoidance of posterior membranes in the stripe assay can be removed by phosphatidylinositol-specific phospholipase C (PI-PLC), which indicates that the relevant molecule is attached to the membrane with a glycolipid anchor. Specific binding of the active molecule to PNA lectin and enhanced incorporation into lipid vesicles allowed the assignment of the activity to a 33 kD glycoprotein (Stahl et al 1990). This protein is more abundant on posterior than on anterior tectal membranes and disappears on or about embryonic day 15 (E15), when the inhibitory activity disappears from the posterior tectal membranes. Interestingly, the establishment of retinal terminal fields in the tectum is completed at that time. After incorporation into lipid vesicles, the purified molecule is active in the stripe assay and induces growth cone collapse (Stahl et al 1990). It is not yet clear how this molecule functions in growth cone guidance when it is present as a gradient. Nevertheless, we do know that growth cones can respond to both step and smooth gradients of the molecule (Baier & Bonhoeffer 1992; Walter et al 1990a,b). Recently, a monoclonal antibody against the 33 kD glycoprotein has been found to stain an increasing anterior to posterior gradient on the tectum (B. Müller et al, in preparation). Molecular cloning of the 33 kD glycoprotein is on the way and will certainly help clarify the role of this molecule in the formation of the retinotectal map in vivo.

Analysis of the behavior of frog retinal growth cones that contact dissociated tectal cells in vitro indicates the presence of a similar neurite growth inhibitory activity on the surface of radial glial cells (Johnston & Gooday 1991). But whether this activity is related to the 33 kD glycoprotein from chick tectum is currently not known.

Molecules Associated with a Neurite Growth Inhibitory Activity on Axonal Surfaces

The surfaces of neighboring axons are generally a good substrate for neurite growth both in vivo and in vitro (Wessels et al 1980). In tissue culture experiments Bray et al (1980) showed that rat sympathetic fibers and rat retinal fibers mix freely when tissue explants of the same type are placed close to each other in a tissue culture dish. However, when a retinal explant is placed close to a sympathetic explant, the outgrowing fibers do not mix and, after some time in culture, form separate territories. Time-lapse video microscopy indicates that this selective fasciculation is the result of a mutual inhibition between retinal and sympathetic fibers (Kapfhammer et al 1986).

Contact of a growth cone with an inhibitory neurite resulted in the shrinkage and, finally, complete collapse of this growth cone within a few minutes (Kapfhammer & Raper 1987a). Throughout the complete sequence of contact and collapse, strong local adhesions between growth cone and neurite were present, which makes it rather unlikely that the observed collapse was simply the result of a reduced or absent adhesion between growth cone and neurite. The observed behavior is much more suggestive of a specific signaling mechanism that induces the collapse of the growth cone. Growth cone collapse is usually reversible, but renewed contact to the inhibitory neurite again leads to collapse. The morphological behavior of collapsing growth cones is similar to that of fibroblasts undergoing contact inhibition of locomotion (Abercrombie 1970; Heaysman 1978).

Later studies clearly showed that the observed inhibition between heteronymous sets of growth cones and neurites is a quite common phenomenon in the vertebrate nervous system. It appears to be present between most combinations of PNS and CNS neurons, but, importantly, is absent between all homonymous pairs studied (Kapfhammer & Raper 1987b). Growth cone collapse specifically induced by heteronymous neuritic surfaces has since been found in several other systems. Examples in the CNS include the avoidance of nasal retinal axons by temporal retinal growth cones (Bonhoeffer & Huf 1986; Raper & Grunewald 1990), the avoidance of tectobulbar neurites by temporal retinal growth cones (Kröger et al 1991), and, in the PNS, the avoidance of DRG axons by preganglionic sympathetic growth cones (Moorman & Hume 1990).

A variety of inhibitory neuronal cell surface molecules can be expected to account for the inhibitory activities associated with different nerve fibers that act only on particular growth cones. As a first step to the purification of these molecules, Raper & Kapfhammer (1990) developed an assay that allows the rapid testing of solubilized biochemical fractions for inhibitory activity. This assay is based on the collapse of the growth cone after its exposure to an inhibitory cell surface molecule. This collapse reaction is very similar to growth cone behavior after contact with an inhibitory neurite. This assay has been used for the purification of the neurite growth inhibitory molecules from posterior somites (Davies et al 1990, see above) and from posterior tectum (Cox et al 1990, see above). Based on the observation that DRG growth cones are inhibited by CNS neurites (Kapfhammer & Raper 1987b), an attempt has been made to purify the inhibitory activity for DRG growth cones from E10 chick brain. The activity can be solubilized by the detergent CHAPS and has been greatly enriched through several steps of column chromatography (Raible & Raper 1990; Raper et al 1992; Raper & Kapfhammer 1990). The purified activity is inhibitory to growth cones of sensory and retinal ganglion cells. It is not quite clear yet how it relates to the inhibitory molecules postulated on the neuronal cell surface and the molecules identified by other groups. These questions will probably only be resolved after complete purification and molecular cloning of the different molecules. A second inhibitory activity from PC12 cells and adrenal membranes that acts rather specifically on sympathetic growth cones has also been identified (Baird & Raper 1991; Raper & Kapfhammer 1990).

Oligodendrocyte- and CNS Myelin-Associated Inhibitors of Neurite Growth

The concept of inhibitors of nerve fiber growth present in the adult CNS of higher vertebrates arose in the context of CNS regeneration. Transplantation experiments have shown that CNS neurons, can regenerate their lesioned neurites in a peripheral nerve environment, but cannot do so within the CNS (Aguayo et al 1990; David & Aguayo 1981; Ramon y Cajal 1928; Tello 1911). In an experiment designed to test the classical "lack of trophic factor hypothesis" (Ramon y Cajal 1928), dissociated perinatal rat sensory, sympathetic, or retinal neurons were cocultured with explants of adult rat sciatic or optic nerves in the presence of NGF or brain-derived neurotrophic factor (BDNF) (Caroni et al 1988; Schwab &

Thoenen 1985). The resulting pattern of fiber growth corresponded exactly to that found in vivo: many axons grew into the sciatic nerves, but none grew into the optic nerves. On the basis of these results, we postulated that there is inhibitory activity present in adult CNS tissue that cannot be overcome by the stimulatory effects of neurotrophic factors (Schwab & Thoenen 1985). Very similar results were obtained by using frozen sections as substrates for cultured neurons or neuroblastoma cells (Carbonetto 1987; Savio & Schwab 1989), and it was shown that this activity is highly enriched in white matter (Crutcher 1989; Khan et al 1990; Sagot et al 1991; Savio & Schwab 1989; Watanabe & Murakami 1989). In co-cultures of dissociated glial cells and neurons, a clear-cut difference emerged between glial cell classes as to their effects on growing neurites: Astrocytes and oligodendrocyte precursor cells were permissive and favorable for neurite growth; mature oligodendrocytes, in contrast, were strictly avoided (Schwab & Caroni 1988). Identical results were obtained by Fawcett et al (1989b) and had, in fact, been noted in an earlier study by Hatten et al (1984). This inhibitory property of oligodendrocytes persisted in the presence of neurotrophic factors and has been observed for a variety of cultured neurons, as well as neuroblastoma and PC12 cells. Spreading and migration of 3T3 fibroblasts is also inhibited (Schwab & Caroni 1988; B. Rubin and M. E. Schwab, unpublished observations).

In a video time-lapse study, DRG growth cones growing on a laminin substrate in the presence of NGF were compared with respect to their interactions with individual astrocytes and oligodendrocytes (Bandtlow et al 1990). Contact with oligodendrocytes always led to rapid arrest of neurite growth, whereas contact with astrocytes resulted in an unchanged or slightly reduced growth velocity. Interestingly, the contact of the tips of the growth cone filopodia with the processes of oligodendrocytes was sufficient to arrest these growth cones, which suggests that second messengers may be involved. Growth arrest was long lasting (hours), but was strictly local, i.e. other neurites arising from the same cells continued to grow normally.

These in vitro experiments also showed that growth inhibition was strictly dependent on physical contact with the oligodendrocytes and that no activity could be recovered from oligodendrocyte culture conditioned medium. When membranes from oligodendrocyte enriched cultures and CNS myelin membranes were used as a substrate for neurons, a strong inhibitory effect was observed (Caroni & Schwab 1988a; Vanselow et al 1990). Separation of spinal cord myelin proteins by SDS-PAGE and subsequent elution of the proteins and their reconstitution into liposomes indicated the presence of two active fractions of 35 and 250 kD molecular weight, respectively (Caroni & Schwab 1988a). These neurite growth inhibitors, called NI-35 and NI-250, have been found in all CNS regions containing white matter, including the optic nerve, but are absent from sciatic nerves. Immunological and recent biochemical data strongly suggest that NI-35 and NI-250 are closely related; NI-250 seems to be a complex containing NI-35 (Bandtlow & Schwab 1991; C. E. Bandtlow et al, in preparation). The partial amino acid sequences available at present from fractions purified by SDS-PAGE and HPLC show no homologies to known proteins. Interestingly, neither activity, nor the corresponding protein bands, can be found in goldfish or trout CNS, a result that accords with the known growth-promoting surface properties of oligodendrocyte-like cells isolated from goldfish optic nerves (Bastmeyer et al 1991; Caroni & Schwab 1988a) and suggests that fish do not possess the relevant inhibitory molecules. This property could explain the capacity for fiber regeneration in the CNS of lower vertebrates.

In the presence of neutralizing antibodies against NI-35 and NI-250, DRG neurites can grow over oligodendrocytes in culture and into optic nerve explants (Bandtlow et al 1990; Caroni & Schwab 1988b). For these in vivo experiments, two independent paradigms were chosen: application of the inhibitor-neutralizing antibody IN-1 (Schnell & Schwab 1990), and elimination of oligodendrocyte precursor cells by x-irradiation of newborn rat spinal cord or optic nerve (Savio & Schwab 1990). In young rats with bilateral thoracic spinal cord transections that completely interrupted the corticospinal tract (CST), the capacity of the CST to regenerate was studied in control animals and in animals treated with the inhibitor neutralizing antibody IN-1 or in oligodendrocyte- and myelin-free rats. In the control animals, CST fibers sprouted for a distance of about 1 mm at the lesion site, but there was no elongation beyond this distance. In contrast, some CST fibers elongated up to 10–20 mm in the experimentally treated animals, and in the majority of cases extended for at least 4–7 mm (Savio & Schwab 1990; Schnell & Schwab 1990; L. Schnell and M. E. Schwab, in preparation). Arborizations of the CST axons in gray matter were observed, but no information is available yet on the detailed anatomy or on the functional capacity of these regenerated connections. Because the number of regenerating fibers in all these experiments is small, and the spontaneous regenerative sprouting of the CST is limited, we are currently trying to enhance CST regeneration by concomitant trophic factor support. Present results indicated that E14 rat spinal cord implants secrete trophic factors that enhance the sprouting of lesioned CST fibers in both newborn (Bregman et al 1989) and adult rats (Schwab et al 1990). A similar enhancement of regeneration by IN-1 antibodies has been obtained for the septo-hippocampal tract (Cadelli & Schwab 1991) and for the adult rat optic nerve (Schwab et al 1990).

Several studies have shown that CST fibers can grow down the spinal cord if lesions are made in the early postnatal period in rats or hamsters or opossum (Bernstein & Stelzner 1983; Bregman et al 1989; Kalil & Reh 1982; Treherne et al 1992). A failure of regrowth occurs at age 4–6 days in rats and hamsters, at the time when axons from DRG neurons cultured on spinal cord frozen sections fail to grow (Sagot et al 1991), and when oligodendrocytes and myelin appear in large parts of the white matter (Schwab & Schnell 1989). Hasan et al (1991) and Shimizu et al (1990) have recently shown a close correlation between failure of regeneration of spinal cord descending tracts and the appearance of myelin in the chick embryo: Successful anatomical and functional regeneration of descending spinal tracts occurs up to E12, but not after E13, which is exactly when myelin formation starts in the spinal cord. By using an antibody against GalC and complement, Keirstead et al (in preparation) recently delayed oligodendrocyte development in chick embryos up to E17. Interestingly, lesions made in these embryos at E15, when regeneration is normally absent, are followed by successful regeneration of the brainstem descending tracts, as assessed anatomically and physiologically.

That the myelin-associated neurite growth inhibitors may play a role in axon guidance has been shown for the developing rat CST. In animals treated with antibody IN-1 or by x-irradiation on the day of birth, CST fibers grow aberrantly into neighboring areas in the rostral half of the spinal cord (Schwab & Schnell 1991), and the extension of side branches into the gray matter occurred earlier and was denser than normal. The absence of oligodendrocytes has to be complete in such experiments, however, because these cells can express NI-35/250 well before they form compact myelin (Caroni & Schwab 1989). Thus, x-irradiation at postnatal day 3 (Pippenger et al 1990) or use of hypomyelinating mice that possess oligodendrocytes at early postnatal stages (Stanfield 1991) are unsuitable models.

After early postnatal ablation of the superior colliculus on one side, together with enucleation of the ipsilateral eye in hamsters, retinal fibers can cross the tectal midline and innervate the remaining superior colliculus (Schneider 1973). Interestingly, they do this in an anatomically aberrant fashion: Instead of growing into the stratum opticum and turning dorsally to end in the superficial gray layer, they extend directly into and terminate within the superficial gray layer. Growth of these fibers occurs when oligodendrocytes start to differentiate in the stratum opticum (Jhaveri et al 1992). IN-1 antibodies were, therefore, applied during this period and resulted in an extended ingrowth of fibers into the stratum opticum, in a manner comparable to that seen in normal animals (Kapfhammer et al 1992). In this system, therefore, the oligodendrocyte-associated components NI-35/250 restrict the access of retinal fibers to a particular layer in the optic tectum shortly after normal innervation is established, which suggests that the presence of NI-35/250 may affect fiber growth in gray matter, as well as in CNS tracts. In the rat optic nerve, inhibitory activity appears at birth coincidently with the first GalC-positive oligodendrocytes, but five to seven days before the first myelin is formed (Caroni & Schwab 1989). Elimination of oligodendrocytes by x-irradiation at birth has allowed us to elicit substantial sprouting in 2-week-old postnatal optic nerves by a single intravitreal injection of fibroblast growth factor (bFGF) at day 10; bFGF had no effect on axon numbers in the presence of oligodendrocytes (Colello et al 1991).

Taken together, these studies suggest that NI-35/250 can exert boundary functions for late growing CNS tracts and, in target areas, can restrict access of fibers to particular regions and layers. Their presence in the adult CNS suggests a second, possibly more important function: the stabilization of the CNS against sprouting in unwanted regions, particularly in the white matter. Whether the restriction of sprouting in gray matter to very short distances is also influenced by these molecules remains to be investigated.

Comparison of Neurite Growth Inhibitory Proteins

The inhibitory proteins described above share several common features (Table 1). The question thus arises whether some of these molecules have similar or even identical structures. No definitive answer to this question can be given, because none of the proteins has been cloned and little sequence information is currently available. The obvious differences in their expression (time, cell type) and in their effects, however, strongly argue against the possibility that only one single inhibitory protein exists. Sequence comparisons may result in the delineation of specific inhibitory domains and allow us to define and extend families of inhibitory membrane proteins.

NEUROTRANSMITTERS AND ELECTRICAL ACTIVITY CAN ARREST GROWTH CONES

Besides membrane-bound molecules, other factors arrest active growth cones by inhibiting their motility and elongation. Recent in vitro studies on a variety of snail and mammalian cell types provide striking examples of growth cone retraction upon exposure to specific soluble molecules. Among the most interesting extrinsic cues that may function in growth cone guidance are neurotransmitters. In vitro, these molecules can indeed act as highly specific regulators of growth cone behavior (see Lipton & Kater 1989). For example, serotonin and dopamine exert a highly selective

Table 1 Comparison of neurite growth inhibitory proteins

	Somite derived	Tectum derived	Brain derived	CNS myelin associated
Activity membrane-bound and extractable by detergents	yes	yes	yes	yes
Active in collapse assay	yes	yes	yes	yes
Heat and trypsin sensitive	yes	yes	yes	yes
Supports neurite out-growth when offered as only substrate	very poorly	yes	n.a.	no
Proposed molecular weight	48/55 kD	33 kD	n.a.	35/250 kD
Binds PNA	yes	yes	no	no
Developmental regulation	present at E3 (chick)	present at E5, disappears at E15 (chick)	present at E10 (chick)	appears post-natally (rat, bovine)
Cellular source	sclerotome cells	n.a. (tectal radial glia?)	n.a.	oligodendrocytes
Sensitive cell types	sensory g.c.'s[a]; others n.a.[b]	temporal retinal g.c.'s; not nasal g.c.'s	sensory and retinal g.c.'s; others n.a.	many neuronal and nonneuronal cells

[a] Growth cone.
[b] Not available.

and cell-specific growth inhibition on specifically identified neurons of the mollusc *Helisoma* (Haydon et al 1984, 1987; McCobb et al 1988b). Similar responses have been described for the growth cones of dendrites of mammalian hippocampal pyramidal cells upon exposure to glutamate, whereas the axonal growth cones of these neurons are relatively unaffected (Mattson et al 1988). The affected growth cones respond by an arrest of filopodial activity, followed by a transient collapse of growth cone structure. Other observations have shown that stimulation of embryonic D_1 dopamine receptors mediates the cessation of filopodial motility and prevents neurite extension in a subset of chick retinal neurons (Lankford et al 1988). Although neurite retraction in young cultures was described, no long-lasting morphological changes of growth cone structures were observed. In cultures of dissociated rat retinal cells, the presence of ACh can be found in the culture medium under certain culture conditions, apparently released from endogenous sources, most likely by amacrine cells (Lipton et al 1988). The pharmacological blockade of the ACh effect by nicotinic antagonists induced the sprouting and re-elongation of the retinal ganglion cells, which suggests that tonic levels of ACh present in the culture medium may exert growth inhibitory effects (Lipton et al 1988).

The simultaneous application of hyperpolarizing neurotransmitters (ACh for the snail, GABA in the case of hippocampal pyramidal cells) can antagonize the growth inhibitory effects exerted by excitatory neuro-transmitters (McCobb et al 1988a; Mattson & Kater 1989). Electrical activity can also act as a stop signal for growth cones. The generation of action potentials abruptly and reversibly halts neurite elongation of *Helisoma* neurons (Cohan et al 1987). Interestingly, different types of neurons require different activity patterns for inhibition (Cohan 1990). Thus, patterned stimulation of mouse sensory neurons causes the immedi-ate withdrawal of filopodia and lamellipodia, in some cases accompanied by a complete retraction of the neurites (Fields et al 1990b). However, neurons can accommodate to chronic stimulation and become insensitive to the stimulus. Although the cellular mechanisms responsible for these responses are largely unknown, the involvement of an as yet unidentified calcium homeostatic buffering system seems plausible (Fields et al 1990a, see below).

EVIDENCE FOR OTHER INHIBITORY SIGNALS AND ACTIVITIES

In addition to the examples described above, in which the inhibition of neurite growth could be assigned to defined molecules, there are more examples in which the presence of signals inhibitory to neurite growth is strongly suggested by experimental evidence, but the molecules involved are not clearly identified yet.

Multifunctional Properties of Tenascin and Related Molecules

Tenascin is a multimeric ECM protein occurring in many organs, including the PNS and CNS. The subunits contain several fibronectin type III, as well as epidermal growth factor (EGF) repeats (Pearson 1988; Reichardt & Tomaselli 1991). Purified tenascin as a substrate for neurons or non-neuronal cells possesses poorly adhesive characteristics (Chiquet-Ehris-mann et al 1988; Halfter et al 1989; Lochter et al 1991). Experiments using recombinant truncated proteins, proteolytic fragments, or domain-specific antibodies have shown a cell-binding domain close to the C-terminal end and the fibronectin repeats, whereas the anti-adhesive effects seem to reside in the EGF-repeat part of the molecule (Friedlander et al 1988; Spring et al 1989). Within the fibronectin repeats, domains with differential effects on neuronal migration and outgrowth seem to exist (Husmann et al 1992). Tenascin can also mediate adhesion of neurons to astrocytes (Grumet et al 1985; Kruse et al 1985). Tenascin as a culture substrate for neurons shows a dual effect: Adhesion of cell bodies and cell spreading is relatively poor; on the other hand, neurite outgrowth is enhanced (Chiquet 1989; Lochter et al 1991). Surprisingly, the addition of soluble tenascin inhibited neurite growth on several types of substrates (Lochter et al 1991).

Tenascin is strongly expressed in the developing and regenerating nervous system (Chiquet-Ehrismann et al 1986; Daniloff et al 1989; Martini & Schachner 1991; Tucker & McKay 1991; Wehrle & Haller 1991). In the periphery, its localization suggests a function that facilitates cell migration and neurite growth. In the developing CNS, tenascin appears early and transiently and seems to be associated mainly with astrocytes (Crossin et al 1989; Grierson et al 1990; Hoffman et al 1988; Steindler et al 1989a). In the barrel field of the somatosensory cortex the presence of tenascin in the forming barrel walls suggested a boundary function associated with astrocytes (Crossin et al 1989; Steindler 1993; Steindler et al 1989b).

The oligodendrocyte-associated, tenascin-related ECM molecules J1 160/180 (Janusin) show similar dual functions as culture substrates for neuronal adhesion and outgrowth, which, in addition, depend on divalent cations and the state of aggregation (Pesheva 1989, 1991). J1 160/180 repulses growth cones at substrate boundaries without, however, leading to growth cone collapse (Schachner 1992).

Glia of the Midline and the Optic Chiasm

The dorsal midline of the spinal cord separates the dorsal columns and is not crossed by any fibers. Snow et al (1990b) have suggested that the glial structure of the roof plate may play an active role in inhibiting the fibers from crossing to the contralateral side. This inhibitory role has been correlated with the immunohistochemical presence of keratan/chondroitin sulfate proteoglycans. Recently a 320 kD proteoglycan has been purified that inhibits neurite outgrowth in vitro and localizes to the midline of the chick spinal cord and hindbrain (Cole & McCabe 1991). There is, however, no direct demonstration of an inhibitory action of the spinal cord roof plate.

The barrier function of midline glial has been more directly demonstrated in the hamster optic tectum, where the optic fibers are confined to one side of the tectum and do not cross to the other side. Here, a population of midline glial cells has been identified that are likely to act as a barrier for the optic axons (D.-Y. Wu 1991; Wu et al 1988). If these midline glial cells are disrupted, either by a heat lesion of the tectum (Wu et al 1989) or by an undercut of the tectal midline (thus leaving the tectal surface and the tectal neuropil intact), the optic fibers cross to the other side of the tectum (D.-Y. Wu 1991; Wu et al 1990). The tectal midline glia also expresses a keratan sulfate epitope (Snow et al 1990b).

A specific inhibitory role of midline glial cells has also been proposed for the confinement of a subpopulation of optic axons to the ipsilateral side of the optic chiasm (Godement et al 1990; Mason & Godement 1992). Because the majority of the optic fibers cross through this glial structure, the inhibitory activity would specifically act on the ipsilateral fibers. Silver et al (1987) have suggested a further glial barrier function for a group of glial cells that separate the optic from the olfactory projections. The molecular basis of this phenomenon is unclear. Kuwada et al (1990) have suggested that in the zebrafish, some, but not all, of the early growth cones in the developing spinal cord are prevented from crossing the midline by an inhibitory activity associated with the floor plate cells.

Peripheral Pathway of Motor Axons

Another system in which inhibitory signals are likely to contribute to axonal pathfinding is the outgrowth of motoneurons in the chick embryo. In a series of experiments, Tosney (1992) has shown that the major pathway of the extending motor axons is bordered by regions that do not support axonal growth. These regions include the posterior somite (Keynes & Stern 1984, 1988, see above), the pelvic girdle (Tosney & Landmesser 1984, 1985), and the perinotochordal mesenchyme (Tosney & Oakley 1990). By surgical manipulations, it was demonstrated that each region actively inhibits fiber ingrowth from motor neurons. Interestingly all these regions show PNA binding and express keratan/chondroitin sulfate epitopes (Oakley & Tosney 1991). Whether the PNA binding inhibitory molecules from posterior somites identified by Keynes and coworkers (see above) are also involved in the inhibitory activity of the other two regions remains to be determined.

Epidermis and Merkel Cells

An inhibitory activity exists in chick epidermis, which is virtually free of sensory nerve fibers (Saxod 1978). Using time-lapse video microscopy, Verna (1985) demonstrated that sensory growth cones avoid epidermal cells and that this avoidance is present at a distance of up to 100 μm from the epidermal explant, which suggests the presence of a secreted inhibitory factor. In a subsequent study, this inhibitory activity was abolished by culture of the epidermal explant in tunicamycin, which suggests the involvement of sugar moieties (Fichard et al 1990). The proteoglycan specific inhibitor β-D-xyloside and antibodies to chondroitin sulfate were also effective in reducing the inhibitory activity from chick epidermis (Fichard et al 1991). These results indicate that a chondroitin sulfate proteoglycan is likely involved in this inhibitory activity.

Merkel cells in the skin are selectively innervated by sensory, but not sympathetic, nerve fibers (Munger 1977). In a recent study, Vos et al (1991) showed that Merkel cells are a source of NGF and that coculture with Merkel cells promotes the survival of sympathetic neurons. However, the reaction of sensory and sympathetic growth cones to direct contact with Merkel cells is quite different. Sensory growth cones make contact and branch onto the surface of the Merkel cells, whereas the sympathetic growth cones collapse and retract after contact to Merkel cell surfaces. This suggests that Merkel cells express on their surface a molecule that is inhibitory for sympathetic, but not sensory, growth cones. This is consistent with the observed innervation pattern in vivo.

Glial Scars at Lesion Sites

Numerous clinical and experimental observations have shown that traumatic CNS lesions result in the formation of scars to which reactive astrocytes make a major contribution (Reier et al 1983; Windle et al 1952). Observations of regenerating fibers at lesion sites and in areas of tissue transplants show that dense astrocytic scars represent a barrier to growing neurites (Jakeman & Reier 1991; Krueger et al 1986; Reier et al 1983). The reasons for this barrier function are not entirely clear. Electron microscopic observations suggest that astrocyte processes are tightly interlinked by junctions and could form a mechanical obstacle. However, frog retinal axons and embryonic DRG neurites regenerate through scars of high density (Fawcett et al 1989a; Reier et al 1983). Alternatively, specific molecules with inhibitory functions could be expressed in these regions. Astrocytes change their growth-promoting properties after prolonged periods in culture and become less supportive for neurite growth (Grierson et al 1990; Smith et al 1990). If nitrocellulose filters are implanted into brains, they are populated by dense astrocyte layers. Again, age plays an important role: Implants from perinatal rats are excellent substrates for in vitro neurite growth, in contrast to implants from older rats, which are much less growth permissive (limited outgrowth does occur, however, which shows that these cells are not totally inhibitory) (Rudge & Silver 1990). Molecular evidence showing a direct, causal relationship to defined constituents expressed by these astrocytes is not available yet; Grierson et al (1990) and McKeon et al (1991) correlate the poor substrate property of aged astrocytes with the appearance of tenascin and specific proteoglycans, but no data on the effect of antibodies against these molecules on neurite growth under these conditions are available.

Proteoglycans

The role of proteoglycans for neurite growth is still unclear. Various proteoglycans are synthesized in the nervous system and by neurons. They differ in their carbohydrate parts and in their protein backbones (Cole & McCabe 1991; Dow et al 1988; Herndon & Lander 1990; Oohira et al 1988). Growth-promoting and inhibitory effects have been reported and seen to be associated with the carbohydrate parts or the core proteins (Dow et al 1988; Iijima et al 1991; Oohira et al 1991; Riopelle & Dow 1990). As described above, keratan and chondroitin sulfate proteoglycans are present in a variety of structures in the CNS and PNS, where neurite growth inhibition seems to occur. A chondroitin-sulfate proteoglycan is colocalized and specifically binds to tenascin (Crossin et al 1989; Hoffman et al 1988; Tan et al 1987). In none of these systems has a detailed biochemical and cell biological analysis of the specific proteoglycans been conducted. For a cartilage-derived keratan, sulfate/chondroitin sulfate proteoglycan coating of culture dishes inhibits outgrowth of DRG and retinal fibers, but only if the proteoglycans are present in relatively high amounts (Snow et al 1990a, 1991). Interestingly, growth cone collapse did not occur. The effect was dependent on the presence of the glycosaminoglycan side chains. Whether these effects are due to a high local concentration of negative charges, for example, or whether they result from a specific, receptor-mediated interaction with these molecules remains to be shown.

CELLULAR MECHANISMS OF INHIBITORY INTERACTIONS

Structural Barriers

Such structures as bone or cartilage in the embryonic chick pelvic girdle, or areas of dense astrocytic processes tightly linked by junctions, e.g. CNS scar areas, are either simply avoided by, or are actively inhibitory for, extending neurites. Whether physical obstruction, in addition to possible specific inhibitory properties of cells forming such regions, also plays a role is still unclear.

Lack of Growth Promoting Signals or Their Receptors Is Not Evidence for an Inhibitor

When given the choice of two different substrates on which to grow, a growth cone will often favor one rather than the other. In some cases, this effect is correlated with adhesiveness, but there are also counter examples: Tenascin promotes neurite outgrowth under certain conditions, but does not support neuronal adhesion (Lochter et al 1991; Wehrle & Chiquet 1990). And, laminin is a preferred substrate for chick sensory neurites, but simultaneously decreases the strength of adhesion between growth cones and substrate (Gundersen 1987). Glass and bovine serum albumin are particularly poor substrates.

Great care should be taken not to interpret such selective growth cone reactions as evidence for inhibitory factors. Genuine inhibitory molecules should be able to counteract positive substrate effects in combination experiments or lead to growth arrest when applied in purified form onto growth cones extending on favorable substrates.

Ligand-Receptor Interactions and Second Messenger Systems

At the moment, no complete amino acid sequences or cDNA information are available for any of the known neurite outgrowth inhibitors. It is therefore impossible to predict whether these molecules are related to each other, thus forming a novel class of negative guidance molecules, or to predict which kind of receptor interactions they may use. However, the observed arrest of growth cones in all cases requires physical contact, which makes a ligand receptor interaction likely. Because NI-35/250 is not found on neurons, a homophilic interaction can be excluded, at least for this molecule. Except for the effects of neurotransmitters, where calcium changes seem to play an important role, the possible second messengers involved for neurite promoting or inhibitory molecules are poorly defined. The paucity of information available limits our discussion to a brief outline of intracellular mechanisms that have been implicated in the control of neurite growth.

CALCIUM   Manipulations affecting levels of intracellular calcium are associated with changes in growth cone motility and behavior (Kater et al 1988). Studies on the cellular mechanisms by which excitatory neurotransmitters mediate their growth inhibitory effects have shown that depolarization and the subsequent opening of voltage-gated calcium channels results in an influx of calcium ions. Direct measurements of intracellular calcium levels within growth cones of responsive *Helisoma* neurons has confirmed that 5-HT exerts a large rise in calcium, which can be blocked by ACh (McCobb & Kater 1988). Similarly, GABA, together with its potentiator diazepam, blocks the calcium increase normally seen upon exposure of hippocampal dendritic growth cones to glutamate (Mattson & Kater 1989). Interestingly, bFGF pretreatment prevents the rise in calcium induced by glutamate (Mattson et al 1989), possibly by the suppression of the expression of a 71 kDa glutamate receptor protein (Michaelis et al 1991). These findings have supported the concept that rises in intracellular calcium levels are inhibitory to growth cone motility (Kater et al 1988). More recently, Kater & Mills (1991) proposed a refined model of calcium as a regulator of growth cone behavior, which implies that there are graded effects of calcium concentrations on growth cone behavior. This model is also supported by a recent study that characterizes the calcium sensitivity of growth cone behavior by titrating intracellular calcium concentrations within the growth cone: When calcium levels fall below or rise above a certain permissive range, motility and outgrowth halts (Lankford & Letourneau 1992). This permissive range for optimal neurite growth may vary between growth cones, depending on the composition of the calcium optima required by different processes that contribute to growth cone behavior. For example, moderate increases of intracellular calcium cause a loss of lamellipodial actin, whereas filopodial actin is spared (Lankford & Letourneau 1992). In addition, calcium-sensitive and -insensitive forms of α-actinin and calspectin are found at different locations of growth cones (Sobue & Kanda 1989). Finally, a clustering of calcium channels in the growth cone may provide a mechanism to produce very local changes in intracellular calcium (Silver et al 1990). These "calcium hotspots" may act on molecules further downstream of the signal transduction cascade, such as calpains or calcium-dependent proteins (Beckerle et al 1987; Heizmann & Hunziker 1991).

In contrast to the well-documented role of calcium for neurotransmitter-mediated effects on growth cones, the evidence that calcium is a signal transducer for inhibitory proteins is unclear. Direct measurements of intracellular calcium in rat DRGs upon exposure to liposome-reconstituted NI-35 showed a fast, transient rise in growth cones and neurites preceding growth cone collapse (Schmidt et al 1991; Bandtlow et al 1992). Although this does not prove that calcium is the transducing agent, the fact that calcium channel blockers prevent the collapse response in a subset of growth cones suggests that calcium may act as an important regulator. A similar rise in calcium was reported when sympathetic preganglionic growth cones contacted DRG neurites (Moorman & Hume 1991). In contrast, Ivins et al (1991) found remarkably stable calcium levels within collapsing chick DRG growth cones upon contact with chick retinal neurites or exposure to the collapsing factor isolated from chick brain (see above). The same growth cones could respond by collapse to high concentrations of intracellular calcium induced by the ionophore A 23187.

cyclicAMP   Like calcium, cAMP is a ubiquitous intracellular regulatory molecule that has been implicated in the control of neurite growth. Elevation of cAMP is inhibitory to neurite growth or growth cone motility in a variety of neuronal cell types that respond to several cues (Bixby 1989; Forscher et al 1987). Effects of dopamine on chicken retinal neurons could be mimicked by forskolin and are consistent with the use of cAMP as a second messenger (Lankford et al 1988). However, it is not entirely clear if cAMP is directly involved in neurite growth inhibition.

GAP-43 AND G-PROTEINS   A potential regulatory protein of growth cone motility is the growth-associated protein GAP-43, which is most prominent in actively extending growth cones. Transfection of nonneuronal cells with GAP-43 induces the formation of processes (Zuber et al 1989). Recent data suggest that GAP-43 may play a role for adhesion, rather than growth cone movement per se; a PC-12 subclone expressing very low levels of GAP-43 responds to NGF with normal neurite outgrowth, but with less adhesive neurites (Baetge & Hammang 1991). In addition, dendritic growth cones lack GAP-43, but often move like axonal growth cones (Goslin & Banker 1990). Although the intracellular mechanisms of GAP-43 action are largely unknown, it has been considered as a "calmodulin sponge," which absorbs free calmodulin and releases it in response to the activation of protein kinase C (PKC) (see Skene 1990). This could be important in modulating cytoskeletal responses to calcium signals. Other possible targets of GAP-43 are G-proteins. Strittmatter et al (1990) demonstrated that GAP-43 can stimulate the binding of GTP to G_0, the predominant G-protein in the brain (Sternweis & Robishaw 1984), which is also highly concentrated in growth cones (Strittmatter et al 1990). Although the relationship of these observations to the control of the neurite outgrowth is unknown, they imply that G_0 could coordinate extracellular signals with the intracellular mechanisms for neurite growth.

PROTEIN KINASE C   Protein kinase C participates in a variety of signal transduction events as a consequence of the receptor-mediated activation of phospholipase C, which leads to the formation of the PKC activator diacylglycerol (see Huang 1989). Recent evidence has shown that dopamine-induced neurite retraction in retinal horizontal cells of catfish is mediated by the activation of PKC via diacylglycerol (Rodrigues & Dowling 1990). However, the growth promoting effects of ECM molecules, like laminin, fibronectin, and collagens, also seem to be dependent on PKC function (Bixby 1989; Bixby & Jhabvala 1990), and antibodies against certain CAMs change the phosphatidylinositol turnover (Schuch et al 1989). Whether PKC is actually a signal transducer for the ECM and

CAM molecules is uncertain, because direct measurements of PKC activation were not obtained. Interpretation of the data is further complicated by the recent suggestion that the function of ECM receptors (integrins) can itself be regulated by PKC (Shaw et al 1990).

TYROSINE KINASES AND PHOSPHATASES   The recent cloning of the *trk* receptor family for the neurotrophins NGF, BDNF, and NT-3 (Klein et al 1991a,b; Lamballe et al 1991; Soppet et al 1991) and axonal receptor-linked tyrosine phosphatases (Yang et al 1991) shed new light on the involvement of both systems in the signal transduction pathways that control outgrowth and guidance. Although the neuron-specific substrates for both activities are currently unknown, they could act on *c-src* or vinculin, proteins that are highly enriched in growth cone membranes and are important regulators of cytoskeletal assembly (Igarashi et al 1990; Maness et al 1988).

SUMMARY AND CONCLUSION

The currently known inhibitory effects on neurite growth reviewed here fall into three groups: inhibitory effects exerted be defined molecules (membrane proteins, neurotransmitters); molecules that are multifunctional in vitro (tenascin, J1-160/180); and descriptive evidence in several systems, which suggests the presence of additional inhibitory or repulsive molecules. In some of the latter examples, correlations exist with the specific localization of ECM components, in particular tenascin and sulfated proteoglycans. Unfortunately, experiments establishing a direct causal link between the presence of these constituents and neurite growth inhibition are sparse or lacking. Because tenascin can exert neurite growth-promoting, as well as inhibitory, effects in vitro, depending on its presence as substrate or in soluble form and probably also on the combination with other ECM molecules, the situation urgently requires experiments in vivo or organ culture using site-specific antibodies for the various functional domains of the tenascin molecule. A slightly different situation exists for proteoglycans, which are a heterogeneous group of molecules where both the sugar parts and the protein backbones can exert inhibitory or growth promoting effects in vitro. Here, a detailed analysis of the proteoglycans actually present at the sites of growth inhibition is needed, including a dissection of the functional domains of these molecules and experiments using activity-neutralizing antibodies in vitro and in vivo. Of particular importance is whether the effects of sulfated proteoglycans are mediated by the high density of negative charges and the peculiar "mucose" substrate properties represented by these molecules, or whether specific molecular interactions, including neuronal receptors, are involved.

Defined molecules with inhibitory or repulsive effects on neurite growth can be further divided into two categories: membrane-bound glycoproteins and neurotransmitters. The effect of transmitters on particular types of neurons can be growth-promoting or inhibitory, depending on the presence of specific transmitter receptor subtypes and their linkage to second messengers. As discussed above, many of these effects seem to be mediated by or influence the calcium homeostasis in the growth cone (Kater & Mills 1991). Because neurons synthesize and secrete transmitters very early in their development, transmitter interactions may play an important role, particularly in shaping axonal and dendritic arborizations within gray matter.

Cell membrane glycoproteins with unique repulsive or inhibitory effects on growth cones have been purified from somites of early chick embryos, embryonic chicken optic tectum or brain, and adult mammalian CNS myelin. As discussed above, important molecular and cell biological differences exist between these molecules, which suggests that they may be different. The full biochemical and molecular characterization of these inhibitory glycoproteins will allow us to address several important questions, including their possible relationships, the existence of a new protein family, and the possible domain structures. In this context, the comparison with the neurite growth-promoting members of the cell adhesion molecule (CAM) families are of particular interest. Many of these molecules possess a typical domain structure, with Ig-domains, fibronectin type III domains, and cadherin domains being especially frequent (Edelman & Crossin 1991; Reichardt & Tomaselli 1991; Takeichi 1991). For tenascin, repeats of EGF-domains have been correlated with anti-adhesive functions (Spring et al 1989). The interaction of cells with each other or with the ECM involves either homophilic interactions (many CAMs) or ligand-receptor interactions (ECM molecules, involving the family of integrins as frequently used receptors). Integrins and many CAMs are transmembrane proteins for which interactions with second messenger mechanisms, as well as with cytoskeletal constituents, have been shown and probably represent a major element in the cellular mechanism of action of these components (Reichardt & Tomaselli 1991; Schuch et al 1989). For inhibitory macromolecules, the existence of receptors is almost completely open at present. However, homophilic interaction is excluded for the myelin-associated inhibitor NI-35, as no such activity could be found in plasma membrane fractions of purified neurons, neuroblastoma cells, or PC12 cells, which makes the presence of a specific receptor very likely.

Except for the neurotransmitters, no clear-cut results exist yet with regard to second messengers involved in inhibitory effects. Growth-promoting and growth-inhibitory signals could act in opposite directions on the same second messenger systems in growth cones, thus influencing the cytoskeleton and the cellular machinery required for movement. Unfortunately, the data on calcium changes in response to inhibitory events of purified neurite growth inhibitors are contradictory at present; therefore, the generality of the model of a critical, growth permissive calcium concentration proposed on the basis of the actions of neurotransmitters on growth cones (Kater & Mills 1991) is unclear at present.

How inhibitory interactions are related to adhesion is another important, largely open question. The possibility exists that molecules, like sulfated proteoglycans, are "inhibitory" by their charge and nonadhesiveness, whereas inhibitors acting via specific receptors would either not, or only secondarily, influence cell adhesion. The situation for inhibitors may be analogous to that for growth promoting substrates: Adhesion of the growth cone is a prerequisite for growth, but the model of guidance by preference for the most adhesive substrate is too simple and is being replaced by an understanding of growth cone reactions to specific signals represented by ECM molecules and CAMs and mediated by specific receptors and their intracellular effects. The presence of inhibitory, in addition to stimulatory, signals significantly increases the regulatory potential of such interactions. Clearly, the exact relation between adhesion and growth cone inhibition has to be addressed as soon as sufficient amounts of highly pure inhibitors or recombinant fragments of their active domains become available.

A final important question is whether there is one or several types of negative reactions of growth cones. For fibroblasts and epithelial cells, two types of contact inhibition of movement have been described: Arrangement of cells alongside each other occurs in transformed rat fibroblasts without arrest of lamellipodial activity at the contact sites ("type II contact inhibition"; Vesely & Weiss 1973). In contrast, primary chick fibroblasts or epithelial cells show paralysis and collapse of their movement organelles upon cell contact ("type I contact inhibition"; Abercrombie 1970; Heaysman 1978; Vesely & Weiss 1973). Interestingly, avoidance without growth cone collapse (resembling "type II" inhibition) has been described for proteoglycans (Snow et al 1991) and J1-160 (Schachner 1991). In contrast, all of the other inhibitory interactions also induce growth cone collapse. This growth cone collapse then can be permanent or transitory, followed by recovery, "desensitization," and continuation of growth. The more subtle repulsive effects of the molecule isolated from embryonic chicken optic tectum fit well with the early appearance and graded distribution of this molecule. This corresponds well to a function as a guidance molecule in the establishment of the tectal map. In contrast, myelin-associated neurite growth inhibitors induce long-lasting effects and appear late in development, and their expression persists throughout life; these characteristics are in line with a stabilizing function for the nervous system. Suppression of side-branch formation in white matter, and a narrow spatial restriction of sprouting in gray matter would, therefore, be main functions of these oligodendrocyte-associated inhibitory molecules, in addition to boundary functions during late phases of CNS development.

As we have reviewed above, molecules exerting inhibitory effects on growing neurites have now been identified in a variety of systems in the developing and adult nervous systems. Inhibitory molecules may prove to be as important for the regulation of neurite growth as the better characterized growth-promoting molecules. We therefore propose the following model for neurite growth and pathfinding: Growth cones interact with their environment—membranes of neighboring cells or axons, ECM, and soluble factors—by complex molecular homophilic and heterophilic (e.g. ligand-receptor) interactions. Among these interactions, some represent specific growth-promoting signals, whereas others exert negative, repulsive, or long-lasting inhibitory effects. The reactions of specific growth cones depends on the particular combination of signals received and on the hierarchy of potencies of the various signals. Different types of neurons or neurites can differ in their receptor systems and, thus, in their responsiveness to various types of growth promoting or inhibitory signals.

Three main directions of future work can be identified: the molecular characterization of the existing and novel neurite growth inhibitory molecules; the identification of presumed receptors for inhibitory proteins and the cellular mechanisms for growth cone repulsion and growth arrest; and in vivo experiments using antibodies or transgenic or gene-ablated animals to show the physiological importance of particular molecules in the context of the combinations of positive and negative signals present in the organism. With the increasing emergence of negative signals as regulators of neurite growth, the further study of inhibitors promises to generate many more unexpected and exciting findings.

ACKNOWLEDGMENT

We thank our colleagues Drs. Ray Colello, Harald Seulberger, Bruno Oesch, and Peter Streit for their helpful comments and criticism of the manuscript, and Mrs. Silvia Kaufmann for her valuable secretarial assistance. The laboratory of the authors is supported by grants from the Swiss National Science Foundation (No. 31-29981.90), the American Paralysis Association (Springfield, NJ), the International Spinal Research Trust (Enfield, UK), the Swiss Multiple Sclerosis Society, and Regeneron Pharmaceuticals (Tarrytown, NY).

Literature Cited

Abercrombie, M. 1970. Contact inhibition in tissue culture. *In Vitro* 6: 128–42

Aguayo, A. J., Carter, D. A., Zwimpfer, T. J., Vidal-Sanz, M., Bray, G. M. 1990. Axonal regeneration and synapse formation in the injured CNS of adult mammals. In *Brain Repair*, ed. A. Björklund, A. Aguayo, D. Ottoson, p. 251. New York: Stockton

Baetge, E. E., Hammang, J. P. 1991. Neurite outgrowth in PC12 cells deficient in GAP-43. *Neuron* 6: 21–30

Baier, H., Bonhoeffer, F. 1992. Axon guidance by gradients of a target-derived component. *Science* 255: 472–75

Baird, J. L., Raper, J. A. 1991. A growth cone collapsing activity from adrenal membranes that affects retinal ganglion growth cones more than dorsal root ganglion growth cones. *Soc. Neurosci.* 17: 741 (Abstr.)

Bandtlow, C. E., Schmidt, M. F., Hassinger, T. D., Schwab, M. E., Kater, S. B. 1992. Role of intracellular calcium in growth cone collapse evoked by the neurite growth inhibitors NI-35. *Science*. In press

Bandtlow, C. E., Schwab, M. E. 1991. Purification and biochemical characterization of rat and bovine CNS myelin associated neurite growth inhibitors NI-35 and NI-250. *Soc. Neurosci.* 17: 1495 (Abstr.)

Bandtlow, C. E., Zachleder, T., Schwab, M. E. 1990. Oligodendrocytes arrest neurite growth by contact inhibition. *J. Neurosci.* 10: 3937–48

Bastmeyer, M., Beckmann, M., Schwab, M. E., Stuermer, C. A. O. 1991. Growth of regenerating goldfish axons is inhibited by rat oligodendrocytes and CNS myelin but not by goldfish optic nerve tract oligodendrocytelike cells and fish CNS myelin. *J. Neurosci.* 11: 626–50

Beckerle, M. C., Burridge, K., DeMartino, G. N., Croall, D. E. 1987. Colocalization of calcium-dependent protease II and one of its substrates at sites of cell adhesion. *Cell* 51: 569–77

Bernstein, D. R., Stelzner, D. J. 1983. Plasticity of the corticospinal tract following midthoracic spinal injury in the postnatal rat. *J. Comp. Neurol.* 221: 382–400

Bixby, J. L. 1989. Protein kinase C is involved in laminin stimulation of neurite outgrowth. *Neuron* 3: 287–97

Bixby, J. L., Jhabvala, P. 1990. Extracellular matrix molecules and cell adhesion molecules induce neurites through different mechanisms. *J. Cell Biol.* 111: 2725–32

Bonhoeffer, F., Huf, J. 1985. Position-dependent properties of retinal axons and their growth cones. *Nature* 315: 409–10

Bonhoeffer, F., Huf, J. 1982. In vitro experiments on axon guidance demonstrating an anterior-posterior gradient on the tectum. *EMBO J.* 1: 427–31

Bonhoeffer, F., Huf, J. 1980. Recognition of cell types by axonal growth cones in vitro. *Nature* 288: 162–64

Bray, D., Wood, P., Bunge, R. P. 1980. Selective fasciculation of nerve fibres in culture. *Exp. Cell Res.* 130: 241–50

Bregman, B. S., Kunkel Bagden, E., McAtee, M., O'Neill, A. 1989. Extension of the critical period for developmental plasticity of the corticospinal pathway. *J. Comp. Neurol.* 282: 355–70

Cadelli, D., Schwab, M. E. 1991. Regeneration of lesioned septohippocampal acetylcholinesterase-positive axons is improved by antibodies against the myelin-associated neurite growth inhibitors NI-35/250. *Eur. J. Neurosci.* 3: 825–32

Carbonetto, S., Evans, D., Cochard, P. 1987. Nerve fiber growth in culture on tissue substrata from central and peripheral nervous systems. *J. Neurosci.* 7: 610–20

Caroni, P., Savio, T., Schwab, M. E. 1988. Central nervous system regeneration: oligodendrocytes and myelin as non-permissive substrates for neurite growth. *Prog. Brain Res.* 78: 363–69

Caroni, P., Schwab, M. E. 1989. Codistribution of neurite growth inhibitors and oligodendrocytes in rat CSN: appearance follows nerve fiber growth and precedes myelination. *Dev. Biol.* 136: 287–95

Caroni, P., Schwab, M. E. 1988a. Two membrane protein fractions from rat central myelin with inhibitory properties for neurite growth and fibroblast spreading. *J. Cell Biol.* 106: 1281–88

Caroni, P., Schwab, M. E. 1988b. Antibody against myelin-associated inhibitor of neurite growth neutralizes nonpermissive substrate properties of CNS white matter. *Neuron* 1: 85–96

Chiquet, M. 1989. Tenascin/J1/Cytotactin: the potential function of hexabrachion proteins in neural development. *Dev. Neurosci.* 11: 266–75

Chiquet-Ehrismann, R., Kalla, P., Pearson, C. A., Beck, K., Chiquet, M. 1988. Tenascin interferes with fibronectin action. *Cell* 53: 383–90

Chiquet-Ehrismann, R., Mackie, E. J., Pearson, C. A., Sakakura, T. 1986. Tenascin: an extracellular matrix protein involved in tissue interactions during fetal development and oncogenesis. *Cell* 47: 131–39

Cohan, C. S. 1990. Frequency-dependent and cell-specific effects of electrical activity on growth cone movements of cultured

Helisoma neurons. *J. Neurobiol.* 21: 400–13

Cohan, C. S., Connor, J. A., Kater, S. B. 1987. Electrically and chemically mediated increase in intracellular calcium in neuronal growth cones. *J. Neurosci.* 7: 3588–99

Cohen, S. 1960. Purification of a nerve-growth promoting protein from the mouse salivary gland and its neurocytotoxic antiserum. *Proc. Natl. Acad. Sci. USA* 46: 302–11

Cole, G. C., McCabe, C. F. 1991. Identification of a developmentally regulated Keratan sulfate proteoglycan that inhibits cell adhesion and neurite outgrowth. *Neuron* 7: 1007–18

Colello, R. J., Kapfhammer, J., Schwab, M. E. 1991. A role for oligodendrocytes in the stabilization of optic nerve axon numbers. *Soc. Neurosci.* 17: 211 (Abstr.)

Cox, E. C., Müller, B., Bonhoeffer, F. 1990. Axonal guidance in the chick visual system: posterior tectal membranes induce collapse of growth cones from the temporal retina. *Neuron* 2: 31–37

Crossin, K. L., Hoffman, S., Tan, S.-S., Edelman, G. M. 1989. Cytotactin and its proteoglycan ligand mark structural and functional boundaries in somatosensory cortex of the early postnatal mouse. *Dev. Biol.* 136: 381–92

Crutcher, K. A. 1989. Tissue sections from the mature rat brain and spinal cord as substrates for neurite outgrowth in vitro: extensive growth on gray matter but little growth on white matter. *Exp. Neurol.* 104: 39–54

Daniloff, J. K., Crossin, K. L., Pinçon-Raymond, M., Murawsky, M., Rieger, F., Edelman, G. M. 1989. Expression of cytoactin in the normal and regenerating neuromuscular system. *J. Cell Biol.* 108: 625–35

David, S., Aguayo, A. J. 1981. Axonal elongation into peripheral nervous system "bridges" after central nervous system injury in adult rats. *Science* 214: 931–33

Davies, A. M., Lumsden, A. 1990. Ontogeny of the somatosensory system: Origins and early development of primary sensory neurons. *Annu. Rev. Neurosci.* 13: 61–73

Davies, J. A., Cook, G. M. W., Stern, C. D., Keynes, R. J. 1990. Isolation from chick somites of a glycoprotein fraction that causes collapse of dorsal root ganglion growth cones. *Neuron* 2: 11–20

Dow, K. E., Mirski, S. E., Roder, J. C., Riopelle, R. J. 1988. Neuronal proteoglycans: biosynthesis and functional interaction with neurons in vitro. *J. Neurosci.* 8: 3278–89

Edelman, G. M., Crossin, K. L. 1991. Cell adhesion molecules: Implication for a molecular histology. *Annu. Rev. Biochem.* 60: 155–90

Fawcett, J. W., Houdson, E., Smith-Thomas, L., Meyer, R. 1989a. The growth of axons in three-dimensional astrocyte cultures. *Dev. Biol.* 135: 449–58

Fawcett, J. W., Rokos, J., Bakst, I. 1989b. Oligodendrocytes repel axons and cause axonal growth cone collapse. *J. Cell Sci.* 92: 93–100

Fichard, A., Verna, J.-M., Olivares, J., Saxod, R. 1991. Involvement of a chondroitin sulfate proteoglycan in the avoidance of chick epidermis by dorsal root ganglia fibers: A study using β-D-xyloside. *Dev. Biol.* 148: 1–9

Fichard, A., Verna, J.-M., Saxod, R. 1990. Effects of tunicamycin on the avoidance reaction of epidermis by sensory neurites in co-cultures. *Int. J. Dev. Neurosci.* 8: 245–54

Fields, R. D., Guthrie, P. B., Kater, S. B. 1990a. Calcium homeostatic capacity is regulated by patterned electrical activity in the growth cones of mouse DRG neurons. *Soc. Neurosci.* 16: 457 (Abstr.)

Fields, R. D., Neale, E. A., Nelson, P. G. 1990b. Effects of patterned electrical activity on neurite outgrowth from mouse sensory neurons. *J. Neurosci.* 10: 2950–64

Forscher, P., Kaczamarek, L. K., Buchamnan, J., Smith, S. J. 1987. Cyclic AMP induces changes in distribution and transport of organelles within growth cones of Aplysia bag neurons. *J. Neurosci.* 7: 3600–11

Friedlander, D. R., Hoffman, S., Edelman, G. M. 1988. Functional mapping of cytotactin: proteolytic fragments active in cell-substrate adhesion. *J. Cell Biol.* 107: 2329–40

Godement, P., Bonhoeffer, F. 1989. Cross-species recognition of tectal cues by retinal fibers in vitro. *Development* 106: 313–20

Godement, P., Salaün, J., Mason, C. A. 1990. Retinal axon pathfinding in the optic chiasm: divergence of crossed and uncrossed fibers. *Neuron* 5: 173–86

Goslin, K., Banker, G. 1990. Rapid changes in the distribution of GAP-43 correlate with the expression of neuronal polarity during normal development and under experimental conditions. *J. Cell Biol.* 110: 1319–31

Grierson, J. P., Petroski, R. E., Ling, D. S. F., Geller, H. M. 1990. Astrocyte topography and tenascin/cytoactin expression: correlation with the ability to support neuritic outgrowth. *Dev. Brain Res.* 55: 11–19

Grumet, M., Hoffman, S., Crossin, K. L., Edelman, G. M. 1985. Cytotactin, an extracellular matrix protein of neural and nonneural tissues that mediates neuron glia interactions. *Proc. Natl. Acad. Sci. USA* 82: 8075–79

Gundersen, R. W. 1987. Response of sensory neurites and growth cones to patterned substrate of laminin and fibronectin in vitro. *Dev. Biol.* 121: 423–31

Halfter, W., Chiquet-Ehrismann, R., Tucker, R. P. 1989. The effect of tenascin and embryonic basal lamina on the behavior and morphology of neural crest cells in vitro. *Dev. Biol.* 132: 14–25

Harrison, R. G. 1910. The outgrowth of the nerve fiber as a mode of protoplasmic movement. *J. Exp. Zool.* 9: 787–846

Hasan, S. J., Nelson, B. H., Valenzuela, J. I., Keirstead, H. S., Shull, S. E., et al. 1991. Functional repair of transected spinal cord in embryonic chick. *Restor. Neurol. Neurosci.* 2: 137–54

Hatten, M. E., Liem, R. K. M., Mason, C. A. 1984. Two forms of cerebellar glial cells interact differentially with neurons in vitro. *J. Cell Biol.* 98: 193–204

Haydon, P. G., McCobb, D. P., Kater, S. B. 1987. The regulation of neurite outgrowth, growth cone motility, and electrical synaptogenesis by serontonin. *J. Neurobiol.* 18: 197–215

Haydon, P. G., McCobb, D. P., Kater, S. B. 1984. Serotonin selectively inhibits growth cone motility and the connection of identified neurons. *Science* 226: 561–64

Heaysman, J. E. M. 1978. Contact inhibition of locomotion: a reappraisal. *Int. Rev. Cytol.* 55: 49–66

Heizmann, C. W., Hunziker, W. 1991. Intracellular calcium-binding proteins: more sites than insights. *Trends Biochem.* 16: 98–103

Herndon, M. E., Lander, A. D. 1990. A diverse set of developmentally regulated proteoglycans is expressed in the rat central nervous system. *Neuron* 4: 949–61

Hoffman, S., Crossin, K. L., Edelman, G. M. 1988. Molecular forms, binding functions, and developmental expression patterns of cytotactin and cytotactin-binding proteoglycan, an interactive pair of extracellular matrix molecules. *J. Cell Biol.* 106: 519–32

Huang, K.-P. 1989. The mechanism of protein kinase C activation. *Trends Neurosci.* 12: 425–31

Husmann, K., Faissner, A., Schachner, M. 1992. Tenascin promotes cerebellar granule cell migration and neurite outgrowth by different domains in the fibronectin Type III repeats. *J. Cell Biol.* 116: 1475–86

Igarashi, M., Saito, S., Komiya, Y. 1990. Vinculin is one of the major endogenous substrates for intrinsic tyrosine kinase in neuronal growth cones isolated from fetal rat brain. *Eur. J. Biochem.* 193: 551–58

Iijima, N., Oohira, A., Mori, T., Kitabatake, K., Kohsaka, S. 1991. Core protein of chondroitin sulfate proteoglycan promotes neurite outgrowth from cultured neocortical neurons. *J. Neurochem.* 56: 706–8

Ivins, J. K., Pittman, R. N. 1989. Growth cone-growth cone interactions in cultures of rat sympathetic neurons. *Dev. Biol.* 135: 147–57

Ivins, J. K., Raper, J. A., Pittman, R. N. 1991. Intracellular calcium levels do not change during contact-mediated collapse of chick DRG growth cone structure. *J. Neurosci.* 11: 1597–1608

Jakeman, L. B., Reier, P. J. 1991. Axonal projections between fetal spinal cord transplants and the adult rat spinal cord: a neuroanatomical tracing study of local interactions. *J. Comp. Neurol.* 307: 311–34

Jessell, T. M. 1988. Adhesion molecules and the hierarchy of neural development. *Neuron* 1: 3–13

Jhaveri, S., Erzurumlu, R. S., Friedman, B., Schneider, G. E. 1992. Oligodendrocytes and myelin formation along the optic tract of the developing hamster: an immunohistochemical study using the Rip antibody. *Glia* 6: 138–48

Johnston, A. R., Gooday, D. J. 1991. Xenopus temporal retinal neurites collapse on contact with glial cells from caudal tectum in vitro. *Development* 113: 409–17

Jones, F. S., Burgoon, M. P., Hoffman, S., Crossin, K. L., Cunningham, B. A., Edelman, G. M. 1988. A cDNA clone for cytotactin contains sequences similar to epidermal growth factor-like repeats and segments of fibronectin and fibrinogen. *Proc. Natl. Acad. Sci. USA* 85: 2186–90

Kalil, K., Reh, T. 1982. A light and electron microscopic study of regrowing pyramidal tract fibers. *J. Comp. Neurol.* 211: 265–75

Kapfhammer, J. P., Grunewald, B. E., Raper, J. A. 1986. The selective inhibition of growth cone extension by specific neurites in culture. *J. Neurosci.* 6: 2527–34

Kapfhammer, J. P., Raper, J. A. 1987a. Collapse of growth cone structure on contact with specific neurites in culture. *J. Neurosci.* 7: 201–12

Kapfhammer, J. P., Raper, J. A. 1987b. Interactions between growth cones and neurites growing from different neural tissues in culture. *J. Neurosci.* 7: 1595–1600

Kapfhammer, J. P., Schwab, M. E., Schneider, G. E. 1992. Antibody neutralization of neurite growth inhibitors from oligodendrocytes results in expanded pattern of postnatally sprouting retinocolicular axons. *J. Neurosci.* 12: 2112–19

Kater, S. B., Mattson, M., Cohan, C. S., Connor, J. 1988. Calcium regulation of the neuronal growth cone. *Trends Neurosci.* 11: 315–21

Kater, S. B., Mills, L. R. 1991. Regulation of growth cone behavior by calcium. *J. Neurosci.* 11: 891–99

Keynes, R. J., Johnson, A. R., Picart, C. J., Dunin-Borkowski, O. M., Cook, G. M. W. 1990. A glycoprotein fraction from adult chicken grey matter causes collapse of CNS and PNS growth cones in vitro. *Soc. Neurosci.* 16: 169 (Abstr.)

Keynes, R. J., Stern, C. D. 1988. The development of neural segmentation in vertebrate embryos. In *The Making of the Nervous System*, ed. J. G. Parnavalas, C. D. Stern, R. V. Stirling, pp. 84–100. Oxford: Oxford Univ. Press

Keynes, R. J., Stern, C. D. 1984. Segmentation in the vertebrate nervous system. *Nature* 310: 786–89

Khan, U., Starega, U., Seeley, P. J. 1990. Selective growth of hippocampal neurites on cryostat sections of rat brain. *Dev. Brain Res.* 54: 87–92

Klein, R., Jing, S., Nanduri, V., O'Rourke, E., Barbacid, M. 1991a. The trk proto-oncogene encodes a receptor for nerve growth factor. *Cell* 65: 189–97

Klein, R., Nanduri, V., Jing, S., Lambelle, F., Tapley, P., et al. 1991b. The trk tyrosine protein kinase is a receptor for brain-derived neurotrophic factor and neurotrophin-3. *Cell* 66: 395–403

Kröger, S., Walter, J. 1991. Molecular mechanisms separating two axonal pathways during embryonic development of the avian optic tectum. *Neuron* 6: 291–303

Krueger, S., Sievers, J., Hansen, C., Sadler, M., Berry, M. 1986. Three morphologically distinct types of interface develop between adult host and fetal brain transplants: implications for scar formation in the adult central nervous system. *J. Comp. Neurol.* 249: 103–16

Kruse, J., Keilhauser, G., Faissner, A., Timpl, R., Schachner, M. 1985. The J1 glycoprotein—a novel nervous system cell adhesion molecule of the L2/HNK-1 family. *Nature* 316: 146–48

Kuwada, J. Y., Bernhardt, R. R., Chitnis, A. B. 1990. Pathfinding by identified growth cones in the spinal cord of zebrafish embryos. *J. Neurosci.* 10: 1299–1308

Lamballe, F., Klein, R., Barbacid, M. 1991. *trkC*, a new member of the trk family of tyrosine protein kinases, is a receptor for neurotrophin-3. *Cell* 66: 967–79

Lankford, K. L., DeMello, F. G., Klein, W. L. 1988. D1-type dopamine receptors

inhibit growth cone motility in cultured retina neurons: evidence that neuro-transmitters act as morphogenic growth regulators in the developing central nervous system. *Proc. Natl. Acad. Sci. USA* 85: 2839–43

Lankford, K. L., Letourneau, P. 1992. Roles of actin filaments and three second messenger systems in short term regulation of chick dorsal root ganglion neurite outgrowth. *Cell Motil. Cytoskel.* 20: 7–9

Letourneau, P. C. 1975. Cell-to-substratum adhesion and guidance of axonal elongation. *Dev. Biol.* 44: 92–101

Levi-Montalcini, R., Hamburger, V. 1951. Selective growth-stimulating effects of mouse sarcoma on the sensory and sympathetic nervous system of the chick embryo. *J. Exp. Zool.* 116: 321–62

Lipton, S. A., Frosch, M. P., Phillips, M. D., Tauck, D. L., Aizenman, E. 1988. Nicotinic antagonists enhance process outgrowth by rat retinal ganglion cells in culture. *Science* 239: 1293–96

Lipton, S. A., Kater, S. B. 1989. Neurotransmitter regulation of neuronal outgrowth, plasticity and survival. *Trends Neurosci.* 12: 265–70

Lochter, A., Vaughan, L., Kaplony, A., Prochiantz, A., Schachner, M., Faissner, A. 1991. J1/tenascin in substrate-bound and soluble form displays contrary effects on neurite outgrowth. *J. Cell Biol.* 113: 1159–71

Maness, P. F., Aubury, M., Shores, C. G., Frame, L., Pfenninger, K. H. 1988. c-src gene product in developing rat brain is enriched in nerve growth cone membranes. *Proc. Natl. Acad. Sci. USA* 85: 5001–5

Martini, R., Schachner, M. 1991. Complex expression pattern of tenascin during innervation of the posterior limb buds of the developing chicken. *J. Neurosci. Res.* 28: 261–79

Mason, C. A., Godement, P. 1992. Growth cone form reflects interactions in visual pathways and cerebellar pathways. In *The Nerve Growth Cone*, ed. P. C. Letourneau, S. B. Kater, E. R. Macagno, pp. 405–23. New York: Raven

Mattson, M. P., Dou, P., Kater, S. B. 1988. Outgrowth-regulating actions of glutamate in isolated hippocampal pyramidal neurons. *J. Neurosci.* 8: 2087–2100

Mattson, M. P., Kater, S. B. 1989. Excitatory and inhibitory neurotransmitters in the generation and degeneration of hippocampal neuroarchitecture. *Brain Res.* 478: 337–48

Mattson, M. P., Murrain, M., Guthrie, P. B., Kater, S. B. 1989. Fibroblast growth factor and glutamate: opposing roles in the generation and degeneration of hippocampal neuroarchitecture. *J. Neurosci.* 9: 3728–40

McCobb, D. P., Cohan, C. S., Connor, J. A., Kater, S. B. 1988a. Interactive effects of serotonin and acetylcholine on neurite elongation. *Neuron* 1: 377–85

McCobb, D. P., Haydon, P. G., Kater, S. B. 1988b. Dopamine and serotonin inhibition of neurite elongation of different identified neurons. *J. Neurosci. Res.* 19: 19–26

McCobb, D. P., Kater, S. B. 1988. Membrane voltage and neurotransmitter regulation of neuronal growth cone motility. *Dev. Biol.* 130: 599–609

McKeon, R. J., Schreiber, R. C., Rudge, J. S., Silver, J. 1991. Reduction of neurite outgrowth in a model of glial scarring following CNS injury is correlated with the expression of inhibitory molecules on reactive astrocytes. *J. Neurosci.* 11: 3398–3411

Michaelis, E. K., Wang, H., Mattson, M. P. 1991. NMDA receptor protein in cultured hippocampal neurons: developmental expression, relation to excitotoxicity, and regulation by basic FGF. *Soc. Neurosci.* 17: 74 (Abstr.)

Moorman, S. J., Hume, R. I. 1991. An increase in internal free calcium concentration coincides with two different growth cone behaviors. *Soc. Neurosci.* 17: 183 (Abstr.)

Moorman, S. J., Hume, R. I. 1990. Growth cones of chick sympathetic preganglionic neurons in vitro interact with other neurons in a cell-specific manner. *J. Neurosci.* 10: 3158–63

Munger, B. L. 1977. Neural-epithelial interactions in sensory receptors. *J. Invest. Dermatol.* 69: 27–40

Oakley, R. A., Tosney, K. W. 1991. Peanut agglutinin and chondroitin-6-sulfate are molecular markers for tissues that act as barriers to axon advance in the avian embryo. *Dev. Biol.* 147: 187–206

Oohira, A., Matsui, F., Katho-Semba, R. 1991. Inhibitory effects of brain chondroitin sulfate proteoglycans on neurite outgrowth from PC12 cells. *J. Neurosci.* 11: 822–27

Oohira, A., Matsui, F., Matsuda, M., Takida, Y., Kuboki, Y. 1988. Occurrence of three distinct molecular species of chondroitin sulfate proteoglycan in the developing rat brain. *J. Biol. Chem.* 263: 10240–46

Pearson, C. A., Pearson, D., Shibahara, S., Hofsteenge, J., Chiquet-Ehrismann, R. 1988. Tenascin: cDNA cloning and induction by TGF-β. *EMBO J.* 7: 2977–81

Pesheva, P., Probstmeier, R., Spiess, E., Schachner, M. 1991. Divalent cations modulate the inhibitory substrate properties of murine glia-derived J1-160 and J1-180 extracellular matrix glycoproteins for neuronal adhesion. *Eur. J. Neurosci.* 3: 356–65

Pesheva, P., Spiess, E., Schachner, M. 1989. J1-160 and J1-180 are oligodendrocyte-secreted nonpermissive substrates for cell adhesion. *J. Cell Biol.* 109: 1765–78

Pippenger, M. A., Sims, T. J., Gilmore, S. A. 1990. Development of the rat corticospinal tract through an altered glial environment. *Dev. Brain Res.* 55: 43–50

Raible, D. W., Raper, J. A. 1990. Additional purification steps of an activity from embryonic chick brain that inhibits neuronal growth cone motility. *Soc. Neurosci.* 16: 313 (Abstr.)

Ramon y Cajal, S. 1928. 1959. *Degeneration and Regeneration of the Nervous System*, ed. and transl. R. M. May. 2 vols. New York: Hafner. 769 pp. (Reprint)

Ramon y Cajal, S. 1929. 1960. *Studies on Vertebrate Neurogenesis*, ed. and transl. L. Guth. Springfield, Ill: Thomas. 432 pp.

Raper, J. A., Chang, S., Raible, D. W. 1992. Interactions between growth cones and axons: Selectively distributed extension-promoting and extension-inhibiting components. In *The Nerve Growth Cone*, ed. P. C. Letourneau, S. B. Kater, E. R. Macagno, pp. 207–17. New York: Raven

Raper, J. A., Grunewald, B. E. 1990. Temporal retinal growth cones collapse on contact with nasal retinal axons. *Exp. Neurol.* 109: 70–74

Raper, J. A., Kapfhammer, J. P. 1990. The enrichment of a neuronal growth cone collapsing activity from embryonic chick brain. *Neuron* 2: 21–29

Reichardt, L. F., Tomaselli, K. J. 1991. Extracellular matrix molecules and their receptors. *Annu. Rev. Neurosci.* 14: 531–70

Reier, P. J., Stensaas, L. J., Guth, L. 1983. The astrocytic scar as an impediment to regeneration in the central nervous system. In *Spinal Cord Reconstruction*, ed. C. C. Kao, R. P. Bunge, P. J. Reier, pp. 163–95. New York: Raven

Riopelle, R. J., Dow, K. E. 1990. Functional interactions of neuronal heparan sulphate proteoglycans with laminin. *Brain Res.* 525: 92–100

Rodrigues, P. S., Dowling, J. E. 1990. Dopamine induces neurite retraction in retinal horizontal cells via diacylglycerol and protein kinase C. *Proc. Natl. Acad. Sci. USA* 87: 9693–97

Rudge, J. S., Silver, J. 1990. Inhibition of

neurite outgrowth on astroglial scars in vitro. *J. Neurosci.* 10: 3594–3603

Sagot, Y., Swerts, J.-P., Cochard, P. 1991. Changes in permissivity for neuronal attachment and neurite outgrowth of spinal cord grey and white matters during development: a study with the "cryo-culture" bioassay. *Brain Res.* 543: 25–35

Sanes, J. R. 1989. Extracellular matrix molecules that influence neural development. *Annu. Rev. Neurosci.* 12: 491–516

Savio, T., Schwab, M. E. 1990. Lesioned corticospinal tract axons regenerate in myelin-free rat spinal cord. *Proc. Natl. Acad. Sci. USA* 87: 4130–33

Savio, T., Schwab, M. E. 1989. Rat CNS white matter, but not gray matter, is non-permissive for neuronal cell adhesion and fiber outgrowth. *J. Neurosci.* 9: 1126–33

Saxod, R. 1978. Development of cutaneous sensory receptors in birds. In *Handbook of Sensory Physiology, Development of Sensory Systems*, ed. M. Jacobson, 9: 337–417. Berlin: Springer-Verlag

Schachner, M. 1992. Neural recognition molecules and their influence on cellular functions. In *The Nerve Growth Cone*, ed. P. C. Letourneau, S. B. Kater, E. R. Macagno, pp. 237–54. New York: Raven

Schmidt, M. F., Bandtlow, C. E., Hassinger, T. D., Schwab, M. E., Kater, S. B. 1991. CNS myelin neurite growth inhibitor NI-35 causes a large transient rise in intracellular calcium which precedes growth arrest and collapse of rat DRG growth cones. *Soc. Neurosci.* 17: 927 (Abstr.)

Schneider, G. E. 1973. Early lesions of the superior colliculus: factors affecting the formation of abnormal projections. *Brain Behav. Evol.* 8: 73–109

Schnell, L., Schwab, M. E. 1990. Axonal regeneration in the rat spinal cord produced by an antibody against myelin-associated neurite growth inhibitors. *Nature* 343: 269–72

Schuch, U., Lohse, M. J., Schachner, M. 1989. Neural cell adhesion molecules influence second messenger systems. *Neuron* 3: 13–20

Schwab, M. E., Caroni, P. 1988. Oligodendrocytes and CNS myelin are non-permissive substrates for neurite growth and fibroblast spreading in vitro. *J. Neurosci.* 8: 2381–93

Schwab, M. E., Schnell, L. 1991. Channelling of developing rat corticospinal tract axons by myelin-associated neurite growth inhibitors. *J. Neurosci.* 11: 709–22

Schwab, M. E., Schnell, L. 1989. Region-specific appearance of myelin constituents in the developing rat spinal cord. *J. Neurocytol.* 18: 161–69

Schwab, M. E., Schnell, L., Cadelli, D. 1990.

Antibodies against myelin-associated neurite growth inhibitors combined with trophic factors and bridges improve regeneration of lesioned nerve fibers in spinal cord and optic nerve. *Soc. Neurosci.* 16: 169 (Abstr.)

Schwab, M. E., Thoenen, H. 1985. Dissociated neurons regenerate into sciatic but not optic nerve explants in culture irrespective of neurotrophic factors. *J. Neurosci.* 5: 2415–23

Shaw, L. M., Messier, J. M., Mercurio, A. M. 1990. The activation-dependent adhesion of macrophages to laminin involves cytoskeletal anchoring and phosphorylation of the $6\beta1$ integrin. *J. Cell Biol.* 110: 2167–74

Shimizu, I., Oppenheim, R. W., O'Brien, M., Shneiderman, A. 1990. Anatomical and functional recovery following spinal cord transection in the chick embryo. *J. Neurobiol.* 21: 918–37

Silver, J., Poston, M., Rutishauser, U. 1987. Axon pathway boundaries in the developing brain. I. Cellular and molecular determinants that separate the optic and olfactory projections. *J. Neurosci.* 7: 2264–72

Silver, R. A., Lamb, A. G., Bolsover, S. R. 1990. Calcium hotspots caused by L-channel clustering promote morphological changes in neuronal growth cones. *Nature* 343: 751–54

Skene, J. H. 1990. GAP-43 as a "calmodulin sponge" and some implications for calcium signalling in axon terminals. *Neurosci. Res.* 13: S112–25 (Suppl.)

Smith, G. M., Rutishauser, U., Silver, J., Miller, R. H. 1990. Maturation of astrocytes in vitro alters the extent and molecular basis of neurite outgrowth. *Dev. Biol.* 138: 377–90

Snow, D. M., Lemmon, V., Carrino, D. A., Caplan, A. I., Silver, J. 1990a. Sulfated proteoglycans in astroglial barriers inhibit neurite outgrowth in vitro. *Exp. Neurol.* 109: 111–30

Snow, D. M., Steindler, D. A., Silver, J. 1990b. Molecular and cellular characterization of the glial roof plate of the spinal cord and optic tectum: a possible role for a proteoglycan in the development of an axon barrier. *Dev. Biol.* 138: 359–76

Snow, D. M., Watanabe, M., Letourneau, P. C., Silver, J. 1991. A chondroitin sulfate proteoglycan may influence the direction of retinal ganglion cell outgrowth. *Development* 113: 1473–85

Sobue, K., Kanda, K. 1989. α-actinins, calspectin (brain spectrin or fodrin), and actin participate in adhesion and movement of growth cones. *Neuron* 3: 311–19

Soppet, D., Escandon, E., Maragos, J., Middlemas, D. S., Reid, S. W., et al. 1991.

The neurotrophic factors brain-derived neurotrophic factor and neurotrophin-3 are ligands for the trkB tyrosine kinase receptor. *Cell* 65: 895–903

Spring, J., Beck, K., Chiquet-Ehrismann, R. 1989. Two contrary functions of tenascin: dissection of the active sites by recombinant tenascin fragments. *Cell* 59: 325–34

Stahl, B., Müller, B., Von Boxberg, Y., Cox, E. C., Bonhoeffer, F. 1990. Biochemical characterization of a putative axonal guidance molecule of the chick visual system. *Neuron* 5: 735–43

Stanfield, B. B. 1991. The corticospinal tract attains a normal configuration in the absence of myelin: observation in Jimpy mutant mice. *Neuron* 7: 249–56

Steindler, D. A. 1993. Glial boundaries in the developing nervous system. *Annu. Rev. Neurosci.* 16: 445–70

Steindler, D. A., Cooper, N. G. F., Faissner, A., Schachner, M. 1989a. Boundaries defined by adhesion molecules during development of the cerebral cortex: the J1/tenascin glycoprotein in the mouse somatosensory cortical barrel field. *Dev. Biol.* 131: 243–60

Steindler, D. A., Faissner, A., Schachner, M. 1989b. Brain "Cordones": transient boundaries of glia and adhesion molecules that define developing functional units. *Comments Dev. Neurobiol.* 1: 29–60

Stern, C. D., Norris, W. E., Bronner-Fraser, M., Carlson, G. J., Faissner, A., et al. 1989. J1/tenascin-related molecules are not responsible for the segmented pattern of neural crest cells or motor axons in the chick embryo. *Development* 107: 309–19

Stern, C. D., Sisodiya, S. M., Keynes, R. J. 1986. Interactions between neurites and somite cells: inhibition and stimulation of nerve growth in the chick embryo. *J. Embryol. Exp. Morphol.* 91: 209–26

Sternweis, P. C., Robishaw, J. D. 1984. Isolation of two proteins with high affinity for guanine nucleotides from membranes of bovine brain. *J. Biol. Chem.* 259: 13806–13

Stirling, V. R. 1991. Molecules, maps and gradients in the retinotectal projection. *Trends Neurosci.* 14: 509–12

Strittmatter, S. M., Valenzuela, D., Kennedy, T. E., Neer, E. J., Fishman, M. C. 1990. Go is a major growth cone protein subject to regulation by GAP-43. *Nature* 344: 836–41

Takeichi, M. 1991. Cadherin cell adhesion receptors as a morphogenetic regulator. *Science* 251: 1451–55

Tan, S.-S., Crossin, K. L., Hoffman, S., Edelman, G. M. 1987. Asymmetric expression in somites of cytoactin and its proteoglycan ligand is correlated with neural crest cell distribution. *Proc. Natl. Acad. Sci. USA* 84: 7977–81

Tello, F. 1911. La influencia del neurotropismo en la regeneracion de los centros nerviosos. *Trab. Lab. Invest. Biol.* 9: 123–59

Thoenen, H. 1991. The changing scene of neurotrophic factors. *Trends Neurosci.* 14: 165–70

Tosney, K. W. 1992. Growth cone navigation in the proximal environment of the chick embryo. In *The Nerve Growth Cone*, ed. P. C. Letourneau, S. B. Kater, E. R. Macagno, pp. 387–403. New York: Raven

Tosney, K. W., Landmesser, L. T. 1985. Development of the major pathways for neurite outgrowth in the chick hindlimb. *Dev. Biol.* 109: 193–214

Tosney, K. W., Landmesser, L. T. 1984. Pattern and specificity of axonal outgrowth following varying degrees of chick limb bud ablation. *J. Neurosci.* 4: 2518–27

Tosney, K. W., Oakley, R. A. 1990. The perinotochordal mesenchyme acts as a barrier to axon advance in the chick embryo: Implications for a general mechanism of axon guidance. *Exp. Neurol.* 109: 75–89

Treherne, J. M., Woodwards, S. K. A., Varga, Z. M., Ritchie, J. M., Nicholss, J. G. 1992. Restoration of conduction and growth of axons through injured spinal cord of neonatal opossum in culture. *Proc. Natl. Acad. Sci. USA* 89: 431–34

Tucker, R. P., McKay, S. E. 1991. The expression of tenascin by neural crest cells and glia. *Development* 112: 1031–39

Udin, S. B., Fawcett, J. W. 1988. Formation of topographic maps. *Annu. Rev. Neurosci.* 11: 289–327

Vanselow, J., Schwab, M. E., Thanos, S. 1990. Responses of regenerating rat retinal ganglion cell axons to contacts with central nervous system myelin in vitro. *Eur. J. Neurosci.* 2: 121–25

Verna, J.-M. 1985. In vitro analysis of interactions between sensory neurons and skin: evidence for selective innervation of dermis and epidermis. *J. Embryol. Exp. Morph.* 86: 53–70

Vesely, P., Weiss, R. A. 1973. Cell locomotion and contact inhibition of normal and neoplastic rat cells. *Int. J. Cancer* 11: 64–76

Vielmetter, J., Stuermer, C. A. O. 1989. Goldfish retinal axons respond to position-specific properties of tectal cell membranes in vitro. *Neuron* 2: 1331–39

Vos, P., Stark, F., Pittman, R. N. 1991. Merkel cells in vitro: Production of nerve growth factor and selective interactions with sensory neurons. *Dev. Biol.* 144: 281–300

Walter, J., Allsopp, T. E., Bonhoeffer, F. 1990a. A common denominator of growth cone guidance and collapse. *Trends Neurosci.* 13: 447–52

Walter, J., Henke-Fahle, S., Bonhoeffer, F. 1987a. Avoidance of posterior tectal membranes by temporal retinal axons. *Development* 101: 909–13

Walter, J., Kern-Veits, B., Huf, J., Stolze, B., Bonhoeffer, F. 1987b. Recognition of position-specific properties of tectal cell membranes by retinal axons in vitro. *Development* 101: 685–96

Walter, J., Müller, B., Bonhoeffer, F. 1990b. Axonal guidance by an avoidance mechanism. *J. Physiol.* 84: 104–10

Watanabe, E., Murakami, F. 1989. Preferential adhesion of chick central neurons to the gray matter of the central nervous system. *Neurosci. Lett.* 97: 69–74

Wehrle, B., Chiquet, M. 1990. Tenascin is accumulated along developing peripheral nerves and allows neurite outgrowth in vitro. *Development* 110: 401–15

Wehrle-Haller, B., Koch, M., Baumgartner, S., Spring, J., Chiquet, M. 1991. Nerve-dependent and -independent tenascin expression in the developing chick limb bud. *Development* 112: 627–37

Wessels, N. K., Letourneau, P. C., Nuttall, R. P., Luduena-Anderson, M., Geiduschek, J. M. 1980. Responses to cell contacts between growth cones, neurites, and ganglionic non-neuronal cells. *J. Neurocytol.* 9: 647–64

Windle, W. F., Clemente, C. D., Chambers, W. W. 1952. Inhibition of formation of a glial barrier as a means of permitting a peripheral nerve to grow into the brain. *J. Comp. Neurol.* 96: 359–69

Wu, D.-Y. 1991. *Radial glia in the developing superior colliculus: evidence for a midline barrier.* PhD thesis. Mass. Inst. Technol., Cambridge, Mass.

Wu, D.-Y., Jhaveri, S., Moya, K. L., Schneider, G. E. 1988. Vimentin and GFAP expression in developing hamster superior colliculus. *Soc. Neurosci.* 14: 1110 (Abstr.)

Wu, D.-Y., Jhaveri, S., Schneider, G. E. 1989. Recrossing of retinal axons after early tectal lesions in hamsters occur only where Vimentin- and GFAP-positive midline cells are damaged. *Soc. Neurosci.* 15: 873 (Abstr.)

Wu, D.-Y., Schneider, G. E., Jhaveri, S. 1990. Retinotectal axons cross to the wrong side following disruption of tectal midline cells in the hamster. *Soc. Neurosci.* 16: 336 (Abstr.)

Yang, X., Seow, K. T., Bahri, S. M., Oon, S. H., Chia, W. 1991. Two *Drosophila* receptor-like tyrosine phosphatase genes are expressed in a subset of developing axons and pioneer neurons in the embryonic CNS. *Cell* 67: 661–73

Zuber, M. X., Goodman, D. W., Karns, L. R., Fishman, M. C. 1989. The neuronal growth-associated protein GAP-43 induces filopodia in non-neuronal cells. *Science* 244: 1193–95

Annu. Rev. Neurosci. 1993. 16:597–623

BEHAVIORALLY BASED MODELING AND COMPUTATIONAL APPROACHES TO NEUROSCIENCE

George N. Reeke, Jr. and Olaf Sporns

The Neurosciences Institute, New York, NY 10021

KEY WORDS: artificial neural networks, synthetic neural modeling, neuronal group selection, emergent behavior, reinforcement learning

INTRODUCTION

In recent years, numerous computer models that attempt to illuminate aspects of behavior have appeared. These models form a highly heterogeneous collection and operate at different levels of description and explanation. Many of them approach the problem from a mathematical or engineering perspective, rather than a biological one. Others are concerned with the nervous system, but only some instantiate a complete, behaving system. In this review, we attempt to describe the assumptions and kinds of results obtained with several of these approaches, illustrating each with a few examples. We make no attempt to be exhaustive.

These different approaches reflect the wide divergence of opinion that exists concerning the proper role of models, the extent to which adaptive behavior can be functionally described independently of its physical basis in the nervous system, and, indeed, the very possibility of neural explanations of behavior. The nature and limitations of models have been widely discussed (e.g. Born 1987; Estes 1975; Gyr et al 1966; Kenny 1971; Lachter & Bever 1988; Neisser 1963; Pinker & Prince 1988; Reeke & Edelman 1988; Searle 1984; Skinner 1990; Winograd & Flores 1986). These discussions have variously argued for the importance of formal rules, of experience in the world, or of evolutionary history in understanding behavior, but little overall agreement has been reached on these issues.

597

0147–006X/93/0301–0597$02.00

Our aim here is to clarify the relationship between theoretical approaches and what has been or might potentially be explained by the corresponding models. We limit ourselves to models that generate some form of observable behavior, because such models are the most likely ones to avoid errors of interpretation that arise from the so-called "homunculus problem" (Reeke & Edelman 1988).

In its most basic form, the homunculus problem occurs when a theory proposed to explain some class of behavior involves the action of an entity whose internal mechanisms are not specified in that theory. This entity generally is invoked to perform some necessary subset of the functions of the brain that is considered by the modeller to be not germane to the theory at hand. It thus constitutes a "homunculus," or "little man," by virtue of having at least some of the cognitive capacities of the "big man" that contains it. Descartes' theory of vision was an early example of a homuncular theory: Once visual impressions were conveyed to the pineal gland, critical elements of analysis, which were not explained by the theory, were postulated to occur there. The theory was thus incomplete.

Many modern "neural network" models of behavior in which the input and output are arrays of simulated neural firing amplitudes may be criticized as homuncular, in that a human observer is required to provide suitable neural inputs and to infer the behavior that would result from the neural outputs obtained from the model. In any particular case, these interpretive steps might inadvertently contain key elements of the mechanisms that underlie the phenomena that the model is attempting to explain. Accordingly, an integrated approach to the modeling of behavior, in which inputs are actual sensors and outputs are actual effectors working in an environment, is most likely to avoid oversimplification of the problem that might otherwise render tests of a particular theory worthless.

Adoption of an integrated approach to behavior still leaves the modeler with a wide range of theoretical and practical choices, not least of which is the level of organization—neural, computational, or psychological—at which explanations for behavior are sought. Although the nervous system is universally agreed to provide a physical substrate for the control of behavior, it is not so widely agreed that one needs to study the nervous system to understand behavior. Ultimately, this choice depends on one's assessment of the causative relevance of activity at the neural level upon higher levels of organization. Choice of the psychological level denies (or postpones) the possibility of relating behavior to neuronal activity; choice of an abstract computational model denies the importance of an organism's evolutionary and developmental history. Yet, these choices are made because study of the nervous system itself is so difficult. If a complete understanding of behavior could be obtained by physiological observation alone, we can be sure that other levels of explanation would disappear, and computer models would cease to appear in neuroscience journals.

Models thus help to distinguish or refine theories in the absence of conclusive experimental data by making predictions of unobvious consequences from given assumptions. They permit theories to be tested against classes of data whose relevance is not otherwise apparent. In fortunate instances, modeling can help bridge between neuronal and higher levels of organization, permitting, for example, statements about neuronal mappings to be transformed into statements about perceptual mappings that can be compared with behavioral data from psychology experiments. Models can also indicate which aspects of variability in a particular experiment are important to a particular theory and which may safely be ignored.

To determine whether a particular model fulfills these functions, one must ask whether it explains some aspect of behavior, or merely describes or predicts it. To have explanatory status, a model must replicate the causal mechanisms of the system being modeled, not merely simulate its output (Webb 1991). Although this distinction is rather obvious, it has been glossed over by many authors. For example, Zipser & Andersen (1988), discussing their model for the response properties of certain posterior parietal neurons, state that "the back propagation paradigm [the approach used in their model] . . . obviously cannot be applied literally to the brain, because information does not travel backwards rapidly through axons" and yet "the striking similarity between model and experimental data certainly supports the conjecture that the cortex and the network generated by back propagation compute in similar ways." Unfortunately, Moore's theorem (discussed in Uttal 1990) guarantees that two finite state automata that behave identically under any finite set of fixed tests can have totally different internal structures and behave differently under tests not in the original set. This theorem undoubtedly applies to more complex systems, such as the brain. Thus, no amount of behavioral simulation can prove that a model is correct, and the ultimate test remains the experimental verification of internal mechanisms.

In the absence of such verification, we suggest some criteria by which the aptness of models can be provisionally assessed. We then present several early models that represent landmarks in the development of behaviorally based modeling. These are followed by a survey of some current models in each of several different classes, with a summary of the salient characteristics we have used to classify and evaluate them.

Criteria of Success for Models

With the understanding that we are concerned with the bases for behavior as it occurs in the biological world, and not with engineering principles for building better behaved artifacts, we suggest that three major criteria must be met by a successful theory: Its assumption must be consistent with available experimental data. It must, in fact, explain behavior, i.e. it must contain no overt or covert homunculi. And, it must make testable predictions that are borne out by experiment.

The third point is obvious. The second has already been discussed, but note that homunculi can occur outside a system as well as inside it. This is true when the model system requires training by an external agent that has any of the behavioral characteristics being modeled, for example, the ability to categorize environmental stimuli. In this case, the trainer must contain, in effect, a system with the same abilities as the system under test or more; the combination of trainer and trainee must be modeled to avoid the homunculus problem. Most models that are trained by "supervised learning" (the paradigm introduced to train the perceptron) (Rosenblatt 1958) contain a hidden homunculus of this kind.

The first point (consistency with biological data) is also worthy of some elaboration, because it has so often been honored in the breach. Some of the general facts most often ignored in the construction of behavioral models are the following (a more detailed discussion is given in Edelman 1987 and in Thorpe & Imbert 1989):

1. The wiring of the nervous system is highly variable from individual to individual and changes during the lifetime of each individual.
2. No single cell in the nervous system uniquely produces signals that have a particular meaning or a reproducible pattern in time. In fact, single cells in the nervous system may die at any time with very little apparent effect.
3. Neuronal processes are relatively slow; behavioral responses to stimuli typically occur in times that are compatible with only a few sequential neuronal firings.
4. Neuronal connections are unidirectional.
5. Connectivity in the nervous system, even in local areas, is very sparse relative to the number that would be required for complete connectivity.
6. No natural mechanism is known by which the output of a neuron can be set directly to a desired value.
7. No mechanism is known by which symbolic representations of information or computational procedures can be encoded in DNA sequences and transmitted between generations.[1]

---

[1] Exceptions to this rule might exist in the form of enzyme cascades capable of carrying out limited sequences of steps. This mechanism is highly inefficient in its use of DNA and hardly capable of representing arbitrary, complex computational procedures of the kind envisioned to exist in computational models of cognitive processing.

Although many proposed models disregard one or more of these criteria, it is possible to construct models that are consistent with all of them. We have proposed a paradigm known as "synthetic neural modeling," which provides a general framework for the construction of such models (Reeke et al 1990a). Keystones of this approach are the formulation of a general theory of nervous system function that considers its evolutionary history and mode of development; the use of biologically realistic elements in a model nervous system that operates according to this theory, simulated together with elements of a particular phenotype and the environment in which that phenotype exists; the simultaneous observation of behavior and low-level neuronal events in this simulated system; and the rigorous testing of the system with basic tasks, such as the ability to categorize objects and events and to carry out simple motor actions. This approach deliberately avoids modeling highly abstract, culturally influenced behaviors (Dehaene & Changeux 1989), such as verbal problem solving, which have been the subject of numerous artificial intelligence (AI) studies. Performance of these tasks must involve more basic skills that are still not well understood. Any appropriate theory can be tested by synthetic neural modeling techniques; the examples we present are based on the theory of neuronal group selection (Edelman 1978, 1987, 1989).

EARLY MODELS

Since the 1940s and 1950s researchers and engineers have attempted to create automata that show "intelligent" or adaptive behavior. Ashby (1940) proposed that adaptive behavior can be considered the behavior of a system at a state of dynamic equilibrium with the environment. This equilibrium was said to be reached and maintained by the action of control loops incorporating negative feedback. Subsequently, Ashby (1952) explored further the adaptive quality of human and animal behavior and, together with Wiener (1948), became a founding father of cybernetics. Cybernetics attempted to explain the goal-seeking behavior of organisms on the basis of mechanical determinism with feedback as its central mechanism. A vital part of the appeal of cybernetics resulted from the design of machines that could show such behavior in ways analogous to actual living organisms. Examples of such devices are Ashby's homeostat (1952); Walter's light-following mechanical tortoise, or "machina speculatrix" (1953); and maze-running electromechanical devices, such as Shannon's "mouse" (1951) or Howard's "rat" (1953). (For more examples, see Kurzweil 1990; Nemes 1970.) At the time, goal-seeking behavior was seen as the essence of intelligence, and the models assumed that such behavior could be explained by a set of mechanical rules (such as feedback). Most researchers did not believe that the actual structure of the nervous systems of humans or animals mattered; it served merely to implement more general schemes and principles. This view has remained very influential in the field of cognitive science and is now known as "functionalism" (Block 1978; Putnam 1960).

With the emergence of computers in the 1950s, the range of mechanical models expanded to include all aspects of human or animal minds. "Thinking" and "problem-solving" were formulated as mechanical processes akin to those carried out inside a computer (Newell et al 1958). In the emerging field of AI, the behavioral capacities of real animals receded into the background, and the developmental aspect of mental and behavioral function was almost entirely ignored. Computer-controlled robotic vehicles, such as "Shakey" (Nilsson 1969; for more recent examples see Cox & Wilfong 1990; Iyengar & Elfes 1991a,b), although ingenious in their design, proved to work reliably only under narrowly defined conditions. These devices were under the control of computer algorithms and did not attempt to incorporate a set of biologically based principles.

CURRENT MODELS

Correlative Models

Although neuroscience has provided many insights into how behavior emerges from brain activity, the direct incorporation of physiological facts and principles into working models of behavior has proven challenging. Correlative models avoid these difficulties by attempting to predict future behavior based on observations of past behavior; their aim is not to provide mechanistic explanations for their predictions. Accordingly, these models do not connect behavior to the nervous system. Nonetheless, they can be of heuristic value for the development of more complete theories.

RESCORLA-WAGNER MODEL   A particularly influential example of this class is the Rescorla-Wagner model, which is aimed at describing relationships between association strengths of stimuli in Pavlovian conditioning (Rescorla & Wagner 1972). As described in Gluck et al (1990), the Rescorla-Wagner model attempts to account for blocking and other conditioning phenomena by proposing that the change (ΔV_i) in the strength of the association between a particular conditioned stimulus (CS_i) and its outcome is proportional to a term that attempts to measure the degree to which that outcome is unexpected, given the stimulus elements present in a particular trial:

$$\Delta V_i = \alpha_i \beta (\lambda - \Sigma_s \varepsilon_s V_s)$$

where α_i is a learning rate parameter applicable to CS_i; β is the corresponding parameter relating to the unconditioned stimulus (US); λ is the maximum possible value of V_i, which is attained when CS_i is the only conditioned stimulus present; and $\Sigma_s \varepsilon_S V_s$ is the sum of the association strengths between all the stimulus elements present in that trial and the US. Thus, as more stimulus elements are added, $\Sigma_s \varepsilon_S V_s$ increases and V_i decreases, consistent with observations of blocking.

This relationship is intuitively reasonable, but suggests no mechanistic basis for its predictions. However, a suggestive connection emerged when Sutton & Barto (1981) showed that the relationship is mathematically equivalent to the so-called "delta rule" of Widrow & Hoff (1960) for adaptive artificial neural networks and is, therefore, equivalent to least-squares error minimization. This convergence suggested a search for neuronal circuits that implement the Rescorla-Wagner rule (Gluck et al 1990). Were such circuits to be found, it would still be necessary, of course, to show experimentally their relevance to the Rescorla-Wagner rule in view of the numerous organizational levels of the nervous system involved in even the simplest conditioned responses. At minimum, an appropriate synthetic neural model would be required to help understand the role of such circuits in the behavioral patterns generated by the system as a whole.

Computational Models (Artificial Intelligence)

Artificial intelligence models are based upon a computational metaphor of brain function that considers data representation and rule-driven manipulation of encoded information to be the enabling elements of cognitive systems. This theory has been most thoughtfully and forcefully set forth by Pylyshyn (1984). Artificial intelligence systems have been most successful in emulating abstract and logical aspects of human behavior, perhaps because some of these are indeed carried out by formal reasoning. However, AI has been less successful in dealing with basic perception, categorization, and movement in the world, which are arguably (Reeke & Edelman 1988) prerequisite to the acquisition and manipulation of formal symbol systems.

SOAR   Rule-based problem solving has a long history in AI (e.g. Newell et al 1958). Early systems clearly suffered from an insufficient knowledge base and tended to fail when tested outside a narrow problem domain. Although some workers believed that insufficient knowledge was the most important deficiency of rule-based systems and attempted to remedy it by constructing ever larger collections of everyday facts expressed as formal production rules (Guha & Lenat 1991), Laird et al (1986) addressed the problem of improving the reasoning ability of rule-based systems. They constructed SOAR, a problem-solving program that attempts to learn from its own efforts at solving problems. The SOAR architecture incorporates elements from ACT* (see below) and contains two key elements derived from an analysis of human problem-solving behavior. The first element, "universal subgoaling," is a process designed to bring the entire power of the system to bear on any new problem that arises. When SOAR does not know how to decide what to do next in a particular situation, it creates a subgoal to decide that question and then attempts to resolve that subgoal. This particularly elegant application of the computer science notion of recursion can succeed in automatically breaking a large problem down into smaller subproblems. The other element, "chunking," is a mechanism for improving the system's efficiency at solving previously encountered problems by linking the result of a chain of reasoning to the initial conditions in such a way that the intermediate reasoning steps are bypassed when a similar set of conditions occurs later. Chunking may be considered a sort of cache for linked problem-solving steps.

Tests carried out with SOAR show good agreement with reaction-time data from human learning experiments. However, like kinetic data in chemistry, reaction time data are not conclusive for deciding internal mechanisms. It is not clear to what extent the essentially inward-looking mechanisms of SOAR can help it increase its "understanding" of relationships between entities in the world, if those relationships are not deducible from its encoded knowledge and rules. It is also not clear how a putative biological system based on SOAR could evolve, obtain its initial quota of symbols and rules, and avoid fatal paralysis when available rules proved inadequate to determine action. Nonetheless, SOAR has proved to be a useful tool for studying aspects of human problem solving.

ACT*   This model, developed by Anderson (1983) (the name signifies "adaptive control of thought"; ACT* is a revised version), attempts to treat cognition as a rule-based system in much the same way as SOAR treats problem-solving. Declarative knowledge acquisition and procedural learning are treated as similarly as possible. The analogy to SOAR extends to the mechanism for rule "compilation," which resembles chunking, and to the extensive tests yielding good agreement with power-law data obtained in human learning experiments. The model is one of few to be applied to human language acquisition. In this task, ACT* learns word meanings and syntax from examples by assuming concepts encoded as rules are already given, in general agreement with the views of Chomsky (1986). In this respect, it shares the assumption of essentially all AI programs, including SOAR, that meaning can be represented entirely within formal symbol systems. This assumption has been strenuously challenged (Bickhard 1991). The need for symbols in formal systems to acquire meaning by interactions with the world outside of the system itself has been well discussed by Putnam (1988) and Harnad (1990), who describes the difficulty as the "symbol grounding" problem.

Emergent Behavior

In recent years, the inability of standard AI to solve fully the problems of planning and controlling movements in the real world has led to a resurgence of interest within the AI community in the behavioral capacities of biological organisms. Much recent work has centered on the design of actual or simulated robots behaving according to biological principles. Common to many of these systems is that their internal control mechanisms neither process explicit representations nor contain a centralized "world model." Instead, these systems, variously called "autonomous agents" (Maes 1990), "animats" (Meyer & Wilson 1991), etc., contain very simple internal architectures, often divided into autonomous modules. The combined action of several modules can produce fairly complex behaviors, such as walking, avoidance, and edge-following. This approach to designing robots is informed by Minsky's (1985) view that intelligence is composed of many mental agents, the "society of mind" (reviewed in Reeke 1991).

SUBSUMPTION ARCHITECTURE   An important line of work is that pursued by Brooks (1989, 1991), who has argued that classical AI systems operate in a way that cannot be transportable to the real world. He attempts to avoid centralized processing and complex internal representations altogether by designing modularized control systems in which each module generates by itself some part of the overall behavior. The modules are arranged in a so-called "subsumption architecture." An important design feature is that direct interconnections between individual modules are hard wired (resembling, for example, the biological concept of lateral inhibition) or are omitted altogether. Another example of this approach is a mobile robot designed by Connell (1990), which collects empty soda cans in an unstructured real-world environment. Connell's robot does not have persistent states; it is directly driven by events in the world at all times. In a related set of simulations inspired by neuroethological observations, Beer (1990) designed a simulation of an artificial insect that shows simple emergent behaviors, such as locomotion, wandering, edge-following, and feeding. Although the behavioral repertoire of the simulation is largely the result of prewired connectivity, the system appears to show adaptive and flexible behavior within its given environment.

Many of the early achievements of systems with "emergent behavior" are quite impressive, and it is encouraging that AI researchers have been inspired by studying animal behavior. However, the insistence on modular architectures and prewired behavioral repertoires is likely to be too restrictive when the approach is scaled up to reproduce or explain the behavior of higher animals. A key structural feature of higher nervous systems is the connectedness of individual neurons within and between functionally specialized regions, providing multimodal memory and a single, coherent picture of the world. Learning and memory are important in shaping an organism's interactions with its environment. In our view, "emergent behavior" is likely to remain limited to the domain of insect-like creatures and fall short of being able to model the behavior of higher vertebrates.

Artificial Neural Networks

The terms "artificial neural network" (ANN) and "connectionist system" encompass a wide variety of systems that contain simple computational elements that are heuristically modeled on neurons and are interconnected in networks. Such systems have a history at least as long as that of AI (e.g. McCulloch & Pitts 1943), but early results were discouraging (Minsky & Papert 1969), and serious development resumed only recently. Artificial neural network systems are extremely diverse, and we refer the reader to other sources for more detailed summaries and criticisms (Graubard 1988; Hanson & Olson 1990; McClelland et al 1986; Pfeifer et al 1989; Poggio 1990; Rumelhart et al 1986; Smolensky 1988; Zornetzer et al 1990). Artificial neural networks have been considered valuable for modeling behavior because they do not require the explicit enumeration of rules commonly required in AI or, alternatively, because they can be used to implement distributed representations of rule-based systems (Touretzky & Hinton 1988). However, ANNs share with AI the fundamental functionalist (Bechtel 1988) assumptions that categories exist a priori in the world independent of a conscious observer and that behavior can be computed by an organism based on detectable regularities in the environment. For example, Sejnowski et al (1988) suggest that a field called "computational neuroscience" can be defined to seek computational explanations for nervous system functions. In their view, "one of the major research objectives of computational neuroscience is to discover the algorithms used in the brain." However, other authors have argued strongly that the brain does not operate by use of algorithms (Edelman 1987), but rather that many elements that cannot be formalized, such as analogy and metaphor, are necessarily involved (Johnson 1987; Lakoff 1987; Searle 1992). To the extent that these arguments are correct, the search for brain algorithms starts from too narrow a premise and is likely to share some of the shortcomings of AI. We describe here a few of the more successful

ANN models that can be interpreted independent of computational descriptions because they generate some form of behavior.

PLACE-FIELD MODELS   Zipser (1986) addressed the problems of place recognition and the ability of animals to navigate toward a remembered goal with the aid of visible landmarks. He began with "place-field" units designed to resemble certain neurons found in rodent hippocampus (O'Keefe & Nadel 1978). The model units responded to a visual scene at a particular location by summing the responses of other units that were activated by the presence of individual landmarks. When these place-field units were modified to respond only when the observer was in a particular (broadly tuned) orientation relative to the scene, their output could be used to guide navigation to a "home" location. However, for reasons involving the method of combining conflicting directional indications from different units, the resulting movement tracks were not as direct as they would be for an actual animal. This problem was eliminated by adopting an alternative, algebraically derived procedure for combining views. Possible neural cognates of this alternative procedure were not identified.

MURPHY   MURPHY (Mel 1990, 1991) is a kinematic controller and path planner that controls a robot arm based on video images. It is perhaps most noteworthy for combining AI planning techniques with a connectionist architecture for low-level control of the arm. MURPHY contains networks with units that respond to objects in the visual scene (object recognition is simplified by careful control of lighting conditions), joint angles and velocity, and hand direction. These networks develop an implicit kinematical model of the camera-arm system, thus avoiding the need for an explicit description of the system's geometry. This system can guide the arm to unobstructed targets; the AI component adds the capability to reach objects in the presence of obstacles. The relative success of this AI-ANN hybrid indicates that each approach can to some extent complement aspects that are lacking in the other.

INFANT   This system (Kuperstein 1988a,b, 1991) has similar goals to those of MURPHY, but they are accomplished entirely with a neural network implementation. The arm functions in a three-dimensional world, and there are two video cameras to provide stereoscopic vision. Visual input maps are routed via arrays of adjustable weights onto target maps, which in turn emit arm motor signals. (Multiple maps for different ranges of joint angles minimize problems caused by the possibility of reaching a target via more than one combination of angles.) The need to solve the difficult "inverse dynamics" problem for the robot arm is avoided by a clever training strategy: A random activity generator is used to activate the target map, and the arm moves to a corresponding random position. The arm is visualized by the cameras, thus producing activity in the input maps. Coefficients between input and target maps are adjusted for best match. After many trials, an object placed anywhere in the visual field evokes activity in the input maps similar to that evoked during training by the image of the arm itself at that location; projected to the target map through the connections that have been adjusted, this activity automatically moves the arm to just the position needed to intercept the object. In the version described, INFANT is not concerned with the categorization of the target as being one that is appropriate for reaching and grasping.

MAVIN   Baloch & Waxman (1991) describe a control system for a mobile robot with a modular distributed architecture based on theoretical proposals of Grossberg (Grossberg 1982; Grossberg & Schmajuk 1987). The robot is designed to respond to simple lighted patterns displayed on CRT screens. It learns to discriminate different patterns independent of their location, size, and orientation. A simple analogue of Pavlovian conditioning is used to train the device to approach some stimuli and avoid others. "Early vision" and motor control for locomotion are implemented via conventional computational techniques; visual object recognition, behavioral conditioning, and eye motion control are handled by ANNs. The authors attempt to deal with the problem, which is ignored in most other models, of controlling multiple behaviors with a single system. This is done by using a simple model of "emotional states."

Along with NOMAD (described below), MAVIN is one of the most ambitious neural network robotic control systems yet attempted. Although certain simplifications have been made in its visual and motor control subsystems, MAVIN generally avoids unbiological assumptions. The degree of behavioral richness that will emerge from further tests with this system will be interesting to observe.

OTHER NEURAL NET MODELS OF BEHAVING SYSTEMS   Alternatives to the models presented above have been proposed and may be instructively compared with them. Nelson's (1991) alternative to Zipser's model of visual homing involves formation of direct associations between learned visual patterns and motor commands. Barhen et al (1988) control a mobile robot by neural network techniques and propose a method for extracting symbolic representations from the neural net for input to an AI navigation routine, a task that was accomplished in a more ad hoc fashion in MURPHY. Eckmiller (1990) proposes a "neural triangular lattice" for remembering and reproducing movement trajectories in a robot controller that is quite different from that of Kuperstein. Verschure et al (1992) have designed a model of an autonomous robot that is completely self-organizing and does not rely on predefined knowledge about the world.

The model can be conditioned to approach targets and avoid obstacles at arbitrary positions in its environment.

"Realistic" Neural Models

In this section, we deal briefly with models that incorporate more or less realistic anatomical and physiological characteristics. Some of these models have been integrated in overall architectures that ultimately produce behavior. Although ANNs use neural principles heuristically to suggest solutions to particular problems, realistic neural models are designed with a different purpose in mind: They help us to understand a specific biological phenomenon, reproduce actual physiological experiments, or generate predictions about possible experiments (e.g. Ambros-Ingerson et al 1990; Miller et al 1989; Pearson et al 1987; Traub et al 1989). Here, we discuss a few examples and their potential relevance for understanding brain function.

NEURAL MODELS OF MOTOR SYSTEMS   The connection between neural organization and behavior is especially close for neural circuitry generating rhythmic output patterns. How rhythmic output can emerge from the activity of a few neurons has been worked out for several systems, e.g. the pyloric network of the crustacean stomatogastric ganglion, the generators of swimming movements in the marine mollusc *Tritonia*, and locomotion in the locust. The neural basis of swimming movements in the lamprey, a lower vertebrate, has been extensively studied both experimentally and in network models by Grillner (Grillner et al 1991; Grillner & Matsushima 1991). In the models, alternating burst activity is generated in segmental spinal networks that contain a relatively small number of well-defined neurons. Several types of conductances, including that produced by voltage-dependent N-methyl-D-aspartate (NMDA) channels, are explicitly modeled, making it possible to observe how specific biochemical modifications influence global network behavior. For example, as in the cognate experiments, the frequency of the rhythmic activity that the network produces depends on the level of bath-applied NMDA (Grillner et al 1988). In addition to modeling the generation of rhythmic movements within one spinal segment, simulations have addressed intersegmental coordination and the role of sensory input (Grillner et al 1990; Williams et al 1990). These studies have shown that it is possible to construct neurally based models that produce a simple behavior, in this case, locomotion.

In a contrasting study, Lockery et al (1989) used a biologically less persuasive back-propagation network to model the system that controls withdrawal-bending movements in the leech. While noting that their model is not physiologically realistic, Lockery et al make the subtle argument that it nonetheless demonstrates that the known anatomical features they have included must be sufficient to perform the task assigned to them, because they can do so even in the less general back-propagation network. This resembles the justification of back propagation given by Zipser & Andersen (1988) and mentioned above.

Because of the vastly greater complexity of the underlying neural mechanisms, there are still very few neurally based models of behavior in higher vertebrates. Some have argued that the study of simple organisms, such as *Aplysia*, will ultimately reveal principles that hold also for higher vertebrates. It appears to us, however, that the behavior of higher vertebrates will not be completely understood by accumulating and elaborating invertebrate circuitry. Adequate neural models of vertebrate behavior will have to be rooted in the full complexity of vertebrate anatomy and physiology.

VISUALLY GUIDED BEHAVIOR IN AMPHIBIA   Several models, collectively called *Rana computatrix*, which address visually guided behavior in frogs and toads, have been constructed by Arbib (Arbib 1989, 1991). Several components deal with the frog's retinotectal system and the organization of tectal columns (Lara et al 1982), as well as with depth perception, detour behavior, and prey-predator discrimination (Cervantes Pérez et al 1985). In a recent neural simulation of the frog's retina, tectum, and thalamic nuclei, Wang & Arbib (1991) examine possible neural strategies for the discrimination of worm-like stimuli. Based on their simulations, these authors predict that pattern discrimination is achieved in the anterior thalamus, which is anatomically an "early" sensory area. Moreover, Wang & Arbib predict certain behavioral characteristics (dishabituation hierarchies) for sets of stimuli that have not yet been tested in the real animal.

Synthetic Neural Models

Realistic models of behaving nervous systems require a theoretical and modeling approach that considers multiple organizational levels, including synaptic, neuronal, area, and global levels of the nervous system; the phenotype of an organism (real or hypothetical); and its history of stimulation and responses in the environment. The synthetic neural modeling approach (Reeke et al 1990a) alluded to earlier was designed to fulfill these requirements. Synthetic neural modeling uses large-scale simulations to study interactions between events at all these levels. Because all relevant variables are represented in the computer, correlations between events in different brain areas, as well as between events in the brain and in the simulated environment, can be readily recorded and analyzed. This approach requires a comprehensive theory of the nervous system; the one employed in the studies we describe is the theory of neuronal group selection (TNGS).

NEURONAL GROUP SELECTION A guiding principle of the TNGS (Edelman 1978, 1987, 1989) is that the nervous system operates by selection on preexisting variance in developmentally generated populations of synapses. According to the theory, such selection is driven by the adaptive value of behavioral acts performed by the organism. It is accomplished by differential modification of synaptic strengths, leading to the formation of strongly interconnected local collectives of neurons known as "neuronal groups." A number of such groups forms a "repertoire," the unit within which selection occurs. Competition between neuronal groups in a repertoire mediated by a similar process of selection leads to enhancement of adaptive behavior by differential strengthening or weakening of synaptic connections into, within, and from those groups whose activity contributes to that behavior.

A central component of the theory is the notion of reentry, which is a form of ongoing, reciprocal exchange of signals between neuronal repertoires in the same or different pathways. Reentry functions to assure that responses are consistent across different neuronal repertoires at any one time and across similar environmental situations occurring at different times. It facilitates the expression of mutually consistent patterns of response in the repertoires it links. It thus integrates disparate responses in repertoires that represent different sensory modalities or submodalities, thus forming the basis of "global mappings" that are key to the classification of stimuli.

Several key predictions of the TNGS have found experimental support. It was predicted (Edelman 1978) that neuronal groups would be found to exist as local populations of strongly interconnected neurons that tend to share their receptive field properties and temporal patterns of discharge. Neuronal groups composed of similarly tuned cells and exhibiting coherent patterns of activity have been demonstrated in the visual cortex (Gray & Singer 1989) and motor cortex (Murthy & Fetz 1992); domains of coactive neurons in neonatal rat cortex (Yuste et al 1992) could be early developmental precursors of neuronal groups. Less direct evidence for the existence of such groups comes from other reports of locally correlated neuronal discharges and frequent observations of clustering of neurons with similar receptive field properties in many areas of the brain. There is also evidence for the occurrence of reentry within and between cortical areas. Interactions along cortico-cortical connections can give rise to temporal correlations between distant neuronal groups (Eckhorn et al 1988; Engel et al 1991; Nelson et al 1992). These findings concerning groups and reentry support fundamental premises of the theory; nonetheless, the theory will require serious revision if important brain functions are shown to depend on the activity of single neurons, or if cortical processing is essentially feedforward and hierarchical with only minor roles for intra- or inter-areal reentry.

Selective mechanisms have been modeled in detail at the level of synapses (Changeux & Dehaene 1989; Finkel & Edelman 1987), local neuronal groups (Finkel & Edelman 1985; Pearson et al 1987; Sporns et al 1989), and the reentrant interaction of multiple cortical maps (Finkel & Edelman 1989; Sporns et al 1991; Tononi et al 1992a,b). Here, we focus on the behavior of a simulated automaton called Darwin III (Reeke et al 1990a,b; Reeke & Sporns 1990).

DARWIN III   Darwin III is a model of a hypothetical simple organism designed to test the main ideas of the TNGS. It has a movable eye and a four-jointed arm with which it can reach out and grasp objects in its two-dimensional environment. These effectors were chosen to permit simple categorizations and motor actions to be modelled in a simulated environment—a more complex system that operates in the real world is described below. The nervous system of Darwin III comprises about 50 inter-connected networks, with a total of 50,000 neurons and 620,000 synapses. The connections, initial connection strengths, and rules for cell responses and synaptic modifications are specified parametrically for each network to establish permissive conditions for the desired behaviors to be selected, but neither prior information about particular stimuli nor explicit algo-rithms for "neural computations" are specified. Training of behavioral responses is unsupervised, i.e. it proceeds in the absence of internal or external representation of desired responses or feedback of detailed error signals into the system by a "teacher." The adaptive value of responses is determined solely by their consequences; criteria for the selection of responses are given neural expression by internal "value systems" that are presumed to arise in animals from evolutionary constraints.

Value systems   In Darwin III, the modification of synapses of certain classes is biased by heterosynaptic inputs from specialized sets of neurons whose responses reflect the automaton's global evaluation of its recent behavior. These neurons have connectivities that allow them to respond to the outcome of adaptive behaviors, and as such they instantiate value systems (Reeke & Edelman 1986). Value systems do not predefine the exact way a behavioral response is executed or determine particular perceptual categories. Rather, they impose biases on synaptic modifications depend-ing on the outcome of previous interactions with the environment. Obvi-ously, value systems must be simple enough to perform their function purely as a consequence of their genetically determined anatomy and cellular physiology. If this were not the case, they would constitute yet another example of hidden homunculi in neural models.

Value systems operate by responding to the consequences of behavior ex post facto. Their output influences large populations of synaptic connections in a diffuse and modulatory fashion; there is no attempt to estimate and correct "errors" at individual synapses, as is the case for the back-propagation training rule used in many ANNs. There is a resemblance between the use of value systems in Darwin III and some variants of more recently described "reinforcement learning" (Mazzoni et al 1991a,b; see also Sutton & Barto 1981). Value systems can be understood as basic evolutionary adaptations that define broad behavioral goals for an organism in terms of their recognizable consequences.

Value systems are based on structures in real nervous systems. Characteristic features of the value repertoires used in Darwin III include the presence of sensory afferents, a relative lack of internal order and topography, and diffuse and widespread efferents that heterosynaptically influence large populations of synapses. Several brainstem nuclei, apparently present in all mammalian species, have widespread efferent connections and presumably modulatory influence on cortical activity (Fallon & Loughlin 1987). The locus coeruleus, for example, sends out axons to large and diverse regions of the cerebral cortex (Foote et al 1983; Gatter & Powell 1977). Several ascending monoaminergic and cholinergic fiber systems specifically influence the responses of cortical target neurons (reviewed in Foote & Morrison 1987). These fiber systems could also be involved in facilitation or amplification of synaptic populations in their target areas and, thus, could serve as biological correlates of value systems.

General properties of Darwin III's units and synaptic rules   A simulation program (the Cortical Network Simulator,"CNS") (Reeke & Edelman 1987) was written to permit the construction of very general neural network systems. In addition to neural repertoires, CNS provides for the simulation of three sensory modalities: vision, touch, and kinesthesia. Signals from the simulated senses are generated in the form of neuronal activity in specialized cells that can be connected to any of the neural repertoires in the model. On the motor side, there may be one or more eyes with lateral and vertical orbital motions, and one or more arms with multiple joints, each controlled by neurons in specified repertoires that are designated as "motor repertoires."

In the instantiation of Darwin III described here, a single eye and a single arm were used. Four sets of interconnected repertoires were constructed: a foveation and fine-tracking oculomotor system, a reaching system using a single multijointed arm, a touch-exploration system using a different set of "muscles" in the same arm, and a reentrant categorizing system. The four systems together form an automaton capable of autonomous behavior involving sensorimotor coordination and simple categorization of stimulus objects.

The responses of neuronal units in Darwin III are determined by a response function that has terms corresponding to a linear summation of thresholded synaptic inputs, noise, output saturation, decay of previous activity, depression, and refractory periods. Training occurs by permanent alterations in synaptic weights, c_{ij}, according to the equation

$$c_{ij}(t+1) = c_{ij}(t) + \delta \cdot \phi(c_{ij}) \cdot (\bar{s}_i - \theta_I) \cdot (m_{ij} - \theta_J) \cdot (v - \theta_V) \cdot R$$

where i and j specify post- and presynaptic cells, respectively; t represents time; δ is a parameter that adjusts the overall rate of synaptic change; $\bar{s}_i$ is the time-averaged activity of cell i; θ_I is an amplification threshold relating to postsynaptic activity; m_{ij} is the average concentration of a hypothetical postsynaptic "modifying substance" that is produced when presynaptic activity is present, and that decays exponentially; θ_J is an amplification threshold relating to presynaptic activity; v is the magnitude of hetero-synaptic input from relevant value scheme neurons; θ_V is an amplification threshold relating to value; and R is a rule selector that may be set to $+1$, 0, or -1 for various combinations of the signs of the three thresholded terms in the equation. Positive values of R lead to enhancement of synapses with correlated pre- and postsynaptic activity (selection) (cf. Hebb 1949, whose proposed rule did not incorporate value systems); negative values of R lead to suppression of such synapses (homeostasis). By choice of the parameters in the synaptic rule, it is possible to simulate any of a wide variety of different kinds of synapses, with properties corresponding, for example, to those of synapses using different neurotransmitters. The result-ing diversity of possibilities has been referred to as "transmitter logic" (Edelman 1987). (For detailed descriptions of the response and ampli-fication functions used in Darwin III, see Reeke et al 1990a.)

Behavior of Darwin III   Darwin III could foveate on objects in its environ-ment, track their movements, reach out with its arm and touch their surfaces, trace their contours, and respond according to whether an object belonged to a particular category (defined by visual and tactile attributes). Its sensorimotor systems (oculomotor and reaching systems) behaved initially at random. Responding to objects in the simulated environment, Darwin III generated exploratory motor activity (random eye movements or "flailing" gestures with its arm). Its internal value systems responded to the consequences of these spontaneous motor acts and continuously influenced the modification of synaptic connections (in most cases, those linking sensory to motor networks). After a relatively short period of time (on the order of a few hundreds to thousands of trials), Darwin III's motor systems functioned with high accuracy; objects presented anywhere within the environment were tracked visually and grasped by the arm (Figure 1).

In its use of random movements to train a kinematic system, the INFANT model discussed above has some features in common with the selective paradigm used to train reaching movements in Darwin III. However, INFANT selects angles, which directly specify the final position or posture of the arm, whereas Darwin III selects gestures, which direct the course of the arm toward its goal. Thus, the dynamics of movement are under nervous system control in Darwin III, but not in INFANT.

Reaching in Darwin III was incorporated into a higher-level behavioral pattern, including stimulus categorization and contingent response. In the categorization system, visual and tactile signals were combined and, as a result of reciprocal, reentrant interactions between visual and touch centers, a behavioral reflex was triggered if an object was visually striped and had "bumps" on its surface. This reflex "flicked" the object away from the vicinity of the automaton. Within limits, Darwin III's behavior could be conditioned—"rewarding" responses to one category of objects gave rise to a marked increase in response frequency, with full generalization to all locations within the environment and to members of the category not part of the training set. Conditioning could be rapidly reversed by switching the reward from one category to another.

In summary, Darwin III offered an opportunity to visualize and relate events at different levels—synapse, cell, group, network, system, and behavior—within a given time period. Darwin III showed sensorimotor coordination in connection with the emergence of relatively rich perceptual categorization. Although Darwin III has made a few steps in the direction of conditioned learning, it cannot yet learn in a fashion that would enable it to change its behavioral responses in any rich fashion toward environmental stimuli as a consequence of its own experience. Learning in this sense would require a greater elaboration of internal value-related states and the ability to couple these into connections between category and motor centers.

DARWIN IV AND NOMAD   Recently, we have begun to explore the applicability of synthetic neural modeling to behavior in a real-world environment (Edelman et al 1992). We have designed a neurally organized multiply adaptive device (NOMAD) controlled by a simulated nervous system. The complete system is called Darwin IV. The nervous system of Darwin IV contains sensory and motor networks analogous to those in Darwin III, in which connection strengths can change according to activity and value. So far, we have demonstrated that NOMAD can be trained to track a light moving in a random path and to approach and segregate red cubes from blue cubes based on value signals related to their respective "taste" (modelled by electrical conductivity). Embedding the system in a real-world environment helps avoid several possible pitfalls of an approach

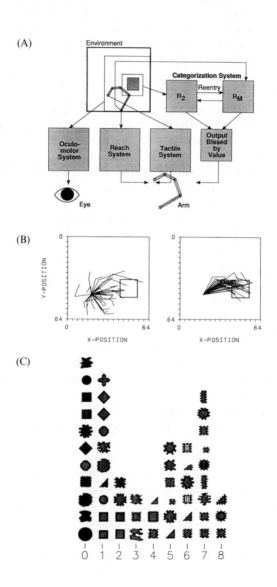

based on simulation only. Real-world stimuli are inherently richer and more variable than simulated ones, and it is harder to bias the system's responses by (inadvertently or deliberately) biasing the stimulation protocol. Synthetic neural modeling as applied to real-world artifacts, such as NOMAD, offers a unique testing ground for behaving neural models.

Taken together, Darwin III and Darwin IV have demonstrated that neural-level models can be coupled to the behavioral level. They show that study of the brain, the phenotype, and the environment as a total system helps clarify many aspects of brain structure that are otherwise inexplicable. They make it possible to understand how the anatomical variance and diversity of the nervous system can be interpreted as key evolved features that lead to its functional capabilities, not as sources of noise and error to be overcome through complex adaptations. Finally, the models show that the TNGS provides a coherent and efficacious description of one way that the brain might acquire its adult functional competence through accepted evolutionary and developmental mechanisms, without a need for homunculi, external teachers, or innate symbolic schemes and formal reasoning.

Figure 1 (*A*) Schematic diagram of the subsystems of Darwin III. Objects (stippled square on right of "Environment" box) move within an "environment," a portion of which is viewed by the eye (large square: limits of peripheral vision; small square: central vision). Movements of the eye and the four-jointed arm (*bottom*) are controlled by the oculomotor and reach systems, respectively. The distal digit contains touch sensors used by the tactile system to trace the edges of objects. The categorization system receives sensory inputs from the central part of the visual field and from joint receptors that signal arm movements. These inputs connect to a higher-order visual center, R_2, and to an area correlating motion signals over time, R_M. Correlation of firing patterns in these repertoires via reentry leads to classification of objects and eventually yields an output biased by value that can activate reflex movements of the arm. (*B*) Traces of paths taken by the tip of the arm before (*left*) and after (*right*) 180 training cycles. Training proceeded as follows: For each of 30 trials, the arm was placed in a standard position with the tip at the point where the trajectories diverge (*lower left*). It was then allowed to move for six cycles, and synaptic changes occurred depending upon the success of the movements relative to a target object whose position is shown by the square at the right. After training, movements that reached the object on a direct path have been selected. (*C*) Distribution of categorization frequencies. Objects are grouped according to the number of flailing responses occurring in eight trials with one version of Darwin III's categorization system. Activation of the flailing response at any time within a 50-cycle time limit on each trial activated the arm and removed the object from the environment. Each such occurrence was counted as a "rejection" response. If no response occurred within the 50-cycle time limit, the trial was ended and a new object was entered. Objects are arranged in nine columns depending on the frequency with which they met with a response. In this version of Darwin III, an intrinsic negative ethological value is attached to "bumpy and striped" objects. (Part *A* and *C* reproduced with permission of The Neurosciences Research Foundation; Part *B* reproduced with permission of the Cornell Theory Center.)

SUMMARY AND CONCLUSIONS

The almost incredible advances that have recently occurred in the power of computers available to scientists in all disciplines have encouraged an explosion of neural network and behavioral models. Some of these have been constrained more by the imagination of the programmer than by rude biological facts. Their relevance for the experimental neuroscientist thus varies from case to case. Some models (e.g. Grillner's model of lamprey swimming movements) are so closely based on known neuro-anatomy and neurophysiology that it becomes possible to generate and test precise experimental predictions. Other models (such as MURPHY and NOMAD) use neurobiological principles in their architectures, but do not portray any particular organism. Although it is harder to relate the study of these models of the study of real animals, they fulfill an important explanatory role. They make possible insights into how behavior is con-trolled by neuronal activity that would be unobtainable in real animals using present methods.

Thus, even the excesses of neural modeling have provided a useful impetus to what is undoubtedly a most promising approach to integrating data from the various disciplines concerned with behavior and the mind. The problems have been pointed out by many authors (see citations in our introduction), and a phase of more critical evaluation appears to have begun. We hope that our brief survey of models based on widely different theoretical approaches, but all aimed at explaining behavior, will encour-age critical comparisons to be made. As in more mature fields, such as thermodynamics, we can expect that more complete models will force an evaluation of theoretical hypotheses against the entire body of available evidence, rather than just a few pertinent test cases. Such evaluation will make possible a much more rigorous exclusion of invalid or inconsistent theoretical ideas. From such studies, a much smaller, but more robust, set of basic principles can be expected to emerge. From the perspective afforded by our own modeling studies, it appears essential that modeling be informed by a general theory of brain function. In this work, the theory of neuronal group selection provides a useful basis for further work by virtue of its consistency with basic evolutionary and physiological prin-ciples and the power of the selection paradigm to shape neural networks in behaviorally adaptive directions.

ACKNOWLEDGMENTS

The authors' research was carried out as part of the theoretical neuro-biology research program at The Neurosciences Institute, which is supported through the Neurosciences Research Foundation. The Foundation receives major support for this research from the John D. and Catherine T. MacArthur Foundation and the Lucille P. Markey Charitable Trust. Olaf Sporns is a W. M. Keck Foundation Fellow. We thank G. M. Edelman for many helpful discussions and critical suggestions.

Literature Cited

Ambros-Ingerson, J., Granger, R., Lynch, G. 1990. Simulation of paleocortex performs hierarchical clustering. *Science* 247: 1344–48

Anderson, J. R. 1983. *The Architecture of Cognition*. Cambridge, Mass: Harvard Univ.

Arbib, M. A. 1991. Neural mechanisms of visuomotor coordination: The evolution of *Rana computatrix*. In *From Animals to Animats. Proc. First Int. Conf. on Simulation of Adaptive Behavior*, ed. J.-A. Meyer, S. W. Wilson, pp. 3–30. Cambridge, Mass: MIT Press

Arbib, M. A. 1989. Visuomotor coordination: Neural models and perceptual robotics. In *Visuomotor Coordination: Amphibians, Comparisons, Models, and Robots*, ed. J.-P. Ewert, M. A. Arbib, pp. 121–71. New York: Plenum

Ashby, W. R. 1952. *Design for a Brain*. London: Chapman & Hall

Ashby, W. R. 1940. Adaptiveness and equilibrium. *J. Ment. Sci.* 86: 478–83

Baloch, A. A., Waxman, A. M. 1991. Visual learning, adaptive expectations, and behavioral conditioning of the mobile robot MAVIN. *Neural Netw.* 4: 271–302

Barhen, J., Dress, W. B., Jorgensen, C. C. 1988. Applications of concurrent neuromorphic algorithms for autonomous robots. In *Neural Computers*, ed. R. Eckmiller, C. von der Malsburg, pp. 321–33. Berlin: Springer-Verlag

Bechtel, W. 1988. *Philosophy of Mind: An Overview for Cognitive Science*. Hillsdale, NJ: Erlbaum

Beer, R. D. 1990. *Intelligence as Adaptive Behavior. An Experiment in Computational Neuroethology*. Boston: Academic

Bickhard, M. H. 1991. Cognitive representation in the brain. *Encycl. Human Biol.* 2: 547–58

Block, N. 1978. Troubles with functionalism. In *Perception and Cognition. Issues in the Foundations of Psychology*, ed. C. W. Savage, pp. 261–325. Minneapolis: Univ. Minn. Press

Born, R., ed. 1987. *Artificial Intelligence: The Case Against*. London: Croom Helm

Brooks, R. A. 1991. New approaches to robotics. *Science* 253: 1227–32

Brooks, R. A. 1989. A robot that walks: emergent behaviors from a carefully evolved network. *Neural Comput.* 1: 253–62

Cervantes Pérez, F., Lara, R., Arbib, M. A. 1985. A neural model of interactions subserving prey-predator discrimination and size preference in anuran amphibia. *J. Theor. Biol.* 113: 117–52

Changeux, J.-P., Dehaene, S. 1989. Neuronal models of cognitive functions. *Cognition* 33: 63–109

Chomsky, N. 1986. *Cartesian Linguistics*. New York: Harper & Row

Connell, J. H. 1990. *Minimalist Mobile Robotics. A Colony-Style Architecture for an Artificial Creature*. Boston: Academic

Cox, I. J., Wilfong, G. T., eds. 1990. *Autonomous Robot Vehicles*. New York: Springer-Verlag

Dehaene, S., Changeux, J.-P. 1989. A simple model of prefrontal cortex function in delayed-response tasks. *J. Cogn. Neurosci.* 1: 244–61

Eckhorn, R., Bauer, R., Jordan, W., Brosch, M., Kruse, W., et al. 1988. Coherent oscillations: a mechanism of feature linking in the visual cortex? Multiple electrode and correlation analyses in the cat. *Biol. Cybern.* 60: 121–30

Eckmiller, R. 1990. The design of intelligent robots as a federation of geometric machines. In *An Introduction to Neural and Electronic Networks*, ed. S. F. Zornetzer, J. L. Davis, C. Lau, pp. 109–28. San Diego: Academic

Edelman, G. M. 1989. *The Remembered Present: A Biological Theory of Consciousness*. New York: Basic Books

Edelman, G. M. 1987. *Neural Darwinism: The Theory of Neuronal Group Selection*. New York: Basic Books

Edelman, G. M. 1978. Group selection and phasic reentrant signaling: a theory of higher brain function. In *The Mindful*

Brain: Cortical Organization and the Group-Selective Theory of Higher Brain Function, ed. G. M. Edelman, V. B. Mountcastle, pp. 51–100. Cambridge, Mass: MIT Press

Edelman, G. M., Reeke, G. N. Jr., Gall, W. E., Tononi, G., Williams, D., et al. 1992. Synthetic neural modeling applied to a real-world artifact. *Proc. Natl. Acad. Sci. USA* 89: 7267–71

Engel, A. K., König, P., Kreiter, A. K., Singer, W. 1991. Interhemispheric synchronization of oscillatory neuronal responses in cat visual cortex. *Science* 252: 1177–79

Estes, W. K. 1975. Some targets for mathematical psychology. *J. Math. Psychol.* 12: 263–82

Fallon, J. H., Loughlin, S. E. 1987. Monoamine innervation of cerebral cortex and a theory of the role of monoamines in cerebral cortex and basal ganglia. In *Cerebral Cortex: Further Aspects of Cortical Function, Including Hippocampus*, ed. E. G. Jones, A. Peters, pp. 41–127. New York: Plenum

Finkel, L. H., Edelman, G. M. 1989. The integration of distributed cortical systems by reentry: a computer simulation of interactive functionally segregated visual areas. *J. Neurosci.* 9: 3188–3208

Finkel, L. H., Edelman, G. M. 1987. Population rules for synapses in networks. In *Synaptic Function*, ed. G. M. Edelman, W. E. Gall, W. M. Cowan, pp. 711–57. New York: Wiley

Finkel, L. H., Edelman, G. M. 1985. Interaction of synaptic modification rules within populations of neurons. *Proc. Natl. Acad. Sci. USA* 82: 1291–95

Foote, S. L., Bloom, F. E., Aston-Jones, G. 1983. The nucleus locus coeruleus: New evidence of anatomical and physiological specificity. *Physiol. Rev.* 63: 844–914

Foote, S. L., Morrison, J. H. 1987. Extrathalamic modulation of cortical function. *Annu. Rev. Neurosci.* 10: 67–95

Gatter, K. C., Powell, T. P. S. 1977. The projection of the locus coeruleus upon the neocortex in the macaque monkey. *Neuroscience* 2: 441–45

Gluck, M. A., Reifsnider, E. S., Thompson, R. F. 1990. Adaptive signal processing and the cerebellum: Models of classical conditioning and VOR adaptation. In *Neuroscience and Connectionist Theory*, ed. M. A. Gluck, D. E. Rumelhart, pp. 131–85. Hillsdale, NJ: Erlbaum Assoc.

Graubard, S. R., ed. 1988. *The Artificial Intelligence Debate: False Starts, Real Foundations*. Cambridge, Mass: MIT Press

Gray, C. M., Singer, W. 1989. Stimulus-specific neuronal oscillations in orientation columns of cat visual cortex. *Proc. Natl. Acad. Sci. USA* 86: 1698–1702

Grillner, S., Buchanan, J. T., Lansner, A. 1988. Simulation of the segmental burst generating network for locomotion in lamprey. *Neurosci. Lett.* 89: 31–35

Grillner, S., Matsushima, T. 1991. The neural network underlying locomotion in lamprey—synaptic and cellular mechanisms. *Neuron* 7: 1–15

Grillner, S., Wallén, P., Brodin, L. 1991. Neuronal network generating locomotor behavior in lamprey: circuitry, transmitters, membrane properties, and simulation. *Annu. Rev. Neurosci.* 14: 169–99

Grillner, S., Wallén, P., Brodin, L., Lansner, A., Ekeberg, Ö., et al. 1990. Neuronal network generating lamprey locomotion—experiments and simulations—supraspinal, intersegmental mechanisms. *Soc. Neurosci. Abstr.* 16: 726

Grossberg, S. 1982. *Studies of Mind and Brain: Neural Principles of Learning, Perception, Development, Cognition, and Motor Control*. Boston: Reidel

Grossberg, S., Schmajuk, N. A. 1987. Neural dynamics of attentionally modulated Pavlovian conditioning: conditioned reinforcement, inhibition, and opponent processing. *Psychobiology* 15: 195–240

Guha, R. V., Lenat, D. B. 1991. Cyc: A midterm report. *Appl. Artif. Intell.* 5: 45–86

Gyr, J. W., Brown, J. S., Willey, R., Zivian, A. 1966. Computer simulation and psychological theories of perception. *Psychol. Bull.* 65: 174–92

Hanson, S. J., Olson, C. R., eds. 1990. *Connectionist Modeling and Brain Function*. Cambridge, Mass: MIT Press

Harnad, S. 1990. The symbol grounding problem. In *Emergent Computation*, ed. S. Forrest, pp. 335–46. Amsterdam: North-Holland

Hebb, D. O. 1949. *The Organization of Behavior: A Neuropsychological Theory*. New York: Wiley

Howard, I. P. 1953. A note on the design of an electro-mechanical maze runner. *Durham Res. Rev.* 3: 54–61

Iyengar, S. S., Elfes, A. 1991a. *Autonomous Mobile Robots: Perception, Mapping and Navigation*. Los Alamitos, Calif: IEEE Comput. Soc. Press

Iyengar, S. S., Elfes, A. 1991b. *Autonomous Mobile Robots: Control, Planning, and Architecture*. Los Alamitos, Calif: IEEE Comput. Soc. Press

Johnson, M. 1987. *The Body in the Mind. The Bodily Basis of Meaning, Imagination, and Reason*. Chicago: Univ. Chicago Press

Kenny, A. J. P. 1971. The homunculus fallacy. In *Interpretations of Life and*

Mind. Essays around the Problem of Reduction, ed. M. Grene, pp. 65–83. New York: Humanities

Kuperstein, M. 1991. INFANT neural controller for adaptive sensory-motor coordination. *Neural Netw.* 4: 131–45

Kuperstein, M. 1988a. Neural model of adaptive hand-eye coordination for single postures. *Science* 239: 1308–11

Kuperstein, M. 1988b. An adaptive neural model for mapping invariant target position. *Behav. Neurosci.* 102: 148–62

Kurzweil, R. 1990. *The Age of Intelligent Machines.* Cambridge, Mass: MIT Press

Lachter, J., Bever, T. G. 1988. The relation between linguistic structure and associative theories of language learning—a constructive critique of some connectionist learning models. *Cognition* 28: 195–247

Laird, J., Rosenbloom, P., Newell, A. 1986. *Universal Subgoaling and Chunking. The Automatic Generation and Learning of Goal Hierarchies.* Boston: Kluwer

Lakoff, G. 1987. *Women, Fire, and Dangerous Things: What Categories Reveal About the Mind.* Chicago: Univ. Chicago Press

Lara, R., Arbib, M. A., Cromarty, A. S. 1982. The role of the tectal column in facilitation of amphibian prey-catching behavior: a neural model. *J. Neurosci.* 2: 521–30

Lockery, S. R., Wittenberg, G., Kristan, W. B., Cottrell, G. W. 1989. Function of identified interneurons in the leech elucidated using neural networks trained by back-propagation. *Nature* 340: 468–71

Maes, P., ed. 1990. *Designing Autonomous Agents: Theory and Practice from Biology to Engineering and Back.* Cambridge, Mass: MIT Press

Mazzoni, P., Andersen, R. A., Jordan, M. I. 1991a. A more biologically plausible learning rule for neural networks. *Proc. Natl. Acad. Sci. USA* 88: 4433–37

Mazzoni, P., Andersen, R. A., Jordan, M. I. 1991b. A more biologically plausible learning rule than backpropagation applied to a network model of cortical area 7a. *Cereb. Cortex* 1: 293–307

McClelland, J. L., Rumelhart, D. E., PDP Res. Group 1986. *Parallel Distributed Processing: Explorations in the Microstructure of Cognition Vol. 2: Psychological and Biological Models.* Cambridge, Mass: MIT Press

McCulloch, W. S., Pitts, W. 1943. A logical calculus of the ideas immanent in nervous activity. *Bull. Math. Biophys.* 5: 115–33

Mel, B. W. 1991. A connectionist model may shed light on neural mechanisms for visually guided reaching. *J. Cogn. Neurosci.* 3: 273–92

Mel, B. W. 1990. *Connectionist Robot Motion Planning. A Neurally-Inspired Approach to Visually-Guided Reaching.* Boston: Academic

Meyer, J.-A., Wilson, S. W., eds. 1991. *From Animals to Animats. Proc. of the First Int. Conf. on Simulation of Adaptive Behavior.* Cambridge, Mass: MIT Press

Miller, K. D., Keller, J. B., Stryker, M. P. 1989. Ocular dominance column development: analysis and simulation. *Science* 245: 605–15

Minsky, M. 1985. *The Society of Mind.* New York: Simon & Schuster

Minsky, M., Papert, S. 1969. *Perceptrons: An Introduction to Computational Geometry.* Cambridge, Mass: MIT Press

Murthy, V. N., Fetz, E. E. 1992. Coherent 25- to 35-Hz oscillations in the sensorimotor cortex of awake behaving monkeys. *Proc. Natl. Acad. Sci. USA* 89: 5670–74

Neisser, U. 1963. The imitation of man by machine. *Science* 139: 193–97

Nelson, J. I., Salin, P. A., Munk, M. H.-J., Arzi, M., Bullier, J. 1992. Spatial and temporal coherence in cortico-cortical connections: a cross-correlation study in areas 17 and 18 in the cat. *Visual Neurosci.* 9: 21–37

Nelson, R. C. 1991. Visual homing using an associative memory. *Biol. Cybern.* 65: 281–91

Nemes, T. 1970. *Cybernetic Machines.* New York: Gordon & Breach

Newell, A., Shaw, J. C., Simon, H. A. 1958. Elements of a theory of human problem solving. *Psychol. Rev.* 65: 151–66

Nilsson, N. J. 1969. A mobile automaton: An application of artificial intelligence techniques. *Proc. Int. Joint Conf. on Artif. Intell.* 233

O'Keefe, J., Nadel, L. 1978. *The Hippocampus as a Cognitive Map.* Oxford: Clarendon

Pearson, J. C., Finkel, L. H., Edelman, G. M. 1987. Plasticity in the organization of adult cortical maps: A computer model, based on neuronal group selection. *J. Neurosci.* 7: 4209–23

Pfeifer, R., Schreter, Z., Fogelman-Soulié, F., Steels, L., eds. 1989. *Connectionism in Perspective.* Amsterdam: Elsevier

Pinker, S., Prince, A. 1988. On language and connectionism. Analysis of a parallel distributed processing model of language acquisition. *Cognition* 28: 73–193

Poggio, T. 1990. A theory of how the brain might work. *Cold Spring Harbor Symp. Quant. Biol.* 55: 899–910

Putnam, H. 1988. *Representation and Reality.* Cambridge, Mass: MIT Press

Putnam, H. 1960. Minds and machines. In

Minds and Machines, ed. A. Anderson, pp. 72–97. Englewood Cliffs, NJ: Prentice-Hall

Pylyshyn, Z. W. 1984. *Computation and Cognition: Toward a Foundation for Cognitive Science*. Cambridge, Mass: MIT Press

Reeke, G. N. Jr. 1991. Book review: Marvin Minsky, *The Society of Mind. Artif. Intell.* 48: 341–48

Reeke, G. N. Jr., Edelman, G. M. 1988. Real brains and artificial intelligence. *Daedalus Proc. Am. Acad. Arts Sci.* 117: 143–73

Reeke, G. N. Jr., Edelman, G. M. 1987. Selective neural networks and their implications for recognition automata. *Int. J. Supercomputer. Appl.* 1: 44–69

Reeke, G. N. Jr., Edelman, G. M. 1986. Recognition automata based on selective neural networks. In *Structure and Dynamics of Nucleic Acids, Proteins, and Membranes*, ed. S. Clementi, S. Chin, pp. 329–53. New York: Plenum

Reeke, G. N. Jr., Finkel, L. H., Sporns, O., Edelman, G. M. 1990a. Synthetic neural modeling: a multilevel approach to the analysis of brain complexity. In *Signal and Sense: Local and Global Order in Perceptual Maps*, ed. G. M. Edelman, W. E. Gall, W. M. Cowan, pp. 607–706. New York: Wiley

Reeke, G. N. Jr., Sporns, O. 1990. Selectionist models of perceptual and motor systems and implications for functionalist theories of brain function. *Physica D* 42: 347–64

Reeke, G. N. Jr., Sporns, O., Edelman, G. M. 1990b. Synthetic neural modeling: the "Darwin" series of automata. *Proc. IEEE* 78: 1498–1530

Rescorla, R. A., Wagner, A. R. 1972. A theory of Pavlovian conditioning: variations in the effectiveness of reinforcement and nonreinforcement. In *Classical Conditioning II: Current Research and Theory*, ed. A. H. Black, W. F. Prokasy, pp. 64–99. New York: Appleton-Century-Crofts

Rosenblatt, F. 1958. The perceptron: A probabilistic model for information storage and organization in the brain. *Psychol. Rev.* 65: 386–408

Rumelhart, D. E., McClelland, J. L., PDP Res. Group 1986. *Parallel Distributed Processing: Explorations in the Microstructure of Cognition. Volume 1: Foundations*. Cambridge, Mass: MIT Press

Searle, J. R. 1992. *The Rediscovery of the Mind*. Cambridge, Mass: MIT Press

Searle, J. 1984. *Minds, Brains, and Science*. Cambridge, Mass: Harvard Univ. Press

Sejnowski, T. J., Koch, C., Churchland, P. S. 1988. Computational neuroscience. *Science* 241: 1299–1306

Shannon, C. E. 1951. Presentation of a maze-solving machine. In *Cybernetics, Trans. Eighth Conf. of the Josiah Macy Found.*, ed. H. von Foerster, pp. 173–80. New York: Macy Found.

Skinner, B. F. 1990. Can psychology be a science of mind? *Am. Psychol.* 45: 1206–10

Smolensky, P. 1988. On the proper treatment of connectionism. *Behav. Brain Sci.* 11: 1–74

Sporns, O., Gally, J. A., Reeke, G. N. Jr., Edelman, G. M. 1989. Reentrant signaling among simulated neuronal groups leads to coherency in their oscillatory activity. *Proc. Natl. Acad. Sci. USA* 86: 7265–69

Sporns, O., Tononi, G., Edelman, G. M. 1991. Modeling perceptual grouping and figure-ground segregation by means of active reentrant conditioning. *Proc. Natl. Acad. Sci. USA* 88: 129–33

Sutton, R. S., Barto, A. G. 1981. Toward a modern theory of adaptive networks: expectation and prediction. *Psychol. Rev.* 88: 135–70

Thorpe, S. J., Imbert, M. 1989. Biological constraints on connectionist modelling. In *Connectionism in Perspective*, ed. R. Pfeifer, Z. Schreter, F. Fogelman-Soulié, L. Steels, pp. 63–92. Amsterdam: Elsevier

Tononi, G., Sporns, O., Edelman, G. M. 1992a. The problem of neural integration: Induced rhythms and short-term correlations. In *Induced Rhythms in the Brain*, ed. E. Başar, T. H. Bullock, pp. 367–95. Boston: Birkhäuser

Tononi, G., Sporns, O., Edelman, G. M. 1992b. Reentry and the problem of integrating multiple cortical areas: simulation of dynamic integration in the visual system. *Cereb. Cortex* 2: 310–35

Touretzky, D. S., Hinton, G. E. 1988. A distributed connectionist production system. *Cogn. Sci.* 12: 423–66

Traub, R. D., Miles, R., Wong, R. K. S. 1989. Model of the origin of rhythmic population oscillations in the hippocampal slice. *Science* 243: 1319–25

Uttal, W. R. 1990. On some two-way barriers between models and mechanisms. *Percep. Psychophys.* 48: 188–203

Verschure, P. F. M. J., Kröse, B. J. A., Pfeifer, R. 1992. Distributed adaptive control: the self-organization of structured behavior. *Robot. Auton. Syst.* 9: 1–15

Walter, W. G. 1953. *The Living Brain*. London: Duckworth

Wang, D. L., Arbib, M. A. 1991. How does the toad's visual system discriminate different worm-like stimuli. *Biol. Cybern.* 64: 251–61

Webb, B. H. 1991. Do computer simulations really cognize? *J. Exp. Theor. Artif. Intell.* 3: 247–54

Widrow, G., Hoff, M. E. 1960. Adaptive switching circuits. *IRE Western Electronic Show and Convention, Convention Record, Part 4* 1960: 96–104

Wiener, N. 1948. *Cybernetics*. Cambridge, Mass: MIT Press

Williams, T. L., Sigvardt, K. A., Kopell, N., Ermentrout, G. B., Remler, M. P. 1990. Forcing of coupled nonlinear oscillators: studies of intersegmental coordination in the lamprey locomotor central pattern generator. *J. Neurophysiol.* 64: 862

Winograd, T., Flores, F. 1986. *Understanding Computers and Cognition. A New Foundation for Design*. Reading, Mass: Addison-Wesley

Yuste, R., Peinado, A., Katz, L. C. 1992. Neuronal domains in developing neocortex. *Science* 257: 665–69

Zipser, D. 1986. Biologically plausible models of place recognition and goal location. In *Parallel Distributed Processing II. Psychological and Biological Models*, ed. J. L. McClelland, D. E. Rumelhart, pp. 432–70. Cambridge, Mass: MIT Press

Zipser, D., Andersen, R. A. 1988. A back-propagation programmed network that simulates response properties of a subset of posterior parietal neurons. *Nature* 331: 679–84

Zornetzer, S. F., Davis, J. L., Lau, C., eds. 1990. *An Introduction to Neural and Electronic Networks*. San Diego: Academic

Annu. Rev. Neurosci. 1993. 16:625–65

LEARNING TO MODULATE TRANSMITTER RELEASE: Themes and Variations in Synaptic Plasticity

Robert D. Hawkins,[1,2] *Eric R. Kandel,*[1-3] *and Steven A. Siegelbaum*[1,3]

[1]Center for Neurobiology and Behavior, Columbia University, College of Physicians and Surgeons; [2]New York State Psychiatric Institute; and [3]Howard Hughes Medical Institute, New York, NY 10032

KEY WORDS: *Aplysia*, presynaptic facilitation, hippocampus, long-term potentiation

INTRODUCTION

Learning is the modification of behavior by experience, and memory is the retention of that modification over time. As learning and memory have become accessible to study with the techniques of cellular and molecular biology, a variety of cellular mechanisms of neuronal plasticity have now been identified that are thought to contribute to different forms of learning in invertebrates and vertebrates (for reviews, see Alkon & Rasmussen 1988; Byrne 1987; Carew & Sahley 1986; Hawkins et al 1987; Ito 1989; Nicoll et al 1988; Thompson 1986). Therefore, researchers can now attempt to define basic principles of learning by comparing these mechanisms and asking: What do they have in common, and how are they different?

Until quite recently, these questions proved surprisingly difficult to answer. Different forms of neuronal plasticity appeared to predominate in different forms of learning, in different regions of the brain, and in different species. In mammals, for example, at least two quite different synaptic mechanisms appeared to operate in different regions of the hippocampus, a structure critical for learning. Moreover, these two mechanisms seemed to differ from those encountered in such invertebrates as *Aplysia*, *Hermissenda*, and *Drosophila*. However, recent cellular studies on hippo-

625

0147–006X/93/0301–0625$02.00

campus have given us new insights that may bring the studies on the different regions of the hippocampus into clearer relationship with one another and also with the findings from invertebrates.

We here describe several common motifs in synaptic plasticity that are emerging from these studies of the hippocampus and relate them to the studies of invertebrates, particularly *Aplysia*. These studies suggest that one common unifying theme in the study of plasticity is the modulation of transmitter release, a modulation that can take surprisingly different forms. Thus, both in hippocampus and in *Aplysia*, modulation of transmitter release during learning can be either nonassociative or associative. Moreover, and perhaps most surprising, different forms of modulation seem to serve as functional building blocks that can be combined to yield more elaborate mechanisms of synaptic plasticity.

Types of Synaptic Plasticity

Because all transformations of neural information in the brain involve only neurons, glia, and their interconnections, neurobiologists since Ramón y Cajal (1911) have long believed that elementary aspects of learning and memory storage are likely to be resolvable at the cellular level. But, which components of the neuron can be changed and how those changes could come about only began to emerge in 1938. Gopfert & Schaefer (1938), while working on the nerve-muscle synapse, discovered the synaptic potential as a discrete physiological link interposed between the action potential in the presynaptic terminal of the neuron and the action potential in the muscle fiber. Subsequently, Fatt & Katz (1951) demonstrated that the synaptic potential resulted from the ionic current in the postsynaptic cell activated by the chemical transmitter acetylcholine released by the presynaptic neurons.

In 1941, Feng discovered that the strength of the synaptic potential is not invariant, but is plastic and varies as a function of activity or use. For example, Feng found that, following a brief period of high frequency (tetanic) activity in the presynaptic neuron, the synaptic potential can be enhanced for minutes to hours. Feng called this form of synaptic plasticity posttetanic potentiation. In 1949, Lloyd, studying the synapse of the IA afferent fibers on motor neurons in the spinal cord, discovered a second and reciprocal type of synaptic plasticity: low frequency, synaptic depression, whereby a low rate of activity produces a depression of synaptic effectiveness. Both of these forms are homosynaptic; the change occurs in the synapses of the activated pathway. About ten years later, Frank & Fuortes (1957), working on the spinal cord, and Dudel & Kuffler (1961), working on the crayfish nerve muscle synapse, provided the first account of heterosynaptic plasticity in which the synaptic strength of one pathway is modified by activity in another pathway. Their discovery of heterosynaptic inhibition demonstrated that activity in one pathway can depress the synaptic potential produced in another. For example, Dudel & Kuffler showed that stimulating the inhibitory axon in the muscle of the lobster decreased the amplitude of the synaptic potential produced by the excitatory motor axons of that muscle. In 1965, a heterosynaptic form of facilitation was described in *Aplysia*, where activity in one pathway enhances activity in another (Kandel & Tauc 1965a,b).

Synaptic Plasticity and Learning

Most chemical synapses show some capability for either homo- or heterosynaptic plasticity. The generality of plastic mechanisms raised this question: To what degree do plastic changes reflect and contribute to the fundamental features we recognize as characteristic of learning in the intact organism?

As it became possible to study learning on the cellular level in the late 1960s and early 1970s, it soon became clear that learning gives rise to plastic changes in neuronal properties that often involve homo- and heterosynaptic changes in the strength of the synaptic connections (Castellucci et al 1970; Krasne 1969; Spencer et al 1966; Zucker 1972). However, these studies also showed that the modification produced by learning in even a simple behavioral circuit involves the activity not of one set of cells and one set of connections; but of many cells and their connections. This finding then raised these questions: What is the correspondence between a given plastic mechanism and a given form of learning? Does the representation of even the most elementary features of learning require the workings of complex circuitry? Or, can the elementary features of learning be resolved in the properties of the individual neurons that participate in learning?

NONASSOCIATIVE LEARNING

One way to address these questions is to see whether distinct forms of learning give rise to distinctive plastic mechanisms that are detectable at the cellular level and that constitute an elementary representation of that learning process. Animal behavior can be modified not simply by one, but by several distinct forms of learning. At the behavioral level, each modification has several defining features. Thus, even simple procedural forms of learning can be subdivided into two broad categories: nonassociative and associative learning. In nonassociative learning, such as habituation or sensitization, the subject learns about the properties of a single stimulus. In associative learning, such as classical or operant conditioning, the subject learns about the relationship between two stimuli (classical conditioning) or between a stimulus and a response (operant conditioning). What are the elementary neural mechanisms of non-associative and associative learning? Do they require complex circuitry, or are they reflected in the plastic properties of individual neurons?

Nonassociative Learning in Invertebrates Involves Alterations in Transmitter Release

Habituation is the simplest form of nonassociative learning. With habituation, an animal learns through repetition to recognize and ignore stimuli that are nonthreatening and, therefore, unimportant (Thorpe 1956). Habituation is involved in most encounters with our environment. For example, a sudden noise commonly triggers an orienting response toward the stimulus accompanied by several autonomic responses, such as an increase in heart rate and respiratory rate. If the noise is repeated, both the orienting and the concomitant autonomic responses abate.

Habituation occurs in all animals. In the marine snail *Aplysia*, for example, a tactile stimulus to the siphon, the respiratory spout of the animal, leads to a brisk reflex withdrawal of its gill. The reflex diminishes, however, with repeated stimulation of the siphon (Pinsker et al 1970). Habituation involves a homosynaptic depression in the synapses between the sensory neurons and the interneurons and motor neurons of this reflex (Castellucci et al 1970). Habituation also leads to a homosynaptic depression in the sensory neuron-to-motor neuron connection in the tail-withdrawal reflex in *Aplysia*, in the sensory neuron-to-interneuron connection of the tail flick escape response in crayfish, and in the connections between interneurons of the flexion reflex of vertebrates (Krasne 1969; Spencer et al 1966; Walters et al 1983; Zucker 1972). Although only a few examples of habituation have been analyzed so far, in each case the excitatory synaptic connections between one or more classes of neurons within the reflex pathway undergo homosynaptic depression (for review, see Hawkins et al 1987). Quantal analysis in invertebrates has shown that homosynaptic depression reflects a decrease in the number of transmitter or quanta released from the presynaptic terminals (Castellucci & Kandel 1974; Zucker 1972).

Sensitization is a slightly more complex form of nonassociative learning, in which an animal learns to strengthen its reflex responses to previously neutral stimuli following the presentation of a potentially threatening stimulus at another site. After hearing a gun shot, a person may react strongly, for a while, to even the most innocuous noise. In *Aplysia*, a noxious stimulus to the neck or tail greatly enhances the gill-withdrawal reflex elicited by a touch to the siphon (Pinsker et al 1970). In the reflexes of *Aplysia* and crayfish and in the spinal reflexes of vertebrates, neuronal connections that are depressed by habituation can also be strengthened by sensitization (Castellucci et al 1970; Spencer et al 1966; Krasne & Glanzman 1986). Thus, a given set of connections can be modified by and contribute to more than one learning process.

In *Aplysia*, the change associated with sensitization in the monosynaptic component between the sensory and motor neurons again involves a presynaptic alteration in transmitter release (Castellucci & Kandel 1976). Sensitizing stimuli from the tail excite a group of modulatory interneurons that enhance transmitter release from sensory neurons, a heterosynaptic process called presynaptic facilitation (Hawkins et al 1981; Hawkins & Schacher 1989; Mackey et al 1989) (Figure 1*A*). Some of the modulatory neurons are serotonergic (Glanzman et al 1989; Mackey et al 1989). Serotonin activates adenylyl cyclase in the sensory neurons, so that sensitizing stimuli increase cAMP and activate the cAMP-dependent protein kinase (kinase A) (Bernier et al 1982; Castellucci et al 1980, 1982). Protein kinase A phosphorylates many substrate proteins in the sensory neurons, including the S-type K^+ channel, or a protein that acts on this channel. Phosphorylation leads to closure of this channel and broadens the action potentials (Klein & Kandel 1978, 1980; Klein et al 1982; Shuster et al 1985;

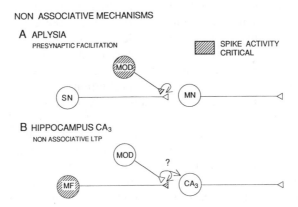

Figure 1   Nonassociative mechanisms, which are thought to contribute to learning in *Aplysia* and the CA_3 region of hippocampus. The shading indicates neurons in which spike activity during training is critical. (*A*) Presynaptic facilitation in *Aplysia*. Activity in a modulatory neuron (MOD) causes facilitation of transmitter release form a sensory neuron (SN), so that the sensory neuron produces a larger EPSP in the motor neuron (MN). (*B*) Nonassociative LTP in the CA_3 region of hippocampus. High frequency firing of the mossy fiber (MF) input to the CA_3 pyramidal cells causes potentiation of the EPSP at the mossy fiber-CA_3 synapses. Activity may also be required in modulatory neurons and/or the CA_3 pyramidal neurons.

Siegelbaum et al 1982). In addition, 5-HT and protein kinase A also modulate the kinetics of a second K^+ channel, a voltage-gated transient K^+ channel, which further enhances broadening of the action potential (Abrams & Goldsmith 1992; Baxter & Byrne 1989; Hochner & Kandel 1992). Broadening of the action potential by 5-HT is thought to contribute to presynaptic facilitation by allowing more Ca^{2+} to enter the presynaptic terminal via the voltage-gated calcium channels.

Aplysia sensory neurons contain two classes of voltage-gated Ca^{2+} channels: a dihydropyridine-sensitive noninactivating channel (L-type) and a dihydropyridine-insensitive inactivating channel (N-type) (Edmonds et al 1990). Serotonin selectively increases the magnitude of the L-type Ca^{2+} current; it has no direct effect on the N-type current. Although, in principle, the direct enhancement of the L-type current might be expected to contribute to presynaptic facilitation, blocking Ca^{2+} influx via the L-type current by dihydropyridines has no effect on either normal transmitter release or presynaptic facilitation (Edmonds et al 1990). Thus, the results of these experiments suggest that 5-HT modulates transmitter release indirectly by increasing Ca^{2+} influx through the dihydropyridine-insensitive (N-type) Ca^{2+} channels as a result of the increase in the duration of the action potential that results from the modulatory decrease in K^+ current.

Can this extra Ca^{2+} influx account for the facilitation? To address this question, Eliot et al (1991) measured directly the Ca^{2+} influx during facilitation, by using fura-2 to image Ca^{2+} in the presynaptic regions of a single sensory neuron innervating a single motor neuron in dissociated cell culture. They first used nitrendipine to block the dihydropyridine-sensitive Ca^{2+} influx into the presynaptic terminals, because, as seen above, the L-type channel does not contribute to facilitation. They then examined Ca^{2+} influx limited to the N-type channel and found that modulation of transmitter release is related by a power of 2.4 to changes in Ca^{2+} influx. Because the N-type Ca^{2+} channel is not regulated by 5-HT, this modulation of Ca^{2+} influx is attributable to the broadening of the action potential by 5-HT, owing to its action on K^+ channels.

Serotonin also enhances transmitter release by a second process that is thought to involve an increase in availability of transmitter vesicles or their mobilization to release sites (Gingrich & Byrne 1985; Hochner et al 1986). This second process involves activation of protein kinase C, in addition to protein kinase A, and is reflected in a substantial increase in spontaneous release that is largely independent of changes in free Ca^{2+} (Braha et al 1990a,b; Ghirardi et al 1992; Sacktor & Schwartz 1990).

Both habituation and sensitization in *Aplysia* show long-term, as well as short-term, forms. Of the two, long-term sensitization has been better analyzed. Whereas a single noxious stimulus to the tail of *Aplysia* produces short-term sensitization lasting minutes, repeated noxious stimulation produces long-term sensitization lasting days to weeks (Frost et al 1985; Pinsker et al 1973). On the cellular level, as on the behavioral level, the long-term process resembles the short-term in several ways: It involves the synaptic connections of the sensory neurons. It is caused by an enhancement of transmitter release. It involves modulation of the S-type K^+ channel. And, it can be induced by the same transmitter (5-HT) and second messenger (cAMP) (Dale et al 1987, 1988; Frost et al 1985; Schacher et al 1988; Scholz & Byrne 1987, 1988). These several similarities between the long- and short-term processes seem to result from the fact that repeated tail stimuli (or repeated application of 5-HT) leads to phosphorylation of at least some of the same substrate proteins as are involved in the short-term process (Sweatt & Kandel 1989). In the long-term, this phosphorylation is persistent and seems to be at least partly caused by a decrease in the amount of regulatory subunit of protein kinase A (Bergold et al 1990, 1992; Greenberg et al 1987). As a result, protein kinase A remains constitutively active long after the level of cAMP has returned to baseline.

Nevertheless, long-term facilitation differs from short-term in two important ways: It requires the expression of genes not involved in the short-term process (Bergold et al 1990; Dash et al 1990; Montarolo et al 1986), and it is associated with morphological changes in the sensory and motor neurons. The number of synaptic terminals of the sensory neurons increases significantly, as does the postsynaptic receptive area of the motor neurons (Bailey & Chen 1988a,b). Morphological changes also occur during long-term plasticity in *Hermissenda* and in crayfish (Alkon et al 1990; Lnenicka et al 1986). In these cases as well, the sites involved in short-term plastic change are also used for long-term change, and the long-term process requires new protein synthesis (Crow & Forrester 1990; Nguyen & Atwood 1990).

In addition to sensitization, *Aplysia* also show short-term behavioral inhibition of the gill-withdrawal reflex upon stimulation of the tail (Mackey et al 1987). The behavioral inhibition precedes sensitization and is thought to result, in part, from presynaptic inhibition of the siphon sensory neurons by inhibitory interneurons that release the neuropeptide FMRFamide (Small et al 1992a). Direct application of FMRFamide to sensory neurons or stimulation of the FMRFamidergic interneurons produces inhibitory electrophysiological actions that are the opposite to those produced by 5-HT (Abrams et al 1984; Mackey et al 1987; Small et al 1992a). Thus, FMRFamide produces presynaptic inhibition of transmitter release, a hyperpolarization of the sensory neuron membrane potential that involves an increase in conductance, and a decrease in the duration of the action potential. Voltage-clamp and single-channel analyses have shown that these effects are caused by an increase in the magnitude of the $S-K^+$ current (Belardetti et al 1987; Brezina et al 1987) and a decrease in the dihydropyridine-insensitive N-type calcium current (Edmonds et al 1990). This dual action to increase K^+ current and decrease Ca^{2+} current is similar to the effects of opioid peptides in producing presynaptic inhibition in vertebrate neurons (MacDonald & Werz 1986; Mudge et al 1979; Williams et al 1982). Presynaptic inhibition with FMRFamide likely results from the synergistic effects of the increase in outward K^+ current and decrease in inward Ca^{2+} current to decrease Ca^{2+} influx during an action potential. In addition, FMRFamide likely exerts a direct inhibitory action on the release process (Dale & Kandel 1990; Man-Son-Hing et al 1989).

The modulatory effects of FMRFamide on the $S-K^+$ current are mediated by a pertussis toxin sensitive G protein (Volterra & Siegelbaum 1988), which leads to the release of arachidonic acid from the membrane (Piomelli et al 1987). The arachidonic acid is metabolized by the 12-lipoxygenase enzymatic pathway to the unstable intermediate compound 12-HPETE (Piomelli et al 1987), which may act directly on the S channel to increase its open probability (Buttner et al 1989). 12-HPETE is further metabolized to several other active compounds, including the epoxy alcohol compound 8-HePeTE (Piomelli et al 1989), which may also have direct modulatory actions on the S channel (Belardetti et al 1989). In addition to modulating the S channel, FMRFamide also antagonizes the increase in protein phosphorylation produced by 5-HT or exogenous application of cAMP, either by activating a phosphatase or inhibiting protein kinase (Ichinose & Byrne 1991; Sweatt et al 1989).

Thus, cellular changes that accompany three forms of nonassociative learning in invertebrates—habituation, sensitization, and inhibition—represent surprisingly clear cellular representations of the behavioral modification. Is there also a compatible representation in vertebrates, and for associative forms of learning?

Nonassociative Long-Term Potentiation in the CA_3 Region of the Hippocampus May Also Utilize Presynaptic Facilitation of Transmitter Release

Since the pioneering work of Scoville & Milner (1957), the hippocampus has been known to be important for the initial storage of declarative memory (memory for people, places, and things) in humans and other mammals (Squire 1992). These studies have shown that the hippocampus may be essential for initially storing long-term memory for a period of days to weeks before the memory trace is consolidated elsewhere, perhaps in different areas of the cerebral cortex (Zola-Morgan & Squire 1990).

The hippocampus has three well-studied synaptic pathways (Figure 2): the perforant pathway synapses onto the granule cells in the dentate gyrus, the granule cells send axons (the mossy fibers) that synapse on the pyramidal cells in the CA_3 region of the hippocampus, and the pyramidal cells in CA_3 send excitatory (Schaffer) collaterals to the pyramidal neurons in the CA_1 region of the hippocampus.

In 1973, Bliss & Lømo first demonstrated that a brief high-frequency train of action potentials in the perforant path produces an increase in the excitatory synaptic potential in the granule cells, which can last for hours, or, under some circumstances, for days or weeks. They called this facilitation long-term potentiation (LTP). Later studies showed that LTP occurs at each of the three major synaptic pathways in the hippocampus, but that the potentiation has different properties (and perhaps different mechanisms) at the mossy fiber synapses in CA_3 than in dentate gyrus or CA_1 (see Zalutsky & Nicoll 1990, 1992). In dentate gyrus and CA_1, the high-frequency stimulation has to be above some threshold level of intensity to produce LTP (a property referred to as cooperativity) (McNaughton et al 1978). However, when weak stimulation of one input pathway (which is itself insufficient to produce LTP) is temporally paired (or associated) with strong stimulation of another input pathway (capable of producing LTP), the weak pathway also undergoes LTP (a property referred to as associ-

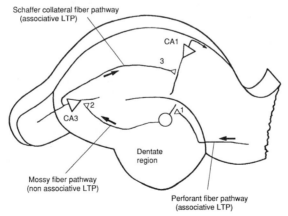

Figure 2 Principal neural circuit connections in the hippocampus. Perforant path fibers synapse on granule cells in the dentate gyrus. The mossy fiber axons of the granule cells synapse on pyramidal cells in the CA_3 region of hippocampus. The Schaffer collateral axons of the CA_3 pyramidal cells synapse on pyramidal cells in the CA_1 region. LTP occurs at each of these synapses. LTP in the dentate gyrus and CA_1 region has associative properties, whereas LTP at the mossy fiber synapses in CA_3 is nonassociative.

ative) (Barrionuevo & Brown 1983; Levy & Steward 1979). By contrast, LTP in the mossy fiber pathway to CA_3 is not associative: When weak stimulation of the mossy fiber pathway is associated with strong stimulation of another input to CA_3, the mossy fiber input does not undergo LTP (Chattarji et al 1989). Moreover, whereas LTP in dentate gyrus and CA_1 is restricted to the tetanized pathway (a property referred to as input specificity) (Andersen et al 1977), strong stimulation of the mossy fibers can produce heterosynaptic LTP of other inputs to CA_3 (Bradler & Barrionuevo 1989).

Hopkins & Johnston (1984, 1988) have found that LTP at the mossy fiber synapses is enhanced by norepinephrine, perhaps acting through a cAMP-dependent mechanism in the postsynaptic cells. Although there is conflicting evidence on whether the induction of LTP at these synapses involves postsynaptic events (Jaffe & Johnston 1990; Williams & Johnston 1989; Zalutsky & Nicoll 1990), the expression of LTP at these synapses apparently involves a presynaptic alteration of transmitter release (Hirata et al 1991; Zalutsky & Nicoll 1990). How this is brought about is still not clear. Specifically, it is not known whether mossy fiber activity is sufficient for enhancement of transmitter release, or whether activation of modulating noradrenergic fibers in hippocampus is also required (Figure 1B). If neuromodulation is required, LTP in CA_3 could have several similarities to presynaptic facilitation in *Aplysia* (Table 1): Both produce long-lasting nonassociative changes in the strength of synaptic connections between specific cells that are thought to contribute to learning, and both may involve heterosynaptic modulation of transmitter release. Norepinephrine, perhaps acting through cAMP, also contributes to aspects of LTP in the CA_1 region (Dunwiddie et al 1992) and in the dentate gyrus (Dahl & Sarvey 1989; Stanton & Sarvey 1985a,b) where it causes phosphorylation of synaptic vesicle proteins (Parfitt et al 1991). Another modulatory transmitter, acetylcholine, acts to suppress LTP in CA_3 (Williams & Johnston 1988), but enhances it in CA_1 (Hirotsu et al 1989; Markram & Segal 1990).

Long-term synaptic changes that resemble LTP also occur in several other brain areas, as well as in sympathetic ganglia, *Aplysia*, and crayfish neuromuscular junction. The changes in sympathetic ganglia, *Aplysia*, and crayfish all involve a presynaptic enhancement of transmitter release (Baxter et al 1985; Briggs et al 1985; Walters & Byrne 1985).

ASSOCIATIVE LEARNING

Classical (Pavlovian) conditioning resembles nonassociative sensitization in that the response of one pathway to stimulation is enhanced by activity in another. Typically, an initially weak or ineffective conditioned stimulus

Table 1 Comparison of nonassociative and associative cellular mechanisms of learning in *Aplysia* and hippocampus

	Nonassociative presynaptic facilitation in *Aplysia*	LTP in the CA$_3$ region of hippocampus	Associative presynaptic facilitation in *Aplysia*	LTP in the CA$_1$ region of hippocampus
Mechanism learning and memory	yes	yes?	yes	yes?
Change synaptic strength	yes	yes	yes	yes
Associative	no	no	activity-dependent Ca^{2+} prime cyclase	Hebbian NMDA receptor
Induction	modulatory	modulatory?	pre-mod coincidence	pre-post coincidence
Maintenance	presynaptic	presynaptic?	presynaptic	presynaptic?
Modulation ion channels	decrease K$^+$ increase Ca^{2+}	increase Ca^{2+}?	decrease K$^+$	
Phosphorylation	kinase A (& C?) persistent	kinase A?	kinase A	kinase B & C? persistent
Protein synthesis-dependent	yes			yes
Morphological changes	increase synaptic boutons			increase vesicles and spine area

(CS), such as a bell, becomes effective in producing a new behavioral response, such as salivation, or in enhancing a previously existing response after it has been paired temporally with a strong unconditioned stimulus (US), such as food. Conditioning is distinguished from sensitization by its requirement for temporal pairing and correlation of the two stimuli during training.

What accounts for the pairing requirement in classical conditioning? A common early assumption underlying studies of learning was that associative changes are properties of complex circuits. Hebb was one of the first to challenge this assumption, by suggesting a cellular mechanism for development and long-term memory that could also account for classical conditioning (Figure 3*B*). Hebb (1949) proposed that associations could be formed by coincident firing of action potentials in the presynaptic and

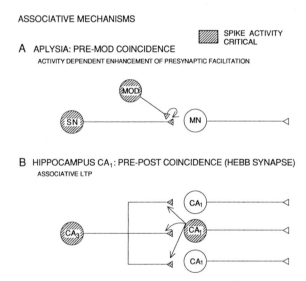

Figure 3 Associative mechanisms, which are thought to contribute to learning in *Aplysia* and the CA$_1$ region of hippocampus. The shading indicates neurons in which spike activity during training is critical. (*A*) Activity-dependent presynaptic facilitation in *Aplysia*. Spike activity in a sensory neuron neuron (SN) just before activity in a modulatory neuron (MOD) causes enhanced facilitation of transmitter release from that sensory neuron. (*B*) Associative (Hebbian) LTP in the CA$_1$ region of hippocampus. Spike activity in a presynaptic neuron (CA$_3$) at the same time as activity in a postsynaptic neuron (CA$_1$) causes potentiation of the EPSP at the synapse between them. This effect may be due to a retrograde message, which is released by the postsynaptic neuron and causes facilitation of transmitter release from the presynaptic neuron. Potentiation may also occur at synapses from the presynaptic neuron to neighboring postsynaptic neurons.

postsynaptic cells (pre-post coincidence): "When an axon of cell A is near enough to excite a cell B and repeatedly or persistently takes part in firing it, some growth process or metabolic change takes place in one or both cells such that A's efficacy, as one of the cells firing B, is increased." According to Hebb's postulate, activity in the postsynaptic neuron is critical. A different mechanism was proposed based on studies of *Aplysia* (Hawkins et al 1983; Kandel & Tauc 1965a,b; Walters & Byrne 1983) (Figure 3*A*). Here, associations are formed when action potentials in a presynaptic cell are coincident with action potentials in a modulatory neuron that synapses on the presynaptic neuron (pre-modulatory coincidence). Recent work has shown that both of these mechanisms are utilized at different synaptic sites; and in the CA_1 region of the hippocampus, both mechanisms may be combined at the same site.

Associative Presynaptic Facilitation and Inhibition in Aplysia *Involve Coincident Activity in a Presynaptic Neuron and a Modulatory Neuron*

The pre-modulatory coincidence mechanism was identified through cellular studies of classical conditioning of the gill-withdrawal reflex in *Aplysia*. This reflex is conditioned by pairing a conditioned stimulus to the siphon with an unconditioned stimulus to the tail (Carew et al 1981). If these two stimuli are repeatedly paired, the siphon stimulation elicits significantly larger gill and siphon withdrawals than if the two stimuli are presented in an unpaired or random fashion. This effect builds up during the training session and is retained for several days. Furthermore, the withdrawal reflex undergoes differential conditioning with stimulation of the siphon and mantle shelf (a region anterior to the gill) or two different sites on the siphon as the discriminative stimuli (Carew et al 1983). Investigations of the interstimulus interval function have shown that there is reliable conditioning when the CS precedes the US by 0.5 seconds (the standard interstimulus interval), but no conditioning when the interval is 2, 5, or 10 seconds, or when the US precedes the CS—that is, there is no backward conditioning (Hawkins et al 1986). These experiments demonstrate stimulus and temporal specificities in conditioning of the *Aplysia* siphon-withdrawal reflex. Recent experiments have shown that conditioning of that reflex also exhibits response specificity. That is, *Aplysia* learn not only to strengthen the magnitude of a previously existing reflex response, they can learn to develop a new type of response to the CS, which resembles the response to the US (Hawkins et al 1989; Walters 1989). Thus, siphon stimulation (the CS) initially produces straight contraction of the siphon, but following conditioning it produces backward bending similar to the response to tail shock (the US). In all of these respects, conditioning of the siphon-withdrawal reflex is similar to many instances of vertebrate conditioning, such as conditioning of the rabbit eye blink response (e.g. Gormezano 1972).

Because of the similarity of sensitization and classical conditioning, it was attractive to think that conditioning might also involve presynaptic facilitation as a mechanism for strengthening the CS pathway. Specifically, the CS and US might converge at the level of individual neurons in the CS pathway, with the US producing greater presynaptic facilitation of those neurons if they fire action potentials just before the US is delivered (as would occur for sensory neurons in the CS pathway). Hawkins et al (1983) tested this possibility by examining the monosynaptic connections from two individual siphon sensory neurons to a siphon motor neuron in a semi-intact preparation before and after a training procedure based on that used in the behavioral experiments. They found that tail shock produced significantly greater facilitation of the monosynaptic EPSP from a sensory neuron to a motor neuron if the shock was preceded by intracellularly produced spike activity in the sensory neuron than if it was either unpaired with spike activity or was presented alone. The results of these experiments were similar both quantitatively and qualitatively with the results of behavioral experiments employing the same protocol and variables, which indicates that this activity-dependent amplification of facilitation could be a cellular mechanism for behavioral conditioning of the siphon-withdrawal reflex. Walters & Byrne (1983) independently demonstrated activity-dependent facilitation of sensory neurons innervating the tail, and Buonomano & Byrne (1990) have shown that the facilitation persists for at least 24 hours. Clark (1984) has investigated the interstimulus interval function for activity-dependent facilitation of the siphon sensory neurons and has found that, as in behavioral conditioning of siphon withdrawal, forward pairing of the spike activity and tail shock is more effective than backward pairing. Activity-dependent facilitation can thus account for aspects of both the stimulus and temporal specificities of conditioning. However, it cannot account for the response specificity of conditioning if the facilitation is assumed to be equal at all the branches of the sensory neuron, because response specificity requires selectively strengthening the synaptic connections onto some postsynaptic cells and not others. Response specificity might be accounted for by branch-specific facilitation (Clark & Kandel 1984; Hawkins et al 1989) or by plasticity at other sites in the reflex circuit, including motor neurons or neuromuscular junctions (Colebrook & Lukowiak 1988; Frost et al 1988; Hawkins et al 1989; Lukowiak 1986; Lukowiak & Colebrook 1988; Walters 1989).

Another mechanism that could account for the response specificity of conditioning is Hebb-type synaptic plasticity at the sensory neuron-motor neuron synapses. Carew et al (1984) tested the Hebb postulate in their system and found that spike activity in the postsynaptic neuron is neither necessary nor sufficient: In some of their experiments, the postsynaptic neuron was held at a hyperpolarized level and did not fire any action potentials in response to the US. And, intracellular stimulation of the postsynaptic neuron did not serve as an effective US. These results indicate that conditioning of the siphon-withdrawal reflex is not caused by Hebb-type synaptic plasticity, but rather is caused in part, by activity-dependent facilitation.

Activity-dependent facilitation itself could result from either a presynaptic or a postsynaptic mechanism. Because facilitation at the sensory neuron-motor neuron synapses underlying behavioral sensitization is presynaptic in origin and involves broadening of the action potential in the sensory neurons, Hawkins et al (1983) investigated the possibility that this mechanism might also be involved in the activity-dependent amplification of facilitation underlying classical conditioning. They used the same experimental protocol as in their synaptic experiments, except that they examined the duration of action potentials in the sensory neurons with the abdominal ganglion bathed in 50 mM tetraethylammonium, which blocks a delayed rectifier K current ($I_{K,v}$) and a calcium-activated K current ($I_{K,Ca}$), thus making the duration of the action potential more sensitive to changes in the remaining currents, especially the S-K^+ current ($I_{K,S}$). The differential training procedure produced a significantly greater increase in spike duration in paired than in unpaired sensory neurons, and this difference was maintained for at least three hours after training. A voltage-clamp analysis showed a pairing-specific decrease in a current with characteristics similar to those of the current that is decreased by serotonin or tail shock during normal presynaptic facilitation (Hawkins & Abrams 1984). Brief application of serotonin can also substitute for tail shock as the US in producing activity-dependent broadening of action potentials in the sensory neurons in the ganglion (Abrams et al 1983) and activity-dependent facilitation of sensory neuron-motor neuron EPSPs in isolated cell culture (Eliot et al 1989b). These experiments provide support for a role of serotonin in activity-dependent facilitation and demonstrate that activity-dependent facilitation occurs at the level of individual sensory cells and does not require additional neuronal circuitry.

The results of these experiments suggest that a mechanism of classical conditioning of the withdrawal reflex is an elaboration of a mechanism of sensitization of the reflex: presynaptic facilitation of the sensory neurons. The pairing specificity characteristic of classical conditioning is thought to result from amplification of presynaptic facilitation by temporally paired spike activity in the sensory neurons. This hypothesis raises two additional questions: Which aspect of the action potential in a sensory neuron interacts with the process of presynaptic facilitation to amplify it? And, which step in the biochemical cascade leading to presynaptic facilitation is sensitive to the action potential? Experiments by Abrams and colleagues suggest that the influx of Ca^{2+} with each action potential is the critical aspect of spike activity and that it primes the serotonin-sensitive adenylyl cyclase in the sensory neurons so that the cyclase subsequently produces more cAMP in response to serotonin. For example, Ca^{2+} must be present in the external medium during training for paired spike activity to enhance the effect of serotonin in producing spike broadening (Abrams 1985; Abrams et al 1983). Furthermore, serotonin produces a greater increase in cAMP levels in sensory cells if it is preceded by spike activity in the sensory cells than if it is not (Kandel et al 1983; Ocorr et al 1985). Abrams and colleagues (Abrams et al 1985, 1991; Eliot et al 1989a; Yovell et al 1986) have begun to analyze cyclase activity in a cell-free membrane homogenate and have found that it is stimulated by Ca^{2+} and calmodulin in addition to being stimulated by serotonin. Moreover, stimulation of the cyclase by both Ca^{2+} and serotonin is synergistic and is most effective when Ca^{2+} precedes the serotonin (Yovell & Abrams 1992). These results suggest that a site of convergence of the CS and the US in conditioning may be the adenylyl cyclase, although other possibilities (including other molecules in the sensory neurons and other sites in the circuit) are not excluded.

A genetic approach in *Drosophila* has produced results that are consistent with these ideas. In particular, adenylyl cyclase activity in the *Drosophila* learning mutant, *rutabaga*, differs from wild-type in that it is not stimulated by Ca^{2+} or calmodulin (Dudai & Zvi 1984; Livingstone et al 1984). Thus, these very different types of evidence from *Aplysia* and *Drosophila* tend to suggest similar molecular mechanisms for associative learning in these different species. Synergistic stimulation of adenylyl cyclase by Ca^{2+} and transmitters has also recently been demonstrated in synaptosomal membranes from rat brain (Natsukari et al 1990).

Activity-dependent neuromodulation is a type of associative cellular mechanism in individual neurons that detects and records that two events have occurred in close temporal contiguity. It has recently been discovered that activity-dependent enhancement occurs for presynaptic inhibition, as well as presynaptic facilitation of the *Aplysia* sensory neurons (Small et al 1989). Facilitation of the sensory neurons is produced by serotonin acting through cAMP, and inhibition is produced by FMRFamide, probably acting through arachidonic acid. Activity-dependent enhancement is, thus, not limited to a single direction of action, transmitter, or second-messenger system. Activity-dependent enhancement has also been reported in experiments in which activity was paired with application of acetylcholine or cGMP in cat motor cortex (Woody et al 1978) and with octopamine at crayfish neuromuscular junction (Breen & Atwood 1983). These results suggest that activity-dependent enhancement of modulatory effects may be a general type of associative mechanism and might occur for modulation involving other transmitters and second messengers, as well.

Associative LTP in the CA₁ Region Involves Coincident Activity in a Presynaptic and Postsynaptic Cell

Unlike facilitation at the sensory neuron-motor neuron synapses in *Aplysia*, LTP in the CA_1 region of the hippocampus shows associative features that have the characteristics of a Hebbian synapse: The induction of LTP in CA_1 requires coincident activity in the postsynaptic pyramidal neuron and presynaptic neurons. Blocking postsynaptic firing blocks the induction of LTP (Malinow & Miller 1986), and intracellularly produced postsynaptic firing can induce LTP when it is paired with weak presynaptic stimulation (Kelso et al 1986; Sastry et al 1986; Wigström et al 1986). Maximal LTP occurs when activation of the pre- and postsynaptic cells is nearly simultaneous (Gustafsson et al 1987; Kelso et al 1986). Because simultaneous pairing is characteristic of declarative learning (e.g. learning facts and spatial relations), whereas forward pairing is characteristic of procedural learning (e.g. basic conditioning), this temporal requirement is consistent with a role of LTP in declarative learning.

The Hebbian property of LTP in CA_1 apparently derives from the properties of the N-methyl-D-aspartate (NMDA) glutamate receptor and its channel. The major excitatory transmitter in the hippocampus is thought to be glutamate, which exerts its action through two major classes of receptors: NMDA and non-NMDA. There is a low density of NMDA receptors in the CA_3 region, but a high density in CA_1 (Monaghan & Cotman 1985). The ion channels associated with the NMDA receptors allow Ca^{2+} influx, which is thought to be essential for the induction of LTP in CA_1 (Lynch et al 1983; Malenka et al 1988, 1992). The NMDA-receptor channel is blocked at negative resting membrane potential by extracellular Mg^{2+} ions. The blockade of this channel is voltage dependent: When the membrane is depolarized, Mg^{2+} is expelled from the channel. Thus, Ca^{2+} influx through the channel requires the coincidence of post-synaptic depolarization and activation of the NMDA receptors by glutamate. Both events are normally produced by strong, high frequency stimulation of presynaptic fibers, which causes sufficient activation of non-NMDA receptor channels to depolarize the postsynaptic cell, thus removing the Mg^{2+} blockade of the NMDA receptor channels and allowing Ca^{2+} to enter the cell.

Blocking activation of the NMDA receptor with selective inhibitors blocks LTP in CA_1, but not in CA_3 (Collingridge et al 1983; Harris & Cotman 1986). This difference may account for some of the differences in properties of LTP in CA_1 and CA_3. Infusion of NMDA receptor blockers into the hippocampus also blocks spatial learning, which suggests that an NMDA receptor mechanism in the hippocampus, perhaps LTP, is involved in that type of learning (Davis et al 1992; Morris et al 1986).

At the molecular level, several lines of evidence suggest that the induction of LTP in CA_1 involves protein phosphorylation initiated by the Ca^{2+} influx into the postsynaptic cell, most likely by Ca^{2+}/calmodulin kinase II (protein kinase B), Ca^{2+} phospholipid-dependent kinase (protein kinase C), and tyrosine kinase (Grant et al 1992; Malenka et al 1989; Malinow et al 1988, 1989; O'Dell et al 1991b; Silva et al 1992). Malinow et al (1988) found that blockers of these kinases block the maintenance, as well as the induction, of LTP, which suggests that maintenance here, as in *Aplysia*, involves a persistent kinase activation. Studies of protein kinase C activation with LTP suggest that the kinase is cleaved and converted to a constitutively active form that functions in the absence of Ca^{2+} (Klann & Sweatt 1990). There is also evidence that the late phases of LTP involve protein synthesis (Frey et al 1988) and morphological changes in presynaptic terminals and dendritic spines (Applegate et al 1987; Meshul & Hopkins 1990), perhaps because of activation of protein kinase A (Chetkovich et al 1991; Frey & Kandel 1992).

LTP in the CA_1 Region May Involve Presynaptic Facilitation

Long-term potentiation in CA_1 is thought to be induced by a postsynaptic mechanism, but the evidence for maintenance is less clear, and the site of expression remains controversial. Bliss et al (1986) initially provided evidence for enhanced presynaptic release of glutamate during LTP. Subsequent studies by Kauer et al (1988) and Muller et al (1988) indicated that expression of LTP in CA_1 might be postsynaptic. They found that during LTP, only the flow of current through the non-NMDA receptor channels was enhanced, whereas if there were a presynaptic increase in transmitter release, one might expect an increase in current through both NMDA and non-NDMA receptor channels.

However, Bashir et al (1991) and Tsien & Malinow (1990) reported an increase in the NMDA and non-NMDA components of the synaptic potential during LTP. This increase in the NMDA component may have been missed in earlier experiments, either because it appears to saturate easily or because low frequency activation of the NMDA receptor may block the formation of LTP (Coan et al 1989; Huang et al 1992). Davies et al (1989) examined the postsynaptic response to iontophoretically applied glutamate agonists and found no change soon after the induction of LTP, but an increase during later stages of LTP, which suggests that expression is initially presynaptic, but becomes postsynaptic. Also suggestive of a presynaptic mechanism, experiments by Hess & Gustafsson (1990) indicate that LTP is associated with a change in shape of the excitatory synaptic potential, perhaps consistent with spike broadening in the presynaptic terminals. However, recent measurements of presynaptic Ca^{2+} signals in CA_1 show no change following induction of LTP (Wu & Saggau 1992)

The most direct way to distinguish between presynaptic and postsynaptic mechanisms is through quantal analysis (del Castillo & Katz 1954; Redman 1990). Several groups have now carried out quantal analysis for LTP (Bekkers & Stevens 1990; Foster & McNaughton 1991; Malinow 1991; Malinow & Tsien 1990). This analysis is based on the assumption that each postsynaptic response (the measure of either a synaptic current or potential), is made up of many integral quantal units. Each quantum is thought to reflect the postsynaptic response elicited by the release of a single synaptic vesicle. The average amplitude of the postsynaptic response, E, is given by $E = m \times q$, where m is the mean number of quanta (number of vesicles released) and q the size of each quantum. Under most circumstances, an increase in m indicates a presynaptic mechanism, whereas a change in q suggests a postsynaptic mechanism.

Although quantal analysis has been a very powerful tool at peripheral synapses, where the methodology was first developed by del Castillo & Katz (1954), it has proven surprisingly difficult to apply to the hippocampus (or to other central synapses) for several reasons (Redman 1990). Bekkers & Stevens (1990) and Malinow & Tsien (1990; Malinow 1991) used whole cell patch clamp recording to circumvent some of these difficulties by improving the signal-to-noise ratio and reported that the expression of LTP is presynaptic. In cultured hippocampal neurons, Bekkers & Stevens found that they could induce LTP by pairing the firing of a presynaptic neuron with depolarization of the postsynaptic cell. Following LTP, the mean synaptic current doubled in amplitude, whereas q, the amplitude of the unitary quantal event, remained unchanged. Direct estimates of q have proved more difficult in intact hippocampal slices. As a result, both groups relied on a more indirect index of presynaptic function, the coefficient of variation (CV) of the evoked response (i.e. the standard deviation divided by the mean). Under the assumption that the number of quanta released per impulse follows a binomial distribution, $1/CV^2$ provides an index of synaptic function that depends solely on presynaptic factors. Both Bekkers & Stevens and Malinow & Tsien found that there was a large increase in $1/CV^2$ following LTP, as would be expected if the increased postsynaptic response was caused by increased transmitter release. Moreover, these groups also report a significant decrease in the number of failures, stimuli that fail to release any quanta, which also suggests a presynaptic change. These results do not rule out a postsynaptic contribution, such as reported by Foster & McNaughton (1991), who found a 30% increase in quantal size (q) during LTP.

However, several of the basic assumptions of quantal analysis, when applied to central nervous system neurons, have recently been challenged (Edwards et al 1990; Korn & Faber 1991; Larkman et al 1991). In particular, it has been suggested that release does not follow a binomial distribution and that the size of the quantal event may reflect quantal clusters of postsynaptic receptors.

Three further studies by Malinow (1991), Malgaroli & Tsien (1992), and Manabe et al (1992) add further data, although they do not as yet settle the point. Malinow has studied transmission between individual presynaptic CA_3 and postsynaptic CA_1 neurons in hippocampal slices. After induction of LTP, the synaptic responses were enhanced by a factor of ten and displayed fewer failures of transmission, and the amplitude distribution was shifted to the right. These changes again are best explained by a presynaptic increase in release. There was also a small increase in quantal size, somewhere around 50%, which could explain some of the postsynaptic modifications. Additional support for presynaptic mechanisms comes from Malgaroli & Tsien (1992). They found an increase in frequency of spontaneous miniature synaptic currents, which parallels a form of LTP in culture induced by glutamate in the presence of low Mg^{2+}. This increase in frequency is not accompanied by an increase in miniature current amplitude.

Finally, Manabe et al (1992) have provided new evidence for a postsynaptic locus by recording spontaneous miniature excitatory synaptic currents from the CA_1 pyramidal cells in hippocampal slices during LTP. They found that LTP produced a clear increase, of about 35%, in the size of the spontaneous synaptic event. This compares with an increase in the evoked synaptic current of 76%. In addition, NMDA application, which causes decrementing potentiation, also caused an increase in miniature synaptic current amplitude that paralleled the increase in the evoked synaptic current. Manabe and colleagues suggest that a substantial portion of the potentiation of the evoked excitatory synaptic current can be accounted for by the increase in the quantal size and by implication of a change in the postsynaptic transmitter sensitivity. However, their data did not rule out a presynaptic contribution to LTP, which involves an increase in the number of quanta released during an evoked EPSP. Moreover, even the observation of an increase in the spontaneous miniature synaptic current amplitude could be consistent with a presynaptic mechanism, which involves cooperative simultaneous release of more than one quantum. Such a possibility has been strongly emphasized by Van der Kloot (1990), based on studies at the vertebrate neuromuscular junction.

Perhaps the safest conclusion at this time is that there is evidence for both pre- and postsynaptic effects during the maintenance of LTP, and both may occur (cf. Kullmann & Nicoll 1992).

LTP May Involve a Retrograde Signal that Communicates Information from the Postsynaptic to the Presynaptic Cell

If the induction of LTP requires postsynaptic activation of NMDA receptors, and maintenance involves a presynaptic increase in transmitter release, some message may be sent from the postsynaptic to the presynaptic neurons. If the retrograde facilitating substance in the hippocampus could diffuse widely, the facilitation should occur at synapses onto postsynaptic cells that did not participate in the induction of LTP. This result has been observed both in monolayer cultures of hippocampus by Bonhoeffer et al (1989) and more recently in hippocampal slices (E. M. Schuman and D. V. Madison, personal communication). In CA_1, induction of LTP in any given cell (by a Hebbian mechanism) leads to the expression of LTP in neighboring postsynaptic cells through a non-Hebbian step in which the neighboring postsynaptic cell does not fire. A similar mechanism has now also been found by Kossel et al (1990) in the visual cortex.

What features should a retrograde signal have? The retrograde messenger is most likely released not from the cell body, but from dendrites of the postsynaptic cell, perhaps from dendritic spines. Because the dendritic spines do not have the conventional machinery for the release of transmitter, the retrograde messenger may be membrane permeable and reach the presynaptic terminals by free diffusion. O'Dell et al (1991a) outlined nine criteria for identifying a retrograde messenger, based on those that have proven useful in the identification of molecules released in normal, anterograde, synaptic transmission, as well as the known properties of LTP in the CA_1 region of hippocampus: 1. The messenger should be synthesized by the postsynaptic CA_1 pyramidal cells. 2. It should be released in response to the activation of NMDA receptors during the induction of LTP. 3. Inhibiting the synthesis of the retrograde messenger should block LTP. 4. There should be a pathway for removing or degrading the active messenger. 5. Exogenous application of the candidate retrograde messenger should mimic LTP. 6. The actions of an exogenously applied candidate messenger should be independent of the NMDA receptor, because during the induction of LTP the release of the endogenous retrograde messenger presumably occurs following NMDA receptor activation.

7. The action of the retrograde messenger should be rapid, because the observed increased release of neurotransmitter during the expression of LTP occurs immediately following its induction (Bekkers & Stevens 1990; Malinow & Tsien 1990). 8. Tetanic LTP should be occluded by the potentiation produced by the retrograde messenger, but other types of synaptic facilitation thought to occur presynaptically (such as paired-pulse facilitation) should not be occluded (Zalutsky & Nicoll 1990). 9. The retrograde messenger should be synapse specific, because during LTP potentiation occurs only at synapses of the stimulated presynaptic fibers and not other fibers. Synapse specificity may be achieved either by spatially restricted diffusion of a locally released messenger, or by activity-dependent actions of the retrograde messenger, whereby activity in the presynaptic terminals renders them receptive to the influence of the retrograde messenger.

Two molecules have been examined that are freely diffusible and might serve as retrograde messengers: arachidonic acid (Piomelli et al 1987; Williams et al 1989) and nitric oxide (Galley et al 1990; Garthwaite et al 1988).

ARACHIDONIC ACID AS A CANDIDATE RETROGRADE MESSENGER IN LTP   Bliss and colleagues first examined the role of arachidonic acid in LTP in the dentate gyrus of the hippocampus. They found that bath application of nordihydroguaiaretic acid, an inhibitor of the lipoxygenase metabolism of arachidonic acid blocked LTP, whereas indomethacin, an inhibitor of cyclo-oxygenase metabolism, did not (Williams & Bliss 1989). Next, they found that tetanic stimulation or activation of NMDA receptors led to the release of arachidonic acid (Lynch et al 1989). Finally, they found that arachidonic acid produced a long-lasting potentiation (Williams et al 1989). However, the potentiation produced by arachidonic acid had a delayed onset, whereas transmitter release increases almost immediately following the normal induction of LTP (Bekkers & Stevens 1990; Malinow & Tsien 1990).

Arachidonic acid was also examined in the CA_1 region by O'Dell et al (1991a). Here, bath application of arachidonic acid had no consistent effect on the magnitude of the extracellularly recorded EPSPs evoked by Schaffer collateral/commissural fiber stimulation. However, weak tetanic stimulation (50 Hz/0.5 sec) that failed to produce long-lasting potentiation when given alone did produce a persistent enhancement of synaptic transmission when delivered in the presence of arachidonic acid, but this potentiation was blocked by 50 μM APV. Because the release of the endogenous retrograde messenger during the induction of LTP must occur at a step subsequent to the activation of the NMDA receptor, the potentiating effect of an exogenously applied candidate retrograde messenger would be expected to be independent of the activation of NMDA receptors.

Thus, although arachidonic acid satisfies some of the criteria for a retrograde messenger at the perforant pathway to granule cell synapses in the dentate gyrus, it does not fulfill as many of these criteria at the Schaffer collateral synapse onto CA_1 pyramidal cells.

NITRIC OXIDE AS A CANDIDATE RETROGRADE MESSENGER   In contrast to arachidonic acid, nitric oxide (NO) seems a better candidate for a diffusible retrograde messenger in CA_1. Nitric oxide is a gas that is generated by the enzyme *NO synthase* from the amino acid l-arginine by splitting off stoichiometric amounts of citrulline. Nitric oxide synthase requires Ca^{2+} calmodulin and the coenzyme NADPH.

The physiological significance of NO was first appreciated in 1980, when Furchgott & Zawalzki discovered endothelial-derived relaxing factor (EDRF), a local hormone released from endothelial cells in response to vasodilators, such as acetylcholine or histamine. Soon thereafter, Moncada and colleagues showed that EDRF was NO (Palmer et al 1987). Nitric oxide diffuses into the underlying smooth muscle, where it causes relaxation and dilation of blood vessels and a rise in cGMP. Garthwaite and collaborators found that activation of NMDA receptors in the brain causes release of NO from cultured cerebellar granule cells (Garthwaite et al 1988) and formation of cGMP through the NO pathway in hippocampus (East & Garthwaite 1991).

O'Dell et al (1991a), Schuman & Madison (1991), Bohme et al (1991), and Haley et al (1992) have investigated the possible role of NO as a retrograde messenger in LTP. All four groups found that specific inhibitors of the enzyme NO synthase block the induction of LTP. This inhibition is reversed by giving excess of the amino acid l-arginine. As would be expected for a retrograde messenger, O'Dell et al (1991a) and Schuman & Madison (1991) found that NO is probably produced postsynaptically, because injecting the inhibitors into the postsynaptic cell blocks LTP. Moreover, perfusion of the slice with NO binding protein, hemoglobin (which does not penetrate cells), also blocks LTP, which indicates that NO must diffuse out of the postsynaptic cell to produce its action, consistent with its possible role as a retrograde messenger in the induction of LTP. To test this idea directly, O'Dell et al (1991a) applied NO to hippocampal neurons in culture and detected an increase in spontaneous release.

The finding of a retrograde messenger that can diffuse to the presynaptic terminals provides a mechanistic explanation for the findings of Bonhoeffer et al (1989) (Figure 3*B*) that induction of LTP in any given cell in CA_1 leads to the expression of LTP in neighboring postsynaptic cells. Similar results have now also been observed by E. M. Schuman and D. V. Madison (personal communication) in the CA_1 region of the hippocampal slice. In addition, Schuman and Madison have found that dialysis of the post-synaptic cell with whole cell recording for 10–20 minutes prevents post-synaptic induction in the cell recorded from. Nevertheless, the dialyzed cell will still show LTP following tetanic stimulation of the presynaptic pathway, presumably because the neighboring cells release a retrograde factor that acts on presynaptic terminals that synapse on the dialyzed cell. Indeed, this effect is blocked by hemoglobin, which binds NO. These studies indicate that NO may diffuse quite widely, perhaps 500 μm or more.

NITRIC OXIDE MAY PRODUCE ACTIVITY-DEPENDENT PRESYNAPTIC FACILITA-TION   In the course of studying the effects of NO on LTP in hippocampal slices, Small et al (1992b,c) found that the effect of NO is greatly enhanced when it is paired with activity in the presynaptic neurons. Thus, activity appears to enhance NO's ability to produce presynaptic facilitation. Nitric oxide produces these effects and initiates LTP even in the presence of APV, which blocks the postsynaptic NMDA receptors and prevents the normal (Hebbian) initiation of LTP through Ca^{2+} influx via the ion channel of that receptor.

These experiments indicate that LTP in CA_1 seems to involve an activity-dependent form of presynaptic facilitation. If so, the synapses in CA_1 use a combination of two independent associative or coincidence-detecting mechanisms. There is the well-characterized Hebbian NMDA receptor mechanism, as well as a non-Hebbian, activity-dependent, presynaptic facilitating mechanism, similar to that seen in *Aplysia* (Figure 3*B*). The associative activation of NMDA receptors in the postsynaptic cells (by glutamate and depolarization) produces a retrograde signal (NO), which initiates an associative activity-dependent (NO and presynaptic activity) facilitation of transmitter release from the presynaptic terminals.

What might be the functional advantage of combining two associative cellular mechanisms, the postsynaptic NMDA receptor and activity-dependent presynaptic facilitation, in this way? One possible advantage of presynaptic plasticity is that it permits an increase in the synaptic signal without increasing synaptic noise, whereas postsynaptic effects, such as changes in receptor sensitivity, would increase both. If presynaptic facili-tation is produced by a diffusible substance, activity dependence of the presynaptic effect could account for the pathway specificity of the poten-tiation, whereby an inactive presynaptic input is not potentiated after induction of LTP by an active input. Conversely, a presynaptic effect could be restricted to synapses where the postsynaptic cell is simultaneously active by using an associative postsynaptic mechanism, such as the NMDA receptor, to cause local release of the facilitating substance. Combining associative pre- and postsynaptic mechanisms in this way would permit the use of presynaptic facilitation in truly Hebbian synapses, with the possible functional advantages, such as output specificity and second-order effects, that Hebb synapses provide (see Hawkins & Kandel 1990).

As these arguments suggest, there may be several similarities between associative presynaptic facilitation in *Aplysia* and associative LTP in CA_1 (Table 1) and between these two mechanism and nonassociative presynaptic facilitation. Both associative mechanisms produce long-lasting changes in synaptic strength that are thought to contribute to associative learning. In both cases there appears to be an important presynaptic component, and in each case the presynaptic component may use a non-Hebbian associative mechanism: an activity-dependent form of presynaptic facilitation.

There are also, however, several important differences between facilitation in *Aplysia* and LTP in CA_1. In hippocampus, the modulatory messenger is evidently released by the postsynaptic cell and not by a diffusely projecting modulatory system, as is the case with 5-HT in *Aplysia*. In addition, different modulatory messengers and kinases appear to be involved. Finally, LTP also involves Hebbian association because of the properties of the NMDA receptor in CA_1 pyramidal neurons.

THERE ARE STILL IMPORTANT QUESTIONS THAT NEED TO BE RESOLVED BEFORE NO CAN BE ACCEPTED AS A RETROGRADE MESSENGER   If the retrograde messenger can diffuse to presynaptic terminals synapsing on neighboring postsynaptic cells, as suggested by the results of Bonhoeffer et al (1989), why does injecting Ca^{2+} chelators or kinase inhibitors in a single postsynaptic cell block tetanic LTP in that cell (Malenka et al 1988, 1989; Malinow et al 1988, 1989; O'Dell et al 1991b), whereas neighboring cells still undergo LTP and presumably release the retrograde message? A possible explanation is that the retrograde message must act both pre- and postsynaptically, and its postsynaptic actions require normal Ca^{2+} levels and kinase activities.

Where is NO located? As yet, little or no brain-specific NO synthase has been found in CA_1 pyramidal cells (Bredt et al 1991), although other isoforms of the enzyme could conceivably be present and generate NO. Thus, the identification of NO as a retrograde messenger must be considered tentative until there are demonstrations that NO can be synthesized in these cells.

Where does NO exert its action? What are the molecular targets for NO in the presynaptic cell? In other tissue, NO acts on one of two targets:

soluble guanylate cyclase (Murad et al 1978) and ADP ribosyltransferase (Brune & Lapetina 1989). Recent evidence suggests that both targets of NO may play a role in hippocampal LTP (Chetkovich & Sweatt 1992; Haley et al 1992; Schumann et al 1992; Small et al 1992c).

POSSIBLE MOLECULAR MECHANISMS FOR CHANGING SYNAPTIC STRENGTH

One of the ultimate goals of learning and memory studies is to identify the molecular mechanisms underlying learning related synaptic plasticity. As discussed above, changes in transmitter release can involve changes in Ca^{2+} influx with each action potential through modulation of either Ca^{2+} or K^+ channels. Alternatively, the changes in release could reflect alterations in the mechanisms for Ca^{2+} homeostasis, such as the *accumulation* of Ca^{2+} within the terminal, because of saturation of Ca^{2+} buffering mechanisms or inhibition of Ca^{2+} transport. This residual Ca^{2+} could drive Ca^{2+}-dependent steps in vesicle mobilization and release. Finally, the changes could be independent of Ca^{2+} and could involve direct changes in proteins important in the release process.

We have already discussed presynaptic facilitation and presynaptic inhibition where there is evidence for increases and decreases, respectively, in the Ca^{2+} influx with each action potential. There also is direct evidence of Ca^{2+} accumulation with posttetanic potentiation (PTP) in *Aplysia* neurons (Connor et al 1986; Kretz et al 1982), in the crayfish nerve muscle synapses (Delaney et al 1989), at the squid giant synapse (Swandulla et al 1991), and in hippocampal mossy fiber terminals (Regehr & Tank 1991) (for earlier evidence, see Erulkar & Rahamimoff 1978). With posttetanic potentiation, there is no change in Ca^{2+} influx; however, because the frequency of firing is high, the Ca^{2+} influx outstrips the Ca^{2+} buffering and Ca^{2+} accumulates in the terminals for a period of minutes. The elevated Ca^{2+} may enhance release directly or activate enzymes that can modulate release.

What happens to Ca^{2+} influx and accumulation with LTP? Regehr & Tank (1991) have examined calcium accumulation at the mossy fiber-CA_3 synapse. Although Ca^{2+} accumulation was associated with PTP, there was no detectable Ca^{2+} accumulation during LTP. However, the authors did not rule out possible changes in presynaptic Ca^{2+} influx. Wu & Saggau (1992) showed that LTP at hippocampal CA_1 synapses was not associated with increased presynaptic Ca^{2+} influx, but they did not address possible Ca^{2+} accumulation. If LTP does not involve changes in presynaptic Ca^{2+} levels, how is release facilitated? To identify possible Ca^{2+}-dependent and independent molecular targets that may participate in presynaptic plasticity, we need to consider the sequence of events at the presynaptic terminal during transmitter release. Transmitter release is a cyclical process involving the transport of vesicles to active zones, the docking of vesicles to their release sites at the active zones, the fusion of the synaptic vesicle membrane with the plasma membrane during exocytosis in response to an increase in intracellular Ca^{2+}, and the retrieval and recycling of vesicle membrane following fusion. Modulation of any one of these steps could regulate transmitter release. Several synaptic vesicle associated proteins that presumably participate in various aspects of the release cycle have recently been cloned and identified. We briefly review the properties of some selected synaptic vesicle proteins as potential sites for plasticity. For more detailed treatments, see reviews by Trimble et al (1991), Sudhof & Jahn (1991), and Smith & Augustine 1988.

Vesicle Anchoring to the Cytoskeleton: The Synapsins

The synapsins are a family of cytoplasmic proteins (synapsins Ia, Ib, IIa, IIb) that bind to synaptic vesicles and actin (see Bahler et al 1990). They are thought to anchor the synaptic vesicles to the cytoskeleton. The synapsins are phosphorylated by both cAMP-dependent protein kinase and calcium/calmodulin-dependent protein kinase II (CaM kinase). Phosphorylation by CaM kinase appears to decrease the binding of synaptic vesicles and actin by synapsin Ia/b. This decrease in binding is thought to play a role in regulating vesicle availability for release. Thus, injection of dephosphorylated synapsin Ia/b into the presynaptic terminal of the squid giant axon inhibits transmitter release (Lin et al 1990; Llinás et al 1985, 1991; Nichols et al 1990), presumably because of the binding and immobilization of synaptic vesicles by synapsin. In contrast, injection of CaM kinase II into the squid axon presynaptic terminal enhances release, presumably by phosphorylating synapsin and mobilizing vesicles for release. Under physiological conditions, the activation of CaM kinase II following presynaptic stimulation could conceivably contribute to certain forms of homosynaptic presynaptic plasticity, including posttetanic potentiation and LTP. However, there is as yet no direct evidence to support such a role.

Vesicle Mobilization and GTP-Binding Proteins

Whereas higher molecular weight GTP-binding proteins (MW ~80 kd) play important roles in signal transduction, low molecular weight GTP-binding proteins (of 20–25 kd), which belong to the p21ras superfamily, are thought to play an important role in the targeting of secretory vesicles to specific membrane compartments and, perhaps, in exocytosis (see Bourne 1988 for review). In the best characterized example, a small GTP-binding protein encoded by the *sec4* gene in yeast is thought to control the docking and fusion of secretory vesicles with the plasma membrane.

Mutations in this gene block vesicle fusion. Two homologous low molecular weight GTP-binding proteins specifically associated with synaptic vesicles—rab3A and rab3B—have been identified in neurons (Fischer von Mollard et al 1991; Matsui et al 1988; Zahraoui et al 1989). By analogy with the *sec4* gene of yeast, these 25 kd proteins have been proposed to play a role in targeting of synaptic vesicles to active zones. Evidence that these proteins may play a role in release comes from the finding that exocytosis causes the dissociation of rab3A from synaptic vesicles (Fischer von Mollard et al 1991). The activity of rab3A can be modulated by a protein found in brain that inhibits the dissociation of GDP and subsequent binding of GTP (Sasaki et al 1990). Although not yet identified, a GTPase activating protein (GAP) might also serve to regulate rab3A function by analogy with the ras GAP.

Although there is currently no direct evidence linking the p21ras superfamily with synaptic plasticity, such proteins are thought to play a role in associative conditioning in *Hermissenda* (Nelson et al 1990). Thus, following associative conditioning of light and rotation, there is an increase in phosphorylation of a 20 kd GTP binding protein (cp20) in the *Hermissenda* retina. Injection of purified cp20 into *Hermissenda* photoreceptors simulates the modulatory effects of associative conditioning on two K currents. Moreover, injection of mammalian ras proteins into the photoreceptors also simulates these actions, which suggests that cp20 is related to ras (Collin et al 1990). However, it is not clear from these experiments whether cp20 corresponds to the *Hermissenda* form of ras or rab, nor whether cp20 is involved in modulating transmitter release from *Hermissenda* neurons.

Exocytosis and Proteins Integral to the Synaptic Vesicle

Two integral membrane synaptic vesicle proteins, synaptophysin and synaptotagmin, are thought to play a more direct role in controlling exocytotic release of transmitter. These two proteins are also potential targets for modulation contributing to synaptic plasticity. Synaptotagmin (or p65) is a 65 kd integral membrane synaptic vesicle protein that has also been suggested to play a role in synaptic vesicle docking and exocytosis (Mathew et al 1981; Perin et al 1990). Synaptotagmin has two internal repeats in its cytoplasmic carboxy terminal tail that are homologous to those domains of protein kinase C, which are thought to be involved in calcium regulation and phospholipid binding. Modulation of the affinity of synaptotagmin for calcium or phospholipid could lead to changes in release. A role for synaptotagmin in release is supported by the finding that this protein binds to the synaptic plasma membrane receptor for α-latrotoxin, a toxin from black widow spider venom that triggers release of synaptic vesicles (Petrenko et al 1991).

Synaptophysin (Mr 34 kd) is another integral membrane synaptic vesicle protein, which is thought to share certain structural features with gap junction ion channels (Buckley et al 1987; Sudhof et al 1987). It contains four putative transmembrane segments and forms a channel when incorporated in planar lipid bilayers (Thomas et al 1988). Almers & Tse (1990) have demonstrated that exocytosis from mast cells involves the transient formation of a fusion pore between the secretory vesicle and plasma membrane that has a conductance similar to gap junction channels. Thus, it has been proposed that synaptophysin is a candidate molecule for forming such a fusion pore. Although it is unclear how a single membrane channel could span both the synaptic vesicle and plasma membrane, synaptophysin does bind to a presynaptic plasma membrane protein, physophilin (Thomas & Betz 1990). Synaptophysin contains a tyrosine rich cytoplasmic domain and is a substrate for the $pp^{60c-src}$ tyrosine kinase (Barnekow et al 1990), suggestive of a potential regulatory site.

OVERALL VIEW

The recent work on LTP suggests the outline of a unified view that encompasses several forms of learning-related plasticity in both vertebrates and invertebrates. Several common themes emerge:

1. Many learning-related changes in neuronal function result from enduring changes in synaptic strength. One mechanism commonly used to change synaptic strength is an alteration in transmitter release. As is the case with behavioral learning, some forms of synaptic plasticity are nonassociative and some are associative.

2. Nonassociative homosynaptic depression and heterosynaptic facilitation and inhibition in *Aplysia* all involve presynaptic modulation of transmitter release. Recent studies of mossy fibers indicate LTP in the CA_3 region of the hippocampus may also have a nonassociative presynaptic component.

3. Two types of associative mechanisms have been described: a Pre-Modulatory mechanism and a Pre-Post (Hebbian) mechanism. The best studied Pre-Modulatory mechanism is activity-dependent presynaptic facilitation, which is thought to contribute to classical conditioning in *Aplysia*. The best studied case of Hebbian plasticity is the induction of LTP in the CA_1 region and in the dentate gyrus of the hippocampus.

4. Long-term potentiation in CA_1 is not restricted to active postsynaptic neurons. It can also spread, presumably by means of a retrograde signal, to presynaptic terminals that end on neurons that are inactive. Thus, whereas LTP is Hebbian at its core, it is non-Hebbian in its surround.

5. The presynaptic actions of NO, a candidate retrograde messenger,

are evidently activity dependent, thus restricting the potentiation to active presynaptic fibers. Therefore, LTP in CA_1 apparently involves two associative mechanisms: the postsynaptic NMDA receptor with Hebbian properties and activity-dependent presynaptic facilitation similar to that seen in *Aplysia*.

These several points illustrate four surprising reductionist possibilities. First, both nonassociative and associative forms of learning are represented in basic cellular processes that do not require complex neural circuitry. Thus, there is even at the cellular level a representation of the computation characteristic of associative classical conditioning. Second, as we have seen in *Aplysia*, *Drosophila*, and hippocampus, nonassociative and associative mechanisms may be related. This suggests that there may be a cellular alphabet for synaptic plasticity where more complex forms of plasticity are based on combinations of components used in simpler forms (Hawkins & Kandel 1984). Third, although we have not considered them here, there are also on the cellular level representations of various short- and long-term stages in memory storage evident on the behavioral level. One particularly interesting finding to emerge from these studies is that long-term memory is associated with structural changes, a finding that draws attention to a possible similarity between memory storage and the late stages of synapse formation. During this phase of development, activity is thought to be important for the pruning and fine tuning of synaptic connections. Moreover, there is preliminary evidence that this fine tuning of connections during development uses activity-dependent coincidence mechanisms similar to those used in learning-related plasticity (Shatz 1992). This raises the possibility that the regulation of synaptic growth in development and learning may share common molecular components and be related in a fundamental way.

The emergence of one set of cellular themes based on presynaptic mechanisms as an important aspect of learning, should not, however, distract attention from other synaptic mechanisms of demonstrated importance. For example, certain types of motor learning in cerebellum involve *postsynaptic* changes in synaptic efficacy (Ito 1989). Moreover, *nonsynaptic* mechanisms, such as changes in excitability, are also important, as revealed in studies of associative learning in *Hermissenda*, rabbit, and cat (Alkon & Rasmussen 1988; Brons & Woody 1980), and by studies of LTP in the hippocampus and sensitization in *Aplysia* (Frost et al 1988; Hess & Gustafsson 1990). Interestingly, these nonsynaptic changes may, in some cases, be caused by the same underlying second messenger mechanisms as the synaptic changes (Klein et al 1986; Schuman & Clark 1990).

The initial success of investigating learning on the cellular level should, therefore, not mislead us about the complexity of the problems still to be faced. Behavioral learning is not only rich and varied, but also mediated by neural circuitry whose computational power greatly enhances the elementary operations evident on the level of single cells. There is parallel processing in memory as there is in perception: In the same animal, several different mechanisms can contribute in parallel to a given instance of learning. Thus, sensitization in *Aplysia* involves plasticity in interneurons and some motor neurons, in addition to presynaptic facilitation of the sensory neurons (Frost et al 1988). Mammalian learning presumably can involve the two different types of LTP in the CA_3 and CA_1 regions of hippocampus. Even at a single synapse, such as the Schaffer collateral synapses on the CA_1 cells, LTP involves changes in postsynaptic excitability, in addition to presynaptic changes in transmitter release (Hess & Gustafsson 1990). Nevertheless, the ability to detect associative mechanisms on the cellular level now invites a molecular analysis of these coincidence-detecting mechanisms. Conversely, the recent renaissance of interest in the aggregate properties of neuronal ensembles is well timed. It is no doubt oversimplistic to think that all forms of learning are represented at the level of single cells and synapses. Some aspects of the representation probably also involve emergent circuit properties. It will therefore be interesting to incorporate the cellular learning rules we have described into neural network models (e.g. Gally et al 1990; Hawkins 1989). Such an analysis may provide a link between the elementary cellular processes and the more complex brain functions involved in remembrance of things past.

ACKNOWLEDGEMENTS

We are grateful to Sarah Mack for preparing the figures and to Harriet Ayers and Andrew Krawetz for typing the manuscript. Preparation of this article was supported by grants from the National Institutes of Health (MH-26212 and MH-45923), the National Institute on Aging (AG08702), and the Howard Hughes Medical Institute.

Literature Cited

Abrams, T. W. 1985. Activity-dependent presynaptic facilitation: An associative mechanism in *Aplysia. Cell. Mol. Neurobiol.* 5: 123–45

Abrams, T. W., Carew, T. J., Hawkins, R. D., Kandel, E. R. 1983. Aspects of the cellular mechanism of temporal specificity in conditioning in *Aplysia*: Preliminary evidence for Ca^{2+} influx as a signal of activity. *Soc. Neurosci. Abstr.* 9: 169

Abrams, T. W., Castellucci, V. F., Camardo, J. S., Kandel, E. R., Lloyd, P. E. 1984. Two endogenous neuropeptides modulate the gill and siphon withdrawal reflex in *Aplysia* by means of presynaptic facilitation involving cyclic AMP-dependent closure of a serotonin-sensitive potassium channel. *Proc. Natl. Acad. Sci. USA* 81: 7956–60

Abrams, T. W., Eliot, L., Dudai, Y., Kandel, E. R. 1985. Activation of adenylate cyclase in *Aplysia* neural tissue by Ca^{2+}/calmodulin, a candidate for an associative mechanism during conditioning. *Soc. Neurosci. Abstr.* 11: 797

Abrams, T. W., Goldsmith, B. A. 1992.

cAMP modulation of multiple K$^+$ currents contributes to both action potential broadening and increased excitability in *Aplysia* sensory neurons. *Soc. Neurosci. Abstr.* 18: 16

Abrams, T. W., Karl, K. A., Kandel, E. R. 1991. Biochemical studies of stimulus convergence during classical conditioning in *Aplysia*: Dual regulation of adenylate cyclase by Ca^{2+}/calmodulin and transmitter. *J. Neurosci.* 11: 2655–65

Alkon, D. L., Rasmussen, H. 1988. A spatial temporal model of cell activation. *Science* 239: 998–1005

Alkon, D. L., Ikeno, H., Dworkin, J., McPhie, D. L., Olds, J. L., et al. 1990. Contraction of neuronal branching volume: An anatomic correlate of Pavlovian conditioning. *Proc. Natl. Acad. Sci. USA* 87: 1611–14

Almers, W., Tse, F. W. 1990. Transmitter release from synapses: Does a preassembled fusion pore initiate exocytosis? *Neuron* 4: 813–18

Andersen, P., Sundberg, S. H., Sveen, O., Wigström, H. 1977. Specific long-lasting potentiation of synaptic transmission in hippocampal slices. *Nature* 266: 736–37

Applegate, M. D., Kerr, D. S., Landfield, P. W. 1987. Redistribution of synaptic vesicles during long-term potentiation in the hippocampus. *Brain Res.* 401: 401–6

Bahler, M., Benfenati, F., Valtorta, F., Greengard, P. 1990. The synapsins and the regulation of synaptic function. *Bioessays* 12(6): 259–63

Bailey, C. H., Chen, M. 1988a. Long-term memory in *Aplysia* modulates the total number of varicosities of single identified sensory neurons. *Proc. Natl. Acad. Sci. USA* 85: 2372–77

Bailey, C. H., Chen, M. 1988b. Long-term sensitization in *Aplysia* increases the number of presynaptic contacts onto the identified gill motor neuron L7. *Proc. Natl. Acad. Sci. USA* 85: 9356–59

Barnekow, A., Jahn, R., Schartl, M. 1990. Synaptophysin: A substrate for the protein tyrosine kinase pp60c-src in intact synaptic vesicles. *Oncogene* 5: 1019–24

Barrionuevo, G., Brown, T. H. 1983. Associative long-term potentiation in hippocampal slices. *Proc. Natl. Acad. Sci. USA* 80: 7347–51

Bashir, Z. I., Alford, S., Davies, S. N., Randall, A. D., Collingridge, G. L. 1991. Long-term potentiation of NMDA receptor-mediated synaptic transmission in the hippocampus. *Nature* 349: 156–58

Baxter, D. A., Bittner, G. D., Brown, T. H. 1985. Quantal mechanisms of long-term synaptic potentiation. *Proc. Natl. Acad. Sci. USA* 82: 5978–82

Baxter, D. A., Byrne, J. H. 1989. Serotonergic modulation of two potassium currents in the pleural sensory neurons of *Aplysia*. *J. Neurophysiol.* 62: 665–79

Bekkers, J. M., Stevens, C. F. 1990. Presynaptic mechanism for long-term potentiation in the hippocampus. *Nature* 346: 724–29

Belardetti, F., Kandel, E. R., Siegelbaum, S. A. 1987. Neuronal inhibition by the peptide FMRFamide involves opening of SK$^+$ channels. *Nature* 325: 153–56

Belardetti, F., Campbell, W. B., Falck, J. R., Demontis, G., Rosolowsky, M. 1989. Products of heme-catalyzed transformation of the arachidonic derivative 12-HPETE open S-type K$^+$ channels in *Aplysia*. *Neuron* 3: 497–505

Bergold, P. J., Beushausen, S. A., Sacktor, T. C., Cheley, S., Bayley, H., Schwartz, J. H. 1992. A regulatory subunit of the cAMP-dependent protein kinase downregulated in Aplysia sensory neurons during long-term sensitization. Neuron 8: 387–97

Bergold, P. J., Sweatt, J. D., Winicov, I., Weiss, K. R., Kandel, E. R., Schwartz, J. H. 1990. Protein synthesis during acquisition of long-term facilitation is needed for the persistent loss of regulatory subunits of the *Aplysia* cAMP-dependent protein kinase. *Proc. Natl. Acad. Sci. USA* 87: 3788–91

Bernier, L., Castellucci, V. F., Kandel, E. R., Schwartz, J. H. 1982. Facilitatory transmitter causes a selective and prolonged increase in adenosine 3':5'-monophosphate in sensory neurons mediating the gill and siphon withdrawal reflex in *Aplysia*. *J. Neurosci.* 2: 1682–91

Bliss, T. V. P., Douglas, R. M., Errington, M. L., Lynch, M. A. 1986. Correlation between long-term potentiation and release of endogenous amino acids from dentate gyrus of anaesthetized rats. *J. Physiol.* (*London*) 377: 391–408

Bliss, T. V. P., Lømo, T. 1973. Long-lasting potentiation of synaptic transmission in the dentate area of the anaesthetized rabbit following stimulation of the perforant path. *J. Physiol.* (*London*) 232: 331–56

Bohme, G. A., Bon, C., Stutzmann, J.-M., Doble, A., Blanchard, J.-C. 1991. Possible involvement of nitric oxide in long-term potentiation. *Eur. J. Pharmacol.* 199: 379–81

Bonhoeffer, T., Staiger, V., Aertsen, A. 1989. Synaptic plasticity in rat hippocampal slice cultures: Local "Hebbian" conjunction of pre- and postsynaptic stimulation leads to distributed synaptic enhancement. *Proc. Natl. Acad. Sci. USA* 86: 8113–17

Bourne, H. R. 1988. Do GTPases direct membrane traffic in secretion? *Cell* 53: 669–71

Bradler, J. E., Barrionuevo, G. 1989. Long-term potentiation in hippocampal CA_3 neurons: Tetanized input regulates heterosynaptic efficacy. *Synapse* 4: 132–42

Braha, O., Dale, N., Hochner, B., Klein, M., Abrams, T. W., Kandel, E. R. 1990a. Second messengers involved in the two processes of presynaptic facilitation that contribute to sensitization and dishabituation in *Aplysia* sensory neurons. *Proc. Natl. Acad. Sci. USA* 87: 2040–44

Braha, O., Dale, N., Klein, M., Kandel, E. R. 1990b. Protein kinase C may contribute to the increase in spontaneous release evoked by 5-HT at cultured *Aplysia* sensory-motor synapses. *Soc. Neurosci. Abstr.* 16: 1013

Bredt, D. S., Glatt, C. E., Huang, P. M., Fotuhi, M., Dawson, T. M., Snyder, S. H. 1991. Nitric oxide synthase protein and mRNA are discretely localized in neuronal populations of the mammalian CNS together with NADPH diaphorase. *Neuron* 7: 615–24

Breen, C. A., Atwood, H. L. 1983. Octopamine—a neurohormone with presynaptic activity-dependent effects at crayfish neuromuscular junctions. *Nature* 30: 716–18

Brezina, V., Eckert, R., Erxleben, C. 1987. Modulation of potassium conductances by an endogenous neuropeptide in neurones of *Aplysia californica*. *J. Physiol. (London)* 382: 267–90

Briggs, C. A., McAfee, D. A., McCaman, R. E. 1985. Long-term potentiation of synaptic acetylcholine release in the superior cervical ganglion of the rat. *J. Physiol. (London)* 363: 181–90

Brons, J. F., Woody, C. D. 1980. Long-term changes in excitability of cortical neurons after Pavlovian conditioning and extinction. *J. Neurophysiol.* 44: 605–15

Brune, B., Lapetina, E. G. 1989. Activation of a cytosolic ADP-ribosyltransferase by nitric oxide-generating agents. *J. Biol. Chem.* 264: 8455–58

Buckley, K. M., Floor, E., Kelly, R. B. 1987. Cloning and sequence analysis of cDNA encoding p38, a major synaptic vesicle protein. *J. Cell. Biol.* 105: 2447–56

Buonomano, D. V., Byrne, J. H. 1990. Long-term synaptic changes produced by a cellular analog of classical conditioning in *Aplysia*. *Science* 249: 420–23

Buttner, N., Siegelbaum, S. A., Volterra, A. 1989. Direct modulation of *Aplysia* S-K^+ channels by a 12-lipoxygenase metabolites of arachidonic acid. *Nature* 347: 553–55

Byrne, J. H. 1987. Cellular analysis of associative learning. *Physiol. Rev.* 67: 329–439

Cajal, S. R. 1911. *Histologie du Système Nerveux de l'Homme et des Vertébrés.* Paris: Maloine. (Republished 1955, *Histologie du Système Nerveux.* Translated by L. Azoulay. Madrid: Inst. Ramón y Cajal)

Carew, T. J., Hawkins, R. D., Abrams, T. W., Kandel, E. R. 1984. A test of Hebb's postulate at identified synapses which mediate classical conditioning in *Aplysia*. *J. Neurosci.* 4: 1217–24

Carew, T. J., Hawkins, R. D., Kandel, E. R. 1983. Differential classical conditioning of a defensive withdrawal reflex in *Aplysia californica*. *Science* 219: 397–400

Carew, T. J., Sahley, C. L. 1986. Invertebrate learning and memory: From behavior to molecules. *Annu. Rev. Neurosci.* 9: 435–87

Carew, T. J., Walters, E. T., Kandel, E. R. 1981. Classical conditioning in a simple withdrawal reflex in *Aplysia californica*. *J. Neurosci.* 1: 1426–37

Castellucci, V. F., Kandel, E. R. 1976. Presynaptic facilitation as a mechanism for behavioral sensitization in *Aplysia*. *Science* 194: 1176–78

Castellucci, V. F., Kandel, E. R. 1974. A quantal analysis of the synaptic depression underlying habituation of the gill-withdrawal reflex in *Aplysia*. *Proc. Natl. Acad. Sci. USA* 71: 5004–8

Castellucci, V. F., Kandel, E. R., Schwartz, J. H., Wilson, F. D., Nairn, A. C., Greengard, P. 1980. Intracellular injection of the catalytic subunit of cyclic AMP-dependent protein kinase simulates facilitation of transmitter release underlying behavioral sensitization in *Aplysia*. *Proc. Natl. Acad. Sci. USA* 77: 7492–96

Castellucci, V. F., Nairn, A., Greengard, P., Schwartz, J. H., Kandel, E. R. 1982. Inhibitor of adenosine 3′:5′-monophosphate-dependent protein kinase blocks presynaptic facilitation in *Aplysia*. *J. Neurosci.* 2: 1673–81.

Castellucci, V. F., Pinsker, H., Kupfermann, I., Kandel, E. R. 1970. Neuronal mechanisms of habituation and dishabituation of the gill-withdrawal reflex in *Aplysia*. *Science* 167: 1745–48

Chattarji, S., Stanton, P. K., Sejnowski, T. J. 1989. Commissural synapses, but not mossy fiber synapses, in hippocampal field CA_3 exhibit associative long-term potentiation and depression. *Brain Res.* 495: 145–50

Chetkovich, D. M., Gray, R., Johnston, D., Sweatt, J. D. 1991. N-methyl-D-aspartate receptor activation increases cAMP levels and voltage-gated Ca^{2+} channel activity in area CA_1 of hippocampus. *Proc. Natl. Acad. Sci. USA* 88: 6467–71

Chetkovich, D. M., Sweatt, J. D. 1992. LTP-inducing tetanic stimulation causes a nitric oxide-mediated increase in cGMP in hippocampal area CA_1. *Soc. Neurosci. Abstr.* 18: 761

Clark, G. A. 1984. A cellular mechanism for the temporal specificity of classical conditioning of the siphon-withdrawal response in *Aplysia*. *Soc. Neurosci. Abstr.* 10: 268

Clark, G. A., Kandel, E. R. 1984. Branch-specific heterosynaptic facilitation in *Aplysia* siphon sensory cells. *Proc. Natl. Acad. Sci. USA* 81: 2577–81

Coan, E. J., Irving, A. J., Collingridge, G. L. 1989. Low-frequency activation of the NMDA receptor system can prevent the induction of LTP. *Neurosci. Lett.* 105: 205–10

Colebrook, E., Lukowiak, K. 1988. Learning by the *Aplysia* model system: Lack of correlation between gill and gill motor neurone responses. *J. Exp. Biol.* 135: 411–29

Collin, C., Papageorge, A. G., Lowy, D. R., Alkon, D. L. 1990. Early enhancement of calcium currents by H-ras oncoproteins injected into *Hermissenda* neurons. *Science* 250: 1743–45

Collingridge, G. L., Kehl, S. J., McLennan, H. 1983. Excitatory amino acids in synaptic transmission in the Schaffer collateral-commissural pathway of the rat hippocampus. *J. Physiol. (London)* 334: 33–46

Connor, J. A., Kretz, R., Shapiro, E. 1986. Calcium levels measured in presynaptic neurone of *Aplysia* under conditions that modulate release. *J. Physiol. (London)* 375: 625–42

Crow, T., Forrester, J. 1990. Inhibition of protein synthesis blocks long-term enhancement of generator potentials produced by one-trial *in vivo* conditioning in *Hermissenda*. *Proc. Natl. Acad. Sci. USA* 87: 4490–94

Dahl, D., Sarvey, J. M. 1989. Norepinephrine induced pathway-specific long-lasting potentiation and depression in the hippocampal dentate gyrus. *Proc. Natl. Acad. Sci. USA* 86: 4776–80

Dale, N., Kandel, E. R. 1990. Facilitatory and inhibitory transmitters modulate spontaneous transmitter release at cultured *Aplysia* sensorimotor synapses. *J. Physiol. (London)* 421: 203–22

Dale, N., Kandel, E. R., Schacher, S. 1987. Serotonin produces long-term changes in the excitability of *Aplysia* sensory neurons in culture that depend on new protein synthesis. *J. Neurosci.* 7: 2232–38

Dale, N., Schacher, S., Kandel, E. R. 1988. Long-term facilitation in *Aplysia* involves increase in transmitter release. *Science* 239: 282–85

Dash, P. K., Hochner, B., Kandel, E. R. 1990. Injection of cAMP-responsive element into the nucleus of *Aplysia* sensory neurons blocks long-term facilitation. *Nature* 345: 718–21

Davies, S. N., Lester, R. A. J., Reymann, K. G., Collingridge, G. L. 1989. Temporally distinct pre- and postsynaptic mechanisms maintain long-term potentiation. *Nature* 338: 500–3

Davis, S., Butcher, S. P., Morris, R. G. M. 1992. The NMDA receptor antagonist D-2-amino-5-phosphonopentanoate (D-AP5) impairs spatial learning and LTP *in vivo* at intracerebral concentrations comparable to those that block LTP *in vitro*. *J. Neurosci.* 12: 21–34

Delaney, K. R., Zucker, R. S., Tank, D. W. 1989. Calcium in motor nerve terminals associated with postsynaptic potentiation. *J. Neurosci.* 9: 3558–67

del Castillo, J., Katz, B. 1954. Quantal components of the end-plate potential. *J. Physiol. (London)* 124: 560–73

Dudai, T., Zvi, S. 1984. Adenylate cyclase in the *Drosophila* memory mutant *rutabaga* displays an altered Ca^{2+} sensitivity. *Neurosci. Lett.* 47: 119–24

Dudel, J., Kuffler, S. W. 1961. Mechanisms of facilitation at the crayfish neuromuscular junction. *J. Physiol. (London)* 155: 530–42

Dunwiddie, T. V., Taylor M., Heginbotham, L. R., Proctor, W. R. 1992. Long-term increases in excitability in the CA_1 region of rat hippocampus induced by β-adrenergic stimulation: Possible mediation by cAMP. *J. Neurosci.* 12: 506–17

East, S. J., Garthwaite, J. 1991. NMDA receptor activation in rat hippocampus induces cyclic GMP formation through the L-arginine-nitric oxide pathway. *Neurosci. Lett.* 123: 17–19

Edmonds, B., Klein, M., Dale, N., Kandel, E. R. 1990. Contribution of two types of calcium channels to synaptic transmission and plasticity. *Science* 250: 1142–47

Edwards, F. A., Konnerth, A., Sakmann, B. 1990. Quantal analysis of inhibitory synaptic transmission in the dentate gyrus of rat hippocampal slices: A patch clamp study. *J. Physiol. (London)* 430: 213–49

Eliot, L. S., Blumenfeld, H., Edmonds, B. W., Kandel, E. R., Siegelbaum, S. A. 1991. Imaging [Ca]$_i$ transients at *Aplysia* sensorimotor synapses: Contribution of direct and indirect modulation to presynaptic facilitation. *Soc. Neurosci. Abstr.* 17: 1485

Eliot, L. S., Dudai, Y., Kandel, E. R., Abrams, T. W. 1989a. Ca^{2+}/calmodulin sensitivity may be common to all forms of neural adenylate cyclase. *Proc. Natl. Acad. Sci. USA* 86: 9564–68

Eliot, L. S., Schacher, S., Kandel, E. R., Hawkins, R. D. 1989b. Pairing-specific, activity-dependent facilitation of *Aplysia* sensory-motor neuron synapses in isolated culture. *Soc. Neurosci. Abstr.* 15: 482

Eruklar, S. D., Rahamimoff, R. 1978. The role of calcium ions in tetanic and post tetanic increase of miniature end plate potential frequency. *J. Physiol. (London)* 278: 501–11

Fatt, P., Katz, B. 1951. An analysis of the end-plate potential recorded with an intracellular electrode. *J. Physiol. (London)* 115: 320–70

Feng, T. P. 1941. Studies on the neuromuscular junction. XXVI. The changes of the end-plate potential during and after prolonged stimulation. *Chin. J. Physiol.* 16: 341–72

Fischer von Mollard, G., Sudhof, T. C., Jahn, R. 1991. A small GTP-binding protein dissociates from synaptic vesicles during exocytosis. *Nature* 349: 79–81

Foster, T. C., McNaughton, B. L. 1991. Long-term enhancement of CA_1 synaptic transmission is due to increased quantal size, not quantal content. *Hippocampus* 1: 79–91

Frank, K., Fuortes, M. G. F. 1957. Presynaptic and postsynaptic inhibition of neurosynaptic reflex. *Fed. Proc.* 16: 39–40

Frey, U., Kandel, E. R. 1992. Protein kinase A induces a protein synthesis-dependent late stage of LTP in hippocampal CA_1 neurons. *Soc. Neurosci. Abstr.* 18: 639

Frey, U., Krug, M., Reymann, K. G., Matthies, H. 1988. Anisomycin, an inhibitor of protein synthesis, blocks late phases of LTP phenomena in the hippocampal CA_1 region *in vitro*. *Brain Res.* 452: 57–65

Frost, W. N., Castellucci, V. F., Hawkins, R. D., Kandel, E. R. 1985. Monosynaptic connections made by the sensory neurons of the gill- and siphon-withdrawal reflex in *Aplysia* participate in the storage of long-term memory for sensitization. *Proc. Natl. Acad. Sci. USA* 82: 8266–69

Frost, W. N., Clark, G. A., Kandel, E. R. 1988. Parallel processing of short-term memory for sensitization in *Aplysia*. *J. Neurobiol.* 19: 297–334

Furchgott, R. F., Zawalzki, J. V. 1980. The obligatory role of endothelial cells in the relaxation of arterial smooth muscle by acetylcholine. *Nature* 288: 373–76

Gally, J. A., Montague, P. R., Reeke, G. N. Jr., Edelman, G. M. 1990. The NO hypothesis: Possible effects of a short-lived, rapidly diffusible signal in the development and function of the nervous system. *Proc. Natl. Acad. Sci. USA* 87: 3547–51

Garthwaite, J., Charles, S. L., Chess-Williams, R. 1988. Endothelium-derived relaxing factor release on activation of NMDA receptors suggests role as intercellular messenger in the brain. *Nature* 336: 385–88

Ghirardi, M., Braha, O., Hochner, B. Montarolo, P. G., Kandel, E. R., Dale, N. E. 1992. The contributions of PKA and PKC to the presynaptic facilitation of evoked and spontaneous transmitter release at depressed and nondepressed synapses in the sensory neurons of *Aplysia*. *Neuron.* 9: 479–89

Gingrich, K. J., Byrne, J. H. 1985. Simulation of synaptic depression, post-tetanic potentiation, and presynaptic facilitation of synaptic potentials from sensory neurons mediating gill-withdrawal reflex in *Aplysia*. *J. Neurophysiol.* 53: 652–69

Glanzman, D. L., Mackey, S. L., Hawkins, R. D., Dyke, A. M., Lloyd, P. E., Kandel, E. R. 1989. Depletion of serotonin in the nervous system of *Aplysia* reduces the behavioral enhancement of gill withdrawal as well as the heterosynaptic facilitation produced by tail shock. *J. Neurosci.* 9: 4200–13

Gopfert, H., Schaefer, H. 1938. Über den direkt und indirekt erregten Aktionsström und die Funktion der Motorischen Endplatte. *Pflügers Arch.* 239: 597–619

Gormezano, I. 1972. Investigation of defense and reward conditioning in the rabbit. In *Classical Conditioning II: Current Research and Theory*, ed. A. H. Black, W. F. Proskasy, pp. 151–81. New York: Appleton-Century-Crofts

Grant, S. G. N., O'Dell, T. J., Karl, K., Stein, P., Soriano, P., Kandel, E. R. 1992. Genetic analysis reveals that *fyn* tyrosine kinase gene is necessary for LTP and learning in mice. *Soc. Neurosci. Abstr.* 18: 638

Greenberg, S. M., Castellucci, V. F., Bayley, H., Schwartz, J. H. 1987. A molecular mechanism for long-term sensitization in *Aplysia*. *Nature* 329: 8266–69

Gustafsson, B., Wigström, H., Abraham, W. C., Huang, Y.-Y. 1987. Long-term potentiation in the hippocampus using depolarizing current pulses as the conditioning stimulus to single volley synaptic potential. *J. Neurosci.* 7: 774–80

Haley, J. E., Wilcox, G. L., Chapman, P. F. 1992. The role of nitric oxide in hippocampal long-term potentiation. *Neuron* 8: 211–16

Harris, E. W., Cotman, C. W. 1986. Long-term potentiation of guinea pig mossy fiber responses is not blocked by N-methyl-D-aspartate antagonists. *Neurosci. Lett.* 70: 132–37

Hawkins, R. D. 1989. A biologically based computational model for several simple forms of learning. In *Computational Models of Learning in Simple Neural Systems*, ed. R. D. Hawkins, G. H. Bower, pp. 65–108. San Diego: Academic

Hawkins, R. D., Abrams, T. W. 1984. Evidence that activity-dependent facilitation underlying classical conditioning in *Aplysia* involves modulation of the same ionic current as normal presynaptic facilitation. *Soc. Neurosci. Abstr.* 10: 268

Hawkins, R. D., Abrams, T. W., Carew, T. J., Kandel, E. R. 1983. A cellular mechanism of classical conditioning in *Aplysia*: Activity-dependent amplification of presynaptic facilitation. *Science* 219: 400–5

Hawkins, R. D., Carew., T. J., Kandel, E. R. 1986. Effects of interstimulus interval and contingency on classical conditioning of the *Aplysia* siphon withdrawal reflex. *J. Neurosci.* 6: 1695–1701

Hawkins, R. D., Castellucci, V. F., Kandel, E. R. 1981. Interneurons involved in mediation and modulation of gill-withdrawal reflex in *Aplysia*. II. Identified neurons produce heterosynaptic facilitation contributing to behavioral sensitization. *J. Neurophysiol.* 45: 315–26

Hawkins, R. D., Clark, G. A., Kandel, E. R. 1987. Cell biological studies of learning in simple vertebrate and invertebrate systems. In *Handbook of Physiology, Section 1: The Nervous System. Vol. V, Higher Functions of the Nervous System*, ed. V. B. Mountcastle, F. Plum, S. R. Geiger, pp. 25–83. Bethesda, Md.: Am. Physiol Soc.

Hawkins, R. D., Kandel, E. R. 1990. Hippocampal LTP and synaptic plasticity in *Aplysia*: Possible relationship of associative cellular mechanisms. *Semin. Neurosci.* 2: 391–401

Hawkins, R. D., Kandel, E. R. 1984. Is there a cell biological alphabet for simple forms of learning? *Psychol. Rev.* 91: 375–91

Hawkins, R. D., Lalevic, N., Clark, G. A., Kandel, E. R. 1989. Classical conditioning of the *Aplysia* siphon-withdrawal reflex exhibits response specificity. *Proc. Natl. Acad. Sci. USA* 86: 7620–24

Hawkins, R. D., Schacher, S. 1989. Identified facilitator neurons L29 and L29 are excited by cutaneous stimuli used in dishabituation, sensitization, and classical conditioning of *Aplysia*. *J. Neurosci.* 9: 4236–45

Hebb, D. O. 1949. *The Organization of Behavior: A Neuropsychological Theory*. New York: Wiley

Hess, G., Gustafsson, B. 1990. Changes in field excitatory postsynaptic potential shape induced by tetanization in the CA_1 region of the guinea-pig hippocampal slice. *Neuroscience* 37: 61–69

Hirata, K., Sawada, S., Yamamoto, C. 1991. Enhancement of transmitter release accompanying with long-term potentiation in synapses between mossy fibers and CA_3 neurons in hippocampus. *Neurosci. Lett.* 123: 73–76

Hirotsu, I., Hori, N., Katsuda, N., Ishihara, I. 1989. Effect of anticholinergic drug on long-term potentiation in rat hippocampal slices. *Brain Res.* 482: 194–97

Hochner, B., Kandel, E. R. 1992. Modulation of a transient K^+ current in the pleural sensory neurons of *Aplysia* by 5-HT and cAMP: Implications for spike broadening. *Proc. Natl. Acad. Sci. USA.* In press.

Hochner, B., Klein, M., Schacher, S., Kandel, E. R. 1986. Additional component in the cellular mechanism of presynaptic facilitation contributes to behavioral dishabituation in *Aplysia*. *Proc. Natl. Acad. Sci. USA* 83: 8794–98

Hopkins, W. F., Johnston, D. 1988. Noradrenergic enhancement of long-term potentiation at mossy fiber synapses in the hippocampus. *J. Neurophysiol.* 59: 667–87

Hopkins, W. F., Johnston, D. 1984. Frequency-dependent noradrenergic modulation of long-term potentiation in hippocampus. *Science* 226: 350–52

Huang, Y.-Y., Colino, A., Selig, D. K., Malenka, R. L. 1992. The influence of prior synaptic activity on the induction of long-term potentiation. *Science* 255: 730–33

Ichinose, M., Byrne, J. H. 1991. Role of protein phosphatase in the modulation of neuronal membrane currents. *Brain Res.* 549: 146–50

Ito, M. 1989. Long-term depression. *Annu. Rev. Neurosci.* 12: 85–102

Jaffe, D., Johnston, D. 1990. Induction of long-term potentiation at hippocampal mossy-fiber synapses follows a Hebbian rule. *J. Neurophysiol.* 64: 948–60

Kandel, E. R., Abrams, T., Bernier, L., Carew, T. J., Hawkins, R. D., Schwartz, J. H. 1983. Classical conditioning and sensitization share aspects of the same molecular cascade in *Aplysia*. *Cold Spring Harbor Symp. Quant. Biol.* 48: 821–30

Kandel, E. R., Tauc, L. 1965a. Heterosynaptic facilitation in neurones of the abdominal ganglion of *Aplysia depilans*. *J. Physiol. (London)* 181: 1–27

Kandel, E. R., Tauc, L. 1965b. Mechanism of heterosynaptic facilitation in the giant cell of the abdominal ganglion of *Aplysia depilans*. *J. Physiol. (London)* 181: 28–47

Kauer, J. A., Malenka, R. C., Nicoll, R. A.

1988. A persistent postsynaptic modification mediates long-term potentiation in the hippocampus. *Neuron* 1: 911–17

Kelso, S. R., Ganong, A. H., Brown, T. H. 1986. Hebbian synapses in hippocampus. *Proc. Natl. Acad. Sci. USA* 83: 5326–30

Klann, E., Sweatt, J. D. 1990. Persistent alteration of protein kinase activity during the maintenance phase of long-term potentiation. *Soc. Neurosci. Abstr.* 16: 144

Klein, M., Camardo, J. S., Kandel, E. R. 1982. Serotonin modulates a specific potassium current in the sensory neurons that show presynaptic facilitation in *Aplysia*. *Proc. Natl. Acad. Sci. USA* 79: 5713–17

Klein, M., Hochner, B., Kandel, E. R. 1986. Facilitatory transmitters and cAMP can modulate accommodation as well as transmitter release in *Aplysia* sensory neurons: Evidence for parallel processing in a single cell. *Proc. Natl. Acad. Sci. USA* 83: 7994–98

Klein, M., Kandel, E. R. 1980. Mechanism of calcium current modulation underlying presynaptic facilitation and behavioral sensitization in *Aplysia*. *Proc. Natl. Acad. Sci. USA* 77: 6912–16

Klein, M., Kandel, E. R. 1978. Presynaptic modulation of voltage-dependent Ca^{2+} current: Mechanism for behavioral sensitization in *Aplysia californica*. *Proc. Natl. Acad. Sci. USA* 75: 3512–16

Korn, H., Faber, D. S. 1991. Quantal analysis and synaptic efficacy in the CNS. *Trends Neurosci.* 14: 439–45

Kossel, A., Bonhoeffer, T., Bolz, T. 1990. Non-Hebbian synapses in rat visual cortex. *Neuroreport* 1: 115–18

Krasne, F. B. 1969. Excitation and habituation of the crayfish escape reflex: The depolarization response in lateral giant fibers of the isolated abdomen. *J. Exp. Biol.* 50: 29–46

Krasne, F. B., Glanzman, D. L. 1986. Sensitization of the crayfish lateral giant escape reaction. *J. Neurosci.* 6: 1013–20

Kretz, R., Shapiro, E., Kandel, E. R. 1982. Post-tetanic potentiation at an identified synapse in *Aplysia* is correlated with a Ca^{2+}-activated K^+ current in the presynaptic neuron: Evidence for Ca^{2+} accumulation. *Proc. Natl. Acad. Sci. USA* 79: 5430–34

Kullmann, D. M., Nicoll, R. A. 1992. Long-term potentiation is associated with increases in quantal content and quantal amplitude. *Nature* 357: 240–44

Larkman, A., Stratford, K., Jack, J. 1991. Quantal analysis of excitatory synaptic action and depression in hippocampal slices. *Nature* 350: 344–47

Levy, W. B., Steward, O. 1979. Synapses as associative memory elements in the hippocampal formation. *Brain Res.* 175: 233–45

Lin, J. W., Sugimori, M., Llinás, R. R., McGuinnes, T. L., Greengard, P. 1990. Effects of synapsin I and calcium-calmodulin-dependent protein kinase II on spontaneous neurotransmitter release in the squid giant synapse. *Proc. Natl. Acad. Sci. USA* 87: 8257–61

Livingstone, M. S., Sziber, P. P., Quinn, W. G. 1984. Loss of calcium/calmodulin responsiveness in adenylate cyclase of *rutabaga*, a *Drosophila* learning mutant. *Cell* 37: 205–15

Llinás, R., Gruner, J. A., Sugimori, M., McGuinnes, T. L., Greengard, P. 1991. Regulation by synapsin I and Ca^{2+}-calmodulin-dependent protein kinase II of the transmitter release in squid giant synapse. *J. Physiol. (London)* 436: 257–82

Llinás, R., McGuinnes, T. L., Leonard, C. S., Sugimori, M., Greengard, P. 1985. Intraterminal injection of synapsin I or calcium-calmodulin-dependent protein kinase II alters neurotransmitter release at the squid giant synapse. *Proc. Natl. Acad. Sci. USA* 82: 3035–39

Lloyd, D. P. C. 1949. Post-tetanic potentiation of response in monosynaptic reflex pathways of the spinal cord. *J. Gen. Physiol.* 33: 147–70

Lnenicka, G. A., Atwood, H. L., Marin, L. 1986. Morphological transformation of synaptic terminals of a phasic motoneuron by long-term tonic stimulation. *J. Neurosci.* 6: 2252–58

Lukowiak, K. 1986. *In vitro* classical conditioning of a gill withdrawal reflex in *Aplysia*: Neural correlates and possible neural mechanisms. *J. Neurobiol.* 17: 83–101

Lukowiak, K., Colebrook, E. 1988. Classical conditioning alters the efficacy of identified gill motor neurons in producing gill withdrawal movement in *Aplysia*. *J. Exp. Biol.* 140: 273–85

Lynch, G., Larson, J., Kelso, S., Barrionuevo, G., Schottler, F. 1983. Intracellular injection of EGTA blocks induction of hippocampal long-term potentiation. *Nature* 305: 719–21

Lynch, M. A., Errington, M. L., Bliss, T. V. P. 1989. Nordihydroguaiaretic acid blocks the synaptic component of long-term potentiation and the associated increase in release of glutamate and arachidonic acid: An *in vivo* study in the dentate gyrus of the rat. *Neuroscience* 30: 693–701

MacDonald, R. L., Werz, M. A. 1986. Dynorphin A decreases voltage-dependent calcium conductance of mouse dorsal root ganglion neurons. *J. Physiol. (London)* 377: 37–249

Mackey, S. L., Glanzman, D. L., Small, S. A., Dyke, A. M., Kandel, E. R., Hawkins, R. D. 1987. Tail shock produces inhibition as well as sensitization of the siphon-withdrawal reflex of *Aplysia*: Possible behavioral role for presynaptic inhibition mediated by the peptide Phe-Met-Arg-Phe-NH$_2$. *Proc. Natl. Acad. Sci. USA* 84: 8730–34

Mackey, S. L., Kandel, E. R., Hawkins, R. D. 1989. Identified serotonergic neurons LCB1 and RCB1 in the cerebral ganglia of *Aplysia* produce presynaptic facilitation of siphon sensory neurons. *J. Neurosci.* 9: 4227–35

Malenka, R. C., Kauer, J. A., Perkel, D. J., Mauk, M. D., Kelly, P. T., et al. 1989. An essential role for postsynaptic calmodulin and protein kinase activity in long-term potentiation. *Nature* 340: 554–57

Malenka, R. C., Kauer, J. A., Zucker, R. S., Nicoll, R. A. 1988. Postsynaptic calcium is sufficient for potentiation of hippocampal synaptic transmission. *Science* 242: 81–84

Malenka, R. C., Lancaster, B., Zucker, R. S. 1992. Temporal limits on the rise in postsynaptic calcium required for the induction of long-term potentiation. *Neuron* 9: 121–28

Malgaroli, A., Tsien, R. W. 1992. Glutamate-induced long-term potentiation of the frequency of miniature synaptic currents in cultured hippocampal neurons. *Nature* 357: 134–39

Malinow, R. 1991 Transmission between pairs of hippocampal slice neurons show quantal levels, oscillations and LTP. *Science* 252: 722–24

Malinow, R., Madison, D. V., Tsien, R. 1988. Persistent protein kinase activity underlying long-term potentiation. *Nature* 335: 820–24

Malinow, R., Miller, J. P. 1986. Postsynaptic hyperpolarization during conditioning reversibly blocks induction of long-term potentiation. *Nature* 320: 529–30

Malinow, R., Schulman, H., Tsien, R. W. 1989. Inhibition of postsynaptic PKC or CaMKII blocks induction but not expression of LTP. *Science* 245: 862–66

Malinow, R., Tsien, R. W. 1990. Presynaptic enhancement shown by whole-cell recordings of long-term potentiation in hippocampal slices. *Nature* 346: 177–80

Manabe, T., Renner, P., Nicoll, R. A. 1992. Postsynaptic contribution to long-term potentiation revealed by the analysis of miniature synaptic currents. *Nature* 355: 50–55

Man-Son-Hing, H., Zoran, M. J., Lukowiak, K., Hadon, P. G. 1989. A neuromodulator of synaptic transmission acts on the secretory apparatus as well as on ion channels. *Nature* 341: 237–39

Markram, H., Segal, M. 1990. Long-lasting facilitation of excitatory postsynaptic potentials in the rat hippocampus by acetylcholine. *J. Physiol. (London)* 427: 381–93

Mathew, W. D., Tsavaler, L., Reichardt, L. F. 1981. Identification of synaptic vesicle-specific membrane protein with a wide distribution in neuronal and neurosecretory tissue. *J. Cell Biol.* 91: 257–69

Matsui, Y., Kikuchi, A., Kondo, J., Hishida, T., Teranishi, T., Takai, Y. 1988. Nucleotide and deduced amino acid sequence of a GTP-binding protein family with molecular weights of 25,000 from bovine brain. *J. Biol. Chem.* 263: 11071–74

McNaughton, B. L., Douglas, R. M., Goddard, G. V. 1978. Synaptic enhancement in fasia dentata: Cooperativity among coactive afferents. *Brain Res.* 157: 277–93

Meshul, C. K., Hopkins, W. F. 1990. Presynaptic ultrastructural correlates of long-term potentiation in the CA$_1$ subfield of the hippocampus. *Brain Res.* 514: 310–19

Monaghan, D. T., Cotman, C. W. 1985. Distribution of N-methyl-D-aspartate-sensitive L-(^{3}H)glutamata-binding sites in rat brain. *J. Neurosci.* 5: 2909–19

Montarolo, P. G., Goelet, P., Castellucci, V. F., Morgan, J., Kandel, E. R., Schacher, S. 1986. A critical period for macromolecular synthesis in long-term heterosynaptic facilitation in *Aplysia*. *Science* 234: 1249–54

Morris, R. G. M., Anderson, E., Lynch, G. S., Baudry, M. 1986. Selective impairment of learning and blockade of long-term potentiation by N-methyl-D-aspartate antagonist, AP-5. *Nature* 319: 774–76

Muller, D., Joly, M., Lynch, G. 1988. Contribution of quisqualate and NMDA receptors to the induction and expression of LTP. *Science* 242: 1694–97

Mudge, A. W., Leeman, S. E., Fischbach, G. D. 1979. Enkephalin inhibits release of substance P from sensory neurones in culture and decreases action potential duration. *Proc. Natl. Acad. Sci. USA* 76: 526–30

Murad, F., Mittal, C. K., Arnold, W. P., Katsuki, S., Kimura, H. 1978. Guanylate cyclase: Activation by azide, nitro compounds, nitric oxide and hydroxyl radical and inhibition by hemoglobin and myoglobin. *Adv. Cyclic Nucleotide Res.* 9: 145–58

Natsukari, N., Hanni, H., Matsunaga, T., Fujita, M. 1990. Synergistic activation of brain adenylate cyclase by calmodulin, and either GTP or catecholamine including dopamine. *Brain Res.* 534: 170–76

Nelson, T. J., Collin, C., Alkon, D. L. 1990. Isolation of a G protein that is modified by learning and reduces potassium currents in *Hermissenda*. *Science* 247: 1479–83

Nichols, R. A., Sihra, T. S., Czernik, A. J., Nairn, A. C., Greengard P. 1990. Calcium/calmodulin-dependent protein kinase II increases glutamate and noradrenaline release from synaptosomes. *Nature* 343: 647–51

Nguyen, P. V., Atwood, H. L. 1990. Expression of long-term adaptation of synaptic transmission requires a critical period of protein synthesis. *J. Neurosci.* 10: 1099–1109

Nicoll, R. A., Kauer, J. A., Malenka, R. C. 1988. The current excitement in long-term potentiation. *Neuron* 1: 97–103

Ocorr, K. A., Walters, E. T., Byrne, J. H. 1985. Associative conditioning analog selectively increases cAMP levels of tail sensory neurons in *Aplysia*. *Proc. Natl. Acad. Sci. USA* 82: 2548–52

O'Dell, T. J., Hawkins, R. D., Kandel, E. R., Arancio, O. 1991a. Tests of the roles of two diffusible substances in long-term potentiation: Evidence for nitric oxide as a possible early retrograde messenger. *Proc. Natl. Acad. Sci. USA* 88: 11285–89

O'Dell, T. J., Kandel, E. R., Grant, S. G. N. 1991b. Long-term potentiation in the hippocampus is blocked by tyrosine kinase inhibitors. *Nature* 353: 558–60

Palmer, R. M. J., Ferrige, A. G., Moncada, S. 1987. Nitric oxide release accounts for the biological activity of endothelium-derived relaxing factor. *Nature* 327: 524–26

Parfitt, K. D., Hoffer, B. J., Browning, M. D. 1991. Norepinephrine and isoproterenol increase the phosphorylation of synapsin I and synapsin II in dentate slices of young but not aged Fisher 344 rats. *Proc. Natl. Acad. Sci. USA* 88: 2361–85

Perin, M. S., Fried, V. A., Mignery, G. A., Jahn, R., Sudhof, T. C. 1990. Phospholipid binding by a synaptic vesicle protein homologous to the regulatory region of protein kinase C. *Nature* 345: 260–63

Petrenko, A. G., Perin, M. S., Davletov, B. A., Ushkaryov, Y. A., Geppert, M., Sudhof, T. C. 1991. Binding of synaptotagmin to the alpha-latrotoxin receptor implicates both in synaptic vesicle exocytosis. *Nature* 353: 65–68

Pinsker, H. M., Hening, W. A., Carew, T. J., Kandel, E. R. 1973. Long-term sensitization of a defensive withdrawal reflex in *Aplysia*. *Science* 182: 1039–42

Pinsker, H., Kupfermann, I., Castellucci, V., Kandel, E. R. 1970. Habituation and dishabituation of the gill-withdrawal reflex in *Aplysia*. *Science* 167: 1740–42

Piomelli, D., Shapiro, E., Zipkin, R., Schwartz, J. H., Feinmark, S. J. 1989. Formation and action of 8-hydroxy-11,12-epoxy-5,9,14-icosatrienoic acid in *Aplysia*: A possible second messenger in neurons. *Proc. Natl. Acad. Sci. USA* 86: 1721–25

Piomelli, D., Volterra, A., Dale, N., Siegelbaum, S. A., Kandel, E. R., et al. 1987. Lipoxygenase metabolites of arachidonic acid as second messengers for presynaptic inhibition of *Aplysia* sensory cells. *Nature* 328: 38–43

Redman, S. 1990. Quantal analysis of synaptic potentials in neurons of the central nervous system. *Physiol. Rev.* 70: 165–98

Regehr, W. D., Tank, D. W. 1991. The maintenance of LTP at hippocampal mossy fiber synapses is independent of sustained presynaptic calcium. *Neuron* 7: 451–59

Sacktor, T. C., Schwartz, J. H. 1990. Sensitizing stimuli cause translocation of protein kinase C in *Aplysia* sensory neurons. *Proc. Natl. Acad. Sci. USA* 87: 2036–39

Sasaki, T., Kikuchi, A., Araki, S., Hata, Y., Isomura, M., et al. 1990. Purification and characterization from bovine brain cytosol of a protein that inhibits the dissociation of GDP from and the subsequent binding of GTP to smg p25A, a ras p21-like GTP-binding protein. *J. Biol. Chem.* 265: 2333–37

Sastry, B. R., Goh, J. W., Auyeung, A. 1986. Associative induction of posttetanic and long-term potentiation in CA_1 of rat hippocampus. *Science* 232: 988–90

Schacher, S., Castellucci, V. F., Kandel, E. R. 1988. cAMP evokes long-term facilitation in *Aplysia* sensory neurons that requires new protein synthesis. *Science* 240: 1667–69

Scholz, K. P., Byrne, J. H. 1988. Intracellular injection of cAMP induces long-term reduction of neuronal K^+ currents. *Science* 240: 1664–66

Scholz, K. P., Byrne, J. H. 1987. Long-term sensitization in *Aplysia*: Biophysical correlates in tal sensory neurons. *Science* 235: 685–87

Schuman, E. M., Clark, G. A. 1990. *Hermissenda* photoreceptors exhibit synaptic facilitation as well as enhanced excitability. *Soc. Neurosci. Abstr.* 16: 21

Schuman, E. M., Madison, D. V. 1991. A requirement for the intracellular messenger nitric oxide in long-term potentiation. *Science* 254: 1503–6

Schuman, E. M., Meffert, M. K., Schulman, H., Madison, D. V. 1992. A potential role for an ADP-ribosyltransferase (ADPRT) in hippocampal long-term potentiation (LTP). *Soc. Neurosci. Abstr.* 18: 761

Scoville, W. B., Milner, B. 1957. Loss of recent memory after bilateral hippocampal lesions. *J. Neurol. Neurosurg. Psychiatry* 20: 11–21

Shatz, C. J. 1992. The developing brain. *Sci. Am.* 267: 60–67

Shuster, M. J., Camardo, J. S., Siegelbaum, S. A., Kandel, E. R. 1985. Cyclic AMP-dependent protein kinase closes the serotonin-sensitive K$^+$ channels of *Aplysia* sensory neurones in cell-free membrane patches. *Nature* 313: 392–95

Siegelbaum, S., Camardo, J. S., Kandel, E. R. 1982. Serotonin and cAMP close single K$^+$ channels in *Aplysia* sensory neurones. *Nature* 299: 413–17

Silva, A. J., Stevens, C. F., Tonegawa, S., Wong, Y. 1992. Deficient hippocampal long-term potentiation in α-calcium-calmodulin kinase II mutant mice. *Science* 257: 201–6

Small, S. A., Cohen, T. E., Kandel, E. R., Hawkins, R. D. 1992a. Identified FMRFamide-immunoreactive neuron LP16 in the left pleural ganglion of *Aplysia* produces presynaptic inhibition of siphon sensory neurons. *J. Neurosci.* 12: 1616–27

Small, S. A., Kandel, E. R., Hawkins, R. D. 1989. Activity-dependent enhancement of presynaptic inhibition in *Aplysia* sensory neuron. *Science* 243: 1603–6

Small, S. A., O'Dell, T. J., Kandel, E. R., Hawkins, R. D. 1992b. Nitric oxide produces long-term enhancement of synaptic transmission in the CA$_1$ region of hippocampus by an activity-dependent mechanism. *Soc. Neurosci. Abstr.* 18: 639

Small, S. A., Zhuo, M., O'Dell, T. J., Kandel, E. R., Hawkins, R. D. 1992c. Nitric oxide produces long-term enhancement of synaptic transmission in the CA$_1$ region of hippocampus by an activity-dependent mechanism. *Cold Spring Harbor Meeting Learning and Memory* (abstr.): 54

Smith, S. J., Augustine, G. J. 1988. Calcium ions, active zones and synaptic transmitter release. *Trends Neurosci.* 11: 458–64

Spencer, W. A., Thompson, R. F., Neilson, D. R. Jr. 1966. Response decrement of the flexion reflex in the acute spinal cat and transient restoration by strong stimuli. *J. Neurophysiol.* 29: 253–73

Squire, L. R. 1992. Memory and the hippocampus: A synthesis from findings with rats, monkeys, and humans. *Psych. Rev.* 99: 195–231

Stanton, P. K., Sarvey, J. M. 1985a. Depletion of norepinephrine, but not serotonin, reduces long-lasting potentiation in the dentate gyrus of rat hippocampal slices. *J. Neurosci.* 5: 2169–76

Stanton, P. K., Sarvey, J. M. 1985b. The effect of high-frequency electrical stimulation and norepinephrine on cyclic AMP levels in normal versus norepinephrine-depleted rat hippocampal slices. *Brain Res.* 358: 343–48

Sudhof, T. C., Jahn, R. 1991. Proteins of synaptic vesicles involved in exocytosis and membrane recycling. *Neuron* 6: 665–77

Sudhof, T. C., Lottspeich, F., Greengard P., Ehrenfreid, M., Jahn, R. 1987. A synaptic vesicle protein with a novel cytoplasmic domain and four transmembrane regions. *Science* 238: 1142–44

Swandulla, D., Hans, M., Zipser, K., Augustine, G. J. 1991. Role of residual calcium in synaptic depression and posttetanic potentiation: Fast and slow calcium signaling in nerve terminals. *Neuron* 7: 915–26

Sweatt, D., Kandel, E. R. 1989. Persistent and transcriptionally-dependent increase in protein phosphorylation in long-term facilitation of *Aplysia* sensory neurons. *Nature* 339: 51–54

Sweatt, J. D., Volterra, A., Edmonds, B., Karl, K. A., Siegelbaum, S. A., Kandel, E. R. 1989. FMRFamide reverses protein phosphorylation produced by 5-HT and cAMP in *Aplysia* sensory neurons. *Nature* 342: 275–78

Thomas, L., Betz, H. 1990. Synaptophysin binds to physophilin, a putative synaptic plasma membrane protein. *J. Cell Biol.* 111: 2041–52

Thomas, L., Hartung, K., Langosch, D., Rehm, H., Bamberg, E., et al. 1988. Identification of synaptophysin as a hexameric channel protein of the synaptic vesicle membrane. *Science* 242: 1050–53

Thompson, R. F. 1986. The neurobiology of learning and memory. *Science* 233: 941–47

Thorpe, W. H. 1956. *Learning and Instinct in Animals.* Cambridge: Harvard Univ. Press

Trimble, W. S., Linial, M., Scheller, R. H. 1991. Cellular and molecular biology of the presynaptic nerve terminal. *Annu. Rev. Neurosci.* 14: 93–122

Tsien, R. W., Malinow, R. 1990. Long-term potentiation: Presynaptic enhancement following postsynaptic activation of Ca^{2+}-dependent protein kinases. *Cold Spring Harbor Symp. Quant. Biol.* 55: 147–59

Van de Kloot, W. 1990. The regulation of quantal size. *Prog. Neurobiol.* 36: 93–130

Volterra, A., Siegelbaum, S. A. 1988. Role of two different guanine nucleotide-binding proteins in the antagonistic modulation of the S-type K$^+$ channel by cAMP and arachidonic acid metabolites in *Aplysia* sensory neurons. *Proc. Natl. Acad. Sci. USA* 85: 7810–14

Walters, E. T. 1989. Transformation of siphon responses during conditioning of *Aplysia* suggests a model of primitive stimulus-response association. *Proc. Natl. Acad. Sci. USA* 86: 7616–19

Walters, E. T., Byrne, J. H. 1985. Long-term enhancement produced by activity-dependent modulation of *Aplysia* sensory neurons. *J. Neurosci.* 5: 662–72

Walters, E. T., Byrne, J. H. 1983. Associative conditioning of single sensory neurons suggests a cellular mechanism for learning. *Science* 219: 405–8

Walters, E. T., Byrne, J. H., Carew, T. J., Kandel, E. R. 1983. Mechanoefferent neurons innervating tail of *Aplysia*: I. Response properties and synaptic connections. *J. Neurophysiol.* 50: 1522–42

Wigström, H., Gustafsson, B., Huang, Y.-Y., Abraham, W. C. 1986. Hippocampal long-lasting potentiation is induced by pairing single afferent volley with intracellularly injected depolarizing current pulses. *Acta Physiol. Scand.* 126: 317–19

Williams, J. H., Bliss, T. V. P. 1989. An *in vitro* study of the effect of lipoxygenase and cyclo-oxygenase inhibitors of arachidonic acid on the induction and maintenance of long-term potentiation in the hippocampus. *Neurosci. Lett.* 107: 301–6

Williams, J. H., Errington, M. L., Lynch, M. A., Bliss, T. V. P. 1989. Arachidonic acid induces a long-term activity-dependent enhancement of synaptic transmission in the hippocampus. *Nature* 341: 739–42

Williams, J. T., Egan, T. M., North, R. A. 1982. Enkephalin opens potassium channels on mammalian central neurons. *Nature* 299: 74–77

Williams, S., Johnston, D. 1989. Long-term potentiation of hippocampal mossy fiber synapses is blocked by postsynaptic injection of calcium chelators. *Neuron* 3: 583–88

Williams, S., Johnston, D. 1988. Muscarinic depression of long-term potentiation in CA_3 hippocampal neurons. *Science* 242: 84–87

Woody, C. D., Swartz, B. E., Gruen, E. 1978. Effects of acetylcholine and cyclic GMP on input resistance of cortical neurons in awake cats. *Brain Res.* 158: 373–95

Wu, L. G., Saggau, P. 1992. The expression of long-term potentiation in hippocampal area CA_1 does not change transients of presynaptic calcium. *Biophys. J.* 61: A509

Yovell, Y., Abrams, T. W. 1992. Temporal asymmetry in activation of *Aplysia* adenylyl cyclase by calcium and transmitter may explain temporal requirements of conditioning. *Proc. Natl. Acad. Sci. USA* 89: 6526–30

Yovell, Y., Dudai, T., Abrams, T. W. 1986. Quantitative analysis of Ca^{2+}/calmodulin activated adenylate cyclase from *Aplysia*, *Drosophila* and rat brain: Possible relevance for associative learning. *Soc. Neurosci. Abstr.* 12: 400

Zahraoui, A., Touchot, N., Chardin, P., Tavitian, A. 1989. The human rab genes encode a family of GTP-binding proteins related to yeast TPT1 and SEC4 products involved in secretion. *J. Biol. Chem.* 264: 12394–12401

Zalutsky, R. A., Nicoll, R. A. 1992. Mossy fiber long-term potentiation shows specificity but no apparent cooperativity. *Neurosci. Lett.* 138: 193–97

Zalutsky, R. A., Nicoll, R. A. 1990. Comparison of two forms of long-term potentiation in single hippocampal neurons. *Science* 248: 1619–24

Zola-Morgan, S. M., Squire, L. R. 1990. The primate hippocampal formation: Evidence for a time-limited role in memory storage. *Science* 250: 288–90

Zucker, R. S. 1972. Crayfish escape behavior and cervical synapses. II. Physiological mechanisms underlying behavioral habituation. *J. Neurophysiol.* 35: 621–37

Annu. Rev. Neurosci. 1993. 16:667–706
Copyright © 1993 by Annual Reviews Inc. All rights reserved

COMPUTATIONAL MODELS OF THE NEURAL BASES OF LEARNING AND MEMORY

Mark A. Gluck

Center for Molecular and Behavioral Neuroscience, Rutgers University, Newark, New Jersey 07102

Richard Granger

Center for the Neurobiology of Learning and Memory, University of California, Irvine, California 92717

KEY WORDS: olfactory cortex, neural network, connectionism, hippocampus, cerebellum

1. INTRODUCTION

Advances in computational analyses of parallel-processing networks have made computer simulation of learning systems an increasingly useful tool for understanding complex aggregate functional effects of changes in neural systems. In this article, we review current efforts to develop computational models of the neural bases of learning and memory, with a focus on the behavioral implications of network-level characterizations of synaptic change in three anatomical regions: olfactory (piriform) cortex, cerebellum, and the hippocampal formation. In each case, the modeling efforts reviewed are based on the assumption that the operation of physiological plasticity rules, as embedded in anatomically defined circuits, gives rise to behavioral properties of learning. Before turning to these network-level effects, we begin by briefly reviewing variants of synaptic plasticity and the learning rules that characterize them.

Forms of Learning and Plasticity

The notion that learning arises from synaptic change dates back to suggestions by Tanzi (1893) and Ramon y Cajal (1909). To the extent that a

667

0147–006X/93/0301–0667$02.00

synaptic plasticity mechanism is proposed as the substrate for a specific form of learning, the characteristics of the plasticity rule may affect the characteristics of memories that arise from its operation. There is increasing evidence for multiple forms of synaptic plasticity (see Granger & Lynch 1992). Behavioral and lesion studies in humans and other species strongly support the intuitive notion that different forms of memory exist (see Zola-Morgan & Squire 1993). Furthermore, these different memory systems may arise from distinct and dissociable mechanisms and anatomical regions within the brain. Studies of the induction and expression of synaptic plasticity, combined with analyses of the anatomical networks within which these synapses are sited, can generate hypotheses about the functional roles that these circuit networks play in observed forms of learning and development (Bear et al 1987; Gluck & Thompson 1989; Miller et al 1990; Thompson & Gluck 1990). This same approach can also be used to address network physiological properties, such as EEG (Freeman 1991), and sometimes to predict previously unanticipated behavioral and memorial properties (Ambros-Ingerson et al 1990). The identification of different forms of plasticity in different brain circuitries suggests that these distinct mechanisms may subserve distinguishable forms of memory. Examples are provided by two major synaptic links inside hippocampus: the mossy-fiber connections from dentate gyrus to field CA3 and the Schaffer-commissural pathway from CA3 to CA1. High-frequency stimulation of either set of fibers results in an enhancement or potentiation of the stimulated synapses, but the forms of potentiation are completely distinct from each other, having different chemistries, longevity, and specificity (Staubli et al 1990; Zalutsky & Nicoll 1990). These significant differences have led to the suggestion that these two forms of plasticity may underlie memories of different type and duration (Lynch & Granger 1990, 1992). We now review arguments that support the hypothesis that long-term potentiation (LTP) may subserve a particular form of memory.

Long-Term Potentiation

Long-term potentiation is synapse-specific, rapidly induced (seconds; Gustafsson & Wigstrom 1990), and long-lasting (weeks; Staubli & Lynch 1987), and exhibits a consolidation period of several minutes during which it is vulnerable to reversal (Arai et al 1990; Staubli et al 1990). A memory system that exhibits corresponding properties can be identified in humans: the ability to encode new information within seconds, in such a way that the encodings become permanent, but apparently disruptable for several minutes after learning. Even rapidly learned and apparently unrehearsed memories are long lasting and are stored with extremely high capacity (see

Standing 1973). Tellingly, LTP is optimally induced by activity patterns that occur in brain during learning (Diamond et al 1986; Larson & Lynch 1986, 1988). Pharmacological blockers of the n-methyl-d-aspartate (NMDA) receptor, a requirement for LTP induction (Collingridge et al 1983), also block spatial learning (Morris et al 1986; Robinson et al 1989) and olfactory learning (Staubli et al 1989). Benzodiazepines, which have known amnesic effects in humans and animals, suppress development of LTP (del Cerro et al 1992). We may hypothesize that LTP is related to the apparently effortless everyday human ability to learn and retain large numbers of minute-to-minute environmental cues. Long-term potentiation has been identified in many telencephalic systems, including the olfactory bulb afferents to olfactory paleocortex (Jung et al 1990; Kanter & Haberly 1990; Roman et al 1987); the cortical perforant path innervation of dentate gyrus, where LTP was first demonstrated (Bliss & Lomo 1973); pyramidal cell systems in hippocampus (Berger 1984); and in neocortex (Lee 1982), including motor (Iriki et al 1989), visual (Artola & Singer 1987), and somatosensory cortices (Lee et al 1991).

The simplest characterization of the requirements for LTP induction resemble those postulated by Hebb (1949), ". . . [when cell A] repeatedly or persistently takes part in firing [cell B], . . . A's efficiency, as one of the cells firing B, is increased." The combination of both ligand and voltage gating of the NMDA receptor conforms nicely to Hebb's implied dual requirement for both pre- and postsynaptic activity at a synapse. Theoretical learning rules based on this correlational or Hebbian postulate yield networks of considerable computational power (Ballard et al 1983; Bienenstock et al 1982; Edelman & Reeke 1982; von der Malsburg 1973). In common among the various computational schemes derived from Hebbian correlational learning rules is the fundamental finding that networks using such rules exhibit a constellation of abilities that center around the phenomenon of data clustering, i.e. the self-organization of learned inputs into groups or clusters based on their similarities. Embedding a simple Hebbian LTP rule into an anatomical network can give rise to additional, and sometimes unexpected, emergent computational properties, as suggested, for example, by the simulation models of the olfactory system.

Long-Term Depression

Ito (1984) and colleagues have provided evidence for a form of heterosynaptic plasticity in the cerebellar cortex: a long-term depression (LTD) of parallel fiber-Purkinje cell synapses following conjoint stimulation of parallel fibers and climbing fibers. More recent studies suggest that some of this plasticity may be related to long-term morphological changes in Purkinje cell arborization and spine density (Anderson et al 1989). In

Section 3, we review recent computational models that incorporate this cerebellar plasticity into a circuit for motor-reflex learning.

Depression of synaptic efficacy has also been demonstrated in telencephalon, typically via so-called "heterosynaptic" depression, which describes a decrease of those synapses whose inputs are silent during postsynaptic activation (for review, see Bindman et al 1991). Recent findings have confirmed the effect of reversal of previously induced LTP (Barrionuevo et al 1980) and have demonstrated synapse-specific (homosynaptic) depression (Artola et al 1990; Dudek & Bear 1992; Stanton & Sejnowski 1989; Stevens 1990).

Abstract network learning models commonly assume mechanisms for both synaptic increase and synaptic decrease, i.e. for LTP and LTD. Indeed, without a way to reduce synaptic strength, it may seem intuitive that all synapses eventually become saturated, thus obscuring already-learned information and preventing further learning. This is not necessarily true; in fact, LTP-only networks have large memory capacity and store significant numbers of inputs with low retrieval error rates (Granger et al 1992). We discuss this work in more depth in Section 4. Other computational implications of combined synaptic increase and decrease have been studied. For example, Bienenstock et al (1982) devised a rule especially suited to the discovery of environmental features. Recent tests of this rule have shown that it provides a potential account of mechanisms that may underlie developmental plasticity in visual cortex (Bear et al 1987; Clothiaux et al 1991).

Error-Correction Rules

Long-term potentiation and LTD share the Hebb-like characteristics of a simple cooccurrence detector: Coactivity of an input and an output cell causes change of the synaptic connection between them. A contrasting family of rules is posited in much artificial neural network research, namely, the family of error-correction rules (Rosenblatt 1962; Rumelhart et al 1986; Widrow & Hoff 1960). In these rules, a new kind of afferent to each target cell is assumed and carries the desired or intended response activity that that cell is supposed to exhibit. The difference between the actual output cell activity and this desired activity determines the change in the strength of connections between the input and output. Architectures supporting the existence of detailed error signals at target neurons are not known in telencephalon, but climbing fibers in the cerebellum may deliver putative error information in a fashion compatible with error-correction algorithms. We present findings from simulation studies of cerebellar circuits instantiating such rules in Section 3.

Error-correction rules in the field of artificial neural networks are principally derived from the perceptron (Rosenblat 1962) or least-mean-squared (LMS) rule of Widrow & Hoff (1960), and have subsequently been generalized to multilayer networks by Werbos (1974), le Cun (1985), Parker (1986), and Rumelhart et al (1986) via backward propagation of error functions through a network. Error-correction rules constitute by far the largest body of work in the field of artificial neural networks, because of their demonstrated computational power. In contrast, the computational characteristics of most telencephalic forms of synaptic plasticity have yet to be determined.

The remainder of this article reviews modeling efforts aimed at studying the relationship between synaptic change and learning behaviors. These models illustrate a range of synaptic plasticity rules, anatomical brain regions (i.e. olfactory cortex, cerebellum, and the hippocampal formation), and learning behaviors. Two different approaches to theory and model development are illustrated by the research reviewed: "bottom-up" models, which attempt to identify circuit function from biological detail, and "top-down" models, which begin with observed behaviors linked to particular anatomical structures and attempt to identify neurobiological mechanisms that implement the behaviors. We first review bottom-up studies of simple Hebb-like plasticity rules based on LTP as it occurs in the olfactory cortex. In Section 3, we review top-down models of error-correction learning rules in the cerebellum. Finally, in Section 4, we compare and contrast bottom-up and top-down approaches to modeling functional characteristics that emerge from plasticity rules in the mammalian hippocampal formation and its overlaying cortical structures.

2. OLFACTORY PALEOCORTEX

The olfactory paleocortex is a phylogenetically old precursor of neocortex, whose simpler anatomy, lack of thalamic relays, and accessibility for physiological and behavioral studies make it an attractive model system for studies of cerebral cortical function. As mentioned earlier, LTP has been shown in the afferents and collaterals of olfactory cortex (Jung 1990; Kanter & Haberly 1990) by using stimulation patterns with characteristics of behaviorally relevant physiological activity (Otto et al 1991). In the following sections, we outline the elements and characteristics of the olfactory paleocortex and its primary input structure, the olfactory bulb, and review computational modeling approaches to understanding their contribution to perceptual memory in the olfactory modality.

Anatomy and Physiology

Figure 1 illustrates the anatomical organization of the major components of the olfactory system. Chemical receptor cells in the nose project via the

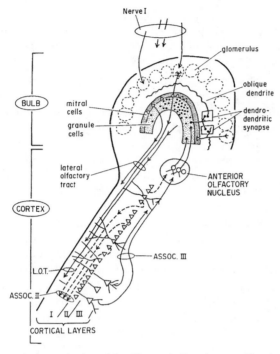

Figure 1   Anatomical organization of the olfactory bulb and cortex. Chemical receptor cells in the nasal epithelium (not shown) send their axons to the olfactory bulb, where they contact the primary excitatory (MT) bulb cells, via specialized structures known as glomeruli, each of which contains the apical dendritic branches from a group of MT cells. The lateral dendrites of MT cells make extensive two-way dendro-dendritic connections with inhibitory granule cells. MT cells sparsely and nontopographically innervate the apical dendrites of layer II/III cells in olfactory cortex, via the LOT. Axons from excitatory cortical cells project back to the inhibitory (granule) layer of the olfactory bulb, both directly and via the anterior olfactory nucleus.

first cranial nerve to the olfactory bulb, where they contact apical dendritic branches of the primary excitatory cells in bulb, the mitral/tufted (MT) cell. In rabbit, roughly 50 million receptors converge onto about 175,000 MT cells, whose dendrites are organized into about 1900 groups or glomeruli, with about 92 MT cells per glomerulus (Mori 1987; Moulton 1976; Shepherd 1972). Mitral/tufted cells give rise to lateral dendrites, which make extensive two-way dendrodendritic connections with inhibitory granule cells (Rall et al 1966; Woolf et al 1991). Inhibitory cells in bulk outnumber the excitatory MT cells by roughly 30 to 1 (Allison 1953).

Mitral/tufted cells project monosynaptically to olfactory cortex, via a bundle of myelinated axons termed the lateral olfactory tract (LOT) that makes nontopographic and sparse ($p < 0.1$) contacts with the apical dendrites of layer II and III cells (Haberly & Price 1977). In contrast to the situation in bulb, cortical inhibitory interneurons are outnumbered by the excitatory cells in layer II by roughly 50 to 1 (Price 1973). The primary efferent projections from cortex are a feedback pathway to the inhibitory granule cells in bulb and axon collaterals to other cortical regions (Price 1973), one of which, the entorhinal cortex, in turn provides the predominant input projection to the hippocampus (Wyss 1981; Zimmer 1971).

Presentation of odors generates spatiotemporal patterns of activity throughout the olfactory system. Olfactory receptors generally fire with increased frequencies as concentration of an odor increases (Gretchell & Shepherd 1978), and characteristic spatial patterns measured by a range of different methods (2-deoxyglucose labeling, optical dyes, and EEG) are robustly associated with particular odors across individuals (Kauer 1987; 1991). Mitral/tufted cells in the olfactory bulb also fire in characteristic spatial patterns for particular odors; again, these patterns are surprisingly constant across individuals (Kauer 1987), although there is evidence that they may change with further learning (Freeman 1991).

Behavior

When performing olfactory discrimination tasks, rats exhibit very rapid learning (within about five trials) and long retention (many months) of odors (Slotnick & Katz 1974; Staubli et al 1987), which makes these memories comparable to the rapid, specific, and enduring memories that characterize much human recognition learning. Rats learn hundreds of distinct olfactory stimuli in a few trials each and retain these memories for months. During olfactory learning behavior, physiological activity throughout the olfactory-hippocampal pathway is relatively tightly synchronized to the 4–8 Hz (theta) rhythm that characterizes sniffing in small mammals (Komisaruk 1970; Macrides 1975; Macrides et al 1982). The finding that this rhythm is optimal for the induction of LTP in olfactory cortex and hippocampus (Larson & Lynch 1986; Pavlides et al 1988) provides the strong suggestion that relationships exist between synchronized rhythmic activity, behavioral learning, and synaptic plasticity (Otto et al 1991).

Computational Modeling

Several researchers have attempted to identify possible links between the distributed representations of the olfactory system and network models (Haberly 1985; Kauer 1991). Computer simulations have been constructed at different levels, ranging from detailed biophysics of dendrodendritic spines (Anton et al 1992), interactions among dendritic elements (Shepherd & Brayton 1987), and collections of physiologically accurate cells (Anton et al 1991; Coultrip et al 1992; Haberly & Bower 1989) to large, hierarchically organized networks that contain simplified models of the anatomical circuitry and physiological rules (Freeman 1991). Most of this work is primarily concerned with modeling physiology, not with learning; nonetheless, we briefly review these physiological models so that their contributions to a larger network model of olfactory learning can be considered.

BIOPHYSICS OF DENDRITIC SPINES   Many proposals have been advanced concerning the possible utility of dendritic spines in neural processing, including their use for control and modulation of information transfer from the input site to the rest of the cell by changes in input resistance (Rall 1970), for concentration of calcium (Gambel & Koch 1987), for active membrane amplification (Rall 1964), and for control of inputs (Shepherd & Brayton 1987). These effects are generally concerned with axodendritic synaptic connections, where the dendritic spine is post-synaptic, and are studied for the postsynaptic cell only. Olfactory bulb contains more complicated synaptic arrangements: Dendro-dendritic connections involve spines on inhibitory interneurons, whose strength is hypothesized to be graded based on the presynaptic voltage (Shepherd 1979).

LOCAL CIRCUITS IN OLFACTORY BULB AND CORTEX   Using these findings as a starting point, Anton et al (1991) have modeled the local circuit effects of these dendrodendritic interactions in the context of the anatomical architecture in which they occur: the glomerular organization of excitatory MT cells and inhibitory granule cells in olfactory bulb. Responsiveness of an MT cell to a given olfactory nerve input is a function of its depth in the MT cell layer (Schneider & Scott 1983). Combined with the graded potentials made possible by dendrodendritic interactions, this range of reponsiveness causes increasing numbers of MT cells to spike in response to increasing frequency of afferent input from the olfactory receptors. The net result is transformation of a frequency-coded signal at the receptors to a spatially coded signal in bulb (Anton et al 1991).

Layer III of olfactory cortex contains roughly 100 times fewer inhibitory interneurons than excitatory pyramidal cells. The axons of these inhibitory cells have a relatively fixed radius of arborization, which gives rise to a patchwork defined by the region of influence of inhibitory cells (see van Hoesen & Pandya 1975). A primary distinction between excitatory and inhibitory cell activity is their relative duration: Excitatory postsynaptic potentials typically last roughly 20 msec, whereas fast inhibitory potentials last at least 100 msec, and other inhibitory currents can last 500–1000 msec. During learning, these circuits operate robustly in synchronous cyclic activity at the 5 Hz theta rhythm (Komisaruk 1970). Local-circuit properties of these physiological characteristics in cortex have been studied via simulations of interactions among excitatory and inhibitory cells engaged in this rhythmic activity (Coultrip et al 1992). In these simulations, neurons receiving the most input activation would be the first to reach spiking threshold, which would excite feedback inhibition, which would in turn inhibit all cells from responding further. The result is the natural generation of a simple competitive or "winner-take-all" mechanism (Coultrip et al 1992), with mathematical characteristics related to those postulated by neural network researchers (Ellias & Grossberg 1975).

LEARNING IN THE OLFACTORY CORTEX-BULB SYSTEM   As mentioned above, LTP induction occurs in olfactory cortex (Jung et al 1990; Kanter & Haberly 1990; Roman et al 1987), with induction requirements (brief bursts at the theta rhythm) matching those found during learning in behaving animals (Otto et al 1991). This argues for investigations of cortical memory function based on this synchronized operating mode, together with the induction and expression rules for LTP. Nonetheless, some computational models of the olfactory system have been constructed based on hypothetical synaptic plasticity mechanisms other than LTP. Haberly & Bower (1989) have described physiological findings in the olfactory cortex, and computational models based on these findings predict that spatial EEG patterns in the cortex should exhibit oscillatory behavior. In an attempt to reproduce measurements of EEG patterns in the olfactory bulb, Freeman (1991) has reported a computational system that exhibits chaotic behavior. This work suggests that multiple specific EEG patterns are associated with an odor, and that different EEG patterns associated with an odor will emerge as other odors are learned. Memorial function in the olfactory system thus might emerge from chaotic behavior (Freeman 1991).

Computer simulations of the olfactory system that incorporate features of LTP induction and expression have been performed by Ambros-Ingerson et al (1990) and have led to findings that suggest not just encoding of sensory cues, but organization of the resulting memories into structures not typically seen in neural network models. In that work, implementation of a repetitive sampling feature meant to represent the cyclic sniffing behavior of mammals (Komisaruk 1970) produced a system that exhibited successively finer-grained encodings of learned cues over sampling cycles. Each sampling cycle includes feedforward activity from bulb to cortex followed by feedback from cortex to bulb. Because this feedback activates long-lasting inhibition in the bulb, the next sample of the input arrives against an inhibitory background in bulk, which effectively masks part of the input, thereby resulting in different activity patterns in bulb and in cortex. Thus, resampling a fixed cue generates different cortical responses with each new sampling cycle.

Learning in this model causes the initial cortical responses to become nearly identical to sufficiently similar inputs from bulb. This phenomenon, called clustering, has been predicted based on Hebbian learning rules by many researchers (Grossberg 1976; Rumelhart & Zipser 1985; von der Malsburg 1973). Essentially, the resulting cortical response corresponds to coarse-grained families or clusters of inputs; a given response signals membership of the input in a given cluster (e.g. fruit odors versus meat odors versus floral odors), thereby partitioning the input space.

The feedback from cortex to the bulb granule cell layer selectively inhibits those portions of bulb response that give rise to the cortical firing pattern; this inhibition in bulb lasts for hundreds of milliseconds (Nicoll 1969). Resampling then causes new bulb and cortex activity against the background of this long-lasting inhibition. The resulting cortical response corresponds to odor components not shared across category members, i.e. the first sample responses to a set of flowers are identical, which signifies that these odors are all members of a single category; subsequent samples correspond to differences among different flowers, thereby effectively distinguishing among subcategories of floral odors. Thus, learning via LTP in the model generates a multilevel hierarchical memory that uncovers statistical relationships inherent in collections of learned cues and, during retrieval, sequentially traverses this hierarchical recognition memory (Ambros-Ingerson et al 1990; see Figure 2A–C).

Predictions from Modeling

These findings from computer simulations and theoretical analyses lead to a hypothesis of paleocortical function: Repetitive sampling and learning combine to cause the cortex to organize its memories into a hierarchical tree that is traversed during recognition. Specific findings from the models give rise to specific and testable predictions at both behavioral and physiological levels: Behaviorally, animals should learn similarity-based categories; physiologically, cell spiking in piriform cortex should be sparse and odor specific during olfactory learning and recognition, in contrast to models with extensive global activity. More specifically, different cortical cells should discharge over successive sampling cycles with progressively more selective tuning.

A recent set of behavioral experiments has tested a prediction of the Ambros-Ingerson et al (1990) model, namely, that rats spontaneously encode and use similarity-based categories. In the model, cortical responses to category members become identical only after several similar cues are learned by the model. This behavior is distinguishable from stimulus generalization, which predicts that choice behavior should, without learning, arise from physical similarity of stimuli alone, rather than from the existence of a sufficient number of similar stimuli. The experimental results supported the computer prediction, thus providing the first evidence that rats build unsupervised, similarity-based categories and that they can do so with widely spaced learning sessions (Granger et al 1991).

Different computational models have generated distinct predictions about the physiological responses of cortical cells during learning of novel olfactory stimuli: Modeling results from Lynch and colleagues have predicted that spiking responses to odors will be sparse (Granger et al 1989; Lynch & Granger 1989), whereas modeling by Freeman and associates predicts extensive excitatory activity (Freeman 1975, 1991). Surprisingly few experimental studies have focused on the responses of single, primary sensory cortical units during acquisition of novel olfactory stimuli in behaving, freely moving animals. Behavioral studies have shown that in olfactory learning tasks mammals are capable of very rapid locomotor responses, within 0.5 seconds, to olfactory cues, even though typical odor delivery systems may take up to 100 msec from odor onset to time of reception at the epithelium (Eichenbaum et al 1987; Staubli et al 1987). Recent experiments were performed to study the processing that occurs within this relatively narrow time period, during which odors are detected and recognized, and appropriate responses are organized. The results indicated that the majority of cells in piriform cortex do not respond to most odors, and few cells exhibited odor specificity of response, i.e. coding is extremely sparse (McCollum et al 1991; see Figure 2E–G). The further prediction from this computational model is that different cortical cells should discharge over successive sampling cycles, with progressively more selective tuning; further experiments are required to test this prediction.

3. THE CEREBELLUM AND MOTOR-REFLEX LEARNING

The cerebellum has long been known to be involved in motor control, movement, and coordination. The anatomical uniformity and seeming simplicity of the cerebellar cortical circuitry (see Figure 3A) has repeatedly proven an attractive target for theorists seeking to understand, and model, neuronal function. Marr (1969) proposed that the two main afferents converging on Purkinje cells in the cerebellar cortex—mossy fibers and climbing fibers—form a simple associative learning system. Mossy fibers, in Marr's theory, provide sensory inputs to the Purkinje cells via parallel fiber inputs whose synaptic efficacy are modified by the presence (or

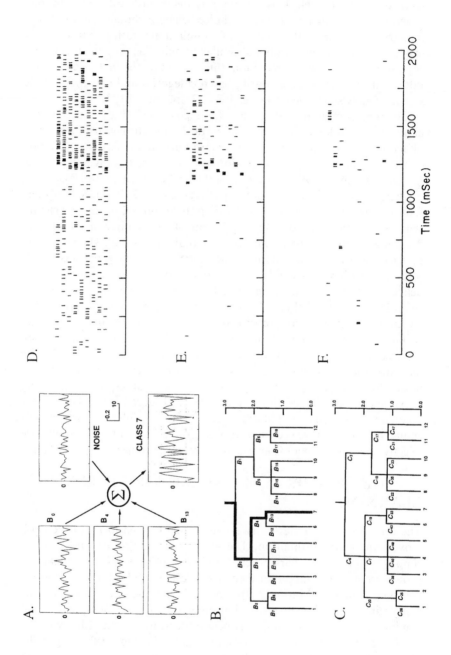

Figure 2   (*A*) A sample member of one group of a constructed hierarchy of simulated cues presented to the olfactory bulb-cortex simulation by Ambros-Ingerson et al (1990). Members of "class 7" are defined as the sum of vectors B0, B4, and B13 plus a "noise" vector; some of these components are shared across members of other categories as shown in (*B*). (*B*) Twelve categories of 50-dimensional real-valued vectors are constructed, each being the sum of four components, some of which are shared across members of different categories. For instance, all members of categories 1–7 contain component B0; all members of categories 6 and 7 contain both B0 and B4; all members of category 7 contain B0, B4, and B13 as shown in (*A*). The height of each node corresponds to the average distance among the means of the data below the node. (*C*) Dendrogram structure created by olfactory cortex-bulb simulation by Ambros-Ingerson et al (1990). After three passes of training on a set of 120 cues chosen randomly from the set constructed in (*B*) above (approximately ten vectors from each of the 12 categories), the network was then tested on a distinct testing set consisting of ten novel instances of each of the 12 categories, and a record was kept of which simulated cortical cells responded on each cue presentation. Analysis of this record showed that cells became tuned to groups of cues that correspond to categories; cell responses are indicated in the dendrogram. For instance, cell C6 spikes on the first simulated "sniff" or cycle in response to all instances of cues from categories 1–7, and not in response to any other cue; cell C19 fires on the second sniff in response to cues from categories 6 and 7 and no other; cell C33 responds on the third sniff only if the cue comes from category 7. The height of each node is given by the average Euclidean distance among the weight vectors represented by each cell at that level (for instance, the average distance between weight vectors C39, C40 and C49 is 1.11). (Reprinted with permission from Ambros-Ingerson et al 1990). (*D*) Percent of animals responding to the positive one of a pair of odors on the first trial (i.e. before any reinforcement) in an olfactory discrimination task. Open squares denote sessions in which one of the two simultaneously presented odors was a member of a created category: odors mixed to be similar to each other, although readily discriminable. Filled squares denote sessions presenting two random (non-category) odors. Sessions alternated between category and non-category presentations. Binomial tests of each score showed responses to the first six sessions, three of which were category sessions, to be at chance. Beginning with the seventh session, category members are approached significantly above chance ($p < 0.05$), whereas approach to noncategory odors remained at chance. (Reprinted with permission from Grange et al 1991.) (*E*) Peristimulus raster plots of unit activity in layer II olfactory (piriform) cortex during odor discrimination sessions for Type II cell. In each panel, rows of dashes indicate cell discharges for a series of trials for one odor. Odor onset begins at 1000 msec. Firing rate decreased during the 1-sec period before odor onset, signaled by a light. Firing increased at odor onset, and then returned to the spontaneous rate later in each trial. Nearly identical response patterns occurred in response to six odors tested. (*F*) Odor-specific type I cell. Odor-specific burst discharges at about 200 msec latency after odor onset. Burst discharges persisted for the majority of trials on this odor. The cell exhibited some odor specificity: No reliable responses were seen for five other odors (not shown) tested on this cell. Six percent of type I cells exhibited this pattern of response. (*G*) Adapting type I cell. Burst discharges occur after odor onset, but responsiveness of this cell is lost after the second trial. Thirty-nine percent of the type I cells tested exhibited this pattern of response. (Reprinted with permission from McCollum et al 1991.)

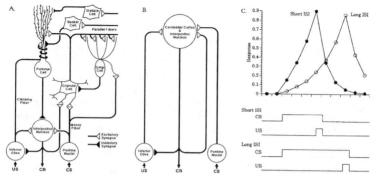

Figure 3 (*A*) Simplified schematic of the neural circuits identified as subserving the conditioned eye-blink response. Depicts the afferents assumed to carry information about CS and US occurrence, the connectivity of the cerebellar cortex and deep nuclei, and the efferents from the cerebellum involved in the generation of the eye-blink CR. Information about the occurrence of the CS comes from mossy fiber projections to the cerebellum via the pontine nuclei. The reinforcement pathway, which carries information about the US, is considered the climbing fibers from the inferior olive. The efferent pathway for the CR consists of projections from the interpositus nucleus that act ultimately upon motor neurons. The efferent response pathway includes two collateral projects. The first is an inhibitory pathway that acts ultimately to inhibit US activation of the inferior olive. The second efferent collateral projects CR information back into the cerebellum as mossy fiber inputs to granule cells. (*B*) Schematic of neural circuit model that abstracts principle pathways in Figure 1*A*. A single higher-order adaptive element corresponds to both the cerebellar cortex and deep nucleus. (Reprinted with permission from Gluck et al 1992.) (*C*) Comparison of response topography for short and long ISI delay conditioning following 500 training trials. From Gluck et al 1992.

absence) of cooccurring climbing fiber input. Thus, the climbing fiber inputs can be characterized as teaching signals that reorganize sensory-motor associations conveyed along mossy and parallel fibers. Albus (1971) modified Marr's thesis slightly by proposing that the efficacy of parallel fiber-Purkinje cell transmission is depressed by this Hebbian process (in contrast to Marr, who proposed an enhancement). Eccles (1977), Calvert & Meno (1972), and Hassul & Daniels (1977) all provided successively more refined and elaborated versions of the Marr-Albus theory.

Recent experimental evidence has provided strong confirmation of these proposals. Two different behavioral paradigms, adaptive modification of the vesitibulo-occular reflex (VOR) (Ito 1982; Robinson 1976) and Pavlovian conditioning of the rabbit eye-blink response (Thompson 1986, 1988), offer converging evidence for the Marr-Albus theory in learning novel motor movements and motor coordination. For both paradigms, new empirical evidence led to the development of subsequent computational models that elaborate and refine the theory. We begin by briefly reviewing theories of the cerebellar substrates of the adaptive VOR. We then focus, in more detail, on computational models of the cerebellar substrates of conditioning, and how these models are leading to new experimental tests and raising new empirical questions.

Adaptation of the Vestibulo-Ocular Reflex

The VOR generates compensatory eye movements to maintain stable retinal images during movement of the head. Robinson (1976, 1977) showed that this VOR may be forced to adapt to changing conditions brought about naturally, by cell loss, disease, and aging, or artificially in the laboratory by special glasses that alter the visual field. Robinson (1977) demonstrated that the vestibulocerebellum is necessary to produce and maintain the large plastic changes that occur in the VOR when it must adapt to changes in the visual field. Similar results have also been obtained with studies of the cerebellar role in responding to unexpected changes in external forces and loads in a motor coordination task (Gilbert & Thach 1972).

More detailed hypotheses of circuit-level mechanisms for the adaptive-VOR were provided by Ito (1982), who argued that the climbing fibers provide retinal slip (error) information, which is used to adapt eye-movement. The precise locus of the neural plasticity that codes for adaption of the VOR is still debatable; Ito (1987) argues for the cerebellar cortex (parallel fiber-Purkinje cell synapses), whereas Lisberger (1987) argues for vestibular nuclei. As described earlier in this review, Ito and colleagues have provided direct evidence for a form of hetero-synaptic plasticity in the cerebellar cortex: a depression of parallel fiber-Purkinje cell synapses following conjoint stimulation of parallel fibers and climbing fibers, precisely the same learning process conjectured by Albus more than ten years earlier. By incorporating Ito's findings and related behavioral work on the adaptive-VOR, Fujita (1982) developed an adaptive filter model of the cerebellar role in the adaptive VOR. Recently, Chapeau-Blondeau & Chauvet (1991) provided a more biologically realistic and physiologically detailed model of the cerebellum, which extends the Fujita model and builds on a related cerebellar theory of Pellionisz & Llinas (1982). This model differs from the others in that it uses explicit propagation delays of neural signals and proposes that the Golgi-granule cell system plays a key role in this timing.

Conditioning of Motor Reflexes

Significant progress in the study of learning systems has been made in the area of classical conditioning (see Bartha et al 1991; Thompson 1986;

Thompson & Gluck 1990). Classical conditioning is a particularly valuable paradigm for analysis of brain substrates of sensory-motor learning, because of the high degree of experimental control possible. In a typical experiment, a neutral stimulus is presented in close succession with a response-evoking stimulus, called the unconditioned stimulus (US). After repeated pairings, the previously neutral stimulus, called the conditioned stimulus (CS), evokes a conditioned response (CR), which is similar to the US-evoked response (UR). The time between the CS onset and the US onset is defined as the interstimulus interval (ISI). By varying the ISI and the intensity, duration, number, and type of stimuli, numerous conditioning phenomena can be explored.

A popular paradigm for studying conditioning behaviors is the rabbit eye-blink task, which has been well characterized by nearly three decades of study in many laboratories (for review, see Gormezano et al 1983). In this preparation, the rabbit is restrained during presentation of stimuli while the movement of the nictitating membrane (third eyelid) is measured. Most commonly, the CS is a light or tone, and the US is a corneal air-puff. Eye-blink conditioning can occur with ISIs from 100 msec to well over 1 second. After about 200 trials, rabbits give CRs on over 90% of the trials. The CR onset initially develops near the US onset and gradually moves earlier in the trial over the course of training. In a well-trained animal, the CR generally peaks near the onset of the US. More complex conditioning phenomena can be observed with compound conditioning paradigms involving the simultaneous presentation of two or more CSs.

The long-term goal of experimental and theoretical work on the rabbit eye-blink preparation is to localize the site(s) of the essential memory trace so that the mechanisms of its formation can be analyzed (for reviews, see Thompson 1986, 1988). As shown in Figure 3A, experimental evidence singles out the pontine nuclei, the inferior olive, the cerebellar cortex, and the interpositus nucleus as crucial contributors to the acquisition of the conditioned response in classical eye-lid conditioning.

CS, US, AND CR PATHWAYS   Information about occurrence of the CS travels along mossy fiber projections to the cerebellum via the pontine nuclei (Steinmentz et al 1986). Lesions of this area and its projections to the cerebellum (middle cerebellar peduncle) have been demonstrated to abolish retention by using peripheral auditory, tactile, and visual CSs, and pontine electrical CS (Knowlton & Thompson 1988). The reinforcement pathway, which carries information about the US, is assumed to be the climbing fibers that originate from the inferior olive (McCormick et al 1985; Yeo et al 1985). Physiological studies of the anterior interpositus nucleus during training show response patterns that match the shape and time course of the learned eye-blink, thus implicating the interpositus as the site of origination of the conditioned response (Foy et al 1984). The interpositus nuclei also regulates the inferior olive via an inhibitory pathway (see Figure 3A), thereby exerting a modulatory influence on the reinforcement (US) pathway.

Computational Modeling

As with the prior example of modeling the olfactory system, the primary goal of cerebellar modeling has been to relate the neural circuitry with observable behavior. One strategy for developing such models is to integrate neural models of acquisition with behavior-level characterizations of classical conditioning phenomena. In this way, the behavioral models can serve as concise embodiments of multiple constraints imposed by the behavioral phenomena on a biological model (Donegan et al 1989). This top-down approach, in which biological theories are motivated by attempts to implement behavioral theories, can be contrasted with the bottom-up approach exemplified by the cortical theories described earlier.

Great theoretical interest in classical conditioning has been motivated by a desire to understand basic behavioral properties and laws of this form of learning. Many models have been specifically applied to the rabbit eye-blink preparation (Donegan & Wagner 1987; Moore & Blazis 1989). Among the simplest and most powerful accounts of conditioning behavior is the Rescorla-Wagner (1972) model. The major goal of this model is to account for stimulus context phenomena that arise with multiple-cue compound training procedures, such as overshadowing, blocking, and conditioned inhibition. This early and influential work is especially relevant to computational network models of learning, because the Rescorla-Wagner model is a special case of the LMS rule (Widrow & Hoff 1960), which is well known in the adaptive network literature (cf. Sutton & Barto 1981) and is closely related to the error backpropagation training procedure found in many recent connectionist network models (Rumelhart et al 1986). As described in Section 1, these error-correcting associative processes have quite different training and associative properties from Hebbian coactivity rules, which characterize most synaptic plasticity found in telencephalon.

One problem in reconciling the data and models from the VOR and conditioning has been the differing temporal parameters that govern synaptic plasticity and behavioral learning. Ito (1987) reported that the temporal parameters necessary for plasticity at the critical cerebellar sites are broad and coarse, i.e. plasticity occurs over a wide range of temporal relationships. As noted above, however, the temporal relationship of the CS, CR, and US are all precisely related at the behavioral level. A possible resolution to this paradox is suggested by the simulation results of Gluck et al (1990). They showed that when the temporal sensitivity of the learning algorithm at the synapses is coarse (broadly tuned), rather than precisely tuned, the model exhibits improved tolerance to noise while still being able to mediate precise timing behaviors, because of the distributed sensory-response coding. These results suggest one possible reconciliation of apparent inconsistencies between imprecise unit responses and the precise timing of behavioral responses (see also Sejnowski 1988, for a related discussion of distributed coarse coding).

Moore and colleagues (Desmond & Moore 1988; Moore & Blazis 1989) have proposed several models for conditioned-response topography that extend the Sutton & Barto (1981) model by using tapped delay lines, assuming that the CS activates different input lines at successively later times throughout the CS period. Confirming experimental evidence on the existence of tapped delay lines in this system has yet to be identified.

Cerebellar Substrates for Timing

An interesting feature of classical conditioning is the behavioral ability of animals to anticipate the temporal interval between the CS and US accurately. A recent cerebellar model described by Gluck et al (1992), analyzed a simplified and reduced version of the circuit in Figure 3A, shown in Figure 3B. Gluck et al (1992) assume that the relevant sites of plasticity occur within a node that corresponds to the aggregate influence of the cerebellar cortex and the interpositus nucleus. All synapses converging on this node are assumed to be modifiable, according to a Hebbian learning rule in which changes are proportional to the correlation between pre- and postsynaptic activity. This rule captures the basic finding that repeated conjoint activation of mossy-parallel fibers and climbing fibers results in LTD, a prolonged decrease in excitability of the parallel fibers synapses on Purkinje cell dendrites (Ito 1989). When filtered through the Purkinje cell's inhibitory action on the interpositus, the net effect of this change is an increase in CS-US associative strength. Purkinje neurons in cerebellar cortex have extensive and elaborate dendritic trees. Each Purkinje cell receives approximately 200,000 parallel fiber synapses, en passant, mostly on distal dendrites, and receives synapses on proximal dendrites from only a single climbing fiber. Purkinje neuron dendrites have active zones, limited regions where synaptic drive can elicit space-limited action potentials. Purkinje neurons illustrate the computational capabilities of dendritic trees proposed by Shepherd (1990), for example, as "and" functions for multiple parallel fibers. In this manner, the inputs to this simplified cerebellar/interpositus unit are presumed to be filtered through a system of higher-order, or "configural" combinations. Each input line interacts multiplicatively with other inputs to this unit. This higher-order filtering is computationally equivalent to assuming the presence of unique configural cues that detect for cooccurrences of stimuli—an assumption concordant with many behavioral models of animal learning (Wagner & Rescorla 1972) and human learning (Gluck 1991; Gluck & Bower 1988; Gluck et al 1989).

TEMPORAL SPECIFICITY OF CONDITIONED RESPONSE   Figure 3C shows response topography from the Gluck et al (1992) model, which illustrates that responses to both short and long ISIs following 500 training trials agree with the qualitative behavioral data on CR topography. The peak CR response occurs at the onset of the US. Furthermore, the increased complexity of the long interval task increases learning time and results in a broader CR shape; again, this is consistent with the behavioral data (Gormezano et al 1983).

BLOCKING AND ERROR-CORRECTION LEARNING   In a blocking experiment (Kamin 1969), training trials pair a CS1–CS2 compound with the US, where CS1 is a stimulus that had been trained to predict the US. If the CS2 is then presented alone, following this compound training, little or no CR results: The associability of CS2 appears to have been blocked by the presence of CS1, already sufficient to generate the CR. The Rescorla-Wagner (1972) equations for conditioning account for this blocking effect via the assumption that associative changes are proportional to the difference between the actual reinforcement (US or no-US) and the expected reinforcement (the CR). The preliminary cerebellar circuit model of Gluck et al (1992) also exhibits blocking because of the error-correcting property of the negative feedback loop from the CR-pathway to the US/reinforcement pathway (Figure 3A,B). This negative feedback loop can be viewed as an implementation of the comparison between expected and actual reinforcement suggested by the Rescorla-Wagner equations (see also Thompson & Gluck 1990).

Functional Role of Other Brain Structures

Hippocampal-region lesions do not impair learning or retention of the CR for simple acquisition of a single conditioned stimulus (see Thompson 1988). However, some more complex forms of conditioning are hippocampally dependent. Logically, a model of cerebellum ought to exhibit deficits on hippocampally dependent tasks, because an isolated cerebellar model presumably lacks processing that would be contributed by the hippocampus. The cerebellar model of Gluck et al (1992) exhibits this property on a set of hippocampally dependent conditioning tasks. For example, as expected from the results of Rudy and Sutherland (Rudy &

Sutherland 1989; Sutherland & Rudy 1989), which implicate the hippo-campus in so-called configural associations, the Gluck et al (1992) model has considerable difficulty learning tasks that require sensitivity to con-figurations, such as in negative patterning (i.e. A+/B+/AB−training, also known as the exclusive-OR or "X-OR" discrimination in logic). Animals with hippocampal-region damage also show marked deficits com-pared with intact animals in several training procedures that demonstrate how preexposure to one or more CSs affects the animal's ability to form associations in later conditioning, including sensory preconditioning (Port & Patterson 1984) and latent inhibition (Solomon & Moore 1975). Con-sistent with these data, the Gluck et al cerebellar model also fails to exhibit these same sensory preexposure behaviors normally observed in intact animals. Animals with hippocampal-region damage are also impaired in training tasks that require a sensitivity to temporal factors, such as long ISI training and trace conditioning (Moyer et al 1990); appropriately, the Gluck et al cerebellar model cannot learn with trace conditioning training or with very long ISIs in a delay conditioning paradigm. Keeping in mind the above results (both empirical and theoretical) on the functional role of the hippocampal region in tasks requiring sensitivity to timing and multiple-cue interactions, we turn now to both physiological and func-tional models of the hippocampal region, and the hippocampus itself.

4. THE HIPPOCAMPAL REGION

The hippocampus and adjacent cortical regions in the medial temporal lobe (see Figure 4) have long been implicated in human learning and memory via lesion data in humans (Scoville & Milner 1957; Squire 1982) and animals (Mishkin 1982; Squire & Zola-Morgan 1983). Although there

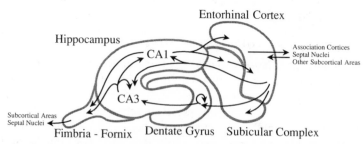

Figure 4 The hippocampal formation and its surrounding cortical environment, including the entorhinal cortex, subicular complex, dentate gyrus, hippocampus (showing areas CA3 and CA1), and the fimbria-fornix. Arrows show major connections between and within structures. Major connections to and from other brain regions are also shown.

is general agreement that this region plays an essential role in many aspects of learning and memory, there is little consensus as to its precise functional role(s).

Some hypotheses of hippocampal-region function have emphasized the critical role of this brain region in specific memorial tasks, notably explicit declarative memories in humans (Squire 1987; Cohen & Squire 1980). As distinct from procedural or implicit memories, these hippocampal-dependent memories are marked by their accessibility to conscious recollection. There has been little or no progress, however, in developing mechanistic or computational models of these disruptions to declarative human memory. Cognitive psychology has yet to provide a detailed computational processing model of intact declarative memory, presumably a necessary precursor to modeling its disruption.

Studies of lower (nonprimate) mammals, especially rats, have established that hippocampal-region lesions interfere with an animal's ability to perform place-learning and spatial navigation tasks (Morris et al 1982; O'Keefe et al 1975). O'Keefe & Nadel (1978) argued that the hippocampal formation plays a key role in these spatial tasks by learning a "spatial map" of the relationship among distal visual cues and related topographical features. Although spatial navigation behavior is not fully characterized at a behavioral (or mechanistic) level, it is more amenable—as a behavioral paradigm—to computational modeling than human declarative memory. Indeed, some preliminary progress has been made in developing mechanistic process theories of spatial navigation and the role of the hippocampal formation in place and route learning (McNaughton 1989; McNaughton & Nadel 1990). Ranck (1984) and McNaughton et al (1991) have argued that visual landmarks become incorporated into the animal's directional system through the association of hippocampal local-view cells with head-direction cells located in the dorsal presubiculum (see also O'Keefe 1991). Detailed computational models for these neurobiological processing theories still remain to be developed and tested as simulation models.

Information-Processing Theories

Another approach to functional theories of hippocampal processing has been to postulate some underlying information-processing role(s) for the hippocampal region and then seek to derive a wider range of task-specific deficits. The goal here is to develop a comprehensive view of hippocampal-region function that illuminates common behavioral functions, and underlying neural mechanisms, in both animal and human memory behaviors. Such a broad characterization has proven elusive, yet some initial progress has been made in modeling the role of the hippocampus in certain complex associative learning paradigms; we review this work in the following section. These modeling efforts have profited from the extensive body of behavioral data that constrains psychological and neurobiological theorizing. In addition, well-understood behavior-level models of classical conditioning (e.g. Mackintosh 1975; Rescorla & Wagner 1972) provide another top-down source of constraints for characterizing the hippocampal-region processing that contributes to this form of learning.

To understand current computational models of hippocampal-region function in associative learning, one must appreciate the theoretical traditions from which they emerge. Thus, before describing these computational models, we review earlier ideas about hippocampal processing, which can be divided into two distinct categories: representational and temporal.

Early researchers viewed the hippocampal region as an attentional control mechanism that altered stimulus representation through processes of stimulus selection. Grastyan et al (1959) proposed that the hippocampus inhibits an animal's orienting response to nonsignificant stimuli. Subsequent elaborations upon this idea were presented by Douglas & Pribram (1966), Kimble (1968), and Douglas (1972), all of whom developed models based on the notion that the hippocampus inhibits orienting (or attentional) processes by identifying stimuli that are correlated with nonreinforcement. Moore (1979) and Solomon (1979) suggested that attenuated stimuli were those with a history of being irrelevant, not just those correlated with nonreinforcement. As an extension of Mackintosh's (1975) attentional hypothesis of stimulus selection, Moore & Stickney (1982) proposed a computational process whereby hippocampal damage would prevent individual cue saliences from being tuned out. Schmajuk & Moore (1985) proposed a real-time variation on this approach, which drew upon the earlier behavioral modeling of Pearce & Hall (1980). Although applicable to a subset of the hippocampal-lesion literature, these models do not address the extensive literature that shows an important role for the hippocampal-region in processing complex multiple-cue stimulus representations and inter-cue relationships.

More recently, researchers have considered a direct representational role for hippocampal function, particularly with respect to processing configurations or relations among different stimulus features. Hirsh (1974, 1980) viewed the hippocampal formation as storing stimulus relations, especially the contextual background within which stimulus-response associations are stored and retrieved. Wickelgren's (1979) "chunking" hypothesis proposed that the hippocampus participates in a process whereby the component features that comprise a stimulus pattern come to be treated as unitary whole or chunk. More recently, Sutherland & Rudy

(1989) have elaborated on this idea: Their configural association theory proposes that the hippocampal formation provides the neural basis for the acquisition and storage of configural associations among stimulus events. In contrast to a simple associative system, which does not rely on the hippocampal formation, Sutherland & Rudy's configural associative system combines representations of elementary stimulus events to construct unique chunked representations, thereby allowing for the formation of associations between these configural representations and other stimulus events. Although most directly applicable to Pavlovian and operant conditioning tasks that involve distinct elementary stimuli (e.g. tones and lights) Sutherland & Rudy suggest a wider range of applications; for example, spatial maps might be viewed as sets of complex configural associations (cf. McNaughton 1989; McNaughton & Nadel 1990). Eichenbaum has similarly emphasized the representational role of hippocampal function, particularly with respect to relational and conjunctive associations among stimuli and the flexible use of these representations in novel situations (Eichenbaum & Buckingham 1991; Eichenbaum et al 1992). Nonetheless, these representational hypotheses have been characterized primarily as verbal or qualitative ideas, not as mechanistic or precise computational simulation models.

Still another class of hypotheses—distinct from those reviewed above—view hippocampal function as buffering or tagging prior representations of stimulus events so that they can be compared with future perceptions. This temporal-processing view can be seen most clearly in the comparator models of Vinogradova (1975) and Gray (1979), in the working memory model of Olton et al (1979), and in the temporary memory-buffer model of hippocampal function of Rawlins (1985). Although these temporal-processing hypotheses fail to address directly the many attentional and representational deficits characterized by the theories reviewed above, they do characterize the temporal nature of many hippocampal-dependent behaviors ranging from delayed nonmatch to sample (Mishkin 1978; Zola-Morgan & Squire 1983, 1990) to ISI and stimulus trace effects in motor-reflex conditioning (see Section 3; also Akase et al 1989).

Computational Models of Hippocampal-System Function

All three classes of hippocampal-lesion deficits (and related hypotheses of hippocampal processing) can characterize subsets of the relevant empirical data. At a qualitative or verbal level, many of these findings can also be related to some aspects of human declarative memory and spatial navigation. What is lacking is a clear mechanistic characterization of hippocampal-region function that simplifies the empirical literature by identifying one or more underlying mechanisms that subserve a wide range of lesion-produced behavioral deficits. As noted earlier, lesions directed at the hippocampus often affect (and disable) adjoining cortical regions in the medial temporal lobe, including the entorhinal, perirhinal, and para-hippocampal cortices. Thus, the multiple functional deficits reviewed above may all be associated with a single underlying mechanism, or they may arise from multiple mechanisms, possibly associated with distinct subregions damaged in these lesions. Discriminating these possibilities based on lesion studies alone is fraught with difficulty, especially in light of the many distinct methods used for so-called hippocampal lesions, ranging from large lesions that eliminate much of the medial temporal area to smaller lesions in different areas, including entorhinal cortex and the fimbria fornix.

In the remainder of this section, we review both functional (top-down) approaches to modeling the hippocampal region and physiological (bottom-up) approaches to modeling the hippocampus proper. We begin by comparing two alternative interpretations of the functional role of the broader hippocampal region, both of which seek to account, primarily, for behavioral data from lesion studies. The first interpretation (Schmajuk & DiCarlo 1991) is based on a detailed circuit-level theory, whereas the other (Gluck & Myers 1992) is posed as an abstract connectionist theory. Following this, we turn to the anatomy and physiology of the hippocampus proper and attempt to derive from these bottom-up constraints an emergent functional processing role for the hippocampus itself. Finally, we compare the functional roles suggested by the aforementioned top-down analyses of the broader hippocampal region with bottom-up theories of the hippocampus proper. In this analysis, we also consider the possible functional roles of the cortical medial temporal regions that overlay the hippocampus, drawing on the previously reviewed models of an analogous cortical structure, the olfactory cortex.

Modeling Cortico-Hippocampal-Cerebellar Function

Schmajuk & Dicarlo (1991) describe a hippocampal model for classical conditioning based on work by Grossberg (1975). They propose that the hippocampus controls self-excitation and competition among sensory representations and stores incentive motivation associations, thereby regulating the contents of a limited short-term memory. Schmajuk & Dicarlo (1992) describe still another—quite different—real-time model, which they propose to map onto regional cerebellar, cortical, and hippocampal circuits. In this model, stimulus configuration occurs in association cortex, and the hippocampus is presumed to compute and broadcast an aggregate prediction of the US to both cortical and cerebellar regions. This model can be traced to earlier attentional and control theories described above.

Although this model can simulate many hippocampal (and cortical) lesion studies, it incorrectly predicts certain specific deficits, such as a hippocampal lesion deficit for conditional inhibition; empirical studies have shown no such deficit (Solomon 1977). Their model also provides no interpretation for important sensory-sensory phenomena, such as the observed lesion deficits in sensory preconditioning and latent inhibition.

By identifying the hippocampus as critical for all forms of stimulus selection, the Schmajuk & Dicarlo (1991, 1992) models depend strongly on early data that show that blocking depends on an intact hippocampus (Solomon 1977); more recent studies, however, have had difficulty replicating this result (Garrud et al 1984). Furthermore, the assumption that these stimulus-selection behaviors are dependent on the hippocampus is inconsistent with data and theory from Thompson and colleagues, who argue for an inherent cerebellar circuit for blocking (and overshadowing) via attenuation of the efficacy of the reinforcement (US) pathway (see Section 3; also Donegan et al 1989; Thompson & Gluck 1990; Gluck et al 1992). Thus, although the hippocampal-region might play a role in blocking and overshadowing through mechanisms of stimulus selection, this region probably does not contain the sole neurobiological substrate for these behaviors.

Connectionist Models of Hippocampal-System Function

Computational models of abstract connectionist networks offer another approach to exploring candidate functional roles of the hippocampal region. Models developed at this abstract level do not directly yield a physiological understanding of the hippocampal region; nevertheless, these models may suggest how some of the behavioral and lesion data might be emergent from some simpler underlying processing role that may be localized in the hippocampal region.

An example of this connectionist-level theorizing is illustrated by the work of Gluck & Myers (1992), who address the question of whether a simple underlying computational function exists that can derive the representational processes subserved by the hippocampal system in associative learning. By developing and testing a connectionist model of hippocampal processing in associative conditioning, their work suggests how learning the hippocampal system might mediate the development of novel stimulus representations in cerebral and cerebellar cortices. The key idea behind their model (shown in Figure 5A) is that the representational (but not temporal) function of the hippocampal region can be approximated by a simple network architecture, called a predictive autoencoder, which develops novel and flexible sensory representations with three key properties: They are distributed, predictive (of future sensory inputs), and compressed (i.e. reduced in size by compressing statistical redundancies among sensory inputs).

This connectionist model is in concordance with a wide variety of associative learning behaviors observed in both intact and hippocampal-lesioned animals. In particular, this model can be interpreted as subsuming—or implementing—many previous characterization of the representational role of the hippocampal region, including contextual representation,

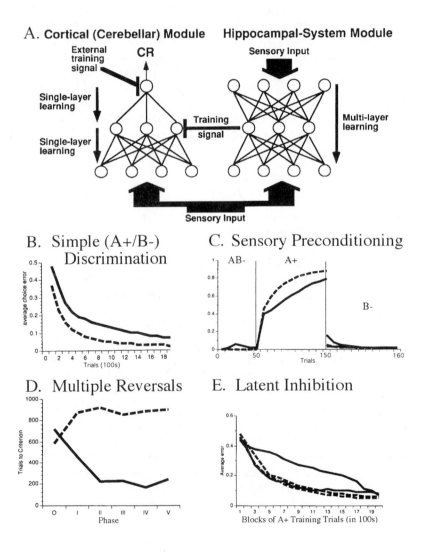

chunking, stimulus selection (attention), and cue configuration. In addition, the model provides a mechanistic interpretation for how the hippocampal region might play a critical role in acquiring certain long-term memories, without itself being the site of these memories.

When the model is applied to experimental data, two versions are considered: the intact system and the lesioned system. With the hippocampal-system module lesioned, the model learns associative stimulus-response (S-R) relationships based on a fixed recoding (representation) of the stimulus inputs. The intact system, however, learns the same S-R relationships based on a flexible recoding of the stimulus inputs in a distributed representation that reflects both the predictive S-R relationships, as well as sensory-sensory correlations in the environment.

To illustrate how the Gluck & Myers model accounts for relevant intact and lesion data, we briefly describe its application to several experiments. Simple discrimination learning $(A+/B-)$ is largely unaffected by hippo-

---

Figure 5 (*A*) The intact Gluck-Myers (1992) cortico-hippocampal model. Learning in the hippocampal system mediates (trains) the development of novel stimulus representations in cerebral and cerebellar cortices (*left*). The hippocampal module has the capacity for multi-layer learning, which results in a novel recoding (or rerepresentation) of its stimulus inputs. The cortical module (*left*) is restricted to using only (single-layer) S-R learning, e.g. the LMS rule of Widrow-Hoff (1960). Hippocampal-lesion experiments are modeled by removing the hippocampal module, which causes the bottom layer of the cortical module to remain fixed (e.g. nonmodifiable). The upper layer of the cortical module, however, can still be trained to learn based on the fixed recoding of the cortical inputs, which occurs in the cortical bottom layer. Thus, learning without the hippocampal module is limited to those discriminations that can be solved without learning a new stimulus representation. (*B*) Simulations of the intact and lesioned cortico-hippocampal model of Gluck & Myers (1992) in *Simple Discrimination*. Training on $CS(A)+/CS(B)-$. Lesioned model learns somewhat faster than the intact model, which must first learn a new representation in the hippocampal module and then transfer this representation to cortical module. (*C*) Simulations of the intact and lesioned cortico-hippocampal model of Gluck & Meyers (1992) in *Sensory Preconditioning*. Pre-exposure to a nonreinforced $CS(A)\&CS(B)$ compound is followed by training to $CS(A)+$; finally, response to $CS(B)$ is tested. Intact and lesioned systems are similar through the first two phases, but only the intact system shows transfer to $CS(B)$ in the third phase. Feedforward backpropagation system likewise shows no transfer in third phase. (*D*) Simulations of the intact and lesioned cortico-hippocampal model of Gluck and Myers (1992). Training on $CS(A)+/CS(B)-$ is followed by reversal training on $CS(A)-/CS(B)+$, and repeatedly reversed. The intact system shows a progressive decrease in the number of trials required to learn each discrimination; in contrast, the lesioned system has difficulty with all but the first discrimination. (*E*) Simulations of the intact and lesioned cortico-hippocampal model of Gluck and Myers (1992). Following a pretraining phase in which $CS(A)$ is repeatedly nonreinforced $(A-$ training), $CS(A)$ is repeatedly reinforced $(A+$ training). The intact, but not the lesioned network, is impaired on $A+$ learning compared with control conditions in which the systems are pretrained with another cue (e.g. $C-$). (Reprinted with permission from Gluck & Myers 1992.)

campal-region lesions; in fact, a faciliation following hippocampal-region lesions has often been reported (e.g. Eichenbaum et al 1987; Schmaltz & Theios 1972). As shown in Figure 5*B*, both the intact and the lesioned network models can solve a simple discrimination task. Furthermore, the lesioned network shows some facilitation because its initial—and sufficient—representation is not altered by hippocampal influence. The Gluck & Myers (1992) model also provides an account of how their proposed hippocampal-system mediates stimulus representations in more complex discrimination paradigms that require an intact hippocampal region. These include sensory preconditioning (Port & Patterson 1984), reversal learning (Berger & Orr 1983), and latent inhibition (Solomon & Moore 1975), as illustrated in Figures 5*C*, 5*D*, and 5*E*, respectively. The model is also consistent with results that show hippocampal involvement in configural learning (Sutherland & Rudy 1989) and contextual-sensitivity (Hirsh 1974; Nadel & Willner 1989; Winocur et al 1987).

In summary, the Gluck & Myers (1992) model shows how a specific computational network architecture can form compact, predictive, and distributed representations of stimuli, which are then made available to other learning systems (such as the cerebellum and cerebral cortex). This model incorporates and refines aspects of many prior, qualitative information-processing theories of hippocampal function. In its current form, the model does not address hippocampal mediation of temporal and sequential processing, functional roles implied by the failure of hippocampal-damaged animals at conditioning with long ISI delays or trace conditioning (Moyer et al 1990). These additional temporal roles either may be interpreted as requiring refinements of the same mechanisms, or they may be localized within different brain structures in the medial temporal lobe. Future efforts will be needed to understand better the interaction between temporal and representational processing in the hippo-campal-region and the precise neurobiological locus (or loci).

We turn now to look more closely at the anatomical and physiological characteristics of the hippocampus itself. We then turn back and see how these characteristics relate to the more global hippocampal-lesion deficits discussed, and modeled, above.

Physiology of Plasticity in Hippocampus

Hippocampus or archicortex has an anatomically simpler design than neocortex, because it largely lacks multiple layers and consists of fewer cell types, which makes it a valuable model system for physiological study of the telencephalon. As shown in Figure 4, signals from the sensory periphery activate subcortical and primary and secondary sensory cortical regions, respectively, and travel eventually to the superficial layers of entorhinal cortex and into hippocampus. Efferent fibers from entorhinal cortex form the primary input to hippocampus, the perforant path, which innervates all three hippocampal structures: the dentate gyrus, field CA3, and field CA1. Dentate gyrus granule cells form the mossy fiber projection to CA3, which in turn gives rise to the Schaffer-commissural afferents to field CA1. Hippocampus is also innervated by the medial septum/diagonal band complex. Efferents from hippocampus contact both the deep layers of entorhinal cortex and the ventral striatum, which in turn give rise to pathways that innervate both hindbrain motor systems and the dorso-medial nucleus of thalamus and thence to frontal cortex.

The phenomenon of LTP was discovered in the hippocampus (Bliss & Lomo 1973) and is now considered a leading candidate for the substrate of memory. Examination of variants of LTP in the hippocampus suggest the possibility that characteristics of LTP might show up as characteristics of memory processes based on LTP, and that variants of LTP may play a role in different forms of memory. In the following sections, we review the current status of findings of the physiological characteristics of LTP in hippocampus and hypotheses concerning the involvement of LTP in hippocampally dependent memory processing.

Much theorizing has occurred on the topic of hippocampal plasticity and its role in memory. Traub & Miles (1991) present a detailed account of their ongoing modeling work on hippocampus, focusing anatomically on field CA3 and functionally on the emergence of spiking and oscillations both in slices and in the intact animal. Their work addresses neither physiological plasticity nor behavioral learning, but nonetheless provides insights into possible processing attributes of field CA3 and, therefore, is of interest with respect to possible hippocampal function. Rolls (1990) has sought to characterize and model unit activity in hippocampus and related structures during behavior. Some of these findings have been incorporated into abstract network models based on hippocampal circuitry. This work, and work from the laboratories of Brown McNaughton, Morris and their colleagues on physiological properties of LTP in hippocampus, has led to several hypotheses about possible relationships between hippocampal function and various well-studied computational phenomena, such as pattern completion, pattern separation, and category learning. This approach relies heavily on hypothetical forms of plasticity based on Hebbian or correlational aspects of LTP (see Brown et al 1990; McNaughton & Nadel 1990).

Non-Hebbian Aspects of LTP in Hippocampal Field CA1

In spite of the general resemblance of NMDA-based LTP induction rules to Hebb's (1949) postulate quoted in Section 1, physiological evidence has clearly shown that the conditions it describes are neither necessary nor sufficient for induction of LTP. The target cell need not be fired, only depolarized for LTP to occur (necessity) (McNaughton et al 1978), and yet even repeated firing will not induce LTP unless the depolarization exceeds the NMDA receptor channel voltage threshold (sufficiency). The trend in the literature has been to embrace the generality of Hebb's proposal and to emphasize its similarities to the requirements for induction of LTP. Yet, the discrepancies between the physiology of LTP and the Hebb correlation requirement raise the question of the relevance of these differences to the computational functions of LTP learning rules. For example, these biological details might yield novel learning rules that confer useful computational abilities to networks. In both hippocampal fields CA1 and CA3, it has been argued that the divergences from simple Hebbian correlational LTP may give rise to learning rules that are not only significantly different from Hebbian rules, but have added computational power that enables them to encode temporal information, absent from simple correlational rules (Granger et al 1992; Lynch & Granger 1992).

A 4–7 Hz EEG rhythm appears throughout the olfactory-hippocampal pathway in animals during exploratory and learning behaviors (Grastyan et al 1959; Komisaruk 1970; Macrides 1975; Macrides et al 1982). Recent studies have established that stimulation patterns based on this theta rhythm can produce robust and stable LTP (Larson & Lynch 1986; Staubli & Lynch 1987). Afferent stimulation arriving in approximate synchrony with the theta rhythm nonetheless exhibits some asynchrony within the envelope of a single peak of the theta rhythm, which gives rise to brief (< 100 msec) sequences of inputs. A Hebbian coactivity rule would predict that as asynchronous afferents arrive, increased depolarization of the target neuron over the staggered input arrival times should cause later inputs to be strengthened more than earlier inputs. However, experiments using three brief inputs stimulated in a staggered sequence over 70 msec show that the earliest-arriving afferents potentiate their synapses the most with subsequently arriving afferents that cause successively less potentiation (Larson & Lynch 1989). Moreover, the predominant effect of potentiation is on responses generated by the AMPA subclass of glutamate receptors, which are active during normal stimulation conditions; potentiation does not have a comparable effect on NMDA receptor dependent potentials (Kauer et al 1988; Muller et al 1988). Thus, subsequent LTP induction episodes are relatively unaffected by prior induction; this effect is not predicted from Hebb's postulate, and contrasts with the development of attractor cells typically found in most network models.

Granger et al (1992) have shown how incorporating simplifications of these non-Hebbian aspects of LTP into a learning rule can give rise to a network that acts as a sequence detector, i.e. a network that selectively responds to specific temporal sequences of inputs. In that work, biophysical simulations of CA1 pyramidal cells predicted a novel LTP expression rule in which cells would preferentially respond to the same sequence on which they were trained. If potentiation were induced on afferents stimulated in the order A-B, then the cell would respond more strongly to subsequent stimulation in that order than to the order B-A. Supporting evidence for the prediction was tested in hippocampal slices. Simulations incorporating simplifications of these LTP induction and expression rules were then tested for their ability to recognize learned sequences; the network performs an accept/reject (or match/mismatch) function, which responds to learned sequences and not to novel sequences. Theoretical analysis of the capacity of the network showed that the number of items learnable by the network was very large and grew approximately linearly as the networks increased in size, thus exhibiting unusually good scaling characteristics. For instance, a network with 500 input lines and 16 target cells exhibited a 50% commission (recognition) error rate after being trained on 10,000 strings of length 4, but increasing the number of cells to 80 reduced the error rate to 4.6% on those 10,000 strings.

Encoding of Temporal Information

As mentioned in Section 1, at least three different forms of plasticity exist in the hippocampus, each with different time duration: Potentiation in cortical projection to dentate gyrus decays steadily over a period of several days (Barnes 1979); mossy fiber potentiation is even more transient, decrementing over the course of a few hours (Staublie et al 1990); LTP in field CA1, however, persists undecremented for at least several weeks (Staubli & Lynch 1987). Evidence that dentate gyrus granule cells fire rapidly for seconds during movement (Deadwyler et al 1979, 1981), combined with the dense recurrent collateral system in field CA3, suggest that CA3 may, when activated by dentate, maintain a cycling firing pattern that could maintain a stimulus trace long enough to bridge temporal delays involved in brief motoric actions (see also Buzsaki 1988). Buzsaki (1989) has proposed a two-stage learning theory that distinguishes between a theta-related exploratory period and a subsequent sharp-wave consolidation period in which CA3 pyramidals teach CA1 cells about associations among recently experienced temporal sequences of stimuli.

Sequential processing by the serially organized component circuits that comprise the cortico-hippocampal pathway (entorhinal cortex → dentate gyrus → field CA3 → CA1) has been hypothesized to be an assembly line of specialized functions, each of which adds a unique aspect to the processing of memories (Lynch & Granger 1991, 1992). Due in part to the different longevity of potentiation in these different anatomical systems, processing via these serial steps not only adds separate qualities to the resultant memory but may also enable memory for sequence or "knowing when." Taken together, the range of temporal durations and dependencies in hippocampal physiology and in the induction and expression of synaptic plasticity suggest that the functionality of the hippocampus itself may be concerned with the encoding, or processing, of time-dependent relationships.

In this regard, many classical conditioning behaviors that are disrupted by hippocampal lesions include real-time, or temporal, effects. As discussed earlier in this section (and in Section 3), these hippocampal-dependent behaviors include trace conditioning, in which the CS terminates before US onset, and other conditioning paradigms in which there is a long interstimulus interval between CS onset and US onset (Akase et al 1989). Furthermore, animal and human studies of tasks that require comparisons across temporal delays (e.g. delayed nonmatching-to-sample) are disrupted by hippocampal damage (Zola-Morgan et al 1982). In summary, convergent evidence from bottom-up physiological modeling and behavioral studies of hippocampal disruption point consistently to a hippocampal role in the processing of temporal information.

Toward Bridging the Physiology-Function Gap

As noted earlier, experimental procedures used to inactivate, or lesion, the hippocampus almost universally have the effect of cutting a link in the cortico-hippocampal-cortical loop: This, therefore, removes the processing contributions of many neural elements, including entorhinal cortex, dentate gyrus, fields CA3 and CA1, and the subicular complex, as well as (possibly) other retrohippocampal regions (see Figure 4). The lines of evidence reviewed above lead to the apparent conclusion that two potentially distinct sources of processing function are disturbed in hippocampal lesions: The first involves the encoding, or representation, of sensory-sensory relationships, whereas the other relates to the spanning of temporal delays in associative learning. Some of the top-down modeling of the broader hippocampal region (reviewed earlier) suggests a hippocampal role in molding novel stimulus representations sensitive to sensory-sensory relationships for use by other cortical systems. In Section 2, we saw that more physiologically based modeling of the superficial layers of olfactory cortex suggest a similar role for that structure, viz. the creation of organized representations encoding similarity relationships among inputs. Given the remarkable similarity between the superficial layers of olfactory and entorhinal cortex, the possibility arises that these hippocampal-region contributions to stimulus representation inferred from lesion data may not be caused by the hippocampus itself, but rather by processing of entorhinal cortex, whose contribution to neocortical processing is also disrupted by hippocampal lesions. This conjecture would leave temporal processing as the purview of hippocampus proper, a hypothesis consistent with the physiological modeling of hippocampal anatomy and physiology reviewed above. This hypothesis is also consistent with the modeling and data on cerebellar function reviewed in Section 3, which indicate a limited capacity for cerebellar processing of complex sensory-sensory relationships and long interstimulus intervals.

In reviewing converging data from bottom-up modeling of cortical and hippocampal circuits and top-down modeling of hippocampal-region and cerebellar function, along with related lesion and physiological data, we see a pattern emerge. We conjecture that a possible synthesis of computational and theoretical interpretations of cortical, cerebellar, and hippocampal processing exists in which representational reformatting for long-term memories ultimately stored in either cerebellar or cerebral cortices may be accomplished by processing of retrohippocampal cortical structures, whereas processing temporal data may be the domain of the hippocampus proper. Other researchers, however, have suggested a different division of labor between the hippocampal and entorhinal systems, based predominantly on lesion studies that involve the hippocampal region (see Otto & Eichenbaum 1992). Unfortunately, confounding such studies is the fact that lesions of either entorhinal cortex or hippocampus affect processing functions of the other structure. Our hypothesis is instead based on primary anatomical and physiological findings in the cortex and hippocampus and modeling studies based on these data. Further integrated interaction between theoretical computational analysis and empirical studies will, of course, be necessary to resolve this issue.

Acknowledgments

For their thoughtful comments and advice on this work, we are indebted to David Amaral, Gyorgy Buszaki, Paul Glauthier, Howard Eichenbaum, Bruce McNaughton, Barbara Knowlton, Catherine Myers, Lynn Nadel, Larry Squire, Richard F. Thompson, and Gary Lynch. This research was supported by grants to M. A. Gluck from the Office of Naval Research (N00014-88-K-0112, with R. F. Thompson, and by a Young Investigator Award) and by grants to R. H. Granger from the Office of Naval Research (N00014-89-J-3179 and N00014-89-J-1255) and the Defense Advanced Research Projects Administration (N00014-92-J-1625).

Literature Cited

Akase, E., Alkon, D., Disterhoff, J. 1989. Hippocampal lesions impair memory of short-delay conditioned eye-blink in rabbits. *Behav. Neurosci.* 103: 935–43

Albus, J. S. 1971. A theory of cerellar function. *Math. Biosci.* 10: 25–61

Allison, A. C. 1953. The morphology of the olfactory system in the vertebrates. *Biol. Rev.* 28: 195–244

Ambros-Ingerson, J., Granger, R., Lynch, G. 1990. Simulation of palecortex performs hierarchial clustering. *Science* 247: 1344–48

Anderson, B. J., Lee, S., Thompson, J., Steinmetz, J. E., Logan, C. G., et al. 1989. Decreased branching of spiny dendrites of rabbit cerebellar Purkinje neurons following associative eyeblink conditioning. *Soc. Neurosci. Abstr.* 640

Anton, P. S., Granger, R., Lynch, G. 1992. Temporal information processing in synapses, cells, and circuits. In *Single Neuron Computation*, ed. T. McKenna, J. Davis, S. Zornetzer. New York: Academic

Anton, P. S., Lynch, G., Granger, R. 1991. Computation of frequency-to-spatial transform by olfactory bulb glomeruli. *Biol. Cybern.* 65: 407–14

Arai, A., Larson, J., Lynch, G. 1990. Anoxia reveals the vulnerable period of the development of long-term potentiation. *Brain Res.* 511: 353–57

Artola, A., Brocher, S., Singer, W. 1990. Different voltage-dependent thresholds for inducing long-term depression and long-term potentiation in slices of rat visual cortex. *Nature* 347: 69–72

Artola, A., Singer, W. 1987. Long-term potentiation and NMDA receptors in rat visual cortex. *Nature* 330: 649–52

Ballard, D. H., Hinton, G. E., Sejnowski, T. 1983. Parallel visual computation. *Nature* 102: 21–26

Barnes, C. A. 1979. Memory deficits associated with senescence: a neurophysiological and behavioral study in the rat. *J. Comp. Physiol. Psychol.* 93: 74–104

Barrionuevo, G., Schottler, F., Lynch, G. 1980. The effects of repetitive low frequency stimulation on control and potentiated synaptic responses in the hippocampus. *Life Sci.* 27: 2385–91

Bartha, G. T., Thompson, R. F., Gluck, M. A. 1991. Sensorimotor learning and the cerebellum. In *Visual Structures and Integrated Function*, ed. M. Arbib, J. Ewer. Berlin: Springer-Verlag

Bear, M., Cooper, L., Ebner, F. 1987. Physiological basis for a theory of synapse modification. *Science* 237: 42–48

Berger, T. W. 1984. Long-term potentiation of hippocampal synaptic transmission affects rate of behavioral learning. *Science* 224: 627–30

Berger, T. W., Orr, W. B. 1983. Hippocampectomy selectively disrupts discrimination reversal learning of the rabbit nictitating membrane response. *Behav. Brain Res.* 8: 49–68

Bienenstock, E., Cooper, L., Munro, P. 1982. Theory for the development of neuron selectivity: orientation specificity and binocular interaction in visual cortex. *J. Neurosci.* 2: 32–48

Bindman, L., Christofi, G., Murphy, K., Nowicky, A. 1991. In *Aspects of Synaptic Transmission*, ed. T. W. Stone, pp. 3–25. London: Taylor & Francis

Bliss, T. V. P., Lomo, T. 1973. Last-lasting potentiation of synaptic transmission in the dentate area of the anesthetized rabbit following stimulation of the perforant path. *J. Physiol.* 232: 334–56

Brown, T., Kairiss, E., Keenan, C. 1990. Hebbian synapses: biophysical mechanisms and algorithms. *Annu. Rev. Neurosci.* 13: 475–511

Buzsaki, G. 1989. A two-stage model of memory trace formation: a role for "noisy" brain states. *Neuroscience* 55: 551–70

Buzsaki, G. 1988. Polysynaptic long-term potentiation: a physiological role of the perforant path—CA3/CA1 pyramidal cell synapse. *Brain Res.* 455: 192–95

Cajal, S. Ramon y. 1909. *Histologie du Systeme Nerveaux de l'Homme et des Vertebres.* Paris: Moloine

Calvert, T. W., Meno, F. 1972. Neural systems modeling applied to the cerebellum. *IEEE Trans. Syst. Man Cybern.* SMC-2: 363–74

Cerro, S. del, Jung, M., Lynch, G. 1992. Benzodiazepines block long-term potentiation in rat hippocampal and piriform cortex slices. *Neuroscience* 49: 1–6

Chapeau-Blondeau, F., Chauvet, G. 1991. A neural network model of the cerebellar cortex performing dynamic associations. *Biol. Cybern.* 65: 267–79

Clothiaux, E. E., Bear, M. F., Cooper, L. N. 1991. Synaptic plasticity in visual cortex: comparison of theory with experiment. *J. Neurophysiol.* 66: 1785–1804

Cohen, N. J., Squire, L. R. 1980. Preserved learning and retention of pattern analysing skill in amnesia: dissociation of knowing how and knowing that. *Science* 210: 207–9

Collingridge, G. L., Kehl, S. L., McLennan, H. 1983. Excitatory amino acids in synaptic transmission in the Schaffer-commissural pathway of the rat hippocampus. *J. Physiol.* (*London*) 334: 33

Coultrip, R., Granger, R., Lynch, G. 1992. A cortical model of winner-take-all computation via lateral inhibition. *Neural Netw.* 5: 47–54

Deadwyler, S. A., West, M., Lynch, G. 1979. Activity of dentate granule cells during learning: differentiation of perforant path input. *Brain Res.* 169: 29–43

Deadwyler, S. A., West, M., Robinson, J. H. 1986. Entorhinal and septal inputs differentially control sensory-evoked responses in rat dentate gyrus. *Science* 211: 1131–83

Desmond, J. E., Moore, J. W. 1988. Adaptive timing in networks: the conditioned response. *Biol. Cybern.* 58: 405–15

Diamond, D. M., Dunwiddie, T. V., Rose, G. M. 1988. Characteristics of hippocampal primed burst potentiation in vitro and in the awake rat. *J. Neurosci.* 8: 4079

Donegan, N. H., Gluck, M. A., Thompson, R. F. 1989. Integrating behavioral and biological models of classical conditioning. *Psychol. Learn. Motivat.* 23: 109–56

Donegan, N. H., Wagner, A. R. 1987. Conditioned dimunition and facilitation of UCR: a sometimes-opponent-process interpretation. In *Classical Conditioning II: Behavioral, Neurophysiological, and Neuro-chemical Studies in the Rabbit*, ed. I. Gormezano, W. Prokasy, R. Thompson. Hillsdale, NJ: Erlbaum Assoc.

Douglas, R. 1972. Pavlovian conditioning and the brain. In *Inhibition and Learning*, ed. R. A. Boakes, M. S. Halliday. London: Academic

Douglas, R., Pribram, K. 1966. Learning and limbic lesions. *Neuropsychologia* 4: 192–220

Dudek, S., Bear, M. 1992. Homosynaptic long-term depression in area CA1 of hippocampus and the effects of NMDA receptor blockade. *Proc. Natl. Acad. Sci. USA* 89: 4363–67

Eccles, J. C. 1977. An instruction selection theory of learning in the cerebellar cortex. *Brain Res.* 127: 327–52

Edelman, G. M., Reeke, G. N. 1982. Selective networks capable for representative transformations, limited generalizations, and associative memory. *Proc. Natl. Acad. Sci. USA* 79: 2091–95

Eichenbaum, H., Buckingham, J. 1991. Studies on hippocampal processing: experiment, theory, and model. In *Neurocomputation and Learning: Foundations of Adaptive Networks*, ed. M. Gabriel, J. Moore. Cambridge, Mass: MIT Press

Eichenbaum, H., Cohen, N. J., Otto, T., Wible, C. 1992. Memory representation in the hippocampus: functional domain and functional organization. In *Memory: Organization and Locus of Change*, ed. L. Squire, G. Lynch, N. Weinberger, J. L. McGaugh. Oxford: Oxford Univ. Press. In press

Eichenbaum, H., Kuperstein, M., Fagan, M., Nagode, J. 1987. Cue-sampling and goal approach correlates of hippocampal unit activity in rats performing in odor-discrimination task. *Neuroscience* 7: 716–32

Ellias, S., Grossberg, S. 1975. Pattern formation, contrast control and oscillations in the short term memory of shunting on-center off-surround networks. *Biol. Cybern.* 20: 69–98

Foy, M. R., Steinmetz, J. E., Thompson, R. F. 1984. Single unit analysis of cerebellum during classically conditioned eyelid response. *Soc. Neurosci. Abstr.* 10: 122

Freeman, W. 1991. The physiology of perception. *Sci. Am.* 264: 78–85

Freeman, W. 1975. *Mass Action in the Nervous System.* New York: Academic

Fujita, M. 1982. Simulation of adaptive modification of the vestibulo-ocular reflex with an adaptive filter model of the cerebellum. *Biol. Cybern.* 45: 207–14

Gambel, E., Koch, C. 1987. The dynamics of free calcium in dendritic spines in response to repetitive synaptic input. *Science* 236: 1311–15

Garrud, P., Rawlins, J. N. P., Mackintosh, N. J., Goodal, G., Cotton, M. M., Feldon, J. 1984. Successful overshadowing and blocking in hippocampectomized rats. *Behav. Brain Res.* 12: 39–53

Getchell, T., Shepherd, G. 1978. Responses of olfactory receptor cells to step pulses of odour at different concentrations in the salamander. *J. Physiol.* 282: 521–40

Gilbert, P. F. C., Thach, W. R. 1972. Purkinje cell activity during motor learning. *Brain Res.* 128: 309–28

Gluck, M. A. 1991. Stimulus generalization and representation in adaptive network models of category learning. *Psychol. Sci.* 2(1): 1–6

Gluck, M. A., Bower, G., Hee, M. 1989. A configural-cue network model of human learning. *11th Annu. Conf. Cogn. Sci. Soc.*

Gluck, M., Bower, G. 1988. Evaluating and adaptive network model of human learning. *J. Mem. Lang.* 27: 166–95

Gluck, M. A., Goren, O., Myers, C., Thompson, R. F. 1992. A higher-order recurrent network model of the cerebellar substrates of response timing in motor-reflex conditioning. *J. Cogn. Neurosci.* In press

Gluck, M. A., Myers, C. E. 1992. Hippocampal function in representation and generalization: a computational theory. *Proc. 1992 Cogn. Sci. Soc. Conf.* Hillsdale, NJ: Erlbaum Assoc.

Gluck, M. A., Reifsnider, E. S., Thompson, R. F. 1990. Adaptive signal processing and the cerebellum: models of classical conditioning and VOR adaptation. In *Neuroscience and Connectionist Theory*, ed. M. A. Gluck, D. E. Rumelhart, pp. 131–85. Hillsdale, NJ: Erlbaum Assoc.

Gluck, M. A., Thompson, R. F. 1989. A biological neural-network analysis of learning and memory. In *An Introduction to Neural and Electronic Networks*, ed. S. Zornetzer, J. Davis, C. Lau, pp. 91–107. New York: Academic

Gormezano, I., Kehoe, E. K., Marshal, B. S. 1983. Twenty years of classical conditioning research with the rabbit. *Prog. Psychobiol. Physiol. Psychol.* 10: 197–275

Granger, R., Ambros-Ingerson, J., Lynch, G. 1989. Derivation of encoding characteristics of layer II cerebral cortex. *J. Cogn. Neurosci.* 1: 61–87

Granger, R., Lynch, G. 1992. Higher olfactory processes: Perceptual learning and memory. *Curr. Biol.* 1: 209–14

Granger, R., Staubli, U., Powers, H., Otto, T., Ambros-Ingerson, J., Lynch, G. 1991. Behavioral tests of a prediction from a cortical network simulation. *Psychol. Sci.* 2: 116–18

Granger, R., Whitson, J., Larson, J., Lynch, G. 1992. Non-Hebbian properties of LTP enable high-capacity encoding of temporal sequences. *Proc. Natl. Acad. Sci. USA*. In press

Grastyan, E., Lissak, K., Madarasz, I., Donhoffer, H. 1959. Hippocampal electrical activity during the development of conditioned reflexes. *Electroencephalogr. Clin. Neurophysiol.* 11: 409–30

Gray, J. A. 1979. *The Neuropsychology of Anxiety.* Oxford: Oxford Univ. Press

Gretchell, T. V., Shepherd, G. M. 1978. Responses of olfactory receptor cells to step pulses of odour at different concentrations in the salamander. *Physiology* 282: 521–40

Grossberg, S. 1976. Adaptive pattern classification and universal recoding: Part I. *Biol. Cybern.* 23: 121–34

Grossberg, S. 1975. A neural model of attention, reinforcement, and discrimination learning. *Int. Rev. Neurobiol.* 18: 263–327

Gustafsson, B., Wigstrom, H. 1990. Long-term potentiation in the CA1 region: its induction and early temporal development. *Prog. Brain Res.* 83: 223–32

Haberly, L. 1985. Neuronal circuitry in olfactory cortex: anatomy and functional implications. *Chem. Senses.* Vol. 10

Haberly, L., Bower, J. 1989. Olfactory cortex: model circuit for study of associative memory. *Trends Neurosci.* 17: 258–64

Haberly, L. B., Price, J. H. 1977. The axonal projection patterns of the mitral and tufted cells of the olfactory bulb in the rat. *Brain Res.* 129: 152–57

Hassul, M., Daniels, P. D. 1977. Cerebellar dynamics: the mossy fiber input. *IEEE Trans. Biomed.* BME-24: 449–56

Hebb, D. 1949. *The Organization of Behavior.* New York: Wiley

Hirsh, R. 1980. The hippocampus, conditional operations and cognition. *Physiol. Psychol.* 8: 175–82

Hirsh, R. 1974. The hippocampus and contextual retrieval of information from memory: a theory. *Behav. Biol.* 12: 42–44

Hoesen, G. van, Pandya, D. 1975. Some connections of the entorhinal (area 28) and pirirhinal (area 35) cortices of the rhesus monkey. I. Temporal lobe afferents. *Brain Res.* 95: 1–24

Iriki, A., Pavlides, C., Keller, A., Asanuma, H. 1989. Long-term potentiation in the motor cortex. *Science* 245: 1385–87

Ito, M. 1989. Long term depression. *Annu. Rev. Neurosci.* 12: 85–102

Ito, M. 1987. Characterization of synaptic plasticity in the cerebellar and cerebral neocortex. In *Life Sciences Research Report 38, The Neural and Molecular Bases of Learning*, ed. J. P. Changeaux, M. Konishi, pp. 263–80

Ito, M. 1984. In *The Cerebellum and Neural Control.* New York: Raven

Ito, M. 1982. Cerebellar control of the vestibulo-flocular reflex around the flocculus hypothesis. *Annu. Rev. Neurosci.* 5: 275–96

Jung, M., Larson, J., Lynch, G. 1990. Long-term potentiation of monosynaptic EPSPs in rat piriform cortex in vitro. *Synapse* 6: 279–83

Kamin, L. J. 1969. Predictability, surprise, attention and conditioning. In *Punishment and Aversive Behavior*, ed. B. Campbell, R. Church, pp. 279–96. New York: Appleton-Century-Crofts

Kanter, E. D., Haberly, L. B. 1990. NMDA-dependent induction of long-term potentiation in afferent and association fiber systems of piriform cortex in vitro. *Brain Res.* 525: 175–79

Kauer, J. S. 1991. Contributions of topography and parallel processing to odor coding in the vertebrate olfactory pathway. *Trends Neurosci.* 14: 79–85

Kauer, J. S. 1987. Coding in the olfactory system. In *The Neurobiology of Taste and Smell*, ed. T. E. Finger, pp. 205–31. New York: Wiley

Kauer, J. S., Malenka, R., Nicoll, R. 1988. A persistent postsynaptic modification mediates long-term potentiation in the hippocampus. *Neuron* 1: 911–17

Kimble, D. P. 1968. Hippocampus and internal inhibition. *Psychol. Bull.* 70: 285–95

Knowlton, B., Thompson, R. F. 1988. Microinjections of local anesthetic into the pontine nuclei reduce the amplitude of the classically conditioned eyeblink response. *Physiol. Behav.* 43: 855–57

Komisaruk, B. R. 1970. Synchrony between limbic system theta activity and rhythmical behavior in rats. *J. Comp. Physiol. Psychol.* 70: 482–92

Larson, J., Lynch, G. 1989. Theta pattern stimulation and the induction of LTP: the sequence in which synapses are stimulated determines the degree to which they potentiate. *Brain Res.* 489: 49–58

Larson, J., Lynch, G. 1988. Role of N-Methyl-D-Aspartate receptors in the induction of synaptic potentiation by burst stimulation patterned after the hippocampal theta rhythm. *Brain Res.* 441: 111–18

Larson, J., Lynch, G. 1986. Induction of synaptic potentiation in hippocampus by patterned stimulation involves two events. *Science* 232: 985–88

le Cun, Y. 1985. Une procedure d'apprentissage pur reseau a seuil assymetrique [A procedure for learning an asymmetric threshold network]. *Proc. Cogn., Paris*, pp. 599–604

Lee, K. 1982. Sustained enhancement of evoked potentials following brief, high-frequency stimulation of the cerebral cortex in vitro. *Brain Res.* 239: 617–23

Lee, S., Weisskopf, M., Ebner, F. 1991. Horizontal long-term potentiation of responses in rat somatosensory cortex. *Brain Res.* 544: 303–10

Lisberger, S. G. 1987. The latency of pathways containing the site of motor learning in the monkey vestibulo-ocular reflex. *Science* 225: 74–76

Lynch, G., Granger, R. 1992. Variations in synaptic plasticity and types of memory in cortico-hippocampal networks. *J. Cogn. Neurosci.* 4: 189–99

Lynch, G., Granger, R. 1991. Serial steps in memory processing: possible clues from studies of plasticity in the olfactory-hippocampal circuit. In *Olfaction As a Model System for Computational Neuroscience*, ed. H. Eichenbaum, J. Davis. Cambridge, Mass: MIT Press

Lynch, G., Granger, R. 1989. Simulation and analysis of a simple cortical network. *Psychol. Learn. Motivat.* 23: 204–41

Mackintosh, N. J. 1975. A theory of attention: variations in the associability of stimuli with reinforcement. *Psychol. Rev.* 82: 276–98

Macrides, F. 1975. Temporal relations between hippocampal slow waves and exploratory sniffing in hamsters. *Behav. Biol.* 14: 295–308

Macrides, F., Eichenbaum, H. B., Forbes, W. B. 1982. Temporal relationship between sniffing and the limbic (theta) rhythm during odor discrimination reversal learning. *J. Neurosci.* 2: 1705–17

Malsberg, C. von der. 1973. Self-organizing of orientation sensitive cells in the striate cortex. *Kybernetik* 14: 85–100

Marr, D. 1969. A theory of cerebellar cortex. *J. Physiol.* 202: 437–70

McCollum, J., Larson, J., Otto, T., Schottler, F., Granger, R., Lynch, G. 1991. Short-latency single-unit processing in olfactory cortex. *J. Cogn. Neurosci.* 3: 293–99

McCormick, D. A., Steinmetz, J. E., Thompson, R. F. 1985. Lesions of the inferior olivary complex cause extinction of the classically conditioned eyeblink response. *Brain Res.* 359: 120–30

McNaughton, B. L. 1989. Neuronal mechanisms for spatial computation and information storage. In *Neural Connections, Mental Computations*, ed. L. Nadel, L. Cover, P. Culicover, R. M. Harnish, pp. 285–350. Cambridge, Mass: MIT Press

McNaughton, B. L., Chen, L. I., Markus, E. J. 1991. Dead reckoning, landmark learning, and the sense of direction: a neurophysiological and computational hypothesis. *J. Cogn. Neurosci.* 3(2): 190–202

McNaughton, B. L., Douglas, R. M., Goddard. 1978. Synaptic enhancement in fascia dentata: Co-operativity among coactive afferents. *Brain Res.* 157: 277–93

McNaughton, B. L., Nadel, L. 1990. Hebb-Marr networks and the neurobiological representation of action in space. In *Neuroscience and Connectionist Theory*, ed. M. A. Gluck, D. E. Rumelhart. Hillsdale, NJ: Erlbaum Assoc.

Miller, K., Keller, J., Stryker, M. 1990. Ocular dominance column development: analysis and simulation. *Science* 245: 605–15

Mishkin, M. 1982. *Philos. Trans. R. Soc. London (Biol.)* 298: 85–92

Mishkin, M. 1978. Memory in monkeys severely impaired by combined but not separate removal of the amygdala and hippocampus. *Nature* 273: 297–98

Moore, J. A. 1979. Brain processes and conditioning. In *Mechanisms of Learning and Behavior*, ed. A. Dickinson, R. A. Boakes. Hillsdale, NJ: Erlbaum Assoc.

Moore, J. W., Blazis, D. E. J. 1989. Simulation of a classically conditioned response: a cerebellar neural network implementation of the Sutton-Barto-Desmond model. In *Neural Models of Plasticity: Experimental and Theoretical Approaches*, ed. J.

H. Byrne, W. O. Berry. New York: Academic

Moore, J. W., Stickney, K. J. 1982. Goal tracking in attentional-associative networks: spatial learning and the hippocampus. *Physiol. Psychol.* 10: 202–8

Mori, K. 1987. Membrane and synaptic properties of identified neurons in the olfactory bulb. *Prog. Neurobiol.* 29: 275–320

Morris, R. G. M., Anderson, E., Lynch, G., Baudry, M. 1986. Selective impairment of learning and blockade of long-term potentiation of NMDA receptor antagonist, AP-5. *Nature* 319: 774–76

Morris, R. G. M., Garrud, P., Rawlins, J. N. P., O'Keefe, J. 1982. Place navigation impaired in rats with hippocampal lesions. *Nature* 297: 681–83

Moulton, D. G. 1976. Spatial patterning of response to odors in the peripheral olfactory system. *Physiol. Rev.* 56: 578–93

Moyer, J. R., Deyo, R. A., Disterhoft, J. F. 1990. Hippocampectomy disrupts trace eye-blink conditioning in rabbits. *Behav. Neurosci.* 104(2): 243–52

Muller, D., Joly, M., Lynch, G. 1988. Contributions of quisqualate and NMDA receptors to the induction and expression of LTP. *Science* 242: 1694–97

Nadel, L., Willner, J. 1989. Context and conditioning: A place for space. *Physiol. Psychol.* 8: 218–28

Nicoll, R. A. 1969. Inhibitory mechanisms in the rabbit olfactory bulb: dendrodendritic mechanisms. *Brain Res.* 14: 157–72

O'Keefe, J. 1991. A computational theory of the hippocampal cognitive map. *Prog. Brain Res.* 83: 301–12

O'Keefe, J., Nadel, L. 1978. *The Hippocampus As a Cognitive Map.* Oxford: Clarendon Univ. Press

O'Keefe, J., Nadel, L., Keightly, S., Kill, D. 1975. Fornix lesions selectively abolish place learning in the rat. *Exp. Neurol.* 48: 152–66

Olton, D. S., Becker, J. T., Handlemann, G. E. 1979. Hippocampus, space, and memory. *Brain Behav. Sci.* 2: 313–65

Otto, T., Eichenbaum, H. 1992. Toward a comprehensive account of hippocampal function: studies of olfactory learning permit an integration of data across multiple levels of neurobiological analysis. In *Neuropsychology of Memory*, ed. N. Butters. New York: Guilford. 2nd ed.

Otto, T., Eichenbaum, H., Wiener, S. I., Wible, C. G. 1991. Learning-related patterns of CA1 spike trains parallel stimulation parameters optimal for inducing hippocampal long-term potentiation. *Hippocampus* 1: 181–92

Parker, D. 1986. A comparison of algorithms

for neuron-like cells. In *Proc. Neural Netw. Comput. Conf.*, ed. J. Denker. New York: Am. Inst. Phys.

Pavlides, C., Greenstein, Y. J., Grudman, M., Winson, J. 1988. Long-term potentiation in the dentate gyrus is induced preferentially on the positive phase of the theta rhythm. *Brain Res.* 439: 383–87

Pearce, J. M., Hall, G. 1980. A model for Pavlovian learning: variations in the effectiveness of conditioned but not of unconditioned stimuli. *Psychol. Rev.* 87: 532–52

Pellionisz, A., Llinas, R. 1982. Space-time representation in the brain. *Neuroscience* 7: 2949–70

Port, R., Patterson, M. 1984. Fimbrial lesions and sensory preconditioning. *Behav. Neurosci.* 98: 584–89

Price, J. L. 1973. An autoradiographic study of complementary laminar patterns of termination of afferent fibers to the olfactory cortex. *J. Comp. Neurol.* 150: 87–108

Rall, W. 1970. Cable properties of dendrites and effects of synaptic location. In *Excitatory Synaptic Mechanisms*, ed. P. Anderson, J. Jansen, pp. 175–87. Oslo: Scand. Univ. Books

Rall, W. 1964. Theoretical significance of dendritic trees for neuronal input-output relations. In *Neural Theory and Modeling*, ed. R. F. Reiss. Stanford, Calif: Stanford Univ. Press

Rall, W., Shepherd, G., Reese, T., Brightman, M. 1966. Dendrodendritic synaptic pathway for inhibition in the olfactory bulb. *Exp. Neurol.* 14: 44–56

Ranck, J. B. 1984. Head direction cells in the deep cell layer of dorsal presubiculum in freely moving rats. *Soc. Neurosci. Abstr.* 19: 599

Rawlins, J. N. P. 1985. Associations across time: the hippocampus as a temporary memory store. *Behav. Brain Sci.* 8: 479–96

Rescorla, R., Wagner, A. 1972. A theory of Pavlovian conditioning: variations in the effectiveness of reinforcement and non-reinforcement. In *Classical Conditioning II: Current Research and Theory*, ed. A. Black, W. Prokasy. New York: Appleton-Century-Crofts

Robinson, D. A. 1977. Vestibular and optokinetic symbiosis: an example of explaining by modelling. *Dev. Neurosci.* 1: 49–58

Robinson, D. A. 1976. Adaptive gain control of vestibulo-ocular reflex by the cerebellum. *J. Neurophysiol.* 36: 954–69

Robinson, G., Crooks, G., Shinkman, P., Gallagher, M. 1989. Behavioral effects of MK-801 mimic defects associated with hippocampal damage. *Psychobiology* 17: 156–64

Rolls, E. 1990. Theoretical and neuro-physiological analysis of the functions of the primate hippocampus in memory. *Cold Spring Harbor Symp. Quant. Biol.* 55: 995–1006

Roman, F., Staubli, U., Lynch, G. 1987. Evidence for synaptic potentiation in a cortical network during learning. *Brain Res.* 418: 221–26

Rosenblatt, F. 1962. *Principles of Neurodynamics: Perceptrons and the Theory of Brain Mechanisms.* Washington, DC: Spartan

Rudy, J., Sutherland, R. 1989. The hippocampal formation is necessary for rats to learn and remember configural discriminations. *Behav. Brain Res.* 84: 97–109

Rumelhart, D. E., Hinton, G., Williams, R. 1986. *Nature* 323: 533–36

Rumelhart, D. E., Zipser, D. 1985. Feature discovery by competitive learning. *Cogn. Sci.* 9: 75–112

Schmajuk, N. A., DiCarlo, J. J. 1992. Stimulus configuration, classical conditioning and hippocampal function. *Psychol. Rev.* 99(2): 268–305

Schmajuk, N. A., DiCarlo, J. J. 1991. Neural dynamics of hippocampal modulation of classical conditioning. In *Neural Network Models of Conditioning and Action*, ed. N. Commons, S. Grossberg, J. E. R. Staddon. Hillsdale, NJ: Erlbaum Assoc.

Schmajuk, N. A., Moore, J. W. 1985. Realtime attentional models for classical conditioning and the hippocampus. *Physiol. Psychol.* 13: 278–90

Schmaltz, L. W., Theios, J. 1972. Acquisition and extinction of a classically conditioned response in hippocampectomized rabbits (*Oryctolagus cuniculus*). *J. Comp. Physiol. Psychol.* 79: 328–33

Schneider, S., Scott, J. 1983. Orthodromic response properties of rat olfactory bulb mitral and tufted cells correlate with their projection patterns. *J. Neurophysiol.* 50: 358–78

Scoville, W. B., Millner, B. 1957. Loss of recent memory after bilateral hippocampal ᴉesions. *J. Neurol. Neurosurg. Psychiatry* 20: 11–21

Sejnowki, T. J. 1988. Neural populations revealed. *Nature* 332: 308

Shepherd, G. M. 1990. The significance of real neuron architectures for neural network simulations. In *Computational Neuroscience*, ed. E. Schwartz, pp. 82–96. Cambridge, Mass: MIT Press

Shepherd, G. M. 1979. *The Synaptic Organization of the Brain.* New York: Oxford Univ. Press

Shepherd, G. M. 1972. Synaptic organization of the mammalian olfactory bulb. *Physiol. Rev.* 52: 864–917

Shepherd, G. M., Brayton, R. K. 1987. Logic operations are properties of computer-simulated interactions between excitable dendritic spines. *Neuroscience* 21: 151–65

Slotnick, B. M., Katz, H. M. 1974. Olfactory learning-set formation in rats. *Science* 185: 796–98

Solomon, P. R., Moore, J. W. 1975. Latent inhibition and stimulus generalization of the classically conditioned nictitating membrane response in rabbits (*Oryctolagus cuniculus*) following dorsal hippocampal ablation. *J. Comp. Physiol. Psychol.* 202: 1192–1203

Solomon, P. R. 1979. Temporal versus spatial information processing theories of hippocampal function. *Psychol. Bull.* 86: 1272–79

Solomon, P. R. 1977. Role of the hippocampus in blocking and conditioned inhibition of rabbit's nictitating membrane response. *J. Comp. Physiol. Psychol.* 91: 407–17

Squire, L. R. 1987. *Memory and Brain.* New York: Oxford Univ. Press

Squire, L. R. 1982. The neuropsychology of human memory. *Annu. Rev. Neurosci.* 5: 241–73

Squire, L. R., Haist, F., Shimamura, A. P. 1989. The neurology of memory: qualitative assessment of retrograde amnesia in two groups of amnesic patients. *J. Neurosci.* 9: 828–39

Squire, L. R., Zola-Morgan, S. 1983. The neurology of memory: the case for correspondence between the findings for man and non-human primate. In *The Physiological Basis of Memory*, ed. J. A. Deutsch. New York: Academic

Standing, L. 1973. Learning 10,000 pictures. *Q. J. Exp. Psychol.* 25: 207–22

Stanton, P. K., Sejnowki, T. J. 1989. Associative long-term depression in the hippocampus induced by Hebbian covariance. *Nature* 339: 215–18

Staubli, U., Fraser, D., Faraday, R., Lynch, G. 1987. Olfaction and the "data" memory system in rats. *Behav. Neurosci.* 101(6): 757–65

Staubli, U., Larson, J., Lynch, G. 1990. Mossy fiber potentiation and long-term potentiation involve different expression mechanisms. *Synapse* 5: 333–35

Staubli, U., Lynch, G. 1987. Stable hippocampal long-term potentiation elicited by "theta" pattern stimulation. *Brain Res.* 435: 227–34

Staubli, U., Thibault, O., DiLorenzo, M., Lynch, G. 1989. Antagonism of NMDA receptors impairs acquisition but not retention of olfactory memory. *Behav. Neurosci.* 103(1): 54–60

Steinmetz, J. E., Rosen, D. J., Chapman,

P. F., Thompson, R. F. 1986. Classical conditioning of the rabbit eyelid response with a mossy fiber stimulation CS. I. Pontine nuclei and middle cerebellar peduncle stimulation. *Behav. Neurosci.* 100: 871–80

Stevens, C. F. 1990. A depression long awaited. *Nature* 347: 16

Sutherland, R., Rudy, J. 1989. Configural association theory: the role of the hippocampal formation in learning, memory and amnesia. *Psychobiology* 17: 129–44

Sutton, R. S., Barto, A. G. 1981. Toward a modern theory of adaptive networks: expectation and prediction. *Psychol. Rev.* 88: 135–70

Tanzi, E. 1893. I fatti e le induzioni nell' odierna istologia del sistema nervoso. *Rev. Sperim. Frenatria Med. Legal.* 19: 419–72

Thompson, R. F. 1988. The neural basis of associative learning of discrete behavioral responses. *Trends Neurosci.* 11(4): 152–55

Thompson, R. F. 1986. The neurobiology of learning and memory. *Science* 233: 941–47

Thompson, R. F., Gluck, M. A. 1990. Brain substrates of basic associative learning and memory. In *Cognitive Neuroscience*, ed. H. J. Weingartner, R. F. Lister. New York: Oxford Univ. Press

Traub, R., Miles, R. 1991. *Neuronal Networks of the Hippocampus.* Cambridge: Cambridge Univ. Press

Vinogradova, O. S. 1975. Functional organization of the limbic system in the process of registration of information: facts and hypotheses. In *The Hippocampus*, ed. K. H. Pribram. New York: Plenum

Werbos, P. 1974. *Beyond Regression: New Tools for Prediction and Analysis in the Behavioral Sciences.* Cambridge: Harvard Univ. Press

Wagner, A. R., Rescorla, R. A. 1972. Inhibition in Pavlovian conditioning: applications of a theory. In *Inhibition and Learning*, ed. R. A. Boakes, S. Halliday, pp. 301–36. New York: Academic

Wickelgren, W. A. 1979. Chunking and consolidation: a theoretical synthesis of semantic networks, configuring in condition-ing, S-R versus cognitive learning, normal forgetting, the amnesic syndrome and the hippocampal arousal system. *Psychol. Rev.* 86: 44–60

Widrow, B., Hoff, M. 1960. Adaptive switching circuits. *Inst. Radio Eng. West. Electron. Show Conv. Conv. Rec.* 4: 96–194

Winocur, G., Rawlins, J., Gray, A. J. 1987. The hippocampus and conditioning to contextual cues. *Behav. Neurosci.* 101: 617–25

Woolf, T. B., Shepherd, G. M., Greer, C. A. 1991. Serial reconstructions of granule cell spines in the mammalian olfactory bulb. *Synapse* 7: 181–92

Wyss, J. 1981. Autoradiographic study of the efferent connections of entorhinal cortex in the rat. *J. Comp. Neurol.* 199: 495–512

Yeo, C. H., Hardiman, M. J., Glickstein, M. 1985. Classical conditioning of the nictitating membrane response of the rabbit: I. Lesions of the cerebellar nuclei. *Exp. Brain Res.* 60: 87–98

Zalutsky, R. A., Nicoll, R. A. 1990. Comparison of two forms of long-term potentiation in single hippocampal neurons. *Science* 248: 1619–24

Zimmer, J. 1971. Ipsilateral afferents to the commissural zone of the fascia dentata demonstrated in decommissurated rats by silver impregnation. *J. Comp. Neurol.* 23: 393–416

Zola-Morgan, S., Squire, L. R. 1993. Neuroanatomy of memory. *Annu. Rev. Neurosci.* 16: 547–63

Zola-Morgan, S., Squire, L. R. 1990. The primate hippocampal formation: evidence for a time-limited role in memory storage. *Science* 250: 288–90

Zola-Morgan, S., Squire, L. R. 1983. The neurology of memory: the case for correspondence between the findings for man and non-human primate. In *The Physiological Basis of Memory*, ed. J. A. Deutsch. New York: Academic

Zola-Morgan, S., Squire, L. R., Mishkin, M. 1982. The neuroanatomy of amnesia: amygdala-hippocampus versus temporal stem. *Science* 218: 1337–39

Annu. Rev. Neurosci. 1993. 16:707–32

PATTERNING THE BRAIN OF THE ZEBRAFISH EMBRYO

Charles B. Kimmel

Institute of Neuroscience, University of Oregon, Eugene, Oregon 97403

KEY WORDS: neuromere, axonal pathfinding, primary neuron, gene expression, pioneer

INTRODUCTION

This review discusses how brain morphogenesis is related to the arrangements of the earliest neurons and their axons in a simple vertebrate, the zebrafish *Brachydanio rerio*. Descriptive work establishes that the early neurons normally arise at stereotyped positions, which are well defined with respect to the overall brain shape. These cells grow their axons without errors or retractions and pioneer pathways, which occur at invariant positions, in a way clearly related to brain regionalization. Experimental analyses reveal that position is, in fact, a key determinant of patterning of both soma and axonal organization. Other possible determinants, such as cell lineage and temporal aspects of development, seem to play less important roles (reviewed by Kimmel & Westerfield 1990). We want to know how position is encoded in the brain and, eventually, we want to understand its molecular genetic basis. Recent analyses reviewed here provide some clues about this issue in two ways: First, understanding the early organization of the system as revealed in morphological studies, and the disorganization that results from discrete perturbation, sheds light on the underlying patterning code, i.e. the "prepattern." Second, initial analyses of early expression patterns of putative regulatory genes have recently been made, which brings analysis of the patterning code to a new level that is clearly closer to revealing the prepattern itself.

For the most part, we only have to be concerned with an interval of development that is about 15 hours long. This interval begins in the late gastrula, about 9 or 10 hours (h) postfertilization, when the brain

707

0147–006X/93/0301–0707$02.00

primordium is still undeveloped. At its end, at 24 h, brain morphogenesis is advanced, and the earliest neuronal groups are interconnected by axons. I call these particular cells primary neurons, defined operationally to mean those cells whose axons are growing by 24 h (Grunwald et al 1988). A definition so simple and arbitrary clearly cannot be correct in every case (see discussions in Kimmel & Westerfield 1990; Wilson & Easter 1992), but it arises naturally from developmental, functional, and some genetic data; the primary neurons seem to be in several ways distinctive from at least some of their followers. They all make long axons (in contrast to short-axon local interneurons that arise later), and, in cases where the same cells can be identified in the adult, the primary neurons are the largest. Limiting our scope to the primary neurons facilitates our analyses, for there are far fewer such cells, and they are more simply arranged than occurs after only one more day of development.

A great deal of work has also recently been done on neuronal development in the spinal cord; however, this is outside the scope of this review (see Eisen 1991a,b; Kuwada et al 1990a,b).

NEUROMERES

In all vertebrates, the neural tube becomes progressively subdivided by constrictions that appear along its length. For example, as described classically, the human brain is first represented by a single "vesicle," then three, and then five. Of course, the regions thus formed are not completely separated one from another as a series of isolated cavities, as may be implied by the term vesicle. A more appropriate term is "neuromere," which means an identifiable swollen region along the wall of the neural anlage that is separated from adjacent ones by constrictions of the wall, the neuromere boundaries, or borders. Sometimes, the term neuromere is used in a more restricted sense (e.g. see Vaage 1969).

Interestingly, even though investigators who study early morphogenesis seem to agree that all vertebrates subdivide their brains similarly, the investigators sometimes disagree substantially about the details of the subdivisions in any single vertebrate (for example in the chick, as reviewed in Puelles et al 1987). The problem clearly merits continued careful consideration and study; swellings and constrictions might be missed or they might arise as artifacts during preparation.

Neuromeres are not evident in the early anlage of the zebrafish brain when one can first recognize it, after about a half-day of development (Figure 1A). During the next six hours, about ten neuromeres form (Figure 1B). The three most rostral neuromeres are substantially larger and become more prominently sculptured than the rest. They correspond to the defi-

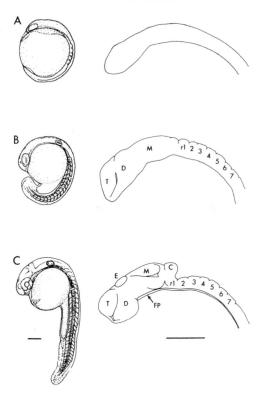

Figure 1 Neuromeres partition the brain during the first embryonic day. Left side views (rostral to the left) of whole embryos (*left*) and their brains (*right*).

(*A*) Subdivisions are not recognized at 12 h, shortly after the neural anlage becomes distinctive. As shown in the whole embryo drawing, the eye rudiment already can be recognized beside the forebrain rudiment. About six somites have formed.

(*B*) About ten neuromeres are present at 18 h, representing the primordia of the telencephalon (T), diencephalon (D), midbrain (M), and seven hindbrain rhombomeres (r1–r7). The ear vesicle lies adjacent to r5. The tail is growing out, and about 18 somites are present. Spontaneous contractions are just beginning in the trunk.

(*C*) Brain morphogenesis is advanced at 24 h. The epiphysis (E) and cerebellum (C) are distinctive. The ventricles are present, and additional sculpturing has delineated the hypothalamus in the ventral diencephalon (D), the midbrain tectum dorsally (M), and tegmentum ventrally. The floor plate (FP), in the ventral midline, extends in an uninterrupted fashion from the rostral midbrain through the hindbrain and spinal cord. Many cell types have arisen by this stage in the embryo, which has formed its full complement of about 30 somites. The eye has developed a lens, and the olfactory placode overlies the telencephalon.

Nomenclature: Generally, this review uses anatomical names as established by their first usage in the zebrafish literature, sometimes as revised by later papers from the same groups. An exception is the designation of rhombomeres, called here r1–r7 to facilitate comparisons with terrestrial vertebrates. These designations correspond, respectively, to the names Ro1–Ro3, Mi1–Mi3, and Ca1 (see Hanneman et al 1988). Scale bars: 200 μm.

nitive midbrain and to the two divisions of the forebrain, the diencephalon and telencephalon. The remaining seven neuromeres subdivide the major part of the hindbrain (Hanneman et al 1988) and are often termed rhombomeres. We have not recognized neuromeres that would subdivide the caudal-most part of the hindbrain (discussed further in Kimmel et al 1991b). Nor do we yet have critical information about the order of appearance of the neuromeres, which is complicated in the chick (Lumsden 1990; Puelles et al 1987) and probably in the fish (C. B. Kimmel 1992, unpublished findings). This early pattern, however it forms, does not persist for very long. For example, during the next two hours, the primordium of the epiphysis appears as a small, well-delineated swelling in the roof of the diencephalon, and the cerebellar primordium forms in the dorso-rostral hindbrain (Figure 1C). Early cerebellar morphogenesis is quite interesting, for a very substantial reorganization, and perhaps a local expansion of tissue occurs within a very few hours, near the end of the first day of development. It could involve only the rostral part of the first rhombomere, but further analysis is required to establish if this is indeed the case.

NEUROMERE BOUNDARY FORMATION

Hatta et al (1991a) characterized the region in which the boundary that separates the midbrain and hindbrain forms, some hours before the boundary is apparent, because, as in other vertebrates, the cells in this region specifically express homeobox-containing genes of the "engrailed" class. The presence of these nuclear and presumably gene-regulatory proteins, called Eng proteins, is detected in an undistinguished region of the neural anlage. Then, at about 14 h, this region forms a slight but recognizable dorsal swelling, bounded rostrally and caudally by shallow constrictions. Neither constriction seems to coincide with the definitive hindbrain-midbrain boundary that appears only later; rather, these early boundaries may be transient. The definitive hindbrain-midbrain boundary then forms in the middle of the Eng-expressing region (Figure 2). The suggestion from this analysis is that an early neuromere transiently arises to produce a boundary between definitive neuromeres. Hatta et al (1991a) speculate that the Eng homeoprotein expression is a critical early step in this process. Clearly, as illustrated by this analysis, neuromere boundary formation can be a dynamic and highly regulated process.

REGIONALIZATION AND DOMAINS OF EXPRESSION OF REGULATORY GENES

Coincidence between where specific regulatory genes are expressed early in development and where neuromeres appear may turn out to be common

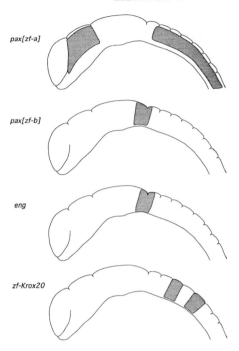

Figure 2 Expression of putative regulatory genes subdivide the neural primordium in distinctive patterns that are clearly related to the neuromeres. Left side views at 18 h, as in Figure 1*B*. The drawings represent my interpretation of the patterns, and not necessarily those of the investigators who did the work. *pax[zf-a]* expression is from Krauss et al (1991a,c) and from Püschel et al (1992), *pax[zf-b]* is from Krauss et al (1991b) and A. Püschel (unpublished), *eng* is from Hatta et al (1991a), and *zf-Krox-20* is from T. Jowett and E. Oxtoby (in preparation).

for many regulatory genes. This is a very exciting prospect, for through this approach we begin to address critically how genes might control early patterning in the brain. Homeobox genes may directly regulate expression of molecules that are crucially important for cell adhesion and shape (Jones et al 1992a,b). When we can combine such observations with perturbation studies, such as mutational analyses, we will begin to learn how genes in fact control patterning.

The first example in vertebrates that clearly relates early gene expression to later neuromere formation was discovered in the mouse embryo by Wilkinson et al (1989). They showed that the expression of a zinc-finger gene, *Krox-20*, which functions as a transcriptional regulator, begins very early in development, eventually focusing into two discrete patches in the neuroepithelium of the hindbrain primordium. These patches later develop as rhombomeres r3 and r5. T. Jowett and E. Oxtoby (in preparation) have now discovered the homologous gene in zebrafish and have found that it is expressed in the fish in the same rhombomere precursor regions as in the mouse (Figure 2). Expression persists through the time when the rhombomeres themselves appear, and the expression borders are sharply coincident with the rhombomere boundaries. The conclusion seems inescapable that, as in the mouse (Wilkinson et al 1989), this gene is part of a regulatory network that establishes the pattern of sub-division of the early hindbrain primordium. One would now also suspect that the regulatory network itself must be highly conserved among all vertebrates.

In the mouse, and probably in the zebrafish, *Krox-20* expression does not include all of the cells of r3 and r5. Specifically, it is absent in the ventral median region, or floor plate, of these rhombomeres (see Figure 1*C*). A more dramatic example of a pattern that is restricted dorsoventrally is the zebrafish paired-box gene *pax[zf-a]*, studied by Krauss et al (1991a,c) and Püschel et al (1992). Expression of this gene begins, at about 9 or 10 h, in a domain that includes the forebrain and optic vesicles. Shortly thereafter, by 14 h, expression begins separately in the hindbrain; thus, there are two domains of *pax[zf-a]* expression along the neuraxis (Figure 2). Especially in the rostral domain, expression is limited to dorsolateral cells and is absent in the brain roof and in a broad expanse of ventral cells. Rostrally, expression may extend into the telencephalon (Püschel et al 1992), and the caudal limit of expression is at or very near the diencephalon-midbrain neuromere boundary. The other expression do-main of *pax[zf-a]* extends along a large expanse of the hindbrain and is specifically absent in most of the first rhombomere.

Krauss et al (1991a,b) have shown that another *pax* gene, *pax[zf-b]*, is expressed in spatially restricted ways in the optic stalk, the eye primordium, the midbrain, and in certain neurons within the hindbrain and spinal cord. Expression in the midbrain is particularly interesting. Within this domain, expression again excludes the floor plate. The caudal border is near the midbrain-hindbrain neuromere boundary, perhaps just crossing this boundary in the fashion more clearly observed for Eng protein expression. The rostral border of *pax[zf-b]* expression is not close to any obvious morphological boundary. It lies well within the midbrain neuromere, subdividing the neuromere's territory into a larger rostral and a smaller caudal part (Figure 2). Later, this border coincides with the caudal end of the tectum.

The patterns considered above all bear on the theme of early regional patterning of the brain, yet they all provide distinctive variations on this theme. This richness of detail in the expression patterns are especially interesting in light of these relationships to the neuromeres.

PRIMARY NEURONS ARISE IN THE NEUROMERE CENTERS

If neuromeres and the genes that regulate their development are important for neuronal patterning, then one would expect that the earliest neurons would have a sensible relationship with the neuromeres. This, in fact, is what we have observed. Along the whole rostrocaudal extent of the neural tube, the first neurons appear in discrete, small clusters, in contrast to the continuous columns of neurons that are present only shortly later along the neuraxis (Chitnis & Kuwada 1990; Hanneman et al 1988; Hanneman & Westerfield 1989; Ross et al 1992; Wilson et al 1990). The early clusters are present bilaterally, and in the brain they lie near the center of each neuromere (Figure 3). The neurons develop within the outermost parts of the ventrolateral walls of the brain, that is, in the basal plates. This generalization probably does not apply to the cluster in the telencephalon (Ross et al 1992), as I discuss later.

The best marker for identifying the primary neurons at early stages of differentiation is the enzyme acetylcholinesterase (Hanneman & Westerfield 1989; Wilson et al 1990). Acetylcholinesterase seems to appear in the young neurons specifically, irrespective of their functional class (Hanneman & Westerfield 1989). As yet, no other marker has been found (e.g. see Chitnis & Kuwada 1990; Hanneman et al 1988) that labels more or different early neurons than those expressing this enzyme. The converse is not true, e.g. expression of the HNK-1 antigen (Metcalfe et al 1990; Wilson et al 1990) begins as early as 14 h and shortly becomes widespread among primary neurons. Yet, not all of them express this cell-surface carbo-

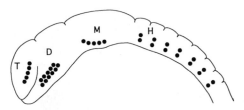

Figure 3   Early neurons arise in single clusters in the center of each neuromere. Cartoon at 18 h, as in Figure 1*B*. Abbreviations are as in Figure 1. Compiled from data in Hanneman et al (1988), Chitnis & Kuwada (1990), and Ross et al (1992).

hydrate, and those that do express it at different levels (Metcalfe et al 1990).

As determined by the acetylcholinesterase marker, the numbers of neurons that initially form each cluster vary according to the brain region and correspond approximately to the sizes of the neuromers themselves. Thus, at about 16 h, the largest clusters are in the forebrain and midbrain (Ross et al 1992); the diencephalic cluster contains about 11 neurons and the telencephalic and midbrain clusters each contain about five. In contrast, only about two neurons develop in each rhombomeric cluster (Hanneman et al 1988). We know that in several of the rhombomeres, the earliest neurons are specific identified reticulospinal neurons: For example, two bilateral pairs of cells always appear to develop first in the center of the fourth rhombomere. On each side, the two cells are immediate neighbors, one of them lying dorsolateral to the other. The lateral cell is the young Mauthner neuron, the medial one is MiM1 (Hanneman et al 1988; Mendelson 1985).

Not all of the primary neurons appear simultaneously. Although the exact temporal order in which neighboring neurons are generated or grow axons is not always critical for their normal development (Mendelson & Kimmel 1986), in some cases it is invariant. For example, the Mauthner cell is consistently generated about three hours before MiM1, and it grows its axon about one hour earlier than MiM1 (Mendelson 1985). Appearance of new neurons during the first day occurs by expansion of the early clusters, as well as by formation of new clusters (Hanneman et al 1988; Wilson et al 1990). Thus, at about the stage at which the epiphysis can be recognized (ca. 20 h), one young neuron, or a very small cluster of young neurons, can be labeled within it (Chitnis & Kuwada 1990; Ross et al 1992; Wilson & Easter 1991a,b). Addition to the early clusters occurs very rapidly; for example, Ross et al (1992) find that the number of neurons in the diencephalic group increases nearly 15-fold between 16 and 24 h.

Both types of addition can be seen in the rhombomeres, where the pattern of neuronal labeling initially restricted to the rhombomere centers gives way, by 24 h, to a pattern of large and small clusters that alternate along the neuraxis (Hanneman et al 1988). The large clusters occupy the central region of the rhombomeres, expanded from the earlier clusters. The small clusters occupy the rhombomere borders and represent new ones. The two zones now present, i.e. the rhombomere center and border regions, develop several distinctive features that suggest their development is separately regulated (Trevarrow et al 1990).

Neurons have not been individually identified in the forebrain and midbrain, as they have been in the hindbrain. However, it seems clear that, as for the Mauthner-MiM1 example in the hindbrain, wherever a cluster is located in the brain, its expansion during the first day is not always simply adding more of the earliest kinds of neurons. To illustrate this point, Metcalfe et al (1990) found that the first neurons to grow axons in the midbrain lie caudally in its prominent basal plate cluster. These neurons project axons caudally, thus forming a major pathway called the medial longitudinal fascicle (MLF). Then, young neurons located rostrally in the same cluster grow axons dorsally to form the posterior commissure (PC). The two neuronal populations later separate into the nucleus MLF and the ventral nucleus PC, respectively. I conclude, therefore, that the addition of new neurons to an older cluster and to a newly formed cluster may be fundamentally similar: Both types of addition increase the complexity of the developing brain.

A SIMPLE SCAFFOLD OF AXONAL BUNDLES

The earliest axons grow between the clusters of cells and form bundles that connect the clusters to one another or connect the brain and spinal cord or the CNS and periphery. Wherever axons grow in the early brain, the pattern is discrete, precise, and stereotyped. Diffuse or incorrect projections that would later be withdrawn are simply not observed (Chitnis & Kuwada 1990; Mendelson 1986; Wilson et al 1990).

The major axonal bundles present by 24 h are shown in Figure 4. By this stage, there is a single prominent longitudinal system of collected bundles that appears continuous along the whole length of the ventral part

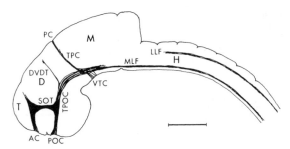

Figure 4 A simple scaffold of axonal bundles arises during the first embryonic day. Left side view, as in Figure 1. Abbreviations: AC, anterior commissure; D, diencephalon; DVDT, dorsoventral diencephalic tract; H, hindbrain; LLF, lateral longitudinal fascicle; M, midbrain; MLF, medial longitudinal fascicle; PC, posterior commissure; POC, postoptic commissure; SOT, supraoptic tract; T, telencephalon; TPC, tract of the posterior commissure; TPOC, tract of the postoptic commissure; VTC, ventral tegmental commissure. Compiled from Wilson et al 1990 and Chitnis & Kuwada 1990. Scale bar: 100 μm.

of the brain. Rostrally, it wraps around the diencephalon as the postoptic commissure (POC) and courses longitudinally through the diencephalon as several bundles, called collectively the tract of the postoptic commissure (TPOC). In the midbrain, some of the TPOC bundles course just dorsal to the MLF, and others may join the MLF (Chitnis & Kuwada 1990). The MLF continues caudally into and through the hindbrain, and then into the spinal cord.

Other axonal bundles join this system. The most rostral one is the supraoptic tract (SOT) from the telencephalon, the region that also forms the anterior commissure. The dorsoventral diencephalic tract (DVDT) joins the TPOC midway through the diencephalon. The tract of the posterior commissure (TPC) joins the TPOC near the diencephalic-midbrain border, and a distributed system of fibers crosses the midbrain floor as the ventral tegmental commissure. Another prominent longitudinal bundle is present in the hindbrain, the lateral longitudinal fascicle (LLF), which does not massively connect to the MLF (but see below for individual axons that project between these bundles).

The LLF arises from axons from primary sensory neurons: Caudally growing axons from neurons in the trigeminal ganglion meet rostrally growing axons from Rohon-Beard cells located in the spinal cord (Metcalfe et al 1990). The other bundles are formed locally by axons growing from neurons in all of the clusters considered above. Some of those in the midbrain and forebrain are illustrated in Figure 5. Contrast this pattern with the growth of a tree. In the case of a real tree, the trunk develops

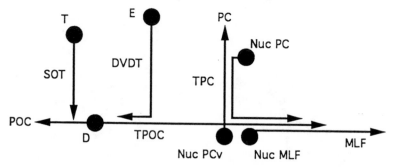

Figure 5 Schematic pattern of early axonal pathfinding in the midbrain and forebrain. The tracts are named in Figure 4. They originate from clusters, shown as single cells, and abbreviated as follows: D, diencephalon basal plate; E, epiphysis; Nuc MLF, nucleus of the MLF; Nuc PC, nucleus of the posterior commissure; Nuc PCv, ventral nucleus of the posterior commissure; T, telencephalon. Compiled from Chitnis & Kuwada (1991) and Wilson & Easter (1991b).

first, it gives rise to the branches, and, in turn, the twigs at the ends of the branches. In the brain we see exactly the opposite pattern of development: The "twigs," i.e. the individual neurons in specific clusters, each grow their axons in specific directions such that the axons meet at stereotyped locations in the brain and collect at these intersections to form the larger "branches," the early axonal bundles, and eventually the "tree-trunk," the ventrolateral longitudinal system in our analogy.

STEREOTYPED AXONAL PATHWAY OF EPIPHYSEAL PIONEERS

I illustrate the precision in axonal pathfinding by considering the early neurons of the epiphysis. Usually, a single epiphyseal neuron pioneers the DVDT, beginning at 19–20 h (Chitnis & Kuwada 1990; Wilson & Easter 1991a,b). I use the word pioneer to mean "first," and not to imply that its presence is required for pathfinding by later axons, an issue considered below. The neuron is located in a stereotyped position, in the caudal epiphysis (Wilson & Easter 1991a). As revealed in studies in which this axon was specifically labeled by applications of DiI, it grows unerringly in a ventral direction along the wall of the diencephalon, thus pioneering the whole length of the DVDT, a distance of about 100 μm, without ever encountering another axon during this part of its journey. Electron microscopy reveals that during this time, the pioneering growth cone courses just beneath the outer surface of the brain wall, thus making contact along the way with processes of neuroepithelial cells (Wilson & Easter 1991a). The neuroepithelial cells along the path of the growth cone do not look specialized or different from their neighbors, before or during the time they contact the leading filopodia from the epiphyseal growth cone. However, the morphology of the epithelial cell processes that directly contact the growth cone itself become more complex. Perhaps local, contact-mediated interactions occur between the growth cone and the epithelium through which it grows.

The pioneering epiphyseal growth cone first encounters other axons upon reaching the ventral diencephalon, probably two or three hours after axogenesis (Wilson & Easter 1991b). These other axons, which represent the early TPOC, grow from the basal plate cluster of neurons in the diencephalon. Invariably, the epiphyseal growth cone turns rostralward at the site of intersection with these axons, thus following the TPOC pathway. It ignores the SOT as it passes by the TPOC-SOT intersection and crosses the midline within the POC.

Pathfinding is exquisitely precise along the whole pathway. Thus, the epiphyseal axon occupies a stereotyped dorsal location within the TPOC and a stereotyped rostral location within the POC. Electron microscopy reveals that restricted positioning of the axon may be caused by a restricted course of the axonal growth cone; for example, at the earliest stages within the TPOC pathway, the epiphyseal growth cone neighbors only the dorsal-most TPOC axons.

The growth cone looks particularly complex along the part of its course within the TPOC, as compared with its earlier course in an axon-free environment. This suggests (Wilson & Easter 1991b) that guidance cues are distributed along the length of the tract, i.e. not just present at the tract intersections. Electron microscopy reveals that epiphyseal growth cones make extensive contacts with processes from neuroepithelial cells, just as they did when pioneering the DVDT part of the pathway. However, the filopodia also make contacts with fascicles of TPOC axons. Thus, the environmental elements with which the epiphyseal neurons interact is more complex within the TPOC part of the pathway than within the DVDT part. This increased complexity might at least partially account for the corresponding change in growth cone morphology: The growth cone may receive guidance cues from both the neuroepithelium and its filopodial contacts with TPOC axons. Interestingly, the epiphyseal growth cone morphology does not change as it passes by the TPOC-SOT intersection, even though its filopodia probably contact SOT axons. Perhaps the growth cone filopodia are less responsive to some of the cellular elements they encounter than to others.

Within a few hours of the time of axonal outgrowth of the epiphyseal pioneer, one or two follower (i.e. later) axons, also from the epiphysis (Wilson & Easter 1991a), grow along the same pathway. Neuroepithelial cells along the pathway envelop the follower axons and pioneer axon alike, to form a single fascicle. As illustrated by Wilson & Easter (1991b), during the initial part of the course, within the DVDT, the growth cones of these early followers do not make particularly close contacts with the epiphyseal pioneer axon, but the follower filopodia may do so. This relationship is similar to that of the epiphyseal pioneer itself within the TPOC, as discussed above. Although Wilson & Easter do not comment on this, the finding seems to suggest that the epiphyseal followers receive guidance cues from both the pioneer fiber and the neuroepithelial cells along the pathway.

Thus, at 24 h, the DVDT is a single bundle of two or three axons. It arises from neurons at a stereotyped location, the epiphysis. It develops without errors being made and connects to the rest of the early axonal scaffold in a highly precise way. During the next day, the few early axons are joined by many others, such that by 48 h, hundreds of axons are present, many of them from sources other than the epiphysis (Wilson & Easter 1991b).

The simple scaffold idea (Chitnis & Kuwada 1990; Wilson et al 1990; see also Kimmel & Westerfield 1990) has emerged as an important one in considering axogenesis in the early zebrafish brain. Interestingly, as Krauss et al (1991a) have pointed out, several of the pathways in the scaffold occur at or near the expression borders of the regulatory genes discussed above. Thus, the TPOC is near the ventral border and the TPC pathway, at or near the caudal border of expression of *pax[zf-a]*; the latter also corresponds to a neuromere boundary. The MLF borders the floor plate, thus coinciding very closely with the ventral expression borders of *pax[zf-b]* in the midbrain and *Krox-20* in specific rhombomeres. Furthermore, beginning at about 25 h, commissures form in rhombomere border regions (Trevarrow et al 1990); these borders again correlate with borders of gene expression. The correlations suggest that the regulatory genes identified in these studies are involved in specifying where the pathways will form (Krauss et al 1991a).

Although the scaffold may be important for pathfinding by later axons, the adult brain has many more axonal tracts, and tract intersections, than can be accounted for by a handful of early neurons that grow unbranched axons to make a scaffold followed slavishly by later-developing axons. For example, the TPOC is not a single bundle, but is present as several bundles at 24 h. Why there are several bundles is unclear. Maybe not one, but several, neuronal populations are present, each one specified to take a slightly different course through the brain; the TPOC represents the collective pathway. Finding out if this is so will require studies at the same level of detail that were carried out for the DVDT, which were somewhat easier to do for the DVDT is a single bundle. As mentioned earlier, the midbrain also provides a clear example of a seemingly simple pattern, which on more detailed analysis turns out to be more complex. In this case, neurons arising at different locations in a single cluster project axons in different directions at different times (Metcalfe et al 1990).

IDIOSYNCRATIC NEURONS IN THE HINDBRAIN

In the cases of pathways that form in the hindbrain, the earliest neurons clearly behave as individuals with respect to the scaffold. Consider the early pathways of identified reticulospinal neurons within rhombomeres r4–r6, diagramed in Figure 6. In many embryos, the Mauthner neuron, in the fourth rhombomere, appears to pioneer a very short segment of the MLF locally (W. K. Metcalfe 1989, unpublished observations). In fact, each Mauthner neuron pioneers the contralateral pathway, which lies on

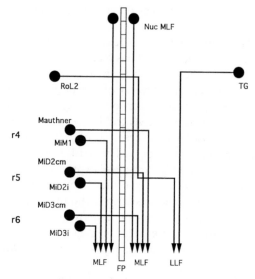

Figure 6 Schematic pattern of early axonal pathfinding by identified reticulospinal neurons within rhombomeres r4–r6. Dorsal view, with rostral to the top. With the exception of the Nuc MLF and the trigeminal ganglion (TG), both of which are cell clusters, the circles represent single identified neurons that are present as bilateral pairs of unique cells, but are shown only on one side of the midline, where the floor plate (FP) is present. Reticulospinal axons run in the MLF, except for cells like RoL2, a cell in r2 whose axon stereotypically switches from the MLF to the LLF. From data in Metcalfe et al 1986.

the other side of the midline: The growing Mauthner axon seems to ignore the position of the ipsilateral MLF, but grows in a medial direction across it. The axon then decussates, crossing at the floor plate at the hindbrain's ventral midline before turning caudally. In contrast, MiM1, the next reticulospinal neuron in rhombomere r4 to generate an axon (Mendelson 1985), chooses the ipsilateral MLF pathway, probably in the company of other reticulospinal axons that grow from cells at more rostral locations in the hind- and midbrains. In the next two rhombomeres, r5 and r6, individual neurons that project contralateral or ipsilateral axons grow out nearly simultaneously; the exact order does not seem to matter (Mendelson & Kimmel 1986). RoL2, a more rostral hindbrain cell, exhibits yet another distinctive pattern (Metcalfe et al 1986). In the middle of the rhombomere r5, the RoL2 axon switches abruptly from the MLF to the LLF. Interestingly, RoL2 does not always make this switch in r5; sometimes, it switches in the center of r6 or r7. This sort of pathway variability suggests that the centers of several rhombomeres possess similar guidance cues to which the RoL2 growth cone can respond. Serially repeating guidance cues characterize neuromeres in insects (Doe et al 1985).

Stereotyped pathfinding by sensory axons in the region of the LLF provides an example in the hindbrain of what might account for the multiple TPOC fascicles in the forebrain. The rostral part of the LLF arises at 18 h from sensory neurons in the trigeminal ganglion. During the next six hours, one or a few sensory axons that arise from two other cranial sensory ganglia, the vestibular and posterior lateral line ganglia, enter the brain and grow longitudinally in it (Kimmel et al 1990). They grow very near the LLF, but do not join it. Rather, a set of three parallel bundles forms, one bundle separated from its neighbors by only about 5 μm (see Figure 7.7 in Metcalfe 1989). Axons appear not to intermingle among these bundles. This degree of precision of early pathfinding is amazing.

There is a strict temporal-spatial patterning of development of these three pathways: The LLF lies most ventrally and arises first, and the lateral line pathway lies most dorsally and arises last (Kimmel et al 1990). Whether the temporal order is important is unknown. However, deleting the neurons of origin of the first pathway, by laser-irradiation before their axons grow, does not appear to disturb the positions of either of the two pathways that develop later, although the courses of individual axons in these pathways may have changed and been undetected (Kimmel et al 1990; 1990, unpublished). The result is negative, but nevertheless suggests that the temporal order of development is unimportant in this system for spatial positioning of specific axonal bundles.

MULTIPLE GUIDANCE CUES

Whereas temporal regulation of development may not generally play an important role in specific pathfinding in the brain, the spatially precise, virtually error-free pathfinding observed in every study to date suggests that position must be crucial. By perturbing the environments through which the axons grow, we might learn more about how position influences pathfinding. Of course, such studies cannot uncover much about the molecular nature of guidance. However, we might learn whether particular cells present in the environments of pioneering axonal growth cones are important for guidance and, later, whether the pioneer axons themselves are important for guidance of those that follow, as suggested by the simple scaffold idea. The first such experiments have recently been carried out.

To address the second issue, the guidance role of the pioneering axons, Chitnis & Kuwada (1991) examined pathfinding in the ventral midbrain by a small set of cells that arises near the PC, they term the nucleus of the posterior commissure (nuc PC). The nuc PC axons first grow ventrally, along the TPC pathway, and probably encounter axons that grow along the same pathway in the opposite direction from nuc PCv (Figure 5). In the tegmentum, the ventrally coursing nuc PC growth cones then encounter TPOC axons, of which about eight have grown into and through this region from their cells of origin in the diencephalon (Chitnis & Kuwada 1990). Normally, where the nuc PC growth cones encounter the TPOC pathway, they turn caudally and follow the TPOC pathway. Thus, with respect to the TPOC pathway, the nuc PC growth cones behave similarly to the epiphyseal growth cones described above, except that they turn in the opposite direction in the pathway (Figure 5).

Chitnis & Kuwada (1991) asked whether TPOC axons guide the nuc PC growth cones by placing a transverse lesion into the caudal diencephalon, which they expected would prevent TPOC axons from reaching the midbrain. Later, they examined the nuc PC axonal courses in these embryos by labeling neurons with Lucifer Yellow or DiI. If TPOC axons provide a unique signal to direct the nuc PC growth cones to turn caudally, then, in the lesioned animals, the growth cones should not turn caudally. However, in more than half of the lesioned embryos, the growth cones turn as usual, by taking the correct caudalwards pathway through the dorsal midbrain tegmentum. Other nuc PC growth cones navigate incorrectly, usually following the wrong axonal pathways. For example, some continue ventrally on the TPC, others turn caudally and follow the MLF, and still others turn rostrally and follow the DVDT. In a few cases, the nuc PC axons grow in brain regions apparently devoid of other axons.

As Chitnis & Kuwada (1991) noted, some of the variability could be caused by TPOC axons being present, but undetected, in some experimental embryos. However, it seems unlikely that all of the many cases in which nuc PC axons grow normally in these animals can be explained away by supposing that invisible TPOC axons are there to guide them. Rather, Chitnis & Kuwada (1991) propose that some other guidance cues, not involving TPOC axons, are available that can direct the caudalwards turn. They further point out that nuc PC growth cones could respond to the same cues that earlier direct the courses of the TPOC axons themselves. I suggested above that this mechanism seems reasonable, considering the nature of the contacts normally made by epiphyseal growth cones and their filopodia (Wilson & Easter 1991b).

To account for the result that not all of the nuc PC axonal courses are normal in the lesioned animals, Chitnis & Kuwada (1991) propose that normally the TPOC axons must also provide guidance information, but that this information is not invariably required for proper pathfinding. At least two sources of guidance cues must be present, one of them being the

TPOC axons and the other being independent of these axons. In the normal embryo, multiple cues would ensure error-free pathfinding.

Hatta (1992) independently arrived at the same conclusion, that error-free pathfinding in the brain requires multiple guidance cues, during the course of investigation of the *cyc* or *"cyclops"* mutation, which, as discussed further below, deletes the floor plate. Consider identified hindbrain reticulospinal axons that normally cross the floor plate as they decussate, or navigate longitudinally beside it, as illustrated in Figure 6. In *cyc* mutants, specific axons in this set make various pathfinding errors, but no single kind of axon consistently makes a particular kind of error. For example, axons that normally decussate by crossing the floor plate sometimes do not cross the midline when the floor plate is missing. Or, more frequently, an axon decussates normally, but then crosses back again to the side of the brain from which it originated (Hatta 1992; see also Kimmel et al 1991a). Collectively, the fibers of the MLF look highly disorganized and fasciculate very poorly. Hatta can rescue the axonal phenotype by supplying a wild-type floor plate to mutants in cell transplantation experiments (as in Hatta et al 1991b); further, he can produce a phenocopy of the axonal disruptions by using a laser to ablate the floor plate in wild-type embryos (Hatta 1992; Hatta et al 1990). Thus, the floor plate must play a role in guidance and must be important for establishing the compact axonal bundles of the normal MLF. But, considering that individual axons can approximate their normal longitudinal courses in mutants, it seems unlikely that the floor plate is the single source of guidance information.

Studies in zebrafish to learn the molecular natures of these guidance cues have only begun. Monoclonal antibodies seem to provide a useful approach to this problem. For example a screen by Trevarrow yielded two antibodies, zn-5 and zn-12, that labeled developing neuronal cell-surface antigens in interesting patterns. The zn-12 antigen turned out to be HNK-1 (Metcalfe et al 1990), already well known in higher vertebrates and implicated in specific cell adhesion or recognition. The zebrafish work implies the same; many types of the primary neurons express high levels of HNK-1 on their axonal growth cones during outgrowth. In a particularly interesting case, axonal growth cones of trigeminal sensory neurons navigate through head mesenchyme to meet the brain wall at the second rhombomere, where they enter the brain to form their central pathway. They express high levels of HNK-1. The brain wall, at r2 specifically, also expresses HNK-1 just before and during the time of the growth cones contact it (Metcalfe et al 1990).

The nature of the antigen recognized by the other antibody, zn-5, is unknown, but several features of the expression pattern suggest it would be worthwhile to learn. It is expressed transiently along specific growing axons, some of which express HNK-1, in a pattern that, like the L1 antigen in the mouse (Fischer et al 1986), suggests it is involved in axonal bundling (B. Trevarrow and C. B. K. 1989, unpublished observations). Furthermore, the zn-5 antigen is also present, again transiently, on cells that seem to form boundaries in the brain (and elsewhere in the body; e.g. see Hatta et al 1991a). In these locations, the antigen may be involved in other cells being able to recognize these boundaries. For example, hindbrain interneurons, whose axons make commissures near the rhombomere boundaries shortly after 24 h, express high levels of the zn-5 antigen as the commissures develop, and expression persists for several days thereafter (Trevarrow et al 1990). The floor plate cells also express this antigen, beginning at about 22 h, and in a most interesting way. During the first several hours of expression, we see high amounts of it on the midline surfaces of these nonneuronal cells, where they contact one another; however, we do not see expression on their lateral surfaces, where the floor plate cells meet other cell types (Hatta et al 1991b). Possibly, this antigen is involved in midline recognition at the floor plate. The midline expression is gone in *cyc* mutants (Hatta et al 1991b), in which neurons fail to identify accurately where the midline is. We note, however, that the zn-5 antigen could not be the only such recognition cue that is normally present, for it seems to come on too late to be used by the earliest neurons. Their axonal growth cones are navigating near the floor plate at least four hours before zn-5 expression can be detected.

SIGNALS THAT ESTABLISH THE CELL GROUPS AND BRAIN REGIONS

Perturbing normal development may not only reveal aspects of axonal patterning, but also regulation of the patterning of brain regions and the specific cells groups within them. Thus, the *cyc* mutation deletes the whole length of floor plate in the brain and trunk spinal cord (Hatta 1992; Hatta et al 1991b). In this respect, it is interesting that most of the regulatory gene expression patterns that have been characterized to date suggest separate genetic control of floor plate and non-floor plate lineages. Mosaic analyses establish that *cyc* acts autonomously in the floor plate, and the same analyses also reveal that mutant cells can differentiate as floor plate; however, they fail to do so in mutants, unless they have wild-type floor plate neighbors. Hatta et al (1991b) speculate that the block is in floor plate specification; that is, the *cyc* gene may lie on the signaling pathway that initially sets up development of the floor plate lineage.

The *cyc* mutation produces a much more severe phenotype in the rostral central nervous system (CNS) than it does elsewhere along the neuraxis.

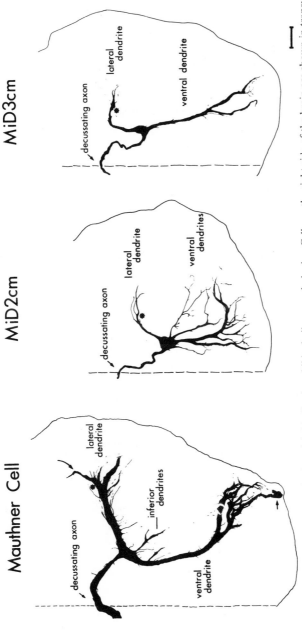

Figure 7 Segmental homologues in the adult hindbrain exhibit similar morphologies. Cells on the right side of the brain are shown in transverse view (dorsal up), as drawn from preparations retrogradely labeled with HRP. The brain wall is also shown, and the midline is dashed. The Mauthner cell develops in r4, MiD2cm in r5, and MiD3cm in r6. Their positions within their segments are all similar, as are their axonal and dendritic projections. Additionally, the cells have the same basic forms as in the newly hatched embryo, but are roughly sevenfold larger in the adult. Reproduced from Lee & Eaton 1991 (with permission). Scale bar: 50 μm.

rostrocaudal, i.e. this particular subdivision is not across the long axis of the body. This interpretation of this boundary seems well supported and is amenable to further examination. For example, fate map studies would reveal the relative positions of the two primordia at earlier stages.

Like the rhombomeres, the midbrain and forebrain neuromeres initially develop single basal plate clusters of neurons. A prominent commissure, the PC, develops between the diencephalic and midbrain neuromeres, thus recalling the intersegmental ventral commissures in the hindbrain (albeit the PC fibers cross dorsally, whereas the hindbrain commissural fibers cross ventrally). Early neurons in both the diencephalon and midbrain contribute caudally growing axons in the ventral longitudinal system of bundles (i.e. in the TPOC and the MLF, respectively) and other axons that make commissures (the POC and the PC). It would be interesting to learn if the commissural axons arise from the more rostral cells in the basal diencephalon, as they do in the midbrain.

Alternatively, the rostral brain may be unsegmented, in a manner analogous to the prostomial region of annelid worms (Weisblat et al 1984) or the acron of insects (Jürgens et al 1986). So far, only general neuronal markers, e.g. acetylcholine esterase, have been used to characterize the early neurons. To address serial homology, one would like to know, in detail, which characteristics, if any, the neurons in forebrain share with their putative homologues in the midbrain. Finding more specific markers would be helpful.

MORE MUTATIONS

In the mouse, there is an interesting gene known as "Small eye" (or *Sey*). Mutational analysis has shown that *Sey* is involved in development of facial structures including the eye, a prominent derivative of the diencephalon. Several independent mutant alleles exist, and all of them produce a similar semidominant phenotype; semidominance itself is somewhat unusual. Heterozygotes have small eyes, and homozygotes have no eyes at all. Semidominance can be interpreted to mean that for normal development, it is crucial that the dosage of the protein coded by *Sey* be correct, as if function were dependent on a defined threshold concentration of the gene product. This interpretation is supported, for *Sey* in particular, by deficiency analysis (Hill et al 1991).

The relevance of *Sey* to neural patterning in the zebrafish is that molecular analysis (Hill et al 1991) shows *Sey* to be the murine homologue of zebrafish *pax[zf-a]*, which is expressed in cells that form the eye, before and during the time that the optic vesicle appears (Krauss et al 1991c; Püschel et al 1992). We can thus imagine, from the mouse work, what *pax*[*zf-a*] might be doing in the fish. A DNA-regulatory protein might well be expected to be required for eye development and to function in the primordium in a dosage-dependent cell autonomous fashion.

However, in zebrafish (Figure 2) and mouse alike, the same expression domain of this gene encompasses the dorsolateral diencephalic wall, where it has been proposed to be "controlling early regionalization in the rostral CNS" (Krauss et al 1991a). If so, *Sey* mutants would be expected to have severe defects in the diencephalon, but no defects at all have been found (Hogan et al 1988). This leaves us with a dilemma that can only be resolved by further analysis. Just how *Sey* acts in the mouse, with respect to the embryonic cells involved, is not very well understood, and molecular identification of the gene provides a major step in this direction.

So far, only a single mutation, *cyc*, has been described that perturbs brain development in zebrafish, but its study has yielded extremely clear information about the cellular nature of patterning interactions. The defects of mutations that affect early development of other zebrafish body parts are similarly interpretable (reviewed in Kimmel et al 1991a; Wilson & Easter 1992), as are the results of the lesion (Chitnis & Kuwada 1991) and teratogen (Holder & Hill 1991) studies reviewed above. Features of the embryo, including its relative simplicity, provide for reasonably unambiguous analysis. Yet, the early patterning, streamlined as it seems to be, is basically the same as in other vertebrates: Extreme evolutionary conservation, which ranges from gene expression patterns to the pattern of the brain's early axonal scaffold, has been emphasized in nearly all of the zebrafish work to date.

Charline Walker picked *cyc* in a screen in which one mutagenizes fish and examined their haploid embryos for interesting phenotypes. She has found, by the same means, other mutations that affect brain patterning, and we are now beginning to study some of these. More importantly, however, because of the early work of Streisinger and coworkers (see Streisinger et al 1981), it is relatively easy to identify and recover many new mutations in zebrafish: A lesson that we learn from *Sey* in the mouse, and especially from neurogenetic studies in *C. elegans* and *Drosophila*, is that we will require many mutations, which represent many gene loci and include separate mutations at individual loci, to unravel the genetic pathways that pattern the early vertebrate brain.

ACKNOWLEDGMENTS

I thank Judith Eisen, Kohei Hatta, Stefan Krauss, Andreas Püschel, Steve Wilson, and Monte Westerfield for their critical comments on the manuscript. Steve Easter, Kohei Hatta, Andreas Püschel, and Trevor Jowett shared unpublished data. Original work from my laboratory was supported by National Institutes of Health grants NS17963 and HD22486.

Literature Cited

Bateson, W. 1894. *Materials for the Study of Variation.* London: Macmillan

Chitnis, A. B., Kuwada, J. Y. 1991. Elimination of a brain tract increases errors in pathfinding by follower growth cones in the zebrafish embryo. *Neuron* 7: 1–20

Chitnis, A. B., Kuwada, J. Y. 1990. Axonogenesis in the brain of zebrofish embryos. *J. Neurosci.* 10: 1892–1905

Doe, C. Q., Kuwada, J. Y., Goodman, C. S. 1985. From epithelium to neuroblasts to neurons: the role of cell interactions and cell lineage during insect neurogenesis. *Philos. Trans. R. Soc. London Ser. B* 312: 67–81

Eisen, J. S. 1991a. Developmental neurobiology of the zebrafish. *J. Neurosci.* 11: 311–17

Eisen, J. S. 1991b. Motoneuronal development in the embryonic zebrafish. *Development* 2(Suppl.): 141–47

Fischer, G., Künemund, V., Schachner, M. 1986. Neurite outgrowth patterns in cerebellar microexplant cultures are affected by antibodies to the cell surface glycoprotein L1. *J. Neurosci.* 6: 605–12

Grunwald, D. J., Kimmel, C. B., Westerfield, M., Walker, C., Streisinger, G. 1988. A neural degeneration mutation that spares primary neurons in the zebrafish. *Dev. Biol.* 126: 115–28

Hanneman, E., Trevarrow, B., Metcalfe, W. K., Kimmel, C. B., Westerfield, M. 1988. Segmental development of the spinal cord and hindbrain of the zebrafish embryo. *Development* 103: 49–58

Hanneman, E., Westerfield, M. 1989. Early expression of acetylcholinesterase activity in functionally distinct neurons of the zebrafish. *J. Comp. Neurol.* 284: 350–61

Hatta, K. 1992. Role of the floor plate in axonal patterning in the zebrafish CNS. *Neuron.* In press

Hatta, K., BreMiller, R. A., Westerfield, M., Kimmel, C. B. 1991a. Diversity of expression of *engrailed* homeoproteins in zebrafish. *Development* 112: 821–32

Hatta, K., Ho, R. K., Walker, C., Kimmel, C. B. 1990. A mutation that deletes the floor plate and disturbs axonal pathfinding in zebrafish. *Soc. Neurosci. Abstr.* 16: 310

Hatta, K., Kimmel, C. B., Ho, R. K.,

Walker, C. 1991b. The cyclops mutation blocks specification of the floor plate of the zebrafish central nervous system. *Nature* 350: 339–41

Hill, R. E., Favor, J., Hogan, B. L. M., Ton, C. C. T., Saunders, G. F., et al. 1991. Mouse *Small eye* results from mutations in a paired-like homeobox-containing gene. *Nature* 354: 522–25

Hogan, B. L. M., Hirst, E. M. A., Horsburgh, G., Hetherington, C. M. 1988. *Small eye* (*Sey*): a mouse model for the genetic analysis of craniofacial abnormalities. *Development* (Suppl.) 103: 115–19

Holder, N., Hill, J. 1991. Retinoic acid modifies development of the midbrain-hindbrain border and affects cranial ganglion formation in zebrafish embryos. *Development* 113: 1159–70

Ingham, P. W. 1991. Segment polarity genes and cell patterning within the *Drosophila* body segment. *Curr. Opin. Gen. Dev.* 1: 261–67

Ingham, P. W. 1990. Genetic control of segmental patterning in the *Drosophila* embryo. In *Genetics of Pattern Formation and Growth Control,* pp. 181–96. New York: Wiley-Liss

Jeffs, P. S., Keynes, R. J. 1990. A brief history of segmentation. *Dev. Biol.* 1: 77–87

Jessell, T. M., Bovolenta, P., Placzek, M., Tessier-Lavigne, M., Dodd, J. 1989. Polarity and patterning in the neural tube: the origin and function of the floor plate. In *Cellular Basis of Morphogenesis,* Ciba Found. Symp., pp. 255–80. Chichester: Wiley

Jones, F. S., Chalepakis, G., Gruss, P., Edelman, G. M. 1992a. Activation of the cytotactin promoter by the homeobox-containing gene *Evx-1. Proc. Natl. Acad. Sci. USA* 89: 2091–95

Jones, F. S., Prediger, E. A., Bittner, D. A., DeRobertis, E. M., Edelman, G. M. 1992b. Cell adhesion molecules as targets for *Hox* genes: neural cell adhesion molecule promoter activity is modulated by cotransfection with *Hox-2.5* and *-2.4. Proc. Natl. Acad. Sci. USA* 89: 2086–90

Jürgens, G., Lehmann, R., Schardin, M., Nüsslein-Volhard, C. 1986. Segmental organisation of the head in the embryo of

Drosophila melanogaster. A blastoderm fate map of the cuticle structures of the larval head. *Roux's Arch. Dev. Biol.* 195: 359–77

Kimmel, C. B., Hatta, K., Eisen, J. S. 1991a. Genetic control of primary neuronal development in zebrafish. *Development* 2(Suppl.): 47–57

Kimmel, C. B., Hatta, K., Metcalfe, W. K. 1990. Early axonal contacts during development of an identified dendrite in the brain of the zebrafish. *Neuron* 4: 535–45

Kimmel, C. B., Schilling, T. F., Hatta, K. 1991b. Patterning of body segments of the zebrafish embryo. *Curr. Top. Dev. Biol.* 25: 77–110

Kimmel, C. B., Westerfield, M. 1990. Primary neurons of the zebrafish. In *Signals and Sense: Local and Global Order in Perceptual Maps*, ed. G. M. Edelman, W. M. Cowan, pp. 561–88. New York: Wiley Interscience

Krauss, S., Johansen, T., Korzh, V., Fjose, A. 1991a. Expression of the zebrafish paired box gene *pax[zf-b]* during early neurogenesis. *Development* 113: 1193–1206

Krauss, S., Johansen, T., Korzh, V., Fjose, A. 1991b. Expression pattern of zebrafish *pax* genes suggests a role in early brain regionalization. *Nature* 353: 267–70

Krauss, S., Johansen, T., Korzh, V., Moens, U., Ericson, J. U., Fjose, A. 1991c. Zebrafish *pas[zf-a]*: a paired box-containing gene expressed in the neural tube. *EMBO J.* 10: 3609–19

Kuwada, J. Y., Bernhardt, R. R., Chitnis, A. B. 1990a. Pathfinding by identified growth cones in the spinal cord of zebrafish embryos. *J. Neurosci.* 10: 1299–1308

Kuwada, J. Y., Bernhardt, R. R., Nguyen, N. 1990b. Development of spinal neurons and tracts in the zebrafish embryo. *J. Comp. Neurol.* 302: 617–28

Lee, R. K. K., Eaton, R. C. 1991. Identifiable reticulospinal neurons of the adult zebrafish, *Brachydanio rerio*. *J. Comp. Neurol.* 304: 34–52

Lumsden, A. 1990. The cellular basis of segmentation in the developing hindbrain. *Trends Neurosci.* 13: 329–39

Lumsden, A., Keynes, R. 1989. Segmental patterns of neuronal development in the chick hindbrain. *Nature* 337: 424–28

Mendelson, B. 1986. Development of reticulospinal neurons of the zebrafish. II. Early axonal outgrowth and cell body position. *J. Comp. Neurol.* 251: 172–84

Mendelson, B. 1985. Soma position is correlated with time of development in three types of identified reticulospinal neurons. *Dev. Biol.* 112: 489–93

Mendelson, B., Kimmel, C. B. 1986. Identified vertebrate neurons that differ in axonal projection develop together. *Dev. Biol.* 118: 309–13

Metcalfe, W. K. 1989. Organization and development of the zebrafish posterior lateral line. In *The Mechanosensory Lateral Line: Neurobiology and Evolution*, ed. S. Coombs, P. Görner, H. Münz, pp. 147–59. New York/Berlin/Heidelberg: Springer-Verlag

Metcalfe, W. K., Mendelson, B., Kimmel, C. B. 1986. Segmental homologies among reticulospinal neurons in the hindbrain of the zebrafish larva. *J. Comp. Neurol.* 251: 147–59

Metcalfe, W. K., Myers, P. Z., Trevarrow, B., Bass, M. B., Kimmel, C. B. 1990. Primary neurons that express the L^2/HNK-1 carbohydrate during early development in the zebrafish. *Development* 110: 491–504

Puelles, L., Amat, J. A., Martinez-de-la-Torre, M. 1987. Segment-related, mosaic neurogenetic pattern in the forebrain and mesencephalon of early chick embryos: I. Topography of AChE-positive neuroblasts up to stage HH18. *J. Comp. Neurol.* 266: 247–68

Püschel, A. W., Gruss, P., Westerfield, M. 1992. Sequence and expression pattern of *pax-6* are highly conserved between zebrafish and mice. *Development* 114: 643–51

Ross, L. S., Parrett, T., Easter, S. S. Jr. 1992. Axonogenesis and morphogenesis in the embryonic zebrafish brain. *J. Neurosci.* 12: 467–82

Streisinger, G., Walker, C., Dower, N., Knauber, D., Singer, F. 1981. Production of clones of homozygous diploid zebrafish (*Brachydanio rerio*). *Nature* 291: 293–96

Trevarrow, B., Marks, D. L., Kimmel, C. B. 1990. Organization of hindbrain segments in the zebrafish embryos. *Neuron* 4: 669–79

Vaage, S. 1969. The segmentation of the primitive neural tube in chick embryos (*Gallus domesticus*). In *Ergebnisse Anatomie Entwicklungsgesch*, 41: 1–88. Berlin/Heidelberg: Springer-Verlag

Weisblat, D. A., Kim, S. Y., Stent, G. S. 1984. Embryonic origins of cells in the leech *Helobdella triserialis*. *Dev. Biol.* 104: 65–85

Weisblat, D. A., Shankland, M. 1985. Cell lineage and segmentation in the leech. *Philos. Trans. R. Soc. London Ser. B* 312: 39–56

Wilkinson, D. G., Bhatt, S., Chavrier, P., Bravo, R., Charnay, P. 1989. Segment-specific expression of a zinc-finger gene in the developing nervous system of the mouse. *Nature* 337: 461–64

Wilson, S. W., Easter, S. S. Jr. 1992. Acquisition of regional and cellular identities in the developing zebrafish nervous system. *Curr. Opin. Neurobiol.* 2: 9–15

Wilson, S. W., Easter, S. S. Jr. 1991a. A pioneering growth cone in the embryonic zebrafish brain. *Proc. Natl. Acad. Sci. USA* 88: 2293–96

Wilson, S. W., Easter, S. S. Jr. 1991b. Stereotyped pathway selection by growth cones of early epiphysial neurons in the embryonic zebrafish. *Development* 112: 723–46

Wilson, S. W., Ross, L. S., Parrett, T., Easter, S. S. Jr. 1990. The development of a simple scaffold of axon tracts in the brain of the embryonic zebrafish, *Brachydanio rerio. Development* 108: 121–45

Wolpert, L. 1969. Positional information and the spatial pattern of cellular differentiation. *J. Theor. Biol.* 25: 1–47

SUBJECT INDEX

A

A23187
dopamine receptor and, 311
Acetylcholine
Ascaris motor neuron and, 51
cellular localization of, 471
endothelial-derived relaxing factor and, 647
intracellular calcium levels and, 581
long-term potentiation and, 634
neuronal nicotinic acetylcholine receptor and, 405
synthesis
primary sympathoadrenal progenitor and, 142
Acetylcholine esterase
clustering in agrin-treated myotubes, 349
sympathetic ganglia and, 474
Acetylcholine receptor
clustering of, 348-56
agrin and, 108, 348-51
extracellular matrix and, 352-53
postsynaptic cytoskeleton and, 353-56
muscarinic, 403
G-proteins and, 22
nicotinic, 403-34
distribution of, 420-23
function of, 423-27
gene expression regulation and, 22
levamisole and, 54
neuromuscular junction and, 348-56
physiological properties of, 404-7
purification of, 411-14
regulation of, 427-32
structure of, 407-14
Acetylcholine receptor gene
nicotinic
heterologous expression of, 415-20
Acetylcholinesterase
Caenorhabditis elegans and, 53

primary neurons and, 713-14
Acrylamide
visual pathway lesions and, 374
ACT*, 604-5
α-Actinin
neurite growth cones and, 582
Adaptin, 97
Adenylate cyclase
dopamine receptor and, 20
G-protein interactions and, 22
neurofibromin and, 200
Adenylyl cyclase
dopamine and, 299-300
dopamine receptor and, 311
Drosophila learning mutant and, 640
ADP ribosyltransferase
nitric oxide and, 650
Adrenaline
See Epinephrine
Adrenergic receptor
G-proteins and, 22
Afferent feedback
reinforcing action of, 280-82
voluntary movement and, 276
Agnosia, 377
Agrammatism
Broca's area and, 522
Agrin
acetylcholine receptor clustering and, 108, 348-52
Akinetic mutism, 524
β-Alanine
gamma-aminobutytric acid transport system and, 77
Alcoholic Korsakoff's syndrome
animal model of, 553
mammillary nuclei in, 552
Alcoholism
Taq 1 restriction fragment length polymorphism and, 304-5
Alzheimer's disease
choline acetyltransferase activity in, 554
dementia due to

impaired word priming in, 169
entorhinal cortex in, 555
motor skill learning in, 175
sympathoadrenal lineage and, 149
Amara, S. G., 73-89
Amines
locomotor activity and, 285-86
Amnesia
animal model of, 547-48
diencephalon and, 552-54
medial temporal lobe and, 548-49
organic
implicit memory tasks and, 160
priming in
longevity of, 165, 167-69
retrograde
medial temporal cortex and, 248
neural basis of, 258-59
short-term memory and, 558
visual object priming in, 171-72
AMPA
rhythmic motor output and, 214
Amphibia
visually guided behavior in, 610
Amygdala
memory and, 551-52
Anderson, D. J., 129-51
Ankyrin
postsynaptic specializations and, 108
sodium channels and, 357
Antibody binding
neurofibromin and, 189-91
Anticholinergics
word fragment completion and, 169
Anticonvulsants
c-*fos* expression and, 18
Antidepressants
mechanism of action of
monoamine transport inhibition and, 74
norepinephrine uptake and, 76
serotonin transporter and, 75-76

733

CUMULATIVE INDEXES

CONTRIBUTING AUTHORS, VOLUMES 11–16

751

CHAPTER TITLES, VOLUMES 11–16

ANNUAL REVIEWS INC.

a nonprofit scientific publisher
4139 El Camino Way
P. O. Box 10139
Palo Alto, CA 94303-0897 • USA

Annual Reviews Inc. publications may be ordered directly from our office; through booksellers and subscription agents, worldwide; and through participating professional societies. **Prices are subject to change without notice.**   California Corp. #161041 • ARI Federal I.D. #94-1156476

- **Individual Buyers:** Prepayment required on new accounts by check or money order (in U.S. dollars, check drawn on U.S. bank) or charge to MasterCard, VISA, or American Express.

- **Institutional Buyers:** Please include purchase order.

- **Students/Recent Graduates:** $10.00 discount from retail price, per volume. Discount does not apply to Special Publications, standing orders, or institutional buyers. **Requirements:** [1] be a degree candidate at, or a graduate within the past three years from, an accredited institution; [2] present proof of status (photocopy of your student I.D. or proof of date of graduation); [3] Order direct from Annual Reviews; [4] prepay.

- **Professional Society Members:** Societies that have a contractual arrangement with Annual Reviews offer our books to members at reduced rates. Check your society for information.

- **California orders** must add applicable sales tax.

- **Canadian orders** must add 7% General Sales Tax. GST Registration #R 121 449-029. Now you can also telephone orders Toll Free from anywhere in Canada (see below).

- **Telephone orders,** paid by credit card, welcomed. **Call Toll Free 1-800-523-8635** from anywhere in USA or Canada. From elsewhere call 415-493-4400, Ext. 1 (not toll free). Monday – Friday, 8:00 am – 4:00 pm, Pacific Time. Students or recent graduates ordering by telephone must supply (by FAX or mail) proof of status if current proof is not on file at Annual Reviews. Written confirmation required on purchase orders from universities before shipment.

- **FAX: 415-855-9815 – 24 hours a day.**

- **Postage paid** by Annual Reviews (4th class bookrate). UPS ground service (within continental U.S.) available at $2.00 extra per book. UPS air service or Airmail also available at cost. UPS requires a street address. P.O. Box, APO, FPO, not acceptable.

- **Regular Orders:** Please list below the volumes you wish to order by volume number.

- **Standing Orders:** New volume in series is sent automatically each year upon publication. Please indicate volume number to begin the standing order. Each year you can save 10% by prepayment of standing-order invoices sent 90 days prior to the publication date. Cancellation may be made at any time.

- **Prepublication Orders:** Volumes not yet published will be shipped in month and year indicated

- **We do not ship on approval.**

ANNUAL REVIEWS SERIES *Volumes not listed are no longer in print*	Prices, postpaid, per volume. USA / other countries (incl. Canada)	Regular Order Please send Volume(s):	Standing Order Begin with Volume:
Annual Review of ANTHROPOLOGY			
Vols. 1-20 (1972-1991)............................$41.00/$46.00			
Vol. 21 (1992).....................................$44.00/$49.00			
Vol. 22 (avail. Oct. 1993)....................$44.00/$49.00		Vol(s). _____	Vol._____
Annual Review of ASTRONOMY AND ASTROPHYSICS			
Vols. 1, 5-14 (1963, 1967-1976)			
16-29 (1978-1991)............................$53.00/$58.00			
Vol. 30 (1992).....................................$57.00/$62.00			
Vol. 31 (avail. Sept. 1993)...................$57.00/$62.00		Vol(s). _____	Vol.____
Annual Review of BIOCHEMISTRY			
Vols. 30-34, 36-60 (1961-1965, 1967-1991) .... $41.00/$47.00			
Vol. 61 (1992) ....$46.00/$52.00			
Vol. 62 (avail. July 1993) ....$46.00/$52.00		Vol(s). _____	Vol.____

ANNUAL REVIEWS SERIES	Prices, postpaid, per volume. USA / other countries (incl. Canada)	Regular Order Please send Volume(s):	Standing Order Begin with Volume:
Volumes not listed are no longer in print			

Annual Review of BIOPHYSICS AND BIOMOLECULAR STRUCTURE

Vols. 1-20	(1972-1991)............................$55.00/$60.00		
Vol. 21	(1992)......................................$59.00/$64.00		
Vol. 22	(avail. June 1993)..................$59.00/$64.00	Vol(s). _____	Vol._____

Annual Review of CELL BIOLOGY

Vols. 1-7	(1985-1991)............................$41.00/$46.00		
Vol. 8	(1992)......................................$46.00/$51.00		
Vol. 9	(avail. Nov. 1993)..................$46.00/$51.00	Vol(s). _____	Vol._____

Annual Review of COMPUTER SCIENCE

Vols. 1-2	(1986-1987)............................$41.00/$46.00		
Vols. 3-4	(1998-1989/1990)...................$47.00/$52.00	Vol(s). _____	Vol._____

Series suspended until further notice. Purchase the complete set for the special promotional price of $100.00 USA / $115.00 other countries, when all four volumes are ordered at the same time. Orders at the special price must be prepaid.

Annual Review of EARTH AND PLANETARY SCIENCES

Vols. 1-19	(1973-1991)............................$55.00/$60.00		
Vol. 20	(1992)......................................$59.00/$64.00		
Vol. 21	(avail. May 1993)...................$59.00/$64.00	Vol(s). _____	Vol._____

Annual Review of ECOLOGY AND SYSTEMATICS

Vols. 2-12, 14-22	(1971-1981, 1983-1991).........$40.00/$45.00		
Vol. 23	(1992)......................................$44.00/$49.00		
Vol. 24	(avail. Nov. 1993)..................$44.00/$49.00	Vol(s). _____	Vol._____

Annual Review of ENERGY AND THE ENVIRONMENT

Vols. 1-16	(1976-1991)............................$64.00/$69.00		
Vol. 17	(1992)......................................$68.00/$73.00		
Vol. 18	(avail. Oct. 1993)...................$68.00/$73.00	Vol(s). _____	Vol._____

Annual Review of ENTOMOLOGY

Vols. 10-16, 18	(1965-1971, 1973)		
20-36	(1975-1991)............................$40.00/$45.00		
Vol. 37	(1992) ....$44.00/$49.00		
Vol. 38	(avail. Jan. 1993) ....$44.00/$49.00	Vol(s). _____	Vol._____

Annual Review of FLUID MECHANICS

Vols. 2-4, 7, 9-11	(1970-1972, 1975, 1977-1979)		
14-23	(1982-1991) ....$40.00/$45.00		
Vol. 24	(1992) ....$44.00/$49.00		
Vol. 25	(avail. Jan. 1993) ....$44.00/$49.00	Vol(s). _____	Vol._____

Annual Review of GENETICS

Vols. 1-12, 14-25	(1967-1978, 1980-1991) ....$40.00/$45.00		
Vol. 26	(1992)......................................$44.00/$49.00		
Vol. 27	(avail. Dec. 1993)..................$44.00/$49.00	Vol(s). _____	Vol._____

Annual Review of IMMUNOLOGY

Vols. 1-9	(1983-1991) ....$41.00/$46.00		
Vol. 10	(1992) ....$45.00/$50.00		
Vol. 11	(avail. April 1993) ....$45.00/$50.00	Vol(s). _____	Vol._____

Annual Review of MATERIALS SCIENCE

Vols. 1, 3-19	(1971, 1973-1989)..................$68.00/$73.00		
Vols. 20-22	(1990-1992) ....$72.00/$77.00		
Vol. 23	(avail. Aug. 1993) ....$72.00/$77.00	Vol(s). _____	Vol._____

ANNUAL REVIEWS SERIES _Volumes not listed are no longer in print_	Prices, postpaid, per volume. USA / other countries (incl. Canada)	Regular Order Please send Volume(s):	Standing Order Begin with Volume:
Annual Review of PUBLIC HEALTH			
Vols. 1-12 (1980-1991) .... $45.00/$50.00			
Vol. 13 (1992) .... $49.00/$54.00			
Vol. 14 (avail. May 1993) .... $49.00/$54.00		Vol(s). _____	Vol._____
Annual Review of SOCIOLOGY			
Vols. 1-17 (1975-1991) .... $45.00/$50.00			
Vol. 18 (1992) .... $49.00/$54.00			
Vol. 19 (avail. Aug. 1993) .... $49.00/$54.00		Vol(s). _____	Vol._____

NEW! Comprehensive Multiyear Index to Annual Review publications on computer disks. Available in the fall of 1992. Price to be announced.

☐ Please send complete information when available.   ☐ DOS   ☐ MAC

SPECIAL PUBLICATIONS	Prices, postpaid, per volume. USA / other countries (incl. Canada)	Regular Order Please send:

The Excitement and Fascination of Science

Volume 1	(1965 softcover) .... $25.00/$29.00	_____ Copy(ies).
Volume 2	(1978 softcover)..................... $25.00/$29.00	_____ Copy(ies).
Volume 3	(1990 hardcover)................... $90.00/$95.00	_____ Copy(ies).

(Volume 3 is published in two parts with complete indexes for Volume 1, 2, and both parts of Volume 3. **Sold as a two-part set only.**)

Intelligence and Affectivity:
Their Relationship During Child Development

(1981 hardcover).................... $8.00/$9.00   _____ Copy(ies).

Send To: **ANNUAL REVIEWS INC., a nonprofit scientific publisher**
4139 El Camino Way   •   P. O. Box 10139
Palo Alto, CA 94303-0897   USA

☐ Please enter my order for publications indicated above. Prices are subject to change without notice.

Date of Order _____   ☐ Proof of student status enclosed

Institutional Purchase Order No. _____   ☐ California order, must add applicable sales tax

Individuals: Prepayment is required in U.S. funds or charge to bank card listed below.   ☐ Canadian order must add 7% GST.

☐ Amount of remittance enclosed: _____   ☐ Optional UPS shipping (domestic ground service except to AK or HI), add $2.00 per volume. UPS requires a street address. No P.O. Box, APO or FPO.

_Or_ charge my   ☐ VISA

☐ MasterCard   ☐ American Express

Account Number _____   Exp. Date ____ / ____

Signature_____

Name _____
     please print

Address _____
     please print

_____ Zip Code _____

_____ Send free copy of current _Prospectus_ ☐

Area(s) of interest   Calif. Corp. No. 161041   ARI Federal I.D. No. 94-1156476

ANNUAL REVIEWS SERIES _Volumes not listed are no longer in print_	Prices, postpaid, per volume. USA / other countries (incl. Canada)	Regular Order Please send Volume(s):	Standing Order Begin with Volume:

Annual Review of MEDICINE
Vols. 9, 11-15   (1958, 1960-1964)
     17-42   (1966-1991) .... $40.00/$45.00
Vol. 43   (1992) .... $44.00/$49.00
Vol. 44   (avail. April 1993) .... $44.00/$49.00   Vol(s). _____   Vol._____

Annual Review of MICROBIOLOGY
Vols. 20-24, 26-45 (1966-1970, 1972-1991) .... $41.00/$46.00
Vol. 46   (1992) .... $45.00/$50.00
Vol. 47   (avail. Oct. 1993) .... $45.00/$50.00   Vol(s). _____   Vol._____

Annual Review of NEUROSCIENCE
Vols. 1-14   (1978-1991) .... $40.00/$45.00
Vol. 15   (1992) .... $44.00/$49.00
Vol. 16   (avail. March 1993) .... $44.00/$49.00   Vol(s). _____   Vol._____

Annual Review of NUCLEAR AND PARTICLE SCIENCE
Vols. 12-41   (1962-1991) .... $55.00/$60.00
Vol. 42   (1992) .... $59.00/$64.00
Vol. 43   (avail. Dec. 1993) .... $59.00/$64.00   Vol(s). _____   Vol._____

Annual Review of NUTRITION
Vols. 1-11   (1981-1991) .... $43.00/$48.00
Vol. 12   (1992) .... $45.00/$50.00
Vol. 13   (avail. July 1993) .... $45.00/$50.00   Vol(s). _____   Vol._____

Annual Review of PHARMACOLOGY AND TOXICOLOGY
Vols. 2-3, 5-31   (1962-1963, 1965-1991) .... $40.00/$45.00
Vol. 32   (1992)..................................... $44.00/$49.00
Vol. 33   (avail. April 1993)................... $44.00/$49.00   Vol(s). _____   Vol._____

Annual Review of PHYSICAL CHEMISTRY
Vols. 12-21, 23-27 (1961-1970, 1972-1976)
     29-42   (1978-1991) .... $44.00/$49.00
Vol. 43   (1992) .... $48.00/$53.00
Vol. 44   (avail. Nov. 1993) .... $48.00/$53.00   Vol(s). _____   Vol._____

Annual Review of PHYSIOLOGY
Vols. 19-53   (1957-1991) .... $42.00/$47.00
Vol. 54   (1992)..................................... $46.00/$51.00
Vol. 55   (avail. March 1993).............. $46.00/$51.00   Vol(s). _____   Vol._____

Annual Review of PHYTOPATHOLOGY
Vols. 3-20, 22-29 (1965-1982, 1984-1991) .... $42.00/$47.00
Vol. 30   (1992) .... $46.00/$51.00
Vol. 31   (avail. Sept. 1993) ....$46.00/$51.00   Vol(s). _____   Vol._____

Annual Review of PLANT PHYSIOLOGY AND PLANT MOLECULAR BIOLOGY
Vols. 16-23, 26-29 (1965-72, 1975-78)
     31-42   (1980-1991) .... $40.00/$45.00
Vol. 43   (1992) .... $44.00/$49.00
Vol. 44   (avail. June 1993) .... $44.00/$49.00   Vol(s). _____   Vol._____

Annual Review of PSYCHOLOGY
Vols. 4, 5, 8, 10, 15, 17 (1953, 1954, 1957, 1959, 1964, 1966)
     20, 22-24, 26-30 (1969, 1971-1973, 1975-1979)
     32-42   (1981-1991) .... $40.00/$45.00
Vol. 43   (1992) .... $43.00/$48.00
Vol. 44   (avail. Feb. 1993) .... $43.00/$48.00   Vol(s). _____   Vol._____